(continued on rear endpapers)

Tenth Edition

College
Algebra

R. David Gustafson
Rock Valley College

Peter D. Frisk
Rock Valley College

Jeffrey D. Hughes
Hinds Community College

BROOKS/COLE
CENGAGE Learning·

Australia · Brazil · Japan · Korea · Mexico · Singapore · Spain · United Kingdom · United States

BROOKS/COLE
CENGAGE Learning™

College Algebra, **Tenth Edition**
R. David Gustafson, Peter D. Frisk,
and Jeffrey D. Hughes

Editor: Gary Whalen

Development Editor: Leslie Lahr

Assistant Editors: Natasha Coats,
Cynthia Ashton

Editorial Assistant: Guanglie Zhang

Media Editor: Lynh Pham

Marketing Manager: Myriah Jasmin FitzGibbon

Marketing Coordinator: Angela Kim

Marketing Communications Manager:
Katherine Malatesta

Project Manager, Editorial Production:
Hal Humphrey

Art Director: Vernon Boes

Print Buyer: Linda Hsu

Permissions Editor: Scott Bragg

Production Service: Ellen Brownstein,
Chapter Two

Text Designer: Kim Rokusek

Photo Researcher: Roman Barnes

Copy Editor: Ellen Brownstein

Illustrations: Lori Heckelman

Cover Designer: Larry Didona

Cover Image: Firefly Productions/CORBIS

Compositor: Graphic World, Inc.

For product information and technology
assistance, contact us at
**Cengage Learning Customer & Sales Support,
1-800-354-9706.**
For permission to use material from this text
or product, submit all requests online at
cengage.com/permissions.
Further permissions questions can be e-mailed to
permissionrequest@cengage.com.

Library of Congress Control Number: 2008928159
ISBN-13: 978-0-495-55888-0
ISBN-10: 0-495-55888-5

Brooks/Cole
10 Davis Drive
Belmont, CA 94002-3098
USA

Cengage Learning is a leading provider of customized learning solutions with office locations around the globe, including Singapore, the United Kingdom, Australia, Mexico, Brazil, and Japan. Locate your local office at **international.cengage.com/region.**

Cengage Learning products are represented in Canada by Nelson Education, Ltd.

For your course and learning solutions, visit **academic.cengage.com.**

Purchase any of our products at your local college store or at our preferred online store **www.ichapters.com.**

Printed in Canada
1 2 3 4 5 6 7 12 11 10 09 08

To our wives, Carol, Martha;
and our children:
Kristy and Steven,
Sarah, Heidi, and David
RDG and PDF

To my mother, Joyce, and
to the memory of my father, Earl
JDH

Contents

Preface

To the Instructor

It is with great delight that we present the tenth edition of *College Algebra*. This edition is enhanced to meet the current expectations of both students and instructors and maintains the same philosophy of the highly successful previous edition. *College Algebra* is designed to prepare students for trigonometry, calculus, statistics, or other disciplines of study. It will also equip students with the necessary skills and intellectual development for employment in our global economy.

Our goal is to write a text that

- includes solid, relevant, and easy to understand mathematics,
- stresses the important concept of function,
- uses real-life applications to motivate student learning and problem solving,
- develops critical-thinking abilities in all students, and
- develops the skills students need to continue their study of mathematics.

We believe that we have accomplished this goal through a successful blending of content and pedagogy. We present a thorough coverage of classic college algebra topics, incorporated into a contemporary framework of tested teaching strategies with carefully selected pedagogical features. In keeping with the spirit of the NCTM and AMATYC standards, this book emphasizes conceptual understanding, problem solving, and the appropriate use of technology.

New Features

1. **Numbered Objectives** *To keep students focused,* objectives are numbered at the beginning of each section and also appear as subheadings within the section.
2. **Section Openers** *To motivate students to read the section,* each section begins with a photo and a real-life situation that is appealing to this generation of students.

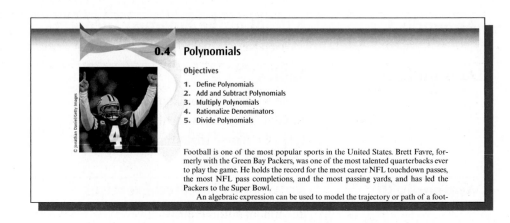

0.4 Polynomials

Objectives

1. Define Polynomials
2. Add and Subtract Polynomials
3. Multiply Polynomials
4. Rationalize Denominators
5. Divide Polynomials

Football is one of the most popular sports in the United States. Brett Favre, formerly with the Green Bay Packers, was one of the most talented quarterbacks ever to play the game. He holds the record for the most career NFL touchdown passes, the most NFL pass completions, and the most passing yards, and has led the Packers to the Super Bowl.

An algebraic expression can be used to model the trajectory or path of a foot-

© Jonathan Daniel/Getty Images

3. **Example Structure** *To help students gain a deeper understanding of how to solve each problem,* solutions begin with a stated approach.

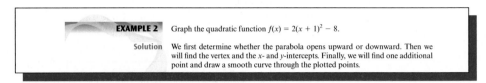

EXAMPLE 2   Graph the quadratic function $f(x) = 2(x + 1)^2 - 8$.

Solution   We first determine whether the parabola opens upward or downward. Then we will find the vertex and the x- and y-intercepts. Finally, we will find one additional point and draw a smooth curve through the plotted points.

4. **Redesigned Chapter Reviews** *To function as a thorough study guide for students,* the chapter reviews are comprehensive and now have three parts: definitions and concepts, examples, and review exercises.

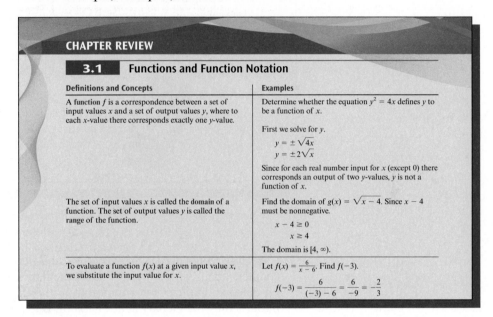

CHAPTER REVIEW

3.1   **Functions and Function Notation**

Definitions and Concepts	Examples
A function f is a correspondence between a set of input values x and a set of output values y, where to each x-value there corresponds exactly one y-value.	Determine whether the equation $y^2 = 4x$ defines y to be a function of x. First we solve for y. $y = \pm\sqrt{4x}$ $y = \pm 2\sqrt{x}$ Since for each real number input for x (except 0) there corresponds an output of two y-values, y is not a function of x.
The set of input values x is called the domain of a function. The set of output values y is called the range of the function.	Find the domain of $g(x) = \sqrt{x - 4}$. Since $x - 4$ must be nonnegative. $x - 4 \geq 0$ $x \geq 4$ The domain is $[4, \infty)$.
To evaluate a function $f(x)$ at a given input value x, we substitute the input value for x.	Let $f(x) = \dfrac{6}{x - 6}$. Find $f(-3)$. $f(-3) = \dfrac{6}{(-3) - 6} = \dfrac{6}{-9} = -\dfrac{2}{3}$

5. **Self Check Answers** *To assist instructors,* the answers to Self Checks are now printed in blue and appear next to the problem in the AIE.

Self Check 11   Solve: $\dfrac{2}{x} < 4$. $(-\infty, 0) \cup \left(\dfrac{1}{2}, \infty\right)$

6. **WebAssign® Exercises Screened in Text** *To help instructors easily identify exercises available for an online assignment,* selected exercises are screened in the AIE. A shaded blue square appears around the exercise number. The exercises are available online at www.webassign.net.

7. **Problems and Projects Available Online** *To increase student's problem-solving skills,* additional problems and projects are now made available online at http://mathematics.brookscole.com.

8. **New Design** *To increase visual interest for students,* we present a new design that is contemporary and student friendly. Color is used to highlight important features.

Updated Features

1. **Applications** *To continue the emphasis on solving problems through realistic applications,* new applications have been added and others updated. All application problems have titles.

65. Maximizing height A ball is thrown straight up from the top of a building 144 ft tall with an initial velocity of 64 ft per second. The distance $s(t)$ (in feet) of the ball from the ground is given by $s(t) = 144 + 64t - 16t^2$. Find the maximum height attained by the ball.

66. Flat-panel television sets A wholesaler of appliances finds that she can sell $(2,400 - p)$ flat-panel television sets each week when the price is p dollars. What price will maximize revenue?

67. Digital cameras A company that produces and sells digital cameras has determined that the total weekly cost C of producing x digital cameras is given by the function

$$C(x) = 1.5x^2 - 144x + 5,856$$

Determine the production level that minimizes the weekly cost for producing the digital cameras and find that weekly minimum cost.

2. **Everyday Connections** *To relate the mathematics in the chapter to student's experiences,* Everyday Connections have been added and existing material updated.

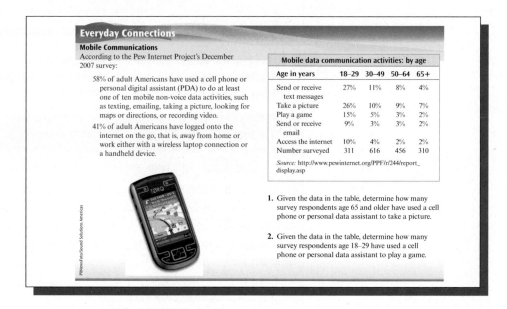

3. **Careers in Mathematics** *To encourage students to explore careers,* each chapters opens with a Careers in Mathematics feature. This content has been up-dated and utilizes information from the *Occupational Outlook Handbook* published by the Department of Labor, Bureau of Statistics. A Web address also is provided to enable students to learn more about the career. (See the beginning of each chapter.)

Content Changes

Each section in the text has been edited to fine-tune the presentation of topics for better flow of ideas and for clarity. There are many new exercises and applications. Some changes made to specific chapters include:

Chapter 0 • **Section 0.1** We expanded the discussion of sets to included subset, union, and intersection.

Chapter 1 • **Section 1.1** Identities, contradictions, and conditional equations are classified in a table.

• **Section 1.3** There is now a subsection explaining how to choose the best method for solving a quadratic equation.

Chapter 2 • **Section 2.4** The topics of symmetry and circles have been rewritten.

Chapter 3 • **Section 3.1** Finding the range of a function defined by an equation has been delayed until the range can be found graphically. Greater emphasis is placed on the difference quotient. The cubing and cube root functions have been included as basic functions.

- **Section 3.2** A table illustrating the characteristics of quadratic functions is given. Emphasis is first placed on parabolas in standard form.
- **Section 3.4** Vertical stretching and shrinking of graphs has been rewritten. Graphing using a combination of translations is now included.
- **Section 3.5** Finding vertical and horizontal asymptotes has been completely rewritten. Steps for graphing a rational function are given and shown throughout the graphing examples.

Chapter 4
- **Section 4.1** More emphasis is now placed on translations of the graphs of exponential functions.
- **Section 4.3** More emphasis is now placed on translations of the natural logarithm functions.
- **Section 4.5** Properties of natural logarithms are now included in the section.

Chapter 5
- **Section 5.1** Synthetic division is introduced earlier in the section.

Chapter 6
- **Section 6.2** Dashed lines are now included in augmented matrices.

Chapter 7
- **Section 7.1** A table showing the characteristics of a parabola in standard form is given. A greater emphasis is placed on finding the vertex, focus, and directrix of a parabola.
- **Section 7.2** Emphasis is first placed on ellipses in standard form.
- **Section 7.3** Steps for graphing an hyperbola are given and shown through examples.

Chapter 8
- **Section 8.2** Finite and infinite sequences are now defined. The definition of series has been expanded.
- **Section 8.3** Sums are clearly referred to as series. Formulas now refer to the sum of a series.
- **Section 8.5** The introduction has been reorganized and revised.

Continued Features

We have kept the pedagogical features that made the previous editions of the book so successful:

1. **Solid Mathematics** The mathematics is sound and not so rigorous that it will confuse students. All exercise sets have Vocabulary and Concept problems, Practice problems, Discovery and Writing problems, and Review problems. The book contains more than 4,000 exercises.

2. **Emphasis on Applications** To show that mathematics is useful, we include a large number of applications. An index of applications is included.

3. **Appealing to Students** The book is written for students to read and understand. The problems within each exercise set are carefully keyed to more than 400 worked examples in which author's notes explain many of the steps used in the problem-solving process. Most examples are followed by a Self Check problem, whose answers appear at the end of each section. Answers to the odd-numbered exercises appear in an appendix in the Student Edition.

4. **Use of Calculators** Throughout the text, there are an abundant number of exercises requiring calculators. These are marked with a calculator icon. Accent on Technology boxes present graphing calculator material by providing clear and concise instructions and examples. Graphing calculators are incorporated into the text, but their use is not required. All of the graphing topics are fully discussed in traditional ways.

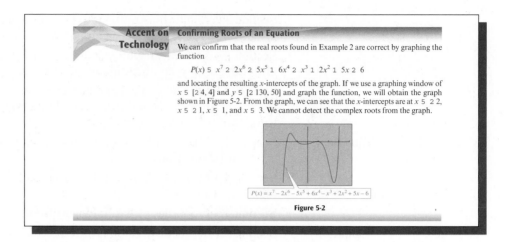

5. **Ample Review Material** A Chapter Review and a Chapter Test appear at the end of each chapter. Cumulative Review Exercises appear after every two chapters.

6. **Historical Notes** Interspersed throughout the text are pictures and captions honoring men and women who have made important contributions to mathematics.

7. **Comments** These appear throughout the text to remind students of important details.

Organization and Coverage

This text can be used in a variety of ways. To maintain optimum flexibility, many chapters are sufficiently independent to allow you to pick and choose topics that are relevant to your students. After teaching Chapters 0-3 in order, you can teach Chapters 4-9 in any order.

Ancillaries for the Instructor

Annotated Instructor's Edition (0-495-55897-4) This instructor's version of the complete student text includes the answers to each exercise printed in blue next to the respective exercise.

Complete Solutions Manual (0-495-55890-7) This manual contains solutions to all exercises from the text, including Chapter Review Exercises, Chapter Tests, and Cumulative Review Exercises.

PowerLecture (0-495-55963-6) This CD-ROM provides the instructor with dynamic media tools for teaching college algebra. PowerPoint® lecture slides and art slides of figures from the text, together with electronic files for the test bank, are available. *Solution Builder* is an electronic version of the *Complete Solutions Manual,* providing instructors with an efficient method for creating solution sets to homework and exams that can be printed or posted. *ExamView* allows you to create, deliver, and customize tests (both print and online) in minutes with this easy-to-use assessment system. *JoinIn* content for electronic response systems is tailored to *College Algebra,* Tenth Edition. Transform your classroom and assess your students' progress with instant in-class quizzes and polls. *TurningPoint®* software lets you pose book-specific questions and display students' answers seamlessly within Microsoft® PowerPoint® slides of your own lecture, in conjunction with the "clicker" hardware of your choice. Enhance your students' interactions with you, your lecture, and each other.

Test Bank (0-495-55891-5) The test bank includes 6 tests per chapter as well as 3 final exams. The tests are made up of a combination of multiple-choice, free-response, true/false, and fill-in-the-blank questions.

Text-Specific Videos (0-495-55892-3) This set of DVDs is available at no charge to qualified adopters of the text. Each tape is broken down into 10- to 20-minute problem-solving lessons that cover each section of the chapter.

Ancillaries for the Student

Enhanced WebAssign® *Enhanced WebAssign* is designed to help students do homework online. This proven and reliable system uses pedagogy and content found in the Gustafson, Frisk, and Hughes' text, and then enhances it to help students learn college algebra more effectively. Automatically graded homework allows students to focus on learning and get interactive study assistance outside of class.

Student Solutions Manual (0-495-55895-8) The *Student Solutions Manual* provides worked-out solutions to all of the odd-numbered problems in the text. It also offers hints and additional problems for practice, similar to those in the text.

Student Companion Website (0-495-55898-2) The *Website* allows students to access book- and course-specific resources, such as tutorial quizzes for each chapter, a glossary of important terms, and links to mathematics sites on the Web.

To the Student

Congratulations! You now own a state-of-the-art textbook that has been written especially for you. We have written a book that you can read and understand, and that you will find interesting. The book includes carefully written sections that will appeal to you because of our use of popular culture. The book also contains real-life applications that will help you see how mathematics is used in our world today. Our goal is for you to use this textbook and study properly so that you will be successful in learning mathematics and reaching your career goal.

The textbook contains an extensive number of worked examples with Self Checks. The answers to the Self Check problems appear at the end of each section. To get the most out of this book, it is important that you read and study the textbook often. We recommend that you work the examples on paper first, and then work the Self Checks. Once you thoroughly understand the concepts taught in the examples, you should then attempt to work the exercises.

So, grab a pencil and paper, begin reading, and dive in. We wish you well.

Acknowledgments

We are grateful to the following people who reviewed previous editions of the text or the current manuscript in its various stages. All of them had valuable suggestions that have been incorporated into this book.

Ebrahim Ahmadizadeh, Northampton Community College

Ricardo Alfaro, University of Michigan–Flint

Richard Andrews, University of Wisconsin

James Arnold, University of Wisconsin

Ronald Atkinson, Tennessee State University

Wilson Banks, Illinois State University

Chad Bemis, Riverside Community College

Anjan Biswas, Tennessee State College

Jerry Bloomberg, Essex Community College

Elaine Bouldin, Middle Tennessee State University

Dale Boye, Schoolcraft College

Eddy Joe Brackin, University of North Alabama

Susan Williams Brown, Gadsden State Community College

Jana Bryant, Manatee Community College

Lee R. Clancy, Golden West College

Krista Blevins Cohlmia, Odessa College

Jan Collins, Embry Riddle College

Cecilia Cooper, William & Harper College

John S. Cross, University of Northern Iowa

Charles D. Cunningham, Jr., James Madison University

M. Hilary Davies, University of Alaska–Anchorage

Elias Deeba, University of Houston–Downtown

Grace DeVelbiss, Sinclair Community College

Lena Dexter, Faulkner State Junior College

Emily Dickinson, University of Arkansas

Mickey P. Dunlap, University of Tennessee, Martin

Gerard G. East, Southwestern Oklahoma State University

Eric Ellis, Essex Community College

Eunice F. Everett, Seminole Community College

Dale Ewen, Parkland College

Harold Farmer, Wallace Community College–Hanceville

Ronald J. Fischer, Evergreen Valley College

Mary Jane Gates, University of Arkansas at Little Rock

Lee R. Gibson, University of Louisville

Marvin Goodman, Monmouth College

Edna Greenwood, Tarrant County College

Jerry Gustafson, Beloit College

Jerome Hahn, Bradley University

Douglas Hall, Michigan State University

Robert Hall, University of Wisconsin

David Hansen, Monterey Peninsula College

Kevin Hastings, University of Delaware

William Hinrichs, Rock Valley College

Arthur M. Hobbs, Texas A & M University

Jack E. Hofer, California Polytechnic State University

Ingrid Holzner, University of Wisconsin

Warren Jaech, Tacoma Community College

Joy St. John Johnson, Alabama A&M University

Nancy Johnson, Broward Community College

Patricia H. Jones, Methodist College

William B. Jones, University of Colorado

Barbara Juister, Elgin Community College

David Kinsey, University of Southern Indiana

Helen Kriegsman, Pittsburg State University

Marjorie O. Labhart, University of Southern Indiana

Betty J. Larson, South Dakota State University

Paul Lauritsen, Brown College

Jaclyn LeFebvre, Illinois Central College

Susan Loveland, University of Alaska–Anchorage

James Mark, Eastern Arizona College

Marcel Maupin, Oklahoma State University, Oklahoma City

Robert O. McCoy, University of Alaska–Anchorage

Judy McKinney, California Polytechnic Institute at Pomona

Sandra McLaurin, University of North Carolina

Marcus McWaters, University of Southern Florida

Donna Menard, University of Massachusetts, Dartmouth

James W. Mettler, Pennsylvania State University

Eldon L. Miller, University of Mississippi

Stuart E. Mills, Louisiana State University, Shreveport

Mila Mogilevskaya, Wichita State University

Gilbert W. Nelson, North Dakota State

Marie Neuberth, Catonsville City College

Charles Odion, Houston Community College

Anthony Peressini, University of Illinois

David L. Phillips, University of Southern Colorado

William H. Price, Middle Tennessee State University

Ronald Putthoff, University of Southern Mississippi

Brooke P. Quinlan, Hillsborough Community College

Leela Rakesh, Carnegie Mellon University

Janet P. Ray, Seattle Central Community College

Robert K. Rhea, J. Sargeant Reynolds Community College

Barbara Riggs, Tennessee Technological University

Minnie Riley, Hinds Community College

Renee Roames, Purdue University

Paul Schaefer, SUNY, Geneseo

Vincent P. Schielack, Jr., Texas A & M University

Robert Sharpton, Miami Dade Community College

L. Thomas Shiflett, Southwest Missouri State University

Richard Slinkman, Bemidji State University

Merreline Smith, California Polytechnic Institute at Pomona

John Snyder, Sinclair Community College

Sandra L. Spain, Thomas Nelson Community College

Warren Strickland, Del Mar College

Paul K. Swets, Angelo State College

Ray Tebbetts, San Antonio College

Faye Thames, Lamar State University

Douglas Tharp, University of Houston–Downtown

Carolyn A. Wailes, University of Alabama, Birmingham

Carol M. Walker, Hinds Community College

William Waller, University of Houston–Downtown

Richard H. Weil, Brown College

Carroll G. Wells, Western Kentucky University

William H. White, University of South Carolina at
 Spartanburg
Clifton Whyburn, University of Houston
Charles R. Williams, Midwestern State University

Harry Wolff, University of Wisconsin
Roger Zarnowski, Angelo State University
Albert Zechmann, University of Nebraska

We want to thank the staff at Brooks/Cole, especially Gary Whalen, Leslie Lahr, Hal Humphrey, Vernon Boes, and Cheryll Linthicum for their support in the production process. We also thank Lori Heckelman for her fine artwork, Mike Welden for his problem checking, and Diane Koenig for her proofreading. Special thanks go to Ellen Brownstein for her patience and copyediting skills, and to Graphic World for their excellent typesetting skills. Finally, we want to thank Mark McComb for contributing the Everyday Connections and Robert McCoy and Mehrdad Simkani for contributing several problems in the Exercise Sets.

R. David Gustafson
Peter D. Frisk
Jeffrey D. Hughes

A Review of Basic Algebra

In this chapter, we review many concepts and skills learned in previous algebra courses. Be sure to master this material now, because it is the basis for the rest of this course.

Careers and Mathematics

Mechanical Engineer   Mechanical engineers research, develop, design, manufacture, and test tools, engines, machines, and other mechanical devices. They also work on power-producing machines such as electrical generators, internal combustion engines, and steam and gas turbines. In addition, they develop power-using machines such as refrigeration and air-conditioning equipment, material-handling systems, elevators and escalators, industrial production equipment, and robots used in manufacturing. Mechanical engineers also design tools that other engineers need for their work and many become administrators or managers. Mechanical engineers held about 226,000 jobs in 2006.

Education   Mechanical engineers typically enter the occupation with a bachelor's degree in mechanical engineering, but some basic research positions may require a graduate degree. Engineers offering their services directly to the public must be licensed. Continuing education to keep current with rapidly changing technology is important for engineers.

Job Outlook   Employment of mechanical engineers is projected to have a 4 percent growth through 2016. The median annual salaries of mechanical engineers were $69,850 in 2006.

For a sample application, see Problem 97 in Section 0.6. For more information, see www.bls.gov/oco/oco027.htm.

© Hugo de Wolf/Alamy

1

5	3			7				
6			1	9	5			
	9	8					6	
8				6				3
4			8		3			1
7				2				6
	6					2	8	
			4	1	9			5
				8			7	9

SUDOKU

0.1 Sets of Real Numbers

Objectives

1. Identify Sets of Real Numbers
2. Identify Properties of Real Numbers
3. Graph Subsets of Real Numbers on the Number Line
4. Graph Intervals on the Number Line
5. Define Absolute Value
6. Find Distances on the Number Line

Sudoku, a game that involves number placement, is very popular. The objective is to fill a 9 by 9 grid so that each column, each row, and each of the 3 by 3 blocks contains the numbers from 1 to 9. A partially completed Sudoku grid is shown above.

To solve Sudoku puzzles, we use logic and the set of numbers, {1, 2, 3, 4, 5, 6, 7, 8, 9}. Sets of numbers are important in mathematics and we begin our study of algebra with this topic.

A **set** is a collection of objects, such as a set of dishes or a set of golf clubs. The set of vowels in the English language can be denoted as {a, e, i, o, u}, where the braces { } are read as "the set of."

If every member of one set B is also a member of another set A, we say that B is a **subset** of A. We can denote this by writing $B \subset A$, where the symbol $\subset$ is read as "is a subset of." If set B equals set A, we can write $B \subseteq A$.

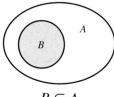
$B \subset A$

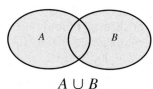
$A \cup B$

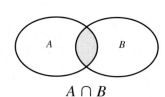
$A \cap B$

If A and B are two sets, we can form a new set consisting of all members that are in set A or set B or both. This set is called the **union** of A and B. We can denote this set by writing $A \cup B$, where the symbol $\cup$ is read as "union."

We can also form the set consisting of all members that are in both set A and set B. This set is called the **intersection** of A and B. We can denote this set by writing $A \cap B$, where the symbol $\cap$ is read as "intersection."

EXAMPLE 1 Let $A = \{a, e, i\}$, $B = \{c, d, e\}$, and $V = \{a, e, i, o, u\}$.
a. Is $A \subset V$?   **b.** Find $A \cup B$.   **c.** Find $A \cap B$.

Solution **a.** Since each member of set A is also a member of set V, $A \subset V$.

b. The union of set A and set B contains the members of set A, set B, or both. Thus, $A \cup B = \{a, c, d, e, i\}$.

c. The intersection of set A and set B contains the members that are in both set A and set B. Thus, $A \cap B = \{e\}$.

Self Check 1 **a.** Is $B \subset V$?   **b.** Find $B \cup V$.
c. Find $A \cap V$.

1. Identify Sets of Real Numbers

There are several sets of numbers that we use in everyday life.

Basic Sets of Numbers	**Natural numbers**
	The numbers that we use for counting: $\{1, 2, 3, 4, 5, 6, \ldots\}$
	Whole numbers
	The set of natural numbers including 0: $\{0, 1, 2, 3, 4, 5, 6, \ldots\}$
	Integers
	The set of whole numbers and their negatives:
	$$\{\ldots, -5, -4, -3, -2, -1, 0, 1, 2, 3, 4, 5, \ldots\}$$

In the definitions above, each group of three dots (called an *ellipsis*) indicates that the numbers continue forever in the indicated direction.

Two important subsets of the natural numbers are the *prime* and *composite* numbers. A **prime number** is a natural number greater than 1 that is divisible only by itself and 1. A **composite number** is a natural number greater than 1 that is not prime.

The set of prime numbers: $\{2, 3, 5, 7, 11, 13, 17, 19, 23, 29, 31, \ldots\}$

The set of composite numbers: $\{4, 6, 8, 9, 10, 12, 14, 15, 16, 18, 20, 21, \ldots\}$

Two important subsets of the set of integers are the *even* and *odd* integers. The **even integers** are the integers that are exactly divisible by 2. The **odd integers** are the integers that are not exactly divisible by 2.

The set of even integers: $\{\ldots, -10, -8, -6, -4, -2, 0, 2, 4, 6, 8, 10, \ldots\}$

The set of odd integers: $\{\ldots, -9, -7, -5, -3, -1, 1, 3, 5, 7, 9, \ldots\}$

So far, we have listed numbers inside braces to specify sets. This method is called the **roster method.** When we give a rule to determine which numbers are in a set, we are using **set-builder notation.** To use set-builder notation to denote the set of prime numbers, we write

$\{x \mid x \text{ is a prime number}\}$ **Read as "the set of all numbers x such that x is a prime number. Recall that when a letter stands for a number, it is called a variable.**

↑ ↑↳   ↑
variable such   rule that determines
     that   membership in the set

The fractions of arithmetic are called *rational numbers.*

Rational Numbers	**Rational numbers** are fractions that have an integer numerator and a nonzero integer denominator. Using set-builder notation, the rational numbers are
	$$\left\{ \frac{a}{b} \mid a \text{ is an integer and } b \text{ is a nonzero integer} \right\}$$

Comment

Remember that the denominator of a fraction can never be 0.

Rational numbers can be written as fractions or decimals. Some examples of rational numbers are

$$5 = \frac{5}{1}, \qquad \frac{3}{4} = 0.75, \qquad -\frac{1}{3} = -0.333\ldots, \qquad \frac{-5}{11} = -0.454545\ldots$$

The $=$ sign indicates that two quantities are equal.

These examples suggest that the decimal forms of all rational numbers are either *terminating decimals* or *repeating decimals*.

EXAMPLE 2   Determine whether the decimal form of each fraction terminates or repeats:
a. $\frac{7}{16}$ **b.** $\frac{65}{99}$

Solution   In each case, we perform a long division and write the quotient as a decimal.

a. To change $\frac{7}{16}$ to a decimal, we perform a long division to get $\frac{7}{16} = 0.4375$. Since 0.4375 terminates, we can write $\frac{7}{16}$ as a terminating decimal.

b. To change $\frac{65}{99}$ to a decimal, we perform a long division to get $\frac{65}{99} = 0.656565\ldots$. Since 0.656565. . . repeats, we can write $\frac{65}{99}$ as a repeating decimal.

We can write repeating decimals in compact form by using an overbar. For example, $0.656565\ldots = 0.\overline{65}$.

Self Check 2   Determine whether the decimal form of each fraction terminates or repeats:
a. $\frac{38}{99}$ **b.** $\frac{7}{8}$

Some numbers have decimal forms that neither terminate nor repeat. These nonterminating, nonrepeating decimals are called **irrational numbers.** Three examples of irrational numbers are

$$1.010010001000010\ldots \qquad \sqrt{2} = 1.414213562\ldots \quad \text{and} \quad \pi = 3.141592654\ldots$$

The union of the set of rational numbers (the terminating and repeating decimals) and the set of irrational numbers (the nonterminating, nonrepeating decimals) is the set of *real numbers* (the set of all decimals).

Real Numbers   A **real number** is any number that is rational or irrational. Using set-builder notation, the set of real numbers is

$$\{x \mid x \text{ is a rational or an irrational number}\}$$

Figure 0-1 shows how many of the previous sets of numbers are related.

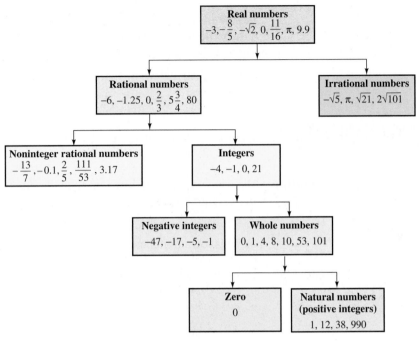

Figure 0-1

EXAMPLE 3   In the set $\{-3, -2, 0, \frac{1}{2}, 1, \sqrt{5}, 2, 4, 5, 6\}$, list all   **a.** even integers   **b.** prime numbers   **c.** rational numbers

Solution   We will check to see whether each number is a member of the set of even integers, the set of prime numbers, and the set of rational numbers.

a. even integers: $-2, 0, 2, 4, 6$

b. prime numbers: $2, 5$

c. rational numbers: $-3, -2, 0, \frac{1}{2}, 1, 2, 4, 5, 6$

Self Check 3   In the set in Example 3, list all   **a.** odd integers,   **b.** composite numbers   **c.** irrational numbers

2.   Identify Properties of Real Numbers

When we work with real numbers, we will use the following properties.

Properties of Real Numbers   If a, b, and c are real numbers,

The Commutative Properties for Addition and Multiplication

$$a + b = b + a \qquad ab = ba$$

The Associative Properties for Addition and Multiplication

$$(a + b) + c = a + (b + c) \qquad (ab)c = a(bc)$$

The Distributive Property of Multiplication over Addition or Subtraction

$$a(b + c) = ab + ac \qquad \text{or} \qquad a(b - c) = ab - ac$$

The Double Negative Rule

$$-(-a) = a$$

Comment

When the associative property is used, the order of the real numbers does not change. The real numbers that occur within the parentheses change.

The distributive property also applies when more than two terms are within parentheses.

EXAMPLE 4   Determine which property of real numbers justifies each statement.
a. $(9 + 2) + 3 = 9 + (2 + 3)$ **b.** $3(x + y + 2) = 3x + 3y + 3 \cdot 2$

Solution   We will compare the form of each statement to the forms listed in the properties of real numbers box.

a. This form matches the associative property of addition.

b. This form matches the distributive property.

Self Check 4   Determine which property of real numbers justifies each statement.
a. $mn = nm$ **b.** $(xy)z = x(yz)$
c. $p + q = q + p$

3. Graph Subsets of Real Numbers on the Number Line

We can graph subsets of real numbers on the **number line.** The number line shown in Figure 0-2 continues forever in both directions. The **positive numbers** are represented by the points to the right of 0, and the **negative numbers** are represented by the points to the left of 0.

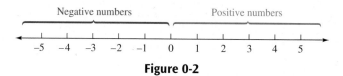

Figure 0-2

Comment

Zero is neither positive nor negative.

Comment

$\sqrt{2}$ can be shown as the diagonal of a square with sides of length 1.

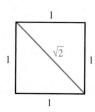

Figure 0-3(a) shows the graph of the natural numbers from 1 to 5. The point associated with each number is called the **graph** of the number, and the number is called the **coordinate** of its point.

Figure 0-3(b) shows the graph of the prime numbers that are less than 10.
Figure 0-3(c) shows the graph of the integers from -4 to 3.
Figure 0-3(d) shows the graph of the real numbers $-\frac{3}{4}$, $-\frac{7}{3}$, $0.\overline{3}$, and $\sqrt{2}$.

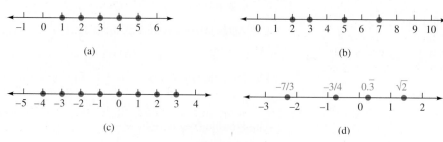

Figure 0-3

The graphs in Figure 0-3 suggest that there is a **one-to-one correspondence** between the set of real numbers and the points on a number line. This means that to each real number there corresponds exactly one point on the number line, and to each point on the number line there corresponds exactly one real-number coordinate.

EXAMPLE 5 Graph the set $\left\{-3, -\frac{4}{3}, 0, \sqrt{5}\right\}$.

Solution We will mark (plot) each number on the number line. To the nearest tenth, $\sqrt{5} = 2.2$.

Self Check 5 Graph the set $\left\{-2, \frac{3}{4}, \sqrt{3}\right\}$. (*Hint:* To the nearest tenth, $\sqrt{3} = 1.7$.)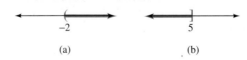

4. Graph Intervals on the Number Line

To show that two quantities are not equal, we can use an **inequality symbol.**

Symbol	Read as	Examples		
$\neq$	"is not equal to"	$5 \neq 8$	and	$0.25 \neq \frac{1}{3}$
$<$	"is less than"	$12 < 20$	and	$0.17 < 1.1$
$>$	"is greater than"	$15 > 9$	and	$\frac{1}{2} > 0.2$
$\leq$	"is less than or equal to"	$25 \leq 25$	and	$1.7 \leq 2.3$
$\geq$	"is greater than or equal to"	$19 \geq 19$	and	$15.2 \geq 13.7$
$\approx$	"is approximately equal to"	$\sqrt{2} \approx 1.414$	and	$\sqrt{3} \approx 1.732$

It is possible to write an inequality with the inequality symbol pointing in the opposite direction. For example,

$12 < 20$	is equivalent to	$20 > 12$
$2.3 \geq -1.7$	is equivalent to	$-1.7 \leq 2.3$

In Figure 0-2, the coordinates of points get larger as we move from left to right on a number line. Thus, if a and b are the coordinates of two points, the one to the right is the greater. This suggests the following facts:

If $a > b$, point a lies to the right of point b on a number line.
If $a < b$, point a lies to the left of point b on a number line.

Figure 0-4(a) shows the graph of the **inequality** $x > -2$ (or $-2 < x$). This graph includes all real numbers x that are greater than -2. The parenthesis at -2 indicates that -2 is not included in the graph. Figure 0-4(b) shows the graph of $x \leq 5$ (or $5 \geq x$). The bracket at 5 indicates that 5 is included in the graph.

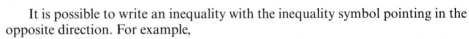

(a) (b)

Figure 0-4

Sometimes two inequalities can be written as a single expression called a **compound inequality.** For example, the compound inequality

$$5 < x < 12$$

Pythagoras of Samos
(569?-475?BC)
Pythagoras is thought to be the world's first pure mathematician. Although he is famous for the theorem that bears his name, he is often called "the father of music," because a society he led discovered some of the fundamentals of musical harmony. This secret society had numerology as its religion. The society is also credited with the discovery of irrational numbers.

is a combination of the inequalities $5 < x$ and $x < 12$. It is read as "5 is less than x, and x is less than 12," and it means that x is between 5 and 12. Its graph is shown in Figure 0-5.

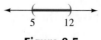

Figure 0-5

The graphs shown in Figures 0-4 and 0-5 are portions of a number line called **intervals.** The interval shown in Figure 0-6(a) is denoted by the inequality $-2 < x < 4$, or in *interval notation* as $(-2, 4)$. The parentheses indicate that the endpoints are not included. The interval shown in Figure 0-6(b) is denoted by the inequality $x > 1$, or as $(1, \infty)$ in interval notation.

$(-2, 4)$
$-2 < x < 4$
x is between −2 and 4.

(a)

$(1, \infty)$
$x > 1$
x is greater than 1.

(b)

Figure 0-6

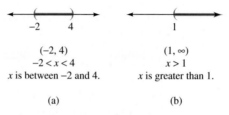

Comment

The symbol ∞ (infinity) is not a real number. It is used to indicate that the graph in Figure 0-6(b) extends infinitely far to the right.

A compound inequality such as $-2 < x < 4$ can be written as two separate inequalities:

$$x > -2 \qquad \text{and} \qquad x < 4$$

This expression represents the **intersection** of two intervals. In interval notation, this expression can be written as

$$(-2, \infty) \cap (-\infty, 4) \quad \text{Read the symbol } \cap \text{ as "intersection."}$$

Since the graph of $-2 < x < 4$ will include all points whose coordinates satisfy both $x > -2$ and $x < 4$ at the same time, its graph will include all points that are larger than -2 but less than 4. This is the interval $(-2, 4)$, whose graph is shown in Figure 0-6(a).

EXAMPLE 6 Write the inequality $-3 < x < 5$ in interval notation and graph it.

Solution This is the interval $(-3, 5)$. Its graph includes all real numbers between -3 and 5, as shown in Figure 0-7.

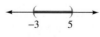

Figure 0-7

Self Check 6 Write the inequality $x < 5$ in interval notation and graph it.

If an interval extends forever in one direction, it is called an **unbounded interval.**

Unbounded Intervals		
Interval	**Inequality**	**Graph**
(a, ∞)	$x > a$	
$(2, \infty)$	$x > 2$	
$[a, \infty)$	$x \geq a$	
$[2, \infty)$	$x \geq 2$	
$(-\infty, a)$	$x < a$	
$(-\infty, 2)$	$x < 2$	
$(-\infty, a]$	$x \leq a$	
$(-\infty, 2]$	$x \leq 2$	
$(-\infty, \infty)$	$-\infty < x > \infty$	

A bounded interval with no endpoints is called an **open interval.** Figure 0-8(a) shows the open interval between -3 and 2. A bounded interval with one endpoint is called a **half-open interval.** Figure 0-8(b) shows the half-open interval between -2 and 3, including -2.

Intervals that include two endpoints are called **closed intervals.** Figure 0-9 shows the graph of a closed interval from -2 to 4.

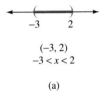

$(-3, 2)$
$-3 < x < 2$

(a)

$[-2, 3)$
$-2 \leq x < 3$

(b)

Figure 0-8

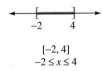

$[-2, 4]$
$-2 \leq x \leq 4$

Figure 0-9

Open Intervals, Half-Open Intervals, and Closed Intervals		
Interval	**Inequality**	**Graph**
Open		
(a, b)	$a < x < b$	
$(-2, 3)$	$-2 < x < 3$	
Half-Open		
$[a, b)$	$a \leq x < b$	
$[-2, 3)$	$-2 \leq x < 3$	
Half-Open		
$(a, b]$	$a < x \leq b$	
$(-2, 3]$	$-2 < x \leq 3$	
Closed		
$[a, b]$	$a \leq x \leq b$	
$[-2, 3]$	$-2 \leq x \leq 3$	

EXAMPLE 7 Write the inequality $3 \leq x$ in interval notation and graph it.

Solution The inequality $3 \leq x$ can be written in the form $x \geq 3$. This is the interval $[3, \infty)$. Its graph includes all real numbers greater than or equal to 3, as shown in Figure 0-10.

3

Figure 0-10

Self Check 7 Write the inequality $-2 < x \leq 5$ in interval notation and graph it.

In Example 8, we will graph a closed interval.

EXAMPLE 8 Write the inequality $5 \geq x \geq -1$ in interval notation and graph it.

Solution The inequality $5 \geq x \geq -1$ can be written in the form

$$-1 \leq x \leq 5$$

This is the interval $[-1, 5]$. Its graph includes all real numbers from -1 to 5. The graph is shown in Figure 0-11.

-1 5

Figure 0-11

Self Check 8 Write the inequality $0 \leq x \leq 3$ in interval notation and graph it.

The expression

$$x < -2 \quad \text{or} \quad x \geq 3 \qquad \text{Read as "}x\text{ is less than }-2\text{ or }x\text{ is greater than or equal to 3."}$$

represents the **union** of two intervals. In interval notation, it is written as

$$(-\infty, -2) \cup [3, \infty) \qquad \text{Read the symbol } \cup \text{ as "union."}$$

Its graph is shown in Figure 0-12.

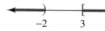

-2 3

Figure 0-12

5. Define Absolute Value

The **absolute value** of a real number x (denoted as $|x|$) is the distance on a number line between 0 and the point with a coordinate of x. For example, points with coordinates of 4 and -4 both lie four units from 0, as shown in Figure 0-13. Therefore, it follows that

$$|-4| = |4| = 4$$

4 units   4 units

$-5 \ -4 \ -3 \ -2 \ -1 \ \ 0 \ \ 1 \ \ 2 \ \ 3 \ \ 4 \ \ 5$

Figure 0-13

In general, for any real number x,

$$|-x| = |x|$$

We can define absolute value algebraically as follows.

Absolute Value If x is a real number, then

$$|x| = x \qquad \text{when } x \geq 0$$
$$|x| = -x \qquad \text{when } x < 0$$

This definition indicates that when x is positive or 0, then x is its own absolute value. However, when x is negative, then $-x$ (which is positive) is its absolute value. Thus, $|x|$ is always nonnegative.

$$|x| \geq 0 \quad \text{for all real numbers } x$$

EXAMPLE 9   Write each number without using absolute value symbols: **a.** $|3|$ **b.** $|-4|$ **c.** $|0|$ **d.** $-|-8|$

Solution   In each case, we will use the definition of absolute value.

a. $|3| = 3$ **b.** $|-4| = 4$ **c.** $|0| = 0$ **d.** $-|-8| = -(8) = -8$

Self Check 9   Write each number without using absolute value symbols: **a.** $|-10|$ **b.** $|12|$ **c.** $-|6|$

In Example 10, we must determine whether the number inside the absolute value symbol is positive or negative.

EXAMPLE 10   Write each number without using absolute value symbols: **a.** $|\pi-1|$ **b.** $|2 - \pi|$ **c.** $|2 - x|$ if $x \geq 5$

Solution   **a.** Since $\pi \approx 3.1416$, $\pi-1$ is positive, and $\pi-1$ is its own absolute value.

$$|\pi - 1| = \pi - 1$$

b. Since $2 - \pi$ is negative, its absolute value is $-(2 - \pi)$.

$$|2 - \pi| = -(2 - \pi) = -2 - (-\pi) = -2 + \pi = \pi - 2$$

c. Since $x \geq 5$, the expression $2 - x$ is negative, and its absolute value is $-(2 - x)$.

$$|2 - x| = -(2 - x) = -2 + x = x - 2 \quad \text{provided } x \geq 5$$

Self Check 10   Write each number without using absolute value symbols. $\left(\textit{Hint: } \sqrt{5} \approx 2.236.\right)$ **a.** $\left|2 - \sqrt{5}\right|$ **b.** $|2 - x|$ if $x \leq 1$

6.   Find Distances on the Number Line

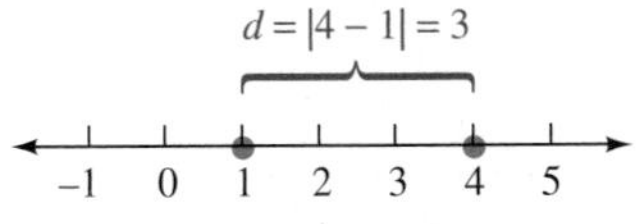

$$d = |4 - 1| = 3$$

Figure 0-14

On the number line shown in Figure 0-14, the distance between the points with coordinates of 1 and 4 is $4 - 1$, or 3 units. However, if the subtraction were done in the other order, the result would be $1 - 4$, or -3 units. To guarantee that the distance d between two points is always positive, we can use absolute value symbols. Thus, the distance d between two points with coordinates of 1 and 4 is

$$d = |4 - 1| = |1 - 4| = 3$$

In general, we have the following definition for the distance between two points on the number line.

Distance between Two Points   If a and b are the coordinates of two points on the number line, the distance between the points is given by the formula

$$d = |b - a|$$

EXAMPLE 11   Find the distance on a number line between points with coordinates of
a. 3 and 5   b. -2 and 3   c. -5 and -1

Solution   We will use the formula for finding the distance between two points on the number line.

a. $d = |5 - 3| = |2| = 2$
b. $d = |3 - (-2)| = |3 + 2| = |5| = 5$
c. $d = |-1 - (-5)| = |-1 + 5| = |4| = 4$

Self Check 11   Find the distance on a number line between points with coordinates of
a. 4 and 10   b. -2 and -7

Self Check Answers   1. a. no   b. {a, c, d, e, i, o, u}   c. {a, e, i}   2. a. repeats   b. terminates
3. a. $-3, 1, 5$ b. $4, 6$ c. $\sqrt{5}$ 4. a. commutative property of
multiplication   b. associative property of multiplication
c. commutative property of addition   5.

6. $(-\infty, 5)$,

7. $(-2, 5]$,

8. $[0, 3]$,

9. a. 10   b. 12   c. -6 10. a. $\sqrt{5} - 2$
b. $2 - x$ 11. a. 6   b. 5

0.1   Exercises

VOCABULARY AND CONCEPTS   *Fill in the blanks.*

1. A ___ is a collection of objects.

2. If every member of one set B is also a member of a second set A, then B is called a _____ of A.

3. If A and B are two sets, the set that contains all members that are in sets A and B or both is called the _____ of A and B.

4. If A and B are two sets, the set that contains all members that are in both sets is called the _____ of A and B.

5. A real number is any number that can be expressed as a _____.

6. A _____ is a letter that is used to represent a number.

7. The smallest prime number is __.

8. All integers that are exactly divisible by 2 are called ____ integers.

9. Natural numbers greater than 1 that are not prime are called _____ numbers.

10. Fractions such as $\frac{2}{3}, \frac{8}{2}$, and $-\frac{7}{9}$ are called _____ numbers.

11. Irrational numbers are _____ that don't terminate and don't repeat.

12. The symbol __ is read as "is less than or equal to."

13. On a number line, the _____ numbers are to the left of 0.

14. The only integer that is neither positive nor negative is __.

15. The associative property of addition states that $(x + y) + z =$ _____.

16. The commutative property of multiplication states that $xy =$ ___.

17. Use the distributive property to complete the statement: $5(m + 2) =$ _____.

18. The statement $(m + n)p = p(m + n)$ illustrates the _____ property of _____.

19. The graph of an _____ is a portion of a number line.

20. The graph of an open interval has ___ endpoints.

21. The graph of a closed interval has ____ endpoints.

22. The graph of a _____ interval has one endpoint.

23. Except for 0, the absolute value of every number is _____.

24. The _____ between two distinct points on a number line is always positive.

Let

$\mathbf{N} =$ *the set of natural numbers*
$\mathbf{W} =$ *the set of whole numbers*
$\mathbf{Z} =$ *the set of integers*
$\mathbf{Q} =$ *the set of rational numbers*
$\mathbf{R} =$ *the set of real numbers*

Determine whether each statement is true or false.
Read the symbol $\subset$ as "is a subset of."

25. $\mathbf{N} \subset \mathbf{W}$ **26.** $\mathbf{Q} \subset \mathbf{R}$
27. $\mathbf{Q} \subset \mathbf{N}$ **28.** $\mathbf{Z} \subset \mathbf{Q}$
29. $\mathbf{W} \subset \mathbf{Z}$ **30.** $\mathbf{R} \subset \mathbf{Z}$

PRACTICE   *Let $A = \{a, b, c, d, e\}$, $B = \{d, e, f, g\}$, and $C = \{a, c, e, f\}$. Find each set.*

31. $A \cup B$ **32.** $A \cap B$
33. $A \cap C$ **34.** $B \cup C$

Consider the following set:
$\left\{-5, -4, -\frac{2}{3}, 0, 1, \sqrt{2}, 2, 2.75, 6, 7\right\}$.

35. Which numbers are natural numbers?
36. Which numbers are whole numbers?
37. Which numbers are integers?
38. Which numbers are rational numbers?

39. Which numbers are irrational numbers?
40. Which numbers are prime numbers?
41. Which numbers are composite numbers?
42. Which numbers are even integers?
43. Which numbers are odd integers?
44. Which numbers are negative numbers?

Graph each subset of the real numbers on a number line.

45. The natural numbers between 1 and 5

46. The composite numbers less than 10

47. The prime numbers between 10 and 20

48. The integers from -2 to 4

49. The integers between -5 and 0

50. The even integers between -9 and -1

51. The odd integers between -6 and 4

52. -0.7, 1.75, and $3\frac{7}{8}$

Write each inequality in interval notation and graph the interval.

53. $x > 2$ **54.** $x < 4$

55. $0 < x < 5$ **56.** $-2 < x < 3$

57. $x > -4$ **58.** $x < 3$

59. $-2 \leq x < 2$ **60.** $-4 < x \leq 1$

61. $x \leq 5$ **62.** $x \geq -1$

63. $-5 < x \leq 0$ **64.** $-3 \leq x < 4$

65. $-2 \leq x \leq 3$ **66.** $-4 \leq x \leq 4$

67. $6 \geq x \geq 2$ **68.** $3 \geq x \geq -2$

Write each pair of inequalities as the intersection of two intervals and graph the result.

69. $x > -5$ and $x < 4$

70. $x \geq -3$ and $x < 6$

71. $x \geq -8$ and $x \leq -3$

72. $x > 1$ and $x \leq 7$

Write each inequality as the union of two intervals and graph the result.

73. $x < -2$ or $x > 2$

74. $x \leq -5$ or $x > 0$

75. $x \leq -1$ or $x \geq 3$

76. $x < -3$ or $x \geq 2$

Write each expression without using absolute value symbols.

77. $|13|$ **78.** $|-17|$

79. $|0|$ **80.** $-|63|$

81. $-|-8|$ **82.** $|-25|$

83. $-|32|$ **84.** $-|-6|$

85. $|\pi - 5|$ **86.** $|8 - \pi|$

87. $|\pi - \pi|$ **88.** $|2\pi|$

89. $|x + 1|$ and $x \geq 2$ **90.** $|x + 1|$ and $x \leq -2$

91. $|x - 4|$ and $x < 0$ **92.** $|x - 7|$ and $x > 10$

Find the distance between each pair of points on the number line.

93. 3 and 8   **94.** -5 and 12

95. -8 and -3 **96.** 6 and -20

APPLICATIONS

97. What subset of the real numbers would you use to describe the populations of several cities?

98. What subset of the real numbers would you use to describe the subdivisions of an inch on a ruler?

99. What subset of the real numbers would you use to report temperatures in several cities?

100. What subset of the real numbers would you use to describe the financial condition of a business?

DISCOVERY AND WRITING

101. Explain why $-x$ could be positive.

102. Explain why every integer is a rational number.

103. Is the statement $|ab| = |a| \cdot |b|$ always true? Explain.

104. Is the statement $\left| \dfrac{a}{b} \right| = \dfrac{|a|}{|b|}$ $(b \neq 0)$ always true? Explain.

105. Is the statement $|a + b| = |a| + |b|$ always true? Explain.

106. Under what conditions will the statement given in Exercise 105 be true?

107. Explain why it is incorrect to write $a < b > c$ if $a < b$ and $b > c$.

108. Explain why $|b - a| = |a - b|$.

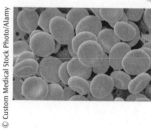

0.2 ## Integer Exponents and Scientific Notation

Objectives

1. Define Natural-Number Exponents
2. Apply the Rules of Exponents
3. Apply the Rules for Order of Operations to Evaluate Expressions
4. Express Numbers in Scientific Notation
5. Use Scientific Notation to Simplify Computations

The number of cells in the human body is approximated to be one hundred trillion or 100,000,000,000,000. One hundred trillion is $(10)(10)(10) \cdot \cdots \cdot (10)$, where ten occurs fourteen times. Fourteen factors of ten can be written as 10^{14}.

In this section, we will use integer exponents to represent repeated multiplication of numbers.

1. Define Natural-Number Exponents

When two or more quantities are multiplied together, each quantity is called a **factor** of the product. The exponential expression x^4 indicates that x is to be used as a factor four times.

$$\overbrace{x^4 = x \cdot x \cdot x \cdot x}^{\text{4 factors of } x}$$

In general, the following is true.

Natural-Number Exponents

For any natural number n,

$$x^n = \overbrace{x \cdot x \cdot x \cdot \,\cdots\, \cdot x}^{n \text{ factors of } x}$$

In the **exponential expression** x^n, x is called the **base**, and n is called the **exponent** or the **power** to which the base is raised. The expression x^n is called a **power of x**. From the definition, we see that a natural-number exponent indicates how many times the base of an exponential expression is to be used as a factor in a product. If an exponent is 1, the 1 is usually not written:

$$x^1 = x$$

EXAMPLE 1 Write each expression without using exponents: **a.** 4^2 **b.** $(-4)^2$ **c.** -5^3 **d.** $(-5)^3$ **e.** $3x^4$ **f.** $(3x)^4$

Solution In each case, we apply the definition of natural-number exponents.

a. $4^2 = 4 \cdot 4 = 16$ Read 4^2 as "four squared."
b. $(-4)^2 = (-4)(-4) = 16$ Read $(-4)^2$ as "negative four squared."
c. $-5^3 = -5(5)(5) = -125$ Read -5^3 as "the negative of five cubed."
d. $(-5)^3 = (-5)(-5)(-5) = -125$ Read $(-5)^3$ as "negative five cubed."
e. $3x^4 = 3 \cdot x \cdot x \cdot x \cdot x$ Read $3x^4$ as "3 times x to the fourth power."
f. $(3x)^4 = (3x)(3x)(3x)(3x) = 81 \cdot x \cdot x \cdot x \cdot x$ Read $(3x)^4$ as "3x to the fourth power."

Self Check 1 Write each expression without using exponents: **a.** 7^3 **b.** $(-3)^2$ **c.** $5a^3$ **d.** $(5a)^4$

Comment

Note the distinction between ax^n and $(ax)^n$:

$$ax^n = a \cdot \overbrace{x \cdot x \cdot x \cdot \,\cdots\, \cdot x}^{n \text{ factors of } x} \qquad\qquad (ax)^n = \overbrace{(ax)(ax)(ax) \cdot \,\cdots\, \cdot (ax)}^{n \text{ factors of } ax}$$

Also note the distinction between $-x^n$ and $(-x)^n$:

$$-x^n = -(\overbrace{x \cdot x \cdot x \cdot \,\cdots\, \cdot x}^{n \text{ factors of } x}) \qquad\qquad (-x)^n = \overbrace{(-x)(-x)(-x) \cdot \,\cdots\, \cdot (-x)}^{n \text{ factors of } -x}$$

Accent on Technology

Using a Calculator to Find Powers

We can use calculators to find powers of numbers. For example, to find 2.35^3 with a scientific calculator, we enter these numbers and press these keys:

$$2.35 \boxed{y^x} \; 3 \; \boxed{=}$$

The display will read $\boxed{12.977875}$.

To find 2.35^3 with a graphing calculator, we enter these numbers and press these keys:

$$2.35 \; \boxed{\wedge} \; 3 \; \boxed{\text{ENTER}}$$

The display will read
$$\begin{array}{l} 2.35\text{^}3 \\ \qquad\qquad 12.977875 \end{array}$$

In either case, $2.35^3 = 12.977875$.

2. Apply the Rules of Exponents

We begin the review of the rules of exponents by considering the product $x^m x^n$. Since x^m indicates that x is to be used as a factor m times, and since x^n indicates that x is to be used as a factor n times, there are $m + n$ factors of x in the product $x^m x^n$.

$$\overset{\substack{m + n \text{ factors of } x}}{\overbrace{x^m x^n = \underbrace{x \cdot x \cdot x \cdot \;\cdots\; \cdot x}_{m \text{ factors of } x} \cdot \underbrace{x \cdot x \cdot x \cdot \;\cdots\; \cdot x}_{n \text{ factors of } x}}} = x^{m+n}$$

This suggests that to multiply exponential expressions with the same base, we *keep the base and add the exponents.*

Product Rule for Exponents

If m and n are natural numbers, then

$$x^m x^n = x^{m+n}$$

Comment

The product rule applies to exponential expressions with the same base. A product of two powers with different bases, such as $x^4 y^3$, cannot be simplified.

To find another property of exponents, we consider the exponential expression $(x^m)^n$. In this expression, the exponent n indicates that x^m is to be used as a factor n times. This implies that x is to be used as a factor mn times.

$$\overset{\substack{mn \text{ factors of } x}}{\overbrace{(x^m)^n = \underbrace{(x^m)(x^m)(x^m) \cdot \;\cdots\; \cdot (x^m)}_{n \text{ factors of } x^m}}} = x^{mn}$$

This suggests that to raise an exponential expression to a power, we *keep the base and multiply the exponents.*

To raise a product to a power, we raise each factor to that power.

$$(xy)^n = \underbrace{(xy)(xy)(xy) \cdot \;\cdots\; \cdot (xy)}_{n \text{ factors of } xy} = (\underbrace{x \cdot x \cdot x \cdot \;\cdots\; \cdot x}_{n \text{ factors of } x})(\underbrace{y \cdot y \cdot y \cdot \;\cdots\; \cdot y}_{n \text{ factors of } y}) = x^n y^n$$

To raise a fraction to a power, we raise both the numerator and the denominator to that power. If $y \neq 0$, then

$$
\left(\frac{x}{y}\right)^n = \overbrace{\left(\frac{x}{y}\right)\left(\frac{x}{y}\right)\left(\frac{x}{y}\right) \cdot \ \cdots \ \cdot \left(\frac{x}{y}\right)}^{n \text{ factors of } \frac{x}{y}}
$$

$$
= \frac{\overbrace{xxx \cdot \ \cdots \ \cdot x}^{n \text{ factors of } x}}{\underbrace{yyy \cdot \ \cdots \ \cdot y}_{n \text{ factors of } y}}
$$

$$
= \frac{x^n}{y^n}
$$

The previous three results are called the **power rules of exponents.**

Power Rules of Exponents   If m and n are natural numbers, then

$$(x^m)^n = x^{mn} \qquad (xy)^n = x^n y^n \qquad \left(\frac{x}{y}\right)^n = \frac{x^n}{y^n} \quad (y \neq 0)$$

EXAMPLE 2   Simplify: **a.** $x^5 x^7$ **b.** $x^2 y^3 x^5 y$ **c.** $(x^4)^9$ **d.** $(x^2 x^5)^3$ **e.** $\left(\dfrac{x}{y^2}\right)^5$ **f.** $\left(\dfrac{5x^2 y}{z^3}\right)^2$

Solution   In each case, we will apply the appropriate rule of exponents.

a. $x^5 x^7 = x^{5+7} = x^{12}$ 　　　　　　　　　　**b.** $x^2 y^3 x^5 y = x^{2+5} y^{3+1} = x^7 y^4$

c. $(x^4)^9 = x^{4 \cdot 9} = x^{36}$ 　　　　　　　　　　**d.** $(x^2 x^5)^3 = (x^7)^3 = x^{21}$

e. $\left(\dfrac{x}{y^2}\right)^5 = \dfrac{x^5}{(y^2)^5} = \dfrac{x^5}{y^{10}} \quad (y \neq 0)$

f. $\left(\dfrac{5x^2 y}{z^3}\right)^2 = \dfrac{5^2 (x^2)^2 y^2}{(z^3)^2} = \dfrac{25 x^4 y^2}{z^6} \quad (z \neq 0)$

Self Check 2   Simplify: **a.** $(y^3)^2$ **b.** $(a^2 a^4)^3$ **c.** $(x^2)^3 (x^3)^2$
d. $\left(\dfrac{3a^3 b^2}{c^3}\right)^3 \quad (c \neq 0)$

If we assume that the rules for natural-number exponents hold for exponents of 0, we can write

$$x^0 x^n = x^{0+n} = x^n = 1x^n$$

Since $x^0 x^n = 1x^n$, it follows that if $x \neq 0$, then $x^0 = 1$.

Zero Exponent $x^0 = 1 \quad (x \neq 0)$

If we assume that the rules for natural-number exponents hold for exponents that are negative integers, we can write

$$x^{-n}x^n = x^{-n+n} = x^0 = 1 \quad (x \neq 0)$$

However, we know that

$$\frac{1}{x^n} \cdot x^n = 1 \quad (x \neq 0) \qquad \tfrac{1}{x^n} \cdot x^n = \tfrac{x^n}{x^n}, \text{ and any nonzero number divided by itself is 1.}$$

Since $x^{-n}x^n = \dfrac{1}{x^n} \cdot x^n$, it follows that $x^{-n} = \dfrac{1}{x^n} \quad (x \neq 0)$.

Negative Exponents   If n is an integer and $x \neq 0$, then

$$x^{-n} = \frac{1}{x^n} \qquad \text{and} \qquad \frac{1}{x^{-n}} = x^n$$

Because of the previous definitions, all of the rules for natural-number exponents will hold for integer exponents.

EXAMPLE 3   Simplify and write all answers without using negative exponents: **a.** $(3x)^0$
b. $3(x^0)$ **c.** x^{-4} **d.** $\dfrac{1}{x^{-6}}$ **e.** $x^{-3}x$ **f.** $(x^{-4}x^8)^{-5}$

Solution   We will use the definitions of zero exponent and negative exponents to simplify each expression.

a. $(3x)^0 = 1$ **b.** $3(x^0) = 3(1) = 3$ **c.** $x^{-4} = \dfrac{1}{x^4}$

d. $\dfrac{1}{x^{-6}} = x^6$ **e.** $x^{-3}x = x^{-3+1}$ **f.** $(x^{-4}x^8)^{-5} = (x^4)^{-5}$
$$= x^{-2} \qquad\qquad = x^{-20}$$
$$= \frac{1}{x^2} \qquad\qquad = \frac{1}{x^{20}}$$

Self Check 3   Simplify and write all answers without using negative exponents: **a.** $7a^0$
b. $3a^{-2}$ **c.** $a^{-4}a^2$ **d.** $(a^3a^{-7})^3$

To develop the quotient rule for exponents, we proceed as follows:

$$\frac{x^m}{x^n} = x^m\left(\frac{1}{x^n}\right) = x^m x^{-n} = x^{m+(-n)} = x^{m-n} \quad (x \neq 0)$$

This suggests that to divide two exponential expressions with the same nonzero base, we *keep the base and subtract the exponent in the denominator from the exponent in the numerator.*

Quotient Rule for Exponents   If m and n are integers, then

$$\frac{x^m}{x^n} = x^{m-n} \quad (x \neq 0)$$

EXAMPLE 4   Simplify and write all answers without using negative exponents:

a. $\dfrac{x^8}{x^5}$ b. $\dfrac{x^2 x^4}{x^{-5}}$

Solution   We will apply the product and quotient rules of exponents.

a. $\dfrac{x^8}{x^5} = x^{8-5}$

  $= x^3$

b. $\dfrac{x^2 x^4}{x^{-5}} = \dfrac{x^6}{x^{-5}}$

  $= x^{6-(-5)}$

  $= x^{11}$

Self Check 4   Simplify and write all answers without using negative exponents:

a. $\dfrac{x^{-6}}{x^2}$ b. $\dfrac{x^4 x^{-3}}{x^2}$

EXAMPLE 5   Simplify and write all answers without using negative exponents:

a. $\left(\dfrac{x^3 y^{-2}}{x^{-2} y^3}\right)^{-2}$ b. $\left(\dfrac{x}{y}\right)^{-n}$

Solution   We will apply the appropriate rules of exponents.

a. $\left(\dfrac{x^3 y^{-2}}{x^{-2} y^3}\right)^{-2} = (x^{3-(-2)} y^{-2-3})^{-2}$

  $= (x^5 y^{-5})^{-2}$

  $= x^{-10} y^{10}$

  $= \dfrac{y^{10}}{x^{10}}$

b. $\left(\dfrac{x}{y}\right)^{-n} = \dfrac{x^{-n}}{y^{-n}}$

  $= \dfrac{x^{-n} x^n y^n}{y^{-n} x^n y^n}$ **Multiply numerator and denominator by 1 in the form $\frac{x^n y^n}{x^n y^n}$.**

  $= \dfrac{x^0 y^n}{y^0 x^n}$ $x^{-n} x^n = x^0$ and $y^{-n} y^n = y^0$.

  $= \dfrac{y^n}{x^n}$ $x^0 = 1$ and $y^0 = 1$.

  $= \left(\dfrac{y}{x}\right)^n$

Self Check 5   Simplify and write all answers without using negative exponents:

a. $\left(\dfrac{x^4 y^{-3}}{x^{-3} y^2}\right)^2$ b. $\left(\dfrac{2a}{3b}\right)^{-3}$

Part b of Example 5 establishes the following rule.

Fractions to a Negative Power	If n is a natural number, then $$\left(\frac{x}{y}\right)^{-n} = \left(\frac{y}{x}\right)^{n} \quad (x \neq 0 \quad \text{and} \quad y \neq 0)$$

3. Apply the Rules for Order of Operations to Evaluate Expressions

When several operations occur in an expression, we must perform the operations in the following order to get the correct result.

Order of Operations	If an expression does not contain grouping symbols such as parentheses or brackets, follow these steps:

1. Find the values of any exponential expressions.
2. Perform all multiplications and/or divisions, working from left to right.
3. Perform all additions and/or subtractions, working from left to right.

If an expression contains grouping symbols such as parentheses, brackets, or braces, use the rules above to perform the calculations within each pair of grouping symbols, working from the innermost pair to the outermost pair.

In a fraction, simplify the numerator and the denominator of the fraction separately. Then simplify the fraction, if possible.

Comment

Many students remember the order of operations rule with the acronym PEMDAS:

  Parentheses
  Exponents
  Multiplication/Division
  Addition/Subtraction

For example, to simplify $\dfrac{3[4 - (6 + 10)]}{2^2 - (6 + 7)}$, we proceed as follows:

$$\frac{3[4 - (6 + 10)]}{2^2 - (6 + 7)} = \frac{3(4 - 16)}{2^2 - (6 + 7)}$$ Simplify within the inner parentheses: $6 + 10 = 16$.

$$= \frac{3(-12)}{2^2 - 13}$$ Simplify within the parentheses: $4 - 16 = -12$ and $6 + 7 = 13$.

$$= \frac{3(-12)}{4 - 13}$$ Evaluate the power: $2^2 = 4$.

$$= \frac{-36}{-9}$$ $3(-12) = -36; 4 - 13 = -9$

$$= 4$$

EXAMPLE 6   If $x = -2$, $y = 3$, and $z = -4$, evaluate:

a. $-x^2 + y^2z$ **b.** $\dfrac{2z^3 - 3y^2}{5x^2}$

Solution   In each part, we will substitute the numbers for the variables, apply the rules of order of operations, and simplify.

a. $-x^2 + y^2z = -(-2)^2 + 3^2(-4)$

$$= -(4) + 9(-4)$$ Evaluate the powers.

$$= -4 + (-36)$$ Do the multiplication.

$$= -40$$ Do the addition.

b. $\dfrac{2z^3 - 3y^2}{5x^2} = \dfrac{2(-4)^3 - 3(3)^2}{5(-2)^2}$

$= \dfrac{2(-64) - 3(9)}{5(4)}$ Evaluate the powers.

$= \dfrac{-128 - 27}{20}$ Do the multiplications.

$= \dfrac{-155}{20}$ Do the subtraction.

$= -\dfrac{31}{4}$ Simplify the fraction.

Self Check 6 If $x = 3$ and $y = -2$, evaluate $\dfrac{2x^2 - 3y^2}{x - y}$.

4. Express Numbers in Scientific Notation

Scientists often work with numbers that are very large or very small. These numbers can be written compactly by expressing them in *scientific notation*.

Scientific Notation A number is written in **scientific notation** when it is written in the form

$$N \times 10^n$$

where $1 \le |N| < 10$ and n is an integer.

Emmy Amalie Noether
(1882-1935)
Emmy Noether is best known for her work in abstract algebra. Her work in the study of invariants led to concepts that Albert Einstein used in his theory of relativity. Einstein described her as the most creative female genius since the beginning of higher education for women. In 1933, she lost her position at the University of Göttingen because of Nazi pressure. She came to the United States and taught at Bryn Mawr College and lectured at the Institute for Advance Study at Princeton.

Light travels 29,980,000,000 centimeters per second. To express this number in scientific notation, we must write it as the product of a number between 1 and 10 and some integer power of 10. The number 2.998 lies between 1 and 10. To get 29,980,000,000, the decimal point in 2.998 must be moved ten places to the right. This is accomplished by multiplying 2.998 by 10^{10}.

Standard notation $\longrightarrow$ 29,980,000,000 = 2.998×10^{10} $\longleftarrow$ Scientific notation

One meter is approximately 0.0006214 mile. To express this number in scientific notation, we must write it as the product of a number between 1 and 10 and some integer power of 10. The number 6.214 lies between 1 and 10. To get 0.0006214, the decimal point in 6.214 must be moved four places to the left. This is accomplished by multiplying 6.214 by $\frac{1}{10^4}$ or by multiplying 6.214 by 10^{-4}.

Standard notation $\longrightarrow$ 0.0006214 = 6.214×10^{-4} $\longleftarrow$ Scientific notation

To write each of the following numbers in scientific notation, we start to the right of the first nonzero digit and count to the decimal point. The exponent gives the number of places the decimal point moves, and the sign of the exponent indicates the direction in which it moves.

a. $372000 = 3.72 \times 10^5$ 5 places to the right

b. $0.000537 = 5.37 \times 10^{-4}$ 4 places to the left

c. $7.36 = 7.36 \times 10^0$ No movement of the decimal point.

Everyday Connections

New Way to the Center of Earth

A NASA technique pinpoints Earth's exact center of mass to understand global climate change.

This spectacular "blue marble" image is the most detailed true-color image of the entire Earth to date. A new NASA-developed technique estimates Earth's center of mass to within 1 millimeter (0.04 inches) a year by using a combination of four space-based techniques. The more accurate frame of reference has applications ranging from improving estimates of changing global sea levels to improving our understanding of earthquakes and volcanoes.

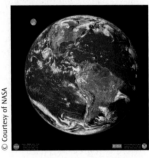

© Courtesy of NASA

Donald Argus of NASA's Jet Propulsion Laboratory, in Pasadena, California, developed the new technique. "Knowing the relative motions of the mass center of Earth's system and the mass center of the solid Earth can help scientists better determine the rate at which ice in Greenland and Antarctica is melting into the ocean," Argus explained. He said the new frame of reference will improve estimates of sea level rise from satellite altimeters like the NASA/French Space Agency Jason satellite, which rely on measurements of the location and motion of the mass center of Earth's system.

The distance from the Earth's center to the North Pole (i.e., the **polar radius**) measures approximately 6,356.750 km, and the distance from the center to the equator (i.e., the **equatorial radius**) measures approximately 6,378.135 km.

1. Express each distance using scientific notation.

2. Given that 1 km is approximately equal to 0.62 miles, use scientific notation to express each distance in miles.

Source: *NASA/GSFC*; http://www.astronomy.com/asy/default.aspx?c=a&id=5671

EXAMPLE 7   Write each number in scientific notation: **a.** 62,000   **b.** −0.0027

Solution   **a.** We must express 62,000 as a product of a number between 1 and 10 and some integer power of 10. This is accomplished by multiplying 6.2 by 10^4.

$$62{,}000 = 6.2 \times 10^4$$

b. We must express −0.0027 as a product of a number whose absolute value is between 1 and 10 and some integer power of 10. This is accomplished by multiplying −2.7 by 10^{-3}.

$$-0.0027 = -2.7 \times 10^{-3}$$

Self Check 7   Write each number in scientific notation:
a. −93,000,000   **b.** 0.0000087

EXAMPLE 8   Write each number in standard notation: **a.** 7.35×10^2 **b.** 3.27×10^{-5}

Solution   **a.** The factor of 10^2 indicates that 7.35 must be multiplied by 2 factors of 10. Because each multiplication by 10 moves the decimal point one place to the right, we have

$$7.35 \times 10^2 = 735$$

b. The factor of 10^{-5} indicates that 3.27 must be divided by 5 factors of 10. Because each division by 10 moves the decimal point one place to the left, we have

$$3.27 \times 10^{-5} = 0.0000327$$

Self Check 8 Write each number in standard notation: **a.** 6.3×10^3
b. 9.1×10^{-4}

5. Use Scientific Notation to Simplify Computations

Another advantage of scientific notation becomes evident when we multiply and divide very large and very small numbers.

EXAMPLE 9 Use scientific notation to calculate $\dfrac{(3,400,000)(0.00002)}{170,000,000}$.

Solution After changing each number to scientific notation, we can do the arithmetic on the numbers and the exponential expressions separately.

$$\frac{(3,400,000)(0.00002)}{170,000,000} = \frac{(3.4 \times 10^6)(2.0 \times 10^{-5})}{1.7 \times 10^8}$$

$$= \frac{6.8}{1.7} \times 10^{6+(-5)-8}$$

$$= 4.0 \times 10^{-7}$$

$$= 0.0000004$$

Self Check 9 Use scientific notation to simplify $\dfrac{(192,000)(0.0015)}{(0.0032)(4,500)}$.

Accent on Technology

Scientific Notation

Calculators often give answers in scientific notation. For example, if we use a calculator to find 21^8, the display will read

On a scientific calculator	*On a graphing calculator*
3.782285936 ^{10}	21^8
	3.782285936E10

In either case, the result is given in scientific notation, and it means $3.782285936 \times 10^{10}$.

We can enter numbers into a calculator in scientific form. For example, to enter 0.000000000061 (which is 6.1×10^{-11}), we enter these numbers and press these keys:

On a scientific calculator	*On a graphing calculator*
6.1 EXP 11 +/−	6.1 EE (−) 11

To use a scientific calculator to simplify $\dfrac{21^8}{0.000000000061}$, we enter the denominator in scientific notation, because there are too many digits to enter it directly. To do the calculation with a scientific calculator, we enter these numbers and press these keys:

21 y^x 8 = ÷ 6.1 EXP 11 +/− =

The display will read 6.200468748 ^{20}. In standard notation, the answer is approximately 620,046,874,800,000,000,000.

The steps are similar using a graphing calculator.

1. **a.** $7 \cdot 7 \cdot 7 = 343$ **b.** $(-3)(-3) = 9$ **c.** $5 \cdot a \cdot a \cdot a$
   d. $(5a)(5a)(5a)(5a) = 625 \cdot a \cdot a \cdot a \cdot a$ 2. **a.** y^6 **b.** a^{18} **c.** x^{12} **d.** $\frac{27a^9b^6}{c^9}$
3. **a.** 7 **b.** $\frac{3}{a^2}$ **c.** $\frac{1}{a^2}$ **d.** $\frac{1}{a^{12}}$ 4. **a.** $\frac{1}{x^8}$ **b.** $\frac{1}{x}$ 5. **a.** $\frac{x^{14}}{y^{10}}$ **b.** $\frac{27b^3}{8a^3}$
6. $\frac{6}{5}$ 7. **a.** -9.3×10^7 **b.** 8.7×10^{-6} 8. **a.** $6{,}300$ **b.** 0.00091 9. 20

0.2 Exercises

Vocabulary and Concepts *Fill in the blanks.*

1. Each quantity in a product is called a _____ of the product.

2. A _____ number exponent tells how many times a base is used as a factor.

3. In the expression $(2x)^3$, __ is the exponent and ___ is the base.

4. The expression x^n is called an _____ expression.

5. A number is in _____ notation when it is written in the form $N \times 10^n$, where $1 \le |N| < 10$ and n is an _____.

6. Unless _____ indicate otherwise, multiplications are performed before additions.

Complete each rule.

7. $x^m x^n = $ _____

8. $(x^m)^n = $ ___

9. $(xy)^n = $ _____

10. $\dfrac{x^m}{x^n} = $ _____

11. $x^0 = $ __

12. $x^{-n} = $ __

Practice *Write each number or expression without using exponents.*

13. 13^2

14. 10^3

15. -5^2

16. $(-5)^2$

17. $4x^3$

18. $(4x)^3$

19. $(-5x)^4$

20. $-6x^2$

21. $-8x^4$

22. $(-8x)^4$

Write each expression using exponents.

23. $7xxx$

24. $-8yyyy$

25. $(-x)(-x)$

26. $(2a)(2a)(2a)$

27. $(3t)(3t)(-3t)$

28. $-(2b)(2b)(2b)(2b)$

29. $xxxyy$

30. $aaabbbb$

Use a calculator to simplify each expression.

31. 2.2^3

32. 7.1^4

33. -0.5^4

34. $(-0.2)^4$

Simplify each expression. Write all answers without using negative exponents. Assume that all variables are restricted to those numbers for which the expression is defined.

35. $x^2 x^3$

36. $y^3 y^4$

37. $(z^2)^3$

38. $(t^6)^7$

39. $(y^5 y^2)^3$

40. $(a^3 a^6)a^4$

41. $(z^2)^3(z^4)^5$

42. $(t^3)^4(t^5)^2$

43. $(a^2)^3(a^4)^2$

44. $(a^2)^4(a^3)^3$

45. $(3x)^3$

46. $(-2y)^4$

47. $(x^2 y)^3$

48. $(x^3 z^4)^6$

49. $\left(\dfrac{a^2}{b}\right)^3$

50. $\left(\dfrac{x}{y^3}\right)^4$

51. $(-x)^0$

52. $4x^0$

53. $(4x)^0$

54. $-2x^0$

55. z^{-4}

56. $\dfrac{1}{t^{-2}}$

57. $y^{-2} y^{-3}$

58. $-m^{-2} m^3$

59. $(x^3 x^{-4})^{-2}$

60. $(y^{-2} y^3)^{-4}$

61. $\dfrac{x^7}{x^3}$

62. $\dfrac{r^5}{r^2}$

63. $\dfrac{a^{21}}{a^{17}}$

64. $\dfrac{t^{13}}{t^4}$

65. $\dfrac{(x^2)^2}{x^2 x}$

66. $\dfrac{s^9 s^3}{(s^2)^2}$

67. $\left(\dfrac{m^3}{n^2}\right)^3$

68. $\left(\dfrac{t^4}{t^3}\right)^3$

69. $\dfrac{(a^3)^{-2}}{aa^2}$

70. $\dfrac{r^9 r^{-3}}{(r^{-2})^3}$

71. $\left(\dfrac{a^{-3}}{b^{-1}}\right)^{-4}$

72. $\left(\dfrac{t^{-4}}{t^{-3}}\right)^{-2}$

73. $\left(\dfrac{r^4 r^{-6}}{r^3 r^{-3}}\right)^2$

74. $\dfrac{(x^{-3}x^2)^2}{(x^2 x^{-5})^{-3}}$

75. $\left(\dfrac{x^5 y^{-2}}{x^{-3} y^2}\right)^4$

76. $\left(\dfrac{x^{-7} y^5}{x^7 y^{-4}}\right)^3$

77. $\left(\dfrac{5x^{-3} y^{-2}}{3x^2 y^{-3}}\right)^{-2}$

78. $\left(\dfrac{3x^2 y^{-5}}{2x^{-2} y^{-6}}\right)^{-3}$

79. $\left(\dfrac{3x^5 y^{-3}}{6x^{-5} y^3}\right)^{-2}$

80. $\left(\dfrac{12x^{-4} y^3 z^{-5}}{4x^4 y^{-3} z^5}\right)^3$

81. $\dfrac{(8^{-2} z^{-3} y)^{-1}}{(5y^2 z^{-2})^3 (5yz^{-2})^{-1}}$

82. $\dfrac{(m^{-2} n^3 p^4)^{-2} (mn^{-2} p^3)^4}{(mn^{-2} p^3)^{-4} (mn^2 p)^{-1}}$

Simplify each expression.

83. $-\dfrac{5[6^2 + (9 - 5)]}{4(2 - 3)^2}$

84. $\dfrac{6[3 - (4 - 7)^2]}{-5(2 - 4^2)}$

Let $x = -2$, $y = 0$, and $z = 3$ and evaluate each expression.

85. x^2

86. $-x^2$

87. x^3

88. $-x^3$

89. $(-xz)^3$

90. $-xz^3$

91. $\dfrac{-(x^2 z^3)}{z^2 - y^2}$

92. $\dfrac{z^2(x^2 - y^2)}{x^3 z}$

93. $5x^2 - 3y^3 z$

94. $3(x - z)^2 + 2(y - z)^3$

95. $\dfrac{-3x^{-3} z^{-2}}{6x^2 z^{-3}}$

96. $\dfrac{(-5x^2 z^{-3})^2}{5xz^{-2}}$

Express each number in scientific notation.

97. 372,000

98. 89,500

99. −177,000,000

100. −23,470,000,000

101. 0.007

102. 0.00052

103. −0.000000693

104. −0.000000089

105. one trillion

106. one millionth

Express each number in standard notation.

107. 9.37×10^5

108. 4.26×10^9

109. 2.21×10^{-5}

110. 2.774×10^{-2}

111. 0.00032×10^4

112. $9,300.0 \times 10^{-4}$

113. -3.2×10^{-3}

114. -7.25×10^3

Use the method of Example 9 to do each calculation. Write all answers in scientific notation.

115. $\dfrac{(65,000)(45,000)}{250,000}$

116. $\dfrac{(0.000000045)(0.00000012)}{45,000,000}$

117. $\dfrac{(0.00000035)(170,000)}{0.00000085}$

118. $\dfrac{(0.0000000144)(12,000)}{600,000}$

119. $\dfrac{(45,000,000,000)(212,000)}{0.00018}$

120. $\dfrac{(0.00000000275)(4,750)}{500,000,000,000}$

Applications *Use scientific notation to compute each answer. Write all answers in scientific notation.*

121. Speed of sound The speed of sound in air is 3.31×10^4 centimeters per second. Compute the speed of sound in <u>meters per minute</u>.

122. Volume of a box Calculate the volume of a box that has dimensions of 6,000 by 9,700 by 4,700 millimeters.

123. Mass of a proton The mass of one proton is 0.0000000000000000000000167248 gram. Find the mass of one billion protons.

124. Speed of light The speed of light in a vacuum is approximately 30,000,000,000 centimeters per second. Find the speed of light in miles per hour. (160,934.4 cm = 1 mile.)

125. Astronomy The distance d, in miles, of the nth planet from the sun is given by the formula

$$d = 9,275,200[3(2^{n-2}) + 4]$$

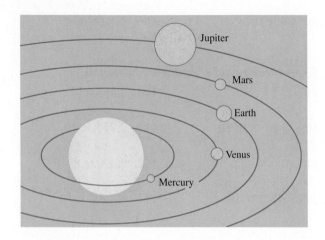

To the nearest million miles, find the distance of Earth and the distance of Mars from the Sun. Give each answer in scientific notation.

126. **License plates** The number of different license plates of the form three digits followed by three letters, as in the illustration, is $10 \cdot 10 \cdot 10 \cdot 26 \cdot 26 \cdot 26$. Write this expression using exponents. Then evaluate it and express the result in scientific notation.

Discovery and Writing   *Write each expression with a single base.*

127. $x^n x^2$

128. $\dfrac{x^m}{x^3}$

129. $\dfrac{x^m x^2}{x^3}$

130. $\dfrac{x^{3m+5}}{x^2}$

131. $x^{m+1} x^3$

132. $a^{n-3} a^3$

133. Explain why $-x^4$ and $(-x)^4$ represent different numbers.

134. Explain why 32×10^2 is not in scientific notation.

Review

135. Graph the interval $(-2, 4)$.

136. Graph the interval $(-\infty, -3] \cup [3, \infty)$.

137. Evaluate $|\pi - 5|$.

138. Find the distance between -7 and -5 on the number line.

0.3   Rational Exponents and Radicals

Objectives

1. Define Rational Exponents Whose Numerators Are 1
2. Define Rational Exponents Whose Numerators Are Not 1
3. Define Radical Expressions
4. Simplify and Combine Radicals
5. Rationalize Denominators and Numerators

"Dead Man's Curve" is a 1964 hit song by the rock and roll duo Jan Berry and Dean Torrence. The song details a teenage drag race that ends in an accident. Today, dead man's curve is a commonly used expression given to dangerous curves on our roads. Every curve has a "critical speed." If we exceed this speed, regardless of how skilled a driver we are, we will lose control of the vehicle.

The radical expression $3.9\sqrt{r}$ gives the critical speed in miles per hour when we travel a curved road with a radius of r feet. A knowledge of square roots and radicals is important and used in the construction of safe highways and roads. We will study the topic of radicals in this section.

1.   Define Rational Exponents Whose Numerators Are 1

If we apply the rule $(x^m)^n = x^{mn}$ to $(25^{1/2})^2$, we obtain

$$(25^{1/2})^2 = 25^{(1/2)\cdot 2} \quad \text{Keep the base and multiply the exponents.}$$
$$= 25^1 \qquad \tfrac{1}{2} \cdot 2 = 1$$
$$= 25$$

Thus, $25^{1/2}$ is a real number whose square is 25. Although both $5^2 = 25$ and $(-5)^2 = 25$, we define $25^{1/2}$ to be the positive real number whose square is 25:

$$25^{1/2} = 5 \quad \text{Read } 25^{1/2} \text{ as "the square root of 25."}$$

In general, we have the following definition.

Rational Exponents   If $a \geq 0$ and n is a natural number, then $a^{1/n}$ (read as "the nth root of a") is the nonnegative real number b such that

$$b^n = a$$

Since $b = a^{1/n}$, we have $b^n = (a^{1/n})^n = a$.

Comment

In the expression $a^{1/n}$, there is no real-number nth root of a when n is even and $a < 0$. For example, $(-64)^{1/2}$ is not a real number, because the square of no real number is -64.

EXAMPLE 1   In each case, we will apply the definition of rational exponents.

a. $16^{1/2} = 4$ Because $4^2 = 16$. Read $16^{1/2}$ as "the square root of 16."

b. $27^{1/3} = 3$ Because $3^3 = 27$. Read $27^{1/3}$ as "the cube root of 27."

c. $\left(\dfrac{1}{81}\right)^{1/4} = \dfrac{1}{3}$ Because $\left(\tfrac{1}{3}\right)^4 = \tfrac{1}{81}$. Read $\left(\tfrac{1}{81}\right)^{1/4}$ as "the fourth root of $\tfrac{1}{81}$."

d. $-32^{1/5} = -(32^{1/5})$ Read $32^{1/5}$ as "the fifth root of 32."
$$= -(2) \qquad \text{Because } 2^5 = 32.$$
$$= -2$$

Self Check 1   Simplify: **a.** $100^{1/2}$ **b.** $243^{1/5}$

If n is even in the expression $a^{1/n}$ and the base contains variables, we often use absolute value symbols to guarantee that an even root is nonnegative.

$(49x^2)^{1/2} = 7|x|$ Because $(7|x|)^2 = 49x^2$. Since x could be negative, absolute value symbols are necessary to guarantee that the square root is nonnegative.

$(16x^4)^{1/4} = 2|x|$ Because $(2|x|)^4 = 16x^4$. Since x could be negative, absolute value symbols are necessary to guarantee that the fourth root is nonnegative.

$(729x^{12})^{1/6} = 3x^2$ Because $(3x^2)^6 = 729x^{12}$. Since x^2 is always nonnegative, no absolute value symbols are necessary. Read $(729x^{12})^{1/6}$ as "the sixth root of $729x^{12}$."

If n is an odd number in the expression $a^{1/n}$, the base a can be negative.

EXAMPLE 2   **a.** $(-8)^{1/3} = -2$ Because $(-2)^3 = -8$.

b. $(-3{,}125)^{1/5} = -5$ Because $(-5)^5 = -3{,}125$.

c. $\left(-\dfrac{1}{1{,}000}\right)^{1/3} = -\dfrac{1}{10}$ Because $\left(-\frac{1}{10}\right)^3 = -\frac{1}{1{,}000}$

Self Check 2   Simplify: **a.** $(-125)^{1/3}$ **b.** $(-100{,}000)^{1/5}$

If n is odd in the expression $a^{1/n}$, we don't need to use absolute value symbols, because odd roots can be negative.

$$(-27x^3)^{1/3} = -3x \quad \text{Because } (-3x)^3 = -27x^3.$$
$$(-128a^7)^{1/7} = -2a \quad \text{Because } (-2a)^7 = -128a^7.$$

We summarize the possibilities concerning $a^{1/n}$ as follows.

If n is a natural number and a is a real number in the expression $a^{1/n}$, then

If $a \geq 0$, then $a^{1/n}$ is the nonnegative real number b such that $b^n = a$.

If $a < 0$ $\begin{cases} \text{and } n \text{ is odd, then } a^{1/n} \text{ is the real number } b \text{ such that } b^n = a. \\ \text{and } n \text{ is even, then } a^{1/n} \text{ is not a real number.} \end{cases}$

The following chart also shows the possibilities that can occur when simplifying $a^{1/n}$.

Simplifying Expressions of the Form $a^{1/n}$			
a **is a real number**	n	$a^{1/n}$	**Examples**
$a = 0$	n is a natural number.	$0^{1/n}$ is the real number 0 because $0^n = 0$.	$0^{1/2} = 0$ because $0^2 = 0$. $0^{1/5} = 0$ because $0^5 = 0$.
$a > 0$	n is a natural number.	$a^{1/n}$ is the nonnegative real number such that $(a^{1/n})^n = a$.	$16^{1/2} = 4$ because $4^2 = 16$. $27^{1/3} = 3$ because $3^3 = 27$.
$a < 0$	n is an odd natural number.	$a^{1/n}$ is the real number such that $(a^{1/n})^n = a$.	$(-32)^{1/5} = -2$ because $(-2)^5 = -32$. $(-125)^{1/3} = -5$ because $(-5)^3 = -125$.
$a < 0$	n is an even natural number.	$a^{1/n}$ is not a real number.	$(-9)^{1/2}$ is not a real number. $(-81)^{1/4}$ is not a real number.

2.   Define Rational Exponents Whose Numerators Are Not 1

The definition of $a^{1/n}$ can be extended to include rational exponents whose numerators are not 1. For example, $4^{3/2}$ can be written as either

$$(4^{1/2})^3 \quad \text{or} \quad (4^3)^{1/2} \quad \text{Because of the power rule, } (x^m)^n = x^{mn}.$$

This suggests the following rule.

Rule for Rational Exponents   If m and n are positive integers, the fraction $\frac{m}{n}$ is in lowest terms, and $a^{1/n}$ is a real number, then

$$a^{m/n} = (a^{1/n})^m = (a^m)^{1/n}$$

In the previous rule, we can view the expression $a^{m/n}$ in two ways:

1. $(a^{1/n})^m$: the mth power of the nth root of a
2. $(a^m)^{1/n}$: the nth root of the mth power of a

For example, $(-27)^{2/3}$ can be simplified in two ways:

$$(-27)^{2/3} = [(-27)^{1/3}]^2 \qquad \text{or} \qquad (-27)^{2/3} = [(-27)^2]^{1/3}$$
$$= (-3)^2 \qquad\qquad\qquad\qquad\qquad = (729)^{1/3}$$
$$= 9 \qquad\qquad\qquad\qquad\qquad\qquad = 9$$

As this example suggests, it is usually easier to take the root of the base first to avoid large numbers.

Comment

It is helpful to think of the phrase *power over root* when we see a rational exponent. The numerator of the fraction represents the *power* and the denominator represents the *root*. Begin with the *root* when simplifying to avoid large numbers.

Negative Rational Exponents

If m and n are positive integers, the fraction $\frac{m}{n}$ is in lowest terms and $a^{1/n}$ is a real number, then

$$a^{-m/n} = \frac{1}{a^{m/n}} \qquad \text{and} \qquad \frac{1}{a^{-m/n}} = a^{m/n} \quad (a \neq 0)$$

EXAMPLE 3

Simplify each expression by applying the rules for rational exponents.

a. $25^{3/2} = (25^{1/2})^3$
$= 5^3$
$= 125$

b. $\left(-\dfrac{x^6}{1{,}000}\right)^{2/3} = \left[\left(-\dfrac{x^6}{1{,}000}\right)^{1/3}\right]^2$
$= \left(-\dfrac{x^2}{10}\right)^2$
$= \dfrac{x^4}{100}$

c. $32^{-2/5} = \dfrac{1}{32^{2/5}}$
$= \dfrac{1}{(32^{1/5})^2}$
$= \dfrac{1}{2^2}$
$= \dfrac{1}{4}$

d. $\dfrac{1}{81^{-3/4}} = 81^{3/4}$
$= (81^{1/4})^3$
$= 3^3$
$= 27$

Self Check 3 Simplify: **a.** $49^{3/2}$ **b.** $16^{-3/4}$ **c.** $\dfrac{1}{(27x^3)^{-2/3}}$

Because of the definition, rational exponents follow the same rules as integer exponents.

EXAMPLE 4   Simplify each expression. Assume that all variables represent positive numbers and write all answers without using negative exponents.

a. $(36x)^{1/2} = 36^{1/2}x^{1/2}$
$$= 6x^{1/2}$$

b. $\dfrac{(a^{1/3}b^{2/3})^6}{(y^3)^2} = \dfrac{a^{6/3}b^{12/3}}{y^6}$
$$= \dfrac{a^2b^4}{y^6}$$

c. $\dfrac{a^{x/2}a^{x/4}}{a^{x/6}} = a^{x/2+x/4-x/6}$
$$= a^{6x/12+3x/12-2x/12}$$
$$= a^{7x/12}$$

d. $\left[\dfrac{-c^{-2/5}}{c^{4/5}}\right]^{5/3} = (-c^{-2/5-4/5})^{5/3}$
$$= [(-1)(c^{-6/5})]^{5/3}$$
$$= (-1)^{5/3}(c^{-6/5})^{5/3}$$
$$= -1c^{-30/15}$$
$$= -c^{-2}$$
$$= -\dfrac{1}{c^2}$$

Self Check 4   Use the directions for Example 4.

a. $\left(\dfrac{y^2}{49}\right)^{1/2}$ **b.** $\dfrac{b^{3/7}b^{2/7}}{b^{4/7}}$ **c.** $\dfrac{(9r^2s)^{1/2}}{rs^{-3/2}}$

3.   Define Radical Expressions

Radical signs can also be used to express roots of numbers.

Definition of $\sqrt[n]{a}$ If n is a natural number greater than 1 and if $a^{1/n}$ is a real number, then
$$\sqrt[n]{a} = a^{1/n}$$

In the **radical expression** $\sqrt[n]{a}$, the symbol $\sqrt{}$ is the **radical sign,** a is the **radicand,** and n is the **index** (or the **order**) of the radical expression. If the order is 2, the expression is a **square root,** and we do not write the index.
$$\sqrt{a} = \sqrt[2]{a}$$

If the index of a radical is 3, we call the radical a **cube root.**

nth Root of a Nonnegative Number   If n is a natural number greater than 1 and $a \geq 0$, then $\sqrt[n]{a}$ is the nonnegative number whose nth power is a.
$$\left(\sqrt[n]{a}\right)^n = a$$

Comment

In the expression $\sqrt[n]{a}$, there is no real-number nth root of a when n is even and $a < 0$. For example, $\sqrt{-64}$ is not a real number, because the square of no real number is -64.

If 2 is substituted for n in the equation $\left(\sqrt[n]{a}\right)^n = a$, we have
$$\left(\sqrt[2]{a}\right)^2 = \left(\sqrt{a}\right)^2 = \sqrt{a}\sqrt{a} = a$$

This shows that if a number a can be factored into two equal factors, either of those factors is a square root of a. Furthermore, if a can be factored into n equal factors, any one of those factors is an nth root of a.

If n is an odd number greater than 1 in the expression $\sqrt[n]{a}$, the radicand can be negative.

EXAMPLE 5   Simplify each expression by applying the definitions of cube roots and fifth roots.

a. $\sqrt[3]{-27} = -3$ Because $(-3)^3 = -27$

b. $\sqrt[3]{-125} = -5$ Because $(-5)^3 = -125$

c. $\sqrt[3]{-\dfrac{27}{1,000}} = -\dfrac{3}{10}$ Because $\left(-\dfrac{3}{10}\right)^3 = -\dfrac{27}{1,000}$

d. $-\sqrt[5]{-243} = -\left(\sqrt[5]{-243}\right)$
$\qquad\qquad\quad = -(-3)$
$\qquad\qquad\quad = 3$

Self Check 5   Find each root: **a.** $\sqrt[3]{216}$ **b.** $\sqrt[5]{-\dfrac{1}{32}}$

We summarize the possibilities concerning $\sqrt[n]{a}$ as follows.

If n is a natural number greater than 1 and a is a real number, then

If $a \geq 0$, then $\sqrt[n]{a}$ is the nonnegative real number such that $\left(\sqrt[n]{a}\right)^n = a$.

If $a < 0$ $\begin{cases} \text{and } n \text{ is odd, then } \sqrt[n]{a} \text{ is the real number such that } \left(\sqrt[n]{a}\right)^n = a. \\ \text{and } n \text{ is even, then } \sqrt[n]{a} \text{ is not a real number.} \end{cases}$

The following chart also shows the possibilities that can occur when simplifying $\sqrt[n]{a}$.

		Simplifying Expressions of the Form $\sqrt[n]{a}$	
a is a real number	n	$\sqrt[n]{a}$	Examples
$a = 0$	n is a natural number.	$\sqrt[n]{0}$ is the real number 0 because $0^n = 0$.	$\sqrt[3]{0} = 0$ because $0^3 = 0$. $\sqrt[5]{0} = 0$ because $0^5 = 0$.
$a > 0$	n is a natural number.	$\sqrt[n]{a}$ is the nonnegative real number such that $\left(\sqrt[n]{a}\right)^n = a$.	$\sqrt{16} = 4$ because $4^2 = 16$. $\sqrt[3]{27} = 3$ because $3^3 = 27$.
$a < 0$	n is an odd natural number.	$\sqrt[n]{a}$ is the real number such that $\left(\sqrt[n]{a}\right)^n = a$.	$\sqrt[5]{-32} = -2$ because $(-2)^5 = -32$. $\sqrt[3]{-125} = -5$ because $(-5)^3 = -125$.
$a < 0$	n is an even natural number.	$\sqrt[n]{a}$ is not a real number.	$\sqrt{-9}$ is not a real number. $\sqrt[4]{-81}$ is not a real number.

We have seen that if $a^{1/m}$ is real, then $a^{m/n} = (a^{1/n})^m = (a^m)^{1/n}$. This same fact can be stated in radical notation.

$$a^{m/n} = \left(\sqrt[n]{a}\right)^m = \sqrt[n]{a^m}$$

Thus, *the mth power of the nth root of a* is the same as *the nth root of the mth power of a.* For example, to find $\sqrt[3]{27^2}$, we can proceed in either of two ways:

$$\sqrt[3]{27^2} = \left(\sqrt[3]{27}\right)^2 = 3^2 = 9 \quad \text{or} \quad \sqrt[3]{27^2} = \sqrt[3]{729} = 9$$

By definition, $\sqrt{a^2}$ represents a nonnegative number. If a could be negative, we must use absolute value symbols to guarantee that $\sqrt{a^2}$ will be nonnegative. Thus, if a is unrestricted,

$$\sqrt{a^2} = |a|$$

A similar argument holds when the index is any even natural number. The symbol $\sqrt[4]{a^4}$, for example, means the *positive* fourth root of a^4. Thus, if a is unrestricted,

$$\sqrt[4]{a^4} = |a|$$

EXAMPLE 6   If x is unrestricted, simplify   **a.** $\sqrt[6]{64x^6}$ **b.** $\sqrt[3]{x^3}$ **c.** $\sqrt{9x^8}$

Solution   We apply the definitions of sixth roots, cube roots, and square roots.

a. $\sqrt[6]{64x^6} = 2|x|$ Use absolute value symbols to guarantee that the result will be nonnegative.

b. $\sqrt[3]{x^3} = x$ Because the index is odd, no absolute value symbols are needed.

c. $\sqrt{9x^8} = 3x^4$ Because $3x^4$ is always nonnegative, no absolute value symbols are needed.

Self Check 6   Use the directions for Example 6:
a. $\sqrt[4]{16x^4}$ **b.** $\sqrt[3]{27y^3}$ **c.** $\sqrt[4]{x^8}$

4.   Simplify and Combine Radicals

Many properties of exponents have counterparts in radical notation. For example, since $a^{1/n}b^{1/n} = (ab)^{1/n}$ and $\frac{a^{1/n}}{b^{1/n}} = \left(\frac{a}{b}\right)^{1/n}$ $(b \neq 0)$, we have the following.

Multiplication and Division Properties of Radicals

If all expressions represent real numbers,

$$\sqrt[n]{a}\sqrt[n]{b} = \sqrt[n]{ab} \qquad \frac{\sqrt[n]{a}}{\sqrt[n]{b}} = \sqrt[n]{\frac{a}{b}} \quad (b \neq 0)$$

In words, we say

The product of two nth roots is equal to the nth root of their product.

The quotient of two nth roots is equal to the nth root of their quotient.

Comment

These properties involve the *n*th root of the product of two numbers or the *n*th root of the quotient of two numbers. There is no such property for sums or differences. For example, $\sqrt{9 + 4} \neq \sqrt{9} + \sqrt{4}$, because

$$\sqrt{9 + 4} = \sqrt{13} \qquad \text{but} \qquad \sqrt{9} + \sqrt{4} = 3 + 2 = 5$$

and $\sqrt{13} \neq 5$. In general,

$$\sqrt{a + b} \neq \sqrt{a} + \sqrt{b} \qquad \text{and} \qquad \sqrt{a - b} \neq \sqrt{a} - \sqrt{b}$$

Numbers that are squares of positive integers, such as 1, 4, 9, 16, 25, and 36 are called **perfect squares.** Expressions such as $4x^2$ and $\frac{1}{9}x^6$ are also perfect squares, because each one is the square of another expression with integer exponents and rational coefficients.

$$4x^2 = (2x)^2 \quad\text{and}\quad \frac{1}{9}x^6 = \left(\frac{1}{3}x^3\right)^2$$

Numbers that are cubes of positive integers, such as 1, 8, 27, 64, 125, and 216 are called **perfect cubes.** Expressions such as $64x^3$ and $\frac{1}{27}x^9$ are also perfect cubes, because each one is the cube of another expression with integer exponents and rational coefficients.

$$64x^3 = (4x)^3 \quad\text{and}\quad \frac{1}{27}x^9 = \left(\frac{1}{3}x^3\right)^3$$

There are also perfect fourth powers, perfect fifth powers, and so on.

We can use perfect powers and the multiplication property of radicals to simplify many radical expressions. For example, to simplify $\sqrt{12x^5}$, we factor $12x^5$ so that one factor is the largest perfect square that divides $12x^5$. In this case, it is $4x^4$. We then rewrite $12x^5$ as $4x^4 \cdot 3x$ and simplify.

$$
\begin{aligned}
\sqrt{12x^5} &= \sqrt{4x^4 \cdot 3x} && \text{Factor } 12x^5 \text{ as } 4x^4 \cdot 3x. \\
&= \sqrt{4x^4}\sqrt{3x} && \text{Use the multiplication property of radicals: } \sqrt{ab} = \sqrt{a}\sqrt{b}. \\
&= 2x^2\sqrt{3x} && \sqrt{4x^4} = 2x^2
\end{aligned}
$$

To simplify $\sqrt[3]{432x^9y}$, we find the largest perfect-cube factor of $432x^9y$ (which is $216x^9$) and proceed as follows:

$$
\begin{aligned}
\sqrt[3]{432x^9y} &= \sqrt[3]{216x^9 \cdot 2y} && \text{Factor } 432x^9y \text{ as } 216x^9 \cdot 2y. \\
&= \sqrt[3]{216x^9}\sqrt[3]{2y} && \text{Use the multiplication property of radicals: } \sqrt[3]{ab} = \sqrt[3]{a}\sqrt[3]{b}. \\
&= 6x^3\sqrt[3]{2y} && \sqrt[3]{216x^9} = 6x^3
\end{aligned}
$$

Radical expressions with the same index and the same radicand are called **like** or **similar radicals.** We can combine the like radicals in $3\sqrt{2} + 2\sqrt{2}$ by using the distributive property.

$$
\begin{aligned}
3\sqrt{2} + 2\sqrt{2} &= (3 + 2)\sqrt{2} \\
&= 5\sqrt{2}
\end{aligned}
$$

This example suggests that to combine like radicals, we *add their numerical coefficients and keep the same radical.*

When radicals have the same index but different radicands, we often can change them to equivalent forms having the same radicand. We then can combine them. For example, to simplify $\sqrt{27} - \sqrt{12}$, we simplify both radicals and combine like radicals.

$$
\begin{aligned}
\sqrt{27} - \sqrt{12} &= \sqrt{9 \cdot 3} - \sqrt{4 \cdot 3} && \text{Factor 27 and 12.} \\
&= \sqrt{9}\sqrt{3} - \sqrt{4}\sqrt{3} && \sqrt{ab} = \sqrt{a}\sqrt{b} \\
&= 3\sqrt{3} - 2\sqrt{3} && \sqrt{9} = 3 \text{ and } \sqrt{4} = 2. \\
&= \sqrt{3} && \text{Combine like radicals.}
\end{aligned}
$$

EXAMPLE 7 Simplify: **a.** $\sqrt{50} + \sqrt{200}$ **b.** $3z\sqrt[5]{64z} - 2\sqrt[5]{2z^6}$

Solution We will simplify each radical expression and then combine like radicals.

a. $\sqrt{50} + \sqrt{200} = \sqrt{25 \cdot 2} + \sqrt{100 \cdot 2}$
$= \sqrt{25}\sqrt{2} + \sqrt{100}\sqrt{2}$
$= 5\sqrt{2} + 10\sqrt{2}$
$= 15\sqrt{2}$

b. $3z\sqrt[5]{64z} - 2\sqrt[5]{2z^6} = 3z\sqrt[5]{32 \cdot 2z} - 2\sqrt[5]{z^5 \cdot 2z}$
$= 3z\sqrt[5]{32}\sqrt[5]{2z} - 2\sqrt[5]{z^5}\sqrt[5]{2z}$
$= 3z(2)\sqrt[5]{2z} - 2z\sqrt[5]{2z}$
$= 6z\sqrt[5]{2z} - 2z\sqrt[5]{2z}$
$= 4z\sqrt[5]{2z}$

Self Check 7 Simplify: **a.** $\sqrt{18} - \sqrt{8}$ **b.** $2\sqrt[3]{81a^4} + a\sqrt[3]{24a}$

5. Rationalize Denominators and Numerators

By **rationalizing the denominator,** we can write a fraction such as

$$\frac{\sqrt{5}}{\sqrt{3}}$$

as a fraction with a rational number in the denominator. All that we must do is multiply both the numerator and the denominator by $\sqrt{3}$. $\left(\text{Note that } \sqrt{3}\sqrt{3} \text{ is the rational number 3.}\right)$

$$\frac{\sqrt{5}}{\sqrt{3}} = \frac{\sqrt{5}\sqrt{3}}{\sqrt{3}\sqrt{3}} = \frac{\sqrt{15}}{3}$$

To rationalize the numerator, we multiply both the numerator and the denominator by $\sqrt{5}$. $\left(\text{Note that } \sqrt{5}\sqrt{5} \text{ is the rational number 5.}\right)$

$$\frac{\sqrt{5}}{\sqrt{3}} = \frac{\sqrt{5}\sqrt{5}}{\sqrt{3}\sqrt{5}} = \frac{5}{\sqrt{15}}$$

EXAMPLE 8 Rationalize each denominator and simplify. Assume that all variables represent positive numbers.

a. $\dfrac{1}{\sqrt{7}}$ **b.** $\sqrt[3]{\dfrac{3}{4}}$ **c.** $\sqrt{\dfrac{3}{x}}$ **d.** $\sqrt{\dfrac{3a^3}{5x^5}}$

Solution We will multiply both the numerator and the denominator by a radical that will make the denominator a rational number.

a. $\dfrac{1}{\sqrt{7}} = \dfrac{1\sqrt{7}}{\sqrt{7}\sqrt{7}}$ **b.** $\sqrt[3]{\dfrac{3}{4}} = \dfrac{\sqrt[3]{3}}{\sqrt[3]{4}}$

$\qquad\qquad = \dfrac{\sqrt{7}}{7}$ $= \dfrac{\sqrt[3]{3}\sqrt[3]{2}}{\sqrt[3]{4}\sqrt[3]{2}}$ **Multiply numerator and denominator by** $\sqrt[3]{2}$, **because** $\sqrt[3]{4}\sqrt[3]{2} = \sqrt[3]{8} = 2$.

$\qquad\qquad\qquad\qquad\qquad\qquad\quad = \dfrac{\sqrt[3]{6}}{\sqrt[3]{8}}$

$\qquad\qquad\qquad\qquad\qquad\qquad\quad = \dfrac{\sqrt[3]{6}}{2}$

c. $\sqrt{\dfrac{3}{x}} = \dfrac{\sqrt{3}}{\sqrt{x}}$ **d.** $\sqrt{\dfrac{3a^3}{5x^5}} = \dfrac{\sqrt{3a^3}}{\sqrt{5x^5}}$

$\qquad\quad = \dfrac{\sqrt{3}\sqrt{x}}{\sqrt{x}\sqrt{x}}$ $= \dfrac{\sqrt{3a^3}\sqrt{5x}}{\sqrt{5x^5}\sqrt{5x}}$ **Multiply numerator and denominator by** $\sqrt{5x}$, **because** $\sqrt{5x^5}\sqrt{5x} = \sqrt{25x^6} = 5x^3$.

$\qquad\quad = \dfrac{\sqrt{3x}}{x}$ $= \dfrac{\sqrt{15a^3x}}{\sqrt{25x^6}}$

$\qquad\qquad\qquad\qquad\qquad\quad = \dfrac{\sqrt{a^2}\sqrt{15ax}}{5x^3}$

$\qquad\qquad\qquad\qquad\qquad\quad = \dfrac{a\sqrt{15ax}}{5x^3}$

Self Check 8   Use the directions for Example 8:   **a.** $\dfrac{6}{\sqrt{6}}$ **b.** $\sqrt[3]{\dfrac{2}{5x}}$

EXAMPLE 9   Rationalize each numerator and simplify. Assume that all variables represent positive numbers: **a.** $\dfrac{\sqrt{x}}{7}$ **b.** $\dfrac{2\sqrt[3]{9x}}{3}$

Solution   We will multiply both the numerator and the denominator by a radical that will make the numerator a rational number.

a. $\dfrac{\sqrt{x}}{7} = \dfrac{\sqrt{x}\cdot\sqrt{x}}{7\sqrt{x}}$ **b.** $\dfrac{2\sqrt[3]{9x}}{3} = \dfrac{2\sqrt[3]{9x}\cdot\sqrt[3]{3x^2}}{3\sqrt[3]{3x^2}}$

$\qquad\quad = \dfrac{x}{7\sqrt{x}}$ $= \dfrac{2\sqrt[3]{27x^3}}{3\sqrt[3]{3x^2}}$

$\qquad\qquad\qquad\qquad\qquad\qquad\qquad\quad = \dfrac{2(3x)}{3\sqrt[3]{3x^2}}$

$\qquad\qquad\qquad\qquad\qquad\qquad\qquad\quad = \dfrac{2x}{\sqrt[3]{3x^2}}$ **Divide out the 3s.**

Self Check 9   Use the directions for Example 9:   **a.** $\dfrac{\sqrt{2x}}{5}$ **b.** $\dfrac{\sqrt[3]{2y^2}}{2}$

After rationalizing denominators, we often can simplify an expression.

EXAMPLE 10 Simplify: $\sqrt{\dfrac{1}{2}} + \sqrt{\dfrac{1}{8}}$.

Solution We will rationalize the denominators of each radical and then combine like radicals.

$$\sqrt{\frac{1}{2}} + \sqrt{\frac{1}{8}} = \frac{1}{\sqrt{2}} + \frac{1}{\sqrt{8}} \qquad \sqrt{\frac{1}{2}} = \frac{\sqrt{1}}{\sqrt{2}} = \frac{1}{\sqrt{2}}; \sqrt{\frac{1}{8}} = \frac{\sqrt{1}}{\sqrt{8}} = \frac{1}{\sqrt{8}}$$

$$= \frac{1\sqrt{2}}{\sqrt{2}\sqrt{2}} + \frac{1\sqrt{2}}{\sqrt{8}\sqrt{2}}$$

$$= \frac{\sqrt{2}}{2} + \frac{\sqrt{2}}{\sqrt{16}}$$

$$= \frac{\sqrt{2}}{2} + \frac{\sqrt{2}}{4}$$

$$= \frac{3\sqrt{2}}{4}$$

Self Check 10 Simplify: $\sqrt[3]{\dfrac{x}{2}} - \sqrt[3]{\dfrac{x}{16}}$.

Another property of radicals can be derived from the properties of exponents. If all of the expressions represent real numbers,

$$\sqrt[n]{\sqrt[m]{x}} = \sqrt[n]{x^{1/m}} = (x^{1/m})^{1/n} = x^{1/(mn)} = \sqrt[mn]{x}$$

$$\sqrt[m]{\sqrt[n]{x}} = \sqrt[m]{x^{1/n}} = (x^{1/n})^{1/m} = x^{1/(nm)} = \sqrt[mn]{x}$$

These results are summarized in the following *theorem* (a fact that can be proved).

Theorem If all of the expressions involved represent real numbers, then

$$\sqrt[m]{\sqrt[n]{x}} = \sqrt[n]{\sqrt[m]{x}} = \sqrt[mn]{x}$$

We can use the previous theorem to simplify many radicals. For example,

$$\sqrt[3]{\sqrt{8}} = \sqrt{\sqrt[3]{8}} = \sqrt{2}$$

Rational exponents can be used to simplify many radical expressions, as shown in the following example.

EXAMPLE 11 Simplify. Assume that x and y are positive numbers.
a. $\sqrt[6]{4}$ **b.** $\sqrt[12]{x^3}$ **c.** $\sqrt[9]{8y^3}$

Solution In each case, we will write the radical as an exponential expression, simplify the resulting expression, and write the final result as a radical.

a. $\sqrt[6]{4} = 4^{1/6} = (2^2)^{1/6} = 2^{2/6} = 2^{1/3} = \sqrt[3]{2}$

b. $\sqrt[12]{x^3} = x^{3/12} = x^{1/4} = \sqrt[4]{x}$

c. $\sqrt[9]{8y^3} = (2^3 y^3)^{1/9} = (2y)^{3/9} = (2y)^{1/3} = \sqrt[3]{2y}$

Self Check 11 Simplify: **a.** $\sqrt[4]{4}$ **b.** $\sqrt[9]{27x^3}$

Self Check Answers   **1. a.** 10 **b.** 3 **2. a.** -5 **b.** -10 **3. a.** 343 **b.** $\frac{1}{8}$ **c.** $9x^2$ **4. a.** $\frac{y}{7}$
b. $b^{1/7}$ **c.** $3s^2$ **5. a.** 6 **b.** $-\frac{1}{2}$ **6. a.** $2|x|$ **b.** $3y$ **c.** x^2
7. a. $\sqrt{2}$ **b.** $8a\sqrt[3]{3a}$ **8. a.** $\sqrt{6}$ **b.** $\frac{\sqrt[3]{50x^2}}{5x}$ **9. a.** $\frac{2x}{5\sqrt{2x}}$ **b.** $\frac{y}{\sqrt[3]{4y}}$
10. $\frac{\sqrt[3]{4x}}{4}$ **11. a.** $\sqrt{2}$ **b.** $\sqrt[3]{3x}$

0.3   Exercises

Vocabulary and Concepts   *Fill in the blanks.*

1. If $a = 0$ and n is a natural number, then $a^{1/n} = $ __.
2. If $a > 0$ and n is a natural number, then $a^{1/n}$ is a _____ number.
3. If $a < 0$ and n is an even number, then $a^{1/n}$ is ___ a real number.
4. $6^{2/3}$ can be written as _____ or _____.
5. $\sqrt[n]{a} = $ ___   6. $\sqrt{a^2} = $ ___
7. $\sqrt[n]{a}\sqrt[n]{b} = $ _____   8. $\sqrt[n]{\dfrac{a}{b}} = $ ___
9. $\sqrt{x+y}$ __ $\sqrt{x} + \sqrt{y}$
10. $\sqrt[m]{\sqrt[n]{x}}$ or $\sqrt[n]{\sqrt[m]{x}}$ can be written as _____.

Practice   *Simplify each expression.*

11. $9^{1/2}$

12. $8^{1/3}$

13. $\left(\dfrac{1}{25}\right)^{1/2}$

14. $\left(\dfrac{16}{625}\right)^{1/4}$

15. $-81^{1/4}$

16. $-\left(\dfrac{8}{27}\right)^{1/3}$

17. $(10,000)^{1/4}$

18. $1,024^{1/5}$

19. $-64^{1/3}$

20. $\left(-\dfrac{27}{8}\right)^{1/3}$

21. $(-64)^{1/2}$

22. $(-125)^{1/3}$

Simplify each expression. Use absolute value symbols when necessary.

23. $(16a^2)^{1/2}$

24. $(25a^4)^{1/2}$

25. $(16a^4)^{1/4}$

26. $(-64a^3)^{1/3}$

27. $(-32a^5)^{1/5}$

28. $(64a^6)^{1/6}$

29. $(-216b^6)^{1/3}$

30. $(256t^8)^{1/4}$

31. $\left(\dfrac{16a^4}{25b^2}\right)^{1/2}$

32. $\left(-\dfrac{a^5}{32b^{10}}\right)^{1/5}$

33. $\left(-\dfrac{1,000x^6}{27y^3}\right)^{1/3}$

34. $\left(\dfrac{49t^2}{100z^4}\right)^{1/2}$

Simplify each expression. Write all answers without using negative exponents.

35. $4^{3/2}$

36. $8^{2/3}$

37. $-16^{3/2}$

38. $(-8)^{2/3}$

39. $-1,000^{2/3}$

40. $100^{3/2}$

41. $64^{-1/2}$

42. $25^{-1/2}$

43. $64^{-3/2}$

44. $49^{-3/2}$

45. $-9^{-3/2}$

46. $(-27)^{-2/3}$

47. $\left(\dfrac{4}{9}\right)^{5/2}$

48. $\left(\dfrac{25}{81}\right)^{3/2}$

49. $\left(-\dfrac{27}{64}\right)^{-2/3}$

50. $\left(\dfrac{125}{8}\right)^{-4/3}$

Simplify each expression. Assume that all variables represent positive numbers. Write all answers without using negative exponents.

51. $(100s^4)^{1/2}$

52. $(64u^6v^3)^{1/3}$

53. $(32y^{10}z^5)^{-1/5}$

54. $(625a^4b^8)^{-1/4}$

55. $(x^{10}y^5)^{3/5}$

56. $(64a^6b^{12})^{5/6}$

57. $(r^8s^{16})^{-3/4}$

58. $(-8x^9y^{12})^{-2/3}$

59. $\left(-\dfrac{8a^6}{125b^9}\right)^{2/3}$

60. $\left(\dfrac{16x^4}{625y^8}\right)^{3/4}$

61. $\left(\dfrac{27r^6}{1,000s^{12}}\right)^{-2/3}$

62. $\left(-\dfrac{32m^{10}}{243n^{15}}\right)^{-2/5}$

63. $\dfrac{a^{2/5}a^{4/5}}{a^{1/5}}$

64. $\dfrac{x^{6/7}x^{3/7}}{x^{2/7}x^{5/7}}$

Simplify each radical expression.

65. $\sqrt{49}$

66. $\sqrt{81}$

67. $\sqrt[3]{125}$

68. $\sqrt[3]{-64}$

69. $\sqrt[3]{-125}$

70. $\sqrt[5]{-243}$

71. $\sqrt[5]{-\dfrac{32}{100,000}}$

72. $\sqrt[4]{\dfrac{256}{625}}$

Simplify each expression, using absolute value symbols when necessary. Write answers without using negative exponents.

73. $\sqrt{36x^2}$

74. $-\sqrt{25y^2}$

75. $\sqrt{9y^4}$

76. $\sqrt{a^4b^8}$

77. $\sqrt[3]{8y^3}$

78. $\sqrt[3]{-27z^9}$

79. $\sqrt[4]{\dfrac{x^4y^8}{z^{12}}}$

80. $\sqrt[5]{\dfrac{a^{10}b^5}{c^{15}}}$

Simplify each expression. Assume that all variables represent positive numbers, so that no absolute value symbols are needed.

81. $\sqrt{8} - \sqrt{2}$

82. $\sqrt{75} - 2\sqrt{27}$

83. $\sqrt{200x^2} + \sqrt{98x^2}$

84. $\sqrt{128a^3} - a\sqrt{162a}$

85. $2\sqrt{48y^5} - 3y\sqrt{12y^3}$

86. $y\sqrt{112y} + 4\sqrt{175y^3}$

87. $2\sqrt[3]{81} + 3\sqrt[3]{24}$

88. $3\sqrt[4]{32} - 2\sqrt[4]{162}$

89. $\sqrt[4]{768z^5} + \sqrt[4]{48z^5}$

90. $-2\sqrt[5]{64y^2} + 3\sqrt[5]{486y^2}$

91. $\sqrt{8x^2y} - x\sqrt{2y} + \sqrt{50x^2y}$

92. $3x\sqrt{18x} + 2\sqrt{2x^3} - \sqrt{72x^3}$

93. $\sqrt[3]{16xy^4} + y\sqrt[3]{2xy} - \sqrt[3]{54xy^4}$

94. $\sqrt[4]{512x^5} - \sqrt[4]{32x^5} + \sqrt[4]{1,250x^5}$

Rationalize each denominator and simplify. Assume that all variables represent positive numbers.

95. $\dfrac{3}{\sqrt{3}}$

96. $\dfrac{2}{\sqrt{x}}$

97. $\dfrac{2}{\sqrt[3]{2}}$

98. $\dfrac{5a}{\sqrt[3]{25a}}$

99. $\dfrac{2b}{\sqrt[4]{3a^2}}$

100. $\sqrt{\dfrac{x}{2y}}$

101. $\sqrt[3]{\dfrac{2u^4}{9v}}$

102. $\sqrt[3]{-\dfrac{3s^5}{4r^2}}$

Rationalize each numerator and simplify. Assume that all variables are positive numbers.

103. $\dfrac{\sqrt{5}}{10}$

104. $\dfrac{\sqrt{y}}{3}$

105. $\dfrac{\sqrt[3]{9}}{3}$

106. $\dfrac{\sqrt[3]{16b^2}}{16}$

107. $\dfrac{\sqrt[5]{16b^3}}{64a}$

108. $\sqrt{\dfrac{3x}{57}}$

Rationalize each denominator and simplify.

109. $\sqrt{\dfrac{1}{3}} - \sqrt{\dfrac{1}{27}}$

110. $\sqrt[3]{\dfrac{1}{2}} + \sqrt[3]{\dfrac{1}{16}}$

111. $\sqrt{\dfrac{x}{8}} - \sqrt{\dfrac{x}{2}} + \sqrt{\dfrac{x}{32}}$

112. $\sqrt[3]{\dfrac{y}{4}} + \sqrt[3]{\dfrac{y}{32}} - \sqrt[3]{\dfrac{y}{500}}$

Simplify each radical expression.

113. $\sqrt[4]{9}$

114. $\sqrt[6]{27}$

115. $\sqrt[10]{16x^6}$

116. $\sqrt[6]{27x^9}$

Discovery and Writing   *We often can multiply and divide radicals with different indices. For example, to multiply $\sqrt{3}$ by $\sqrt[3]{5}$, we first write each radical as a sixth root*

$$\sqrt{3} = 3^{1/2} = 3^{3/6} = \sqrt[6]{3^3} = \sqrt[6]{27}$$
$$\sqrt[3]{5} = 5^{1/3} = 5^{2/6} = \sqrt[6]{5^2} = \sqrt[6]{25}$$

and then multiply the sixth roots.

$$\sqrt{3}\sqrt[3]{5} = \sqrt[6]{27}\sqrt[6]{25} = \sqrt[6]{(27)(25)} = \sqrt[6]{675}$$

Division is similar. Use this idea to write each of the following expressions as a single radical.

117. $\sqrt{2}\sqrt[3]{2}$

118. $\sqrt{3}\sqrt[3]{5}$

119. $\dfrac{\sqrt[4]{3}}{\sqrt{2}}$

120. $\dfrac{\sqrt[3]{2}}{\sqrt{5}}$

121. For what values of x does $\sqrt[4]{x^4} = x$? Explain.

122. If all of the radicals involved represent real numbers and $y \neq 0$, explain why

$$\sqrt[n]{\dfrac{x}{y}} = \dfrac{\sqrt[n]{x}}{\sqrt[n]{y}}$$

123. If all of the radicals involved represent real numbers and there is no division by 0, explain why

$$\left(\dfrac{x}{y}\right)^{-m/n} = \sqrt[n]{\dfrac{y^m}{x^m}}$$

124. The definition of $x^{m/n}$ requires that $\sqrt[n]{x}$ be a real number. Explain why this is important. (*Hint:* Consider what happens when n is even, m is odd, and x is negative.)

Review

125. Write $-2 < x \le 5$ using interval notation.

126. Write the expression $|3 - x|$ without using absolute value symbols. Assume that $x > 4$.

Evaluate each expression when $x = -2$ and $y = 3$.

127. $x^2 - y^2$

128. $\dfrac{xy + 4y}{x}$

129. Write 617,000,000 in scientific notation.

130. Write 0.00235×10^4 in standard notation.

0.4 Polynomials

Objectives

1. Define Polynomials
2. Add and Subtract Polynomials
3. Multiply Polynomials
4. Rationalize Denominators
5. Divide Polynomials

Football is one of the most popular sports in the United States. Brett Favre, formerly with the Green Bay Packers, was one of the most talented quarterbacks ever to play the game. He holds the record for the most career NFL touchdown passes, the most NFL pass completions, and the most passing yards, and has led the Packers to the Super Bowl.

An algebraic expression can be used to model the trajectory or path of a football when passed by Brett Favre. Suppose the height in feet, t seconds after the football leaves Favre's hand, is given by the algebraic expression

$$-0.1t^2 + t + 5.5$$

At a time of $t = 3$ seconds, we see that the height of the football is 7.6 ft because

$$-0.1(3)^2 + 3 + 5.5 = 7.6$$

Algebraic expressions such as $-0.1t^2 + t + 5.5$ are called **polynomials,** and we will study them in this section.

1. Define Polynomials

A **monomial** is a real number or the product of a real number and one or more variables with whole-number exponents. The number is called the **coefficient** of the variables. Some examples of monomials are

$$3x, \qquad 7ab^2, \qquad -5ab^2c^4, \qquad x^3, \qquad \text{and} \qquad -12$$

with coefficients of 3, 7, -5, 1, and -12, respectively.

The **degree** of a monomial is the sum of the exponents of its variables. All nonzero constants (except 0) have a degree of 0.

The degree of $3x$ is 1.	The degree of $7ab^2$ is 3.
The degree of $-5ab^2c^4$ is 7.	The degree of x^3 is 3.
The degree of -12 is 0 (since $-12 = -12x^0$).	0 has no defined degree.

A monomial or a sum of monomials is called a **polynomial.** Each monomial in that sum is called a **term of the polynomial.** A polynomial with two terms is called a **binomial,** and a polynomial with three terms is called a **trinomial.**

Monomials	Binomials	Trinomials
$3x^2$	$2a + 3b$	$x^2 + 7x - 4$
$-25xy$	$4x^3 - 3x^2$	$4y^4 - 2y + 12$
a^2b^3c	$-2x^3 - 4y^2$	$12x^3y^2 - 8xy - 24$

The **degree of a polynomial** is the degree of the term in the polynomial with highest degree. The only polynomial with no defined degree is 0, which is called the **zero polynomial.** Here are some examples.

- $3x^2y^3 + 5xy^2 + 7$ is a trinomial of 5th degree, because its term with highest degree (the first term) is 5.

- $3ab + 5a^2b$ is a binomial of degree 3.

- $5x + 3y^2 + \sqrt[4]{3}z^4 - \sqrt{7}$ is a polynomial, because its variables have whole-number exponents. It is of degree 4.

- $-7y^{1/2} + 3y^2 + \sqrt[5]{3}z$ is not a polynomial, because one of its variables (y in the first term) does not have a whole-number exponent.

If two terms of a polynomial have the same variables with the same exponents, they are **like** or **similar** terms. To combine the like terms in the sum $3x^2y + 5x^2y$ or the difference $7xy^2 - 2xy^2$, we use the distributive property:

$$3x^2y + 5x^2y = (3 + 5)x^2y \qquad 7xy^2 - 2xy^2 = (7 - 2)xy^2$$
$$= 8x^2y \qquad\qquad\qquad = 5xy^2$$

This illustrates that *to combine like terms, we add (or subtract) their coefficients and keep the same variables and the same exponents.*

2. Add and Subtract Polynomials

Recall that we can use the distributive property to remove parentheses enclosing the terms of a polynomial. When the sign preceding the parentheses is +, we simply drop the parentheses:

$$+(a + b - c) = +1(a + b - c)$$
$$= 1a + 1b - 1c$$
$$= a + b - c$$

When the sign preceding the parentheses is −, we drop the parentheses and the − sign and change the sign of each term within the parentheses.

$$-(a + b - c) = -1(a + b - c)$$
$$= -1a + (-1)b - (-1)c$$
$$= -a - b + c$$

We can use these facts to add and subtract polynomials. *To add (or subtract) polynomials, we remove parentheses (if necessary) and combine like terms.*

EXAMPLE 1   Add: $(3x^3y + 5x^2 - 2y) + (2x^3y - 5x^2 + 3x)$.

Solution   To add the polynomials, we remove parentheses and combine like terms.

$$(3x^3y + 5x^2 - 2y) + (2x^3y - 5x^2 + 3x)$$
$$= 3x^3y + 5x^2 - 2y + 2x^3y - 5x^2 + 3x$$
$$= 3x^3y + 2x^3y + 5x^2 - 5x^2 - 2y + 3x \qquad \text{Use the commutative property to rearrange terms.}$$
$$= 5x^3y - 2y + 3x \qquad \text{Combine like terms.}$$

We can add the polynomials in a vertical format by writing like terms in a column and adding the like terms, column by column.

$$\begin{array}{l} 3x^3y + 5x^2 - 2y \\ \underline{2x^3y - 5x^2 \qquad + 3x} \\ 5x^3y \qquad - 2y + 3x \end{array}$$

Self Check 1   Add: $(4x^2 + 3x - 5) + (3x^2 - 5x + 7)$.

EXAMPLE 2   Subtract: $(2x^2 + 3y^2) - (x^2 - 2y^2 + 7)$.

Solution   To subtract the polynomials, we remove parentheses and combine like terms.

$$(2x^2 + 3y^2) - (x^2 - 2y^2 + 7)$$
$$= 2x^2 + 3y^2 - x^2 + 2y^2 - 7$$
$$= 2x^2 - x^2 + 3y^2 + 2y^2 - 7 \qquad \text{Use the commutative property to rearrange terms.}$$
$$= x^2 + 5y^2 - 7 \qquad \text{Combine like terms.}$$

We can subtract the polynomials in a vertical format by writing like terms in a column and subtracting the like terms, column by column.

$$\begin{array}{l} 2x^2 + 3y^2 \\ \underline{-(\ x^2 - 2y^2 + 7)} \\ x^2 + 5y^2 - 7 \end{array} \qquad \begin{array}{l} 2x^2 - x^2 = (2 - 1)x^2 = 1x^2 = x^2 \\ 3y^2 - (-2)y^2 = 3y^2 + 2y^2 = (3 + 2)y^2 = 5y^2 \\ 0 - 7 = -7 \end{array}$$

Self Check 2   Subtract: $(4x^2 + 3x - 5) - (3x^2 - 5x + 7)$.

We can also use the distributive property to remove parentheses enclosing several terms that are multiplied by a constant. For example,

$$4(3x^2 - 2x + 6) = 4(3x^2) - 4(2x) + 4(6)$$
$$= 12x^2 - 8x + 24$$

This example suggests that *to add multiples of one polynomial to another, or to subtract multiples of one polynomial from another, we remove parentheses and combine like terms.*

EXAMPLE 3 Simplify: $7x(2y^2 + 13x^2) - 5(xy^2 - 13x^3)$.

Solution
$$7x(2y^2 + 13x^2) - 5(xy^2 - 13x^3)$$
$$= 14xy^2 + 91x^3 - 5xy^2 + 65x^3 \quad \text{Use the distributive property to remove parentheses.}$$
$$= 14xy^2 - 5xy^2 + 91x^3 + 65x^3 \quad \text{Use the commutative property to rearrange terms.}$$
$$= 9xy^2 + 156x^3 \quad \text{Combine like terms.}$$

Self Check 3 Simplify: $3(2b^2 - 3a^2b) + 2b(b + a^2)$.

3. Multiply Polynomials

To find the product of $3x^2y^3z$ and $5xyz^2$, we proceed as follows:
$$(3x^2y^3z)(5xyz^2) = 3 \cdot x^2 \cdot y^3 \cdot z \cdot 5 \cdot x \cdot y \cdot z^2$$
$$= 3 \cdot 5 \cdot x^2 \cdot x \cdot y^3 \cdot y \cdot z \cdot z^2 \quad \text{Use the commutative property to rearrange terms.}$$
$$= 15x^3y^4z^3$$

This illustrates that *to multiply two monomials, we multiply the coefficients and then multiply the variables.*

To find the product of a monomial and a polynomial, we use the distributive property.

$$3xy^2(2xy + x^2 - 7yz) = 3xy^2(2xy) + (3xy^2)(x^2) - (3xy^2)(7yz)$$
$$= 6x^2y^3 + 3x^3y^2 - 21xy^3z$$

This illustrates that *to multiply a polynomial by a monomial, we multiply each term of the polynomial by the monomial.*

To multiply one binomial by another, we use the distributive property twice.

EXAMPLE 4 Multiply: **a.** $(x + y)(x + y)$ **b.** $(x - y)(x - y)$ **c.** $(x + y)(x - y)$

Solution **a.** $(x + y)(x + y) = (x + y)x + (x + y)y$
$$= x^2 + xy + xy + y^2$$
$$= x^2 + 2xy + y^2$$

b. $(x - y)(x - y) = (x - y)x - (x - y)y$
$$= x^2 - xy - xy + y^2$$
$$= x^2 - 2xy + y^2$$

c. $(x + y)(x - y) = (x + y)x - (x + y)y$
$$= x^2 + xy - xy - y^2$$
$$= x^2 - y^2$$

Self Check 4 Multiply: **a.** $(x + 2)(x + 2)$ **b.** $(x - 3)(x - 3)$
c. $(x + 4)(x - 4)$

The products in Example 4 are called **special products.** Because they occur so often, it is worthwhile to learn their forms.

Special Product Formulas

$$(x + y)^2 = (x + y)(x + y) = x^2 + 2xy + y^2$$
$$(x - y)^2 = (x - y)(x - y) = x^2 - 2xy + y^2$$
$$(x + y)(x - y) = x^2 - y^2$$

Comment

Remember that $(x + y)^2$ and $(x - y)^2$ have trinomials for their products and that

$$(x + y)^2 \neq x^2 + y^2 \quad \text{and} \quad (x - y)^2 \neq x^2 - y^2$$

For example,

$$(3 + 5)^2 \neq 3^2 + 5^2 \quad \text{and} \quad (3 - 5)^2 \neq 3^2 - 5^2$$
$$8^2 \neq 9 + 25 \qquad\qquad\qquad (-2)^2 \neq 9 - 25$$
$$64 \neq 34 \qquad\qquad\qquad\qquad 4 \neq -16$$

We can use the **FOIL method** to multiply one binomial by another. The word FOIL is an acronym for **F**irst terms, **O**uter terms, **I**nner terms, and **L**ast terms. To use this method to multiply $3x - 4$ by $2x + 5$, we write

First terms   Last terms

$$(3x - 4)(2x + 5) = 3x(2x) + 3x(5) - 4(2x) - 4(5)$$

Inner terms
$$= 6x^2 + 15x - 8x - 20$$
Outer terms
$$= 6x^2 + 7x - 20$$

In this example,

- the product of the first terms is $6x^2$,
- the product of the outer terms is $15x$,
- the product of the inner terms is $-8x$, and
- the product of the last terms is -20.

The resulting like terms of the product are then combined.

EXAMPLE 5   Use the FOIL method to multiply: $\left(\sqrt{3} + x\right)\left(2 - \sqrt{3}x\right)$.

Solution
$$\left(\sqrt{3} + x\right)\left(2 - \sqrt{3}x\right) = 2\sqrt{3} - \sqrt{3}\sqrt{3}x + 2x - x\sqrt{3}x$$
$$= 2\sqrt{3} - 3x + 2x - \sqrt{3}x^2$$
$$= 2\sqrt{3} - x - \sqrt{3}x^2$$

Self Check 5   Multiply: $\left(2x + \sqrt{3}\right)\left(x - \sqrt{3}\right)$.

To multiply a polynomial with more than two terms by another polynomial, we multiply each term of one polynomial by each term of the other polynomial and combine like terms whenever possible.

EXAMPLE 6 Multiply: **a.** $(x + y)(x^2 - xy + y^2)$ **b.** $(x + 3)^3$

Solution **a.** $(x + y)(x^2 - xy + y^2) = x^3 - x^2y + xy^2 + yx^2 - xy^2 + y^3$
$$= x^3 + y^3$$

b. $(x + 3)^3 = (x + 3)(x + 3)^2$
$$= (x + 3)(x^2 + 6x + 9)$$
$$= x^3 + 6x^2 + 9x + 3x^2 + 18x + 27$$
$$= x^3 + 9x^2 + 27x + 27$$

Self Check 6 Multiply: $(x + 2)(2x^2 + 3x - 1)$.

We can use a vertical format to multiply two polynomials, such as the polynomials given in Self Check 6. We first write the polynomials as follows and draw a line beneath them. We then multiply each term of the upper polynomial by each term of the lower polynomial and write the results so that like terms appear in each column. Finally, we combine like terms, column by column.

$$2x^2 + 3x - 1$$
$$\underline{ x + 2}$$
$4x^2 + 6x - 2$ Multiply $2x^2 + 3x - 1$ by 2.
$\underline{2x^3 + 3x^2 - x}$ Multiply $2x^2 + 3x - 1$ by x.
$2x^3 + 7x^2 + 5x - 2$ In each column, combine like terms.

If n is a whole number, the expressions $a^n + 1$ and $2a^n - 3$ are polynomials and we can multiply them as follows:

$$(a^n + 1)(2a^n - 3) = 2a^{2n} - 3a^n + 2a^n - 3$$
$$= 2a^{2n} - a^n - 3 \qquad \text{Combine like terms.}$$

We also can use the methods previously discussed to multiply expressions that are not polynomials, such as $x^{-2} + y$ and $x^2 - y^{-1}$.

$$(x^{-2} + y)(x^2 - y^{-1}) = x^{-2+2} - x^{-2}y^{-1} + x^2y - y^{1-1}$$
$$= x^0 - \frac{1}{x^2y} + x^2y - y^0$$
$$= 1 - \frac{1}{x^2y} + x^2y - 1 \qquad x^0 = 1 \text{ and } y^0 = 1.$$
$$= x^2y - \frac{1}{x^2y}$$

4. Rationalize Denominators

If the denominator of a fraction is a binomial containing square roots, we can use the product formula $(x + y)(x - y)$ to rationalize the denominator. For example, to rationalize the denominator of

$$\frac{6}{\sqrt{7} + 2}$$ To rationalize a denominator means to change the denominator into a rational number.

we multiply the numerator and denominator by $\sqrt{7} - 2$ and simplify.

$$\frac{6}{\sqrt{7} + 2} = \frac{6\left(\sqrt{7} - 2\right)}{\left(\sqrt{7} + 2\right)\left(\sqrt{7} - 2\right)} \qquad \frac{\sqrt{7} - 2}{\sqrt{7} - 2} = 1$$

$$= \frac{6\left(\sqrt{7} - 2\right)}{7 - 4}$$

$$= \frac{6\left(\sqrt{7} - 2\right)}{3} \qquad \text{Here the denominator is a rational number.}$$

$$= 2\left(\sqrt{7} - 2\right)$$

In this example, we multiplied both the numerator and the denominator of the given fraction by $\sqrt{7} - 2$. This binomial is the same as the denominator of the given fraction $\sqrt{7} + 2$, except for the sign between the terms. Such binomials are called *conjugate binomials* or *radical conjugates*.

Conjugate Binomials | **Conjugate binomials** are binomials that are the same except for the sign between their terms. The conjugate of $a + b$ is $a - b$, and the conjugate of $a - b$ is $a + b$.

EXAMPLE 7 Rationalize the denominator: $\dfrac{\sqrt{3x} - \sqrt{2}}{\sqrt{3x} + \sqrt{2}}$ $(x > 0)$.

Solution We multiply the numerator and the denominator by $\sqrt{3x} - \sqrt{2}$ (the conjugate of $\sqrt{3x} + \sqrt{2}$) and simplify.

$$\frac{\sqrt{3x} - \sqrt{2}}{\sqrt{3x} + \sqrt{2}} = \frac{\left(\sqrt{3x} - \sqrt{2}\right)\left(\sqrt{3x} - \sqrt{2}\right)}{\left(\sqrt{3x} + \sqrt{2}\right)\left(\sqrt{3x} - \sqrt{2}\right)} \qquad \frac{\sqrt{3x} - \sqrt{2}}{\sqrt{3x} - \sqrt{2}} = 1$$

$$= \frac{\sqrt{3x}\sqrt{3x} - \sqrt{3x}\sqrt{2} - \sqrt{2}\sqrt{3x} + \sqrt{2}\sqrt{2}}{\left(\sqrt{3x}\right)^2 - \left(\sqrt{2}\right)^2}$$

$$= \frac{3x - \sqrt{6x} - \sqrt{6x} + 2}{3x - 2}$$

$$= \frac{3x - 2\sqrt{6x} + 2}{3x - 2}$$

Self Check 7 Rationalize the denominator: $\dfrac{\sqrt{x} + 2}{\sqrt{x} - 2}$.

In calculus, we often rationalize a numerator.

EXAMPLE 8 Rationalize the numerator: $\dfrac{\sqrt{x + h} - \sqrt{x}}{h}$.

Solution To rid the numerator of radicals, we multiply the numerator and the denominator by the conjugate of the numerator and simplify.

$$\frac{\sqrt{x+h}-\sqrt{x}}{h}=\frac{\left(\sqrt{x+h}-\sqrt{x}\right)\left(\sqrt{x+h}+\sqrt{x}\right)}{h\left(\sqrt{x+h}+\sqrt{x}\right)} \qquad \frac{\sqrt{x+h}+\sqrt{x}}{\sqrt{x+h}+\sqrt{x}}=1$$

$$=\frac{x+h-x}{h\left(\sqrt{x+h}+\sqrt{x}\right)} \qquad \text{Here the numerator has no radicals.}$$

$$=\frac{h}{h\left(\sqrt{x+h}+\sqrt{x}\right)}$$

$$=\frac{1}{\sqrt{x+h}+\sqrt{x}} \qquad \text{Divide out the common factor of } h.$$

Self Check 8   Rationalize the numerator: $\dfrac{\sqrt{4+h}-2}{h}$.

5.   Divide Polynomials

To divide monomials, we write the quotient as a fraction and simplify by using the rules of exponents. For example,

$$\frac{6x^2y^3}{-2x^3y}=-3x^{2-3}y^{3-1}$$

$$=-3x^{-1}y^2$$

$$=-\frac{3y^2}{x}$$

To divide a polynomial by a monomial, we write the quotient as a fraction, write the fraction as a sum of separate fractions, and simplify each one. For example, to divide $8x^5y^4+12x^2y^5-16x^2y^3$ by $4x^3y^4$, we proceed as follows:

$$\frac{8x^5y^4+12x^2y^5-16x^2y^3}{4x^3y^4}=\frac{8x^5y^4}{4x^3y^4}+\frac{12x^2y^5}{4x^3y^4}+\frac{-16x^2y^3}{4x^3y^4}$$

$$=2x^2+\frac{3y}{x}-\frac{4}{xy}$$

Gottfried Wilhelm Leibniz
(1646-1716)
Leibniz, a German philosopher and logician, is best known as one of the inventors of calculus, along with Isaac Newton. He also developed the binary numeration system that is basic to modern computers. He even invented a calculating machine that would add, subtract, multiply, divide, and find roots of numbers.

To divide two polynomials, we can use long division. To illustrate, we consider the division

$$\frac{2x^2+11x-30}{x+7}$$

which can be written in long division form as

$$x+7\overline{)2x^2+11x-30}$$

The binomial $x+7$ is called the **divisor,** and the trinomial $2x^2+11x-30$ is called the **dividend.** The final answer, called the **quotient,** will appear above the long division symbol.

We begin the division by asking "What monomial, when multiplied by x, gives $2x^2$?" Because $x\cdot 2x=2x^2$, the answer is $2x$. We place $2x$ in the quotient, multiply each term of the divisor by $2x$, subtract, and bring down the -30.

$$\begin{array}{r} 2x \\ x+7\overline{)2x^2+11x-30} \\ \underline{2x^2+14x} \\ -3x-30 \end{array}$$

We continue the division by asking "What monomial, when multiplied by x, gives $-3x$?" We place the answer, -3, in the quotient, multiply each term of the divisor by -3, and subtract. This time, there is no number to bring down.

$$
\begin{array}{r}
2x \ - \ \ 3 \\
x + 7 \overline{)\ 2x^2 + 11x - 30} \\
\underline{2x^2 + 14x } \\
-\ 3x - 30 \\
\underline{-\ 3x - 21} \\
-\ 9
\end{array}
$$

Because the degree of the remainder, -9, is less than the degree of the divisor, the division process stops, and we can express the result in the form

$$
\text{quotient} + \frac{\text{remainder}}{\text{divisor}}
$$

Thus,

$$
\frac{2x^2 + 11x - 30}{x + 7} = 2x - 3 + \frac{-9}{x + 7}
$$

EXAMPLE 9 Divide $6x^3 - 11$ by $2x + 2$.

Solution We set up the division, leaving spaces for the missing powers of x in the dividend.

$$
2x + 2 \overline{)\ 6x^3 - 11}
$$

The division process continues as usual, with the following results:

$$
\begin{array}{r}
3x^2 - 3x + 3 \\
2x + 2 \overline{)\ 6x^3 - 11} \\
\underline{6x^3 + 6x^2 } \\
-\ 6x^2 \\
\underline{-\ 6x^2 - 6x } \\
+\ 6x - 11 \\
\underline{+\ 6x + \ \ 6} \\
-\ 17
\end{array}
$$

Thus, $\dfrac{6x^3 - 11}{2x + 2} = 3x^2 - 3x + 3 + \dfrac{-17}{2x + 2}$.

Comment

In Example 9, we could write the missing powers of x using coefficients of 0:

$$
2x + 2 \overline{)\ 6x^3 + 0x^2 + 0x - 11}
$$

Self Check 9 Divide: $3x + 1 \overline{)\ 9x^2 - 1}$.

EXAMPLE 10 Divide $-3x^3 - 3 + x^5 + 4x^2 - x^4$ by $x^2 - 3$.

Solution The division process works best when the terms in the divisor and dividend are written with their exponents in descending order.

$$
\begin{array}{r}
x^3 - x^2 + 1 \\
x^2 - 3\overline{)x^5 - x^4 - 3x^3 + 4x^2 - 3} \\
\underline{x^5 \qquad - 3x^3} \\
-x^4 \qquad\quad + 4x^2 \\
\underline{-x^4 \qquad\quad + 3x^2} \\
x^2 - 3 \\
\underline{x^2 - 3} \\
0
\end{array}
$$

Thus, $\dfrac{-3x^3 - 3 + x^5 + 4x^2 - x^4}{x^2 - 3} = x^3 - x^2 + 1.$

Self Check 10 Divide: $x^2 + 1\overline{)3x^2 - x + 1 - 2x^3 + 3x^4}.$

Self Check Answers
1. $7x^2 - 2x + 2$ **2.** $x^2 + 8x - 12$ **3.** $8b^2 - 7a^2b$ **4. a.** $x^2 + 4x + 4$
b. $x^2 - 6x + 9$ **c.** $x^2 - 16$ **5.** $2x^2 - x\sqrt{3} - 3$ **6.** $2x^3 + 7x^2 + 5x - 2$
7. $\dfrac{x + 4\sqrt{x} + 4}{x - 4}$ **8.** $\dfrac{1}{\sqrt{4 + h} + 2}$ **9.** $3x - 1$ **10.** $3x^2 - 2x + \dfrac{x + 1}{x^2 + 1}$

0.4 Exercises

Vocabulary and Concepts   *Fill in the blanks.*

1. A _____ is a real number or the product of a real number and one or more _____.

2. The _____ of a monomial is the sum of the exponents of its _____.

3. A _____ is a polynomial with three terms.

4. A _____ is a polynomial with two terms.

5. A monomial is a polynomial with ___ term.

6. The constant 0 is called the ___ polynomial.

7. Terms with the same variables with the same exponents are called ___ terms.

8. The _____ of a polynomial is the same as the degree of its term of highest degree.

9. To combine like terms, we add their _____ and keep the same _____ and the same exponents.

10. The conjugate of $3\sqrt{x} + 2$ is _____.

Determine whether the given expression is a polynomial. If so, tell whether it is a monomial, a binomial, or a trinomial, and give its degree.

11. $x^2 + 3x + 4$

12. $5xy - x^3$

13. $x^3 + y^{1/2}$

14. $x^{-3} - 5y^{-2}$

15. $4x^2 - \sqrt{5x^3}$

16. x^2y^3

17. $\sqrt{15}$

18. $\dfrac{5}{x} + \dfrac{x}{5} + 5$

19. 0

20. $3y^3 - 4y^2 + 2y + 2$

Practice   *Perform the operations and simplify.*

21. $(x^3 - 3x^2) + (5x^3 - 8x)$

22. $(2x^4 - 5x^3) + (7x^3 - x^4 + 2x)$

23. $(y^5 + 2y^3 + 7) - (y^5 - 2y^3 - 7)$

24. $(3t^7 - 7t^3 + 3) - (7t^7 - 3t^3 + 7)$

25. $2(x^2 + 3x - 1) - 3(x^2 + 2x - 4) + 4$

26. $5(x^3 - 8x + 3) + 2(3x^2 + 5x) - 7$

27. $8(t^2 - 2t + 5) + 4(t^2 - 3t + 2) - 6(2t^2 - 8)$

28. $-3(x^3 - x) + 2(x^2 + x) + 3(x^3 - 2x)$

29. $y(y^3 - 1) - y^2(y^2 + 2) - y(2y^2 - 2)$

30. $-4a^2(a + 1) + 3a(a^2 - 4) - a^2(a + 2)$

$(x^2y - 4y^2x) - (yx^2 + 3xy^2) + (2x^2y + 3xy^2)$

31. $xy(x - 4y) - y(x^2 + 3xy) + xy(2x + 3y)$

32. $3mn(m + 2n) - 6m(3mn + 1) - 2n(4mn - 1)$

33. $2x^2y^3(4xy^4)$

34. $-15a^3b(-2a^2b^3)$

35. $-3m^2n(2mn^2)\left(-\dfrac{mn}{12}\right)$

36. $-\dfrac{3r^2s^3}{5}\left(\dfrac{2r^2s}{3}\right)\left(\dfrac{15rs^2}{2}\right)$

37. $-4rs(r^2 + s^2)$

38. $6u^2v(2uv^2 - y)$

39. $6ab^2c(2ac + 3bc^2 - 4ab^2c)$

$12a^2b^2c^2 + 18ab^3c^3 - 24a^2b^4c^2$

40. $-\dfrac{mn^2}{2}(4mn - 6m^2 - 8)$

41. $(a + 2)(a + 2)$ **42.** $(y - 5)(y - 5)$

43. $(a - 6)^2$ **44.** $(t + 9)^2$

45. $(x + 4)(x - 4)$ **46.** $(z + 7)(z - 7)$

47. $(x - 3)(x + 5)$ **48.** $(z + 4)(z - 6)$

49. $(u + 2)(3u - 2)$ **50.** $(4x + 1)(2x - 3)$

51. $(5x - 1)(2x + 3)$ **52.** $(4x - 1)(2x - 7)$

53. $(3a - 2b)^2$ **54.** $(4a + 5b)(4a - 5b)$

55. $(3m + 4n)(3m - 4n)$ **56.** $(4r + 3s)^2$

57. $(2y - 4x)(3y - 2x)$ **58.** $(-2x + 3y)(3x + y)$

59. $(9x - y)(x^2 - 3y)$ **60.** $(8a^2 + b)(a + 2b)$

61. $(5z + 2t)(z^2 - t)$ **62.** $(y - 2x^2)(x^2 + 3y)$

63. $(3x - 1)^3$ **64.** $(2x - 3)^3$

65. $(3x + 1)(2x^2 + 4x - 3)$

66. $(2x - 5)(x^2 - 3x + 2)$

67. $(3x + 2y)(2x^2 - 3xy + 4y^2)$

68. $(4r - 3s)(2r^2 + 4rs - 2s^2)$

Multiply the expressions as you would multiply polynomials.

69. $2y^n(3y^n + y^{-n})$

70. $3a^{-n}(2a^n + 3a^{n-1})$

71. $-5x^{2n}y^n(2x^{2n}y^{-n} + 3x^{-2n}y^n)$

72. $-2a^{3n}b^{2n}(5a^{-3n}b - ab^{-2n})$

73. $(x^n + 3)(x^n - 4)$

74. $(a^n - 5)(a^n - 3)$

75. $(2r^n - 7)(3r^n - 2)$

76. $(4z^n + 3)(3z^n + 1)$

77. $x^{1/2}(x^{1/2}y + xy^{1/2})$

78. $ab^{1/2}(a^{1/2}b^{1/2} + b^{1/2})$

79. $(a^{1/2} + b^{1/2})(a^{1/2} - b^{1/2})$

80. $(x^{3/2} + y^{1/2})^2$

Rationalize each denominator.

81. $\dfrac{2}{\sqrt{3} - 1}$ **82.** $\dfrac{1}{\sqrt{5} + 2}$

83. $\dfrac{3x}{\sqrt{7} + 2}$ **84.** $\dfrac{14y}{\sqrt{2} - 3}$

85. $\dfrac{x}{x - \sqrt{3}}$ **86.** $\dfrac{y}{2y + \sqrt{7}}$

87. $\dfrac{y + \sqrt{2}}{y - \sqrt{2}}$ **88.** $\dfrac{x - \sqrt{3}}{x + \sqrt{3}}$

89. $\dfrac{\sqrt{2} - \sqrt{3}}{1 - \sqrt{3}}$ **90.** $\dfrac{\sqrt{3} - \sqrt{2}}{1 + \sqrt{2}}$

91. $\dfrac{\sqrt{x} - \sqrt{y}}{\sqrt{x} + \sqrt{y}}$ **92.** $\dfrac{\sqrt{2x} + y}{\sqrt{2x} - y}$

Rationalize each numerator.

93. $\dfrac{\sqrt{2} + 1}{2}$ **94.** $\dfrac{\sqrt{x} - 3}{3}$

95. $\dfrac{y - \sqrt{3}}{y + \sqrt{3}}$ **96.** $\dfrac{\sqrt{a} - \sqrt{b}}{\sqrt{a} + \sqrt{b}}$

97. $\dfrac{\sqrt{x+3} - \sqrt{x}}{3}$ **98.** $\dfrac{\sqrt{2+h} - \sqrt{2}}{h}$

Perform each division and write all answers without using negative exponents.

99. $\dfrac{36a^2b^3}{18ab^6}$

100. $\dfrac{-45r^2s^5t^3}{27r^6s^2t^8}$

101. $\dfrac{16x^6y^4z^9}{-24x^9y^6z^0}$

102. $\dfrac{32m^6n^4p^2}{26m^6n^7p^2}$

103. $\dfrac{5x^3y^2 + 15x^3y^4}{10x^2y^3}$

104. $\dfrac{9m^4n^9 - 6m^3n^4}{12m^3n^3}$

105. $\dfrac{24x^5y^7 - 36x^2y^5 + 12xy}{60x^5y^4}$

106. $\dfrac{9a^3b^4 + 27a^2b^4 - 18a^2b^3}{18a^2b^7}$

Perform each division. If there is a nonzero remainder, write the answer in quotient + $\frac{remainder}{divisor}$ form.

107. $x + 3\overline{)3x^2 + 11x + 6}$

108. $3x + 2\overline{)3x^2 + 11x + 6}$

109. $2x - 5\overline{)2x^2 - 19x + 37}$

110. $x - 7\overline{)2x^2 - 19x + 35}$

111. $x^2 + x - 1\overline{)x^3 - 2x^2 - 4x + 3}$

112. $x^2 - 3\overline{)x^3 - 2x^2 - 4x + 5}$

113. $\dfrac{x^5 - 2x^3 - 3x^2 + 9}{x^3 - 2}$

114. $\dfrac{x^5 - 2x^3 - 3x^2 + 9}{x^3 - 3}$

115. $\dfrac{x^5 - 32}{x - 2}$

116. $\dfrac{x^4 - 1}{x + 1}$

117. $11x - 10 + 6x^2\overline{)36x^4 - 121x^2 + 120 + 72x^3 - 142x}$

118. $x + 6x^2 - 12\overline{)-121x^2 + 72x^3 - 142x + 120 + 36x^4}$

Applications

119. Geometry Find an expression that represents the area of the brick wall.

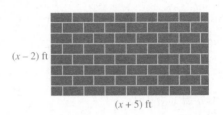

$(x - 2)$ ft

$(x + 5)$ ft

120. Geometry The area of the triangle shown in the illustration is represented as $(x^2 + 3x - 40)$ square feet. Find an expression that represents its height.

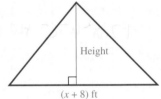

Height

$(x + 8)$ ft

121. Gift boxes The corners of a 12 in.-by-12 in. piece of cardboard are folded inward and glued to make a box. Write a polynomial that represents the volume of the resulting box shown below.

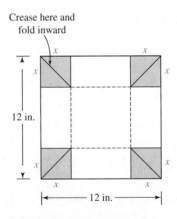

Crease here and fold inward

x

x

12 in.

12 in.

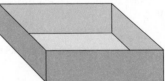

122. Travel Complete the following table, which shows the rate (mph), time traveled (hr), and distance traveled (mi) by a family on vacation.

r	$\cdot$	t	$=$	d
$3x + 4$				$3x^2 + 19x + 20$

Discovery and Writing

123. Show that a trinomial can be squared by using the formula $(a + b + c)^2 = a^2 + b^2 + c^2 + 2ab + 2bc + 2ac$.

124. Show that $(a + b + c + d)^2 = a^2 + b^2 + c^2 + d^2 + 2ab + 2ac + 2ad + 2bc + 2bd + 2cd$.

125. Explain the FOIL method.

126. Explain how to rationalize the numerator of $\dfrac{\sqrt{x} + 2}{x}$.

127. Explain why $(a + b)^2 \neq a^2 + b^2$.

128. Explain why $\sqrt{a^2 + b^2} \neq \sqrt{a^2} + \sqrt{b^2}$.

Review *Simplify each expression. Assume that all variables represent positive numbers.*

129. $9^{3/2}$

130. $\left(\dfrac{8}{125}\right)^{-2/3}$

131. $\left(\dfrac{625x^4}{16y^8}\right)^{3/4}$

132. $\sqrt{80x^4}$

133. $\sqrt[3]{16ab^4} - b\sqrt[3]{54ab}$

134. $x\sqrt[4]{1{,}280x} + \sqrt[4]{80x^5}$

Juan Mabromata/AFP/Getty Images

0.5 Factoring Polynomials

Objectives

1. Factor Out a Common Monomial
2. Factor by Grouping
3. Factor the Difference of Two Squares
4. Factor Trinomials
5. Factor Trinomials by Grouping
6. Factor the Sum and Difference of Two Cubes
7. Factor Miscellaneous Polynomials

The television network series *CSI: Crime Scene Investigation* and its spinoffs, *CSI: NY* and *CSI: Miami,* are favorite television shows of many people. Its fans enjoy watching a team of forensic scientists uncover the circumstances that led to an unusual death or crime. These mysteries are captivating because many viewers are interested in the field of forensic science.

In this section, we will investigate a mathematics mystery. We will be given a polynomial and asked to unveil or uncover the two or more polynomials that were multiplied together to obtain the given polynomial. The process that we will use to solve this mystery is called *factoring.*

When two or more polynomials are multiplied together, each one is called a **factor** of the resulting product. For example, the factors of $7(x + 2)(x + 3)$ are

$$7, \quad x + 2, \quad \text{and} \quad x + 3$$

The process of writing a polynomial as the product of several factors is called *factoring.*

In this section, we will discuss factoring where the coefficients of the polynomial and the polynomial factors are integers. If a polynomial cannot be factored by using integers only, we call it a **prime polynomial.**

1. Factor Out a Common Monomial

The simplest type of factoring occurs when we see a common monomial factor in each term of the polynomial. In this case, our strategy is to use the distributive property and factor out the greatest common factor.

4. Factor Trinomials

Trinomials that are squares of binomials can be factored by using the following formulas.

Factoring Trinomial Squares	(1)	$x^2 + 2xy + y^2 = (x + y)(x + y) = (x + y)^2$
	(2)	$x^2 - 2xy + y^2 = (x - y)(x - y) = (x - y)^2$

For example, to factor $a^2 - 6a + 9$, we note that it can be written in the form

$$a^2 - 2(3a) + 3^2 \quad x = a \text{ and } y = 3.$$

which matches the left side of Equation 2 above. Thus,

$$\begin{aligned} a^2 - 6a + 9 &= a^2 - 2(3a) + 3^2 \\ &= (a - 3)(a - 3) \\ &= (a - 3)^2 \quad \text{You can check by multiplying.} \end{aligned}$$

Factoring trinomials that are not squares of binomials often requires some guesswork. If a trinomial with no common factors is factorable, it will factor into the product of two binomials.

EXAMPLE 7 Factor: $x^2 + 3x - 10$.

Solution To factor $x^2 + 3x - 10$, we must find two binomials $x + a$ and $x + b$ such that

$$x^2 + 3x - 10 = (x + a)(x + b)$$

where the product of a and b is -10 and the sum of a and b is 3.

$$ab = -10 \qquad \text{and} \qquad a + b = 3$$

To find such numbers, we list the possible factorizations of -10:

$$10(-1) \qquad 5(-2) \qquad -10(1) \qquad -5(2)$$

Only in the factorization $5(-2)$ do the factors have a sum of 3. Thus, $a = 5$ and $b = -2$, and

$$x^2 + 3x - 10 = (x + a)(x + b)$$

$$(3) \qquad x^2 + 3x - 10 = (x + 5)(x-2) \quad \text{You can check by multiplying.}$$

Because of the commutative property of multiplication, the order of the factors in Equation 3 is not important. Equation 3 also can be written as

$$x^2 + 3x - 10 = (x - 2)(x + 5)$$

Self Check 7 Factor: $p^2 - 5p - 6$.

EXAMPLE 8 Factor: $2x^2 - x - 6$.

Solution Since the first term is $2x^2$, the first terms of the binomial factors must be $2x$ and x:

$$2x^2 - x - 6 = \left(2x \quad\right)\left(x \quad\right)$$

The product of the last terms must be -6, and the sum of the products of the outer terms and the inner terms must be $-x$. Since the only factorization of -6 that will cause this to happen is $3(-2)$, we have

$$2x^2 - x - 6 = (2x + 3)(x - 2) \quad \text{You can check by multiplying.}$$

Self Check 8   Factor: $6x^2 - x - 2$.

It is not easy to give specific rules for factoring trinomials, because some guesswork is often necessary. However, the following hints are helpful.

Strategy for Factoring a General Trinomial with Integer Coefficients

1. Write the trinomial in descending powers of one variable.

2. Factor out any greatest common factor, including -1 if that is necessary to make the coefficient of the first term positive.

3. When the sign of the first term of a trinomial is $+$ and the sign of the third term is $+$, the sign between the terms of each binomial factor is the same as the sign of the middle term of the trinomial.
   When the sign of the first term is $+$ and the sign of the third term is $-$, one of the signs between the terms of the binomial factors is $+$ and the other is $-$.

4. Try various combinations of first terms and last terms until you find one that works. If no possibilities work, the trinomial is prime.

5. Check the factorization by multiplication.

EXAMPLE 9   Factor: $10xy + 24y^2 - 6x^2$.

Solution   We write the trinomial in descending powers of x and then factor out the common factor of -2.

$$10xy + 24y^2 - 6x^2 = -6x^2 + 10xy + 24y^2$$
$$= -2(3x^2 - 5xy - 12y^2)$$

Since the sign of the third term of $3x^2 - 5xy - 12y^2$ is $-$, the signs between the binomial factors will be opposite. Since the first term is $3x^2$, the first terms of the binomial factors must be $3x$ and x:

$$-2(3x^2 - 5xy - 12y^2) = -2\left(3x \quad \right)\left(x \quad \right)$$

The product of the last terms must be $-12y^2$, and the sum of the outer terms and the inner terms must be $-5xy$. Of the many factorizations of $-12y^2$, only $4y(-3y)$ leads to a middle term of $-5xy$. So we have

$$10xy + 24y^2 - 6x^2 = -6x^2 + 10xy + 24y^2$$
$$= -2(3x^2 - 5xy - 12y^2)$$
$$= -2(3x + 4y)(x - 3y) \qquad \text{You can check by multiplying.}$$

Self Check 9   Factor: $-6x^2 - 15xy - 6y^2$.

5.   Factor Trinomials by Grouping

Another way of factoring trinomials involves factoring by grouping. This method can be used to factor trinomials of the form $ax^2 + bx + c$. For example, to factor $6x^2 + 5x - 6$, we note that $a = 6$, $b = 5$, and $c = -6$, and proceed as follows:

1. Find the product ac: $6(-6) = -36$. This number is called the **key number.**

2. Find two factors of the key number (-36) whose sum is $b = 5$. Two such numbers are 9 and -4.

$$9(-4) = -36 \qquad \text{and} \qquad 9 + (-4) = 5$$

3. Use the factors 9 and -4 as coefficients of two terms to be placed between $6x^2$ and -6.

$$6x^2 + 5x - 6 = 6x^2 + 9x - 4x - 6$$

4. Factor by grouping:

$$6x^2 + 9x - 4x - 6 = 3x(2x + 3) - 2(2x + 3)$$
$$= (2x + 3)(3x - 2) \qquad \text{Factor out } 2x + 3.$$

EXAMPLE 10   Factor by grouping: $15x^2 + x - 2$.

Solution   Since $a = 15$ and $c = -2$ in the trinomial, $ac = -30$. We now find factors of -30 whose sum is $b = 1$. Such factors are 6 and -5. We use these factors as coefficients of two terms to be placed between $15x^2$ and -2.

$$15x^2 + 6x - 5x - 2$$

Finally, we factor by grouping.

$$3x(5x + 2) - (5x + 2) = (5x + 2)(3x - 1) \qquad \text{You can check by multiplying.}$$

Self Check 10   Factor by grouping: $15a^2 + 17a - 4$.

We often can factor polynomials with variable exponents. For example, if n is a natural number,

$$a^{2n} - 5a^n - 6 = (a^n + 1)(a^n - 6)$$

because

$$(a^n + 1)(a^n - 6) = a^{2n} - 6a^n + a^n - 6$$
$$= a^{2n} - 5a^n - 6 \qquad \text{Combine like terms.}$$

6. Factor the Sum and Difference of Two Cubes

Two other types of factoring involve binomials that are the sum or the difference of two cubes. Like the difference of two squares, they can be factored by using a formula.

Factoring the Sum and Difference of Two Cubes

$$x^3 + y^3 = (x + y)(x^2 - xy + y^2)$$
$$x^3 - y^3 = (x - y)(x^2 + xy + y^2)$$

EXAMPLE 11   Factor: $x^3 - 8$.

Solution   This binomial can be written as $x^3 - 2^3$, which is the difference of two cubes. Substituting into the formula for the difference of two cubes gives

$$x^3 - 2^3 = (x - 2)(x^2 + 2x + 2^2)$$
$$= (x - 2)(x^2 + 2x + 4) \qquad \text{You can check by multiplying.}$$

Self Check 11   Factor: $p^3 - 64$.

EXAMPLE 12   Factor: $27x^6 + 64y^3$.

Solution   We can write this expression as the sum of two cubes and factor it as follows:
$$27x^6 + 64y^3 = (3x^2)^3 + (4y)^3$$
$$= (3x^2 + 4y)[(3x^2)^2 - (3x^2)(4y) + (4y)^2]$$
$$= (3x^2 + 4y)(9x^4 - 12x^2y + 16y^2) \quad \text{You can check by multiplying.}$$

Self Check 12   Factor: $8a^3 + 1{,}000b^6$.

7.   Factor Miscellaneous Polynomials

EXAMPLE 13   Factor: $x^2 - y^2 + 6x + 9$.

Solution   Here we will factor a trinomial and a difference of two squares.
$$x^2 - y^2 + 6x + 9 = x^2 + 6x + 9 - y^2 \qquad \text{Use the commutative property to}$$
$$\text{rearrange the terms.}$$
$$= (x + 3)^2 - y^2 \qquad \text{Factor } x^2 + 6x + 9.$$
$$= (x + 3 + y)(x + 3 - y) \qquad \text{Factor the difference of two}$$
$$\text{squares.}$$

We could try to factor this expression in another way.
$$x^2 - y^2 + 6x + 9 = (x + y)(x - y) + 3(2x + 3) \qquad \text{Factor } x^2 - y^2 \text{ and } 6x + 9.$$

However, we are unable to finish the factorization. If grouping in one way doesn't work, try various other ways.

Self Check 13   Factor: $a^2 + 8a - b^2 + 16$.

EXAMPLE 14   Factor: $z^4 - 3z^2 + 1$.

Solution   This trinomial cannot be factored as the product of two binomials, because no combination will give a middle term of $-3z^2$. However, if the middle term were $-2z^2$, the trinomial would be a perfect square, and the factorization would be easy:
$$z^4 - 2z^2 + 1 = (z^2 - 1)(z^2 - 1)$$
$$= (z^2 - 1)^2$$

We can change the middle term in $z^4 - 3z^2 + 1$ to $-2z^2$ by adding z^2 to it. However, to make sure that adding z^2 does not change the value of the trinomial, we must also subtract z^2. We can then proceed as follows.
$$z^4 - 3z^2 + 1 = z^4 - 3z^2 + z^2 + 1 - z^2 \qquad \text{Add and subtract } z^2.$$
$$= z^4 - 2z^2 + 1 - z^2 \qquad \text{Combine } -3z^2 \text{ and } z^2.$$
$$= (z^2 - 1)^2 - z^2 \qquad \text{Factor } z^4 - 2z^2 + 1.$$
$$= (z^2 - 1 + z)(z^2 - 1 - z) \qquad \text{Factor the difference of two squares.}$$

In this type of problem, we will always try to add and subtract a perfect square in hopes of making a perfect-square trinomial that will lead to factoring a difference of two squares.

Self Check 14   Factor: $x^4 + 3x^2 + 4$.

It is helpful to identify the problem type when we must factor polynomials that are given in random order.

Identifying Factoring Problem Types

1. Factor out all common monomial factors.
2. If an expression has two terms, check whether the problem type is
   a. **The difference of two squares:** $x^2 - y^2 = (x + y)(x - y)$
   b. **The sum of two cubes:** $x^3 + y^3 = (x + y)(x^2 - xy + y^2)$
   c. **The difference of two cubes:** $x^3 - y^3 = (x - y)(x^2 + xy + y^2)$
3. If an expression has three terms, try to factor it as a **trinomial.**
4. If an expression has four or more terms, try factoring by **grouping.**
5. Continue until each individual factor is prime.
6. Check the results by multiplying.

Self Check Answers

1. $4a(a - 2b)$ 2. $a^2b^2c^2(1 + abc)$ 3. $(x + 2)(x + y)$
4. $(3a + 4b)(3a - 4b)$ 5. $(a^2 + 9b^2)(a + 3b)(a - 3b)$ 6. $-3(x + 2)(x - 2)$
7. $(p - 6)(p + 1)$ 8. $(3x - 2)(2x + 1)$ 9. $-3(x + 2y)(2x + y)$
10. $(3a + 4)(5a - 1)$ 11. $(p - 4)(p^2 + 4p + 16)$
12. $8(a + 5b^2)(a^2 - 5ab^2 + 25b^4)$ 13. $(a + 4 + b)(a + 4 - b)$
14. $(x^2 + 2 + x)(x^2 + 2 - x)$

0.5 Exercises

Vocabulary and Concepts   *Fill in the blanks.*

1. When polynomials are multiplied together, each polynomial is a _____ of the product.
2. If a polynomial cannot be factored using _____ coefficients, it is called a _____ polynomial.

Complete each factoring formula.

3. $ax + bx =$ _____
4. $x^2 - y^2 =$ _____
5. $x^2 + 2xy + y^2 =$ _____
6. $x^2 - 2xy + y^2 =$ _____
7. $x^3 + y^3 =$ _____
8. $x^3 - y^3 =$ _____

Practice   *In each expression, factor out the greatest common monomial.*

9. $3x - 6$
10. $5y - 15$
11. $8x^2 + 4x^3$
12. $9y^3 + 6y^2$
13. $7x^2y^2 + 14x^3y^2$
14. $25y^2z - 15yz^2$

In each expression, factor by grouping.

15. $a(x + y) + b(x + y)$
16. $b(x - y) + a(x - y)$
17. $4a + b - 12a^2 - 3ab$
18. $x^2 + 4x + xy + 4y$

Factor each expression, if possible.

19. $4x^2 - 9$
20. $36z^2 - 49$
21. $4 - 9r^2$
22. $16 - 49x^2$
23. $25x^4 + 1$
24. $36t^4 + 121$
25. $(x + z)^2 - 25$
26. $(x - y)^2 - 9$

In each expression, factor the trinomial.

27. $x^2 + 8x + 16$
28. $a^2 - 12a + 36$
29. $b^2 - 10b + 25$
30. $y^2 + 14y + 49$
31. $m^2 + 4mn + 4n^2$
32. $r^2 - 8rs + 16s^2$

33. $12x^2 - xy - 6y^2$ **34.** $8x^2 - 10xy - 3y^2$

In each expression, factor the trinomial by grouping.

35. $x^2 + 10x + 21$ **36.** $x^2 + 7x + 10$

37. $x^2 - 4x - 12$ **38.** $x^2 - 2x - 63$

39. $6p^2 + 7p - 3$ **40.** $4q^2 - 19q + 12$

In each expression, factor the sum of two cubes.

41. $t^3 + 343$ **42.** $r^3 + 8s^3$

In each expression, factor the difference of two cubes.

43. $8z^3 - 27$ **44.** $125a^3 - 64$

Factor each expression completely. If an expression is prime, so indicate.

45. $3a^2bc + 6ab^2c + 9abc^2$
46. $5x^3y^3z^3 + 25x^2y^2z^2 - 125xyz$
47. $3x^3 + 3x^2 - x - 1$
48. $4x + 6xy - 9y - 6$
49. $2txy + 2ctx - 3ty - 3ct$
50. $2ax + 4ay - bx - 2by$
51. $ax + bx + ay + by + az + bz$
52. $6x^2y^3 + 18xy + 3x^2y^2 + 9x$
53. $x^2 - (y - z)^2$ **54.** $z^2 - (y + 3)^2$

55. $(x - y)^2 - (x + y)^2$ **56.** $(2a + 3)^2 - (2a - 3)^2$

57. $x^4 - y^4$ **58.** $z^4 - 81$

59. $3x^2 - 12$ **60.** $3x^3y - 3xy$

61. $18xy^2 - 8x$ **62.** $27x^2 - 12$

63. $x^2 - 2x + 15$ **64.** $x^2 + x + 2$

65. $-15 + 2a + 24a^2$ **66.** $-32 - 68x + 9x^2$

67. $6x^2 + 29xy + 35y^2$ **68.** $10x^2 - 17xy + 6y^2$

69. $12p^2 - 58pq - 70q^2$ **70.** $3x^2 - 6xy - 9y^2$

71. $-6m^2 + 47mn - 35n^2$ **72.** $-14r^2 - 11rs + 15s^2$

73. $-6x^3 + 23x^2 + 35x$ **74.** $-y^3 - y^2 + 90y$

75. $6x^4 - 11x^3 - 35x^2$ **76.** $12x + 17x^2 - 7x^3$

77. $x^4 + 2x^2 - 15$ **78.** $x^4 - x^2 - 6$

79. $a^{2n} - 2a^n - 3$ **80.** $a^{2n} + 6a^n + 8$

81. $6x^{2n} - 7x^n + 2$ **82.** $9x^{2n} + 9x^n + 2$

83. $4x^{2n} - 9y^{2n}$ **84.** $8x^{2n} - 2x^n - 3$

85. $10y^{2n} - 11y^n - 6$ **86.** $16y^{4n} - 25y^{2n}$

87. $2x^3 + 2,000$ **88.** $3y^3 + 648$

89. $(x + y)^3 - 64$

90. $(x - y)^3 + 27$

91. $64a^6 - y^6$

92. $a^6 + b^6$
93. $a^3 - b^3 + a - b$
94. $(a^2 - y^2) - 5(a + y)$
95. $64x^6 + y^6$
96. $z^2 + 6z + 9 - 225y^2$
97. $x^2 - 6x + 9 - 144y^2$
98. $x^2 + 2x - 9y^2 + 1$
99. $(a + b)^2 - 3(a + b) - 10$
100. $2(a + b)^2 - 5(a + b) - 3$
101. $x^6 + 7x^3 - 8$
102. $x^6 - 13x^4 + 36x^2$
103. $x^4 + 3x^2 + 4$
104. $x^4 + x^2 + 1$
105. $x^4 + 7x^2 + 16$
106. $y^4 + 2y^2 + 9$
107. $4a^4 + 1 + 3a^2$
108. $x^4 + 25 + 6x^2$

Applications

109. Candy To find the amount of chocolate used in the outer coating of the malted-milk ball shown on the next page, we can find the volume V of the chocolate shell using the formula $V = \frac{4}{3}\pi r_1{}^3 - \frac{4}{3}\pi r_2{}^3$. Factor the expression on the right side of the formula.

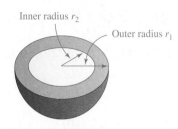

Inner radius r_2

Outer radius r_1

110. Movie stunts   The formula that gives the distance a stuntwoman is above the ground t seconds after she falls over the side of a 144-foot tall building is $f = 144 - 16t^2$. Factor the right side.

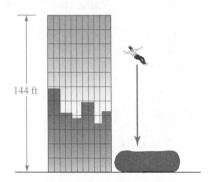

144 ft

Discovery and Writing

111. Explain how to factor the difference of two squares.

112. Explain how to factor the difference of two cubes.

113. Explain how to factor $a^2 - b^2 + a + b$.

114. Explain how to factor $x^2 + 2x + 1$.

Factor the indicated monomial from the given expression.

115. $3x + 2; 2$ **116.** $5x - 3; 5$

117. $x^2 + 2x + 4; 2$ **118.** $3x^2 - 2x - 5; 3$

119. $a + b; a$ **120.** $a - b; b$

121. $x + x^{1/2}; x^{1/2}$ **122.** $x^{3/2} - x^{1/2}; x^{1/2}$

123. $2x + \sqrt{2}y; \sqrt{2}$ **124.** $\sqrt{3}a - 3b; \sqrt{3}$

125. $ab^{3/2} - a^{3/2}b; ab$ **126.** $ab^2 + b; b^{-1}$

Factor each expression by grouping three terms and two terms.

127. $x^2 + x - 6 + xy - 2y$
128. $2x^2 + 5x + 2 - xy - 2y$
129. $a^4 + 2a^3 + a^2 + a + 1$
130. $a^4 + a^3 - 2a^2 + a - 1$

Review

131. Which natural number is neither prime nor composite?

132. Graph the interval $[-2, 3)$.

133. Simplify: $(x^3x^2)^4$.

134. Simplify: $\dfrac{(a^3)^3(a^2)^4}{(a^2a^3)^3}$.

135. $\left(\dfrac{3x^4x^3}{6x^{-2}x^4}\right)^0$

136. Simplify: $\sqrt{20x^5}$.

137. Simplify: $\sqrt{20x} - \sqrt{125x}$.

138. Rationalize the denominator: $\dfrac{3}{\sqrt[3]{3}}$.

0.6   Rational Expressions

Objectives

1. Define Rational Expressions
2. Simplify Rational Expressions
3. Multiply and Divide Rational Expressions
4. Add and Subtract Rational Expressions
5. Simplify Complex Fractions

Abercrombie and Fitch (A&F) is a successful American clothing company founded in 1892 by David Abercrombie and Ezra Fitch. Today the stores are popular shopping destinations for university students wanting to keep up with the latest styles and trends.

Suppose that a clothing manufacturer finds that the cost in dollars of producing x fleece vintage shirts is given by the algebraic expression $13x + 1,000$. The average cost of producing each shirt could be obtained by dividing the production cost, $13x + 1,000$, by the number of shirts produced, x. The algebraic fraction

$$\frac{13x + 1,000}{x}$$

represents the average cost per shirt. We see that the average cost of producing 200 shirts would be $18.

$$\frac{13(200) + 1,000}{200} = 18$$

An understanding of algebraic fractions is important in solving many real-life problems.

1.   Define Rational Expressions

If x and y are real numbers, the quotient $\frac{x}{y}$ $(y \neq 0)$ is called a **fraction.** The number x is called the **numerator,** and the number y is called the **denominator.**

Algebraic fractions are quotients of algebraic expressions. If the expressions are polynomials, the fraction is called a **rational expression.** The first two of the following algebraic fractions are rational expressions. The third is not, because the numerator and denominator are not polynomials.

$$\frac{5y^2 + 2y}{y^2 - 3y - 7} \qquad \frac{8ab^2 - 16c^3}{2x + 3} \qquad \frac{x^{1/2} + 4x}{x^{3/2} - x^{1/2}}$$

Remember that the denominator of a fraction cannot be zero.

We summarize some of the properties of fractions as follows:

Properties of Fractions If a, b, c, and d are real numbers and no denominators are 0, then

Equality of Fractions

$$\frac{a}{b} = \frac{c}{d} \quad \text{if and only if} \quad ad = bc$$

Fundamental Property of Fractions

$$\frac{ax}{bx} = \frac{a}{b}$$

Multiplication and Division of Fractions

$$\frac{a}{b} \cdot \frac{c}{d} = \frac{ac}{bd} \quad \text{and} \quad \frac{a}{b} \div \frac{c}{d} = \frac{a}{b} \cdot \frac{d}{c} = \frac{ad}{bc}$$

Addition and Subtraction of Fractions

$$\frac{a}{b} + \frac{c}{b} = \frac{a+c}{b} \quad \text{and} \quad \frac{a}{b} - \frac{c}{b} = \frac{a-c}{b}$$

The first two examples illustrate each of the previous properties of fractions.

EXAMPLE 1 Assume that no denominators are 0.

a. $\dfrac{2a}{3} = \dfrac{4a}{6}$ Because $2a(6) = 3(4a)$.

b. $\dfrac{6xy}{10xy} = \dfrac{3(2xy)}{5(2xy)}$ Factor the numerator and denominator and divide out the common factors.

$$= \frac{3}{5}$$

Self Check 1 **a.** Is $\dfrac{3y}{5} = \dfrac{15z}{25}$?   **b.** Simplify: $\dfrac{15a^2b}{25ab^2}$.

EXAMPLE 2 Assume that no denominators are 0.

a. $\dfrac{2r}{7s} \cdot \dfrac{3r}{5s} = \dfrac{2r \cdot 3r}{7s \cdot 5s}$

$$= \frac{6r^2}{35s^2}$$

b. $\dfrac{3mn}{4pq} \div \dfrac{2pq}{7mn} = \dfrac{3mn}{4pq} \cdot \dfrac{7mn}{2pq}$

$$= \frac{21m^2n^2}{8p^2q^2}$$

c. $\dfrac{2ab}{5xy} + \dfrac{ab}{5xy} = \dfrac{2ab + ab}{5xy}$

$$= \frac{3ab}{5xy}$$

d. $\dfrac{6uv^2}{7w^2} - \dfrac{3uv^2}{7w^2} = \dfrac{6uv^2 - 3uv^2}{7w^2}$

$$= \frac{3uv^2}{7w^2}$$

Self Check 2 Perform each operation:

a. $\dfrac{3a}{5b} \cdot \dfrac{2a}{7b}$ **b.** $\dfrac{2ab}{3rs} \div \dfrac{2rs}{4ab}$

c. $\dfrac{5pq}{3t} + \dfrac{3pq}{3t}$ **d.** $\dfrac{5mn^2}{3w} - \dfrac{mn^2}{3w}$

To add or subtract rational expressions with unlike denominators, we write each expression as an equivalent expression with a common denominator. We then can add or subtract the expressions. For example,

$$\frac{3x}{5} + \frac{2x}{7} = \frac{3x(7)}{5(7)} + \frac{2x(5)}{7(5)}$$

$$= \frac{21x}{35} + \frac{10x}{35}$$

$$= \frac{21x + 10x}{35}$$

$$= \frac{31x}{35}$$

$$\frac{4a^2}{15} - \frac{3a^2}{10} = \frac{4a^2(2)}{15(2)} - \frac{3a^2(3)}{10(3)}$$

$$= \frac{8a^2}{30} - \frac{9a^2}{30}$$

$$= \frac{8a^2 - 9a^2}{30}$$

$$= \frac{-a^2}{30}$$

$$= -\frac{a^2}{30}$$

A rational expression is **in lowest terms** if all factors common to the numerator and the denominator have been removed. To **simplify a rational expression** means to write it in lowest terms.

2.   Simplify Rational Expressions

To simplify rational expressions, we use the fundamental property of fractions. This enables us to divide out all factors that are common to the numerator and the denominator.

EXAMPLE 3   Simplify: $\dfrac{x^2 - 9}{x^2 - 3x}$ $(x \neq 0, 3)$.

Solution   We factor the difference of two squares in the numerator, factor out x in the denominator, and divide out the common factor of $x - 3$.

$$\frac{x^2 - 9}{x^2 - 3x} = \frac{(x + 3)(x - 3)}{x(x - 3)} \qquad \frac{x - 3}{x - 3} = 1$$

$$= \frac{x + 3}{x}$$

Self Check 3   Simplify: $\dfrac{a^2 - 4a}{a^2 - a - 12}$ $(a \neq 4, -3)$.

We will encounter the following properties of fractions in the next examples.

Properties of Fractions   If a and b represent real numbers and there are no divisions by 0, then

$$\frac{a}{1} = a \qquad\qquad\qquad \frac{a}{a} = 1$$

$$\frac{a}{b} = \frac{-a}{-b} = -\frac{a}{-b} = -\frac{-a}{b} \qquad -\frac{a}{b} = \frac{a}{-b} = \frac{-a}{b} = -\frac{-a}{-b}$$

EXAMPLE 4 Simplify: $\dfrac{x^2 - 2xy + y^2}{y - x}$ $(x \neq y)$.

Solution We factor the trinomial in the numerator, factor -1 from the denominator, and divide out the common factor of $x - y$.

$$\dfrac{x^2 - 2xy + y^2}{y - x} = \dfrac{(x - y)(x - y)}{-(x - y)} \qquad \dfrac{x - y}{x - y} = 1$$

$$= \dfrac{x - y}{-1}$$

$$= -\dfrac{x - y}{1}$$

$$= -(x - y)$$

Self Check 4 Simplify: $\dfrac{a^2 - ab - 2b^2}{2b - a}$ $(2b - a \neq 0)$.

EXAMPLE 5 Simplify: $\dfrac{x^2 - 3x + 2}{x^2 - x - 2}$ $(x \neq 2, -1)$.

Solution We factor the numerator and denominator and divide out the common factor of $x - 2$.

$$\dfrac{x^2 - 3x + 2}{x^2 - x - 2} = \dfrac{(x - 1)(x - 2)}{(x + 1)(x - 2)} \qquad \dfrac{x - 2}{x - 2} = 1$$

$$= \dfrac{x - 1}{x + 1}$$

Self Check 5 Simplify: $\dfrac{a^2 + 3a - 4}{a^2 + 2a - 3}$ $(a \neq 1, -3)$.

3. Multiply and Divide Rational Expressions

EXAMPLE 6 Multiply: $\dfrac{x^2 - x - 2}{x^2 - 1} \cdot \dfrac{x^2 + 2x - 3}{x - 2}$ $(x \neq 1, -1, 2)$.

Solution To multiply the rational expressions, we multiply the numerators, multiply the denominators, factor, and divide out the common factors.

$$\dfrac{x^2 - x - 2}{x^2 - 1} \cdot \dfrac{x^2 + 2x - 3}{x - 2} = \dfrac{(x^2 - x - 2)(x^2 + 2x - 3)}{(x^2 - 1)(x - 2)}$$

$$= \dfrac{(x - 2)(x + 1)(x - 1)(x + 3)}{(x + 1)(x - 1)(x - 2)} \qquad \dfrac{x - 2}{x - 2} = 1, \dfrac{x + 1}{x + 1} = 1, \dfrac{x - 1}{x - 1} = 1$$

$$= x + 3$$

Self Check 6 Simplify: $\dfrac{x^2 - 9}{x^2 - x} \cdot \dfrac{x - 1}{x^2 - 3x}$ $(x \neq 0, 1, 3)$.

EXAMPLE 7   Divide: $\dfrac{x^2 - 2x - 3}{x^2 - 4} \div \dfrac{x^2 + 2x - 15}{x^2 + 3x - 10}$ $(x \neq 2, -2, 3, -5)$.

Solution   To divide the rational expressions, we multiply by the reciprocal of the second rational expression. We then simplify by factoring the numerator and denominator and dividing out the common factors.

$$\dfrac{x^2 - 2x - 3}{x^2 - 4} \div \dfrac{x^2 + 2x - 15}{x^2 + 3x - 10} = \dfrac{x^2 - 2x - 3}{x^2 - 4} \cdot \dfrac{x^2 + 3x - 10}{x^2 + 2x - 15}$$

$$= \dfrac{(x^2 - 2x - 3)(x^2 + 3x - 10)}{(x^2 - 4)(x^2 + 2x - 15)}$$

$$= \dfrac{(x - 3)(x + 1)(x - 2)(x + 5)}{(x + 2)(x - 2)(x + 5)(x - 3)} \qquad \dfrac{x - 3}{x - 3} = 1, \dfrac{x - 2}{x - 2} = 1, \dfrac{x + 5}{x + 5} = 1$$

$$= \dfrac{x + 1}{x + 2}$$

Self Check 7   Simplify: $\dfrac{a^2 - a}{a + 2} \div \dfrac{a^2 - 2a}{a^2 - 4}$ $(a \neq 0, 2, -2)$.

EXAMPLE 8   Simplify: $\dfrac{2x^2 - 5x - 3}{3x - 1} \cdot \dfrac{3x^2 + 2x - 1}{x^2 - 2x - 3} \div \dfrac{2x^2 + x}{3x}$ $\left(x \neq \dfrac{1}{3}, -1, 3, 0, -\dfrac{1}{2}\right)$.

Solution   We can change the division to a multiplication, factor, and simplify.

$$\dfrac{2x^2 - 5x - 3}{3x - 1} \cdot \dfrac{3x^2 + 2x - 1}{x^2 - 2x - 3} \div \dfrac{2x^2 + x}{3x}$$

$$= \dfrac{2x^2 - 5x - 3}{3x - 1} \cdot \dfrac{3x^2 + 2x - 1}{x^2 - 2x - 3} \cdot \dfrac{3x}{2x^2 + x}$$

$$= \dfrac{(2x^2 - 5x - 3)(3x^2 + 2x - 1)(3x)}{(3x - 1)(x^2 - 2x - 3)(2x^2 + x)}$$

$$= \dfrac{(x - 3)(2x + 1)(3x - 1)(x + 1)3x}{(3x - 1)(x + 1)(x - 3)x(2x + 1)} \qquad \dfrac{x - 3}{x - 3} = 1, \dfrac{2x + 1}{2x + 1} = 1, \dfrac{3x - 1}{3x - 1} = 1, \dfrac{x + 1}{x + 1} = 1, \dfrac{x}{x} = 1$$

$$= 3$$

Self Check 8   Simplify: $\dfrac{x^2 - 25}{x - 2} \div \dfrac{x^2 - 5x}{x^2 - 2x} \cdot \dfrac{x^2 + 2x}{x^2 + 5x}$ $(x \neq 0, 2, 5, -5)$.

4.   Add and Subtract Rational Expressions

To add (or subtract) rational expressions with like denominators, we add (or subtract) the numerators and keep the common denominator.

EXAMPLE 9   Add: $\dfrac{2x + 5}{x + 5} + \dfrac{3x + 20}{x + 5}$ $(x \neq -5)$.

Solution $\dfrac{2x + 5}{x + 5} + \dfrac{3x + 20}{x + 5} = \dfrac{5x + 25}{x + 5}$ **Add the numerators and keep the common denominator.**

$$= \dfrac{5(x + 5)}{1(x + 5)} \qquad \text{Factor out 5 and divide out the common factor of } x + 5.$$
$$\dfrac{x + 5}{x + 5} = 1$$

$$= 5$$

Add: $\dfrac{3x - 2}{x - 2} + \dfrac{x - 6}{x - 2}$ $(x \neq 2)$

To add (or subtract) rational expressions with unlike denominators, we must find a common denominator, called the **least** (or lowest) **common denominator (LCD).** Suppose the unlike denominators of three rational expressions are 12, 20, and 35. To find the LCD, we first find the prime factorization of each number.

$$
\begin{array}{lll}
12 = 4 \cdot 3 & 20 = 4 \cdot 5 & 35 = 5 \cdot 7 \\
 = 2^2 \cdot 3 & = 2^2 \cdot 5 &
\end{array}
$$

Because the LCD is the smallest number that can be divided by 12, 20, and 35, it must contain factors of 2^2, 3, 5, and 7. Thus,

$$LCD = 2^2 \cdot 3 \cdot 5 \cdot 7 = 420$$

That is, 420 is the smallest number that can be divided without remainder by 12, 20, and 35.

When finding an LCD, we always factor each denominator and then create the LCD by using each factor the greatest number of times that it appears in any one denominator. The product of these factors is the LCD.

This rule also applies if the unlike denominators of the rational expressions contain variables. Suppose the unlike denominators are $x^2(x - 5)$ and $x(x - 5)^3$. To find the LCD, we use x^2 and $(x - 5)^3$. Thus, the LCD is the product. $x^2(x - 5)^3$.

Comment

Remember that to find the *least* common denominator, use each factor the *greatest* number of times that it occurs.

EXAMPLE 10 Add: $\dfrac{1}{x^2 - 4} + \dfrac{2}{x^2 - 4x + 4}$ $(x \neq 2, -2)$.

Solution We factor each denominator and find the LCD.

$$
\begin{array}{l}
x^2 - 4 = (x + 2)(x - 2) \\
x^2 - 4x + 4 = (x - 2)(x - 2) = (x - 2)^2
\end{array}
$$

The LCD is $(x + 2)(x - 2)^2$. We then write each rational expression with its denominator in factored form, convert each rational expression into an equivalent expression with a denominator of $(x + 2)(x - 2)^2$, add the expressions, and simplify.

$$
\begin{aligned}
\dfrac{1}{x^2 - 4} + \dfrac{2}{x^2 - 4x + 4} &= \dfrac{1}{(x + 2)(x - 2)} + \dfrac{2}{(x - 2)(x - 2)} \\
&= \dfrac{1(x - 2)}{(x + 2)(x - 2)(x - 2)} + \dfrac{2(x + 2)}{(x - 2)(x - 2)(x + 2)} \qquad \tfrac{x-2}{x-2} = 1, \tfrac{x+2}{x+2} = 1 \\
&= \dfrac{1(x - 2) + 2(x + 2)}{(x + 2)(x - 2)(x - 2)} \\
&= \dfrac{x - 2 + 2x + 4}{(x + 2)(x - 2)(x - 2)} \\
&= \dfrac{3x + 2}{(x + 2)(x - 2)^2}
\end{aligned}
$$

Comment

Always attempt to simplify the final result. In this case, the final fraction is already in lowest terms.

Self Check 10 Add: $\dfrac{3}{x^2 - 6x + 9} + \dfrac{1}{x^2 - 9}$ $(x \neq 3, -3)$.

EXAMPLE 11 Simplify: $\dfrac{x-2}{x^2-1} - \dfrac{x+3}{x^2+3x+2} + \dfrac{3}{x^2+x-2}$ $(x \neq 1, -1, -2)$.

Solution We factor the denominators to find the LCD.

$$x^2 - 1 = (x+1)(x-1)$$
$$x^2 + 3x + 2 = (x+2)(x+1)$$
$$x^2 + x - 2 = (x+2)(x-1)$$

The LCD is $(x+1)(x+2)(x-1)$. We now write each rational expression as an equivalent expression with this LCD, and proceed as follows:

$$\frac{x-2}{x^2-1} - \frac{x+3}{x^2+3x+2} + \frac{3}{x^2+x-2}$$

$$= \frac{x-2}{(x+1)(x-1)} - \frac{x+3}{(x+1)(x+2)} + \frac{3}{(x-1)(x+2)}$$

$$= \frac{(x-2)(x+2)}{(x+1)(x-1)(x+2)} - \frac{(x+3)(x-1)}{(x+1)(x+2)(x-1)} + \frac{3(x+1)}{(x-1)(x+2)(x+1)} \qquad \frac{x+2}{x+2}=1, \frac{x-1}{x-1}=1, \frac{x+1}{x+1}=1$$

$$= \frac{(x^2-4) - (x^2+2x-3) + (3x+3)}{(x+1)(x+2)(x-1)}$$

$$= \frac{x^2 - 4 - x^2 - 2x + 3 + 3x + 3}{(x+1)(x+2)(x-1)}$$

$$= \frac{x+2}{(x+1)(x+2)(x-1)}$$

$$= \frac{1}{(x+1)(x-1)} \qquad \text{Divide out the common factor of } x+2. \ \tfrac{x+2}{x+2}=1$$

Self Check 11 Simplify: $\dfrac{4y}{y^2-1} - \dfrac{2}{y+1} + 2$ $(y \neq 1, -1)$.

5. Simplify Complex Fractions

A **complex fraction** is a fraction that has a fraction in its numerator or a fraction in its denominator.

EXAMPLE 12 Simplify: $\dfrac{\dfrac{1}{x} + \dfrac{1}{y}}{\dfrac{x}{y}}$ $(x, y \neq 0)$.

Method 1. We note that the LCD of the three fractions in the complex fraction is xy. So we multiply the numerator and denominator of the complex fraction by xy and simplify:

$$\frac{\dfrac{1}{x} + \dfrac{1}{y}}{\dfrac{x}{y}} = \frac{xy\left(\dfrac{1}{x} + \dfrac{1}{y}\right)}{xy\left(\dfrac{x}{y}\right)} = \frac{\dfrac{xy}{x} + \dfrac{xy}{y}}{\dfrac{xxy}{y}} = \frac{y+x}{x^2}$$

Method 2. We combine the fractions in the numerator of the complex fraction to obtain a single fraction over a single fraction.

$$\frac{\dfrac{1}{x}+\dfrac{1}{y}}{\dfrac{x}{y}}=\frac{\dfrac{1(y)}{x(y)}+\dfrac{1(x)}{y(x)}}{\dfrac{x}{y}}=\frac{\dfrac{y+x}{xy}}{\dfrac{x}{y}}$$

Then we use the fact that any fraction indicates a division:

$$\frac{\dfrac{y+x}{xy}}{\dfrac{x}{y}}=\frac{y+x}{xy}\div\frac{x}{y}=\frac{y+x}{xy}\cdot\frac{y}{x}=\frac{(y+x)y}{xyx}=\frac{y+x}{x^2}$$

Self Check 12   Simplify: $\dfrac{\dfrac{1}{x}-\dfrac{1}{y}}{\dfrac{1}{x}+\dfrac{1}{y}}$ $(x, y \neq 0)$.

Self Check Answers   **1. a.** no   **b.** $\dfrac{3a}{5b}$ **2. a.** $\dfrac{6a^2}{35b^2}$ **b.** $\dfrac{4a^2b^2}{3r^2s^2}$ **c.** $\dfrac{8pq}{3t}$ **d.** $\dfrac{4mn^2}{3w}$ **3.** $\dfrac{a}{a+3}$
4. $-(a+b)$ **5.** $\dfrac{a+4}{a+3}$ **6.** $\dfrac{x+3}{x^2}$ **7.** $a-1$ **8.** $x+2$ **9.** 4
10. $\dfrac{4x+6}{(x+3)(x-3)^2}$ **11.** $\dfrac{2y}{y-1}$ **12.** $\dfrac{y-x}{y+x}$

0.6 Exercises

Vocabulary and Concepts   *Fill in the blanks.*

1. In the fraction $\frac{a}{b}$, a is called the _____.

2. In the fraction $\frac{a}{b}$, b is called the _____.

3. $\frac{a}{b}=\frac{c}{d}$ if and only if _____.

4. The denominator of a fraction can never be _____.

Complete each formula.

5. $\dfrac{a}{b}\cdot\dfrac{c}{d}=$ ____

6. $\dfrac{a}{b}\div\dfrac{c}{d}=$ ____

7. $\dfrac{a}{b}+\dfrac{c}{b}=$ _____

8. $\dfrac{a}{b}-\dfrac{c}{b}=$ _____

Determine whether the fractions are equal. Assume that no denominators are 0.

9. $\dfrac{8x}{3y},\dfrac{16x}{6y}$

10. $\dfrac{3x^2}{4y^2},\dfrac{12y^2}{16x^2}$

11. $\dfrac{25xyz}{12ab^2c},\dfrac{50a^2bc}{24xyz}$

12. $\dfrac{15rs^2}{4rs^2},\dfrac{37.5a^3}{10a^3}$

Practice   *Simplify each rational expression. Assume that no denominators are 0.*

13. $\dfrac{7a^2b}{21ab^2}$

14. $\dfrac{35p^3q^2}{49p^4q}$

Perform the operations and simplify, whenever possible. Assume that no denominators are 0.

15. $\dfrac{4x}{7}\cdot\dfrac{2}{5a}$

16. $\dfrac{-5y}{2z}\cdot\dfrac{4}{y^2}$

17. $\dfrac{8m}{5n}\div\dfrac{3m}{10n}$

18. $\dfrac{15p}{8q}\div\dfrac{-5p}{16q^2}$

19. $\dfrac{3z}{5c}+\dfrac{2z}{5c}$

20. $\dfrac{7a}{4b}-\dfrac{3a}{4b}$

21. $\dfrac{15x^2y}{7a^2b^3}-\dfrac{x^2y}{7a^2b^3}$

22. $\dfrac{8rst^2}{15m^4t^2}+\dfrac{7rst^2}{15m^4t^2}$

Simplify each fraction. Assume that no denominators are 0.

23. $\dfrac{2x-4}{x^2-4}$

24. $\dfrac{x^2-16}{x^2-8x+16}$

25. $\dfrac{25 - x^2}{x^2 + 10x + 25}$ **26.** $\dfrac{4 - x^2}{x^2 - 5x + 6}$

27. $\dfrac{6x^3 + x^2 - 12x}{4x^3 + 4x^2 - 3x}$ **28.** $\dfrac{6x^4 - 5x^3 - 6x^2}{2x^3 - 7x^2 - 15x}$

29. $\dfrac{x^3 - 8}{x^2 + ax - 2x - 2a}$ **30.** $\dfrac{xy + 2x + 3y + 6}{x^3 + 27}$

Perform the operations and simplify, whenever possible.
Assume that no denominators are 0.

31. $\dfrac{x^2 - 1}{x} \cdot \dfrac{x^2}{x^2 + 2x + 1}$

32. $\dfrac{y^2 - 2y + 1}{y} \cdot \dfrac{y + 2}{y^2 + y - 2}$

33. $\dfrac{3x^2 + 7x + 2}{x^2 + 2x} \cdot \dfrac{x^2 - x}{3x^2 + x}$

34. $\dfrac{x^2 + x}{2x^2 + 3x} \cdot \dfrac{2x^2 + x - 3}{x^2 - 1}$

35. $\dfrac{x^2 + x}{x - 1} \cdot \dfrac{x^2 - 1}{x + 2}$

36. $\dfrac{x^2 + 5x + 6}{x^2 + 6x + 9} \cdot \dfrac{x + 2}{x^2 - 4}$

37. $\dfrac{2x^2 + 32}{8} \div \dfrac{x^2 + 16}{2}$

38. $\dfrac{x^2 + x - 6}{x^2 - 6x + 9} \div \dfrac{x^2 - 4}{x^2 - 9}$

39. $\dfrac{z^2 + z - 20}{z^2 - 4} \div \dfrac{z^2 - 25}{z - 5}$

40. $\dfrac{ax + bx + a + b}{a^2 + 2ab + b^2} \div \dfrac{x^2 - 1}{x^2 - 2x + 1}$

41. $\dfrac{3x^2 + 5x - 2}{x^3 + 2x^2} \div \dfrac{6x^2 + 13x - 5}{2x^3 + 5x^2}$

42. $\dfrac{x^2 + 13x + 12}{8x^2 - 6x - 5} \div \dfrac{2x^2 - x - 3}{8x^2 - 14x + 5}$

43. $\dfrac{x^2 + 7x + 12}{x^3 - x^2 - 6x} \cdot \dfrac{x^2 - 3x - 10}{x^2 + 2x - 3} \cdot \dfrac{x^3 - 4x^2 + 3x}{x^2 - x - 20}$

44. $\dfrac{x^2 - 2x - 3}{21x^2 - 50x - 16} \cdot \dfrac{3x - 8}{x - 3} \div \dfrac{x^2 + 6x + 5}{7x^2 - 33x - 10}$

45. $\dfrac{x^3 + 27}{x^2 - 4} \div \left(\dfrac{x^2 + 4x + 3}{x^2 + 2x} \div \dfrac{x^2 + x - 6}{x^2 - 3x + 9} \right)$

46. $\dfrac{x(x - 2) - 3}{x(x + 7) - 3(x - 1)} \cdot \dfrac{x(x + 1) - 2}{x(x - 7) + 3(x + 1)}$

47. $\dfrac{3}{x + 3} + \dfrac{x + 2}{x + 3}$

48. $\dfrac{3}{x + 1} + \dfrac{x + 2}{x + 1}$

49. $\dfrac{4x}{x - 1} - \dfrac{4}{x - 1}$

50. $\dfrac{6x}{x - 2} - \dfrac{3}{x - 2}$

51. $\dfrac{2}{5 - x} + \dfrac{1}{x - 5}$

52. $\dfrac{3}{x - 6} - \dfrac{2}{6 - x}$

53. $\dfrac{3}{x + 1} + \dfrac{2}{x - 1}$

54. $\dfrac{3}{x + 4} + \dfrac{x}{x - 4}$

55. $\dfrac{a + 3}{a^2 + 7a + 12} + \dfrac{a}{a^2 - 16}$

56. $\dfrac{a}{a^2 + a - 2} + \dfrac{2}{a^2 - 5a + 4}$

57. $\dfrac{x}{x^2 - 4} - \dfrac{1}{x + 2}$

58. $\dfrac{b^2}{b^2 - 4} - \dfrac{4}{b^2 + 2b}$

59. $\dfrac{3x - 2}{x^2 + 2x + 1} - \dfrac{x}{x^2 - 1}$

60. $\dfrac{2t}{t^2 - 25} - \dfrac{t + 1}{t^2 + 5t}$

61. $\dfrac{2}{y^2 - 1} + 3 + \dfrac{1}{y + 1}$

62. $2 + \dfrac{4}{t^2 - 4} - \dfrac{1}{t - 2}$

63. $\dfrac{1}{x - 2} + \dfrac{3}{x + 2} - \dfrac{3x - 2}{x^2 - 4}$

64. $\dfrac{x}{x - 3} - \dfrac{5}{x + 3} + \dfrac{3(3x - 1)}{x^2 - 9}$

65. $\left(\dfrac{1}{x - 2} + \dfrac{1}{x - 3} \right) \cdot \dfrac{x - 3}{2x}$

66. $\left(\dfrac{1}{x + 1} - \dfrac{1}{x - 2} \right) \div \dfrac{1}{x - 2}$

67. $\dfrac{3x}{x - 4} - \dfrac{x}{x + 4} - \dfrac{3x + 1}{16 - x^2}$

68. $\dfrac{7x}{x-5} + \dfrac{3x}{5-x} + \dfrac{3x-1}{x^2-25}$

69. $\dfrac{1}{x^2+3x+2} - \dfrac{2}{x^2+4x+3} + \dfrac{1}{x^2+5x+6}$

70. $\dfrac{-2}{x-y} + \dfrac{2}{x-z} - \dfrac{2z-2y}{(y-x)(z-x)}$

71. $\dfrac{3x-2}{x^2+x-20} - \dfrac{4x^2+2}{x^2-25} + \dfrac{3x^2-25}{x^2-16}$

72. $\dfrac{3x+2}{8x^2-10x-3} + \dfrac{x+4}{6x^2-11x+3} - \dfrac{1}{4x+1}$

87. $\dfrac{3xy}{1-\dfrac{1}{xy}}$

88. $\dfrac{x-3+\dfrac{1}{x}}{-\dfrac{1}{x}-x+3}$

89. $\dfrac{3x}{x+\dfrac{1}{x}}$

90. $\dfrac{2x^2+4}{2+\dfrac{4x}{5}}$

91. $\dfrac{\dfrac{x}{x+2}-\dfrac{2}{x-1}}{\dfrac{3}{x+2}+\dfrac{x}{x-1}}$

92. $\dfrac{\dfrac{2x}{x-3}+\dfrac{1}{x-2}}{\dfrac{3}{x-3}-\dfrac{x}{x-2}}$

Simplify each complex fraction. Assume that no denominators are 0.

73. $\dfrac{\dfrac{3a}{b}}{\dfrac{6ac}{b^2}}$

74. $\dfrac{\dfrac{3t^2}{9x}}{\dfrac{t}{18x}}$

75. $\dfrac{\dfrac{3a^2b}{ab}}{27}$

76. $\dfrac{\dfrac{3u^2v}{4t}}{3uv}$

77. $\dfrac{\dfrac{x-y}{ab}}{\dfrac{y-x}{ab}}$

78. $\dfrac{\dfrac{x^2-5x+6}{2x^2y}}{\dfrac{x^2-9}{2x^2y}}$

79. $\dfrac{\dfrac{1}{x}+\dfrac{1}{y}}{xy}$

80. $\dfrac{xy}{\dfrac{11}{x}+\dfrac{11}{y}}$

81. $\dfrac{\dfrac{1}{x}+\dfrac{1}{y}}{\dfrac{1}{x}-\dfrac{1}{y}}$

82. $\dfrac{\dfrac{1}{x}-\dfrac{1}{y}}{\dfrac{1}{x}+\dfrac{1}{y}}$

83. $\dfrac{\dfrac{3a}{b}-\dfrac{4a^2}{x}}{\dfrac{1}{b}+\dfrac{1}{ax}}$

84. $\dfrac{1-\dfrac{x}{y}}{\dfrac{x^2}{y^2}-1}$

85. $\dfrac{x+1-\dfrac{6}{x}}{x+5+\dfrac{6}{x}}$

86. $\dfrac{2z}{1-\dfrac{3}{z}}$

Write each expression without using negative exponents, and simplify the resulting complex fraction. Assume that no denominators are 0.

93. $\dfrac{1}{1+x^{-1}}$

94. $\dfrac{y^{-1}}{x^{-1}+y^{-1}}$

95. $\dfrac{3(x+2)^{-1}+2(x-1)^{-1}}{(x+2)^{-1}}$

96. $\dfrac{2x(x-3)^{-1}-3(x+2)^{-1}}{(x-3)^{-1}(x+2)^{-1}}$

Applications

97. Engineering The stiffness k of the shaft shown in the illustration is given by the following formula where k_1 and k_2 are the individual stiffnesses of each section. Simplify the complex fraction on the right side of the formula.

$$k = \dfrac{1}{\dfrac{1}{k_1}+\dfrac{1}{k_2}}$$

Section 1   Section 2

98. Electronics The combined resistance R of three resistors with resistances of R_1, R_2, and R_3 is given by the following formula. Simplify the complex fraction on the right side of the formula.

$$R = \dfrac{1}{\dfrac{1}{R_1}+\dfrac{1}{R_2}+\dfrac{1}{R_3}}$$

Discovery and Writing *Simplify each complex fraction. Assume that no denominators are 0.*

99. $\dfrac{x}{1 + \dfrac{1}{3x^{-1}}}$

100. $\dfrac{ab}{2 + \dfrac{3}{2a^{-1}}}$

101. $\dfrac{1}{1 + \dfrac{1}{1 + \dfrac{1}{x}}}$

102. $\dfrac{y}{2 + \dfrac{2}{2 + \dfrac{2}{y}}}$

103. Explain why the formula $\dfrac{a}{b} + \dfrac{c}{d} = \dfrac{ad + bc}{bd}$ is valid.

104. Explain why the formula $\dfrac{a}{b} \div \dfrac{c}{d} = \dfrac{a}{b} \cdot \dfrac{d}{c}$ is valid.

105. Explain the commutative property of addition and explain why it is useful.

106. Explain the distributive property and explain why it is useful.

Review   *Write each expression without using absolute value symbols.*

107. $|-6|$

108. $|5 - x|$, given that $x < 0$

Simplify each expression.

109. $\left(\dfrac{x^3 y^{-2}}{x^{-1} y}\right)^{-3}$

110. $(27x^6)^{2/3}$

111. $\sqrt{20} - \sqrt{45}$

112. $2(x^2 + 4) - 3(2x^2 + 5)$

CHAPTER REVIEW

0.1   Sets of Real Numbers

Definitions and Concepts	Examples
Natural numbers: The numbers that we count with.	$1, 2, 3, 4, 5, 6, 7, 8, 9, 10, \ldots$
Whole numbers: The natural numbers and 0.	$0, 1, 2, 3, 4, 5, 6, 7, 8, 9, 10, \ldots$
Integers: The whole numbers and the negatives of the natural numbers.	$\ldots, -5, -4, -3, -2, -1, 0, 1, 2, 3, 4, 5, \ldots$
Rational numbers: $\{x \mid x$ can be written in the form $\frac{a}{b}$ $(b \neq 0)$, where a and b are integers.$\}$ All decimals that either terminate or repeat.	$2 = \dfrac{2}{1}, -5 = -\dfrac{5}{1}, \dfrac{2}{3}, -\dfrac{7}{3}, 0.25 = \dfrac{3}{4}$ $\dfrac{1}{4} = 0.25 \quad \dfrac{13}{5} = 2.6 \quad \dfrac{2}{3} = 0.666\ldots = 0.\overline{6} \quad \dfrac{7}{11} = 0.\overline{63}$
Irrational numbers: Nonrational real numbers. All decimals that neither terminate nor repeat.	$\sqrt{2}, \quad -\sqrt{26}, \quad \pi, \quad 0.232232223\ldots$
Real numbers: Any number that can be expressed as a decimal.	$-6, -\dfrac{9}{13}, 0, \pi, \sqrt{31}, 10.73$
Prime numbers: A natural number greater than 1 that is divisible only by itself and 1.	$2, 3, 5, 7, 11, 13, 17, \ldots$
Composite numbers: A natural number greater than 1 that is not prime.	$4, 6, 8, 9, 10, 12, 14, 15, \ldots$
Even integers: The integers that are exactly divisible by 2.	$\ldots, -6, -4, -2, 0, 2, 4, 6, \ldots$
Odd integers: The integers that are not exactly divisible by 2.	$\ldots, -5, -3, -1, 1, 3, 5, \ldots$

Associative properties: of addition $(a + b) + c = a + (b + c)$ of multiplication $(ab)c = a(bc)$	$(5 + 4) + 7 = 5 + (4 + 7)$ $(5 \cdot 4) \cdot 7 = 5(4 \cdot 7)$
Commutative properties: of addition $a + b = b + a$ of multiplication $ab = ba$	$4 + 7 = 7 + 4$ $1.7 + 2.5 = 2.5 + 1.7$ $4 \cdot 7 = 7 \cdot 4$ $1.7(2.5) = 2.5(1.7)$
Distributive property: $a(b + c) = ab + ac$	$3(x + 6) = 3x + 3 \cdot 6$ $0.2(y - 10) = 0.2y - 0.2(10)$
Double negative rule: $-(-a) = a$	$-(-5) = 5$ $-(-8) = 8$
Open intervals have no endpoints.	$(-3, 2)$
Closed intervals have two endpoints.	$[-3, 2]$
Half-open intervals have one endpoint.	$(-3, 2]$
Absolute value: If $x \geq 0$, then $|x| = x$. If $x < 0$, then $|x| = -x$.	$\|7\| = 7$ $\|3.5\| = 3.5$ $\left\|\dfrac{7}{2}\right\| = \dfrac{7}{2}$ $\|0\| = 0$ $\|-7\| = 7$ $\|-3.5\| = 3.5$ $\left\|-\dfrac{7}{2}\right\| = \dfrac{7}{2}$
The distance d between points a and b on the number line is $d = \|b - a\|$.	The distance d on the number line between points with coordinates of -3 and 2, is $d = \|2 - (-3)\| = \|5\| = 5$.

Exercises

Consider the set $\{-6, -3, 0, \frac{1}{2}, 3, \pi, \sqrt{5}, 6, 8\}$. List the numbers in this set that are

1. natural numbers
2. whole numbers
3. integers
4. rational numbers
5. irrational numbers
6. real numbers

Consider the set $\{-6, -3, 0, \frac{1}{2}, 3, \pi, \sqrt{5}, 6, 8\}$. List the numbers in this set that are

7. prime numbers
8. composite numbers
9. even integers
10. odd integers

Determine which property of real numbers justifies each statement.

11. $(a + b) + 2 = a + (b + 2)$
12. $a + 7 = 7 + a$
13. $4(2x) = (4 \cdot 2)x$
14. $3(a + b) = 3a + 3b$
15. $(5a)7 = 7(5a)$
16. $(2x + y) + z = (y + 2x) + z$
17. $-(-6) = 6$

Graph each subset of the real numbers:

18. the prime numbers between 10 and 20
19. the even integers from 6 to 14

Graph each interval on the number line.

20. $-3 < x \leq 5$

21. $x \geq 0$ or $x < -1$

22. $(-2, 4]$

23. $(-\infty, 2) \cap (-5, \infty)$

24. $(-\infty, -4) \cup [6, \infty)$

Write each expression without absolute value symbols.

25. $|6|$ **26.** $|-25|$

27. $|1 - \sqrt{2}|$ **28.** $|\sqrt{3} - 1|$

29. On the number line, find the distance between points with coordinates of -5 and 7.

0.2   Integer Exponents and Scientific Notation

Definitions and Concepts	Examples
Natural-number exponents: $$x^n = \overbrace{x \cdot x \cdot x \cdot \cdots \cdot x}^{n \text{ factors of } x}$$	 $$x^5 = x \cdot x \cdot x \cdot x \cdot x$$
Rules of exponents: If there are no divisions by 0, $x^m x^n = x^{m+n}$ $(x^m)^n = x^{mn}$ $(xy)^n = x^n y^n$ $\left(\dfrac{x}{y}\right)^n = \dfrac{x^n}{y^n}$ $x^0 = 1 \quad (x \neq 0)$ $x^{-n} = \dfrac{1}{x^n}$ $\dfrac{x^m}{x^n} = x^{m-n}$ $\left(\dfrac{x}{y}\right)^{-n} = \left(\dfrac{y}{x}\right)^n$	$x^2 x^5 = x^{2+5} = x^7$ $(x^2)^7 = x^{2 \cdot 7} = x^{14}$ $(xy)^5 = x^5 y^5$ $\left(\dfrac{x}{y}\right)^5 = \dfrac{x^5}{y^5}$ $6^0 = 1$ $6^{-2} = \dfrac{1}{6^2} = \dfrac{1}{36}$ $\dfrac{x^6}{x^4} = x^{6-4} = x^2$ $\left(\dfrac{x}{6}\right)^{-3} = \left(\dfrac{6}{x}\right)^3 = \dfrac{6^3}{x^3} = \dfrac{216}{x^3}$
A number is written in **scientific notation** when it is written in the form $N \times 10^n$, where $1 \leq |N| < 10$.	Write each number in scientific notation. $386{,}000 = 3.86 \times 10^5$ $0.0025 = 2.5 \times 10^{-3}$ Write each number in standard form. $7.3 \times 10^3 = 7{,}300$ $5.25 \times 10^{-4} = 0.000525$

Exercises

Write each expression without using exponents.

30. $-5a^3$ **31.** $(-5a)^2$

Write each expression using exponents.

32. $3ttt$ **33.** $(-2b)(3b)$

Simplify each expression.

34. $n^2 n^4$ **35.** $(p^3)^2$

36. $(x^3 y^2)^4$ **37.** $\left(\dfrac{a^4}{b^2}\right)^3$

38. $(m^{-3} n^0)^2$ **39.** $\left(\dfrac{p^{-2} q^2}{2}\right)^3$

40. $\dfrac{a^5}{a^8}$ **41.** $\left(\dfrac{a^2}{b^3}\right)^{-2}$

42. $\left(\dfrac{3x^2 y^{-2}}{x^2 y^2}\right)^{-2}$ **43.** $\left(\dfrac{a^{-3} b^2}{ab^{-3}}\right)^{-2}$

44. $\left(\dfrac{-3x^3 y}{xy^3}\right)^{-2}$ **45.** $\left(-\dfrac{2m^{-2} n^0}{4m^2 n^{-1}}\right)^{-3}$

46. If $x = -3$ and $y = 3$, evaluate $-x^2 - xy^2$

Write each number in scientific notation.

47. $6{,}750$ **48.** 0.00023

Write each number in standard notation.

49. 4.8×10^2 **50.** 0.25×10^{-3}

51. Use scientific notation to simplify
$$\dfrac{(45{,}000)(350{,}000)}{0.000105}.$$

0.3 Rational Exponents and Radicals

Definitions and Concepts	Examples
Summary of $a^{1/n}$ definitions: If $a \geq 0$, then $a^{1/n}$ is the nonnegative number b such that $b^n = a$.	$16^{1/2} = 4$ because $4^2 = 16$.
If $a < 0$ and n is odd, then $a^{1/n}$ is the real number b such that $b^n = a$.	$(-27)^{1/3} = -3$ because $(-3)^3 = -27$.
If $a < 0$ and n is even, then $a^{1/n}$ is not a real number.	$(-16)^{1/2}$ is not a real number because no number squared is -16.
Rule for rational exponents: If m and n are positive integers, $\frac{m}{n}$ is in lowest terms, and $a^{1/n}$ is a real number, then $$a^{m/n} = (a^{1/n})^m = (a^m)^{1/n}$$ **Definition of $\sqrt[n]{a}$:** $$\sqrt[n]{a} = a^{1/n}$$	$8^{2/3} = (8^{1/3})^2 = 2^2 = 4 \quad$ or $\quad (8^2)^{1/3} = 64^{1/3} = 4$ $\sqrt[3]{125} = 125^{1/3} = 5$
Properties of radicals: If all radicals are real numbers and there are no divisions by 0, then $$\sqrt[n]{ab} = \sqrt[n]{a}\sqrt[n]{b}$$ $$\sqrt[n]{\frac{a}{b}} = \frac{\sqrt[n]{a}}{\sqrt[n]{b}}$$ $$\sqrt[m]{\sqrt[n]{a}} = \sqrt[n]{\sqrt[m]{a}} = \sqrt[mn]{a}$$	$\sqrt[5]{32x^{10}} = \sqrt[5]{32}\sqrt[5]{x^{10}} = 2x^2$ $\sqrt[4]{\frac{x^{12}}{625}} = \frac{\sqrt[4]{x^{12}}}{\sqrt[4]{625}} = \frac{x^{12/4}}{5} = \frac{x^3}{5}$ $\sqrt[3]{\sqrt{64}} = \sqrt[3]{8} = 2 \qquad \sqrt{\sqrt[3]{64}} = \sqrt{4} = 2$ $\sqrt[3\cdot2]{64} = \sqrt[6]{64} = 2$

Exercises

Simplify each expression, if possible.

52. $121^{1/2}$

53. $\left(\dfrac{27}{125}\right)^{1/3}$

54. $(32x^5)^{1/5}$

55. $(81a^4)^{1/4}$

56. $(-1{,}000x^6)^{1/3}$

57. $(-25x^2)^{1/2}$

58. $(x^{12}y^2)^{1/2}$

59. $\left(\dfrac{x^{12}}{y^4}\right)^{-1/2}$

60. $\left(\dfrac{-c^{2/3}c^{5/3}}{c^{-2/3}}\right)^{1/3}$

61. $\left(\dfrac{a^{-1/4}a^{3/4}}{a^{9/2}}\right)^{-1/2}$

Simplify each expression.

62. $64^{2/3}$

63. $32^{-3/5}$

64. $\left(\dfrac{16}{81}\right)^{3/4}$

65. $\left(\dfrac{32}{243}\right)^{2/5}$

66. $\left(\dfrac{8}{27}\right)^{-2/3}$

67. $\left(\dfrac{16}{625}\right)^{-3/4}$

68. $(-216x^3)^{2/3}$

69. $\dfrac{p^{a/2}p^{a/3}}{p^{a/6}}$

Simplify each expression.

70. $\sqrt{36}$

71. $-\sqrt{49}$

72. $\sqrt{\dfrac{9}{25}}$

73. $\sqrt[3]{\dfrac{27}{125}}$

74. $\sqrt{x^2y^4}$

75. $\sqrt[3]{x^3}$

76. $\sqrt[4]{\dfrac{m^8n^4}{p^{16}}}$

77. $\sqrt[5]{\dfrac{a^{15}b^{10}}{c^5}}$

Simplify and combine terms.

78. $\sqrt{50} + \sqrt{8}$

79. $\sqrt{12} + \sqrt{3} - \sqrt{27}$

80. $\sqrt[3]{24x^4} - \sqrt[3]{3x^4}$

Rationalize each denominator.

81. $\dfrac{\sqrt{7}}{\sqrt{5}}$

82. $\dfrac{8}{\sqrt{8}}$

83. $\dfrac{1}{\sqrt[3]{2}}$

84. $\dfrac{2}{\sqrt[3]{25}}$

Rationalize each numerator.

85. $\dfrac{\sqrt{2}}{5}$

86. $\dfrac{\sqrt{5}}{5}$

87. $\dfrac{\sqrt{2x}}{3}$

88. $\dfrac{3\sqrt[3]{7x}}{2}$

0.4   Polynomials

Definitions and Concepts	Examples
Monomial: A polynomial with one term.	$2, \quad 3x, \quad 4x^2y, \quad -x^3y^2z$
Binomial: A polynomial with two terms.	$2t + 3, \quad 3r^2 - 6r, \quad 4m + 5n$
Trinomial: A polynomial with three terms.	$3p^2 - 7p + 8, \quad 3m^2 + 2n - p$
The **degree of a monomial** is the sum of the exponents on its variables.	The degree of $4x^2y^3$ is $2 + 3 = 5$.
The **degree of a polynomial** is the degree of the term in the polynomial with highest degree.	The degree of the first term of $3p^2q^3 - 6p^3q^4 + 9pq^2$ is $2 + 3 = 5$. The degree of the second term is $3 + 4 = 7$. The degree of the third term is $1 + 2 = 3$. The degree of the polynomial is the largest of these. It is 7.
$a(b + c + d + \cdots) = ab + ac + ad + \cdots$	$3(x + 2) = 3x + 6$ $2x(3x^2 - 2y + 3) = 6x^3 - 4xy + 6x$
To add or subtract polynomials, remove parentheses and combine like terms.	Add: $(3x^2 + 5x) + (2x^2 - 2x)$. $\begin{aligned}(3x^2 + 5x) + (2x^2 - 2x) &= 3x^2 + 5x + 2x^2 - 2x\\ &= 3x^2 + 2x^2 + 5x - 2x\\ &= 5x^2 + 3x\end{aligned}$ Subtract: $(4a^2 - 5b) - (3a^2 - 7b)$. $\begin{aligned}(4a^2 - 5b) - (3a^2 - 7b) &= 4a^2 - 5b - 3a^2 + 7b\\ &= 4a^2 - 3a^2 - 5b + 7b\\ &= a^2 + 2b\end{aligned}$
Special products: $\quad (x + y)^2 = x^2 + 2xy + y^2$ $\quad (x - y)^2 = x^2 - 2xy + y^2$ $\quad (x + y)(x - y) = x^2 - y^2$	$(2m + 3)^2 = 4m^2 + 12m + 9$ $(4t - 3s)^2 = 16t^2 - 24ts + 9s^2$ $\begin{aligned}(2m + n)(2m - n) &= 4m^2 - 2mn + 2mn - n^2\\ &= 4m^2 - n^2\end{aligned}$
To multiply a binomial by another, use the FOIL method.	$\begin{aligned}(2a + b)(a - b) &= 2a(a) + 2a(-b) + ba + b(-b)\\ &= 2a^2 - 2ab + ab - b^2\\ &= 2a^2 - ab - b^2\end{aligned}$

The **conjugate** of $a + b$ is $a - b$.

Rationalize the denominator: $\dfrac{x}{\sqrt{x} + 2}$.

$$\frac{x}{\sqrt{x} + 2} = \frac{x\left(\sqrt{x} - 2\right)}{\left(\sqrt{x} + 2\right)\left(\sqrt{x} - 2\right)}$$

Multiply the numerator and denominator by the conjugate of $\sqrt{x} + 2$.

$$= \frac{x\sqrt{x} - 2x}{x - 4}$$

Simplify.

To divide a polynomial by a monomial, write the division as the sum of several fractions, and simplify each one.

$$\frac{3x^2 - 2x}{3x} = \frac{3x^2}{3x} - \frac{2x}{3x} = x - \frac{2}{3}$$

To divide polynomials, use long division.

Divide: $2x + 3 \overline{)6x^2 + 7x - x + 3}$.

$$
\begin{array}{r}
3x^2 - x + 1 \\
2x + 3 \overline{)6x^3 + 7x^2 - x + 3} \\
\underline{6x^3 + 9x^2} \\
-2x^2 - x \\
\underline{-2x^2 - 3x} \\
+2x + 3 \\
\underline{+2x + 3} \\
0
\end{array}
$$

Exercises

Give the degree of each polynomial and tell whether the polynomial is a monomial, a binomial, or a trinomial.

89. $x^3 - 8$

90. $8x - 8x^2 - 8$

91. $\sqrt{3}x^2$

92. $4x^4 - 12x^2 + 1$

Perform the operations and simplify.

93. $2(x + 3) + 3(x - 4)$

94. $3x^2(x - 1) - 2x(x + 3) - x^2(x + 2)$

95. $(3x + 2)(3x + 2)$

96. $(3x + y)(2x - 3y)$

97. $(4a + 2b)(2a - 3b)$

98. $(z + 3)(3z^2 + z - 1)$

99. $(a^n + 2)(a^n - 1)$

100. $\left(\sqrt{2} + x\right)^2$

101. $\left(\sqrt{2} + 1\right)\left(\sqrt{3} + 1\right)$

102. $\left(\sqrt[3]{3} - 2\right)\left(\sqrt[3]{9} + 2\sqrt[3]{3} + 4\right)$

Rationalize each denominator.

103. $\dfrac{2}{\sqrt{3} - 1}$

104. $\dfrac{-2}{\sqrt{3} - \sqrt{2}}$

105. $\dfrac{2x}{\sqrt{x} - 2}$

106. $\dfrac{\sqrt{x} - \sqrt{y}}{\sqrt{x} + \sqrt{y}}$

Rationalize each numerator.

107. $\dfrac{\sqrt{x} + 2}{5}$

108. $\dfrac{1 - \sqrt{a}}{a}$

Perform each division.

109. $\dfrac{3x^2y^2}{6x^3y}$

110. $\dfrac{4a^2b^3 + 6ab^4}{2b^2}$

111. $2x + 3 \overline{)2x^3 + 7x^2 + 8x + 3}$

112. $x^2 - 1 \overline{)x^5 + x^3 - 2x - 3x^2 - 3}$

0.5   Factoring Polynomials

Definitions and Concepts	Examples
Factoring out a common monomial: $$ab + ac = a(b + c)$$	**Factor:** $$3p^3 - 6p^2q + 9p = 3p(p^2 - 2pq + 3)$$
Factoring the difference of two squares: $$x^2 - y^2 = (x + y)(x - y)$$	$$4x^2 - 9 = (2x + 3)(2x - 3)$$
Factoring trinomial squares: $$x^2 + 2xy + y^2 = (x + y)^2$$ $$x^2 - 2xy + y^2 = (x - y)^2$$	$$9a^2 + 12ab + 4b^2 = (3a + 2b)(3a + 2b) = (3a + 2b)^2$$ $$r^2 - 4rs + 4s^2 = (r - 2s)(r - 2s) = (r - 2s)^2$$
Factoring the sum and difference of two cubes: $$x^3 + y^3 = (x + y)(x^2 - xy + y^2)$$ $$x^3 - y^3 = (x - y)(x^2 + xy + y^2)$$	$$r^3 + 8 = (r + 2)(r^2 - 2r + 4)$$ $$27a^3 - 8b^3 = (3a - 2b)(9a^2 + 6ab + 4b^2)$$

Exercises

Factor each expression completely, if possible.

113. $3t^3 - 3t$ **114.** $5r^3 - 5$

115. $6x^2 + 7x - 24$ **116.** $3a^2 + ax - 3a - x$

117. $8x^3 - 125$ **118.** $6x^2 - 20x - 16$

119. $x^2 + 6x + 9 - t^2$ **120.** $3x^2 - 1 + 5x$

121. $8z^3 + 343$

122. $1 + 14b + 49b^2$

123. $121z^2 + 4 - 44z$

124. $64y^3 - 1,000$

125. $2xy - 4zx - wy + 2zw$

126. $x^8 + x^4 + 1$

0.6   Rational Expressions

Definitions and Concepts	Examples
Properties of fractions: If there are no divisions by 0, then	
$\dfrac{a}{b} = \dfrac{c}{d}$ if and only if $ad = bc$.	$\dfrac{3x}{4} = \dfrac{6x}{8}$ because $(3x)8$ and $4(6x)$ both equal $24x$.
$\dfrac{a}{b} = \dfrac{ax}{bx}$	$\dfrac{6x^2}{8x^3} = \dfrac{3 \cdot \cancel{2} \cdot \cancel{x} \cdot \cancel{x}}{\cancel{2} \cdot 4 \cdot \cancel{x} \cdot \cancel{x} \cdot x} = \dfrac{3}{4x}$
$\dfrac{a}{b} \cdot \dfrac{c}{d} = \dfrac{ac}{bd}$	$\dfrac{3p}{2q} \cdot \dfrac{2p}{6q} = \dfrac{3p \cdot 2p}{2q \cdot 6q} = \dfrac{\cancel{3}p \cdot 2p}{\cancel{2}q \cdot \cancel{3} \cdot 2q} = \dfrac{p^2}{2q^2}$
$\dfrac{a}{b} \div \dfrac{c}{d} = \dfrac{ad}{bc}$	$\dfrac{2t}{3s} \div \dfrac{2t}{6s} = \dfrac{2t}{3s} \cdot \dfrac{6s}{2t} = \dfrac{2t \cdot 6s}{3s \cdot 2t} = \dfrac{\cancel{2t} \cdot 2 \cdot \cancel{3s}}{\cancel{3s}\cancel{2t}} = 2$
$\dfrac{a}{b} + \dfrac{c}{b} = \dfrac{a + c}{b}$ $\dfrac{a}{b} - \dfrac{c}{b} = \dfrac{a - c}{b}$	$\dfrac{x}{4} + \dfrac{y}{4} = \dfrac{x + y}{4}$ $\dfrac{3p}{2q} - \dfrac{p}{2q} = \dfrac{3p - p}{2q} = \dfrac{2p}{2q} = \dfrac{p}{q}$
$a \cdot 1 = a$ $\dfrac{a}{1} = a$ $\dfrac{a}{a} = 1$	$7 \cdot 1 = 7$ $\dfrac{7}{1} = 7$ $\dfrac{7}{7} = 1$

$\dfrac{a}{b} = \dfrac{-a}{-b} = -\dfrac{a}{-b} = -\dfrac{-a}{b}$	$\dfrac{7}{2} = \dfrac{-7}{-2} = -\dfrac{7}{-2} = -\dfrac{-7}{2}$
$-\dfrac{a}{b} = \dfrac{a}{-b} = \dfrac{-a}{b} = -\dfrac{-a}{-b}$	$-\dfrac{7}{2} = \dfrac{7}{-2} = \dfrac{-7}{2} = -\dfrac{-7}{-2}$
To simplify a rational expression, factor the numerator and denominator, if possible, and divide out factors that are common to the numerator and denominator.	To simplify $\dfrac{2 - x}{2x + 4}$, we factor -1 from the numerator and 2 from the denominator to get $$\frac{2 - x}{2x - 4} = \frac{-(-2 + x)}{2(x - 2)} = \frac{-(x - 2)}{2(x - 2)} = \frac{-1}{2} = -\frac{1}{2}$$
To add or subtract rational expressions with unlike denominators, we find the LCD of the expressions, write each expression with a denominator that is the LCD, add or subtract the expressions, and simplify the result, if possible.	$\dfrac{2x}{x + 2} - \dfrac{2x}{x - 3} = \dfrac{2x(x - 3)}{(x + 2)(x - 3)} - \dfrac{2x(x + 2)}{(x - 3)(x + 2)}$ $= \dfrac{2x(x - 3) - 2x(x + 2)}{(x + 2)(x - 3)}$ $= \dfrac{2x^2 - 6x - 2x^2 - 4x}{(x + 2)(x - 3)}$ $= \dfrac{-10x}{(x + 2)(x - 3)}$
Use the rules for rational expressions to simplify complex fractions.	$\dfrac{1 - \dfrac{y}{2}}{\dfrac{1}{y} + \dfrac{1}{2}} = \dfrac{2y\left(1 - \dfrac{y}{2}\right)}{2y\left(\dfrac{1}{y} + \dfrac{1}{2}\right)} = \dfrac{2y - y^2}{2 + y} = \dfrac{y(2 - y)}{2 + y}$

Exercises

Simplify each rational expression.

127. $\dfrac{2 - x}{x^2 - 4x + 4}$

128. $\dfrac{a^2 - 9}{a^2 - 6a + 9}$

Perform each operation and simplify. Assume that no denominators are 0.

129. $\dfrac{x^2 - 4x + 4}{x + 2} \cdot \dfrac{x^2 + 5x + 6}{x - 2}$

130. $\dfrac{2y^2 - 11y + 15}{y^2 - 6y + 8} \cdot \dfrac{y^2 - 2y - 8}{y^2 - y - 6}$

131. $\dfrac{2t^2 + t - 3}{3t^2 - 7t + 4} \div \dfrac{10t + 15}{3t^2 - t - 4}$

132. $\dfrac{p^2 + 7p + 12}{p^3 + 8p^2 + 4p} \div \dfrac{p^2 - 9}{p^2}$

133. $\dfrac{x^2 + x - 6}{x^2 - x - 6} \cdot \dfrac{x^2 - x - 6}{x^2 + x - 2} \div \dfrac{x^2 - 4}{x^2 - 5x + 6}$

134. $\left(\dfrac{2x + 6}{x + 5} \div \dfrac{2x^2 - 2x - 4}{x^2 - 25}\right)\dfrac{x^2 - x - 2}{x^2 - 2x - 15}$

135. $\dfrac{2}{x - 4} + \dfrac{3x}{x + 5}$

136. $\dfrac{5x}{x - 2} - \dfrac{3x + 7}{x + 2} + \dfrac{2x + 1}{x + 2}$

137. $\dfrac{x}{x - 1} + \dfrac{x}{x - 2} + \dfrac{x}{x - 3}$

138. $\dfrac{x}{x + 1} - \dfrac{3x + 7}{x + 2} + \dfrac{2x + 1}{x + 2}$

139. $\dfrac{3(x + 1)}{x} - \dfrac{5(x^2 + 3)}{x^2} + \dfrac{x}{x + 1}$

140. $\dfrac{3x}{x + 1} + \dfrac{x^2 + 4x + 3}{x^2 + 3x + 2} - \dfrac{x^2 + x - 6}{x^2 - 4}$

Simplify each complex fraction. Assume that no denominators are 0.

141. $\dfrac{\dfrac{5x}{2}}{\dfrac{3x^2}{8}}$

142. $\dfrac{\dfrac{3x}{y}}{\dfrac{6x}{y^2}}$

143. $\dfrac{\dfrac{1}{x} + \dfrac{1}{y}}{x - y}$

144. $\dfrac{x^{-1} + y^{-1}}{y^{-1} - x^{-1}}$

CHAPTER TEST

Consider the set $\left\{-7, -\frac{2}{3}, 0, 1, 3, \sqrt{10}, 4\right\}$.

1. List the numbers in the set that are odd integers.
2. List the numbers in the set that are prime numbers.

Determine which property justifies each statement.

3. $(a + b) + c = (b + a) + c$
4. $a(b + c) = ab + ac$

Graph each interval on a number line.

5. $-4 < x \le 2$

6. $(-\infty, -3) \cup [6, \infty)$

Write each expression without using absolute value symbols.

7. $|-17|$
8. $|x - 7|$, when $x < 0$.

Find the distance on a number line between points with the following coordinates.

9. -4 and 12 10. -20 and -12

Simplify each expression. Assume that all variables represent positive numbers, and write all answers without using negative exponents.

11. $x^4 x^5 x^2$ 12. $\dfrac{r^2 r^3 s}{r^4 s^2}$

13. $\dfrac{(a^{-1}a^2)^{-2}}{a^{-3}}$ 14. $\left(\dfrac{x^0 x^2}{x^{-2}}\right)^6$

Write each number in scientific notation.

15. $450{,}000$ 16. 0.000345

Write each number in standard notation.

17. 3.7×10^3 18. 1.2×10^{-3}

Simplify each expression. Assume that all variables represent positive numbers, and write all answers without using negative exponents.

19. $(25a^4)^{1/2}$ 20. $\left(\dfrac{36}{81}\right)^{3/2}$

21. $\left(\dfrac{8t^6}{27s^9}\right)^{-2/3}$ 22. $\sqrt[3]{27a^6}$

23. $\sqrt{12} + \sqrt{27}$

24. $2\sqrt[3]{3x^4} - 3x\sqrt[3]{24x}$

25. Rationalize the denominator:

$$\dfrac{x}{\sqrt{x} - 2}.$$

26. Rationalize the numerator:

$$\dfrac{\sqrt{x} - \sqrt{y}}{\sqrt{x} + \sqrt{y}}.$$

Perform each operation.

27. $(a^2 + 3) - (2a^2 - 4)$
28. $(3a^3 b^2)(-2a^3 b^4)$
29. $(3x - 4)(2x + 7)$
30. $(a^n + 2)(a^n - 3)$
31. $(x^2 + 4)(x^2 - 4)$
32. $(x^2 - x + 2)(2x - 3)$
33. $x - 3\overline{)6x^2 + x - 23}$
34. $2x - 1\overline{)2x^3 + 3x^2 - 1}$

Factor each polynomial.

35. $3x + 6y$
36. $x^2 - 100$
37. $10t^2 - 19tw + 6w^2$
38. $3a^3 - 648$
39. $x^4 - x^2 - 12$
40. $6x^4 + 11x^2 - 10$

Perform each operation and simplify if possible. Assume that no denominators are 0.

41. $\dfrac{x}{x + 2} + \dfrac{2}{x + 2}$

42. $\dfrac{x}{x + 1} - \dfrac{x}{x - 1}$

43. $\dfrac{x^2 + x - 20}{x^2 - 16} \cdot \dfrac{x^2 - 25}{x - 5}$

44. $\dfrac{x + 2}{x^2 + 2x + 1} \div \dfrac{x^2 - 4}{x + 1}$

Simplify each complex fraction. Assume that no denominators are 0.

45. $\dfrac{\dfrac{1}{a} + \dfrac{1}{b}}{\dfrac{1}{b}}$

46. $\dfrac{x^{-1}}{x^{-1} + y^{-1}}$

Equations and Inequalities

Careers and Mathematics

Registered Nurse   Registered nurses (RNs), regardless of specialty or work setting, treat patients, educate patients and the public about various medical conditions, and provide advice and emotional support to patients' family members. RNs record patients' medical histories and symptoms, help perform diagnostic tests and analyze results, operate medical machinery, administer treatment and medications, and help with patient follow-up and rehabilitation.

RNs also teach patients and their families how to manage their illnesses or injuries. Some RNs work to promote general health by educating the public about warning signs and symptoms of disease. RNs also may run general health screening or immunization clinics, blood drives, and public seminars on various conditions.

Registered nurses held about 2.5 million jobs in 2006. About 3 in 5 jobs were in hospitals.

Education   There are three major educational paths to registered nursing—a bachelor's of science degree in nursing (BSN), an associate degree in nursing (ADN), and a diploma. BSN programs, offered by colleges and universities, take about 4 years to complete. ADN programs, offered by community and junior colleges, take about 2 to 3 years to complete. Diploma programs, administered in hospitals, last about 3 years. Individuals then must complete a national licensing examination in order to obtain a nursing license.

Job Outlook   Job opportunities for RNs are expected to be very good. Employment is projected to grow 23 percent through 2016. The median annual salaries of registered nurses were $57,280 in 2006.

For a sample application, see Problem 41 in Section 1.2. For more information go to www.bls.gov/oco/ocos083.htm.

The topic of this chapter is equations—one of the most important concepts in algebra. Equations are used in almost every academic discipline and vocational area, especially in chemistry, physics, medicine, economics, and business.

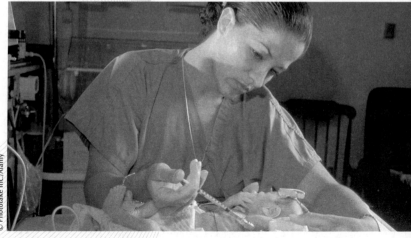

© Phototake Inc./Alamy

© avatra image /Alamy

1.1 Equations

Objectives

1. Review the Properties of Equality
2. Solve Linear Equations
3. Solve Rational Equations
4. Solve Formulas for a Specific Variable

It's been said that "weddings today are as beautiful as they are expensive." Suppose a couple budgets $3,000 for their wedding reception at an historic home. If $600 is charged for renting the home and there is a $24-per-person fee for food and beverages, how many guests can the couple accommodate at their reception?

If we let the **variable** x represent the number of guests, the expression $600 + 24x$ represents the cost for the reception. That is, $600 for the home rental plus $24 times the number of guests, x. We want to know the value of x that makes the expression equal $3,000.

We can write the statement $600 + 24x = 3,000$ to indicate that the two quantities are equal. A statement indicating that two quantities are equal is called an **equation.** In this section, we will review how to solve equations of this type.

If $x = 100$, the equation is true because when we substitute 100 for x, we obtain a true statement.

$$600 + 24(\mathbf{100}) = 3,000$$
$$600 + 2,400 = 3,000$$
$$3,000 = 3,000$$

The couple can accommodate 100 guests at their wedding reception.

An equation can be either true or false. For example, the equation $2 + 2 = 4$ is true, and the equation $2 + 3 = 6$ is false. An equation such as $3x - 2 = 10$ can be true or false depending on the value of x. If $x = 4$, the equation is true, because 4 satisfies the equation.

$$3x - 2 = 10$$
$$3(4) - 2 \overset{?}{=} 10 \qquad \text{Substitute 4 for } x. \text{ Read } \overset{?}{=} \text{ as "is possibly equal to."}$$
$$12 - 2 \overset{?}{=} 10$$
$$10 = 10$$

François Vieta (Viête)
(1540–1603)
By using letters in place of unknown numbers, Vieta simplified algebra and brought its notation closer to the notation that we use today. The one symbol he didn't use was the equal sign.

This equation is false for all other values of x.

Any number that satisfies an equation is called a **solution** or **root** of the equation. The set of all solutions of an equation is called its **solution set.** We have seen that the solution set of $3x - 2 = 10$ is $\{4\}$. To **solve** an equation means to find its solution set.

There can be restrictions on the values of a variable. For example, in the fraction

$$\frac{x^2 + 4}{x - 2}$$

we cannot replace x with 2, because that would make the denominator equal to 0.

EXAMPLE 1   Find the restrictions on the values of b in the equation: $\sqrt{b} = \dfrac{2}{b - 1}$.

Solution   For $\sqrt{b}$ to be a real number, b must be nonnegative, and for $\dfrac{2}{b-1}$ to be a real number, b cannot be 1. Thus, the values of b are restricted to the set of all nonnegative real numbers except 1.

Self Check 1   Find the restrictions on a: $\sqrt{a} = \dfrac{3}{a - 2}$.

There are three types of equations: identities, contradictions, and conditional equations. These are defined and illustrated in the following table.

Type of equation	Definition	Example
Identity	Every acceptable real number replacement for the variable is a solution.	$x^2 - 9 = (x + 3)(x - 3)$ Every real number x is a solution.
Contradiction	No real number is a solution.	$x = x + 1$ The equation has no solution. No real number can be 1 greater than itself.
Conditional Equation	Solution set contains some but not all real numbers.	$3x - 2 = 10$ The equation has one solution; the number 4.

Two equations with the same solution set are called **equivalent equations.**

1.   Review the Properties of Equality

There are certain properties of equality that we can use to transform equations into equivalent but less complicated equations. If we use these properties, the resulting equations will be equivalent and will have the same solution set.

Properties of Equality   **The Addition and Subtraction Properties**
If a, b, and c are real numbers and $a = b$, then

$$a + c = b + c \qquad \text{and} \qquad a - c = b - c$$

The Multiplication and Division Properties
If a, b, and c are real numbers and $a = b$, then

$$ac = bc \qquad \text{and} \qquad \frac{a}{c} = \frac{b}{c} \quad (c \neq 0)$$

The Substitution Property
In an equation, a quantity may be substituted for its equal without changing the truth of the equation.

2.   Solve Linear Equations

The easiest equations to solve are the **first-degree** or **linear equations.** Since these equations involve first-degree polynomials, they also are called **first-degree polynomial equations.**

Linear Equations A **linear equation in one variable** (say, x) is any equation that can be written in the form

$$ax + b = 0 \qquad (a \text{ and } b \text{ are real numbers and } a \neq 0)$$

To solve the linear equation $2x + 3 = 0$, we subtract 3 from both sides of the equation and divide both sides by 2.

$$2x + 3 = 0$$
$$2x + 3 - 3 = 0 - 3 \qquad \text{To undo the addition of 3, subtract 3 from both sides.}$$
$$2x = -3$$
$$\frac{2x}{2} = -\frac{3}{2} \qquad \text{To undo the multiplication by 2, divide both sides by 2.}$$
$$x = -\frac{3}{2}$$

To verify that $-\frac{3}{2}$ satisfies the equation, we substitute $-\frac{3}{2}$ for x and simplify:

$$2x + 3 = 0$$
$$2\left(-\frac{3}{2}\right) + 3 \overset{?}{=} 0 \qquad \text{Substitute } -\frac{3}{2} \text{ for } x.$$
$$-3 + 3 \overset{?}{=} 0 \qquad 2\left(-\frac{3}{2}\right) = -3$$
$$0 = 0$$

Since both sides of the equation are equal, the solution checks.

In Exercise 74, you will be asked to solve the general linear equation $ax + b = 0$ for x, thereby showing that every conditional linear equation has exactly one solution.

EXAMPLE 2 Find the solution set: $3(x + 2) = 5x + 2$.

Solution We proceed as follows:

$$3(x + 2) = 5x + 2$$
$$3x + 6 = 5x + 2 \qquad \text{Use the distributive property and remove parentheses.}$$
$$3x + 6 - 3x = 5x - 3x + 2 \qquad \text{Subtract } 3x \text{ from both sides.}$$
$$6 = 2x + 2 \qquad \text{Combine like terms.}$$
$$6 - 2 = 2x + 2 - 2 \qquad \text{Subtract 2 from both sides.}$$
$$4 = 2x \qquad \text{Simplify.}$$
$$\frac{4}{2} = \frac{2x}{2} \qquad \text{Divide both sides by 2.}$$
$$2 = x \qquad \text{Simplify.}$$

Since all of the above equations are equivalent, the solution set of the original equation is $\{2\}$. Verify that 2 satisfies the equation.

Self Check 2 Find the solution set: $4(x - 3) = 7x - 3$.

Everyday Connections

Mobile Communications

According to the Pew Internet Project's December 2007 survey:

- 58% of adult Americans have used a cell phone or personal digital assistant (PDA) to do at least one of ten mobile non-voice data activities, such as texting, emailing, taking a picture, looking for maps or directions, or recording video.

- 41% of adult Americans have logged onto the internet on the go, that is, away from home or work either with a wireless laptop connection or a handheld device.

Mobile data communication activities: by age				
Age in years	18–29	30–49	50–64	65+
Send or receive text messages	27%	11%	8%	4%
Take a picture	26%	10%	9%	7%
Play a game	15%	5%	3%	2%
Send or receive email	9%	3%	3%	2%
Access the internet	10%	4%	2%	2%
Number surveyed	311	616	456	310

Source: http://www.pewinternet.org/PPF/r/244/report_display.asp

PRNewsFoto/Sound Solutions Americas

1. Given the data in the table, determine how many survey respondents age 65 and older have used a cell phone or personal data assistant to take a picture.

2. Given the data in the table, determine how many survey respondents age 18–29 have used a cell phone or personal data assistant to play a game.

EXAMPLE 3   Find the solution set: $\dfrac{3}{2}y - \dfrac{2}{3} = \dfrac{1}{5}y$.

Solution   To clear the equation of fractions, we multiply both sides by the LCD of the three fractions and proceed as follows:

$$\frac{3}{2}y - \frac{2}{3} = \frac{1}{5}y$$

$$30\left(\frac{3}{2}y - \frac{2}{3}\right) = 30\left(\frac{1}{5}y\right) \qquad \text{Multiply both sides by 30, the LCD of } \tfrac{3}{2}, \tfrac{2}{3}, \text{ and } \tfrac{1}{5}.$$

$$45y - 20 = 6y \qquad \text{Remove parentheses and simplify.}$$

$$45y - 20 + 20 = 6y + 20 \qquad \text{Add 20 to both sides.}$$

$$45y = 6y + 20 \qquad \text{Simplify.}$$

$$45y - 6y = 6y - 6y + 20 \qquad \text{Subtract } 6y \text{ from both sides.}$$

$$39y = 20 \qquad \text{Combine like terms.}$$

$$\frac{39y}{39} = \frac{20}{39} \qquad \text{Divide both sides by 39.}$$

$$y = \frac{20}{39} \qquad \text{Simplify.}$$

The solution set is $\left\{\dfrac{20}{39}\right\}$. Verify that $\dfrac{20}{39}$ satisfies the equation.

Self Check 3 Find the solution set: $\frac{2}{3}p - 3 = \frac{p}{6}$.

EXAMPLE 4 Solve: **a.** $3(x + 5) = 3(1 + x)$ **b.** $5 + 5(x + 2) - 2x = 3x + 15$

Solution **a.**
$$3(x + 5) = 3(1 + x)$$
$$3x + 15 = 3 + 3x \qquad \text{Remove parentheses.}$$
$$3x - 3x + 15 = 3 + 3x - 3x \qquad \text{Subtract } 3x \text{ from both sides.}$$
$$15 = 3 \qquad \text{Combine like terms.}$$

Since $15 = 3$ is false, the equation has no roots. Its solution set is the empty set, which is denoted as $\varnothing$. This equation is a contradiction.

b. $5 + 5(x + 2) - 2x = 3x + 15$
$$5 + 5x + 10 - 2x = 3x + 15 \qquad \text{Remove parentheses.}$$
$$3x + 15 = 3x + 15 \qquad \text{Simplify.}$$

Because both sides of the final equation are identical, every value of x will make the equation true. The solution set is the set of all real numbers. This equation is an identity.

Self Check 4 Solve: **a.** $-2(x - 4) + 6x = 4(x + 1)$
b. $2(x + 1) + 4 = 2(x + 3)$

3. Solve Rational Equations

Rational equations are equations that contain rational expressions. Some examples of rational equations are

$$\frac{2}{x - 3} = 7, \qquad \frac{x + 1}{x - 2} = \frac{3}{x - 2}, \qquad \text{and} \qquad \frac{x + 2}{x + 3} + \frac{1}{x^2 + 2x - 3} = 1$$

When solving these equations, we will multiply both sides by a quantity containing a variable. When we do this, we could inadvertently multiply both sides of an equation by 0 and obtain a solution that makes the denominator of a fraction 0. In this case, we have found a false solution, called an **extraneous solution.** These solutions do not satisfy the equation and must be discarded.

The following equation has an extraneous solution.

$$\frac{x + 1}{x - 2} = \frac{3}{x - 2}$$
$$(x - 2)\left(\frac{x + 1}{x - 2}\right) = (x - 2)\left(\frac{3}{x - 2}\right) \qquad \text{Multiply both sides by } x - 2.$$
$$x + 1 = 3 \qquad \frac{x - 2}{x - 2} = 1$$
$$x = 2 \qquad \text{Subtract 1 from both sides.}$$

If we check by substituting 2 for x, we obtain 0's in the denominator. Thus, 2 is not a root. The solution set is $\varnothing$.

EXAMPLE 5 Solve: $\dfrac{x + 2}{x + 3} + \dfrac{1}{x^2 + 2x - 3} = 1$.

Solution Note that x cannot be -3, because that would cause the denominator of the first fraction to be 0. To find other restrictions, we factor the trinomial in the denominator of the second fraction.

$$x^2 + 2x - 3 = (x + 3)(x - 1)$$

Since the denominator will be 0 when $x = -3$ or $x = 1$, x cannot be -3 or 1.

$$\frac{x + 2}{x + 3} + \frac{1}{x^2 + 2x - 3} = 1$$

$$\frac{x + 2}{x + 3} + \frac{1}{(x + 3)(x - 1)} = 1 \qquad \text{Factor } x^2 + 2x - 3.$$

$$(x + 3)(x - 1)\left[\frac{x + 2}{x + 3} + \frac{1}{(x + 3)(x - 1)}\right] = (x + 3)(x - 1)1 \qquad \text{Multiply both sides by } (x + 3)(x - 1).$$

$$\cancel{(x + 3)}(x - 1)\left(\frac{x + 2}{\cancel{x + 3}}\right) + \cancel{(x + 3)(x - 1)}\frac{1}{\cancel{(x + 3)(x - 1)}} = (x + 3)(x - 1)1 \qquad \text{Remove brackets.}$$

$$(x - 1)(x + 2) + 1 = (x + 3)(x - 1) \qquad \text{Simplify.}$$

$$x^2 + x - 2 + 1 = x^2 + 2x - 3 \qquad \text{Multiply the binomials.}$$

$$x - 1 = 2x - 3 \qquad \text{Subtract } x^2 \text{ from both sides and combine like terms.}$$

$$2 = x \qquad \text{Add 3 and subtract } x \text{ from both sides.}$$

Because 2 is a meaningful replacement for x, it is a root. However, it is a good idea to check it.

$$\frac{x + 2}{x + 3} + \frac{1}{x^2 + 2x - 3} = 1$$

$$\frac{2 + 2}{2 + 3} + \frac{1}{2^2 + 2(2) - 3} \stackrel{?}{=} 1 \qquad \text{Substitute 2 for } x.$$

$$\frac{4}{5} + \frac{1}{5} \stackrel{?}{=} 1$$

$$1 = 1$$

Since 2 satisfies the equation, it is a root.

Self Check 5   Solve: $\frac{3}{5} + \frac{7}{x + 2} = 2$.

4.   Solve Formulas for a Specific Variable

Many equations, called **formulas,** contain several variables. For example, the formula that converts degrees Celsius to degrees Fahrenheit is $F = \frac{9}{5}C + 32$. If we want to change a large number of Fahrenheit readings to degrees Celsius, it would be tedious to substitute each value of F into the formula and then repeatedly solve it for C. It is better to solve the formula for C, substitute the values for F, and evaluate C directly.

EXAMPLE 6   Solve $F = \frac{9}{5}C + 32$ for C.

Solution   We use the same methods as for solving linear equations.

$$F = \frac{9}{5}C + 32$$

$$F - 32 = \frac{9}{5}C \qquad \text{Subtract 32 from both sides.}$$

$$\frac{5}{9}(F - 32) = \frac{5}{9}\left(\frac{9}{5}C\right) \qquad \text{Multiply both sides by } \frac{5}{9}.$$

$$\frac{5}{9}(F - 32) = C \qquad \text{Simplify.}$$

This result also can be written in the form $C = \frac{5F - 160}{9}$.

Self Check 6   Solve $C = \frac{5}{9}(F - 32)$ for F.

EXAMPLE 7   The formula $A = P + Prt$ is used to find the amount of money in a savings account at the end of a specified time. A represents the amount, P represents the principal (the original deposit), r represents the rate of simple interest per unit of time, and t represents the number of units of time. Solve this formula for P.

Solution   We factor P from both terms on the right side of the equation and proceed as follows:

$$A = P + Prt$$
$$A = P(1 + rt) \qquad \text{Factor out } P.$$
$$\frac{A}{1 + rt} = P \qquad \text{Divide both sides by } 1 + rt.$$
$$P = \frac{A}{1 + rt}$$

Self Check 7   Solve $pq = fq + fp$ for f.

Self Check Answers   **1.** all nonnegative real numbers but 2   **2.** $\{-3\}$ **3.** $\{6\}$ **4. a.** no solution   **b.** all real numbers   **5.** 3   **6.** $F = \frac{9}{5}C + 32$ **7.** $f = \frac{pq}{q + p}$

1.1   Exercises

Vocabulary and Concepts   *Fill in the blanks.*

1. If a number satisfies an equation, it is called a _____ or a _____ of the equation.
2. If an equation is true for all values of its variable, it is called an _____.
3. A contradiction is an equation that is true for ___ values of its variable.
4. A _____ equation is true for some values of its variable and is not true for others.
5. An equation of the form $ax + b = 0$ is called a _____ equation.
6. If an equation contains rational expressions, it is called a _____ equation.
7. A conditional linear equation has ___ root.
8. The _____ of a fraction can never be 0.

Practice   *Each quantity represents a real number. Find any restrictions on x.*

9. $x + 3 = 1$

10. $\frac{1}{2}x - 7 = 14$

11. $\frac{1}{x} = 12$

12. $\frac{3}{x - 2}$

13. $\sqrt{x} = 4$

14. $\sqrt[3]{x} = 64$

15. $\frac{1}{x - 3} = \frac{5}{x + 2}$

16. $\frac{24}{\sqrt{x - 3}}$

Solve each equation, if possible. Classify each one as an identity, a conditional equation, or an equation with no solutions.

17. $2x + 5 = 15$

18. $3x + 2 = x + 8$

19. $2(n + 2) - 5 = 2n$

20. $3(m + 2) = 2(m + 3) + m$

21. $\frac{x + 7}{2} = 7$

22. $\frac{x}{2} - 7 = 14$

23. $2(a + 1) = 3(a - 2) - a$

24. $x^2 = (x + 4)(x - 4) + 16$

25. $3(x - 3) = \dfrac{6x - 18}{2}$

26. $x(x + 2) = (x + 1)^2$

27. $\dfrac{3}{b - 3} = 1$

28. $x^2 - 8x + 15 = (x - 3)(x + 5)$

29. $2x^2 + 5x - 3 = (2x - 1)(x + 3)$

30. $2x^2 + 5x - 3 = 2x\left(x + \dfrac{19}{2}\right)$

Solve each equation. If an equation has no solution, so indicate.

31. $2x + 7 = 10 - x$

32. $9a - 3 = 15 + 3a$

33. $\dfrac{5}{3}z - 8 = 7$

34. $\dfrac{4}{3}y + 12 = -4$

35. $\dfrac{z}{5} + 2 = 4$

36. $\dfrac{3p}{7} - p = -4$

37. $\dfrac{3x - 2}{3} = 6x + 7$

38. $\dfrac{7}{2}x + 5 = x + \dfrac{15}{2}$

39. $5(x - 2) = 2x + 8$

40. $5(r - 4) = -5(r - 4)$

41. $2(2x + 1) - \dfrac{3x}{2} = \dfrac{-3(4 + x)}{2}$

42. $(x - 2)(x - 3) = (x + 3)(x + 4)$

43. $7(2x + 5) - 6(x + 8) = 7$

44. $(t + 1)(t - 1) = (t + 2)(t - 3) + 4$

45. $(x - 2)(x + 5) = (x - 3)(x + 2)$

46. $\dfrac{3x + 1}{20} = \dfrac{1}{2}$

47. $\dfrac{3}{2}(3x - 2) - 10x - 4 = 0$

48. $a(a - 3) + 5 = (a - 1)^2$

49. $x(x + 2) = (x + 1)^2 - 1$

50. $\dfrac{3 + x}{3} + \dfrac{x + 7}{2} = 4x + 1$

51. $\dfrac{(y + 2)^2}{3} = y + 2 + \dfrac{y^2}{3}$

52. $2x - \dfrac{7}{6} + \dfrac{x}{6} = \dfrac{4x + 3}{6}$

53. $2(s + 2) + (s + 3)^2 = s(s + 5) + 2\left(\dfrac{17}{2} + s\right)$

54. $\dfrac{3}{x} + \dfrac{1}{2} = \dfrac{4}{x}$

55. $\dfrac{2}{x + 1} + \dfrac{1}{3} = \dfrac{1}{x + 1}$

56. $\dfrac{3}{x - 2} + \dfrac{1}{x} = \dfrac{3}{x - 2}$

57. $\dfrac{9(t + 3)}{t(t + 3)} = \dfrac{7}{t + 3}$

58. $x + \dfrac{2(-2x + 1)}{3x + 5} = \dfrac{3x^2}{3x + 5}$

59. $\dfrac{2}{(a - 7)(a + 2)} = \dfrac{4}{(a + 3)(a + 2)}$

60. $\dfrac{2}{n - 2} + \dfrac{1}{n + 1} = \dfrac{1}{n^2 - n - 2}$

61. $\dfrac{2x + 3}{x^2 + 5x + 6} + \dfrac{3x - 2}{x^2 + x - 6} = \dfrac{5x - 2}{x^2 - 4}$

62. $\dfrac{3x}{x^2 + x} - \dfrac{2x}{x^2 + 5x} = \dfrac{x + 2}{x^2 + 6x + 5}$

63. $\dfrac{3x + 5}{x^3 + 8} + \dfrac{3}{x^2 - 4} = \dfrac{2(3x - 2)}{(x - 2)(x^2 - 2x + 4)}$

64. $\dfrac{1}{n + 8} - \dfrac{3n - 4}{5n^2 + 42n + 16} = \dfrac{1}{5n + 2}$

65. $\dfrac{1}{11 - n} - \dfrac{2(3n - 1)}{-7n^2 + 74n + 33} = \dfrac{1}{7n + 3}$

66. $\dfrac{4}{a^2 - 13a - 48} - \dfrac{2}{a^2 - 18a + 32} = \dfrac{1}{a^2 + a - 6}$

67. $\dfrac{5}{y + 4} + \dfrac{2}{y + 2} = \dfrac{6}{y + 2} - \dfrac{1}{y^2 + 6y + 8}$

68. $\dfrac{6}{2a - 6} - \dfrac{3}{3 - 3a} = \dfrac{1}{a^2 - 4a + 3}$

69. $\dfrac{3y}{6 - 3y} + \dfrac{2y}{2y + 4} = \dfrac{8}{4 - y^2}$

70. $\dfrac{3 + 2a}{a^2 + 6 + 5a} - \dfrac{2 - 3a}{a^2 - 6 + a} = \dfrac{5a - 2}{a^2 - 4}$

71. $\dfrac{a}{a + 2} - 1 = -\dfrac{3a + 2}{a^2 + 4a + 4}$

72. $\dfrac{x - 1}{x + 3} + \dfrac{x - 2}{x - 3} = \dfrac{1 - 2x}{3 - x}$

Solve each formula for the specified variable.

73. $k = 2.2p$; p

74. $ax + b = 0$; x

75. $p = 2l + 2w$; w

76. $V = \dfrac{1}{3}\pi r^2 h$; h

77. $V = \dfrac{1}{3}\pi r^2 h;\ r^2$

78. $z = \dfrac{x - \mu}{\sigma};\ \mu$

79. $P_n = L + \dfrac{si}{f};\ s$

80. $P_n = L + \dfrac{si}{f};\ f$

81. $F = \dfrac{mMg}{r^2};\ m$

82. $\dfrac{1}{f} = \dfrac{1}{p} + \dfrac{1}{q};\ f$

83. $\dfrac{x}{a} + \dfrac{y}{b} = 1;\ y$

84. $\dfrac{x}{a} - \dfrac{y}{b} = 1;\ a$

85. $\dfrac{1}{r} = \dfrac{1}{r_1} + \dfrac{1}{r_2};\ r$

86. $\dfrac{1}{r} = \dfrac{1}{r_1} + \dfrac{1}{r_2};\ r_1$

87. $l = a + (n - 1)d;\ n$

88. $l = a + (n - 1)d;\ d$

89. $a = (n - 2)\dfrac{180}{n};\ n$

90. $S = \dfrac{a - lr}{1 - r};\ a$

91. $R = \dfrac{1}{\dfrac{1}{r_1} + \dfrac{1}{r_2} + \dfrac{1}{r_3}};\ r_1$

92. $R = \dfrac{1}{\dfrac{1}{r_1} + \dfrac{1}{r_2} + \dfrac{1}{r_3}};\ r_3$

Discovery and Writing

93. Explain why a conditional linear equation always has exactly one root.

94. Define an extraneous solution and explain how such a solution occurs.

Review   *Simplify each expression. Use absolute value symbols when necessary.*

95. $(25x^2)^{1/2}$

96. $\left(\dfrac{25p^2}{16q^4}\right)^{1/2}$

97. $\left(\dfrac{125x^3}{8y^6}\right)^{-2/3}$

98. $\left(-\dfrac{27y^3}{1{,}000x^6}\right)^{1/3}$

99. $\sqrt{25y^2}$

100. $\sqrt[3]{-125y^9}$

101. $\sqrt[4]{\dfrac{a^4 b^{12}}{z^8}}$

102. $\sqrt[5]{\dfrac{x^{10}y^5}{z^{15}}}$

1.2   Applications of Linear Equations

Objectives

1. Solve Number Problems
2. Solve Geometric Problems
3. Solve Investment Problems
4. Solve Break-Point Analysis Problems
5. Solve Shared-Work Problems
6. Solve Mixture Problems
7. Solve Uniform Motion Problems

John Shearer/WireImage/Getty Images

Tim McGraw is one of the most successful country music singers. He is married to country singer Faith Hill and is the son of former baseball player Tug McGraw. His album sales have exceeded 40 million copies and his *Soul to Soul* concert tours have been attended by thousands of fans.

Suppose we hear that Borders bookstore has Tim McGraw's CD *Let It Go* on sale for 30% off the original price. Knowing this is a bargain, we immediately drive to Borders, quickly pay the $12.99 selling price for the CD, and listen to the great music as we drive. Later that day, this question comes to mind: What was the original price of the CD?

A linear equation can be used to model this problem. We can let x represent the original price of the CD and subtract 30% of x (the discount) and we will get the selling price $12.99. The linear equation is $x - 0.3x = 12.99$. We can solve this equation to determine the original price of the CD.

$$x - 0.3x = 12.99$$
$$0.7x = 12.99 \quad \text{Combine like terms.}$$
$$x = \frac{12.99}{0.7} \quad \text{Divide both sides by 0.7.}$$
$$x = 18.56$$

The original price of the CD is $18.56.

In this section, we will use the equation-solving techniques discussed in the previous section to solve applied problems (often called *word problems*). To solve these problems, we must translate the verbal description of the problem into an equation. The process of finding the equation that describes the words of the problem is called **mathematical modeling.** The equation itself is often called a **mathematical model** of the situation described in the word problem.

The following list of steps provides a strategy to follow when we try to find the equation that models an applied problem.

Strategy for Modeling with Equations

1. Analyze the problem to see what you are to find. Often, drawing a diagram or making a table will help you visualize the facts.

2. Pick a variable to represent the quantity that is to be found, and write a sentence telling what that variable represents. Express all other quantities mentioned in the problem as expressions involving this single variable.

3. Find a way to express a quantity in two different ways. This might involve a formula from geometry, finance, or physics.

4. Form an equation indicating that the two quantities found in Step 3 are equal.

5. Solve the equation.

6. Answer the questions asked in the problem.

7. Check the answers in the words of the problem.

This list does not apply to all situations, but it can be used for a wide range of problems with only slight modifications.

1. Solve Number Problems

EXAMPLE 1   A student has scores of 74%, 78%, and 70% on three exams. What score is needed on a fourth exam for the student to earn an average grade of 80%?

Solution   To find an equation that models the problem, we can let x represent the required grade on the fourth exam. The average grade will be one-fourth of the sum of the four grades. We know this average is to be 80.

The average of the four grades	equals	the required average grade.
$\dfrac{74 + 78 + 70 + x}{4}$	$=$	80

We can solve this equation for x.

$\dfrac{222 + x}{4} = 80$ **74 + 78 + 70 = 222**

$222 + x = 320$ **Multiply both sides by 4.**

$x = 98$ **Subtract 222 from both sides.**

To earn an average of 80%, the student must score 98% on the fourth exam.

2.   Solve Geometric Problems

EXAMPLE 2   A city ordinance requires a man to install a fence around the swimming pool shown in Figure 1-1. He wants the border around the pool to be of uniform width. If he has 154 feet of fencing, find the width of the border.

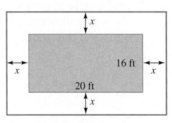

Figure 1-1

Solution   We can let x represent the width of the border. The distance around the large rectangle, called its **perimeter,** is given by the formula $P = 2l + 2w$, where l is the length, $20 + 2x$, and w is the width, $16 + 2x$. Since the man has 154 feet of fencing, the perimeter will be 154 feet. To find an equation that models the problem, we substitute these values into the formula for perimeter.

$P = 2l + 2w$ **This is the formula for the perimeter of a rectangle.**

$154 = 2(20 + 2x) + 2(16 + 2x)$ **Substitute 154 for P, 20 + 2x for l, and 16 + 2x for w.**

$154 = 40 + 4x + 32 + 4x$ **Use the distributive property to remove parentheses.**

$154 = 72 + 8x$ **Combine like terms.**

$82 = 8x$ **Subtract 72 from both sides.**

$10\dfrac{1}{4} = x$ **Divide both sides by 8.**

This border will be $10\frac{1}{4}$ feet wide.

3.   Solve Investment Problems

EXAMPLE 3   A woman invested $10,000, part at 9% and the rest at 14%. If the annual income from these investments is $1,275, how much did she invest at each rate?

Solution We can let x represent the amount invested at 9%. Then $10,000 - x$ represents the amount invested at 14%. Since the annual income from any simple-interest investment is the product of the interest rate and the amount invested, we have the following information.

Type of investment	Rate	Amount invested	Interest earned
9% investment	0.09	x	$0.09x$
14% investment	0.14	$10,000 - x$	$0.14(10,000 - x)$

The total income from these two investments can be expressed in two ways: as $1,275 and as the sum of the incomes from the two investments.

The income from the 9% investment	plus	the income from the 14% investment	equals	the total income.
$0.09x$	$+$	$0.14(10,000 - x)$	$=$	$1,275$

We can solve this equation for x.

$$0.09x + 0.14(10,000 - x) = 1,275$$
$$9x + 14(10,000 - x) = 127,500 \qquad \text{To eliminate the decimal points, multiply both sides by 100.}$$
$$9x + 140,000 - 14x = 127,500 \qquad \text{Use the distributive property to remove parentheses.}$$
$$-5x + 140,000 = 127,500 \qquad \text{Combine like terms.}$$
$$-5x = -12,500 \qquad \text{Subtract 140,000 from both sides.}$$
$$x = 2,500 \qquad \text{Divide both sides by } -5.$$

The amount invested at 9% was $2,500, and the amount invested at 14% was $7,500 ($10,000 - $2,500). These amounts are correct, because 9% of $2,500 is $225, 14% of $7,500 is $1,050, and the sum of these amounts is $1,275.

4. Solve Break-Point Analysis Problems

Running a machine involves two costs—**setup costs** and **unit costs.** Setup costs include the cost of installing a machine and preparing it to do a job. Unit cost is the cost to manufacture one item, which includes the costs of material and labor.

EXAMPLE 4 Suppose that one machine has a setup cost of $400 and a unit cost of $1.50, and a second machine has a setup cost of $500 and a unit cost of $1.25. Find the **break point** (the number of units manufactured at which the cost on each machine is the same).

Solution We can let x represent the number of items to be manufactured. The cost c_1 of using machine 1 is

$$c_1 = 400 + 1.5x$$

and the cost c_2 of using machine 2 is

$$c_2 = 500 + 1.25x$$

The break point occurs when these two costs are equal.

The cost of using machine 1	equals	the cost of using machine 2.
$400 + 1.5x$	$=$	$500 + 1.25x$

We can solve this equation for x.

$$
\begin{aligned}
400 + 1.5x &= 500 + 1.25x & \\
1.5x &= 100 + 1.25x & \text{Subtract 400 from both sides.} \\
0.25x &= 100 & \text{Subtract } 1.25x \text{ from both sides.} \\
x &= 400 & \text{Divide both sides by 0.25.}
\end{aligned}
$$

The break point is 400 units. This result is correct, because it will cost the same amount to manufacture 400 units with either machine.

$$c_1 = \$400 + \$1.5(400) = \$1,000 \quad \text{and} \quad c_2 = \$500 + \$1.25(400) = \$1,000 \quad \blacksquare$$

5. Solve Shared-Work Problems

EXAMPLE 5 The Toll Way Authority needs to pave 100 miles of interstate highway before freezing temperatures come in about 60 days. Sjostrom and Sons has estimated that it can do the job in 110 days. Scandroli and Sons has estimated that it can do the job in 140 days. If the authority hires both contractors, will the job get done in time?

Solution Since Sjostrom can do the job in 110 days, they can do $\frac{1}{110}$ of the job in one day, and since Scandroli can do the job in 140 days, they can do $\frac{1}{140}$ of the job in one day. If we let n represent the number of days it will take to pave the highway if both contractors work together, they can do $\frac{1}{n}$ of the job in one day. The work that they can do together in one day is the sum of what each can do in one day.

The part Sjostrom can pave in one day	plus	the part Scandroli can pave in one day	equals	the part they can pave together in one day.
$\dfrac{1}{110}$	$+$	$\dfrac{1}{140}$	$=$	$\dfrac{1}{n}$

We can solve this equation for n.

$$
\begin{aligned}
\frac{1}{110} + \frac{1}{140} &= \frac{1}{n} & \\
(110)(140)n\left(\frac{1}{110} + \frac{1}{140}\right) &= (110)(140)n\left(\frac{1}{n}\right) & \text{\textbf{Multiply both sides by (110)(140)}}n \\
& & \text{\textbf{to eliminate the fractions.}} \\
\frac{(110)(140)n}{110} + \frac{(110)(140)n}{140} &= \frac{(110)(140)n}{n} & \text{\textbf{Use the distributive property to}} \\
& & \text{\textbf{remove parentheses.}} \\
140n + 110n &= 15,400 & \tfrac{110}{110} = 1, \tfrac{140}{140} = 1, \text{ and } \tfrac{n}{n} = 1. \\
250n &= 15,400 & \text{\textbf{Combine like terms.}} \\
n &= 61.6 & \text{\textbf{Divide both sides by 250.}}
\end{aligned}
$$

It will take the contractors about 62 days to pave the highway. With any luck, the job will be done in time. $\blacksquare$

6.   Solve Mixture Problems

EXAMPLE 6   A container is partially filled with 20 liters of whole milk containing 4% butterfat. How much 1% milk must be added to obtain a mixture that is 2% butterfat?

Solution   Since the first container shown in Figure 1-2(a) contains 20 liters of 4% milk, it contains 0.04(20) liters of butterfat. To this amount, we will add the contents of the second container, which holds 0.01(*l*) liters of butterfat.
    The sum of these two amounts will equal the number of liters of butterfat in the third container, which is 0.02(20 + *l*) liters of butterfat. This information is presented in table form in Figure 1-2(b).

The butterfat in the 4% milk	plus	the butterfat in the 1% milk	equals	the butterfat in the 2% milk.
4% of 20 liters	+	1% of *l* liters	=	2% of (20 + *l*) liters

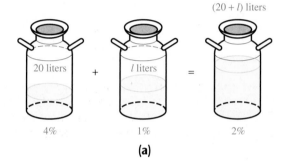

(20 + *l*) liters

20 liters   +   *l* liters   =

4%   1%   2%

(a)

	Percentage · of butterfat	Amount of milk	= Amount of butterfat
4% milk	0.04	20	0.04(20)
1% milk	0.01	*l*	0.01(*l*)
2% milk	0.02	20 + *l*	0.02(20 + *l*)

(b)

Figure 1-2

We can solve this equation for *l*.

$$0.04(20) + 0.01(l) = 0.02(20 + l)$$

$$4(20) + l = 2(20 + l) \qquad \text{Multiply both sides by 100.}$$

$$80 + l = 40 + 2l \qquad \text{Remove parentheses.}$$

$$40 = l \qquad \text{Subtract 40 and } l \text{ from both sides.}$$

    To dilute the 20 liters of 4% milk to a 2% mixture, 40 liters of 1% milk must be added. To check, we note that the final mixture contains 0.02(60) = 1.2 liters of pure butterfat, and that this is equal to the amount of pure butterfat in the 4% milk and the 1% milk; 0.04(20) + 0.01(40) = 1.2 liters.

7.   Solve Uniform Motion Problems

EXAMPLE 7   A man leaves home driving at the rate of 50 mph. When his daughter discovers that he has forgotten his wallet, she drives after him at the rate of 65 mph. How long will it take her to catch her dad if he had a 15-minute head start?

Solution   Uniform motion problems are based on the formula $d = rt$, where d is the distance, r is the rate, and t is the time. We can organize the information given in the problem in a chart like the one shown in Figure 1-3 on the next page. In the chart, t represents the number of hours the daughter must drive to overtake her father. Because the father has a 15-minute, or $\frac{1}{4}$ hour, head start, he has been on the road for $\left(t + \frac{1}{4}\right)$ hours.

	d	**=**	**r**	**·**	**t**
Man	$50\left(t + \frac{1}{4}\right)$		50		$t + \frac{1}{4}$
Daughter	$65t$		65		t

(a)

50 mph

65 mph

(b)

Figure 1-3

We can set up the following equation and solve it for t.

The distance the man drives	equals	the distance the daughter drives.
$50\left(t + \dfrac{1}{4}\right)$	$=$	$65t$

We can solve this equation for t.

$$50\left(t + \frac{1}{4}\right) = 65t$$

$$50t + \frac{25}{2} = 65t \qquad \text{Remove parentheses.}$$

$$\frac{25}{2} = 15t \qquad \text{Subtract } 50t \text{ from both sides.}$$

$$\frac{5}{6} = t \qquad \text{Divide both sides by 15 and simplify.}$$

It will take the daughter $\frac{5}{6}$ hours, or 50 minutes, to overtake her dad.

George Polya
(1888–1985)
Polya, a Hungarian, became a professor of mathematics at Stanford University. His approach to problem solving made him very popular with faculty and students. Polya's book *How to Solve It* became a best-seller. His problem-solving approach involves four steps:

1. Understand the problem.
2. Devise a plan.
3. Carry out the plan.
4. Check back.

1.2   Exercises

Vocabulary and Concepts   *Fill in the blanks.*

1. To average n scores, ____ the scores and divide by n.

2. The formula for the _____ of a rectangle is $P = 2l + 2w$.

3. The simple annual interest earned on an investment is the product of the interest rate and the _____ invested.

4. The number of units manufactured at which the cost on two machines is equal is called the _____.

5. Distance traveled is the product of the ____ and the ____.

6. 5% of 30 liters is ____ liters.

Practice   *Solve each problem.*

7. **Test scores** Juan scored 5 points higher on his midterm and 13 points higher on his final than he did on his first exam. If his mean (average) score was 90, what was his score on the first exam?

8. **Test scores** Sally took four tests in science class. On each successive test, her score improved by 3 points. If her mean score was 69.5%, what did she score on the first test?

9. **Teacher certification** On the Illinois certification test for teachers specializing in learning disabilities, a teacher earned the scores shown in the accompanying table. What was the teacher's score in program development?

Human development with special needs	82
Assessment	90
Program development and instruction	?
Professional knowledge and legal issues	78
AVERAGE SCORE	86

10. **Golfing** Par on a golf course is 72. If a golfer shot rounds of 76, 68, and 70 in a tournament, what will she need to shoot on the final round to average par?

11. **Replacing locks**   A locksmith charges $40 plus $28 for each lock installed. How many locks can be replaced for $236?

12. **Delivering ads**   A college student earns $20 per day delivering advertising brochures door-to-door, plus 75¢ for each person he interviews. How many people did he interview on a day when he earned $56?

13. **Width of a picture frame**   The picture frame with the dimensions shown in the illustration was built with 14 feet of framing material. Find its width.

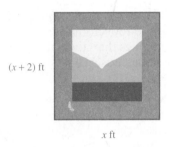

$(x + 2)$ ft

x ft

14. **Fencing a garden**   If a gardener fences in the total rectangular area shown in the illustration instead of just the square area, he will need twice as much fencing to enclose the garden. How much fencing will he need?

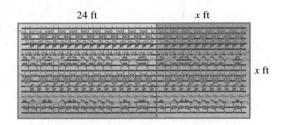

24 ft x ft

x ft

15. **Wading pool dimensions**   The area of the triangular swimming pool shown in the illustration is doubled by adding a rectangular wading pool. Find the dimensions of the pool. (*Hint:* The area of a triangle $= \frac{1}{2}bh$, and the area of a rectangle $= lw$.)

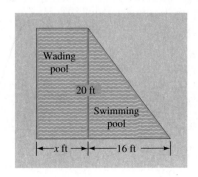

Wading pool

20 ft

Swimming pool

x ft   16 ft

16. **House construction**   A builder wants to install a triangular window with the angles shown in the illustration. What angles will he have to cut to make the window fit? (*Hint:* The sum of the angles in a triangle equals 180°.)

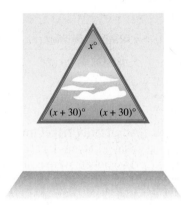

$x°$

$(x + 30)°$ $(x + 30)°$

17. **Length of a living room**   If a carpenter adds a porch with dimensions shown in the illustration to the living room, the living area will be increased by 50%. Find the length of the living room.

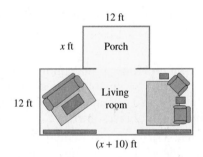

12 ft

x ft   Porch

12 ft   Living room

$(x + 10)$ ft

18. **Depth of water in a trough**   The trough in the illustration has a cross-sectional area of 54 square inches. Find the depth, d, of the trough. (*Hint:* Area of a trapezoid $= \frac{1}{2}h(b_1 + b_2)$.)

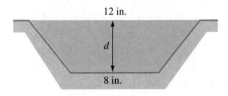

12 in.

d

8 in.

19. **Investment problem**   An executive invests $22,000, some at 7% and some at 6% annual interest. If he receives an annual return of $1,420, how much is invested at each rate?

20. **Financial planning** After inheriting some money, a woman wants to invest enough to have an annual income of $5,000. If she can invest $20,000 at 9% annual interest, how much more will she have to invest at 7% to achieve her goal? (See the table.)

Type	Rate	Amount	Income
9% investment	0.09	20,000	.09(20,000)
7% investment	0.07	x	.07x

21. **Ticket sales** A full-price ticket for a college basketball game costs $2.50, and a student ticket costs $1.75. If 585 tickets were sold, and the total receipts were $1,217.25, how many tickets were student tickets?

22. **Ticket sales** Of the 800 tickets sold to a movie, 480 were full-price tickets costing $7 each. If the gate receipts were $4,960, what did a student ticket cost?

23. **Investment problem** A woman invests $37,000, part at 8% and the rest at $9\frac{1}{2}$% annual interest. If the $9\frac{1}{2}$% investment provides $452.50 more income than the 8% investment, how much is invested at each rate?

24. **Investment problem** Equal amounts are invested at 6%, 7%, and 8% annual interest. If the three investments yield a total of $2,037 annual interest, find the total investment.

25. **Discounts** After being discounted 20%, a radio sells for $63.96. Find the original price.

26. **Markups** A merchant increases the wholesale cost of a washing machine by 30% to determine the selling price. If the washer sells for $588.90, find the wholesale cost.

27. **Break-point analysis** A machine to mill a brass plate has a setup cost of $600 and a unit cost of $3 for each plate manufactured. A bigger machine has a setup cost of $800 but a unit cost of only $2 for each plate manufactured. Find the break point.

28. **Break-point analysis** A machine to manufacture fasteners has a setup cost of $1,200 and a unit cost of $0.005 for each fastener manufactured. A newer machine has a setup cost of $1,500 but a unit cost of only $0.0015 for each fastener manufactured. Find the break point.

29. **Computer sales** A computer store has fixed costs of $8,925 per month and a unit cost of $850 for every computer it sells. If the store can sell all the computers it can get for $1,275 each, how many must be sold for the store to break even? (*Hint:* The *break-even point* occurs when costs equal income.)

30. **Restaurant management** A restaurant has fixed costs of $137.50 per day and an average unit cost of $4.75 for each meal served. If a typical meal costs $6, how many customers must eat at the restaurant each day for the owner to make a profit?

31. **Roofing houses** A man estimates that it will take him 7 days to roof his house. A professional roofer estimates that it will take him 4 days to roof the same house. How long will it take if they work together?

32. **Sealing asphalt** One crew can seal a parking lot in 8 hours and another in 10 hours. How long will it take to seal the parking lot if the two crews work together?

33. **Mowing lawns** If a woman can mow a lawn with a lawn tractor in 2 hours, and her husband can mow the same lawn with a push mower in 4 hours, how long will it take to mow the lawn if they work together?

34. **Filling swimming pools** A garden hose can fill a swimming pool in 3 days, and a larger hose can fill the pool in 2 days. How long will it take to fill the pool if both hoses are used?

35. **Filling swimming pools** An empty swimming pool can be filled in 10 hours. When full, the pool can be drained in 19 hours. How long will it take to fill the empty pool if the drain is left open?

36. **Preparing seafood** Sam stuffs shrimp in his job as a seafood chef. He can stuff 1,000 shrimp in 6 hours. When his sister helps him, they can stuff 1,000 shrimp in 4 hours. If Sam gets sick, how long will it take his sister to stuff 500 shrimp?

37. **Diluting solutions** How much water should be added to 20 ounces of a 15% solution of alcohol to dilute it to a 10% solution?

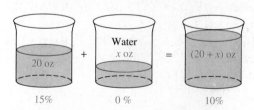

38. Increasing concentrations   The beaker shown below contains a 2% saltwater solution.

   a. How much water must be boiled away to increase the concentration of the salt solution from 2% to 3%?

   b. Where on the beaker would the new water level be?

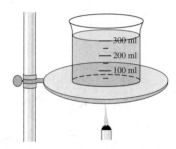

39. Winterizing cars   A car radiator has a 6-liter capacity. If the liquid in the radiator is 40% antifreeze, how much liquid must be replaced with pure antifreeze to bring the mixture up to a 50% solution?

40. Mixing milk   If a bottle holding 3 liters of milk contains $3\frac{1}{2}$% butterfat, how much skimmed milk must be added to dilute the milk to 2% butterfat?

41. Preparing solutions   A nurse has 1 liter of a solution that is 20% alcohol. How much pure alcohol must she add to bring the solution up to a 25% concentration?

42. Diluting solutions   If there are 400 cubic centimeters of a chemical in 1 liter of solution, how many cubic centimeters of water must be added to dilute it to a 25% solution?   (*Hint:* 1,000 cc = 1 liter.)

43. Cleaning swimming pools   A swimming pool contains 15,000 gallons of water. How many gallons of chlorine must be added to "shock the pool" and bring the water to a $\frac{3}{100}$% solution?

44. Mixing fuels   An automobile engine can run on a mixture of gasoline and a substitute fuel. If gas costs $3.50 per gallon and the substitute fuel costs $2 per gallon, what percent of a mixture must be substitute fuel to bring the cost down to $2.75 per gallon?

45. Evaporation   How many liters of water must evaporate to turn 12 liters of a 24% salt solution into a 36% solution?

46. Increasing concentrations   A beaker contains 320 ml of a 5% saltwater solution. How much water should be boiled away to increase the concentration to 6%?

47. Lowering fat   How many pounds of extra-lean hamburger that is 7% fat must be mixed with 30 pounds of hamburger that is 15% fat to obtain a mixture that is 10% fat?

48. Dairy foods   How many gallons of cream that is 22% butterfat must be mixed with milk that is 2% butterfat to get 20 gallons of milk containing 4% butterfat?

49. Mixing solutions   How many gallons of a 5% alcohol solution must be mixed with 90 gallons of 1% solution to obtain a 2% solution?

50. Preparing medicines   A doctor prescribes an ointment that is 2% hydrocortisone. A pharmacist has 1% and 5% concentrations in stock. How much of each should the pharmacist use to make a 1-ounce tube?

51. Driving rates   John drove to a distant city in 5 hours. When he returned, there was less traffic, and the trip took only 3 hours. If John averaged 26 mph faster on the return trip, how fast did he drive each way?

52. Distance problem   Suzi drove home at 60 mph, but her brother Jim, who left at the same time, could drive at only 48 mph. When Suzi arrived, Jim still had 60 miles to go. How far did Suzi drive?

53. Distance problem   Two cars leave Pima Community College traveling in opposite directions. One car travels at 60 mph and the other at 64 mph. In how many hours will they be 310 miles apart?

54. Bank robberies   Some bank robbers leave town, speeding at 70 mph. Ten minutes later, the police give chase, traveling at 78 mph. How long will it take the police to overtake the robbers?

55. Jogging problem   Two cross-country runners are 440 yards apart and are running toward each other, one at 8 mph and the other at 10 mph. In how many seconds will they meet?

56. Driving rates   One morning, John drove 5 hours before stopping to eat. After lunch, he increased his speed by 10 mph. If he completed a 430-mile trip in 8 hours of driving time, how fast did he drive in the morning?

57. Boating problem   A motorboat goes 5 miles upstream in the same time it requires to go 7 miles downstream. If the river flows at 2 mph, find the speed of the boat in still water.

58. Wind velocity A plane can fly 340 mph in still air. If it can fly 200 miles downwind in the same amount of time it can fly 140 miles upwind, find the velocity of the wind.

59. Feeding cattle A farmer wants to mix 2,400 pounds of cattle feed that is to be 14% protein. Barley (11.7% protein) will make up 25% of the mixture. The remaining 75% will be made up of oats (11.8% protein) and soybean meal (44.5% protein). How many pounds of each will he use?

60. Feeding cattle If the farmer in Exercise 59 wants only 20% of the mixture to be barley, how many pounds of each should he use?

 Use a calculator to help solve each problem.

61. Machine tool design 712.51 cubic millimeters of material was removed by drilling the blind hole as shown in the illustration. Find the depth of the hole. (*Hint:* The volume of a cylinder is given by $V = \pi r^2 h$.)

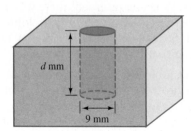

d mm

9 mm

62. Architecture The Norman window with dimensions as shown is a rectangle topped by a semicircle. If the area of the window is 68.2 square feet, find its height h.

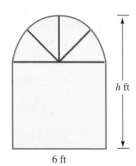

h ft

6 ft

Discovery and Writing

63. Which type of problem was easiest for you to solve? Why?

64. Which type of problem was hardest for you to solve? Why?

Review *Factor each expression.*

65. $x^2 - 2x - 63$ **66.** $2x^2 + 11x - 21$

67. $9x^2 - 12x - 5$ **68.** $9x^2 - 2x - 7$

69. $x^2 + 6x + 9$ **70.** $x^2 - 10x + 25$

71. $x^3 + 8$ **72.** $27a^3 - 64$

1.3 Quadratic Equations

Objectives

1. Solve Quadratic Equations Using Factoring and the Square Root Property
2. Solve Quadratic Equations Using Completing the Square
3. Solve Quadratic Equations Using the Quadratic Formula
4. Determine the Easiest Method to Solve a Quadratic Equation
5. Solve Formulas for Indicated Variables
6. Define and Use the Discriminant
7. Write Rational Equations in Quadratic Form and Solve the Equations

Fenway Park, often called America's Most Beloved Ballpark, is home to the Boston Red Sox baseball club. The park opened in 1912 and is currently the oldest major league baseball stadium.

In baseball, the distance between home plate and first base is 90 feet and the distance between first base and second base is 90 feet. To find the distance between home plate and second base, we can use the *Pythagorean theorem,* which states:

The sum of the squares of the two legs of a right triangle is equal to the square of its hypotenuse.

Because home plate, first base, and second base form a right triangle, we can let x represent the distance between home plate and second base (the **hypotenuse** of the right triangle) and 90 feet represent the length of each leg. We then can apply the Pythagorean theorem and write the equation $90^2 + 90^2 = x^2$. To find the distance between home plate and second base, we must solve this equation.

$$90^2 + 90^2 = x^2$$
$$8{,}100 + 8{,}100 = x^2 \quad \text{Square 90 two times.}$$
$$16{,}200 = x^2 \quad \text{Simplify.}$$

To find x, we must determine what positive number squared gives 16,200. From Chapter 0, we know that this number is the square root of 16,200.

$$\sqrt{16{,}200} \approx 127.3 \quad \text{Use a calculator and round to the nearest tenth.}$$

To the nearest tenth, the distance between home plate and second base is 127.3 feet.

Since this equation contains the term x^2, it is an example of a new type of equation, called a *quadratic* equation. In this section, we will learn several strategies for solving these equations.

1. Solve Quadratic Equations Using Factoring and the Square Root Property

Polynomial equations such as $2x^2 - 11x - 21 = 0$ and $3x^2 - x - 2 = 0$ are called *quadratic* or *second-degree* equations.

Quadratic Equations | A **quadratic equation** is an equation that can be written in the form $ax^2 + bx + c = 0$, where a, b, and c are real numbers and $a \neq 0$.

To solve quadratic equations by factoring, we can use the following theorem.

Zero-Factor Theorem | If a and b are real numbers, and if $ab = 0$, then

$$a = 0 \qquad \text{or} \qquad b = 0$$

Proof Suppose that $ab = 0$. If $a = 0$, we are finished, because at least one of a or b is 0.

If $a \neq 0$, then a has a reciprocal $\frac{1}{a}$, and we can multiply both sides of the equation $ab = 0$ by $\frac{1}{a}$ to obtain

$$ab = 0$$
$$\frac{1}{a}(ab) = \frac{1}{a}(0) \quad \text{Multiply both sides by } \frac{1}{a}.$$
$$\left(\frac{1}{a} \cdot a\right)b = 0 \quad \text{Use the associative property to group } \frac{1}{a} \text{ and } a \text{ together.}$$
$$1b = 0 \quad \frac{1}{a} \cdot a = 1$$
$$b = 0$$

Thus, if $a \neq 0$, then b must be 0, and the theorem is proved.

EXAMPLE 1 Solve by factoring: $2x^2 - 9x - 35 = 0$.

Solution The left side can be factored and written as

$$(2x + 5)(x - 7) = 0$$

Comment

The zero-factor theorem can be used only when there is a constant of 0 on the right side of the equation.

This product can be 0 if and only if one of the factors is 0. So we can use the zero-factor theorem and set each factor equal to 0. We then can solve each equation for x.

$$2x + 5 = 0 \quad \text{or} \quad x - 7 = 0$$
$$2x = -5 \qquad\qquad x = 7$$
$$x = -\frac{5}{2}$$

Because $(2x + 5)(x - 7) = 0$ only if one of its factors is zero, $-\frac{5}{2}$ and 7 are the only solutions of the equation.

Verify that each one satisfies the equation.

Self Check 1 Solve by factoring: $6x^2 + 7x - 3 = 0$.

In many quadratic equations, the quadratic expression does not factor over the set of integers. For example, the left side of $x^2 - 5x + 3 = 0$ is a prime polynomial and cannot be factored over the set of integers.

To develop a method to solve these equations, we consider the equation $x^2 = c$. If c is positive, it has two real roots that can be found by subtracting c from both sides, factoring $x^2 - c$ over the set of real numbers, setting each factor equal to 0, and solving for x.

$$x^2 = c$$
$$x^2 - c = 0 \qquad\qquad \text{Subtract } c \text{ from both sides.}$$
$$x^2 - \left(\sqrt{c}\right)^2 = 0 \qquad\qquad \left(\sqrt{c}\right)^2 = c$$
$$\left(x - \sqrt{c}\right)\left(x + \sqrt{c}\right) = 0 \qquad\qquad \text{Factor the difference of two squares.}$$
$$x - \sqrt{c} = 0 \quad \text{or} \quad x + \sqrt{c} = 0 \qquad \text{Set each factor equal to 0.}$$
$$x = \sqrt{c} \qquad\qquad x = -\sqrt{c}$$

The roots of $x^2 = c$ are $x = \sqrt{c}$ and $x = -\sqrt{c}$. This fact is summarized in the **square root property.**

Square Root Property If $c > 0$, the equation $x^2 = c$ has two real roots:

$$x = \sqrt{c} \quad \text{or} \quad x = -\sqrt{c}$$

EXAMPLE 2 Solve using the square root property: $x^2 - 8 = 0$.

Solution We solve for x^2 and use the square root property.

$$x^2 - 8 = 0$$
$$x^2 = 8$$
$$x = \sqrt{8} \quad \text{or} \quad x = -\sqrt{8}$$
$$x = 2\sqrt{2} \qquad x = -2\sqrt{2} \quad \sqrt{8} = \sqrt{4}\sqrt{2} = 2\sqrt{2}$$

Verify that each root satisfies the equation.

Self Check 2 Solve using the square root property: $x^2 - 12 = 0$.

EXAMPLE 3 Solve: $(x + 4)^2 = 1$.

Solution Again, we will use the square root property.

$$(x + 4)^2 = 1$$

$$x + 4 = \sqrt{1} \quad \text{or} \quad x + 4 = -\sqrt{1}$$

$$x + 4 = 1 \quad\quad\quad x + 4 = -1$$

$$x = -3 \quad\quad\quad\quad x = -5$$

Verify that each root satisfies the equation.

Self Check 3 Solve: $(x + 5)^2 = 4$.

2. Solve Quadratic Equations Using Completing the Square

Another way to solve quadratic equations is called **completing the square.** This method is based on the following products:

$$x^2 + 2ax + a^2 = (x + a)^2 \quad \text{and} \quad x^2 - 2ax + a^2 = (x - a)^2$$

The trinomials $x^2 + 2ax + a^2$ and $x^2 - 2ax + a^2$ are perfect-square trinomials, because each one factors as the square of a binomial. In each case, the coefficient of the first term is 1. If we take one-half of the coefficient of x in the middle term and square it, we obtain the third term.

$$\left[\frac{1}{2}(2a)\right]^2 = a^2 \quad \text{and} \quad \left[\frac{1}{2}(-2a)\right]^2 = (-a)^2 = a^2$$

This suggests that to make $x^2 + bx$ a perfect-square trinomial, we find one-half of b, square it, and add the result to the binomial. For example, to make $x^2 + 10x$ a perfect-square trinomial, we find one-half of 10 to get 5, square 5 to get 25, and add 25 to $x^2 + 10x$.

$$x^2 + 10x + \left[\frac{1}{2}(10)\right]^2 = x^2 + 10x + (5)^2$$

$$= x^2 + 10x + 25 \quad \text{Note that } x^2 + 10x + 25 = (x + 5)^2.$$

To make $x^2 - 11x$ a perfect-square trinomial, we find one-half of -11 to get $-\frac{11}{2}$, square $-\frac{11}{2}$ to get $\frac{121}{4}$, and add $\frac{121}{4}$ to $x^2 - 11x$.

$$x^2 - 11x + \left[\frac{1}{2}(-11)\right]^2 = x^2 - 11x + \left(-\frac{11}{2}\right)^2$$

$$= x^2 - 11x + \frac{121}{4} \quad \begin{array}{l}\text{Note that}\\ x^2 - 11x + \frac{121}{4} = \left(x - \frac{11}{2}\right)^2.\end{array}$$

To solve a quadratic equation in x by completing the square, we follow these steps.

Completing the Square
1. If the coefficient of x^2 is not 1, make it 1 by dividing both sides of the equation by the coefficient of x^2.
2. If necessary, add a number to both sides of the equation to get the constant on the right side of the equation.
3. Complete the square on x:
   a. Identify the coefficient of x, take one-half of it, and square the result.
   b. Add the number found in part a to both sides of the equation.
4. Factor the perfect-square trinomial and combine like terms.
5. Solve the resulting quadratic equation by using the square root property.

To use completing the square to solve $x^2 - 10x + 24 = 0$, we note that the coefficient of x^2 is 1. We move on to Step 2 and subtract 24 from both sides to get the constant term on the right side of the equal sign.

$$x^2 - 10x = -24$$

We then can complete the square by adding $\left[\frac{1}{2}(-10)\right]^2 = 25$ to both sides.

$$x^2 - 10x + 25 = -24 + 25$$
$$x^2 - 10x + 25 = 1 \qquad \text{Simplify on the right side.}$$

We then factor the perfect-square trinomial on the left side.

$$(x - 5)^2 = 1$$

Finally, we use the square root property to solve this equation.

$$x - 5 = 1 \quad \text{or} \quad x - 5 = -1$$
$$x = 6 \qquad\qquad x = 4$$

EXAMPLE 4 Use completing the square to solve $x^2 + 4x - 6 = 0$.

Solution Here the coefficient of x^2 is already 1. We move to Step 2 and add 6 to both sides to isolate the binomial $x^2 + 4x$.

$$x^2 + 4x = 6$$

We then find the number to add to both sides by completing the square. Since one-half of 4 (the coefficient of x) is 2 and $2^2 = 4$, we add 4 to both sides.

$$x^2 + 4x + 4 = 6 + 4 \qquad \text{Add 4 to both sides.}$$
$$x^2 + 4x + 4 = 10$$
$$(x + 2)^2 = 10 \qquad \text{Factor } x^2 + 4x + 4.$$
$$x + 2 = \sqrt{10} \quad \text{or} \quad x + 2 = -\sqrt{10} \qquad \text{Use the square root property.}$$
$$x = -2 + \sqrt{10} \qquad\qquad x = -2 - \sqrt{10}$$

Verify that each root satisfies the original equation.

Self Check 4 Solve by completing the square: $x^2 - 2x - 9 = 0$.

EXAMPLE 5 Solve by completing the square: $x(x + 3) = 2$.

Solution We remove parentheses to get

$$x^2 + 3x = 2$$

Since the coefficient of x^2 is 1 and the constant is on the right side, we move to Step 3 and find the number to be added to both sides to complete the square. Since one-half of 3 (the coefficient of x) is $\frac{3}{2}$ and the square of $\frac{3}{2}$ is $\frac{9}{4}$, we add $\frac{9}{4}$ to both sides.

$$x^2 + 3x + \frac{9}{4} = 2 + \frac{9}{4} \qquad \text{Add } \frac{9}{4} \text{ to both sides.}$$

$$\left(x + \frac{3}{2}\right)^2 = \frac{17}{4} \qquad \text{Factor } x^2 + 3x + \frac{9}{4}.$$

$$x + \frac{3}{2} = \frac{\sqrt{17}}{2} \quad \text{or} \quad x + \frac{3}{2} = -\frac{\sqrt{17}}{2} \qquad \text{Use the square root property.}$$

$$x = \frac{-3 + \sqrt{17}}{2} \qquad\qquad x = \frac{-3 - \sqrt{17}}{2} \qquad \text{Subtract } \frac{3}{2} \text{ from both sides.}$$

Verify that each root satisfies the original equation.

Self Check 5   Solve by completing the square: $x(x + 5) = 1$.

EXAMPLE 6   Solve by completing the square: $6x^2 + 5x - 6 = 0$.

Solution   We begin by dividing both sides of the equation by 6 to make the coefficient of x^2 equal to 1. Then we proceed as follows:

$$6x^2 + 5x - 6 = 0$$

$$x^2 + \frac{5}{6}x - 1 = 0 \qquad \text{Divide both sides by 6.}$$

$$x^2 + \frac{5}{6}x = 1 \qquad \text{Add 1 to both sides.}$$

$$x^2 + \frac{5}{6}x + \frac{25}{144} = 1 + \frac{25}{144} \qquad \text{Add } \left(\frac{1}{2} \cdot \frac{5}{6}\right)^2, \text{ or } \frac{25}{144}, \text{ to both sides.}$$

$$\left(x + \frac{5}{12}\right)^2 = \frac{169}{144} \qquad \text{Factor } x^2 + \frac{5}{6}x + \frac{25}{144}.$$

We now apply the square root property.

$$x + \frac{5}{12} = \sqrt{\frac{169}{144}} \quad \text{or} \quad x + \frac{5}{12} = -\sqrt{\frac{169}{144}}$$

$$x + \frac{5}{12} = \frac{13}{12} \qquad\qquad x + \frac{5}{12} = -\frac{13}{12}$$

$$x = \frac{8}{12} \qquad\qquad\qquad x = -\frac{18}{12}$$

$$x = \frac{2}{3} \qquad\qquad\qquad x = -\frac{3}{2}$$

Verify that each root satisfies the original equation.

Self Check 6   Solve by completing the square: $2x^2 - 5x - 3 = 0$.

3.   Solve Quadratic Equations Using the Quadratic Formula

We can solve the equation $ax^2 + bx + c = 0$ $(a \neq 0)$ by completing the square. The result will be a formula that we can use to solve quadratic equations.

$$ax^2 + bx + c = 0$$

$$\frac{ax^2}{a} + \frac{b}{a}x + \frac{c}{a} = \frac{0}{a} \qquad \text{Divide both sides by } a.$$

$$x^2 + \frac{b}{a}x = -\frac{c}{a} \qquad \text{Simplify and subtract } \frac{c}{a} \text{ from both sides.}$$

$$x^2 + \frac{b}{a}x + \frac{b^2}{4a^2} = \frac{b^2}{4a^2} - \frac{4ac}{4aa} \qquad \begin{array}{l} \text{Add } \frac{b^2}{4a^2} \text{ to both sides and multiply the numerator and} \\ \text{denominator of } \frac{c}{a} \text{ by } 4a. \end{array}$$

$$\left(x + \frac{b}{2a}\right)^2 = \frac{b^2 - 4ac}{4a^2} \qquad \text{Factor the left side and add the fractions on the right side.}$$

We can now use the square root property.

$$x + \frac{b}{2a} = \sqrt{\frac{b^2 - 4ac}{4a^2}} \qquad \text{or} \quad x + \frac{b}{2a} = -\sqrt{\frac{b^2 - 4ac}{4a^2}}$$

$$x = -\frac{b}{2a} + \frac{\sqrt{b^2 - 4ac}}{2a} \qquad\qquad x = -\frac{b}{2a} - \frac{\sqrt{b^2 - 4ac}}{2a}$$

$$x = \frac{-b + \sqrt{b^2 - 4ac}}{2a} \qquad\qquad x = \frac{-b - \sqrt{b^2 - 4ac}}{2a}$$

These values of x are the two roots of the equation $ax^2 + bx + c = 0$. They usually are combined into a single expression, called the **quadratic formula.**

Quadratic Formula   The solutions of the general quadratic equation, $ax^2 + bx + c = 0$, are

$$x = \frac{-b \pm \sqrt{b^2 - 4ac}}{2a} \quad (a \neq 0)$$

Comment

Be sure to write the quadratic formula correctly. Do not write it as

$$x = -b \pm \frac{\sqrt{b^2 - 4ac}}{2a}$$

The quadratic formula should be read twice, once using the $+$ sign and once using the $-$ sign. The quadratic formula implies that

$$x = \frac{-b + \sqrt{b^2 - 4ac}}{2a} \qquad \text{or} \qquad x = \frac{-b - \sqrt{b^2 - 4ac}}{2a}$$

EXAMPLE 7   Use the quadratic formula to solve $x^2 - 5x + 3 = 0$.

Solution   In this equation $a = 1$, $b = -5$, and $c = 3$. We will substitute these values into the quadratic formula.

$$x = \frac{-b \pm \sqrt{b^2 - 4ac}}{2a} \qquad \text{This is the quadratic formula.}$$

$$x = \frac{-(-5) \pm \sqrt{(-5)^2 - 4(1)(3)}}{2(1)} \qquad \text{Substitute 1 for } a, -5 \text{ for } b, \text{ and 3 for } c.$$

$$x = \frac{5 \pm \sqrt{13}}{2} \qquad (-5)^2 - 4(1)(3) = 25 - 12 = 13$$

Both values satisfy the original equation.

Self Check 7   Solve using the quadratic formula: $3x^2 - 5x + 1 = 0$.

EXAMPLE 8   Use the quadratic formula to solve $2x^2 + 8x - 7 = 0$.

Solution   In this equation, $a = 2$, $b = 8$, and $c = -7$. We will substitute these values into the quadratic formula.

$$x = \frac{-b \pm \sqrt{b^2 - 4ac}}{2a}$$ This is the quadratic formula.

$$x = \frac{-8 \pm \sqrt{8^2 - 4(2)(-7)}}{2(2)}$$ Substitute 2 for a, 8 for b, and -7 for c.

$$x = \frac{-8 \pm \sqrt{120}}{4}$$ $8^2 - 4(2)(-7) = 64 + 56 = 120$

$$x = \frac{-8 \pm 2\sqrt{30}}{4}$$ $\sqrt{120} = \sqrt{4 \cdot 30} = 2\sqrt{30}$

$$x = \frac{-4 \pm \sqrt{30}}{2} \quad \text{or} \quad x = -2 \pm \frac{\sqrt{30}}{2}$$

Both values satisfy the original equation.

Self Check 8   Solve using the quadratic formula: $4x^2 + 16x - 13 = 0$.

4.   Determine the Easiest Method to Solve a Quadratic Equation

So far, we have solved quadratic equations by *factoring*, by the *square root method*, by *completing the square*, and by the *quadratic formula*. With so many methods available, it is useful to think about which one will be the easiest way to solve a specific quadratic equation. Although we have used completing the square to develop the quadratic formula, it is usually the most complicated way to solve a quadratic equation. Therefore, unless specified, we usually will not use this method. However, we will complete the square again later in the book to write certain equations in specific forms.

The following chart summarizes the different types of quadratic equations that can occur, a suggested method for solving them, and an example of each type.

Type of quadratic equation	Easiest method to solve it	Example
Equations of the form $ax^2 + bx + c = 0$ where the left side factors easily.	Use factoring and the zero-factor theorem.	Solve: $6x^2 - 11x + 3 = 0$ $(2x - 3)(3x - 1) = 0$ $2x - 3 = 0$ or $3x - 1 = 0$ $x = \dfrac{3}{2}$ $\bigg\vert$ $x = \dfrac{1}{3}$
Equations of the form $ax^2 + bx = 0$ where the constant term is missing.	Use factoring and the zero-factor theorem.	Solve: $9x^2 + 6x = 0$ $3x(3x + 2) = 0$ $3x = 0$ or $3x + 2 = 0$ $x = 0$ $\bigg\vert$ $x = -\dfrac{2}{3}$
Equations of the form $ax^2 - c = 0$ where the term involving x is missing and the left side factors easily.	Use factoring and the zero-factor theorem.	Solve: $4x^2 - 9 = 0$ $(2x + 3)(2x - 3) = 0$ $2x + 3 = 0$ or $2x - 3 = 0$ $x = -\dfrac{3}{2}$ $\bigg\vert$ $x = \dfrac{3}{2}$

(continued)

Type of quadratic equation	Easiest method to solve it	Example
Equations of the form $ax^2 - c = 0$ or $x^2 = k$ where k is a contant.	Use the square root method.	Solve: $2x^2 - 5 = 0$ $$x^2 = \frac{5}{2}$$ $$x = \pm\sqrt{\frac{5}{2}}$$ $$x = \pm\frac{\sqrt{10}}{2}$$
Equations of the form $ax^2 + bx + c = 0$ where the left side cannot be factored easily or cannot be factored at all.	Use the quadratic formula.	Solve: $3x^2 - x - 5 = 0$. $$x = \frac{-b \pm \sqrt{b^2 - 4ac}}{2a}$$ $$x = \frac{-(-1) \pm \sqrt{(-1)^2 - 4(3)(-5)}}{2(3)} \quad \begin{array}{l} a = 3, \\ b = 1, \\ c = -5 \end{array}$$ $$x = \frac{1 \pm \sqrt{1 + 60}}{6}$$ $$x = \frac{1 \pm \sqrt{61}}{6}$$

5. Solve Formulas for Indicated Variables

Many formulas involve quadratic equations. For example, if an object is fired straight up into the air with an initial velocity of 88 feet per second, its height is given by the formula $h = 88t - 16t^2$, where h represents its height (in feet) and t represents the elapsed time (in seconds) since it was fired.

To solve this formula for t, we use the quadratic formula.

$$h = 88t - 16t^2$$

$16t^2 - 88t + h = 0$ Add $16t^2$ and $-88t$ to both sides.

$$t = \frac{-(-88) \pm \sqrt{(-88)^2 - 4(16)(h)}}{2(16)}$$ Substitute into the quadratic formula.

$$t = \frac{88 \pm \sqrt{7{,}744 - 64h}}{32}$$ Simplify.

6. Define and Use the Discriminant

We can predict what type of roots a quadratic equation will have before we solve it. Suppose that the coefficients a, b, and c in the equation $ax^2 + bx + c = 0$ $(a \neq 0)$ are real numbers. Then the two roots of the equation are given by the quadratic formula

$$x = \frac{-b \pm \sqrt{b^2 - 4ac}}{2a} \quad (a \neq 0)$$

The value of $b^2 - 4ac$, called the **discriminant,** determines the nature of the roots. The possibilities are summarized as follows.

Nature of the Roots of a Quadratic Equation

If a, b, and c are real numbers and $b^2 - 4ac$ is . . .	then the two solutions are
positive	unequal real numbers
0	equal real numbers
negative	not real numbers

If a, b, and c are rational numbers and $b^2 - 4ac$ is . . .	then the two solutions are
0	equal rational numbers
a nonzero perfect square	unequal rational numbers
a positive nonperfect square	unequal irrational numbers

EXAMPLE 9   Determine the nature of the roots of $3x^2 + 4x + 1 = 0$.

Solution   We calculate the discriminant $b^2 - 4ac$.

$$b^2 - 4ac = 4^2 - 4(3)(1) \quad \text{Substitute 4 for } b\text{, 3 for } a\text{, and 1 for } c.$$
$$= 16 - 12$$
$$= 4$$

Since a, b, and c are rational numbers and the discriminant is a nonzero perfect square, the two roots will be unequal rational numbers.

Self Check 9   Determine the nature of the roots of $4x^2 - 3x - 2 = 0$.

EXAMPLE 10   If k is a constant, many quadratic equations are represented by the equation

$$(k - 2)x^2 + (k + 1)x + 4 = 0$$

Find the values of k that will give an equation with roots that are equal real numbers.

Solution   We calculate the discriminant $b^2 - 4ac$ and set it equal to 0.

$$b^2 - 4ac = (k + 1)^2 - 4(k - 2)(4)$$
$$0 = k^2 + 2k + 1 - 16k + 32$$
$$0 = k^2 - 14k + 33$$
$$0 = (k - 3)(k - 11)$$
$$k - 3 = 0 \quad \text{or} \quad k - 11 = 0$$
$$k = 3 \quad \mid \quad k = 11$$

When $k = 3$ or $k = 11$, the equation will have equal roots. As a check, we let $k = 3$ and note that the equation $(k - 2)x^2 + (k + 1)x + 4 = 0$ becomes

$$(3 - 2)x^2 + (3 + 1)x + 4 = 0$$
$$x^2 + 4x + 4 = 0$$

The roots of this equation are equal real numbers, as expected:

$$x^2 + 4x + 4 = 0$$
$$(x + 2)(x + 2) = 0$$
$$x + 2 = 0 \quad \text{or} \quad x + 2 = 0$$
$$x = -2 \quad \mid \quad x = -2$$

Similarly, $k = 11$ will give an equation with equal real roots.

Self Check 10   Find k such that $(k - 2)x^2 - (k + 3)x + 9 = 0$ will have equal roots.

7. Write Rational Equations in Quadratic Form and Solve the Equations

If an equation can be written in quadratic form, it can be solved with the techniques used for solving quadratic equations.

EXAMPLE 11   Solve: $\dfrac{1}{x - 1} + \dfrac{3}{x + 1} = 2$.

Solution   Since neither denominator can be zero, $x \neq 1$ and $x \neq -1$. If either number appears as a root, it must be discarded.

$$\frac{1}{x - 1} + \frac{3}{x + 1} = 2$$

$$(x - 1)(x + 1)\left[\frac{1}{x - 1} + \frac{3}{x + 1}\right] = (x - 1)(x + 1)2 \qquad \text{Multiply both sides by } (x - 1)(x + 1).$$

$$(x + 1) + 3(x - 1) = 2(x^2 - 1) \qquad \text{Remove brackets and simplify.}$$

$$4x - 2 = 2x^2 - 2 \qquad \text{Remove parentheses and simplify.}$$

$$0 = 2x^2 - 4x \qquad \text{Add } 2 - 4x \text{ to both sides.}$$

The resulting equation is a quadratic equation that we can solve by factoring.

$$2x^2 - 4x = 0$$
$$2x(x - 2) = 0 \qquad \text{Factor } 2x^2 - 4x.$$
$$2x = 0 \quad \text{or} \quad x - 2 = 0$$
$$x = 0 \quad \big| \quad \qquad x = 2$$

Verify these results by checking each root in the original equation.

Self Check 11   Solve: $\dfrac{1}{x - 1} + \dfrac{2}{x + 1} = 1$.

Self Check Answers   **1.** $\frac{1}{3}, -\frac{3}{2}$ **2.** $2\sqrt{3}, -2\sqrt{3}$ **3.** $-3, -7$ **4.** $1 \pm \sqrt{10}$ **5.** $\frac{-5 \pm \sqrt{29}}{2}$
6. $3, -\frac{1}{2}$ **7.** $\frac{5 \pm \sqrt{13}}{6}$ **8.** $-2 \pm \frac{\sqrt{29}}{2}$ **9.** unequal and irrational
10. $3, 27$ **11.** $0, 3$

1.3   Exercises

Vocabulary and Concepts   *Fill in the blanks.*

1. A quadratic equation is an equation that can be written in the form _____, where $a \neq 0$.

2. If a and b are real numbers and _____, then $a = 0$ or $b = 0$.

3. If $c > 0$, the equation $x^2 = c$ has two roots. They are $x =$ ___ and $x =$ ____.

4. The quadratic formula is _____ ($a \neq 0$).

5. If a, b, and c are real numbers and if $b^2 - 4ac = 0$, the roots of the quadratic equation are _____.

6. If a, b, and c are real numbers and $b^2 - 4ac < 0$, the roots of the quadratic equation are _____.

Practice   *Solve each equation by factoring. Check all answers.*

7. $x^2 - x - 6 = 0$

8. $x^2 + 8x + 15 = 0$

9. $x^2 - 144 = 0$

10. $x^2 + 4x = 0$

11. $2x^2 + x - 10 = 0$

12. $3x^2 + 4x - 4 = 0$

13. $5x^2 - 13x + 6 = 0$

14. $2x^2 + 5x - 12 = 0$

15. $15x^2 + 16x = 15$

16. $6x^2 - 25x = -25$

17. $12x^2 + 9 = 24x$

18. $24x^2 + 6 = 24x$

Use the square root property to solve each equation. You may need to factor an expression.

19. $x^2 = 9$

20. $x^2 = 20$

21. $y^2 - 50 = 0$

22. $x^2 - 75 = 0$

23. $(x - 1)^2 = 4$

24. $(y + 2)^2 - 49 = 0$

25. $a^2 + 2a + 1 = 9$

26. $x^2 - 6x + 9 = 25$

Complete the square to make each binomial a perfect-square trinomial.

27. $x^2 + 6x$

28. $x^2 + 8x$

29. $x^2 - 4x$

30. $x^2 - 12x$

31. $a^2 + 5a$

32. $t^2 + 9t$

33. $r^2 - 11r$

34. $s^2 - 7s$

35. $y^2 + \dfrac{3}{4}y$

36. $p^2 + \dfrac{3}{2}p$

37. $q^2 - \dfrac{1}{5}q$

38. $m^2 - \dfrac{2}{3}m$

Solve each equation by completing the square.

39. $x^2 - 8x + 15 = 0$

40. $x^2 + 10x + 21 = 0$

41. $x^2 + x - 6 = 0$

42. $x^2 - 9x + 20 = 0$

43. $x^2 - 25x = 0$

44. $x^2 + x = 0$

45. $3x^2 + 4x = 4$

46. $2x^2 + 5x = 12$

47. $x^2 + 5 = -5x$

48. $x^2 + 1 = -4x$

49. $3x^2 = 1 - 4x$

50. $2x^2 = 3x + 1$

Use the quadratic formula to solve each equation.

51. $x^2 - 12 = 0$

52. $x^2 - 20 = 0$

53. $2x^2 - x - 15 = 0$

54. $6x^2 + x - 2 = 0$

55. $5x^2 - 9x - 2 = 0$

56. $4x^2 - 4x - 3 = 0$

57. $2x^2 + 2x - 4 = 0$

58. $3x^2 + 18x + 15 = 0$

59. $-3x^2 = 5x + 1$

60. $2x(x + 3) = -1$

61. $5x\left(x + \dfrac{1}{5}\right) = 3$

62. $7x^2 = 2x + 2$

Solve each formula for the indicated variable.

63. $h = \dfrac{1}{2}gt^2$; t

64. $x^2 + y^2 = r^2$; x

65. $h = 64t - 16t^2$; t

66. $y = 16x^2 - 4$; x

67. $\dfrac{x^2}{a^2} + \dfrac{y^2}{b^2} = 1; y$

68. $\dfrac{x^2}{a^2} - \dfrac{y^2}{b^2} = 1; x$

69. $\dfrac{x^2}{a^2} - \dfrac{y^2}{b^2} = 1; a$

70. $\dfrac{x^2}{a^2} - \dfrac{y^2}{b^2} = 1; b$

71. $x^2 + xy - y^2 = 0; x$

72. $x^2 - 3xy + y^2 = 0; y$

Use the discriminant to determine the nature of the roots of each equation. **Do not solve the equation.**

73. $x^2 + 6x + 9 = 0$

74. $x^2 - 5x + 2 = 0$

75. $3x^2 - 2x + 5 = 0$

76. $9x^2 + 42x + 49 = 0$

77. $10x^2 + 29x = 21$

78. $10x^2 + x = 21$

79. $-3x^2 + 2x = 21$

80. $-8x^2 - 2x = 13$

81. Does $1{,}492x^2 + 1{,}984x - 1{,}776 = 0$ have any roots that are real numbers?

82. Does $2{,}004x^2 + 10x + 1{,}994 = 0$ have any roots that are real numbers?

83. Find two values of k such that $x^2 + kx + 3k - 5 = 0$ will have two roots that are equal.

84. For what value(s) of b will the solutions of $x^2 - 2bx + b^2 = 0$ be equal?

Change each equation to quadratic form and solve it by the most efficient method.

85. $x + 1 = \dfrac{12}{x}$

86. $x - 2 = \dfrac{15}{x}$

87. $8x - \dfrac{3}{x} = 10$

88. $15x - \dfrac{4}{x} = 4$

89. $\dfrac{5}{x} = \dfrac{4}{x^2} - 6$

90. $\dfrac{6}{x^2} + \dfrac{1}{x} = 12$

91. $x\left(30 - \dfrac{13}{x}\right) = \dfrac{10}{x}$

92. $x\left(20 - \dfrac{17}{x}\right) = \dfrac{10}{x}$

93. $(a - 2)(a + 4) = 2a(a - 3)$

94. $\dfrac{4 + a}{2a} = \dfrac{a - 2}{3}$

95. $\dfrac{1}{x} + \dfrac{3}{x + 2} = 2$

96. $\dfrac{1}{x - 1} + \dfrac{1}{x - 4} = \dfrac{5}{4}$

97. $\dfrac{1}{x + 1} + \dfrac{5}{2x - 4} = 1$

98. $\dfrac{x(2x + 1)}{x - 2} = \dfrac{10}{x - 2}$

99. $x + 1 + \dfrac{x + 2}{x - 1} = \dfrac{3}{x - 1}$

100. $\dfrac{1}{4 - y} = \dfrac{1}{4} + \dfrac{1}{y + 2}$

101. $\dfrac{24}{a} - 11 = \dfrac{-12}{a + 1}$

102. $\dfrac{36}{b} - 17 = \dfrac{-24}{b + 1}$

Discovery and Writing

103. If r_1 and r_2 are the roots of $ax^2 + bx + c = 0$, show that $r_1 + r_2 = -\dfrac{b}{a}$.

104. If r_1 and r_2 are the roots of $ax^2 + bx + c = 0$, show that $r_1 r_2 = \dfrac{c}{a}$.

In Exercises 105 and 106, a stone is thrown straight upward, higher than the top of a tree. The stone is even with the top of the tree at time t_1 on the way up and at time t_2 on the way down. If the height of the tree is h feet, both t_1 and t_2 are solutions of $h = v_0 t - 16t^2$.

105. Show that the tree is $16t_1 t_2$ feet tall.

106. Show that v_0 is $16(t_1 + t_2)$ feet per second.

107. Explain why the zero-factor theorem is true.

108. Explain how to complete the square on $x^2 - 17x$.

Quadratic equations can be solved automatically by using a computer program, such as **Excel.**

109. Solve: $2x^2 - 3x - 4 = 0$.

  a. Open an Excel spreadsheet. In cell B1, enter the left side of the equation as $= 2*A1^2 - 3*A1 - 4$. (Cell A1 is reserved for the value of x.)

  b. After pressing ENTER , you will see the number -4 in cell B1. Since cell A1 is empty, its value is considered to be 0 and the value of the quadratic is -4 when $x = 0$.

  c. To solve the equation, enter a guess for the solution in cell A1. To find a positive solution, enter 1 as a guess in cell A1.

d. Click inside cell B1. On the Menu bar, look under the Tools menu for SOLVER. (If there is no SOLVER, choose Add-Ins and then, inside the dialog box, select Solver Add-In and click OK.) After choosing SOLVER, a parameters window will appear. Inside the window, in front of Equal To, select Value Of, and be sure it is followed by the number 0. In the GUESS box, enter A1. Then click Solve. When the Solver Results dialog box opens, be sure that Keep Solver Solution is selected, and click OK. The solution of the equation appears in cell A1.

e. Find the negative solution of the equation.

110. Use Excel to solve $6x^2 + 13x - 5 = 0$.

Review   *Simplify each expression.*

111. $5x(x - 2) - x(3x - 2)$

112. $(x + 3)(x - 9) - x(x - 5)$

113. $(m + 3)^2 - (m - 3)^2$

114. $[(y + z)(y - z)]^2$

115. $\sqrt{50x^3} - x\sqrt{8x}$

116. $\dfrac{2x}{\sqrt{5} - 2}$

1.4   Applications of Quadratic Equations

Objectives

1. Solve Geometric Problems
2. Solve Uniform Motion Problems
3. Solve Falling Body Problems
4. Solve Business Problems
5. Solve Shared-Work Problems

The Grand Canyon Skywalk is a tourist attraction located along the Colorado River in the state of Arizona. The glass walkway is shaped like a horseshoe and is 4,000 feet above the floor of the canyon. The Grand Canyon, known for its overwhelming size and beauty, is awe-inspiring and one of our nation's most astounding natural wonders.

If a Clif energy bar is accidentally dropped over the side of the skywalk, how long will it take it to hit the canyon floor?

If t represents the time in seconds, the quadratic equation $-16t^2 + 4,000 = 0$ models the time it takes the energy bar to fall to the canyon floor. We can solve this equation by using the square root property

$$-16t^2 + 4,000 = 0$$
$$-16t^2 = -4,000 \qquad \text{Subtract 4,000 from both sides.}$$
$$t^2 = 250 \qquad \text{Divide both sides by } -16.$$
$$t = \pm\sqrt{250} \qquad \text{Use the square root property.}$$
$$\approx \pm15.8 \qquad \text{Round to the nearest tenth.}$$

Because time cannot be negative, we disregard the negative answer. The time it will take the energy bar to reach the canyon floor is about 15.8 seconds.

As this example illustrates, the solutions of many problems involve quadratic equations.

1. Solve Geometric Problems

EXAMPLE 1 The length of a rectangle exceeds its width by 3 feet. If its area is 40 square feet, find its dimensions.

Solution To find an equation that models the problem, we can let w represent the width of the rectangle. Then, $w + 3$ will represent its length (see Figure 1-4). Since the formula for the area of a rectangle is $A = lw$ (area = length × width), the area of the rectangle is $(w + 3)w$, which is equal to 40.

The length of the rectangle	times	the width of the rectangle	equals	the area of the retangle.
$(w + 3)$	·	w	=	40

We can solve this equation for w.

$$(w + 3)w = 40$$
$$w^2 + 3w = 40$$
$$w^2 + 3w - 40 = 0 \quad \text{Subtract 40 from both sides.}$$
$$(w - 5)(w + 8) = 0 \quad \text{Factor.}$$
$$w - 5 = 0 \quad \text{or} \quad w + 8 = 0$$
$$w = 5 \quad \quad \quad w = -8$$

When $w = 5$, the length is $w + 3 = 8$. The solution -8 must be discarded, because a rectangle cannot have a negative width.

We can verify that this solution is correct by observing that a rectangle with dimensions of 5 feet by 8 feet has an area of 40 square feet.

w ft

$(w + 3)$ ft

Figure 1-4

EXAMPLE 2 On a college campus, a sidewalk 85 meters long (represented by the red lines in Figure 1-5) joins a dormitory building D with the student center C. However, the students prefer to walk directly from D to C. If segment DC is 65 meters long, how long is each piece of the existing sidewalk?

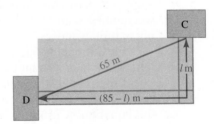

C

65 m

l m

D

$(85 - l)$ m

Figure 1-5

Solution We note that the triangle shown in the figure is a right triangle, with a **hypotenuse** that is 65 meters long. If we let the shorter leg of the triangle be l meters long, the length of the longer leg will be $(85 - l)$ meters. By the **Pythagorean theorem**, we know that the sum of the squares of the two legs of a right triangle is equal to the square of the hypotenuse. Thus, we can form the equation

$$l^2 + (85 - l)^2 = 65^2 \quad \text{In a right triangle, } a^2 + b^2 = c^2.$$

which we can solve as follows:

$$l^2 + 7{,}225 - 170l + l^2 = 4{,}225 \qquad \text{Expand } (85 - l)^2.$$
$$2l^2 - 170l + 3{,}000 = 0 \qquad \text{Combine like terms and subtract 4,225 from both sides.}$$
$$l^2 - 85l + 1{,}500 = 0 \qquad \text{Divide both sides by 2.}$$

Since the left side is difficult to factor, we will solve this equation using the quadratic formula.

$$l = \frac{-b \pm \sqrt{b^2 - 4ac}}{2a}$$

$$l = \frac{-(-85) \pm \sqrt{(-85)^2 - 4(1)(1{,}500)}}{2(1)}$$

$$l = \frac{85 \pm \sqrt{1{,}225}}{2}$$

$$l = \frac{85 \pm 35}{2}$$

$$l = \frac{85 + 35}{2} \quad \text{or} \quad l = \frac{85 - 35}{2}$$
$$= 60 \qquad\qquad\qquad = 25$$

The length of the shorter leg is 25 meters. The length of the longer leg is $(85 - 25)$ meters, or 60 meters.

2. Solve Uniform Motion Problems

EXAMPLE 3

A man drives 600 miles to a convention. On the return trip, he is able to increase his speed by 10 mph and save 2 hours of driving time. How fast did he drive in each direction?

Solution

We can let s represent the car's speed (in mph) driving to the convention. On the return trip, his speed was $s + 10$ mph. Recall that the distance traveled by an object moving at a constant rate for a certain time is given by the formula $d = rt$. If we divide both sides of this formula by r, we will have a formula for time.

$$t = \frac{d}{r}$$

We can organize the information given in this problem as in Figure 1-6.

	d	$=$	r	$\cdot$	t
Outbound trip	600		s		$\dfrac{600}{s}$
Return trip	600		$s + 10$		$\dfrac{600}{s + 10}$

Figure 1-6

Although neither the outbound nor the return travel time is given, we know the difference of those times.

Sofia Kovalevskaya
(1850–1891)

In addition to being a great mathematician, Kovalevskaya was also an early advocate for women's rights. She hoped to study mathematics at the University of Berlin, but strict rules prohibited women from attending lectures there. Undaunted, she studied with Karl Weierstrauss, who taught at the university. In 1874, she was granted a Ph.D. from the University of Göttingen. Her work produced the basis for future discoveries by other mathematicians.

The longer time of the outbound trip	minus	the shorter time of the return trip	equals	the difference in travel times.
$\dfrac{600}{s}$	$-$	$\dfrac{600}{s+10}$	$=$	2

We can solve this equation for s.

$$\frac{600}{s} - \frac{600}{s+10} = 2$$

$$s(s+10)\left(\frac{600}{s} - \frac{600}{s+10}\right) = s(s+10)2 \qquad \text{Multiply both sides by } s(s+10) \text{ to clear the equation of fractions.}$$

$$600(s+10) - 600s = 2s(s+10) \qquad \text{Simplify.}$$

$$600s + 6{,}000 - 600s = 2s^2 + 20s \qquad \text{Remove parentheses.}$$

$$6{,}000 = 2s^2 + 20s \qquad \text{Combine like terms.}$$

$$0 = 2s^2 + 20s - 6{,}000 \qquad \text{Subtract 6,000 from both sides.}$$

$$0 = s^2 + 10s - 3{,}000 \qquad \text{Divide both sides by 2.}$$

$$0 = (s - 50)(s + 60) \qquad \text{Factor.}$$

$$s - 50 = 0 \quad \text{or} \quad s + 60 = 0 \qquad \text{Set each factor equal to 0.}$$

$$s = 50 \qquad\qquad s = -60$$

The solution $s = -60$ must be discarded. The man drove 50 mph to the convention and $50 + 10$, or 60 mph, on the return trip.

These answers are correct, because a 600-mile trip at 50 mph would take $\frac{600}{50}$, or 12 hours. At 60 mph, the same trip would take only 10 hours, which is 2 hours less time.

3. Solve Falling Body Problems

EXAMPLE 4   If an object is thrown straight up into the air with an initial velocity of 144 feet per second, its height is given by the formula $h = 144t - 16t^2$, where h represents its height (in feet) and t represents the time (in seconds) since it was thrown. How long will it take for the object to return to the point from which it was thrown?

Solution   When the object returns to its starting point, its height is again 0. Thus, we can set h equal to 0 and solve for t.

$$h = 144t - 16t^2$$

$$0 = 144t - 16t^2 \qquad \text{Let } h = 0.$$

$$0 = 16t(9 - t) \qquad \text{Factor.}$$

$$16t = 0 \quad \text{or} \quad 9 - t = 0 \qquad \text{Set each factor equal to 0.}$$

$$t = 0 \qquad\qquad t = 9$$

At $t = 0$, the object's height is 0, because it was just released. When $t = 9$, the height is again 0, and the object has returned to its starting point.

4. Solve Business Problems

EXAMPLE 5   A bus company shuttles 1,120 passengers daily between Rockford, Illinois and O'Hare airport. The current one-way fare is $10. For each 25¢ increase in the fare, the company predicts that it will lose 48 passengers. What increase in fare will produce daily revenue of $10,208?

Solution Let q represent the number of quarters the fare will be increased. Then the new fare will be $(10 + 0.25q)$. Since the company will lose 48 passengers for each 25¢ increase, $48q$ passengers will be lost when the rate increases by q quarters. The passenger load will then be $(1,120 - 48q)$ passengers.

Since the daily revenue of $10,208 will be the product of the rate and the number of passengers, we have

$$(10 + 0.25q)(1,120 - 48q) = 10,208$$

$$11,200 - 480q + 280q - 12q^2 = 10,208 \qquad \text{Remove parentheses.}$$

$$-12q^2 - 200q + 992 = 0 \qquad \begin{array}{l}\text{Combine like terms and subtract 10,208} \\ \text{from both sides.}\end{array}$$

$$3q^2 + 50q - 248 = 0 \qquad \text{Divide both sides by } -4.$$

Since the left side is difficult to factor, we will solve this equation with the quadratic formula.

$$q = \frac{-b \pm \sqrt{b^2 - 4ac}}{2a}$$

$$q = \frac{-50 \pm \sqrt{50^2 - 4(3)(-248)}}{2(3)} \qquad \text{Substitute 3 for } a, \text{ 50 for } b, \text{ and } -248 \text{ for } c.$$

$$q = \frac{-50 \pm \sqrt{2,500 + 2,976}}{6}$$

$$q = \frac{-50 \pm \sqrt{5,476}}{6}$$

$$q = \frac{-50 \pm 74}{6}$$

$$q = \frac{-50 + 74}{6} \qquad \text{or} \qquad q = \frac{-50 - 74}{6}$$

$$= \frac{24}{6} \qquad\qquad\qquad = \frac{-124}{6}$$

$$= 4 \qquad\qquad\qquad\quad = -\frac{62}{3}$$

Since the number of riders cannot be negative, the result of $-\frac{62}{3}$ must be discarded. To generate $10,208 in daily revenues, the company should raise the fare by 4 quarters, or $1, to $11. ▬▬

5. Solve Shared-Work Problems

EXAMPLE 6 One environmental company can clean up an oil spill on a beach in 2 days less time than its competitor. Working together, they were able to clean up the spill in 10 days. How long would it have taken the first company to clean up the spill if it worked alone?

Solution Suppose the first company can clean up the spill in x days. Then the first company can do $\frac{1}{x}$ of the job each day. Because the first company can do the work in 2 days less than its competitor, it will take the competitor $(x + 2)$ days to clean up the spill. The competitor can do $\frac{1}{x + 2}$ of the job each day.

Working together, they can clean up the spill in 10 days. So together they can do $\frac{1}{10}$ of the job each day. The sum of the work each can do in one day is equal to the work that they can do together in one day.

The part the first company can clean up in one day	plus	the part the second company can clean up in one day	equals	the part they can clean up together in one day.
$\dfrac{1}{x}$	$+$	$\dfrac{1}{x+2}$	$=$	$\dfrac{1}{10}$

We can solve this equation for x.

$$\frac{1}{x} + \frac{1}{x+2} = \frac{1}{10}$$

$$10x(x+2)\left(\frac{1}{x} + \frac{1}{x+2}\right) = 10x(x+2)\left(\frac{1}{10}\right)$$

Multiply both sides by $10x(x+2)$ to eliminate the fractions.

$$\frac{10x(x+2)}{x} + \frac{10x(x+2)}{x+2} = \frac{10x(x+2)}{10}$$

Distribute the multiplication by $10x(x+2)$.

$$10(x+2) + 10x = x(x+2)$$

$\frac{x}{x} = 1$, $\frac{x+2}{x+2} = 1$, and $\frac{10}{10} = 1$.

$$10x + 20 + 10x = x^2 + 2x$$

Use the distributive property to remove parentheses.

$$0 = x^2 - 18x - 20$$

Subtract $20x$ and 20 from both sides.

Since the right side cannot be factored over the integers, we will solve the equation with the quadratic formula.

$$x = \frac{-b \pm \sqrt{b^2 - 4ac}}{2a}$$

$$x = \frac{-(-18) \pm \sqrt{(-18)^2 - 4(1)(-20)}}{2(1)}$$

Substitute 1 for a, -18 for b, and -20 for c.

$$x = \frac{18 \pm \sqrt{324 + 80}}{2}$$

$$x = \frac{18 \pm \sqrt{404}}{2}$$

$$x = \frac{18 \pm 20.09975124}{2}$$

$$x = \frac{18 + 20.09975124}{2} \quad \text{or} \quad x = \frac{18 - 20.09975124}{2}$$

$$\approx 19.05 \qquad\qquad\qquad \approx -1.05$$

Since the work cannot be completed in a negative number of days, we discard the solution of -1.05. Thus, the first company can complete the job working alone in a little over 19 days.

1.4 Exercises

Vocabulary and Concepts   *Fill in the blanks.*

1. The formula for the area of a rectangle is _____.
2. The formula that relates distance, rate, and time is _____.

Practice   *Solve each problem.*

3. **Geometric problem**   A rectangle is 4 feet longer than it is wide. If its area is 32 square feet, find its dimensions.
4. **Geometric problem**   A rectangle is 5 times as long as it is wide. If the area is 125 square feet, find its perimeter.

5. Geometric problem   The side of a square is 4 centimeters shorter than the side of a second square. If the sum of their areas is 106 square centimeters, find the length of one side of the larger square.

6. Geometric problem   The base of a triangle is one-third as long as its height. If the area of the triangle is 24 square meters, how long is its base?

7. Flags   In 1912, an order by President Taft fixed the width and length of the U.S. flag in the ratio 1 to 1.9. If 100 square feet of cloth are to be used to make a U.S. flag, estimate its dimensions to the nearest $\frac{1}{4}$ foot.

8. Imax screens   A large movie screen is in the Panasonic Imax theater at Darling Harbor, Sydney, Australia. The rectangular screen has an area of 11,349 square feet. Find the dimensions of the screen if it is 20 feet longer than it is wide.

9. Metal fabrication   A piece of tin, 12 inches on a side, is to have four equal squares cut from its corners, as in the illustration. If the edges are then to be folded up to make a box with a floor area of 64 square inches, find the depth of the box.

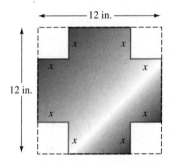

10. Making gutters   A piece of sheet metal, 18 inches wide, is bent to form the gutter shown in the illustration. If the cross-sectional area is 36 square inches, find the depth of the gutter.

11. Manufacturing   A manufacturer of television sets received an order for sets with a 46-inch screen (measured along the diagonal), as shown in the illustration. If the screens are to be rectangular in shape and $17\frac{1}{2}$ inches wider than they are high, find the dimensions of the screen to the nearest tenth of an inch.

12. Finding dimensions   The oriental rug shown is 2 feet longer than it is wide. To the nearest tenth of a foot, find its dimensions.

13. Cycling rates   A cyclist rides from DeKalb to Rockford, a distance of 40 miles. His return trip takes 2 hours longer, because his speed decreases by 10 mph. How fast does he ride each way?

14. Travel times   A farmer drives a tractor from one town to another, a distance of 120 kilometers. He drives 10 kilometers per hour faster on the return trip, cutting 1 hour off the time. How fast does he drive each way?

15. Uniform motion problem   If the speed were increased by 10 mph, a 420-mile trip would take 1 hour less time. How long will the trip take at the slower speed?

16. Uniform motion problem   By increasing her usual speed by 25 kilometers per hour, a bus driver decreases the time on a 25-kilometer trip by 10 minutes. Find the usual speed.

17. Ballistics   The height of a projectile fired upward with an initial velocity of 400 feet per second is given by the formula $h = -16t^2 + 400t$, where h is the height in feet and t is the time in seconds. Find the time required for the projectile to return to earth.

18. **Ballistics**   The height of an object tossed upward with an initial velocity of 104 feet per second is given by the formula $h = -16t^2 + 104t$, where h is the height in feet and t is the time in seconds. Find the time required for the object to return to its point of departure.

19. **Falling coins**   An object will fall s feet in t seconds, where $s = 16t^2$. How long will it take for a penny to hit the ground if it is dropped from the top of the Sears Tower in Chicago?   (*Hint:* The tower is 1,454 feet tall.)

20. **Movie stunts**   According to the *Guinness Book of World Records, 1998,* stuntman Dan Koko fell a distance of 312 feet into an airbag after jumping from the Vegas World Hotel and Casino. The distance d in feet traveled by a free-falling object in t seconds is given by the formula $d = 16t^2$. To the nearest tenth of a second, how long did the fall last?

21. **Accidents**   The height h (in feet) of an object that is dropped from a height of s feet is given by the formula $h = s - 16t^2$, where t is the time the object has been falling. A 5-foot-tall woman on a sidewalk looks directly overhead and sees a window washer drop a bottle from 4 stories up. How long does she have to get out of the way? Round to the nearest tenth. (A story is 12 feet.)

22. **Ballistics**   The height of an object thrown upward with an initial velocity of 32 feet per second is given by the formula $h = -16t^2 + 32t$, where t is the time in seconds. How long will it take the object to reach a height of 16 feet?

23. **Setting fares**   A bus company has 3,000 passengers daily, paying a 25¢ fare. For each nickel increase in fare, the company projects that it will lose 80 passengers. What fare increase will produce $994 in daily revenue?

24. **Jazz concerts**   A jazz group on tour has been drawing average crowds of 500 people. It is projected that for every $1 increase in the $12 ticket price, the average attendance will decrease by 50. At what ticket price will nightly receipts be $5,600?

25. **Concert receipts**   Tickets for the annual symphony orchestra pops concert cost $15, and the average attendance at the concerts has been 1,200 people. Management projects that for each 50¢ decrease in ticket price, 40 more patrons will attend. How many people attended the concert if the receipts were $17,280?

26. **Projecting demand**   The *Vilas County News* earns a profit of $20 per year for each of its 3,000 subscribers. Management projects that the profit per subscriber would increase by 1¢ for each additional subscriber over the current 3,000. How many subscribers are needed to bring a total profit of $120,000?

27. **Architecture**   A **golden rectangle** is one of the most visually appealing of all geometric forms. The front of the Parthenon, built in Athens in the 5th century B.C. and shown in the illustration, is a golden rectangle. In a golden rectangle, the length l and the height h of the rectangle must satisfy the following equation.

$$\frac{l}{h} = \frac{h}{l - h}$$

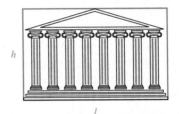

If a rectangular billboard is to have a height of 15 feet, how long should it be if it is to form a golden rectangle? Round to the nearest tenth of a foot.

28. **Golden ratio**   Rectangle $ABCD$, shown here, will be a **golden rectangle** if $\frac{AB}{AD} = \frac{BC}{BE}$ where $AE = AD$. Let $AE = 1$ and find the ratio of AB to AD.

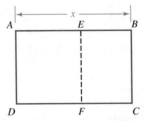

29. **Filling storage tanks**   Two pipes are used to fill a water storage tank. The first pipe can fill the tank in 4 hours, and the two pipes together can fill the tank in 2 hours less time than the second pipe alone. How long would it take for the second pipe to fill the tank?

30. Filling swimming pools A hose can fill a swimming pool in 6 hours. Another hose needs 3 more hours to fill the pool than the two hoses combined. How long would it take the second hose to fill the pool?

31. Mowing lawns Kristy can mow a lawn in 1 hour less time than her brother Steven. Together they can finish the job in 5 hours. How long would it take Kristy if she worked alone?

32. Milking cows Working together, Sarah and Heidi can milk the cows in 2 hours. If they work alone, it takes Heidi 3 hours longer than it takes Sarah. How long would it take Heidi to milk the cows alone?

33. Geometric problem Is it possible for a rectangle to have a width that is 3 units shorter than its diagonal and a length that is 4 units longer than its diagonal?

34. Geometric problem If two opposite sides of a square are increased by 10 meters and the other sides are decreased by 8 meters, the area of the rectangle that is formed is 63 square meters. Find the area of the original square.

35. Investment problems Maude and Matilda each have a bank CD. Maude's is $1,000 larger than Matilda's, but the interest rate is 1% less. Last year Maude received interest of $280, and Matilda received $240. Find the rate of interest for each CD.

36. Investment problem Scott and Laura have both invested some money. Scott invested $3,000 more than Laura and at a 2% higher interest rate. If Scott received $800 annual interest and Laura received $400, how much did Scott invest?

37. Buying microwave ovens Some mathematics professors would like to purchase a $150 microwave oven for the department workroom. If four of the professors don't contribute, everyone's share will increase by $10. How many professors are in the department?

38. Planting windscreens A farmer intends to construct a windscreen by planting trees in a quarter-mile row. His daughter points out that 44 fewer trees will be needed if they are planted 1 foot farther apart. If her dad takes her advice, how many trees will be needed? A row starts and ends with a tree. (*Hint:* 1 mile = 5,280 feet.)

39. Puzzle problem If a wagon wheel had 10 more spokes, the angle between spokes would decrease by 6°. How many spokes does the wheel have?

40. Puzzle problem A merchant could sell one model of digital cameras at list price for $180. If he had 3 more cameras, he could sell each one for $10 less and still receive $180. Find the list price of each camera.

41. Geometric problem If one leg of a right triangle is 14 meters shorter than the other leg, and the hypotenuse is 26 meters, find the length of the two legs.

42. Geometric problem Find the dimensions of a rectangle whose area is 180 cm^2 and whose perimeter is 54 cm.

43. Automobile engines As the piston shown moves upward, it pushes a cylinder of a gasoline/air mixture that is ignited by the spark plug. The formula that gives the volume of a cylinder is $V = \pi r^2 h$, where r is the radius and h the height. Find the radius of the piston (to the nearest hundredth of an inch) if it displaces 47.75 cubic inches of gasoline/air mixture as it moves from its lowest to its highest point.

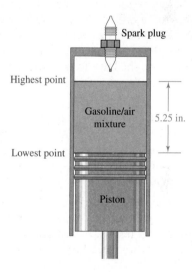

44. History One of the important cities of the ancient world was Babylon. Greek historians wrote that the city was square-shaped. Its area numerically exceeded its perimeter by about 124. Find its dimensions in miles. (Round to the nearest tenth.)

Discovery and Writing

45. Which of the preceding problems did you find the hardest? Why?

46. Which of the preceding problems did you find the easiest? Why?

Review *Perform the operations and simplify.*

47. $\dfrac{2}{x} - \dfrac{1}{x-3}$

48. $\dfrac{1}{x} \cdot \dfrac{x^2 - 5x}{x-3}$

49. $\dfrac{x+3}{x^2 - x - 6} \div \dfrac{x^2 + 3x}{x^2 - 9}$

50. $\dfrac{\dfrac{1}{x} - \dfrac{1}{2}}{x-2}$

51. $\dfrac{\dfrac{1}{x} + \dfrac{1}{y}}{\dfrac{1}{x} - \dfrac{1}{y}}$

52. $\dfrac{x}{x+1} + \dfrac{x+1}{x} \cdot \dfrac{x^2 - x}{3}$

1.5 Complex Numbers

Objectives

1. Define and Simplify Imaginary Numbers
2. Define and Perform Operations on Complex Numbers
3. Find Powers of i
4. Define the Absolute Value of a Complex Number
5. Solve Quadratic Equations with Complex Roots
6. Factor the Sum of Two Squares

The Blue Man Group is a highly successful group of creative performers who cover themselves in blue grease paint, wear latex bald caps, and dress in black. They combine rock music, comedy, multimedia theatrics, and sophisticated lighting to entertain their audiences.

There are a variety of themes used in Blue Man performances and the group uses fractals in their shows. Fractals are beautiful art designs that are computer generated from numbers called *complex numbers*. We will study these numbers in this section. The fractal that is shown is called a *Julia set*.

All of the quadratic equations that we considered in Section 1.3 had roots that were real numbers. However, the solutions of many quadratic equations are not real numbers. For example, if we use the quadratic formula to solve $x^2 + x + 2 = 0$, we get solutions that are not real.

$$x = \frac{-b \pm \sqrt{b^2 - 4ac}}{2a}$$

$$x = \frac{-1 \pm \sqrt{1^2 - 4(1)(2)}}{2(1)}$$ Substitute 1 for a, 1 for b, and 2 for c.

$$x = \frac{-1 \pm \sqrt{1 - 8}}{2}$$

$$x = \frac{-1 \pm \sqrt{-7}}{2}$$

Each solution involves $\sqrt{-7}$. This is not a real number, because the square of no real number is -7.

1.   Define and Simplify Imaginary Numbers

For years, mathematicians believed that numbers such as

$$\sqrt{-1}, \qquad \sqrt{-4}, \qquad \sqrt{-5}, \qquad \text{and} \qquad \sqrt{-7}$$

were nonsense. Even the great English mathematician Sir Isaac Newton (1642–1727) called them "impossible numbers." In the 17th century, these symbols were called **imaginary numbers** by René Descartes. Today, they have important uses, such as describing the behavior of alternating current in electronics.

The imaginary numbers are based on the **imaginary unit *i*,** where

$$i^2 = -1$$

Because i represents the square root of -1, we also write

$$i = \sqrt{-1}$$

Because imaginary numbers follow the rules for exponents, we have

$$(3i)^2 = 3^2 i^2 = 9(-1) = -9 \quad i^2 = -1$$

Since $(3i)^2 = -9$, $3i$ is a square root of -9, and we can write

$$\sqrt{-9} = 3i$$

This result also can be obtained by using the multiplication property of radicals.

$$\sqrt{-9} = \sqrt{9(-1)}$$
$$= \sqrt{9}\sqrt{-1} \quad \sqrt{ab} = \sqrt{a}\sqrt{b}$$
$$= 3i \quad \sqrt{9} = 3 \text{ and } \sqrt{-1} = i$$

We can use the multiplication property of radicals to simplify imaginary numbers.

$$\sqrt{-25} = \sqrt{25(-1)} = \sqrt{25}\sqrt{-1} = 5i$$
$$\sqrt{-7} = \sqrt{7(-1)} = \sqrt{7}\sqrt{-1} = \sqrt{7}i$$
$$\sqrt{\frac{-100}{49}} = \sqrt{\frac{100}{49}(-1)} = \sqrt{\frac{100}{49}}\sqrt{-1} = \frac{10}{7}i$$

Comment

If a and b are both negative, then $\sqrt{ab} \neq \sqrt{a}\sqrt{b}$. For example, the correct simplification of $\sqrt{-16}\sqrt{-4}$ is

$$\sqrt{-16}\sqrt{-4} = (4i)(2i) = 8i^2 = 8(-1) = -8$$

The following simplification is incorrect, because we get a different result.

$$\sqrt{-16}\sqrt{-4} = \sqrt{(-16)(-4)} = \sqrt{64} = 8$$ Here, a and b are both negative, and the multiplication property of radicals does not apply.

2. Define and Perform Operations on Complex Numbers

Numbers that are the sum or difference of a real number and an imaginary number, such as $3 + 4i$, $-5 + 7i$, and $-1 - 9i$, are called *complex numbers*.

Complex Numbers   A **complex number** is a number that can be written in the form $a + bi$, where a and b are real numbers and $i = \sqrt{-1}$.
The number a is called the **real part,** and b is called the **imaginary part.**

If $b = 0$, the complex number $a + bi$ is the real number a. If $a = 0$ and $b \neq 0$, the complex number $a + bi$ is the imaginary number bi. It follows that the set of real numbers and the set of imaginary numbers are subsets of the set of complex numbers.

Figure 1-7 illustrates how the various sets of numbers are related.

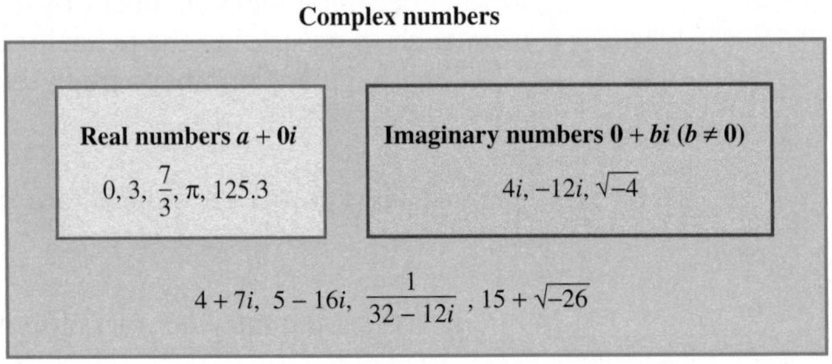

Complex numbers

Real numbers $a + 0i$	Imaginary numbers $0 + bi$ $(b \neq 0)$
$0, 3, \dfrac{7}{3}, \pi, 125.3$	$4i, -12i, \sqrt{-4}$

$$4 + 7i, \ 5 - 16i, \ \frac{1}{32 - 12i}, \ 15 + \sqrt{-26}$$

Figure 1-7

To determine whether two complex numbers are equal, we can use the following definition.

Equality of Complex Numbers   Two complex numbers are equal if their real parts are equal and their imaginary parts are equal. If $a + bi$ and $c + di$ are two complex numbers, then

$$a + bi = c + di \quad \text{if and only if} \quad a = c \text{ and } b = d$$

EXAMPLE 1   For what numbers x and y is $3x + 4i = (2y + x) + xi$?

Solution   Since the numbers are equal, their imaginary parts must be equal: $x = 4$. Since their real parts are equal, $3x = 2y + x$. We can solve the system

$$\begin{cases} x = 4 \\ 3x = 2y + x \end{cases}$$

by substituting 4 for x in the second equation and solving for y. We find that $y = 4$. The solution is $x = 4$ and $y = 4$.

Self Check 1   Find x: $a + (x + 3)i = a - (2x - 1)i$.

Complex numbers can be added and subtracted as if they were binomials.

Adding and Subtracting Complex Numbers

Two complex numbers such as $a + bi$ and $c + di$ are added and subtracted as if they were binomials:

$$(a + bi) + (c + di) = (a + c) + (b + d)i$$
$$(a + bi) - (c + di) = (a - c) + (b - d)i$$

Because of the preceding definition, the sum or difference of two complex numbers is another complex number.

EXAMPLE 2   Simplify: **a.** $(3 + 4i) + (2 + 7i)$ **b.** $(-5 + 8i) - (2 - 12i)$

Solution   **a.** $(3 + 4i) + (2 + 7i) = 3 + 4i + 2 + 7i$
$$= 3 + 2 + 4i + 7i$$
$$= 5 + 11i$$

b. $(-5 + 8i) - (2 - 12i) = -5 + 8i - 2 + 12i$
$$= -5 - 2 + 8i + 12i$$
$$= -7 + 20i$$

Self Check 2   Simplify: **a.** $(5 - 2i) + (-3 + 9i)$
b. $(2 + 5i) - (6 + 7i)$

Complex numbers can also be multiplied as if they were binomials.

Multiplying Complex Numbers

The numbers $a + bi$ and $c + di$ are multiplied as if they were binomials, with $i^2 = -1$:

$$(a + bi)(c + di) = (ac - bd) + (ad + bc)i$$

Because of this definition, the product of two complex numbers is another complex number.

EXAMPLE 3   Multiply: **a.** $(3 + 4i)(2 + 7i)$ **b.** $(5 - 7i)(1 + 3i)$.

Solution   **a.** $(3 + 4i)(2 + 7i) = 6 + 21i + 8i + 28i^2$
$$= 6 + 21i + 8i + 28(-1) \quad i^2 = -1$$
$$= 6 - 28 + 29i$$
$$= -22 + 29i$$

b. $(5 - 7i)(1 + 3i) = 5 + 15i - 7i - 21i^2$
$$= 5 + 15i - 7i - 21(-1) \quad i^2 = -1$$
$$= 5 + 21 + 8i$$
$$= 26 + 8i$$

Self Check 3   Multiply: $(2 - 5i)(3 + 2i)$

To avoid errors in determining the sign of the result, always express numbers in $a + bi$ form before attempting any algebraic manipulations.

EXAMPLE 4 Multiply: $\left(-2 + \sqrt{-16}\right)\left(4 - \sqrt{-9}\right)$.

Solution We change each number to $a + bi$ form:

$$-2 + \sqrt{-16} = -2 + \sqrt{16}\sqrt{-1} = -2 + 4i$$
$$4 - \sqrt{-9} = 4 - \sqrt{9}\sqrt{-1} = 4 - 3i$$

and then find the product.

$$(-2 + 4i)(4 - 3i) = -8 + 6i + 16i - 12i^2$$
$$= -8 + 6i + 16i - 12(-1) \quad i^2 = -1$$
$$= -8 + 12 + 22i$$
$$= 4 + 22i$$

Self Check 4 Multiply: $\left(3 + \sqrt{-25}\right)\left(2 - \sqrt{-9}\right)$.

Before we discuss the division of complex numbers, we introduce the concept of a complex conjugate.

Complex Conjugates The complex numbers $a + bi$ and $a - bi$ are called **complex conjugates** of each other.

For example,

$2 + 5i$ and $2 - 5i$ are complex conjugates.

$-\dfrac{1}{2} + 4i$ and $-\dfrac{1}{2} - 4i$ are complex conjugates.

What makes this concept important is the fact that the product of two complex conjugates is always a real number. For example,

$$(2 + 5i)(2 - 5i) = 4 - 10i + 10i - 25i^2$$
$$= 4 - 25(-1) \qquad i^2 = -1$$
$$= 4 + 25$$
$$= 29$$

In general, we have

$$(a + bi)(a - bi) = a^2 - abi + abi - b^2i^2$$
$$= a^2 - b^2(-1) \qquad i^2 = -1$$
$$= a^2 + b^2$$

To divide complex numbers, we use the concept of complex conjugates to rationalize the denominator.

EXAMPLE 5 Divide and write the result in $a + bi$ form: $\dfrac{3}{2 + i}$.

Solution To divide, we rationalize the denominator and simplify.

$$\frac{3}{2+i} = \frac{3(2-i)}{(2+i)(2-i)}$$

To make the denominator a real number, multiply the numerator and denominator by the complex conjugate of $2+i$, which is $2-i$.

$$= \frac{6-3i}{4-2i+2i-i^2}$$

Multiply.

$$= \frac{6-3i}{4+1}$$

Simplify the denominator.

$$= \frac{6-3i}{5}$$

$$= \frac{6}{5} - \frac{3}{5}i$$

It is common to accept $\dfrac{6}{5} - \dfrac{3}{5}i$ as a substitute for $\dfrac{6}{5} + \left(-\dfrac{3}{5}\right)i$.

Self Check 5   Divide and write the result in $a+bi$ form: $\dfrac{3}{3-i}$.

EXAMPLE 6   Divide and write the result in $a+bi$ form: $\dfrac{2-\sqrt{-16}}{3+\sqrt{-1}}$.

Solution

$$\frac{2-\sqrt{-16}}{3+\sqrt{-1}} = \frac{2-4i}{3+i}$$

Change each number to $a+bi$ form.

$$= \frac{(2-4i)(3-i)}{(3+i)(3-i)}$$

To make the denominator a real number, multiply the numerator and denominator by $3-i$.

$$= \frac{6-2i-12i+4i^2}{9-3i+3i-i^2}$$

Remove parentheses.

$$= \frac{2-14i}{9+1}$$

Combine like terms; $i^2 = -1$.

$$= \frac{2}{10} - \frac{14i}{10}$$

$$= \frac{1}{5} - \frac{7}{5}i$$

Self Check 6   Divide and write the result in $a+bi$ form: $\dfrac{3+\sqrt{-25}}{2-\sqrt{-1}}$.

Examples 5 and 6 illustrate that the quotient of two complex numbers is another complex number.

3.   Find Powers of i

The powers of i with natural number exponents produce an interesting pattern.

$$i^1 = \sqrt{-1} = i \qquad\qquad i^5 = i^4 i = 1i = i$$
$$i^2 = \left(\sqrt{-1}\right)^2 = -1 \qquad i^6 = i^4 i^2 = 1(-1) = -1$$
$$i^3 = i^2 i = -1i = -i \qquad i^7 = i^4 i^3 = 1(-i) = -i$$
$$i^4 = i^2 i^2 = (-1)(-1) = 1 \qquad i^8 = i^4 i^4 = 1(1) = 1$$

Comment
$i^0 = 1$ because any number except 0 raised to the 0th power is 1.

The pattern continues: $i, -1, -i, 1, \ldots$.

EXAMPLE 7   Simplify: i^{365}.

Solution   Since $i^4 = 1$, each occurrence of i^4 is a factor of 1. To determine how many factors of i^4 are in i^{365}, we divide 365 by 4. The quotient is 91, and the remainder is 1.

$$i^{365} = (i^4)^{91} \cdot i^1$$
$$= 1^{91} \cdot i^1 \qquad i^4 = 1$$
$$= i \qquad\qquad 1^{91} = 1 \text{ and } 1 \cdot i = i$$

Self Check 7   Simplify: $i^{1,999}$

The result of Example 7 illustrates the following theorem.

Powers of i   If n is a natural number that has a remainder of r when divided by 4, then
$$i^n = i^r$$
When n is divisible by 4, the remainder r is 0 and $i^0 = 1$.

We also can simplify powers of i that involve negative integer exponents.

$$i^{-1} = \frac{1}{i} = \frac{1 \cdot i}{i \cdot i} = \frac{i}{-1} = -i \qquad\qquad i^{-2} = \frac{1}{i^2} = \frac{1}{-1} = -1$$

$$i^{-3} = \frac{1}{i^3} = \frac{1 \cdot i}{i^3 \cdot i} = \frac{i}{i^4} = \frac{i}{1} = i \qquad\qquad i^{-4} = \frac{1}{i^4} = \frac{1}{1} = 1$$

4.   Define the Absolute Value of a Complex Number

Absolute Value of a Complex Number   If $a + bi$ is a complex number, then
$$|a + bi| = \sqrt{a^2 + b^2}$$

Because of the previous definition, the absolute value of a complex number is a real number. For this reason, i does not appear in the result.

EXAMPLE 8   Write without absolute value symbols: **a.** $|3 + 4i|$ **b.** $|4 - 6i|$

Solution   In each case, we apply the definition of absolute value of a complex number.

a. $|3 + 4i| = \sqrt{3^2 + 4^2}$ **b.** $|4 - 6i| = \sqrt{4^2 + (-6)^2}$
$\qquad\quad = \sqrt{9 + 16}$ $\qquad\qquad = \sqrt{16 + 36}$
$\qquad\quad = \sqrt{25}$ $\qquad\qquad = \sqrt{52}$
$\qquad\quad = 5$ $\qquad\qquad = \sqrt{4 \cdot 13}$
$\qquad\qquad\qquad\qquad\qquad\qquad\qquad = \sqrt{4}\sqrt{13}$
$\qquad\qquad\qquad\qquad\qquad\qquad\qquad = 2\sqrt{13}$

Self Check 8   Write without absolute value symbols: $|2 - 5i|$.

EXAMPLE 9 Write without absolute value symbols: **a.** $\left|\dfrac{2i}{3+i}\right|$ **b.** $|a + 0i|$

Solution **a.** We first write $\dfrac{2i}{3+i}$ in $a + bi$ form:

$$\frac{2i}{3+i} = \frac{2i(3-i)}{(3+i)(3-i)} = \frac{6i - 2i^2}{9 - i^2} = \frac{6i + 2}{10} = \frac{1}{5} + \frac{3}{5}i$$

and then find the absolute value of $\frac{1}{5} + \frac{3}{5}i$.

$$\left|\frac{2i}{3+i}\right| = \left|\frac{1}{5} + \frac{3}{5}i\right| = \sqrt{\left(\frac{1}{5}\right)^2 + \left(\frac{3}{5}\right)^2} = \sqrt{\frac{10}{25}} = \frac{\sqrt{10}}{5}$$

b. $|a + 0i| = \sqrt{a^2 + 0^2} = \sqrt{a^2} = |a|$

From part b, we see that $|a| = \sqrt{a^2}$.

Self Check 9 Write without absolute value symbols: $\left|\dfrac{3i}{2-i}\right|$.

5. Solve Quadratic Equations with Complex Roots

The roots of many quadratic equations are complex numbers, as the following example shows.

EXAMPLE 10 Solve: $x^2 - 4x + 5 = 0$.

Solution In this equation, $a = 1$, $b = -4$, and $c = 5$.

$$x = \frac{-b \pm \sqrt{b^2 - 4ac}}{2a}$$

$$= \frac{-(-4) \pm \sqrt{(-4)^2 - 4(1)(5)}}{2(1)} \qquad \text{Substitute 1 for } a, -4 \text{ for } b, \text{ and 5 for } c.$$

$$= \frac{4 \pm \sqrt{16 - 20}}{2}$$

$$= \frac{4 \pm \sqrt{-4}}{2}$$

$$= \frac{4 \pm 2i}{2} \qquad\qquad \sqrt{-4} = \sqrt{4}\sqrt{-1} = 2i$$

$$= 2 \pm i \qquad\qquad \frac{4 \pm 2i}{2} = \frac{2(2 \pm i)}{2} = 2 \pm i$$

The roots $x = 2 + i$ and $x = 2 - i$ both satisfy the equation. Note that the roots are complex conjugates.

Self Check 10 Solve: $x^2 + 3x + 4 = 0$.

6. Factor the Sum of Two Squares

We have seen that if the sum of two squares has no common factor, it cannot be factored over the set of integers. However, it is possible to factor the sum of two squares over the set of complex numbers. For example, to factor $9x^2 + 16y^2$, we proceed as follows:

$$
\begin{aligned}
9x^2 + 16y^2 &= 9x^2 - (-1)16y^2 \\
&= 9x^2 - i^2(16y^2) \qquad i^2 = -1 \\
&= 9x^2 - 16y^2i^2 \\
&= (3x + 4yi)(3x - 4yi) \quad \text{Factor the difference of two squares.}
\end{aligned}
$$

Self Check Answers   **1.** $-\frac{2}{3}$ **2. a.** $2 + 7i$ **b.** $-4 - 2i$ **3.** $16 - 11i$ **4.** $21 + i$ **5.** $\frac{9}{10} + \frac{3}{10}i$ **6.** $\frac{1}{5} + \frac{13}{5}i$ **7.** $-i$ **8.** $\sqrt{29}$ **9.** $\frac{3\sqrt{5}}{5}$ **10.** $-\frac{3}{2} \pm \frac{\sqrt{7}}{2}i$

1.5 Exercises

Vocabulary and Concepts   *Fill in the blanks.*

1. $\sqrt{-3}$, $\sqrt{-9}$ and $\sqrt{-12}$ are examples of _____ numbers.
2. In the complex number $a + bi$, a is the ____ part, and b is the _____ part.
3. If $a = 0$ and $b \neq 0$ in the complex number $a + bi$, the number is an _____ number.
4. If $b = 0$ in the complex number $a + bi$, the number is a ____ number.
5. The complex conjugate of $2 + 5i$ is _____.
6. By definition, $|a + bi| =$ _____.
7. The absolute value of a complex number is a ____ number.
8. The product of two complex conjugates is a ____ number.

Practice   *Find the values of x and y.*

9. $x + (x + y)i = 3 + 8i$
10. $x + 5i = y - yi$
11. $3x - 2yi = 2 + (x + y)i$
12. $\begin{cases} 2 + (x + y)i = 2 - i \\ x + 3i = 2 + 3i \end{cases}$

Perform all operations. Give all answers in $a + bi$ form.

13. $(2 - 7i) + (3 + i)$
14. $(-7 + 2i) + (2 - 8i)$
15. $(5 - 6i) - (7 + 4i)$
16. $(11 + 2i) - (13 - 5i)$

17. $(14i + 2) + (2 - \sqrt{-16})$
18. $\left(5 + \sqrt{-64}\right) - (23i - 32)$
19. $\left(3 + \sqrt{-4}\right) - \left(2 + \sqrt{-9}\right)$
20. $\left(7 - \sqrt{-25}\right) + \left(-8 + \sqrt{-1}\right)$
21. $(2 + 3i)(3 + 5i)$
22. $(5 - 7i)(2 + i)$
23. $(2 + 3i)^2$
24. $(3 - 4i)^2$
25. $\left(11 + \sqrt{-25}\right)\left(2 - \sqrt{-36}\right)$
26. $\left(6 + \sqrt{-49}\right)\left(6 - \sqrt{-49}\right)$
27. $\left(\sqrt{-16} + 3\right)\left(2 + \sqrt{-9}\right)$
28. $\left(12 - \sqrt{-4}\right)\left(-7 + \sqrt{-25}\right)$

29. $\dfrac{1}{i^3}$
30. $\dfrac{3}{i^5}$
31. $\dfrac{-4}{i^{10}}$
32. $\dfrac{-10}{i^{24}}$
33. $\dfrac{1}{2 + i}$
34. $\dfrac{-2}{3 - i}$
35. $\dfrac{2i}{7 + i}$
36. $\dfrac{-3i}{2 + 5i}$
37. $\dfrac{2 + i}{3 - i}$
38. $\dfrac{3 - i}{1 + i}$
39. $\dfrac{4 - 5i}{2 + 3i}$
40. $\dfrac{34 + 2i}{2 - 4i}$
41. $\dfrac{5 - \sqrt{-16}}{-8 + \sqrt{-4}}$
42. $\dfrac{3 - \sqrt{-9}}{2 - \sqrt{-1}}$

43. $\dfrac{2 + i\sqrt{3}}{3 + i}$

44. $\dfrac{3 + i}{4 - i\sqrt{2}}$

Simplify each expression.

45. i^9
46. i^{27}
47. i^{38}
48. i^{99}
49. i^{-6}
50. i^0
51. i^{-10}
52. i^{-31}

Write without absolute value symbols.

53. $|3 + 4i|$
54. $|5 + 12i|$
55. $|2 + 3i|$
56. $|5 - i|$
57. $\left| -7 + \sqrt{-49} \right|$
58. $\left| -2 - \sqrt{-16} \right|$
59. $\left| \dfrac{1}{2} + \dfrac{1}{2}i \right|$
60. $\left| \dfrac{1}{2} - \dfrac{1}{4}i \right|$
61. $|-6i|$
62. $|5i|$
63. $\left| \dfrac{2}{1 + i} \right|$
64. $\left| \dfrac{3}{3 + i} \right|$
65. $\left| \dfrac{-3i}{2 + i} \right|$
66. $\left| \dfrac{5i}{i - 2} \right|$
67. $\left| \dfrac{i + 2}{i - 2} \right|$
68. $\left| \dfrac{2 + i}{2 - i} \right|$

Use the quadratic formula to solve each equation. Simplify all solutions and write them in $a + bi$ form.

69. $x^2 + 2x + 2 = 0$
70. $a^2 + 4a + 8 = 0$
71. $y^2 + 4y + 5 = 0$
72. $x^2 + 2x + 5 = 0$
73. $x^2 - 2x = -5$
74. $z^2 - 3z = -8$
75. $x^2 - \dfrac{2}{3}x = -\dfrac{2}{9}$
76. $x^2 + \dfrac{5}{4} = x$

Factor each expression over the set of complex numbers.

77. $x^2 + 4$
78. $16a^2 + 9$
79. $25p^2 + 36q^2$
80. $100r^2 + 49s^2$
81. $2y^2 + 8z^2$
82. $12b^2 + 75c^2$
83. $50m^2 + 2n^2$
84. $64a^4 + 4b^2$

Applications *In electronics, the formula $V = IR$ is called* Ohm's law. *It gives the relationship in a circuit between the voltage V (in volts), the current I (in amperes), and the resistance R (in ohms).*

85. **Electronics** Find V when $I = 3 - 2i$ amperes and $R = 3 + 6i$ ohms.

86. **Electronics** Find R when $I = 2 - 3i$ amperes and $V = 21 + i$ volts.

87. **Electronics** The impedance Z in an AC (alternating current) circuit is a measure of how much the circuit impedes (hinders) the flow of current through it. The impedance is related to the voltage V and the current I by the following formula.

$$V = IZ$$

If a circuit has a current of $(0.5 + 2.0i)$ amps and an impedance of $(0.4 - 3.0i)$ ohms, find the voltage.

88. **Fractals** Complex numbers are fundamental in the creation of the intricate geometric shape shown below, called a *fractal*. The process of creating this image is based on the following sequence of steps, which begins by picking any complex number, which we will call z.

1. Square z, and then add that result to z.
2. Square the result from Step 1, and then add it to z.
3. Square the result from Step 2, and then add it to z.

If we begin with the complex number i, what is the result after performing Steps 1, 2, and 3?

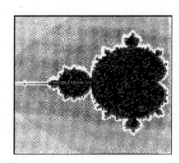

Discovery and Writing

89. Show that the addition of two complex numbers is commutative by adding the complex numbers $a + bi$ and $c + di$ in both orders and observing that the sums are equal.

90. Show that the multiplication of two complex numbers is commutative by multiplying the complex numbers $a + bi$ and $c + di$ in both orders and observing that the products are equal.

91. Show that the addition of complex numbers is associative.

92. Find three examples of complex numbers that are reciprocals of their own conjugates.

93. Explain how to determine whether two complex numbers are equal.

94. Define the complex conjugate of a complex number.

Review *Simplify each expression. Assume that all variables represent nonnegative numbers.*

95. $\sqrt{8x^3}\sqrt{4x}$

96. $\left(\sqrt{x} - 5\right)^2$

97. $\left(\sqrt{x + 1} - 2\right)^2$

98. $\left(-3\sqrt{2x + 1}\right)^2$

99. $\dfrac{4}{\sqrt{5} - 1}$

100. $\dfrac{x - 4}{\sqrt{x} + 2}$

1.6 Polynomial and Radical Equations

Objectives

1. Solve Polynomial Equations by Factoring
2. Solve Other Equations by Factoring
3. Solve Radical Equations

© Greg Vaughn/Alamy

Pikes Peak is in the Rocky Mountains, near Colorado Springs, Colorado. Standing at its summit, a person can see for miles.

The distance a person can see from the peak of the mountain is called the *horizon distance.* If this distance, d, is measured in miles and the height of the observer, h, is measured in feet, d and h are related by the formula $d = \sqrt{1.5h}$.

Since the height of Pike's Peak is approximately 14,000 feet, we can substitute 14,000 for h into the formula and simplify.

$$d = \sqrt{1.5h}$$
$$= \sqrt{1.5(\mathbf{14{,}000})} \quad \text{Substitute 14,000 for } h.$$
$$= \sqrt{21{,}000}$$
$$\approx 144.9137675$$

From the top of Pikes Peak, a person can see about 145 miles.

Since this equation contains a radical, it is called a *radical equation,* one of the topics of the section.

1. Solve Polynomial Equations by Factoring

The equation $ax^2 + bx + c = 0$ is a polynomial equation of second degree, because its left side contains a second-degree polynomial. Many polynomial equations of higher degree can be solved by factoring.

EXAMPLE 1   Solve: **a.** $6x^3 - x^2 - 2x = 0$ **b.** $x^4 - 5x^2 + 4 = 0$

Solution   We will solve each equation by factoring.

a.
$$6x^3 - x^2 - 2x = 0$$
$$x(6x^2 - x - 2) = 0 \qquad \text{Factor out } x.$$
$$x(3x - 2)(2x + 1) = 0 \qquad \text{Factor out } 6x^2 - x - 2.$$

We set each factor equal to 0.

$$x = 0 \quad \text{or} \quad 3x - 2 = 0 \quad \text{or} \quad 2x + 1 = 0$$
$$x = \frac{2}{3} \qquad\qquad x = -\frac{1}{2}$$

Verify that each solution satisfies the original equation.

b.
$$x^4 - 5x^2 + 4 = 0$$
$$(x^2 - 4)(x^2 - 1) = 0 \qquad \text{Factor } x^4 - 5x^2 + 4.$$
$$(x + 2)(x - 2)(x + 1)(x - 1) = 0 \qquad \text{Factor each difference of two squares.}$$

We set each factor equal to 0.

$$x + 2 = 0 \quad \text{or} \quad x - 2 = 0 \quad \text{or} \quad x + 1 = 0 \quad \text{or} \quad x - 1 = 0$$
$$x = -2 \qquad\quad x = 2 \qquad\quad x = -1 \qquad\quad x = 1$$

Verify that each solution satisfies the original equation.

Self Check 1   Solve: $2x^3 + 3x^2 - 2x = 0$.

2. Solve Other Equations by Factoring

To solve another type of equation by factoring, we use a property that states that equal powers of equal numbers are equal.

Power Property of Real Numbers   If a and b are numbers, n is an integer, and $a = b$, then
$$a^n = b^n$$

When we raise both sides of an equation to the same power, the resulting equation might not be equivalent to the original one. For example, if we raise both sides of

(1) $x = 4$ with a solution set of $\{4\}$

to the second power, we obtain

(2) $x^2 = 16$ with a solution set of $\{4, -4\}$

Equations 1 and 2 have different solution sets, and the solution -4 of Equation 2 does not satisfy Equation 1. Because raising both sides of an equation to the same power often introduces **extraneous solutions** (false solutions that don't satisfy the original equation), we must check all suspected roots to be certain that they satisfy the original equation.

The following equation has an extraneous solution.

$$x - x^{1/2} - 6 = 0$$
$$(x^{1/2} - 3)(x^{1/2} + 2) = 0 \qquad\qquad \text{Factor } x - x^{1/2} - 6.$$
$$x^{1/2} - 3 = 0 \quad \text{or} \quad x^{1/2} + 2 = 0 \qquad \text{Set each factor equal to 0.}$$
$$x^{1/2} = 3 \qquad\qquad x^{1/2} = -2$$

Because equal powers of equal numbers are equal, we can square both sides of the previous equations to get

$$(x^{1/2})^2 = (3)^2 \quad \text{or} \quad (x^{1/2})^2 = (-2)^2$$
$$x = 9 \quad \quad \quad \quad x = 4$$

The number 9 satisfies the equation $x - x^{1/2} - 6 = 0$ but 4 does not, as the following check shows:

If x = 9	*If x = 4*
$x - x^{1/2} - 6 = 0$	$x - x^{1/2} - 6 = 0$
$9 - 9^{1/2} - 6 \overset{?}{=} 0$	$4 - 4^{1/2} - 6 \overset{?}{=} 0$
$9 - 3 - 6 \overset{?}{=} 0$	$4 - 2 - 6 \overset{?}{=} 0$
$0 = 0$	$-4 \neq 0$

The number 9 is the only root.

EXAMPLE 2 Solve: **a.** $2x^{2/5} - 5x^{1/5} - 3 = 0$ **b.** $3(3x - 2x^{1/2}) = -1$

Solution We will solve each equation by factoring.

a.
$$2x^{2/5} - 5x^{1/5} - 3 = 0$$
$$(2x^{1/5} + 1)(x^{1/5} - 3) = 0 \qquad \text{Factor } 2x^{2/5} - 5x^{1/5} - 3.$$
$$2x^{1/5} + 1 = 0 \quad \text{or} \quad x^{1/5} - 3 = 0 \qquad \text{Set each factor equal to 0.}$$
$$2x^{1/5} = -1 \quad \quad \quad \quad x^{1/5} = 3$$
$$x^{1/5} = -\frac{1}{2}$$

We can raise both sides of each of the previous equations to the fifth power to obtain

$$(x^{1/5})^5 = \left(-\frac{1}{2}\right)^5 \quad \text{or} \quad (x^{1/5})^5 = (3)^5$$
$$x = -\frac{1}{32} \quad \quad \quad \quad x = 243$$

Verify that each solution satisfies the original equation.

b.
$$3(3x - 2x^{1/2}) = -1$$
$$9x - 6x^{1/2} + 1 = 0 \qquad \text{Remove parentheses and add 1 to both sides.}$$
$$(3x^{1/2} - 1)(3x^{1/2} - 1) = 0 \qquad \text{Factor.}$$
$$3x^{1/2} - 1 = 0 \quad \text{or} \quad 3x^{1/2} - 1 = 0 \qquad \text{Set each factor equal to 0.}$$
$$x^{1/2} = \frac{1}{3} \quad \quad \quad \quad x^{1/2} = \frac{1}{3} \qquad \text{Solve each equation.}$$

We can square both sides of each of the previous equations to obtain

$$(x^{1/2})^2 = \left(\frac{1}{3}\right)^2 \quad \text{or} \quad (x^{1/2})^2 = \left(\frac{1}{3}\right)^2$$
$$x = \frac{1}{9} \quad \quad \quad \quad x = \frac{1}{9}$$

Here, the solutions are the same. Verify that each one satisfies the equation.

Self Check 2 Solve: $x^{2/5} - x^{1/5} - 2 = 0$.

3. Solve Radical Equations

Radical equations are equations containing radicals with variables in the radicand. To solve such equations, we use the power property of real numbers.

EXAMPLE 3   Solve: $\sqrt{x + 3} - 4 = 7$.

Solution   We will isolate the radical on the left side and then square both sides.

$$\sqrt{x + 3} - 4 = 7$$
$$\sqrt{x + 3} = 11 \qquad \text{Add 4 to both sides to isolate the radical.}$$
$$\left(\sqrt{x + 3}\right)^2 = (11)^2 \qquad \text{Square both sides.}$$
$$x + 3 = 121 \qquad \text{Simplify.}$$
$$x = 118 \qquad \text{Subtract 3 from both sides.}$$

Comment

Remember to check all roots when solving radical equations, because raising both sides of an equation to a power can introduce extraneous roots.

Since squaring both sides might introduce extraneous roots, we must check the result of 118.

$$\sqrt{x + 3} - 4 = 7$$
$$\sqrt{118 + 3} - 4 \overset{?}{=} 7 \qquad \text{Substitute 118 for } x.$$
$$\sqrt{121} - 4 \overset{?}{=} 7$$
$$11 - 4 \overset{?}{=} 7$$
$$7 = 7$$

Because it checks, 118 is a root of the equation.

Self Check 3   Solve: $\sqrt{x - 3} + 4 = 7$.

EXAMPLE 4   Solve: $\sqrt{x + 3} = 3x - 1$.

Solution   We will square both sides of the equation to eliminate the radicals and solve the resulting equations by factoring.

$$\sqrt{x + 3} = 3x - 1$$
$$\left(\sqrt{x + 3}\right)^2 = (3x - 1)^2 \qquad \text{Square both sides.}$$
$$x + 3 = 9x^2 - 6x + 1 \qquad \text{Remove parentheses.}$$
$$0 = 9x^2 - 7x - 2 \qquad \text{Add } -x - 3 \text{ to both sides.}$$
$$0 = (9x + 2)(x - 1) \qquad \text{Factor } 9x^2 - 7x - 2.$$
$$9x + 2 = 0 \quad \text{or} \quad x - 1 = 0 \qquad \text{Set each factor equal to 0.}$$
$$x = -\frac{2}{9} \qquad\qquad x = 1$$

Since squaring both sides can introduce extraneous roots, we must check each result.

$$\sqrt{x+3} = 3x - 1 \qquad \text{or} \qquad \sqrt{x+3} = 3x - 1$$

$$\sqrt{-\frac{2}{9} + 3} \overset{?}{=} 3\left(-\frac{2}{9}\right) - 1 \qquad\qquad \sqrt{1+3} \overset{?}{=} 3(1) - 1$$

$$\sqrt{\frac{25}{9}} \overset{?}{=} -\frac{2}{3} - 1 \qquad\qquad \sqrt{4} \overset{?}{=} 3 - 1$$

$$\frac{5}{3} \neq -\frac{5}{3} \qquad\qquad\qquad 2 = 2$$

Since $-\frac{2}{9}$ does not satisfy the equation, it is extraneous. Since 1 checks, it is the only solution.

Self Check 4   Solve: $\sqrt{x-2} = 2x - 10$.

EXAMPLE 5   Solve: $\sqrt[3]{x^3 + 56} = x + 2$.

Solution   To eliminate the radical, we cube both sides of the equation.

$$\sqrt[3]{x^3 + 56} = x + 2$$

$$\left(\sqrt[3]{x^3 + 56}\right)^3 = (x+2)^3 \qquad \text{Cube both sides.}$$

$$x^3 + 56 = x^3 + 6x^2 + 12x + 8 \qquad \text{Remove parentheses.}$$

$$0 = 6x^2 + 12x - 48 \qquad \text{Simplify.}$$

$$0 = x^2 + 2x - 8 \qquad \text{Divide both sides by 6.}$$

$$0 = (x+4)(x-2) \qquad \text{Factor } x^2 + 2x - 8.$$

$$x + 4 = 0 \quad \text{or} \quad x - 2 = 0 \qquad \text{Set each factor equal to 0.}$$

$$x = -4 \qquad\qquad x = 2$$

We check each suspected solution to see whether either is extraneous.

For x = −4	*For x = 2*
$\sqrt[3]{x^3 + 56} = x + 2$	$\sqrt[3]{x^3 + 56} = x + 2$
$\sqrt[3]{(-4)^3 + 56} \overset{?}{=} -4 + 2$	$\sqrt[3]{2^3 + 56} \overset{?}{=} 2 + 2$
$\sqrt[3]{-64 + 56} \overset{?}{=} -2$	$\sqrt[3]{8 + 56} \overset{?}{=} 4$
$\sqrt[3]{-8} \overset{?}{=} -2$	$\sqrt[3]{64} \overset{?}{=} 4$
$-2 = -2$	$4 = 4$

Since both values satisfy the equation, -4 and 2 are roots.

Self Check 5   Solve: $\sqrt[3]{x^3 + 7} = x + 1$.

EXAMPLE 6   Solve: $\sqrt{2x + 3} + \sqrt{x - 2} = 4$.

Solution   We can write the equation in the form

$$\sqrt{2x + 3} = 4 - \sqrt{x - 2} \qquad \text{Subtract } \sqrt{x-2} \text{ from both sides.}$$

so that the left side contains one radical. We then square both sides to get

$$\left(\sqrt{2x + 3}\right)^2 = \left(4 - \sqrt{x - 2}\right)^2$$

$$2x + 3 = 16 - 8\sqrt{x - 2} + x - 2$$

$$2x + 3 = 14 - 8\sqrt{x - 2} + x \qquad \text{Combine like terms.}$$

$$x - 11 = -8\sqrt{x - 2} \qquad \text{Subtract 14 and } x \text{ from both sides.}$$

We then square both sides again to eliminate the radical.

$$(x - 11)^2 = \left(-8\sqrt{x - 2}\right)^2 \qquad \text{Square both sides.}$$

$$x^2 - 22x + 121 = 64(x - 2) \qquad \text{Remove parentheses.}$$

$$x^2 - 22x + 121 = 64x - 128 \qquad \text{Remove parentheses.}$$

$$x^2 - 86x + 249 = 0 \qquad \text{Subtract } 64x, \text{ add 128 to both sides, and combine terms.}$$

$$(x - 3)(x - 83) = 0 \qquad \text{Factor.}$$

$$x - 3 = 0 \quad \text{or} \quad x - 83 = 0 \qquad \text{Set each factor equal to 0.}$$

$$x = 3 \qquad\qquad x = 83$$

Substituting these results into the equation will show that 83 doesn't check; it is extraneous. However, 3 does satisfy the equation and is a root.

Self Check 6 Solve: $\sqrt{2x + 1} + \sqrt{x + 5} = 6$.

Accent on Technology

Highway Engineering

A highway curve banked at 8° will accommodate traffic traveling s mph if the radius of the curve is r feet, according to the formula $s = 1.45\sqrt{r}$. To find what radius is necessary to accommodate 70-mph traffic, highway engineers substitute 70 for s in the formula and solve for r. See Figure 1-8.

$$s = 1.45\sqrt{r}$$

$$70 = 1.45\sqrt{r} \qquad \text{Substitute 70 for } s.$$

$$\frac{70}{1.45} = \sqrt{r} \qquad \text{Divide both sides by 1.45.}$$

$$\left(\frac{70}{1.45}\right)^2 = r \qquad \text{Square both sides.}$$

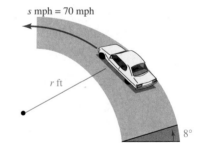

s mph = 70 mph

r ft

8°

Figure 1-8

We can use a calculator to find r by entering these numbers and pressing these keys:

Scientific calculator   *Graphing calculator*

(70 ÷ 1.45) x^2 (70 ÷ 1.45) x^2 ENTER

The display will read 2330.558859 . The radius of the curve should be 2,331 feet.

Self Check Answers   **1.** $0, \frac{1}{2}, -2$ **2.** $-1, 32$ **3.** 12   **4.** 6   **5.** $1, -2$ **6.** 4

1.6   Exercises

Vocabulary and Concepts   *Fill in the blanks.*

1. Equal powers of equal real numbers are _____.
2. If a and b are real numbers and $a = b$, then $a^2 = $ ___.
3. False solutions that don't satisfy the equation are called _____ solutions.
4. _____ equations contain radicals with variables in their _____.

Practice   *Use factoring to solve each equation for real values of the variable.*

5. $x^3 + 9x^2 + 20x = 0$
6. $x^3 + 4x^2 - 21x = 0$
7. $6a^3 - 5a^2 - 4a = 0$
8. $8b^3 - 10b^2 + 3b = 0$
9. $y^4 - 26y^2 + 25 = 0$
10. $y^4 - 13y^2 + 36 = 0$
11. $x^4 - 37x^2 + 36 = 0$
12. $x^4 - 50x^2 + 49 = 0$
13. $2y^4 - 46y^2 = -180$
14. $2x^4 - 102x^2 = -196$
15. $z^{3/2} - z^{1/2} = 0$
16. $r^{5/2} - r^{3/2} = 0$
17. $2m^{2/3} + 3m^{1/3} - 2 = 0$
18. $6t^{2/5} + 11t^{1/5} + 3 = 0$
19. $x - 13x^{1/2} + 12 = 0$
20. $p + p^{1/2} - 20 = 0$
21. $2t^{1/3} + 3t^{1/6} - 2 = 0$
22. $z^3 - 7z^{3/2} - 8 = 0$
23. $6p + p^{1/2} - 1 = 0$
24. $3r - r^{1/2} - 2 = 0$

Find all real solutions of each equation.

25. $\sqrt{x - 2} = 5$
26. $\sqrt{a - 3} - 5 = 0$
27. $3\sqrt{x + 1} = \sqrt{6}$
28. $\sqrt{x + 3} = 2\sqrt{x}$
29. $\sqrt{5a - 2} = \sqrt{a + 6}$
30. $\sqrt{16x + 4} = \sqrt{x + 4}$
31. $2\sqrt{x^2 + 3} = \sqrt{-16x - 3}$
32. $\sqrt{x^2 + 1} = \dfrac{\sqrt{-7x + 11}}{\sqrt{6}}$

33. $\sqrt[3]{7x + 1} = 4$
34. $\sqrt[3]{11a - 40} = 5$
35. $\sqrt[4]{30t + 25} = 5$
36. $\sqrt[4]{3z + 1} = 2$
37. $\sqrt{x^2 + 21} = x + 3$
38. $\sqrt{5 - x^2} = -(x + 1)$
39. $\sqrt{y + 2} = 4 - y$
40. $\sqrt{3z + 1} = z - 1$
41. $x - \sqrt{7x - 12} = 0$
42. $x - \sqrt{4x - 4} = 0$
43. $x + 4 = \sqrt{\dfrac{6x + 6}{5}} + 3$
44. $\sqrt{\dfrac{8x + 43}{3}} - 1 = x$
45. $\sqrt{\dfrac{x^2 - 1}{x - 2}} = 2\sqrt{2}$
46. $\dfrac{\sqrt{x^2 - 1}}{\sqrt{3x - 5}} = \sqrt{2}$
47. $\sqrt[3]{x^3 + 7} = x + 1$
48. $\sqrt[3]{x^3 - 7} + 1 = x$
49. $\sqrt[3]{8x^3 + 61} = 2x + 1$
50. $\sqrt[3]{8x^3 - 37} = 2x - 1$
51. $\sqrt{2p + 1} - 1 = \sqrt{p}$
52. $\sqrt{r} + \sqrt{r + 2} = 2$
53. $\sqrt{x + 3} = \sqrt{2x + 8} - 1$
54. $\sqrt{x + 2} + 1 = \sqrt{2x + 5}$
55. $\sqrt{y + 8} - \sqrt{y - 4} = -2$
56. $\sqrt{z + 5} - 2 = \sqrt{z - 3}$
57. $\sqrt{2b + 3} - \sqrt{b + 1} = \sqrt{b - 2}$
58. $\sqrt{a + 1} + \sqrt{3a} = \sqrt{5a + 1}$
59. $\sqrt{\sqrt{b} + \sqrt{b + 8}} = 2$
60. $\sqrt{\sqrt{x + 19} - \sqrt{x - 2}} = \sqrt{3}$

Applications

61. **Height of a bridge**   The distance d (in feet) that an object will fall in t seconds is given by the following formula. To find the height of a bridge above a river, a man drops a stone into the water (see the illustration). If it takes the stone 5 seconds to hit the water, how high is the bridge?

$$t = \sqrt{\dfrac{d}{16}}$$

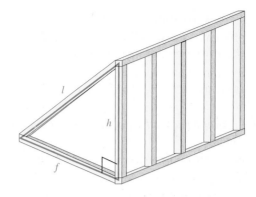

62. Horizon distance The higher a lookout tower, the farther an observer can see. (See the illustration.) The distance d (called the **horizon distance**, measured in miles) is related to the height h of the observer (measured in feet) by the following formula.

$$d = \sqrt{1.5h}$$

How tall must a tower be for the observer to see 30 miles?

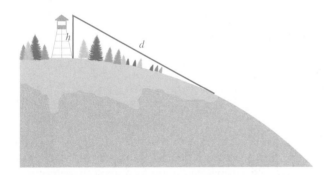

63. Carpentry During construction, carpenters often brace walls, as shown in the illustration. The appropriate length of the brace is given by the following formula.

$$l = \sqrt{f^2 + h^2}$$

If a carpenter nails a 10-foot brace to the wall 6 feet above the floor, how far from the base of the wall should he nail the brace to the floor?

64. Windmills The power generated by a windmill is related to the velocity of the wind by the following formula

$$v = \sqrt[3]{\frac{P}{0.02}}$$

where P is the power (in watts) and v is the velocity of the wind (in mph). To the nearest 10 watts, find the power generated when the velocity of the wind is 31 mph.

65. Diamonds The *effective rate of interest r* earned by an investment is given by the following formula

$$r = \sqrt[n]{\frac{A}{P}} - 1$$

where P is the initial investment that grows to value A after n years. If a diamond buyer got $4,000 for a 1.03-carat diamond that he had purchased 4 years earlier, and earned an annual rate of return of 6.5% on the investment, what did he originally pay for the diamond?

66. Theater productions The ropes, pulleys, and sandbags shown in the illustration are part of a mechanical system used to raise and lower scenery for a stage play. For the scenery to be in the proper position, the following formula must apply:

$$w_2 = \sqrt{w_1^2 + w_3^2}$$

If $w_2 = 12.5$ lb and $w_3 = 7.5$ lb, find w_1.

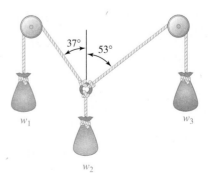

Discovery and Writing

67. Explain why squaring both sides of an equation might introduce extraneous roots.

68. Can cubing both sides of an equation introduce extraneous roots? Explain.

Review *Graph each subset of the real numbers on the number line.*

69. The natural numbers between −4 and 4

70. The integers between −4 and 4

Write each inequality in interval notation and graph the interval.

71. $x \geq 3$

72. $x < -5$

73. $-2 \leq x < 1$

74. $-3 \leq x \leq 3$

75. $x < 1$ or $x \geq 2$

76. $x \leq 1$ or $x > 2$

© Arco Images GmbH/Alamy

1.7 Inequalities

Objectives

1. Understand the Properties of Inequalities
2. Solve Linear Inequalities
3. Solve Compound Inequalities
4. Solve Quadratic Inequalities
5. Solve Rational Inequalities

Studying abroad is a wonderful opportunity for students. Suppose you read about a program to study Spanish in Costa Rica for several weeks during the summer.

The cost of the program includes $900 for round trip airfare plus $350 per week, which covers tuition, meals, and living accommodations. If you have $3,000, how many weeks can you afford to spend in Costa Rica?

If x represents the number of weeks you can spend in Costa Rica, we can write an inequality that represents the cost.

Airfare	plus	cost of $350 per week for x weeks	is less than or equal to	$3,000.
900	+	350x	≤	3,000

To solve this inequality, we can proceed as follows:

$$900 + 350x \leq 3,000$$
$$350x \leq 2,100 \quad \text{Subtract 900 from both sides.}$$
$$x \leq 6 \quad \text{Divide both sides by 350.}$$

Since x can be equal to 6, there is enough money to spend up to 6 weeks in Costa Rica.

In this section, we will review the inequality symbols, learn inequality properties, and learn strategies that can be applied to solve several types of inequalities.

We previously introduced the following symbols.

Symbol	Read as	Examples
$\neq$	"is not equal to"	$8 \neq 10$ and $25 \neq 12$
$<$	"is less than"	$8 < 10$ and $12 < 25$
$>$	"is greater than"	$30 > 10$ and $100 > -5$
$\leq$	"is less than or equal to"	$-6 \leq 12$ and $-8 \leq -8$
$\geq$	"is greater than or equal to"	$12 \geq -5$ and $9 \geq 9$
$\approx$	"is approximately equal to"	$7.49 \approx 7.5$ and $\frac{1}{3} \approx 0.33$

Since the coordinates of points get larger as we move from left to right on the number line,

$a > b$ if point a lies to the right of point b on a number line.

$a < b$ if point a lies to the left of point b on a number line.

EXAMPLE 1

a. $5 > -3$, because 5 lies to the right of -3 on the number line.

b. $-7 < -2$, because -7 lies to the left of -2 on the number line.

c. $3 \leq 3$, because $3 = 3$.

d. $2 \leq 3$, because $2 < 3$.

e. $x + 1 > x$, because $x + 1$ lies one unit to the right of x on the number line.

Self Check 1

Write an inequality symbol to make a true statement:

a. $25 \quad\rule{1cm}{0.4pt}\quad 12$ 　　　b. $-5 \quad\rule{1cm}{0.4pt}\quad -5$

c. $-12 \quad\rule{1cm}{0.4pt}\quad -20$

1. Understand the Properties of Inequalities

The Trichotomy Property　For any real numbers a and b, one of the following statements is true.

$$a < b, \quad a = b, \quad \text{or} \quad a > b$$

The trichotomy property indicates that one of the following statements is true about two real numbers. Either the first is less than the second, or the first is equal to the second, or the first is greater than the second.

The Transitive Property　If a, b, and c are real numbers, then

if $a < b$ and $b < c$, then $a < c$.

if $a > b$ and $b > c$, then $a > c$.

The first part of the transitive property indicates that if a first number is less than a second and the second number is less than a third, then the first number is less than the third.

The second part of the transitive property is similar, with the words "is greater than" substituted for "is less than."

Addition and Subtraction Properties of Inequality

Let a, b, and c represent real numbers.

If $a < b$, then $a + c < b + c$.

If $a < b$, then $a - c < b - c$.

Similar properties exist for $>$, $\leq$, and $\geq$.

This property states that *any real number can be added to (or subtracted from) both sides of an inequality to obtain another inequality with the same order (direction).*

For example, if we add 4 to (or subtract 4 from) both sides of $8 < 12$, we get

$$8 < 12 \qquad\qquad 8 < 12$$
$$8 + 4 < 12 + 4 \qquad 8 - 4 < 12 - 4$$
$$12 < 16 \qquad\qquad 4 < 8$$

and the $<$ symbol is unchanged.

Multiplication and Division Properties of Inequality

Let a, b, and c represent real numbers.

Part 1: If $a < b$ and $c > 0$, then $ca < cb$.

If $a < b$ and $c > 0$, then $\dfrac{a}{c} < \dfrac{b}{c}$.

Part 2: If $a < b$ and $c < 0$, then $ca > cb$.

If $a < b$ and $c < 0$, then $\dfrac{a}{c} > \dfrac{b}{c}$.

Similar properties exist for $>$, $\leq$, and $\geq$.

This property has two parts. Part 1 states that *both sides of an inequality can be multiplied (or divided) by the same positive number to obtain another inequality with the same order.*

For example, if we multiply (or divide) both sides of $8 < 12$ by 4, we get

$$8 < 12 \qquad\qquad 8 < 12$$
$$8(4) < 12(4) \qquad \frac{8}{4} < \frac{12}{4}$$
$$32 < 48 \qquad\qquad 2 < 3$$

and the $<$ symbol is unchanged.

Part 2 states that *both sides of an inequality can be multiplied (or divided) by the same negative number to obtain another inequality with the opposite order.*

For example, if we multiply (or divide) both sides of $8 < 12$ by -4, we get

$$8 < 12 \qquad\qquad 8 < 12$$
$$8(-4) > 12(-4) \qquad \frac{8}{-4} > \frac{12}{-4}$$
$$-32 > -48 \qquad\qquad -2 > -3$$

and the $<$ symbol is changed to a $>$ symbol.

Comment

Unless we are multiplying or dividing by a negative number, the properties of inequality are the same as the properties of equality.

2.   Solve Linear Inequalities

Linear inequalities are inequalities such as $ax + c < 0$ or $ax - c \geq 0$ where $a \neq 0$. Numbers that make an inequality true when substituted for the variable are solutions of the inequality. Inequalities with the same solution set are called **equivalent inequalities.** Because of the previous properties, we can solve inequalities as we do equations. However, we must always remember to change the order of an inequality when multiplying (or dividing) both sides by a negative number.

EXAMPLE 2   Solve: $3(x + 2) < 8$.

Solution   We proceed as with equations.

$$3(x + 2) < 8$$

$3x + 6 < 8$ Remove parentheses.

$3x < 2$ Subtract 6 from both sides.

$x < \dfrac{2}{3}$ Divide both sides by 3.

Figure 1-9

All numbers that are less than $\frac{2}{3}$ are solutions of the inequality. The solution set can be expressed in interval notation as $\left(-\infty, \frac{2}{3}\right)$ and be graphed as in Figure 1-9.

Self Check 2   Solve: $5(p - 4) > 25$.

EXAMPLE 3   Solve: $-5(x - 2) \leq 20 + x$.

Solution   We proceed as with equations.

$$-5(x - 2) \leq 20 + x$$

$-5x + 10 \leq 20 + x$ Remove parentheses.

$-6x + 10 \leq 20$ Subtract x from both sides.

$-6x \leq 10$ Subtract 10 from both sides.

We now divide both sides of the inequality by -6, which changes the order of the inequality.

$x \geq \dfrac{10}{-6}$ Divide both sides by -6.

$x \geq -\dfrac{5}{3}$ Simplify the fraction.

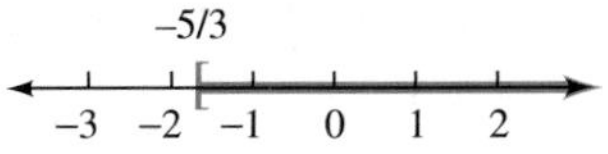

Figure 1-10

The graph of the solution set is shown in Figure 1-10. It is the interval $\left[-\frac{5}{3}, \infty\right)$.

Self Check 3   Solve: $-4(x + 3) \geq 16$.

EXAMPLE 4   An empty truck with driver weighs 4,350 pounds. It is loaded with feed corn weighing 31 pounds per bushel. Between farm and market is a bridge with a 10,000-pound load limit. How many bushels can the truck legally carry?

Solution   The empty truck with driver weighs 4,350 pounds, and corn weighs 31 pounds per bushel. If we let b represent the number of bushels in a legal load, the weight of the corn will be $31b$ pounds. Since the combined weight of the truck, driver, and cargo cannot exceed 10,000 pounds, we can form the following inequality.

The weight of the empty truck with driver	plus	the weight of the corn	must be less than or equal to	10,000 pounds.
4,350	+	31b	≤	10,000

We can solve the inequality as follows.

$$4{,}350 + 31b \le 10{,}000$$
$$31b \le 5{,}650 \qquad \text{Subtract 4,350 from each side.}$$
$$b \le 182.2580645 \qquad \text{Divide both sides by 31.}$$

The truck can legally carry $182\frac{1}{4}$ bushels or less.

3. Solve Compound Inequalities

The statement that x is between 2 and 5 implies two inequalities,

$$x > 2 \quad \text{and} \quad x < 5$$

It is customary to write both inequalities as one **compound inequality:**

$$2 < x < 5 \qquad \text{Read as "2 is less than } x \text{ and } x \text{ is less than 5."}$$

To express that x is not between 2 and 5, we must convey the idea that either x is greater than or equal to 5, or that x is less than or equal to 2. This is equivalent to the statement

$$x \ge 5 \quad \text{or} \quad x \le 2$$

This inequality is satisfied by all numbers x that satisfy one or both of its parts.

Comment

Remember that $2 < x < 5$ means that $x > 2$ and $x < 5$. The word *and* indicates that both inequalities must be true at the same time.

Comment

It is incorrect to write $x \ge 5$ or $x \le 2$ as $2 \ge x \ge 5$ because this would mean that $2 \ge 5$, which is false.

EXAMPLE 5   Solve: $5 < 3x - 7 \le 8$.

Solution   We can isolate x between the inequality symbols by adding 7 to each part of the inequality to get

$$5 + 7 < 3x - 7 + 7 \le 8 + 7 \qquad \text{Add 7 to each part.}$$
$$12 < 3x \le 15 \qquad \text{Do the additions.}$$

and dividing all parts by 3 to get

$$4 < x \le 5$$

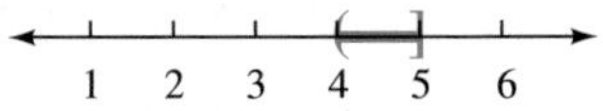

The solution set is the interval (4, 5], whose graph appears in Figure 1-11.

Figure 1-11

Self Check 5   Solve: $-5 \le 2x + 1 < 9$.

EXAMPLE 6   Solve: $3 + x \le 3x + 1 < 7x - 2$.

Solution   Because it is impossible to isolate x between the inequality symbols, we must solve each inequality separately.

$$
\begin{array}{c|c}
3 + x \le 3x + 1 & \text{and} \quad 3x + 1 < 7x - 2 \\
3 \le 2x + 1 & 1 < 4x - 2 \\
2 \le 2x & 3 < 4x \\
1 \le x & \dfrac{3}{4} < x \\
x \ge 1 & x > \dfrac{3}{4}
\end{array}
$$

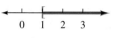

Figure 1-12

Since the connective in this inequality is *and,* the solution set is the intersection (or overlap) of the intervals $[1, \infty)$ and $\left(\frac{3}{4}, \infty\right)$, which is $[1, \infty)$. The graph is shown in Figure 1-12.

Self Check 6   Solve: $x + 1 < 2x - 3 \le 3x - 5$.

It is possible for an inequality to be true for all values of its variable. It is also possible for an inequality to have no solutions. For example,

$x < x + 1$ is true for all numbers x.
$x > x + 1$ is true for no numbers x.

4.   Solve Quadratic Inequalities

If $a \ne 0$, inequalities like $ax^2 + bx + c < 0$ and $ax^2 + bx + c > 0$ are called **quadratic inequalities.** We will begin by discussing two methods for solving the quadratic inequality $x^2 - x - 6 > 0$.

EXAMPLE 7   Solve: $x^2 - x - 6 > 0$.

Method 1   First we solve the equation $x^2 - x - 6 = 0$.

$$x^2 - x - 6 = 0$$
$$(x + 2)(x - 3) = 0$$
$$x + 2 = 0 \quad \text{or} \quad x - 3 = 0$$
$$x = -2 \quad | \quad x = 3$$

Figure 1-13

The graphs of these solutions establish the three intervals shown in Figure 1-13. To determine which intervals are solutions, we test a number in each interval and see whether it satisfies the inequality.

Interval	Test value	Inequality $x^2 - x - 6 > 0$	Result
$(-\infty, -2)$	-6	$(-6)^2 - (-6) - 6 \overset{?}{>} 0$ $36 > 0$ true	The numbers in this interval are solutions.
$(-2, 3)$	0	$0^2 - 0 - 6 \overset{?}{>} 0$ $-6 > 0$ false	The numbers in this interval are not solutions.
$(3, \infty)$	5	$5^2 - 5 - 6 \overset{?}{>} 0$ $14 > 0$ true	The numbers in this interval are solutions.

Figure 1-14

The solutions are in the intervals $(-\infty, -2)$ or $(3, \infty)$, as shown in Figure 1-14.

Method 2   A second method relies on the number line and a notation that keeps track of the signs of the factors of $x^2 - x - 6$, which are $(x - 3)(x + 2)$. (See Figure 1-15 on the next page.) First, we consider the factor $x - 3$.

If $x = 3$, then $x - 3 = 0$.
If $x < 3$, then $x - 3$ is negative.
If $x > 3$, then $x - 3$ is positive.

Then we consider the factor $x + 2$.

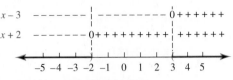

If $x = -2$, then $x + 2 = 0$.

If $x < -2$, then $x + 2$ is negative.

If $x > -2$, then $x + 2$ is positive.

Figure 1-15

We place this information on the **sign graph** shown in Figure 1-15 by using $+$ and $-$ signs. Only to the left of -2 and to the right of 3 do the signs of both factors agree. Only there is the product positive.

Self Check 7   Solve: $x(x + 1) - 6 \geq 0$.

EXAMPLE 8   Solve: $x(x + 3) < -2$.

We remove parentheses and add 2 to both sides to make the right side of the equation equal to 0 and solve $x^2 + 3x + 2 < 0$.

Method 1   First we solve the equation $x^2 + 3x + 2 = 0$.

$$x^2 + 3x + 2 = 0$$
$$(x + 2)(x + 1) = 0$$
$$x + 2 = 0 \quad \text{or} \quad x + 1 = 0$$
$$x = -2 \quad\quad\quad x = -1$$

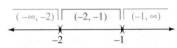

Figure 1-16

These solutions establish the intervals shown in Figure 1-16.

The solutions of $x^2 + 3x + 2 < 0$ will be the numbers in one or more of these intervals. To determine which intervals are solutions, we test a number in each interval to see whether it satisfies the inequality.

Interval	Test value	Inequality $x^2 + 3x + 2 < 0$	Result
$(-\infty, -2)$	-7	$(-7)^2 + 3(-7) + 2 \overset{?}{<} 0$ $30 < 0$ **false**	The numbers in this interval are not solutions.
$(-2, -1)$	$-\dfrac{3}{2}$	$\left(-\dfrac{3}{2}\right)^2 + 3\left(-\dfrac{3}{2}\right) + 2 \overset{?}{<} 0$ $-\dfrac{1}{4} < 0$ **true**	The numbers in this interval are solutions.
$(-1, \infty)$	0	$0^2 + 0 + 2 \overset{?}{<} 0$ $2 < 0$ **false**	The numbers in this interval are not solutions.

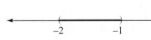

Figure 1-17

The solution is the interval $(-2, -1)$ whose graph appears in Figure 1-17.

Method 2   We construct the sign graph shown in Figure 1-18. Only between -2 and -1 do the factors have opposite signs. Here, the product is negative.

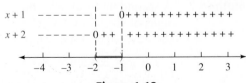

Figure 1-18

Self Check 8   Solve: $x^2 - 5x - 6 < 0$.

EXAMPLE 9   Solve: $x^2 - 5 \geq 0$.

Solution   This equation will not factor using only integers. However, we can solve it by adding 5 to both sides and using the square-root method.

$$x^2 - 5 = 0$$
$$x^2 - 5 + 5 = 0 + 5 \quad \text{Add 5 to both sides.}$$
$$x^2 = 5$$
$$x = \sqrt{5} \quad \text{or} \quad x = -\sqrt{5}$$

These solutions establish the intervals shown in the table and Figure 1-19. To decide which ones are solutions, we test a number in each interval to see whether it is a solution.

Interval	Test value	Inequality $x^2 - 5 \geq 0$	Result
$(-\infty, -\sqrt{5})$	-3	$(-3)^2 - 5 \overset{?}{\geq} 0$ $4 \geq 0$ true	The numbers in this interval are solutions.
$\left(-\sqrt{5}, \sqrt{5}\right)$	0	$(0)^2 - 5 \overset{?}{\geq} 0$ $-5 \geq 0$ false	The numbers in this interval are not solutions.
$\left(\sqrt{5}, \infty\right)$	3	$(3)^2 - 5 \overset{?}{\geq} 0$ $4 \geq 0$ true	The numbers in this interval are solutions.

Figure 1-19

As shown in Figure 1-19, the solution is the union of two intervals: $\left(-\infty, -\sqrt{5}\right] \cup \left[\sqrt{5}, \infty\right)$.

Self Check 9   Solve: $x^2 - 7 < 0$.

4.   Solve Rational Inequalities

Inequalities that contain fractions with polynomial numerators and denominators are called **rational inequalities.** To solve them, we can use the same techniques that we use to solve quadratic inequalities.

EXAMPLE 10   Solve the rational inequality: $\dfrac{x^2 - x - 2}{x^2 - 4x + 3} \leq 0$.

Method 1   The intervals are found by solving $x^2 - x - 2 = 0$ and $x^2 - 4x + 3 = 0$. The solutions of the first equation are -1 and 2, and the solutions of the second equation are 1 and 3. These solutions establish the five intervals shown in Figure 1-20.

Figure 1-20

The solutions of $\dfrac{x^2 - x - 2}{x^2 - 4x + 3} \leq 0$ will be the numbers in one or more of these intervals. To determine which intervals are solutions, we test a number in each interval to see whether it satisfies the inequality.

Interval	Test value	Inequality $\dfrac{x^2 - x - 2}{x^2 - 4x + 3} \le 0$	Result
$(-\infty, -1)$	-2	$\dfrac{(-2)^2 - (-2) - 2}{(-2)^2 - 4(-2) + 3} \le 0$ $\dfrac{4}{15} \le 0$ false	The numbers in this interval are not solutions.
$(-1, 1)$	0	$\dfrac{(0)^2 - (0) - 2}{(0)^2 - 4(0) + 3} \le 0$ $-\dfrac{2}{3} \le 0$ true	The numbers in this interval are solutions.
$(1, 2)$	1.5	$\dfrac{(1.5)^2 - (1.5) - 2}{(1.5)^2 - 4(1.5) + 3} \le 0$ $\dfrac{5}{3} \le 0$ false	The numbers in this interval are not solutions.
$(2, 3)$	2.5	$\dfrac{(2.5)^2 - (2.5) - 2}{(2.5)^2 - 4(2.5) + 3} \le 0$ $-\dfrac{7}{3} \le 0$ true	The numbers in this interval are solutions.
$(3, \infty)$	4	$\dfrac{(4)^2 - (4) - 2}{(4)^2 - 4(4) + 3} \le 0$ $\dfrac{10}{3} \le 0$ false	The numbers in this interval are not solutions.

The numbers in the intervals $(-1, 1)$ and $(2, 3)$ satisfy the inequality, but numbers in the intervals $(-\infty, -1)$, $(1, 2)$, and $(3, \infty)$ do not. The graph of the solution set is shown in Figure 1-21. The solution set is $[-1, 1) \cup [2, 3)$.

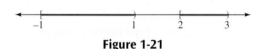

Figure 1-21

Because $x = -1$ and $x = 2$ make the numerator 0, they satisfy the inequality. Thus, their graphs are drawn with brackets to show that -1 and 2 are included. Because 1 and 3 give 0's in the denominator, the parentheses at $x = 1$ and $x = 3$ show that 1 and 3 are not in the solution set.

Method 2   We factor each trinomial and write the inequality in the form

$$\frac{(x - 2)(x + 1)}{(x - 3)(x - 1)} \le 0$$

We then construct the sign graph shown in Figure 1-22. The value of the fraction will be 0 when $x = 2$ and $x = -1$. The value will be negative when there is an odd number of negative factors. This happens between -1 and 1 and between 2 and 3.

The graph of the solution set also appears in Figure 1-22. The brackets at -1 and 2 show that these numbers are in the solution set. The parentheses at 1 and 3 show that these numbers are not in the solution set.

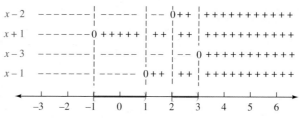

Figure 1-22

Self Check 10   Solve: $\dfrac{x^2 + 2x - 3}{x^2 + 4x + 3} > 0$.

EXAMPLE 11   Solve: $\dfrac{6}{x} > 2$.

Solution   To get a 0 on the right side, we subtract 2 from both sides. We then combine like terms on the left side.

$$\frac{6}{x} > 2$$

$$\frac{6}{x} - 2 > 0 \qquad \text{Subtract 2 from both sides.}$$

$$\frac{6}{x} - \frac{2x}{x} > 0 \qquad \frac{x}{x} = 1$$

$$\frac{6 - 2x}{x} > 0 \qquad \text{Add the numerators and keep the common denominator.}$$

The inequality now has the form of a rational inequality. The intervals are found by solving $6 - 2x = 0$ and $x = 0$. The solution of the first equation is 3, and the solution of the second equation is 0. This determines the intervals $(-\infty, 0)$, $(0, 3)$, and $(3, \infty)$. Because only the numbers in the interval $(0, 3)$ satisfy the original inequality, the solution set is $(0, 3)$. The graph is shown in Figure 1-23.

We could construct a sign graph as in Figure 1-24 and obtain the same solution set.

Figure 1-23

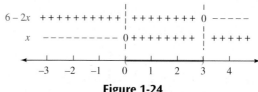

Figure 1-24

Self Check 11   Solve: $\dfrac{2}{x} < 4$.

Comment

It is tempting to solve Example 11 by multiplying both sides by x and solving the inequality $6 > 2x$. However, multiplying both sides by x gives $6 > 2x$ only when x is positive. If x is negative, multiplying both sides by x will reverse the direction of the $>$ symbol, and the inequality $\dfrac{6}{x} > 2$ will be equivalent to $6 < 2x$. If you fail to consider both cases, you will get a wrong answer.

1. a. $>$ or $\geq$ **b.** $\leq$ or $\geq$ **c.** $>$ or $\geq$ **2.** $(9, \infty)$ ⟵————(——→
                                                                                                            9

3. $(-\infty, -7]$ ⟵————┤————→   **5.** $[-3, 4)$ ⟵—┤———┤→
                        -7 -3 4

6. $(4, \infty)$ ⟵———(———→   **7.** $(-\infty, -3] \cup [2, \infty)$ ⟵—┤———┤—→
                  4 -3 2

8. $(-1, 6)$ ⟵—(———)——→   **9.** $\left(-\sqrt{7}, \sqrt{7}\right)$ ⟵——(———)——→
              -1 6 $-\sqrt{7}$ $\sqrt{7}$

10. $(-\infty, -3) \cup (-3, -1) \cup (1, \infty)$ ⟵——)(——)——(——→
                                              -3 -1 1

11. $(-\infty, 0) \cup \left(\frac{1}{2}, \infty\right)$ ⟵———)——(———→
                                        0   1/2

1.7 Exercises

Vocabulary and Concepts   *Fill in the blanks. Assume that all variables represent real numbers.*

1. If $x > y$, then x lies to the _____ of y on a number line.

2. $a < b$, _____, or $a > b$.

3. If $a < b$ and $b < c$, then _____.

4. If $a < b$, then $a + c <$ _____.

5. If $a < b$, then $a - c <$ _____.

6. If $a < b$ and $c > 0$, then ac __ bc.

7. If $a < b$ and $c < 0$, then ac __ bc.

8. If $a < b$ and $c < 0$, then $\frac{a}{c}$ __ $\frac{b}{c}$.

9. $3x - 5 < 12$ and $ax + c > 0$ $(a \neq 0)$ are examples of _____ inequalities.

10. $ax^2 + bx - c \geq 0$ and $3x^2 - 6x < 0$ are examples of _____ inequalities.

11. If two inequalities have the same solution set, they are called _____ inequalities.

12. An inequality that contains a fraction with a polynomial numerator and denominator is called a _____ inequality.

Practice   *Solve each inequality, graph the solution set, and write the answer in interval notation. Do not worry about drawing your graphs exactly to scale.*

13. $3x + 2 < 5$

14. $-2x + 4 < 6$

15. $3x + 2 \geq 5$

16. $-2x + 4 \geq 6$

17. $-5x + 3 > -2$

18. $4x - 3 > -4$

19. $-5x + 3 \leq -2$

20. $4x - 3 \leq -4$

21. $2(x - 3) \leq -2(x - 3)$

22. $3(x + 2) \leq 2(x + 5)$

23. $\frac{3}{5}x + 4 > 2$

24. $\frac{1}{4}x - 3 > 5$

25. $\frac{x + 3}{4} < \frac{2x - 4}{3}$

26. $\frac{x + 2}{5} > \frac{x - 1}{2}$

27. $\frac{6(x - 4)}{5} \geq \frac{3(x + 2)}{4}$

28. $\frac{3(x + 3)}{2} < \frac{2(x + 7)}{3}$

29. $\frac{5}{9}(a + 3) - a \geq \frac{4}{3}(a - 3) - 1$

30. $\frac{2}{3}y - y \leq -\frac{3}{2}(y - 5)$

31. $\frac{2}{3}a - \frac{3}{4}a < \frac{3}{5}\left(a + \frac{2}{3}\right) + \frac{1}{3}$

32. $\dfrac{1}{4}b + \dfrac{2}{3}b - \dfrac{1}{2} > \dfrac{1}{2}(b + 1) + b$

33. $4 < 2x - 8 \le 10$ **34.** $3 \le 2x + 2 < 6$

35. $9 \ge \dfrac{x - 4}{2} > 2$ **36.** $5 < \dfrac{x - 2}{6} < 6$

37. $0 \le \dfrac{4 - x}{3} \le 5$ **38.** $0 \ge \dfrac{5 - x}{2} \ge -10$

39. $-2 \ge \dfrac{1 - x}{2} \ge -10$ **40.** $-2 \le \dfrac{1 - x}{2} < 10$

41. $-3x > -2x > -x$ **42.** $-3x < -2x < -x$

43. $x < 2x < 3x$ **44.** $x > 2x > 3x$

45. $2x + 1 < 3x - 2 < 12$

46. $2 - x < 3x + 5 < 18$

47. $2 + x < 3x - 2 < 5x + 2$

48. $x > 2x + 3 > 4x - 7$

49. $3 + x > 7x - 2 > 5x - 10$

50. $2 - x < 3x + 1 < 10x$

51. $x \le x + 1 \le 2x + 3$

52. $-x \ge -2x + 1 \ge -3x + 1$

53. $x^2 + 7x + 12 < 0$ **54.** $x^2 - 13x + 12 \le 0$

55. $x^2 - 5x + 6 \ge 0$ **56.** $6x^2 + 5x - 6 > 0$

57. $x^2 + 5x + 6 < 0$ **58.** $x^2 + 9x + 20 \ge 0$

59. $6x^2 + 5x + 1 \ge 0$ **60.** $x^2 + 9x + 20 < 0$

61. $6x^2 - 5x < -1$ **62.** $9x^2 + 24x > -16$

63. $2x^2 \ge 3 - x$ **64.** $9x^2 \le 24x - 16$

65. $x^2 - 3 \ge 0$ **66.** $x^2 - 7 \le 0$

67. $x^2 - 11 < 0$ **68.** $x^2 - 20 > 0$

69. $\dfrac{x + 3}{x - 2} < 0$ **70.** $\dfrac{x + 3}{x - 2} > 0$

71. $\dfrac{x^2 + x}{x^2 - 1} > 0$ **72.** $\dfrac{x^2 - 4}{x^2 - 9} < 0$

73. $\dfrac{x^2 + 5x + 6}{x^2 + x - 6} \ge 0$ **74.** $\dfrac{x^2 + 10x + 25}{x^2 - x - 12} \le 0$

75. $\dfrac{6x^2 - x - 1}{x^2 + 4x + 4} > 0$ **76.** $\dfrac{6x^2 - 3x - 3}{x^2 - 2x - 8} < 0$

77. $\dfrac{3}{x} > 2$

78. $\dfrac{3}{x} < 2$

79. $\dfrac{6}{x} < 4$

80. $\dfrac{6}{x} > 4$

81. $\dfrac{3}{x - 2} \le 5$

82. $\dfrac{3}{x + 2} \le 4$

83. $\dfrac{6}{x^2 - 1} < 1$

84. $\dfrac{6}{x^2 - 1} > 1$

Applications *Solve each problem.*

85. Long distance A long-distance telephone call costs 36¢ for the first three minutes and 11¢ for each additional minute. How long can a person talk for less than $2?

86. Buying a computer A student who can afford to spend up to $2,000 sees the ad shown in the illustration. If she buys a computer, how many games can she buy?

Big Sale!!!!

$1,695.95

Games $19.95

87. Buying CDs Andy can spend up to $275 on a CD player and some CDs. If he can buy a disk player for $150 and disks for $9.75, what is the greatest number of disks that he can buy?

88. Buying DVDs Mary wants to spend less than $600 for a DVD recorder and some DVDs. If the recorder of her choice costs $425 and DVDs cost $7.50 each, how many DVDs can she buy?

89. Buying a refrigerator A woman who has $1,200 to spend wants to buy a refrigerator. Refer to the following table and write an inequality that shows how much she can pay for the refrigerator.

State sales tax	6.5%
City sales tax	0.25%

90. Renting a rototiller The cost of renting a rototiller is $17.50 for the first hour and $8.95 for each additional hour. How long can a person have the rototiller if the cost must be less than $75?

91. Real estate taxes A city council has proposed the following two methods of taxing real estate:

Method 1	$2,200 + 4% of assessed value
Method 2	$1,200 + 6% of assessed value

For what range of assessments *a* would the first method benefit the taxpayer?

92. Medical plans A college provides its employees with a choice of the two medical plans shown in the following table. For what size hospital bills is Plan 2 better for the employee than Plan 1? (*Hint:* The cost to the employee includes both the deductible payment and the employee's coinsurance payment.)

Plan 1	Plan 2
Employee pays $100	Employee pays $200
Plan pays 70% of the rest	Plan pays 80% of the rest

93. Medical plans To save costs, the college in Exercise 92 raised the employee deductible, as shown in the following table. For what size hospital bills is Plan 2 better for the employee than Plan 1? (*Hint:* The cost to the employee includes both the deductible payment and the employee's coinsurance payment.)

Plan 1	Plan 2
Employee pays $200	Employee pays $400
Plan pays 70% of the rest	Plan pays 80% of the rest

94. Geometry The perimeter of a rectangle is to be between 180 inches and 200 inches. Find the range of values for its length when its width is 40 inches.

95. Geometry The perimeter of an equilateral triangle is to be between 50 centimeters and 60 centimeters. Find the range of lengths of one side.

96. Geometry The perimeter of a square is to be from 25 meters to 60 meters. Find the range of values for its area.

Discovery and Writing

97. Express the relationship $20 < l < 30$ in terms of P, where $P = 2l + 2w$.

98. Express the relationship $10 < C < 20$ in terms of F, where $F = \frac{9}{5}C + 32$.

99. The techniques used for solving linear equations and linear inequalities are similar, yet different. Explain.

100. Explain why the relation $\geq$ is transitive.

Review *In Exercises 101–106,*
$$A = \left\{-9, -\pi, -2, -\frac{1}{2}, 0, 1, 2, \sqrt{7}, \frac{21}{2}\right\}.$$

101. Which numbers are even integers?

102. Which numbers are natural numbers?

103. Which numbers are prime numbers?

104. Which numbers are irrational numbers?

105. Which numbers are real numbers?

106. Which numbers are rational numbers?

1.8 Absolute Value

Objectives

1. Review Absolute Value
2. Solve Equations of the Form $|x| = k$
3. Solve Equations with Two Absolute Values
4. Solve Inequalities of the Form $|x| < k$
5. Solve Inequalities of the Form $|x| > k$
6. Solve Inequalities with Two Absolute Values

Cancun, Mexico is a popular spring break destination for college students. The beautiful beaches, exciting nightlife, and tropical weather make it a pleasurable experience.

The average annual temperature in Cancun is 78° F with fluctuations of approximately 7 degrees. We can represent this temperature range using absolute value notation. If we let x represent the temperature at a given time, the absolute value of the difference between x and 78 is less than or equal to 7. We can write this as $|x - 78| \leq 7$.

In this section, we will review absolute value and examine its consequences in greater detail.

1. Review Absolute Value

Absolute Value   The **absolute value** of the real number x, denoted by $|x|$, is defined as follows:

If $x \geq 0$, then $|x| = x$.
If $x < 0$, then $|x| = -x$.

This definition provides a way to associate a nonnegative real number with any real number.

- If $x \geq 0$, then x (which is positive or 0) is its own absolute value.
- If $x < 0$, then $-x$ (which is positive) is the absolute value.

Either way, $|x|$ is positive or 0:

$$|x| \geq 0 \quad \text{for all real numbers } x$$

EXAMPLE 1 Write each expression without using absolute value symbols:
a. $|7|$ **b.** $|-3|$ **c.** $-|-7|$ **d.** $|x - 2|$

Solution In each case, we will apply the definition of absolute value.

a. Because 7 is positive, $|7| = 7$.
b. Because -3 is negative, $|-3| = -(-3) = 3$.
c. The expression $-|-7|$ means "the negative of the absolute value of -7." Thus, $-|-7| = -(7) = -7$.
d. To denote the absolute value of a variable quantity, we must give a conditional answer.

$$\text{If } x - 2 \geq 0, \text{ then } |x - 2| = x - 2.$$
$$\text{If } x - 2 < 0, \text{ then } |x - 2| = -(x - 2) = -x + 2 = 2 - x.$$

Self Check 1 Write each expression without using absolute value symbols:
a. $|0|$ **b.** $|-17|$
c. $|x + 5|$

2. Solve Equations of the Form $|x| = k$

In the equation $|x| = 8$, x can be either 8 or -8, because $|8| = 8$ and $|-8| = 8$. In general, the following is true.

Absolute Value Equations If $k \geq 0$, then

$$|x| = k \quad \text{is equivalent to} \quad x = k \quad \text{or} \quad x = -k$$

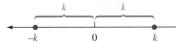

Figure 1-25

Comment

If $k < 0$, then $|x| = k$ has no solution.

The absolute value of a number represents the distance on the number line from a point to the origin. The solutions of $|x|$ are the coordinates of the two points that lie exactly k units from the origin. (See Figure 1-25.)

The equation $|x - 3| = 7$ indicates that a point on the number line with a coordinate of $x - 3$ is 7 units from the origin. Thus, $|x - 3|$ can be 7 or -7.

$$\begin{array}{lcl} x - 3 = 7 & \text{or} & x - 3 = -7 \\ x = 10 & & x = -4 \end{array}$$

The solutions of 10 and -4 are shown in Figure 1-26. Both of these numbers satisfy the equation.

$$\begin{array}{ccc} |x - 3| = 7 & \text{and} & |x - 3| = 7 \\ |10 - 3| = 7 & & |-4 - 3| = 7 \\ |7| = 7 & & |-7| = 7 \\ 7 = 7 & & 7 = 7 \end{array}$$

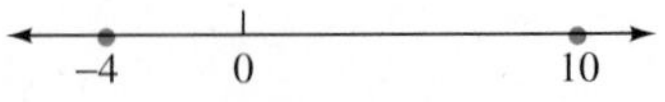

Figure 1-26

EXAMPLE 2   Solve: $|3x - 5| = 7$.

Solution   The equation $|3x - 5| = 7$ is equivalent to two equations

$$3x - 5 = 7 \qquad \text{or} \qquad 3x - 5 = -7$$

which can be solved separately:

$$
\begin{array}{c|c}
3x - 5 = 7 \quad \text{or} & 3x - 5 = -7 \\
3x = 12 & 3x = -2 \\
x = 4 & x = -\dfrac{2}{3}
\end{array}
$$

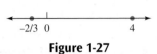

Figure 1-27

The solution set consists of the points shown in Figure 1-27.

Self Check 2   Solve: $|2x + 3| = 7$.

3.   Solve Equations with Two Absolute Values

The equation $|a| = |b|$ is true when $a = b$ or when $a = -b$. For example,

$$
\begin{array}{ccc}
|3| = |3| & \text{or} & |3| = |-3| \\
3 = 3 & & 3 = 3
\end{array}
$$

In general, the following is true.

Equations with Two Absolute Values	If a and b represent algebraic expressions, the equation $|a| = |b|$ is equivalent to $$a = b \qquad \text{or} \qquad a = -b$$

EXAMPLE 3   Solve: $|2x| = |x - 3|$.

Solution   The equation $|2x| = |x - 3|$ will be true when $2x$ and $x - 3$ are equal or when they are negatives. This gives two equations, which can be solved separately:

$$
\begin{array}{c|c}
2x = x - 3 \quad \text{or} & 2x = -(x - 3) \\
x = -3 & 2x = -x + 3 \\
 & 3x = 3 \\
 & x = 1
\end{array}
$$

Verify that -3 and 1 satisfy the equation.

Self Check 3   Solve: $|3x + 1| = |5x - 3|$.

4.   Solve Inequalities of the Form $|x| < k$

The inequality $|x| < 5$ indicates that a point with coordinate x is less than 5 units from the origin. (See Figure 1-28 on the next page.) Thus, x is between -5 and 5, and

$$|x| < 5 \qquad \text{is equivalent to} \qquad -5 < x < 5$$

In general, the inequality $|x| < k$ $(k > 0)$ indicates that a point with coordinate x is less than k units from the origin. (See Figure 1-29 on the next page.)

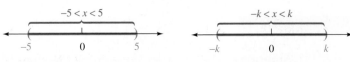

Figure 1-28   **Figure 1-29**

Inequalities of the Form $\|x\| < k$	If $k > 0$, then $\|x\| < k$	is equivalent to	$-k < x < k$.
	If $k > 0$, $\|x\| \le k$	is equivalent to	$-k \le x \le k$.

EXAMPLE 4   Solve: $\|x - 2\| < 7$.

Solution   The inequality $\|x - 2\| < 7$ is equivalent to

$$-7 < x - 2 < 7$$

We can add 2 to each part of this inequality to get

$$-5 < x < 9$$

The solution set is the interval $(-5, 9)$, shown in Figure 1-30.

Figure 1-30

Self Check 4   Solve: $\|x + 3\| < 9$.

5.   Solve Inequalities of the Form $\|x\| > k$

The inequality $\|x\| > 5$ indicates that a point with coordinate x is more than 5 units from the origin. (See Figure 1-31.) Thus, $x < -5$ or $x > 5$.

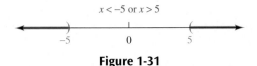

Figure 1-31

In general, the inequality $\|x\| > k$ $(k > 0)$ indicates that a point with coordinate x is more than k units from the origin. (See Figure 1-32.)

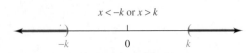

Figure 1-32

Inequalities of the Form $\|x\| > k$	If $k > 0$, then		
	$\|x\| > k$	is equivalent to	$x < -k$ or $x > k$
	$\|x\| \ge k$	is equivalent to	$x \le -k$ or $x \ge k$

EXAMPLE 5   Solve: $\left| \dfrac{2x + 3}{2} \right| + 7 \ge 12$.

Solution   We begin by subtracting 7 from both sides of the inequality to isolate the absolute value on the left side.

$$\left|\frac{2x+3}{2}\right| \geq 5$$

This result is equivalent to two inequalities that can be solved separately.

$$\begin{array}{rcl|rcl} \dfrac{2x+3}{2} &\leq& -5 \quad \text{or} & \dfrac{2x+3}{2} &\geq& 5 \\ 2x+3 &\leq& -10 & 2x+3 &\geq& 10 \\ 2x &\leq& -13 & 2x &\geq& 7 \\ x &\leq& -\dfrac{13}{2} & x &\geq& \dfrac{7}{2} \end{array}$$

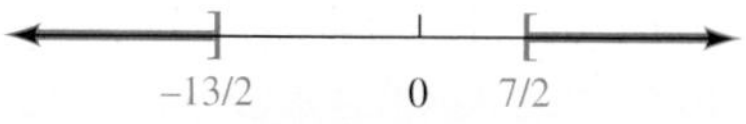

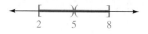

Figure 1-33

The solution set is the union of the intervals $\left(-\infty, -\frac{13}{2}\right]$ and $\left[\frac{7}{2}, \infty\right)$. Its graph appears in Figure 1-33.

Self Check 5 Solve: $\left|\dfrac{3x-6}{3}\right| + 2 \geq 12.$

EXAMPLE 6 Solve: $0 < |x-5| \leq 3$.

Solution The inequality $0 < |x-5| \leq 3$ consists of two inequalities that can be solved separately. The solution will be the intersection of the inequalities

$$0 < |x-5| \qquad \text{and} \qquad |x-5| \leq 3$$

The inequality $0 < |x-5|$ is true for all x except 5. The inequality $|x-5| \leq 3$ is equivalent to the inequality

$$-3 \leq x-5 \leq 3$$
$$2 \leq x \leq 8 \qquad \text{Add 5 to each part.}$$

Figure 1-34

The solution set is the intersection of these two solutions, which is the interval [2, 8], except 5. This is the union of the intervals [2, 5) and (5, 8], as shown in Figure 1-34.

Self Check 6 Solve: $0 < |x+2| \leq 5$.

5. Solve Inequalities with Two Absolute Values

In Example 9 of Section 1.5, we saw that $|a|$ could be defined as

$$|a| = \sqrt{a^2}$$

We will use this fact in the next example.

EXAMPLE 7 Solve $|x+2| > |x+1|$ and give the result in interval notation.

Solution

$$\begin{array}{rcll} |x+2| &>& |x+1| & \\ \sqrt{(x+2)^2} &>& \sqrt{(x+1)^2} & \text{Use } |a| = \sqrt{a^2}. \\ (x+2)^2 &>& (x+1)^2 & \text{Square both sides.} \\ x^2+4x+4 &>& x^2+2x+1 & \text{Expand each binomial.} \\ 4x &>& 2x-3 & \text{Subtract } x^2 \text{ and 4 from both sides.} \\ 2x &>& -3 & \text{Subtract } 2x \text{ from both sides.} \\ x &>& -\dfrac{3}{2} & \text{Divide both sides by 2.} \end{array}$$

The solution set is the interval $\left(-\frac{3}{2}, \infty\right)$. Check several numbers in this interval to verify that this interval is the solution.

Self Check 7 Solve $|x - 3| \leq |x + 2|$ and give the result in interval notation.

Three other properties of absolute value are sometimes useful.

Properties of Absolute Value

If a and b are real numbers, then

1. $|ab| = |a||b|$ **2.** $\left|\dfrac{a}{b}\right| = \dfrac{|a|}{|b|}$ $(b \neq 0)$ **3.** $|a + b| \leq |a| + |b|$

Properties 1 and 2 above indicate that the absolute value of a product (or a quotient) is the product (or the quotient) of the absolute values.

Property 3 indicates that the absolute value of a sum is either equal to or less than the sum of the absolute values.

Self Check Answers

1. a. 0   **b.** 17   **c.** if $x + 5 \geq 0$, $|x + 5| = x + 5$; if $x + 5 < 0$,
$|x + 5| = -x - 5$ **2.** $-5, 2$ **3.** $2, \frac{1}{4}$

4. $(-12, 6)$

5. $(-\infty, -8] \cup [12, \infty)$

6. $[-7, -2) \cup (-2, 3]$ **7.** $\left[\frac{1}{2}, \infty\right)$

1.8 Exercises

Vocabulary and Concepts   *Fill in the blanks.*

1. If $x \geq 0$, then $|x| = \_\_$.

2. If $x < 0$, then $|x| = \_\_\_$.

3. $|x| = k$ is equivalent to _____.

4. $|a| = |b|$ is equivalent to $a = b$ or _____.

5. $|x| < k$ is equivalent to _____.

6. $|x| > k$ is equivalent to _____.

7. $|x| \geq k$ is equivalent to _____.

8. $\sqrt{a^2} = \_\_\_$.

Practice   *Write each expression without absolute value symbols.*

9. $|7|$ **10.** $|-9|$

11. $|0|$ **12.** $|3 - 5|$

13. $|5| - |-3|$ **14.** $|-3| + |5|$

15. $|\pi - 2|$ **16.** $|\pi - 4|$

17. $|x - 5|$ and $x \geq 5$ **18.** $|x - 5|$ and $x \leq 5$

19. $|x^3|$ **20.** $|2x|$

Solve for x.

21. $|x + 2| = 2$ **22.** $|2x + 5| = 3$

23. $|3x - 1| = 5$ **24.** $|7x - 5| = 3$

25. $\left|\dfrac{3x - 4}{2}\right| = 5$ **26.** $\left|\dfrac{10x + 1}{2}\right| = \dfrac{9}{2}$

27. $\left|\dfrac{2x - 4}{5}\right| = 2$ **28.** $\left|\dfrac{3x + 11}{7}\right| = 1$

29. $\left|\dfrac{x - 3}{4}\right| = -2$ **30.** $\left|\dfrac{x - 5}{3}\right| = 0$

31. $\left| \dfrac{4x - 2}{x} \right| = 3$ **32.** $\left| \dfrac{2(x - 3)}{3x} \right| = 6$

57. $3 \left| \dfrac{3x - 1}{2} \right| > 5$ **58.** $2 \left| \dfrac{8x + 2}{5} \right| \le 1$

33. $|x| = x$

34. $|x| + x = 2$

35. $|x + 3| = |x|$

36. $|x + 5| = |5 - x|$

59. $\dfrac{|x - 1|}{-2} > -3$ **60.** $\dfrac{|2x - 3|}{-3} < -1$

37. $|x - 3| = |2x + 3|$

38. $|x - 2| = |3x + 8|$

39. $|x + 2| = |x - 2|$ **40.** $|2x - 3| = |3x - 5|$

61. $0 < |2x + 1| < 3$

41. $\left| \dfrac{x + 3}{2} \right| = |2x - 3|$ **42.** $\left| \dfrac{x - 2}{3} \right| = |6 - x|$

62. $0 < |2x - 3| < 1$

63. $8 > |3x - 1| > 3$

43. $\left| \dfrac{3x - 1}{2} \right| = \left| \dfrac{2x + 3}{3} \right|$

64. $8 > |4x - 1| > 5$

44. $\left| \dfrac{5x + 2}{3} \right| = \left| \dfrac{x - 1}{4} \right|$

Solve each inequality, express the solution set in interval notation, and graph it.

65. $2 < \left| \dfrac{x - 5}{3} \right| < 4$

45. $|x - 3| < 6$ **46.** $|x - 2| \ge 4$

66. $3 < \left| \dfrac{x - 3}{2} \right| < 5$

67. $10 > \left| \dfrac{x - 2}{2} \right| > 4$

47. $|x + 3| > 6$ **48.** $|x + 2| \le 4$

68. $5 \ge \left| \dfrac{x + 2}{3} \right| > 1$

69. $2 \le \left| \dfrac{x + 1}{3} \right| < 3$

49. $|2x + 4| \ge 10$ **50.** $|5x - 2| < 7$

70. $8 > \left| \dfrac{3x + 1}{2} \right| > 2$

51. $|3x + 5| + 1 \le 9$ **52.** $|2x - 7| - 3 > 2$

Solve each inequality and express the solution using interval notation.

53. $|x + 3| > 0$ **54.** $|x - 3| \le 0$

71. $|x + 1| \ge |x|$ **72.** $|x + 1| < |x + 2|$

73. $|2x + 1| < |2x - 1|$ **74.** $|3x - 2| \ge |3x + 1|$

55. $\left| \dfrac{5x + 2}{3} \right| < 1$ **56.** $\left| \dfrac{3x + 2}{4} \right| > 2$

75. $|x + 1| < |x|$ **76.** $|x + 2| \le |x + 1|$

77. $|2x + 1| \ge |2x - 1|$ **78.** $|3x - 2| < |3x + 1|$

Applications

79. Finding temperature ranges The temperatures on a summer day satisfy the inequality $|t - 78°| \leq 8°$, where t is the temperature in degrees Fahrenheit. Express this range without using absolute value symbols.

80. Finding operating temperatures A car CD player has an operating temperature of $|t - 40°| < 80°$, where t is the temperature in degrees Fahrenheit. Express this range without using absolute value symbols.

81. Range of camber angles The specifications for a certain car state that the camber angle c of its wheels should be $0.6° \pm 0.5°$. Express this range with an inequality containing an absolute value.

82. Tolerance of a sheet of steel A sheet of steel is to be 0.25 inch thick, with a tolerance of 0.015 inch. Express this specification with an inequality containing an absolute value.

83. Humidity level A Steinway piano should be placed in an environment in which the relative humidity h is between 38% and 72%. Express this range with an inequality containing an absolute value.

84. Light bulbs A light bulb is expected to last h hours, where $|h - 1{,}500| \leq 200$. Express this range without using absolute value symbols.

85. Error analysis In a lab, students measured the percent of copper p in a sample of copper sulfate. The students know that copper sulfate is actually 25.46% copper by mass. They are to compare their results to the actual value and find the amount of *experimental error*.

   a. Which measurements shown in the following illustration satisfy the absolute value inequality $|p - 25.46| \leq 1.00$?

   b. What can be said about the amount of error for each of the trials listed in part a?

Lab 4	Section A
Title:	
"Percent copper (CU) in copper sulfate ($CuSO_4 \cdot 5H_2O$)"	

Results

	% Copper
Trial #1:	22.91%
Trial #2:	26.45%
Trial #3	26.49%
Trial #4:	24.76%

86. Error analysis See Exercise 85.

   a. Which measurements satisfy the absolute value inequality $|p - 25.46| > 1.00$?

   b. What can be said about the amount of error for each of the trials listed in part a?

Discovery and Writing

87. Explain how to find the absolute value of a number.

88. Explain why the equation $|x| + 9 = 0$ has no solution.

89. Explain the use of parentheses and brackets when graphing inequalities.

90. If $k > 0$, explain the differences between the solution sets of $|x| < k$ and $|x| > k$.

Review *Write each number in scientific notation.*

91. 37,250

92. 0.0003725

Write each number in standard notation.

93. 5.23×10^5 **94.** 7.9×10^{-4}

Simplify each expression.

95. $(x - y)^2 - (x + y)^2$

96. $(p + q)^2 + (p - q)^2$

1.1 Equations

Definitions and Concepts	Examples
An **equation** is a statement indicating that two quantities are equal.	Equations: $2x - 5 = 0,$ $\dfrac{2(x-2)}{x-3} = \dfrac{7x+3}{x+2}$
There can be restrictions on the variable in an equation.	In the equation $2x - 5 = 10$, x can be any real number. In the equation $\dfrac{2(x-2)}{x-3} = \dfrac{7x+3}{x+2}$, x cannot be 3 or -2, because this would give a 0 in the denominator.
Properties of equality: If $a = b$ and c is a number, then $\quad a + c = b + c \quad$ and $\quad a - c = b - c$ $\quad ac = bc \qquad$ and $\quad \dfrac{a}{c} = \dfrac{b}{c} \quad (c \neq 0)$	If $a = b$, then $\quad a + 7 = b + 7 \quad$ and $\quad a - 7 = b - 7$ $\quad 7a = 7b \qquad$ and $\quad \dfrac{a}{7} = \dfrac{b}{7}$
A **linear equation** is an equation that can be written in the form $ax + b = 0 \quad (a \neq 0)$. To solve a linear equation, use the properties of equality to isolate x on one side of the equation.	Solve $3x - 5 = 4$. $\qquad 3x - 5 = 4$ $3x - 5 + 5 = 4 + 5 \qquad$ Add 5 to both sides. $\qquad\qquad 3x = 9 \qquad$ Combine like terms. $\qquad\qquad \dfrac{3x}{3} = \dfrac{9}{3} \qquad$ Divide both sides by 3. $\qquad\qquad x = 3$
An **identity** is an equation that is true for all acceptable replacements for its variable. A **contradiction** is an equation that is false for all acceptable replacements for its variable.	**Identities:** $x + x = 2x, \quad 2(x + 1) = 2x + 2$ **Contradictions:** $x + 1 = x, \quad 2(x + 1) = 2x + 3$
Rational equations are equations that contain rational expressions. To solve rational equations, multiply both sides of the equation by an expression that will remove the denominators and solve the resulting equation. Be sure to check the answers to identify any **extraneous solutions.**	Solve $\dfrac{2x}{x-3} = \dfrac{6}{x-3}$. $\qquad \dfrac{2x}{x-3} = \dfrac{6}{x-3}$ $(x-3)\dfrac{2x}{x-3} = (x-3)\dfrac{6}{x-3} \qquad$ Multiply both sides by $x - 3$. $\qquad\qquad 2x = 6 \qquad$ Simplify. $\qquad\qquad x = 3 \qquad$ Divide both sides by 2 and simplify. The result of 3 is extraneous because when you substitute 3 into the original equation, you get a denominator of 0.

Formulas can be solved for a specific variable.	Solve $A = \frac{1}{2}bh$ for h.

$$A = \frac{1}{2}bh$$

$$2A = bh \qquad \text{Multiply both sides by 2.}$$

$$\frac{2A}{b} = \frac{bh}{b} \qquad \text{Divide both sides by } b.$$

$$\frac{2A}{b} = h \qquad \text{Simplify.}$$

Exercises

Find the restrictions on x, if any.

1. $3x + 7 = 4$

2. $x + \frac{1}{x} = 2$

3. $\sqrt{x} = 4$

4. $\frac{1}{x-2} = \frac{2}{x-3}$

Solve each equation and classify it as an identity, a conditional equation, or an equation with no solution.

5. $3(9x + 4) = 28$

6. $\frac{3}{2}a = 7(a + 11)$

7. $8(3x - 5) - 4(x + 3) = 12$

8. $\frac{x+3}{x+4} + \frac{x+3}{x+2} = 2$

9. $\frac{3}{x-1} = \frac{1}{2}$

10. $\frac{8x^2 + 72x}{9 + x} = 8x$

11. $\frac{3x}{x-1} - \frac{5}{x+3} = 3$

12. $x + \frac{1}{2x-3} = \frac{2x^2}{2x-3}$

13. $\frac{4}{x^2 - 13x - 48} - \frac{1}{x^2 + x - 6} = \frac{2}{x^2 - 18x + 32}$

14. $\frac{a-1}{a+3} + \frac{2a-1}{3-a} = \frac{2-a}{a-3}$

Solve each formula for the indicated variable.

15. $C = \frac{5}{9}(F - 32); F$

16. $P_n = l + \frac{si}{f}; f$

17. $\frac{1}{f} = \frac{1}{f_1} + \frac{1}{f_2}; f_1$

18. $S = \frac{a - lr}{1 - r}; l$

1.2 Applications of Linear Equations

Definitions and Concepts	**Examples**
Use the following steps to solve an application problem:	Two students leave their dorm in two cars traveling in opposite directions. If one student drives at a rate of 55 mph and the other at a rate of 50 mph, how long will it take for them to be 210 miles apart?
1. Analyze the problem.	
2. Pick a variable to represent the quantity to be found.	
3. Form an equation.	
4. Solve the equation.	
5. Check the solution in the words of the problem.	

Analyze the problem and pick a variable.   We can organize the facts of the problem in the following chart. Since each student drives the same amount of time, let t represent that time.

	d	$=$	r	$\cdot$	t
Student 1	$55t$		55		t
Student 2	$50t$		50		t

Form and solve an equation.   Since the students are driving in opposite directions, the distance they are apart in t hours is the sum of the distances they drive, a total of 210 miles. We can form and solve the following equation:

$$55t + 50t = 210$$
$$105t = 210 \qquad \text{Combine like terms.}$$
$$t = 2 \qquad \text{Divide both sides by 105.}$$

Check.   In 2 hours, student 1 drives 55(2) miles and student 2 drives 50(2) miles, or 110 miles plus 100 miles. At this time, they will be 210 miles apart.

Exercises

19. Test scores   Carlos took four tests in an English class. On each successive test, his score improved by 4 points. If his mean score was 66%, what did he score on the first test?

20. Fencing a garden   A homeowner has 100 ft of fencing to enclose a rectangular garden. If the garden is to be 5 ft longer than it is wide, find its dimensions.

21. Travel   Two women leave a shopping center by car, traveling in opposite directions. If one car averages 45 mph and the other 50 mph, how long will it take for the cars to be 285 apart?

22. Travel   Two taxis leave an airport and travel in the same direction. If the average speed of one taxi is 40 mph and the average speed of the other taxi is 46 mph, how long will it take before the cars are 3 miles apart?

23. Preparing a solution   A liter of fluid is 50% alcohol. How much water must be added to dilute it to a 20% solution?

24. Washing windows   Scott can wash 37 windows in 3 hours, and Bill can wash 27 windows in 2 hours. How long will it take the two of them to wash 100 windows?

25. Filling a tank   A tank can be filled in 9 hours by one pipe and in 12 hours by another. How long will it take both pipes to fill the empty tank?

26. Producing brass   How many ounces of pure zinc must be alloyed with 20 ounces of brass that is 30% zinc and 70% copper to produce brass that is 40% zinc?

27. Lending money   A bank lends $10,000, part of it at 11% annual interest and the rest at 14%. If the annual income is $1,265, how much was lent at each rate?

28. Producing oriental rugs   An oriental rug manufacturer can use one loom with a setup cost of $750 that can weave a rug for $115. Another loom, with a setup cost of $950, can produce a rug for $95. How many rugs are produced if the costs are the same on each loom?

1.3 Quadratic Equations

Definitions and Concepts	Examples
A **quadratic equation** is an equation that can be written in the form $ax^2 + bx + c = 0$, where a, b, and c are real numbers and $a \neq 0$.	$3x^2 - 5x - 7 = 0$, $\quad 5x^2 - 25 = 0$, $\quad 7x^2 + 14x = 0$

Zero-factor theorem:
If $ab = 0$, then $a = 0$ or $b = 0$.

Solve $x^2 - x - 6 = 0$ using the zero-factor theorem.

$$x^2 - x - 6 = 0$$
$$(x + 2)(x - 3) = 0 \quad \text{Factor } x^2 - x - 6.$$
$$x + 2 = 0 \quad \text{or} \quad x - 3 = 0$$
$$x = -2 \quad | \quad x = 3$$

Square root property:
If $c > 0$, $x^2 = c$ has two real roots:
$$x = \sqrt{c} \quad \text{or} \quad x = -\sqrt{c}$$

If $x^2 = 32$, then
$$
\begin{aligned}
x &= \sqrt{32} & \text{or} \quad x &= -\sqrt{32} \\
&= \sqrt{16 \cdot 2} & &= -\sqrt{16 \cdot 2} \\
&= 4\sqrt{2} & &= -4\sqrt{2}
\end{aligned}
$$

To complete the square:
1. Make the coefficient of x^2 equal to 1.
2. Get the constant on the right side of the equation.
3. Complete the square on x.
   Take one-half the coefficient of x, square it, and add it to both sides of the equation.
4. Factor the resulting perfect-square trinomial and combine like terms.
5. Solve the resulting quadratic equation by using the square root property.

Solve $x^2 - x - 6 = 0$ by completing the square:
1. Since the coefficient of x^2 is 1, we go to Step 2.
2. Add 6 to both sides to get the constant on the right side: $x^2 - x = 6$.
3. $x^2 - x + \left(-\dfrac{1}{2}\right)^2 = 6 + \left(-\dfrac{1}{2}\right)^2$
4. $\quad \left(x - \dfrac{1}{2}\right)^2 = \dfrac{25}{4}$
5. $\qquad x - \dfrac{1}{2} = \pm\sqrt{\dfrac{25}{4}}$
$$x = \dfrac{1}{2} \pm \dfrac{5}{2}$$
$$x = \dfrac{6}{2} = 3 \quad \text{or} \quad x = -\dfrac{4}{2} = -2$$

Quadratic formula:
$$x = \frac{-b \pm \sqrt{b^2 - 4ac}}{2a} \quad (a \neq 0)$$

If $3x^2 - 5x + 1 = 0$, then $a = 3$, $b = -5$, and $c = 1$, so
$$x = \frac{-b \pm \sqrt{b^2 - 4ac}}{2a} = \frac{-(-5) \pm \sqrt{(-5)^2 - 4(3)(1)}}{2(3)}$$
$$= \frac{5 \pm \sqrt{25 - 12}}{6} = \frac{5 \pm \sqrt{13}}{6}$$

Discriminant: The value $b^2 - 4ac$ is called the *discriminant*.

If $b^2 - 4ac = 0$, the roots of $ax^2 + bx + c = 0$ are equal rational numbers.

In $4x^2 - 12x + 9 = 0$, the discriminant is
$$b^2 - 4ac = (-12)^2 - 4(4)(9) = 0$$

The roots are equal rational numbers.

If $b^2 - 4ac$ is a nonzero perfect square, then the roots are unequal rational numbers.

If $b^2 - 4ac$ is a positive nonperfect square, then the roots are unequal irrational numbers.

If $b^2 - 4ac$ is negative, then the roots are not real numbers.

In $x^2 - x - 6 = 0$, the discriminant is
$$b^2 - 4ac = (-1)^2 - 4(1)(-6) = 25$$
Because 25 is a nonzero perfect square, the roots are unequal rational numbers.

In $x^2 - x - 7 = 0$, the discriminant is
$$b^2 - 4ac = (-1)^2 - 4(1)(-7) = 29$$
Because 29 is a positive nonperfect square, the roots are unequal irrational numbers.

In $3x^2 - 2x - 1 = 0$, the discriminant is
$$b^2 - 4ac = (-2)^2 - 4(3)(1) = -8$$
The roots are not real numbers.

Exercises

Solve each equation by factoring.

29. $2x^2 - x - 6 = 0$

30. $12x^2 + 13x = 4$

31. $5x^2 - 8x = 0$

32. $27x^2 = 30x - 8$

Solve each equation by completing the square.

33. $x^2 - 8x + 15 = 0$

34. $3x^2 + 18x = -24$

35. $5x^2 - x - 1 = 0$

36. $5x^2 - x = 0$

Use the quadratic formula to solve each equation.

37. $x^2 + 5x - 14 = 0$

38. $3x^2 - 25x = 18$

39. $5x^2 = 1 - x$

40. $-5 = a^2 + 2a$

41. Calculate the discriminant associated with the equation $6x^2 + 5x + 1 = 0$.

42. Determine the nature of the roots of the equation in Exercise 41.

43. Find the value of k that will make the roots of $kx^2 + 4x + 12 = 0$ equal.

44. Find the values of k that will make the roots of $4y^2 + (k + 2)y = 1 - k$ equal.

45. Solve: $\dfrac{1}{a} - \dfrac{1}{5} = \dfrac{3}{2a}$

46. Solve: $\dfrac{4}{a - 4} + \dfrac{4}{a - 1} = 5$.

1.4   Applications of Quadratic Equations

Definitions and Concepts

Many real-life problems are modeled by quadratic equations.

Examples

If a missile is launched straight up into the air with an initial velocity of 128 feet per second, its height will be given by the formula $h = -16t^2 + 128t$, where h represents its height (in feet) and t represents the time (in seconds) since it was launched. How long will it take the missile to return to its starting point?

When the missile returns to its starting point, its height will again be 0. So we let $h = 0$ and solve for t.

$$0 = -16t^2 + 128t$$
$$0 = -16t(t - 8)$$
$$-16t = 0 \quad \text{or} \quad t - 8 = 0$$
$$t = 0 \quad | \quad t = 8$$

The missile will leave its starting point at 0 seconds and return at 8 seconds.

Exercises

47. Fencing a field A farmer wishes to enclose a rectangular garden with 300 yards of fencing. A river runs along one side of the garden, so no fencing is needed there. Find the dimensions of the rectangle if the area is 10,450 square yards.

48. Flying rates A jet plane, flying 120 mph faster than a propeller-driven plane, travels 3,520 miles in 3 hours less time than the propeller plane requires to fly the same distance. How fast does each plane fly?

49. Flight of a ball A ball thrown into the air reaches a height h (in feet) according to the formula $h = -16t^2 + 64t$, where t is the time elapsed since the ball was thrown. Find the shortest time it will take the ball to reach a height of 48 feet.

50. Width of a walk A man built a walk of uniform width around a rectangular pool. If the area of the walk is 117 square feet and the dimensions of the pool are 16 feet by 20 feet, how wide is the walk?

1.5 Complex Numbers

Definitions and Concepts	Examples
$a + bi = c + di$ if and only if $a = c$ and $b = d$	$3 + \sqrt{4}i = \dfrac{6}{2} + 2i$ because $3 = \dfrac{6}{2}$ and $\sqrt{4} = 2$
$(a + bi) + (c + di) = (a + c) + (b + d)i$	$(-3 + 4i) + (2 + 7i) = (-3 + 2) + (4 + 7)i$ $= -1 + 11i$
$(a + bi) - (c + di) = (a - c) + (b - d)i$	$(-3 + 4i) - (2 + 7i) = (-3 - 2) + (4 - 7)i$ $= -5 - 3i$
$(a + bi)(c + di) = (ac - bd) + (ad + bc)i$ or multiply them as if they were binomials. $i = \sqrt{-1}$	$(-3 + 4i)(2 + 7i) = -3(2) - 3(7i) + 4i(2) + 4i(7i)$ $= -6 - 21i + 8i + 28i^2$ $= -6 - 21i + 8i - 28$ $= -34 - 13i$
The **complex conjugate** of $a + bi$ is $a - bi$.	$3 + 4i$ and $3 - 4i$ are complex conjugates.
To divide complex numbers, rationalize the denominator.	Divide $2 + i$ by $2 - i$. $\dfrac{2 + i}{2 - i} = \dfrac{(2 + i)(2 + i)}{(2 - i)(2 + i)} \qquad \dfrac{2 + i}{2 + i} = 1$ $= \dfrac{4 + 2i + 2i + i^2}{4 + 2i - 2i - i^2}$ $= \dfrac{4 + 4i - 1}{4 - (-1)}$

$$= \frac{3 + 4i}{5}$$

$$= \frac{3}{5} + \frac{4}{5}i$$

If n is a natural number that has a remainder of r when divided by 4, then $i^n = i^r$.	$i^2 = -1, i^3 = -i, i^4 = 1, i^5 = i, i^6 = -1, i^7 = -i,$ $i^8 = 1, \ldots$
$\lvert a + bi \rvert = \sqrt{a^2 + b^2}$	$\lvert 5 - 7i \rvert = \sqrt{5^2 + (-7)^2} = \sqrt{25 + 49} = \sqrt{74}$

Exercises

Perform all operations and express all answers in a + bi form.

51. $(2 - 3i) + (-4 + 2i)$ **52.** $(2 - 3i) - (4 + 2i)$

53. $\left(3 - \sqrt{-36}\right) + \left(\sqrt{-16} + 2\right)$

54. $\left(3 + \sqrt{-9}\right)\left(2 - \sqrt{-25}\right)$

55. $\dfrac{3}{i}$ **56.** $-\dfrac{2}{i^3}$

57. $\dfrac{3}{1 + i}$ **58.** $\dfrac{2i}{2 - i}$

59. $\dfrac{3 + i}{3 - i}$ **60.** $\dfrac{3 - 2i}{1 + i}$

61. Simplify: i^{53}.   **62.** Simplify: i^{103}.

63. $\lvert 3 - i \rvert$ **64.** $\left\lvert \dfrac{1 + i}{1 - i} \right\rvert$

65. Solve: $3x^2 - 2x + 1 = 0$.

66. Solve: $3x^2 + 4 = 2x$.

1.6   Polynomial and Radical Equations

Definitions and Concepts	Examples
Many polynomial equations of higher degree can be solved by factoring.	Solve: $x^3 - 5x^2 + 6x = 0$. $x^3 - 5x^2 + 6x = 0$ $x(x^2 - 5x + 6) = 0$ Factor out x. $x(x - 2)(x - 3) = 0$ Factor $x^2 - 5x + 6$. $x = 0$ or $x - 2 = 0$ or $x - 3 = 0$ $x = 2$ $x = 3$ The solution set is $\{0, 2, 3\}$.
Power property of real numbers: If $a = b$, then $a^2 = b^2$.	If $x = 5$, then $x^2 = 5^2$ or $x^2 = 25$.
Factoring can be used to solve certain nonpolynomial equations.	Solve: $2x - 5x^{1/2} + 3 = 0$. $2x - 5x^{1/2} + 3 = 0$ $(2x^{1/2} - 3)(x^{1/2} - 1) = 0$ $2x^{1/2} - 3 = 0$ or $x^{1/2} - 1 = 0$ $x^{1/2} = \dfrac{3}{2}$ $x^{1/2} = 1$ $x = \dfrac{9}{4}$ $x = 1$ Both roots check.

To solve radical equations, use the power property of real numbers. Check all roots because extraneous roots can be introduced.

Solve: $\sqrt{2x - 3} = x - 1$.

$$\sqrt{2x - 3} = x - 1$$
$$\left(\sqrt{2x - 3}\right)^2 = (x - 1)^2 \qquad \text{Square both sides.}$$
$$2x - 3 = x^2 - 2x + 1$$
$$0 = x^2 - 4x + 4$$
$$0 = (x - 2)(x - 2)$$
$$x - 2 = 0 \quad \text{or} \quad x - 2 = 0$$
$$x = 2 \qquad\qquad x = 2$$

Since 2 checks, it is a root.

Exercises

Solve each equation.

67. $\dfrac{3x}{2} - \dfrac{2x}{x - 1} = x - 3$ **68.** $\dfrac{12}{x} - \dfrac{x}{2} = x - 3$

69. $x^4 - 2x^2 + 1 = 0$ **70.** $x^4 + 36 = 37x^2$

71. $a - a^{1/2} - 6 = 0$ **72.** $x^{2/3} + x^{1/3} - 6 = 0$

73. $\sqrt{x - 1} + x = 7$ **74.** $\sqrt{a + 9} - \sqrt{a} = 3$

75. $\sqrt{5 - x} + \sqrt{5 + x} = 4$

76. $\sqrt{y + 5} + \sqrt{y} = 1$

1.7   Inequalities

Definitions and Concepts	Examples
If a, b, and c are real numbers:	
If $a < b$, $a + c < b + c$ and $a - c < b - c$.	If $x < 10$, then $x + 6 < 10 + 6$ and $x - 6 < 10 - 6$.
If $a < b$ and $c > 0$, $ac < bc$ and $\dfrac{a}{c} < \dfrac{b}{c}$.	If $x < 12$, then $3x < 3(12)$ and $\dfrac{x}{3} < \dfrac{12}{3}$.
If $a < b$ and $c < 0$, $ac > bc$ and $\dfrac{a}{c} > \dfrac{b}{c}$.	If $x \leq 12$, then $-3x \geq -3(12)$ and $\dfrac{x}{-3} \geq \dfrac{12}{-3}$.
Trichotomy property: $\qquad a < b, \qquad a = b, \qquad \text{or} \qquad a > b$	Either $x < 3$, $x = 3$, or $x > 3$.
Transitive property: If $a < b$ and $b < c$, then $a < c$.	If $x < 4$ and $4 < y$, then $x < y$.
Solving inequalities: Use the same steps to solve inequalities as you would use to solve equations. However, remember to reverse the order of the inequality when you multiply (or divide) both sides of an inequality by a negative number.	Solve: $-4(x - 3) \geq 7$. $\begin{aligned} -4(x - 3) &\geq 7 \\ -4x + 12 &\geq 7 \qquad \text{Remove parentheses.} \\ -4x &\geq -5 \qquad \text{Subtract 12 from both sides.} \\ x &\leq \frac{5}{4} \qquad \text{Divide both sides by } -4. \end{aligned}$ In interval notation, the solution is $\left(-\infty, \frac{5}{4}\right]$.

Solve: $-5 \leq 2x + 1 < 3$.

$-5 \leq 2x + 1 < 3$

$-6 \leq 2x < 2$ **Subtract 1 from all three parts.**

$-3 \leq x < 1$ **Divide each part by 2.**

In interval notation, the solution is $[-3, 1)$.

Solve: $\dfrac{x + 1}{x - 2} > 0$.

First note that the solution of $x + 1 = 0$ $(x = -1)$ and the solution of $x - 2 = 0$ $(x = 2)$ form three intervals: $(-\infty, -1)$, $(-1, 2)$, and $(2, \infty)$. To determine which intervals are solutions, we test a number in each interval to see whether it satisfies the inequality.

Interval	Test value	Inequality $\dfrac{x + 1}{x - 2} > 0$	Result
$(-\infty, -1)$	-2	$\dfrac{-2 + 1}{-2 - 2} = \dfrac{1}{4} > 0$	The numbers in this interval are solutions.
$(-1, 2)$	0	$\dfrac{0 + 1}{0 - 2} = -\dfrac{1}{2} < 0$	The numbers in this interval are not solutions.
$(2, \infty)$	3	$\dfrac{3 + 1}{3 - 2} = 4 > 0$	The numbers in this interval are solutions.

The solution is the union of two intervals:

$(-\infty, -1) \cup (2, \infty)$

Exercises

Solve each inequality.

77. $2x - 9 < 5$

78. $5x + 3 \geq 2$

79. $\dfrac{5(x - 1)}{2} < x$

80. $\dfrac{1}{4}x + \dfrac{2}{3}x - x > \dfrac{1}{2} + \dfrac{1}{2}(x + 1)$

81. $0 \leq \dfrac{3 + x}{2} < 4$

82. $2 + a < 3a - 2 \leq 5a + 2$

83. $(x + 2)(x - 4) > 0$ **84.** $(x - 1)(x + 4) < 0$

85. $x^2 - 2x - 3 < 0$ **86.** $2x^2 + x - 3 > 0$

89. $\dfrac{x^2 + x - 2}{x - 3} \geq 0$ **90.** $\dfrac{5}{x} < 2$

87. $\dfrac{x + 2}{x - 3} \geq 0$ **88.** $\dfrac{x - 1}{x + 4} \leq 0$

1.8   Absolute Value

Definitions and Concepts	Examples
$|x| = \begin{cases} x \text{ when } x \geq 0 \\ -x \text{ when } x < 0 \end{cases}$	$|5| = 5 \qquad |-7| = 7 \qquad -|-10| = -10$
If $k \geq 0$, then $|x| = k$ is equivalent to $\quad x = k \quad$ or $\quad x = -k$ If a and b are algebraic expressions, $|a| = |b|$ is equivalent to $a = b$ or $a = -b$.	Solve: $|x - 2| = 6$. $\quad x - 2 = 6 \quad$ or $\quad x - 2 = -6$ $\qquad x = 8 \qquad\qquad x = -4$ Solve: $|3x| = |x - 2|$. $\quad 3x = x - 2 \quad$ or $\quad 3x = -(x - 2)$ $\quad 2x = -2 \qquad\qquad 3x = -x + 2$ $\quad\; x = -1 \qquad\qquad\; 4x = 2$ $\qquad\qquad\qquad\qquad\quad x = \dfrac{1}{2}$ Both solutions check.
If $k > 0$, then $|x| < k$ is equivalent to $\quad -k < x < k$ If $k > 0$, then $|x| > k$ is equivalent to $\quad x > k \quad$ or $\quad x < -k$	Solve: $|x - 2| < 6$. $\quad -6 < x - 2 < 6$ $\quad -4 < x < 8 \quad$ **Add 2 to all three parts.** Solve: $|x - 2| > 6$. $\quad x - 2 > 6 \quad$ or $\quad x - 2 < -6$ $\qquad x > 8 \qquad\qquad x < -4 \quad$ **Add 2 to both parts.** Thus, $x < -4$ or $x > 8$.
Properties of absolute value: **1.** $|ab| = |a||b|$ **2.** $\left|\dfrac{a}{b}\right| = \dfrac{|a|}{|b|} \quad (b \neq 0)$ **3.** $|a + b| \leq |a| + |b|$	$|-3x| = |-3||x| = 3|x|$ $\left|\dfrac{-3}{x}\right| = \dfrac{|-3|}{|x|} = \dfrac{3}{|x|} \quad (x \neq 0)$ $|x + 3| \leq |x| + |3|$

Exercises

Solve each equation or inequality.

91. $|x + 1| = 6$

92. $|2x - 1| = |2x + 1|$

93. $\left|\dfrac{3x + 11}{7}\right| - 1 = 0$

94. $\left|\dfrac{2a - 6}{3a}\right| - 6 = 0$

95. $|x + 3| < 3$ **96.** $|3x - 7| \geq 1$ **99.** $1 < |2x + 3| < 4$ **100.** $0 < |3x - 4| < 7$

97. $\left| \dfrac{x + 2}{3} \right| < 1$ **98.** $\left| \dfrac{x - 3}{4} \right| > 8$

CHAPTER TEST

Find all restrictions on x.

1. $\dfrac{x}{x(x - 1)}$ **2.** $\sqrt{x}$

Solve each equation.

3. $7(2a + 5) - 7 = 6(a + 8)$

4. $\dfrac{3}{x^2 - 5x - 14} = \dfrac{4}{x^2 + 5x + 6}$

5. Solve for x: $z = \dfrac{x - \mu}{\sigma}$.

6. Solve for a: $\dfrac{1}{a} = \dfrac{1}{b} + \dfrac{1}{c}$.

7. A student's average on three tests is 75. If the final is to count as two one-hour tests, what grade must the student make to bring the average up to 80?

8. A woman invested part of $20,000 at 6% interest and the rest at 7%. If her annual interest is $1,260, how much did she invest at 6%?

Solve each equation.

9. $4x^2 - 8x + 3 = 0$

10. $2b^2 - 12 = -5b$

11. Write the quadratic formula.

12. Use the quadratic formula to solve $3x^2 - 5x - 9 = 0$.

13. Find k such that $x^2 + (k + 1)x + k + 4 = 0$ will have two equal roots.

14. The height of a projectile shot up into the air is given by the formula $h = -16t^2 + 128t$. Find the time t required for the projectile to return to its starting point.

Perform each operation and write all answers in a + bi form.

15. $(4 - 5i) - (-3 + 7i)$

16. $(4 - 5i)(3 - 7i)$

17. $\dfrac{2}{2 - i}$ **18.** $\dfrac{1 + i}{1 - i}$

Simplify each expression.

19. i^{13} **20.** i^0

Find each absolute value.

21. $|5 - 12i|$ **22.** $\left| \dfrac{1}{3 + i} \right|$

Solve each equation.

23. $z^4 - 13z^2 + 36 = 0$

24. $2p^{2/5} - p^{1/5} - 1 = 0$

25. $\sqrt{x + 5} = 12$

26. $\sqrt{2z + 3} = 1 - \sqrt{z + 1}$

Solve each inequality.

27. $5x - 3 \leq 7$

28. $\dfrac{x + 3}{4} > \dfrac{2x - 4}{3}$

29. $1 + x < 3x - 3 < 4x - 2$

30. $\dfrac{x + 2}{x - 1} \leq 0$

Solve each equation.

31. $\left| \dfrac{3x + 2}{2} \right| = 4$ **32.** $|x + 3| = |x - 3|$

Solve each inequality and graph the solution set.

33. $|2x - 5| > 2$

34. $\left| \dfrac{2x + 3}{3} \right| \leq 5$

CUMULATIVE REVIEW EXERCISES

Consider the set $\left\{ -5, -3, -2, 0, 1, \sqrt{2}, 2, \frac{5}{2}, 5, 6, 11 \right\}$.

1. Which numbers are even integers?
2. Which numbers are prime numbers?

Write each inequality as an interval and graph it.

3. $-4 \leq x < 7$

4. $x \geq 2$ or $x < 0$

Determine which property of the real numbers justifies each expression.

5. $(a + b) + c = c + (a + b)$
6. If $x < 3$ and $3 < y$, then $x < y$.

Simplify each expression. Assume that all variables represent positive numbers. Give all answers with positive exponents.

7. $(81a^4)^{1/2}$ **8.** $81(a^4)^{1/2}$

9. $(a^{-3}b^{-2})^{-2}$ **10.** $\left(\dfrac{4x^4}{12x^2y} \right)^{-2}$

11. $\left(\dfrac{4x^0y^2}{x^2y} \right)^{-2}$ **12.** $\left(\dfrac{4x^{-5}y^2}{6x^{-2}y^{-3}} \right)^2$

13. $(a^{1/2}b)^2(ab^{1/2})^2$ **14.** $(a^{1/2}b^{1/2}c)^2$

Rationalize each denominator and simplify.

15. $\dfrac{3}{\sqrt{3}}$ **16.** $\dfrac{2}{\sqrt[3]{4x}}$

17. $\dfrac{3}{y - \sqrt{3}}$ **18.** $\dfrac{3x}{\sqrt{x} - 1}$

Simplify each expression and combine like terms.

19. $\sqrt{75} - 3\sqrt{5}$ **20.** $\sqrt{18} + \sqrt{8} - 2\sqrt{2}$

21. $\left(\sqrt{2} - \sqrt{3} \right)^2$ **22.** $\left(3 - \sqrt{5} \right)\left(3 + \sqrt{5} \right)$

Perform the operations and simplify when necessary.

23. $(3x^2 - 2x + 5) - 3(x^2 + 2x - 1)$
24. $5x^2(2x^2 - x) + x(x^2 - x^3)$
25. $(3x - 5)(2x + 7)$
26. $(z + 2)(z^2 - z + 2)$
27. $3x + 2 \overline{)6x^3 + x^2 + x + 2}$
28. $x^2 + 2 \overline{)3x^4 + 7x^2 - x + 2}$

Factor each polynomial.

29. $3t^2 - 6t$
30. $3x^2 - 10x - 8$
31. $x^8 - 2x^4 + 1$

32. $x^6 - 1$

Perform the operations and simplify.

33. $\dfrac{x^2 - 4}{x^2 + 5x + 6} \cdot \dfrac{x^2 - 2x - 15}{x^2 + 3x - 10}$

34. $\dfrac{6x^3 + x^2 - x}{x + 2} \div \dfrac{3x^2 - x}{x^2 + 4x + 4}$

35. $\dfrac{2}{x + 3} + \dfrac{5x}{x - 3}$

36. $\dfrac{x - 2}{x + 3}\left(\dfrac{x + 3}{x^2 - 4} - 1\right)$

37. $\dfrac{\dfrac{1}{a} + \dfrac{1}{b}}{\dfrac{1}{ab}}$

38. $\dfrac{x^{-1} - y^{-1}}{x - y}$

Solve each equation.

39. $\dfrac{3x}{x + 5} = \dfrac{x}{x - 5}$

40. $8(2x - 3) - 3(5x + 2) = 4$

Solve each formula for the indicated variable.

41. $\dfrac{1}{R} = \dfrac{1}{R_1} + \dfrac{1}{R_2}; \; R$

42. $S = \dfrac{a - lr}{1 - r}; \; r$

43. Gardening A gardener wishes to enclose her rectangular raspberry patch with 40 feet of fencing. The raspberry bushes are planted along the garage, so no fencing is needed on that side. Find the dimensions if the total area is to be 192 square feet.

44. Financial planning A college student invested part of a $25,000 inheritance at 7% interest and the rest at 6%. If his annual interest is $1,670, how much did he invest at 6%?

Perform the operations. If the result is not real, express the answer in a + bi form.

45. $\dfrac{2 + i}{2 - i}$

46. $\dfrac{i(3 - i)}{(1 + i)(1 + i)}$

47. $|3 + 4i|$

48. $\dfrac{5}{i^7} + 5i$

Solve each equation.

49. $\dfrac{x + 3}{x - 1} - \dfrac{6}{x} = 1$

50. $x^4 + 36 = 13x^2$

51. $\sqrt{y + 2} + \sqrt{11 - y} = 5$

52. $z^{2/3} - 13z^{1/3} + 36 = 0$

Solve each inequality and graph the solution set.

53. $5x - 7 \le 4$

54. $x^2 - 8x + 15 > 0$

55. $\dfrac{x^2 + 4x + 3}{x - 2} \ge 0$

56. $\dfrac{9}{x} > x$

57. $|2x - 3| \ge 5$

58. $\left|\dfrac{3x - 5}{2}\right| < 2$

The Rectangular Coordinate System and Graphs of Equations

Mathematical expressions often indicate relationships between two variables. To visualize these relationships, we draw graphs of their equations.

Careers and Mathematics

Economist   Economists, such as Alan Greenspan, former Chairman of the Federal Reserve Board, study how society distributes scarce resources such as land, labor, raw materials, and machinery to produce goods and services. They also research issues such as energy costs, inflation, interest rates, imports, and employment levels.

Economists use mathematical models to help predict answers to questions such as the nature and length of business cycles, the effects of a specific rate of inflation on the economy, and the effects of tax legislation on unemployment levels.

Economists held about 15,000 jobs in 2006.

Education   A master's or Ph.D. degree in economics is required for many private-sector economist jobs and for advancement to more responsible positions. In the federal government, candidates for entry-level economist positions must have a bachelor's degree with a minimum of 21 semester hours of economics and 3 hours of statistics, accounting, or calculus.

Job Outlook   Employment of economists is expected to grow about 7 percent through 2016. The median annual earnings of economists were $77,010 in 2006.

For a sample application, see Example 10 in Section 2.3. For more information, go to www.bls.gov/ocos055.htm.

Stephen Jaffe/IMF via Getty Images

174

2.1 The Rectangular Coordinate System

Objectives

1. Plot Points in the Rectangular Coordinate System
2. Graph Linear Equations
3. Graph Horizontal and Vertical Lines
4. Solve Problems Using Linear Equations
5. Find the Distance Between Two Points
6. Find the Midpoint of a Line Segment

We often say that a picture is worth a thousand words. In fact, pictures and graphs are an effective way to present information. For this reason, they appear frequently in newspapers and magazines. For example, the graph shown in Figure 2-1 provides a visual representation of the manatee population in Florida after the year 2000.

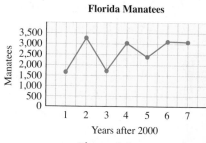

Figure 2-1

From the graph we can determine many facts about manatees. Among them are:

- The number of manatees in Florida declined from 2002 to 2003.
- In 2002, the population was about 3,300 animals.
- In this time period, the lowest population of manatees occurred in 2001.

In mathematics, graphs are also an effective way to present information. In this chapter, we will draw graphs of equations containing two variables and then discuss the information that we can derive from graphs.

The solutions of an equation with variables x and y such as $y = -\frac{1}{2}x + 4$ are ordered pairs of real numbers (x, y) that satisfy the equation. To find some ordered pairs that satisfy the equation, we substitute **input values** of x into the equation and find the corresponding **output values** of y. For example, if we substitute 2 for x, we obtain

$$y = -\frac{1}{2}x + 4$$

$$y = -\frac{1}{2}(2) + 4 \quad \text{Substitute 2 for } x.$$

$$= -1 + 4$$

$$= 3$$

Since $y = 3$ when $x = 2$, the ordered pair $(2, 3)$ is a solution of the equation. The first coordinate, 2, of the ordered pair is usually called the **x-coordinate.** The second coordinate, 3, is usually called the **y-coordinate.** The solution (2, 3) and several other solutions are listed in the table of values shown in Figure 2-2.

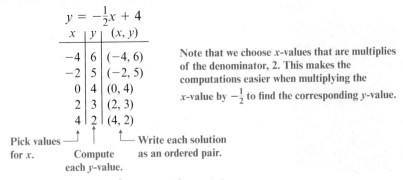

$$y = -\frac{1}{2}x + 4$$

x	y	(x, y)
−4	6	(−4, 6)
−2	5	(−2, 5)
0	4	(0, 4)
2	3	(2, 3)
4	2	(4, 2)

Note that we choose *x*-values that are multiplies of the denominator, 2. This makes the computations easier when multiplying the *x*-value by $-\frac{1}{2}$ to find the corresponding *y*-value.

Pick values for *x*. Compute each *y*-value. Write each solution as an ordered pair.

Figure 2-2

Accent on Technology

Generating Tables with a Graphing Calculator

If an equation in *x* and *y* is solved for *y*, we can use a graphing calculator to generate a table of solutions. The instructions in this discussion are for a TI-84 Plus graphing calculator. For details about other brands, please consult the owner's manual.

To construct a table of solutions for $x + 2y = 8$, we first solve the equation for *y*.

$$x + 2y = 8$$
$$2y = -x + 8 \qquad \text{Subtract } x \text{ from both sides.}$$
$$y = -\frac{1}{2}x + 4 \qquad \text{Divide both sides by 2 and simplify.}$$

Note that this is the equation shown in Figure 2-2.

To construct a table of values for $y = -\frac{1}{2}x + 4$, we press 2nd TBLSET and enter one value for *x* on the line labeled TblStart=. In Figure 2-3(a), −4 has been entered on this line. Other values for *x* that will appear in the table are determined by setting an **increment value** on the line labeled ΔTbl =. In Figure 2-3(a), an increment value of 2 has been entered. This means that each *x*-value in the table will be 2 units larger than the previous one.

To enter the equation, we press y = and enter −(1/2)x + 4, as shown in Figure 2-3(b). Finally, we press 2nd TABLE to obtain the table of values shown in Figure 2-3(c). This table contains all of the solutions listed in Figure 2-2, plus the two additional solutions (6, 1) and (8, 0). To see other values, we simply scroll up and down the screen by pressing the up and down arrow keys.

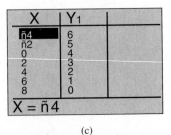

```
TABLE SETUP
 TblStart = −4
 ΔTbl = 2■
  Indpnt:  Auto  Ask
  Depend:  Auto  Ask
```

(a)

```
Plot1  Plot2  Plot3
\Y1 ≡ ñ(1/2)X + 4■
\Y2 =
\Y3 =
\Y4 =
\Y5 =
\Y6 =
\Y7 =
```

(b)

X	Y1
ñ4	6
ñ2	5
0	4
2	3
4	2
6	1
8	0

X = ñ4

(c)

Figure 2-3

Before we can present the table of solutions shown in Figure 2-2 in graphical form, we need to discuss the rectangular coordinate system.

1. Plot Points in the Rectangular Coordinate System

The **rectangular coordinate system** consists of two perpendicular number lines that divide the plane into four **quadrants,** numbered as shown in Figure 2-4. The horizontal number line is called the **x-axis,** and the vertical number line is called the **y-axis.** These axes intersect at a point called the **origin,** which is the 0 point on each axis. The positive direction on the x-axis is to the right, the positive direction on the y-axis is upward, and the same unit distance is used on both axes, unless otherwise indicated.

To plot (or graph) the point associated with the pair $x = 2$ and $y = 3$, denoted as (2, 3), we start at the origin, count 2 units to the right, and then count 3 units up. (See Figure 2-5.) Point P (which lies in the first quadrant) is the graph of the pair (2, 3). The pair (2, 3) gives the **coordinates** of point P.

To plot point Q with coordinates (−4, 6), we start at the origin, count 4 units to the left, and then count 6 units up. Point Q lies in the second quadrant. Point R with coordinates (6, −4) lies in the fourth quadrant.

Comment

The pairs (−4, 6) and (6, −4) represent different points. (−4, 6) is in the second quadrant and (6, −4) is in the fourth quadrant.

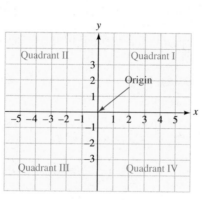

Figure 2-4

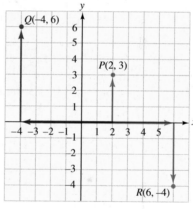

Figure 2-5

2. Graph Linear Equations

The **graph of the equation** $y = -\frac{1}{2}x + 4$ is the graph of all points (x, y) on the rectangular coordinate system whose coordinates satisfy the equation. To graph $y = -\frac{1}{2}x + 4$, we plot the pairs listed in the table of solutions shown in Figure 2-6. These points lie on the line shown in the figure. This line is the graph of the equation.

René Descartes
(1596–1650)
Descartes is famous for his work in philosophy as well as for his work in mathematics. His philosophy is expressed in the words "I think, therefore I am." He is best known in mathematics for his invention of a coordinate system and his work with conic sections.

$$y = -\frac{1}{2}x + 4$$

x	y	(x, y)
−4	6	(−4, 6)
−2	5	(−2, 5)
0	4	(0, 4)
2	3	(2, 3)
4	2	(4, 2)

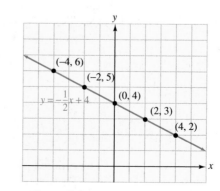

Figure 2-6

Comment

When we say that the graph of an equation is a line, we imply two things:

1. Every point with coordinates that satisfy the equation will lie on the line.
2. Every point on the line will have coordinates that satisfy the equation.

When the graph of an equation is a line, we call the equation a **linear equation.** These equations often are written in **standard form** as $Ax + By = C$, where A, B, and C are specific numbers (called **constants**) and x and y are variables. Either A or B can be 0, but A and B cannot both be 0. Here are four examples of linear equations written in standard form.

Linear Equation	*Values of A, B, and C*
$3x + 2y = 6$	$A = 3, B = 2, C = 6$
$5x - 2y = -10$	$A = 5, B = -2, C = -10$
$2y = 7$	$A = 0, B = 2, C = 7$
$x = -4$	$A = 1, B = 0, C = -4$

EXAMPLE 1   Graph: $x + 2y = 5$.

Solution   We will solve the equation for y and form a table of solutions by picking values for x, substituting them into the equation, and solving for the other variable y. We will plot the points represented in the table of solutions and draw a line through the points.

Solve the equation for y.

$$x + 2y = 5$$
$$x - x + 2y = 5 - x \qquad \text{Subtract } x \text{ from both sides.}$$
$$2y = -x + 5 \qquad \text{Simplify.}$$
$$y = -\frac{1}{2}x + \frac{5}{2} \qquad \text{Divide both sides by 2.}$$

Pick values for x and solve for y.

If we pick $x = 0$, we can find y as follows:

$$y = -\frac{1}{2}x + \frac{5}{2}$$
$$y = -\frac{1}{2}(0) + \frac{5}{2} \qquad \text{Substitute 0 in for } x.$$
$$y = \frac{5}{2} \qquad \text{Simplify.}$$

The ordered pair $\left(0, \frac{5}{2}\right)$ satisfies the equation.

To find another ordered pair, we pick $x = 1$ and find y.

$$y = -\frac{1}{2}x + \frac{5}{2}$$
$$y = -\frac{1}{2}(1) + \frac{5}{2} \qquad \text{Substitute 1 in for } x.$$
$$y = 2 \qquad \text{Simplify.}$$

The ordered pair $(1, 2)$ satisfies the equation.

These pairs and others that satisfy the equation are shown in Figure 2-7. We plot the points and join them with a line to get the graph of the equation.

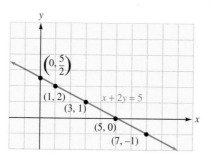

$$x + 2y = 5$$

x	y	(x, y)
0	$\frac{5}{2}$	$\left(0, \frac{5}{2}\right)$
1	2	$(1, 2)$
3	1	$(3, 1)$
5	0	$(5, 0)$
7	-1	$(7, -1)$

Figure 2-7

Self Check 1   Graph: $3x - 2y = 6$.

Comment
Although there are infinitely many points that lie on a line, only two are required to graph a line. However, it is a good idea to find a third point as a check.

EXAMPLE 2   Graph: $3(y + 2) = 2x - 3$.

Solution   We will solve the equation for y and find pairs (x, y) that satisfy the equation. Then, we will plot the points and graph the line.

Solve the equation for y.

$$3(y + 2) = 2x - 3$$

$$3y + 6 = 2x - 3 \qquad \text{Use the distributive property to remove parentheses.}$$

$$3y = 2x - 9 \qquad \text{Subtract 6 from both sides.}$$

$$y = \frac{2}{3}x - 3 \qquad \text{Divide both sides by 3.}$$

Pick values for x and solve for y.
We now substitute numbers for x to find the corresponding values of y. If we let $x = 0$ and find y, we get

$$y = \frac{2}{3}x - 3$$

$$y = \frac{2}{3}(0) - 3 \qquad \text{Substitute 0 for } x.$$

$$y = -3 \qquad \text{Simplify.}$$

The point $(0, -3)$ lies on the graph.
   If we let $x = 3$, we get

$$y = \frac{2}{3}x - 3$$

$$y = \frac{2}{3}(3) - 3 \qquad \text{Substitute 3 for } x.$$

$$y = 2 - 3 \qquad \text{Simplify.}$$

$$y = -1$$

The point $(3, -1)$ lies on the graph.
   We plot these points and others, as in Figure 2-8, and draw the line that passes through the points.

$$3(y + 2) = 2x - 3$$

x	y	(x, y)
-3	-5	$(-3, -5)$
0	-3	$(0, -3)$
3	-1	$(3, -1)$
6	1	$(6, 1)$

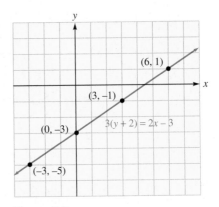

Figure 2-8

Self Check 2   Graph: $2(x - 1) = 6 - 8y$.

In Figure 2-7 in Example 1, the graph intersects the y-axis at the point $\left(0, \frac{5}{2}\right)$, which is called the **$y$-intercept.** It intersects the x-axis at the point $(5, 0)$, which is called the **x-intercept.**

Intercepts of a Line   The **y-intercept** of a line is the point $(0, b)$, where the line intersects the y-axis. To find b, substitute 0 for x in the equation of the line and solve for y.

The **x-intercept** of a line is the point $(a, 0)$, where the line intersects the x-axis. To find a, substitute 0 for y in the equation of the line and solve for x.

EXAMPLE 3   Use the x- and y-intercepts to graph the equation $3x + 2y = 12$.

Solution   To find the y-intercept, we substitute 0 for x and solve for y. To find the x-intercept, we substitute 0 for y and solve for x. We also will find a third point as a check and then plot the points and draw the graph.

Find the y-intercept.
To find the y-intercept, we substitute 0 for x and solve for y.

$$3x + 2y = 12$$
$$3(0) + 2y = 12 \qquad \text{Substitute 0 for } x.$$
$$2y = 12 \qquad \text{Simplify.}$$
$$y = 6 \qquad \text{Divide both sides by 2.}$$

The y-intercept is the point $(0, 6)$.

Find the x-intercept.
To find the x-intercept, we substitute 0 for y and solve for x.

$$3x + 2y = 12$$
$$3x + 2(0) = 12 \qquad \text{Substitute 0 for } y.$$
$$3x = 12 \qquad \text{Simplify.}$$
$$x = 4 \qquad \text{Divide both sides by 3.}$$

The x-intercept is the point $(4, 0)$.

Find a third point as a check.
If we let $x = 2$, we will find that $y = 3$.

$$3x + 2y = 12$$
$$3(2) + 2y = 12 \qquad \text{Substitute 2 for } x.$$
$$6 + 2y = 12 \qquad \text{Simplify.}$$
$$2y = 6 \qquad \text{Subtract 6 from both sides.}$$
$$y = 3 \qquad \text{Divide both sides by 2.}$$

The point (2, 3) satisfies the equation.
    We plot each pair (as in Figure 2-9) and join them with a line to get the graph of the equation.

$3x + 2y = 12$

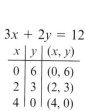

x	y	(x, y)
0	6	(0, 6)
2	3	(2, 3)
4	0	(4, 0)

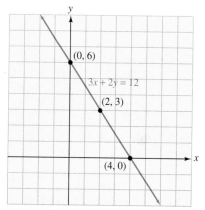

Figure 2-9

Self Check 3   Graph: $2x - 3y = 12$.

3.   Graph Horizontal and Vertical Lines

In the next example, we will graph a horizontal and a vertical line.

EXAMPLE 4   Graph: **a.** $y = 2$ **b.** $x = -3$

Solution   In each case, we will plot a few ordered pairs that satisfy the equation and then draw the graph of the line.

a. In the equation $y = 2$, the value of y is always 2. Any value can be used for x. If we pick x-values of $-3, 0, 2$, and 4, we get the ordered pairs: $(-3, 2), (0, 2),$ $(2, 2),$ and $(4, 2)$. Plotting the pairs shown in Figure 2-10, we see that the graph is a horizontal line, parallel to the x-axis and having a y-intercept of $(0, 2)$. The line has no x-intercept.

b. In the equation $x = -3$, the value of x is always -3. Any value can be used for y. If we pick y-values of $-2, 0, 2$ and 3, we get the ordered pairs: $(-3, -2),$ $(-3, 0), (-3, 2)$ and $(-3, 3)$. After plotting the pairs shown in Figure 2-10, we see that the graph is a vertical line, parallel to the y-axis and having an x-intercept of $(-3, 0)$. The line has no y-intercept.

$y = 2$

x	y	(x, y)
-3	2	$(-3, 2)$
0	2	$(0, 2)$
2	2	$(2, 2)$
4	2	$(4, 2)$

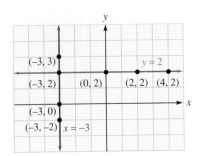

$x = -3$

x	y	(x, y)
-3	-2	$(-3, -2)$
-3	0	$(-3, 0)$
-3	2	$(-3, 2)$
-3	3	$(-3, 3)$

Figure 2-10

Self Check 4   Graph: **a.** $x = 2$ **b.** $y = -3$

Example 4 suggests the following facts.

Equations of Vertical and Horizontal Lines

If a and b are real numbers, then

- The graph of the equation $x = a$ is a vertical line with x-intercept of $(a, 0)$. If $a = 0$, the line $x = 0$ is the y-axis.
- The graph of the equation $y = b$ is a horizontal line with y-intercept of $(0, b)$. If $b = 0$, the line $y = 0$ is the x-axis.

Accent on Technology

Grace Murray Hopper
(1906–1992)
Grace Hopper graduated from Vassar College in 1928 and obtained a master's degree from Yale in 1930. In 1943, she entered the U.S. Naval Reserve. While in the Navy, she became a programmer of the Mark I, the world's first large computer. She is credited for first using the word "bug" to refer to a computer problem. The first bug was actually a moth that flew into one of the relays of the Mark II. From then on, locating computer problems was called "debugging" the system.

Graphing Calculators

We can graph equations with a graphing calculator. To see a graph, we must choose the minimum and maximum values of the x- and y-coordinates that will appear on the calculator's window. A window with standard settings of

$$\text{Xmin} = -10 \qquad \text{Xmax} = 10 \qquad \text{Ymin} = -10 \qquad \text{Ymax} = 10$$

will produce a graph in which the value of x is in the interval $[-10, 10]$, and the value of y is in the interval $[-10, 10]$.

To use a graphing calculator to graph $3x + 2y = 12$, we first solve the equation for y.

$$3x + 2y = 12$$
$$2y = -3x + 12 \qquad \text{Subtract } 3x \text{ from both sides.}$$
$$y = -\frac{3}{2}x + 6 \qquad \text{Divide both sides by 2.}$$

After we enter the right side, the screen should look something like this:

$$Y_1 = -(3/2)X + 6 \qquad \text{or} \qquad f(x) = -(3/2)X + 6$$

We then press GRAPH to obtain the graph shown in Figure 2-11(a). To show more detail, we can draw the graph in a different window. A window with settings of $[-1, 5]$ for x and $[-2, 7]$ for y will give the graph shown in Figure 2-11(b).

Finding intercepts of a graph

We can trace to find the coordinates of any point on a graph. After pressing TRACE, a flashing cursor will appear on the screen. The coordinates of the cursor also will appear at the bottom of the screen.

Continued

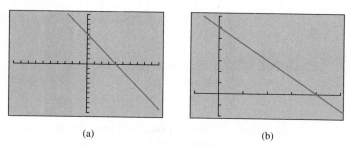

(a)   (b)

Figure 2-11

To find the x-intercept of the graph of $2y = -5x - 7$ $\left(\text{or } y = -\frac{5}{2}x - \frac{7}{2}\right)$, we graph the equation, using $[-10, 10]$ for x and $[-10, 10]$ for y, and press TRACE to get Figure 2-12(a). We then can move the cursor along the line toward the x-intercept until we arrive at a point with the coordinates shown in Figure 2-12(b).

To get better results, we can zoom in to get a magnified picture, trace again, and move the cursor to the point with coordinates shown in Figure 2-12(c). Since the y-coordinate is almost 0, this point is nearly the x-intercept.

We can achieve better results with repeated zooms.

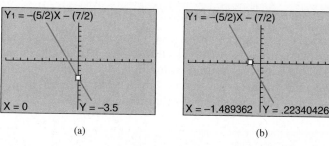
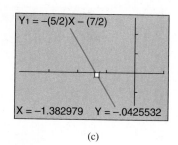

(a)   (b)   (c)

Figure 2-12

We also can find the coordinates of the x-intercept of the graph of $y = -\left(\frac{5}{2}\right)x - \left(\frac{7}{2}\right)$ by using the ZERO command, found under the CALC menu. After we guess left and right bounds and press ENTER after the prompt GUESS, the cursor automatically moves to the x-intercept of the graph, and the coordinates of that point are displayed on the screen. See Figure 2-13(a).

To find the y-intercept of the graph, we can use the VALUE command, which is also found under the CALC menu. With this option, we enter an x-value of 0, as shown in Figure 2-13(b). After we press ENTER, the cursor highlights the y-intercept, and its coordinates are displayed. See Figure 2-13(c).

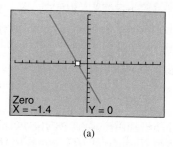
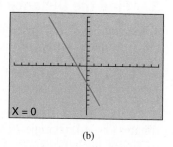
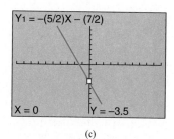

(a)   (b)   (c)

Figure 2-13

4. Solve Problems Using Linear Equations

EXAMPLE 5   A computer purchased for \$2,750 is expected to depreciate according to the formula $y = -550x + \$2,750$, where y is the value of the computer after x years. When will the computer be worth nothing?

Solution   The computer will have no value when its value (y) is 0. To find x when $y = 0$, we substitute 0 for y and solve for x.

$$y = -550x + 2,750$$
$$0 = -550x + 2,750$$
$$-2,750 = -550x \qquad \text{Subtract 2,750 from both sides.}$$
$$5 = x \qquad \text{Divide both sides by } -550.$$

The computer will have no value in 5 years.

Self Check 5   When will the value of the computer be \$1,650?

Accent on Technology

Depreciation

To solve Example 5 with a graphing calculator, we graph $y = -550x + 2,750$ in the window $X = [-10, 10]$ and $Y = [-10, 3,000]$, as shown in Figure 2-14(a). We chose the maximum y-value of 3,000 because y is almost 3,000 when $x = 0$. We then trace to get Figure 2-14(b). We then zoom in and trace again to get Figure 2-14(c), which shows that y is nearly 0 when $x = 5$.

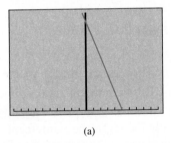
(a)
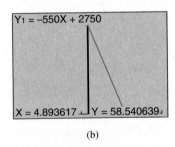
(b)
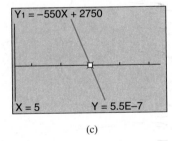
(c)

Figure 2-14

We can obtain the same result using the ZERO command.

---

5. Find the Distance between Two Points

To derive the formula used to find the distance between two points on a rectangular coordinate system, we use **subscript notation** and denote the points as

$P(x_1, y_1)$ Read as "point P with coordinates of x sub 1 and y sub 1."

$Q(x_2, y_2)$ Read as "point Q with coordinates of x sub 2 and y sub 2."

If $P(x_1, y_1)$ and $Q(x_2, y_2)$ are two points in Figure 2-15 and point R has coordinates (x_2, y_1), triangle PQR is a right triangle. By the Pythagorean theorem, the square of the hypotenuse of right triangle PQR is equal to the sum of the squares of the two legs. Because leg RQ is vertical, the square of its length is $(y_2 - y_1)^2$. Since leg PR is horizontal, the square of its length is $(x_2 - x_1)^2$. Thus, we have

(1) $d^2 = (x_2 - x_1)^2 + (y_2 - y_1)^2$

Because equal positive numbers have equal positive square roots, we can take the positive square root of both sides of Equation 1 to obtain the **distance formula**.

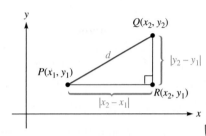

Figure 2-15

$d = \sqrt{(x_2 - x_1)^2 + (y_2 - y_1)^2}$

The Distance Formula The distance d between points (x_1, y_1) and (x_2, y_2) is given by

$$d = \sqrt{(x_2 - x_1)^2 + (y_2 - y_1)^2}$$

EXAMPLE 6 Find the distance between $P(-1, -2)$ and $Q(-7, 8)$.

Solution We use the distance formula, $d = \sqrt{(x_2 - x_1)^2 + (y_2 - y_1)^2}$, to find the distance between $P(-1, -2)$ and $Q(-7, 8)$.

If we let $P(-1, -2) = P(x_1, y_1)$ and $Q(-7, 8) = Q(x_2, y_2)$, we can substitute -1 for x_1, -2 for y_1, -7 for x_2, and 8 for y_2 into the formula and simplify.

$$d(PQ) = \sqrt{(x_2 - x_1)^2 + (y_2 - y_1)^2} \quad \text{Read } d(PQ) \text{ as "the length of segment } PQ\text{."}$$
$$d(PQ) = \sqrt{[-7 - (-1)]^2 + [8 - (-2)]^2}$$
$$= \sqrt{(-6)^2 + (10)^2}$$
$$= \sqrt{36 + 100}$$
$$= \sqrt{136}$$
$$= \sqrt{4 \cdot 34}$$
$$= 2\sqrt{34} \qquad\qquad \sqrt{4 \cdot 34} = \sqrt{4}\sqrt{34} = 2\sqrt{34}$$

Self Check 6 Find the distance between $P(-2, -5)$ and $Q(3, 7)$.

6. Find the Midpoint of a Line Segment

If point M in Figure 2-16 lies midway between points $P(x_1, y_1)$ and $Q(x_2, y_2)$, point M is called the **midpoint** of segment PQ. To find the coordinates of M, we find the average of the x-coordinates and the average of the y-coordinates of P and Q.

The Midpoint Formula The midpoint of the line segment with endpoints at $P(x_1, y_1)$ and $Q(x_2, y_2)$ is the point M with coordinates of

$$M = \left(\frac{x_1 + x_2}{2}, \frac{y_1 + y_2}{2} \right)$$

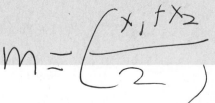

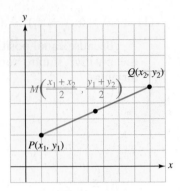

Figure 2-16

You will be asked to prove this formula in Exercise 107 by using the distance formula to show that $d(PM) + d(MQ) = d(PQ)$.

EXAMPLE 7   Find the midpoint of the segment joining $P(-7, 2)$ and $Q(1, -4)$.

Solution   We use the midpoint formula, $M = \left(\dfrac{x_1 + x_2}{2}, \dfrac{y_1 + y_2}{2}\right)$ to find the midpoint of the line segment joining $P(-7, 2)$ and $Q(1, -4)$. To do so, we substitute $P(-7, 2)$ for $P(x_1, y_1)$ and $Q(1, -4)$ for $Q(x_2, y_2)$ into the midpoint formula to get

$$x_M = \frac{x_1 + x_2}{2} \qquad \text{and} \qquad y_M = \frac{y_1 + y_2}{2}$$

$$= \frac{-7 + 1}{2} \qquad\qquad\qquad = \frac{2 + (-4)}{2}$$

$$= \frac{-6}{2} \qquad\qquad\qquad\quad = \frac{-2}{2}$$

$$= -3 \qquad\qquad\qquad\qquad = -1$$

The midpoint is $M(-3, -1)$.

Self Check 7   Find the midpoint of the segment joining $P(-7, -8)$ and $Q(-2, 10)$.

EXAMPLE 8   The midpoint of the segment joining $P(-3, 2)$ and $Q(x_2, y_2)$ is $M(1, 4)$. Find the coordinates of Q.

Solution   We can let $P(x_1, y_1) = P(-3, 2)$ and $M(x_M, y_M) = M(1, 4)$, and then find the coordinates x_2 and y_2 of point $Q(x_2, y_2)$.

$$x_M = \frac{x_1 + x_2}{2} \qquad \text{and} \qquad y_M = \frac{y_1 + y_2}{2}$$

$$1 = \frac{-3 + x_2}{2} \qquad\qquad\qquad 4 = \frac{2 + y_2}{2}$$

$$2 = -3 + x_2 \qquad\qquad\quad 8 = 2 + y_2 \qquad \textbf{Multiply both sides by 2.}$$

$$5 = x_2 \qquad\qquad\qquad\qquad 6 = y_2$$

The coordinates of point Q are $(5, 6)$.

Self Check 8   If the midpoint of a segment PQ is $M(2, -5)$ and one endpoint is $Q(6, 9)$, find P.

Self Check Answers

1.
2.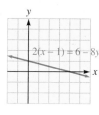
3.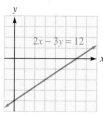

4.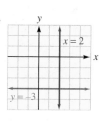

5. 2 years   6. 13   7. $M\left(-\frac{9}{2}, 1\right)$ 8. $(-2, -19)$

2.1 Exercises

Vocabulary and Concepts   *Fill in the blanks.*

1. The coordinate axes divide the plane into four _____.

2. The coordinate axes intersect at the _____.

3. The positive direction on the x-axis is _____.

4. The positive direction on the y-axis is _____.

5. The x-coordinate is the ____ coordinate in an ordered pair.

6. The y-coordinate is the _____ coordinate in an ordered pair.

7. A _____ equation is an equation whose graph is a line.

8. The point where a line intersects the _____ is called the y-intercept.

9. The point where a line intersects the x-axis is called the _____.

10. The graph of the equation $x = a$ will be a _____ line.

11. The graph of the equation $y = b$ will be a _____ line.

12. Complete the distance formula:

    $d = $ _____

13. If a point divides a segment into two equal segments, the point is called the _____ of the segment.

14. The midpoint of the segment joining $P(x_1, y_1)$ and $Q(x_2, y_2)$ is _____.

Practice   *Refer to the illustration and determine the coordinates of each point.*

15. A
16. B
17. C
18. D
19. E
20. F
21. G
22. H

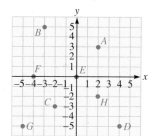

Graph each point. Indicate the quadrant in which the point lies, or the axis on which it lies.

23. $(2, 5)$ 24. $(-3, 4)$
25. $(-4, -5)$ 26. $(6, 2)$
27. $(5, 2)$ 28. $(3, -4)$
29. $(4, 0)$ 30. $(0, 2)$

Solve each equation for y and graph the equation. Then check your graph with a graphing calculator.

31. $y - 2x = 7$ 32. $y + 3 = -4x$

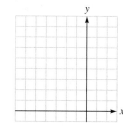
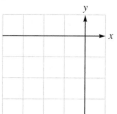

33. $y + 5x = 5$

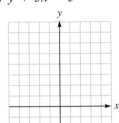

34. $y - 3x = 6$

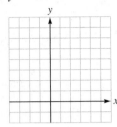

43. $2x - y = 4$

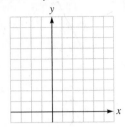

44. $3x + y = 9$

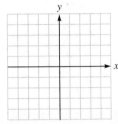

35. $6x - 3y = 10$

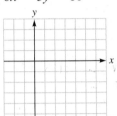

36. $4x + 8y - 1 = 0$

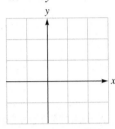

45. $3x + 2y = 6$

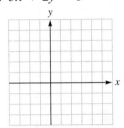

46. $2x - 3y = 6$

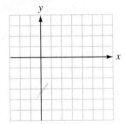

37. $3x = 6y - 1$

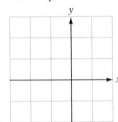

38. $2x + 1 = 4y$

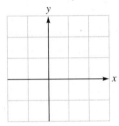

47. $4x - 5y = 20$

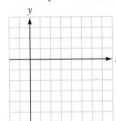

48. $3x - 5y = 15$

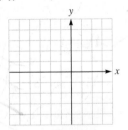

39. $2(x + y + 1) = x + 2$

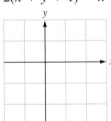

40. $5(x + 2) = 3y - x$

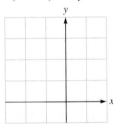

Graph each equation.

49. $y = 3$

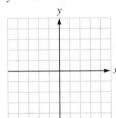

50. $x = -4$

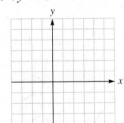

Find the x- and y-intercepts and use them to graph each equation.

41. $x + y = 5$

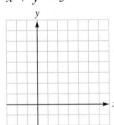

42. $x - y = 3$

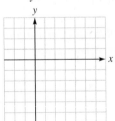

51. $3x + 5 = -1$

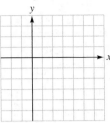

52. $7y - 1 = 6$

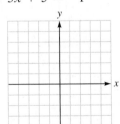

53. $3(y + 2) = y$

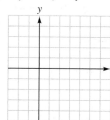

54. $4 + 3y = 3(x + y)$

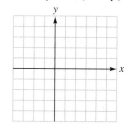

55. $3(y + 2x) = 6x + y$

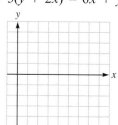

56. $5(y - x) = x + 5y$

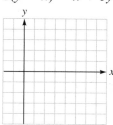

Use a graphing calculator to graph each equation and then find the x-coordinate of the x-intercept to the nearest hundredth.

57. $y = 3.7x - 4.5$

58. $y = \dfrac{3}{5}x + \dfrac{5}{4}$

59. $1.5x - 3y = 7$

60. $0.3x + y = 7.5$

Find the distance between P and O(0, 0).

61. $P(4, -3)$ **62.** $P(-5, 12)$

63. $P(-3, 2)$ **64.** $P(5, 0)$

65. $P(1, 1)$ **66.** $P(6, -8)$

67. $P\left(\sqrt{3}, 1\right)$ **68.** $P\left(\sqrt{7}, \sqrt{2}\right)$

Find the distance between P and Q.

69. $P(3, 7)$; $Q(6, 3)$ **70.** $P(4, 9)$; $Q(9, 21)$

71. $P(4, -6)$; $Q(-1, 6)$ **72.** $P(0, 5)$; $Q(6, -3)$

73. $P(-2, -15)$; $Q(-9, -39)$

74. $P(-7, 11)$; $Q(3, -13)$

75. $P(3, -3)$; $Q(-5, 5)$

76. $P(6, -3)$; $Q(-3, 2)$

77. $P(\pi, -2)$; $Q(\pi, 5)$

78. $P\left(\sqrt{5}, 0\right)$; $Q(0, 2)$

Find the midpoint of the line segment PQ.

79. $P(2, 4)$; $Q(6, 8)$ **80.** $P(3, -6)$; $Q(-1, -6)$

81. $P(2, -5)$; $Q(-2, 7)$ **82.** $P(0, 3)$; $Q(-10, -13)$

83. $P(-8, 5)$; $Q(8, -5)$ **84.** $P(3, -2)$; $Q(2, -3)$

85. $P(0, 0)$; $Q\left(\sqrt{5}, \sqrt{5}\right)$

86. $P\left(\sqrt{3}, 0\right)$; $Q\left(0, -\sqrt{5}\right)$

One endpoint P and the midpoint M of line segment PQ are given. Find the coordinates of the other endpoint, Q.

87. $P(1, 4)$; $M(3, 5)$ **88.** $P(2, -7)$; $M(-5, 6)$

89. $P(5, -5)$; $M(5, 5)$ **90.** $P(-7, 3)$; $M(0, 0)$

91. Show that a triangle with vertices at $(13, -2)$, $(9, -8)$, and $(5, -2)$ is isosceles.

92. Show that a triangle with vertices at $(-1, 2)$, $(3, 1)$, and $(4, 5)$ is isosceles.

93. In the illustration, points M and N are the midpoints of AC and BC, respectively. Find the length of MN.

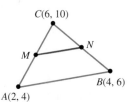

94. In the illustration, points M and N are the midpoints of AC and BC, respectively. Show that $d(MN) = \frac{1}{2}[d(AB)]$.

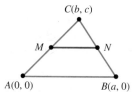

95. In the illustration, point M is the midpoint of the hypotenuse of right triangle AOB. Show that the area of rectangle $OLMN$ is one-half of the area of triangle AOB.

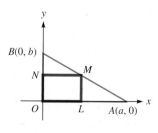

96. Rectangle $ABCD$ in the illustration is twice as long as it is wide, and its sides are parallel to the coordinate axes. If the perimeter is 42, find the coordinates of point C.

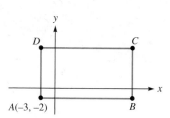

Applications

97. House appreciation A house purchased for $225,000 is expected to appreciate according to the formula $y = 17,500x + 225,000$, where y is the value of the house after x years. Find the value of the house 5 years later.

98. Car depreciation A car purchased for $17,000 is expected to depreciate according to the formula $y = -1,360x + 17,000$. When will the car be worthless?

99. Demand equations The number of photo scanners that consumers buy depends on price. The higher the price, the fewer photo scanners people will buy. The equation that relates price to the number of photo scanners sold at that price is called a **demand equation.** If the demand equation for a photo scanner is $p = -\frac{1}{10}q + 170$, where p is the price and q is the number of photo scanners sold at that price, how many photo scanners will be sold at a price of $150?

100. Supply equations The number of television sets that manufacturers produce depends on price. The higher the price, the more TVs manufacturers will produce. The equation that relates price to the number of TVs produced at that price is called a **supply equation.** If the supply equation for a 13-inch LCD TV is $p = \frac{1}{10}q + 130$, where p is the price and q is the number of TVs produced for sale at that price, how many TVs will be produced if the price is $150?

101. Meshing gears The rotational speed V of a large gear (with N teeth) is related to the speed v of the smaller gear (with n teeth) by the equation $V = \frac{nv}{N}$. If the larger gear in the illustration is making 60 revolutions per minute, how fast is the smaller gear spinning?

102. Crime prevention The number n of incidents of family violence requiring police response appears to be related to d, the money spent on crisis intervention, by the equation $n = 430 - 0.005d$. What expenditure would reduce the number of incidents to 350?

103. Navigation See the illustration. An ocean liner is located 23 miles east and 72 miles north of Pigeon Cove Lighthouse, and its home port is 47 miles west and 84 miles south of the lighthouse. How far is the ship from port?

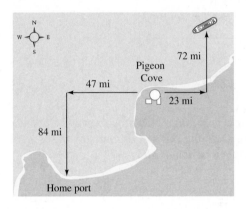

104. Engineering Two holes are to be drilled at locations specified by the engineering drawing shown in the illustration. Find the distance between the centers of the holes.

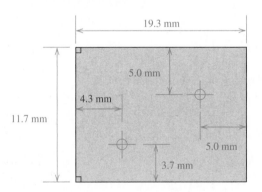

Discovery and Writing

105. Explain how to graph a line using the intercept method.

106. Explain how to determine the quadrant in which the point $P(a, b)$ lies.

107. In Figure 2-16, show that $d(PM) + d(MQ) = d(PQ)$.

108. Use the result of Exercise 107 to explain why point M is the midpoint of segment PQ.

Review   *Graph each interval on the number line.*

109. $[-3, 2) \cup (-2, 3]$ **110.** $(-1, 4) \cap [-2, 2]$

111. $[-3, -2) \cap (2, 3]$ **112.** $[-4, -3) \cup (2, 3]$

Solve each equation.

113. $\dfrac{3}{y + 6} = \dfrac{4}{y + 4}$

114. $\dfrac{z + 4}{z^2 + z} - \dfrac{z + 1}{z^2 + 2z} = \dfrac{8}{z^2 + 3z + 2}$

2.2   The Slope of a Nonvertical Line

Objectives

1. Find the Slope of a Line
2. Use Slope to Solve Problems
3. Find Slopes of Horizontal and Vertical Lines
4. Find Slopes of Parallel and Perpendicular Lines

The world's steepest passenger railway is the Lookout Mountain Incline Railway in Chattanooga, Tennessee. Passengers experience breathtaking views of the city and surrounding mountains as the trolley-style railcars travel up Lookout Mountain. The grade or steepness of the track is 72.7% near the top.

Mathematicians use the term **slope** to represent the measure of the steepness of a line. We will explore the topic of slope in this section because it has many real-life applications.

1.   Find the Slope of a Line

Suppose that a college student rents a room for $300 per month, plus a $200 non-refundable deposit. The table shown in Figure 2-17(a) gives the cost (y) for several numbers of months (x). If we construct a graph from these data, we get the line shown in Figure 2-17(b).

Time in months (x)	Total cost (y)
0	200
1	500
2	800
3	1,100
4	1,400

(a)

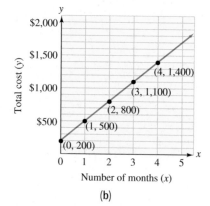

(b)

Figure 2-17

From the graph, we can see that if x changes from 0 to 1, y changes from 200 to 500. As x changes from 1 to 2, y changes from 500 to 800, and so on. The ratio of the change in y divided by the change in x is the constant 300.

$$\frac{\text{Change in } y}{\text{Change in } x} = \frac{500 - 200}{1 - 0} = \frac{800 - 500}{2 - 1} = \frac{1,100 - 800}{3 - 2} = \frac{1,400 - 1,100}{4 - 3} = \frac{300}{1} = 300$$

The ratio of the change in y divided by the change in x between any two points on any line is always a constant. This constant rate of change is called the **slope** of the line.

The Slope of a Nonvertical Line

The **slope of the nonvertical line** (see Figure 2-18) passing through points $P(x_1, y_1)$ and $Q(x_2, y_2)$ is

$$m = \frac{\text{change in } y}{\text{change in } x} = \frac{y_2 - y_1}{x_2 - x_1} \quad (x_2 \neq x_1)$$

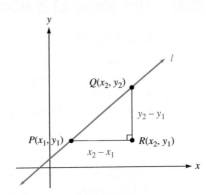

Figure 2-18

Comment

Slope is often considered to be a measure of the steepness or tilt of a line. Note that you can use the coordinates of any two points on a line to compute the slope of the line.

EXAMPLE 1

Find the slope of the line passing through $P(-1, -2)$ and $Q(7, 8)$. (See Figure 2-19.)

Solution

We will substitute the points $P(-1, -2)$ and $Q(7, 8)$ into the slope formula, $m = \frac{\text{change in } y}{\text{change in } x} = \frac{y_2 - y_1}{x_2 - x_1}$, to find the slope of the line.

Let $P(x_1, y_1) = P(-1, -2)$ and $Q(x_2, y_2) = Q(7, 8)$. Then we substitute -1 for x_1, -2 for y_1, 7 for x_2, and 8 for y_2 to get

$$m = \frac{\text{change in } y}{\text{change in } x}$$

$$m = \frac{y_2 - y_1}{x_2 - x_1}$$

$$= \frac{8 - (-2)}{7 - (-1)}$$

$$= \frac{10}{8}$$

$$= \frac{5}{4}$$

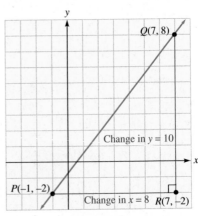

Figure 2-19

The slope of the line is $\frac{5}{4}$. We would have obtained the same result if we had let $P(x_1, y_1) = P(7, 8)$ and $Q(x_2, y_2) = Q(-1, -2)$.

Self Check 1   Find the slope of the line passing through $P(-3, -4)$ and $Q(5, 9)$.

Comment

When calculating slope, always subtract the y-values and the x-values in the same order.

$$m = \frac{y_2 - y_1}{x_2 - x_1} \quad \text{or} \quad m = \frac{y_1 - y_2}{x_1 - x_2}$$

Otherwise, you will obtain an incorrect result.

A slope can be a positive real number, 0, or a negative real number. If the denominator of the slope formula is 0, slope is not defined.

The change in y (often denoted as Δy) is the **rise** of the line between points P and Q. The change in x (often denoted as Δx) is the **run.** Using this terminology, we can define slope to be the ratio of the rise to the run:

$$m = \frac{y_2 - y_1}{x_2 - x_1} = \frac{\Delta y}{\Delta x} = \frac{\text{rise}}{\text{run}} \quad (\Delta x \neq 0)$$

EXAMPLE 2   Find the slope of the line determined by $5x + 2y = 10$. (See Figure 2-20.)

Solution   We will find the coordinates of the y- and x-intercepts and substitute into the slope formula, $m = \frac{\text{change in } y}{\text{change in } x} = \frac{y_2 - y_1}{x_2 - x_1}$, to find the slope of the line.

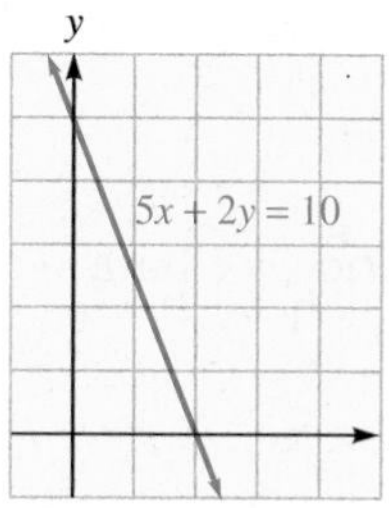

- If $y = 0$, then $x = 2$, and the point $(2, 0)$ lies on the line.
- If $x = 0$, then $y = 5$, and the point $(0, 5)$ lies on the line.

We then find the slope of the line between $P(2, 0)$ and $Q(0, 5)$.

$$m = \frac{\text{change in } y}{\text{change in } x}$$

$$m = \frac{y_2 - y_1}{x_2 - x_1}$$

$$= \frac{5 - 0}{0 - 2}$$

$$= -\frac{5}{2}$$

The slope is $-\frac{5}{2}$.

Figure 2-20

Self Check 2   Find the slope of the line determined by $3x - 2y = 9$.

2. Use Slope to Solve Problems

EXAMPLE 3   If carpet costs $25 per square yard plus a delivery charge of $30, the total cost c of n square yards is given by the formula

Total cost	equals	cost per square yard	times	the number of square yards purchased	plus	the delivery charge.
c	$=$	25	$\cdot$	n	$+$	30

Graph the equation $c = 25n + 30$ and interpret the slope of the line.

Solution   We will complete a table of solutions and graph the equation on a coordinate system with a vertical c-axis and a horizontal n-axis. Figure 2-21 shows a table of ordered pairs and the graph.

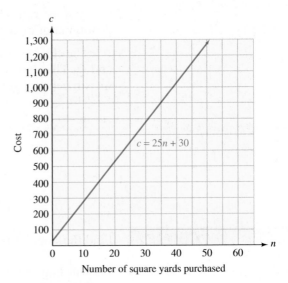

$$c = 25n + 30$$

n	c	(n, c)
10	280	(10, 280)
20	530	(20, 530)
30	780	(30, 780)
40	1,030	(40, 1,030)
50	1,280	(50, 1,280)

Figure 2-21

If we pick the points (30, 780) and (50, 1,280) to find the slope, we have

$$m = \frac{\Delta c}{\Delta n}$$

$$= \frac{c_2 - c_1}{n_2 - n_1}$$

$$= \frac{1{,}280 - 780}{50 - 30} \quad \text{Substitute 1,280 for } c_2\text{, 780 for } c_1\text{, 50 for } n_2\text{, and 30 for } n_1.$$

$$= \frac{500}{20}$$

$$= 25$$

The slope of 25 (in dollars/square yard) is the cost per square yard of the carpet.

Everyday Connections

Sales of Digital and Non-Digital Media

"It's important to note that a lot of people are beholden to the old media world and, by definition, the new world changes things and the old world is always going to resist that change. I think that there is skepticism from people and that's OK. I accept that as a challenge. Hopefully, that makes us stronger."

Michael Robertson, founder of MP3.com

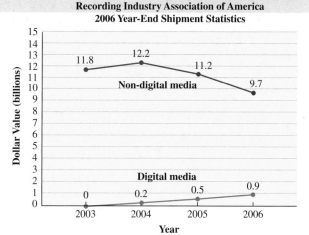

Source: www.riaa.com

Since 2003, the sales of non-digital media have declined, and the sales of digital media have increased, as shown in the following graph.

We can approximate the average **rate of growth (or decrease)** of a quantity during a given time interval by calculating the slope of the line segment that connects the endpoints of the graph on the given interval.

Use the data from the graph to compute the average rate of growth (or decrease) of the following:

1. Shipped digital media value from 2004–2006.

2. Shipped non-digital media value from 2004–2006.

EXAMPLE 4 It takes a skier 25 minutes to complete the course shown in Figure 2-22. Find his average rate of descent in feet per minute.

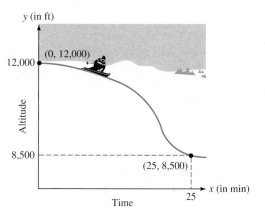

Figure 2-22

Solution To find the average rate of descent, we will find the ratio of the change in altitude to the change in time. To find this ratio, we will calculate the slope of the line passing through the points $(0, 12{,}000)$ and $(25, 8{,}500)$.

$$\begin{aligned}
\text{Average rate} &= \frac{12{,}000 - 8{,}500}{0 - 25} \\
\text{of descent} &\\
&= \frac{3{,}500}{-25} \\
&= -140
\end{aligned}$$

The average rate of descent is -140 ft/min.

3. Find Slopes of Horizontal and Vertical Lines

If $P(x_1, y_1)$ and $Q(x_2, y_2)$ are points on the horizontal line shown in Figure 2-23(a), then $y_1 = y_2$, and the numerator of the fraction

$$\frac{y_2 - y_1}{x_2 - x_1} \quad \text{On a horizontal line, } x_2 \neq x_1.$$

is 0. Thus, the value of the fraction is 0, and the slope of the horizontal line is 0.

If $P(x_1, y_1)$ and $Q(x_2, y_2)$ are points on the vertical line shown in Figure 2-23(b), then $x_1 = x_2$, and the denominator of the fraction

$$\frac{y_2 - y_1}{x_2 - x_1} \quad \text{On a vertical line, } y_2 \neq y_1.$$

is 0. Since the denominator of a fraction cannot be 0, the slope of a vertical line is not defined.

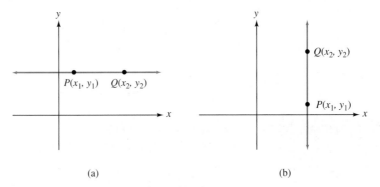

(a)   (b)

Figure 2-23

Slopes of Horizontal and Vertical Lines

The slope of a horizontal line (a line with an equation of the form $y = b$) is 0.

The slope of a vertical line (a line with an equation of the form $x = a$) is not defined.

If a line rises as we follow it from left to right, as in Figure 2-24(a), its slope is positive. If a line drops as we follow it from left to right, as in Figure 2-24(b), its slope is negative.

If a line is horizontal, as in Figure 2-24(c), its slope is 0. If a line is vertical, as in Figure 2-24(d), it has no defined slope.

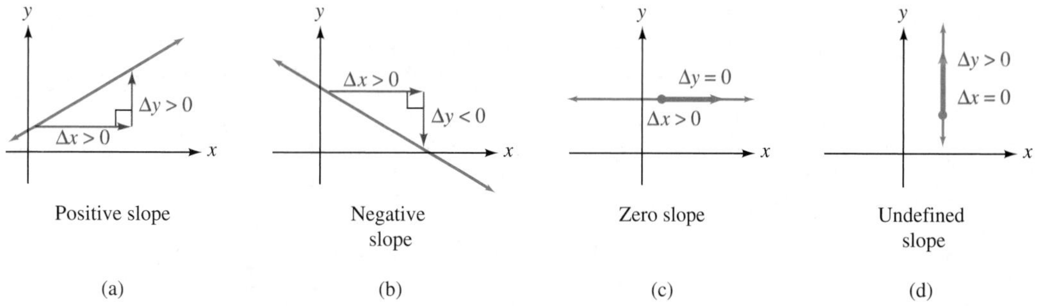

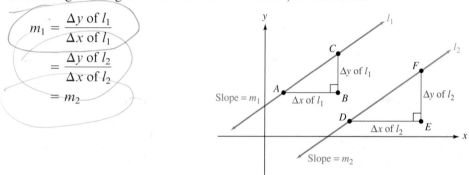

Figure 2-24

4.   Find Slopes of Parallel and Perpendicular Lines

To see a relationship between parallel lines and their slopes, we refer to the parallel lines l_1 and l_2 shown in Figure 2-25, with slopes of m_1 and m_2, respectively. Because right triangles ABC and DEF are similar, it follows that

$$m_1 = \frac{\Delta y \text{ of } l_1}{\Delta x \text{ of } l_1}$$

$$= \frac{\Delta y \text{ of } l_2}{\Delta x \text{ of } l_2}$$

$$= m_2$$

Figure 2-25

This shows that if two nonvertical lines are parallel, they have the same slope. It also is true that when two lines have the same slope, they are parallel.

Slopes of Parallel Lines   Nonvertical parallel lines have the same slope, and lines having the same slope are parallel.

Since vertical lines are parallel, lines with undefined slopes are parallel.

EXAMPLE 5   The lines in Figure 2-26 are parallel. Find y.

Solution   Since the lines are parallel, their slopes are equal. To find y, we will find the slope of each line, set them equal, and solve the resulting equation.

slope of PQ = slope of RS

$$\frac{-2 - 4}{1 - (-3)} = \frac{y - 5}{3 - (-2)}$$

$$\frac{-6}{4} = \frac{y - 5}{5} \qquad \text{Simplify.}$$

$$-30 = 4(y - 5) \qquad \text{Multiply both sides by 20.}$$

$$-30 = 4y - 20 \qquad \text{Remove parentheses and simplify.}$$

$$-10 = 4y \qquad \text{Add 20 to both sides.}$$

$$-\frac{5}{2} = y \qquad \text{Divide both sides by 4 and simplify.}$$

Thus, $y = -\frac{5}{2}$.

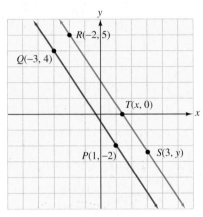

Figure 2-26

Self Check 5   Find x in Figure 2-26.

The following theorem relates perpendicular lines and their slopes.

Slopes of Perpendicular Lines	If two nonvertical lines are perpendicular, the product of their slopes is -1.
	If the product of the slopes of two lines is -1, the lines are perpendicular.

Proof   Suppose l_1 and l_2 are lines with slopes of m_1 and m_2 that intersect at some point. See Figure 2-27. Then superimpose a coordinate system over the lines so that the intersection point is the origin. Let $P(a, b)$ be a point on l_1, and let $Q(c, d)$ be a point on l_2. Neither point P nor point Q can be the origin.

Comment

If the product of two numbers is -1, the numbers are called **negative reciprocals.**

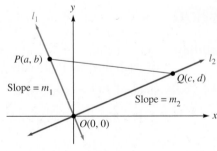

Figure 2-27

First, we suppose that l_1 and l_2 are perpendicular. Then triangle POQ is a right triangle with its right angle at O. By the Pythagorean theorem,

$$d(OP)^2 + d(OQ)^2 = d(PQ)^2$$
$$(a - 0)^2 + (b - 0)^2 + (c - 0)^2 + (d - 0)^2 = (a - c)^2 + (b - d)^2$$
$$a^2 + b^2 + c^2 + d^2 = a^2 - 2ac + c^2 + b^2 - 2bd + d^2$$
$$0 = -2ac - 2bd$$
$$bd = -ac$$

(1)
$$\frac{b}{a} \cdot \frac{d}{c} = -1 \qquad \text{Divide both sides by } ac.$$

The coordinates of P are (a, b), and the coordinates of O are $(0, 0)$. Using the definition of slope, we have

$$m_1 = \frac{b - 0}{a - 0} = \frac{b}{a}$$

Similarly, we have

$$m_2 = \frac{d}{c}$$

We substitute m_1 for $\frac{b}{a}$ and m_2 for $\frac{d}{c}$ in Equation 1 to obtain

$$m_1 m_2 = -1$$

Hence, if lines l_1 and l_2 are perpendicular, the product of their slopes is -1.

Conversely, we suppose that the product of the slopes of lines l_1 and l_2 is -1. Because the steps in the previous discussion are reversible, we have $d(OP)^2 + d(OQ)^2 = d(PQ)^2$. By the Pythagorean theorem, triangle POQ is a right triangle. Thus, l_1 and l_2 are perpendicular.

It is also true that a horizontal line is perpendicular to a vertical line.

EXAMPLE 6   Are the lines shown in Figure 2-28 perpendicular?

Solution   We will determine the slopes of the lines and see whether their product is -1.

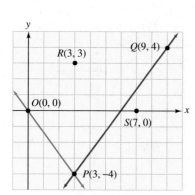

$$\text{Slope of } OP = \frac{\Delta y}{\Delta x} \qquad\qquad \text{Slope of } PQ = \frac{\Delta y}{\Delta x}$$

$$= \frac{y_2 - y_1}{x_2 - x_1} \qquad\qquad = \frac{y_2 - y_1}{x_2 - x_1}$$

$$= \frac{-4 - 0}{3 - 0} \qquad\qquad = \frac{4 - (-4)}{9 - 3}$$

$$= -\frac{4}{3} \qquad\qquad = \frac{8}{6}$$

$$\qquad\qquad\qquad\qquad = \frac{4}{3}$$

Figure 2-28

Since the product of the slopes is $-\frac{16}{9}$ and not -1, the lines are not perpendicular.

Self Check 6   Is either line in Figure 2-28 perpendicular to the line passing through R and S?

Self Check Answers   **1.** $\frac{13}{8}$ **2.** $\frac{3}{2}$ **5.** $\frac{4}{3}$ **6.** yes

2.2 Exercises

Vocabulary and Concepts   *Fill in the blanks.*

1. The slope of a nonvertical line is defined to be the change in y _____ by the change in x.
2. The change in __ is often called the rise.
3. The change in x is often called the ___.
4. When computing the slope from the coordinates of two points, always subtract the y-values and the x-values in the _____.
5. The symbol Δy means _____ y.
6. The slope of a _____ line is 0.
7. The slope of a _____ line is undefined.
8. If the slopes of two lines are equal, the lines are _____.
9. If the product of the slopes of two lines is -1, the lines are _____.
10. If two lines are perpendicular, the product of their slopes is ___.

Practice   *Find the slope of the line passing through each pair of points, if possible.*

11.

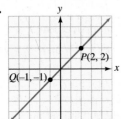

12.

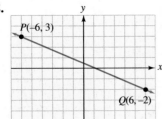

13. $P(2, 5)$; $Q(3, 10)$

14. $P(3, -1)$; $Q(5, 3)$

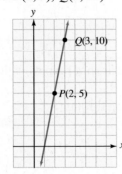

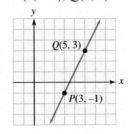

15. $P(3, -2)$; $Q(-1, 5)$
16. $P(3, 7)$; $Q(6, 16)$
17. $P(8, -7)$; $Q(4, 1)$
18. $P(5, 17)$; $Q(17, 17)$
19. $P(-4, 3)$; $Q(-4, -3)$
20. $P\left(2, \sqrt{7}\right)$; $Q\left(\sqrt{7}, 2\right)$
21. $P\left(\dfrac{3}{2}, \dfrac{2}{3}\right)$; $Q\left(\dfrac{5}{2}, \dfrac{7}{3}\right)$
22. $P\left(-\dfrac{2}{5}, \dfrac{1}{3}\right)$; $Q\left(\dfrac{3}{5}, -\dfrac{5}{3}\right)$
23. $P(a + b, c)$; $Q(b + c, a)$ assume $c \neq a$
24. $P(b, 0)$; $Q(a + b, a)$ assume $a \neq 0$

Find two points on the line and find the slope of the line.

25. $y = 3x + 2$ 26. $y = 5x - 8$
27. $5x - 10y = 3$ 28. $8y + 2x = 5$
29. $3(y + 2) = 2x - 3$ 30. $4(x - 2) = 3y + 2$
31. $3(y + x) = 3(x - 1)$ 32. $2x + 5 = 2(y + x)$

Find the slope of the line, if possible.

33. $y = 7$ 34. $2y = 5$
35. $x = -\dfrac{1}{2}$ 36. $x - 7 = 0$

Determine whether the slope of the line is positive, negative, 0, or undefined.

37.

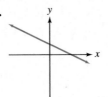

38.

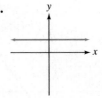

39.

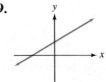

40.

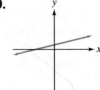

41.

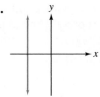

42.

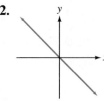

Determine whether the lines with the given slopes are parallel, perpendicular, or neither.

43. $m_1 = 3$; $m_2 = -\dfrac{1}{3}$ **44.** $m_1 = \dfrac{2}{3}$; $m_2 = \dfrac{3}{2}$

45. $m_1 = \sqrt{8}$; $m_2 = 2\sqrt{2}$ **46.** $m_1 = 1$; $m_2 = -1$

47. $m_1 = -\sqrt{2}$; $m_2 = \dfrac{\sqrt{2}}{2}$

48. $m_1 = 2\sqrt{7}$; $m_2 = \sqrt{28}$

49. $m_1 = -0.125$; $m_2 = 8$

50. $m_1 = 0.125$; $m_2 = \dfrac{1}{8}$

51. $m_1 = ab^{-1}$; $m_2 = -a^{-1}b$ $(a \neq 0, b \neq 0)$

52. $m_1 = \left(\dfrac{a}{b}\right)^{-1}$; $m_2 = -\dfrac{b}{a}$ $(a \neq 0, b \neq 0, a \neq b)$

Determine whether the line through the given points and the line through $R(-3, 5)$ and $S(2, 7)$ are parallel, perpendicular, or neither.

53. $P(2, 4)$; $Q(7, 6)$ **54.** $P(-3, 8)$; $Q(-13, 4)$

55. $P(-4, 6)$; $Q(-2, 1)$ **56.** $P(0, -9)$; $Q(4, 1)$

57. $P(a, a)$; $Q(3a, 6a)$ $(a \neq 0)$
58. $P(b, b)$; $Q(-b, 6b)$ $(a \neq 0)$

Find the slopes of lines PQ and PR, and determine whether points P, Q, and R lie on the same line.

59. $P(-2, 8)$; $Q(-6, 9)$; $R(2, 5)$
60. $P(1, -1)$; $Q(3, -2)$; $R(-3, 0)$
61. $P(-a, a)$; $Q(0, 0)$; $R(a, -a)$
62. $P(a, a + b)$; $Q(a + b, b)$; $R(a - b, a)$

Determine which, if any, of the three lines PQ, PR, and QR are perpendicular.

63. $P(5, 4)$; $Q(2, -5)$; $R(8, -3)$

64. $P(8, -2)$; $Q(4, 6)$; $R(6, 7)$

65. $P(1, 3)$; $Q(1, 9)$; $R(7, 3)$

66. $P(2, -3)$; $Q(-3, 2)$; $R(3, 8)$

67. $P(0, 0)$; $Q(a, b)$; $R(-b, a)$

68. $P(a, b)$; $Q(-b, a)$; $R(a - b, a + b)$

69. Right triangles Show that the points $A(-1, -1)$, $B(-3, 4)$, and $C(4, 1)$ are the vertices of a right triangle.

70. Right triangles Show that the points $D(0, 1)$, $E(-1, 3)$, and $F(3, 5)$ are the vertices of a right triangle.

71. Squares Show that the points $A(1, -1)$, $B(3, 0)$, $C(2, 2)$, and $D(0, 1)$ are the vertices of a square.

72. Squares Show that the points $E(-1, -1)$, $F(3, 0)$, $G(2, 4)$, and $H(-2, 3)$ are the vertices of a square.

73. Parallelograms Show that the points $A(-2, -2)$, $B(3, 3)$, $C(2, 6)$, and $D(-3, 1)$ are the vertices of a parallelogram. (Show that both pairs of opposite sides are parallel.)

74. Trapezoids Show that points $E(1, -2)$, $F(5, 1)$, $G(3, 4)$, and $H(-3, 4)$ are the vertices of a trapezoid. (Show that only one pair of opposite sides is parallel.)

75. Geometry In the illustration, points M and N are midpoints of CB and BA, respectively. Show that MN is parallel to AC.

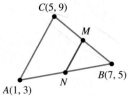

76. Geometry In the illustration, $d(AB) = d(AC)$. Show that AD is perpendicular to BC.

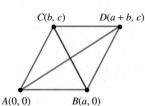

Applications

77. Rate of growth When a college started an aviation program, the administration agreed to predict enrollments using a straight-line method. If the enrollment during the first year was 12, and the enrollment during the fifth year was 26, find the rate of growth per year (the slope of the line). See the illustration.

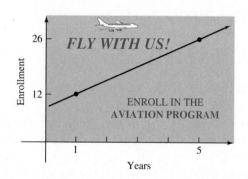

78. Rate of growth A small business predicts sales according to a straight-line method. If sales were $50,000 in the first year and $110,000 in the third year, find the rate of growth in dollars per year (the slope of the line).

79. Rate of decrease The price of computers has been dropping steadily for the past ten years. If a desktop PC cost $6,700 ten years ago, and the same computing power cost $2,200 three years ago, find the rate of decrease per year. (Assume a straight-line model.)

80. Hospital costs The table shows the changing mean daily cost for a hospital room. For the ten-year period, find the rate of change per year of the portion of the room cost that is absorbed by the hospital.

Year	Total cost to the hospital	Amount passed on to patient
1995	$459	$214
2000	670	295
2005	812	307

81. Charting temperature changes The following Fahrenheit temperature readings were recorded over a four-hour period.

Time	12:00	1:00	2:00	3:00	4:00
Temperature	47°	53°	59°	65°	71°

Let t represent the time (in hours), with 12:00 corresponding to $t = 0$. Let T represent the temperature. Plot the points (t, T), and draw the line through those points. Explain the meaning of $\frac{\Delta T}{\Delta t}$.

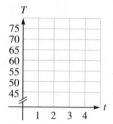

82. Tracking the Dow The Dow Jones Industrial Averages at the close of trade on three consecutive days were as follows:

Day	Monday	Tuesday	Wednesday
Close	12,981	12,964	12,947

Let d represent the day, with $d = 0$ corresponding to Monday, and let D represent the Dow Jones average. Plot the points (d, D), and draw the graph. Explain the meaning of $\frac{\Delta D}{\Delta d}$.

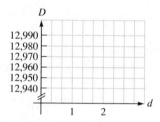

83. Speed of an airplane A pilot files a flight plan indicating her intention to fly at a constant speed of 590 mph. Write an equation that expresses the distance traveled in terms of the flying time. Then graph the equation and interpret the slope of the line. (*Hint: d = rt.*)

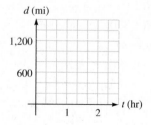

84. Growth of savings A student deposits $25 each month in a Holiday Club account at her bank. The account pays no interest. Write an equation that expresses the amount in her account in terms of the number of deposits. Then graph the line, and interpret the slope of the line.

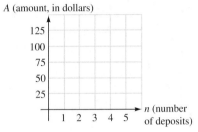

A (amount, in dollars)

125
100
75
50
25

1 2 3 4 5   n (number of deposits)

Discovery and Writing

85. Explain why the slope of a vertical line is undefined.

86. Explain how to determine whether two lines are parallel, perpendicular, or neither.

Review   *Solve each equation for y and simplify.*

87. $3x + 7y = 21$

88. $y - 3 = 5(x + 2)$

89. $\dfrac{x}{5} + \dfrac{y}{2} = 1$

90. $x - 5y = 15$

Factor each expression.

91. $6p^2 + p - 12$

92. $b^3 - 27$

93. $mp + mq + np + nq$

94. $x^4 + x^2 - 2$

2.3 # Writing Equations of Lines

Objectives

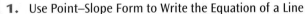

1. Use Point–Slope Form to Write the Equation of a Line
2. Use Slope–Intercept Form to Write the Equation of a Line
3. Graph Linear Equations Using the Slope and *y*-Intercept
4. Determine Whether Graphs of Linear Equations Are Parallel, Perpendicular, or Neither
5. Write Equations of Parallel and Perpendicular Lines
6. Recognize and Use the Standard Form of the Equation of a Line
7. Write the Equation of a Line that Models a Real-Life Problem
8. Use Linear Curve Fitting to Solve Problems

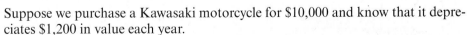

Suppose we purchase a Kawasaki motorcycle for $10,000 and know that it depreciates $1,200 in value each year.

We can use the facts given to write the linear equation that represents the value *y* of the motorcycle *x* years after it was purchased. Because the motorcycle's value decreases $1,200 each year, the slope of its line graph is $-1,200$. Because the purchase price is $10,000, we know that when we let $x = 0$, the value of *y* will equal 10,000. The linear equation that satisfies these two conditions is $y = -1,200x + 10,000$. This equation represents the straight-line depreciation of the motorcycle.

In this section, we will write the equations of lines given specific characteristics or features of the line.

1.   Use Point–Slope Form to Write the Equation of a Line

Suppose that line l in Figure 2-29 has a slope of m and passes through the point $P(x_1, y_1)$. If $Q(x, y)$ is any other point on line l, we have

$$m = \frac{y - y_1}{x - x_1}$$

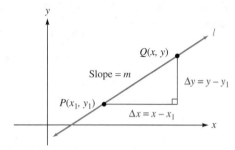

Figure 2-29

If we multiply both sides by $x - x_1$, we have

(1) $y - y_1 = m(x - x_1)$

Since Equation 1 displays the coordinates of the point (x_1, y_1) on the line and the slope m of the line, it is called the **point–slope form** of the equation of a line.

Point–Slope Form of the Equation of a Line	The equation of the line passing through $P(x_1, y_1)$ and with slope m is $$y - y_1 = m(x - x_1)$$

EXAMPLE 1   Write the equation of the line with slope $-\frac{5}{3}$ and passing through $P(3, -1)$.

Solution   We will substitute $-\frac{5}{3}$ for m, 3 for x_1, and -1 for y_1 in the point–slope form and simplify.

$$y - y_1 = m(x - x_1) \qquad \text{This is the point–slope form.}$$

$$y - (-1) = -\frac{5}{3}(x - 3) \qquad \text{Substitute } -\frac{5}{3} \text{ for } m, \text{ 3 for } x_1, \text{ and } -1 \text{ for } y_1.$$

$$y + 1 = -\frac{5}{3}x + 5 \qquad \text{Remove parentheses.}$$

$$y = -\frac{5}{3}x + 4 \qquad \text{Subtract 1 from both sides.}$$

The equation of the line is $y = -\frac{5}{3}x + 4$.

Self Check 1   Write the equation of the line with slope $-\frac{2}{3}$ and passing through $P(-4, 5)$.

EXAMPLE 2   Find the equation of the line passing through $P(3, 7)$ and $Q(-5, 3)$.

Solution   We will find the slope of the line and then choose either point P or point Q and substitute both the slope and coordinates of the point into the point–slope form. First we find the slope of the line.

$$m = \frac{y_2 - y_1}{x_2 - x_1} \quad \text{This is the slope formula.}$$

$$= \frac{3 - 7}{-5 - 3} \quad \text{Substitute 3 for } y_2, \text{ 7 for } y_1, -5 \text{ for } x_2, \text{ and 3 for } x_1.$$

$$= \frac{-4}{-8}$$

$$= \frac{1}{2}$$

We can choose either point P or point Q and substitute its coordinates into the point–slope form. If we choose $P(3, 7)$, we substitute $\frac{1}{2}$ for m, 3 for x_1, and 7 for y_1.

$$y - y_1 = m(x - x_1) \quad \text{This is the point–slope form.}$$

$$y - 7 = \frac{1}{2}(x - 3) \quad \text{Substitute } \frac{1}{2} \text{ for } m, \text{ 3 for } x_1, \text{ and 7 for } y_1.$$

$$y = \frac{1}{2}x - \frac{3}{2} + 7 \quad \text{Remove parentheses and add 7 to both sides.}$$

$$y = \frac{1}{2}x + \frac{11}{2} \qquad -\frac{3}{2} + 7 = -\frac{3}{2} + \frac{14}{2} = \frac{11}{2}$$

The equation of the line is $y = \frac{1}{2}x + \frac{11}{2}$.

Self Check 2   Find the equation of the line passing through $P(-5, 4)$ and $Q(8, -6)$.

2.   Use Slope–Intercept Form to Write the Equation of a Line

Since the y-intercept of the line shown in Figure 2-30 is the point $P(0, b)$, we can write the equation of the line by substituting 0 for x_1 and b for y_1 into the point–slope form and simplifying.

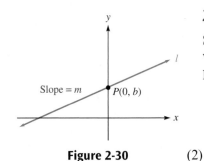

Figure 2-30

$$y - y_1 = m(x - x_1) \quad \text{This is the point–slope form.}$$

$$y - b = m(x - 0) \quad \text{Substitute 0 for } x_1 \text{ and } b \text{ for } y_1.$$

$$y - b = mx \qquad x - 0 = x$$

(2) $\qquad y = mx + b \qquad \text{Add } b \text{ to both sides.}$

Because Equation 2 displays the slope m and the y-coordinate b of the y-intercept, it is called the **slope–intercept form** of the equation of a line.

Slope–Intercept Form of the Equation of a Line   The equation of the line with slope m and y-intercept $(0, b)$ is

$$y = mx + b$$

Three examples of linear equations written in slope–intercept form are shown below.

Comment

Note that when a line is written in slope–intercept form, the coefficient of x is the slope and the constant term is the y-intercept.

Example	Slope	y-intercept
$y = 2x + 7$	$m = 2$	$b = 7, (0, 7)$
$y = \frac{2}{3}x - 5$	$m = \frac{2}{3}$	$b = -5, (0, -5)$
$y = -4x + \frac{1}{5}$	$m = -4$	$b = \frac{1}{5}, \left(0, \frac{1}{5}\right)$

EXAMPLE 3 Use slope–intercept form to write the equation of the line with slope 4 that passes through $P(5, 9)$.

Solution Since we know that $m = 4$ and that the ordered pair $(5, 9)$ satisfies the equation, we substitute 4 for m, 5 for x, and 9 for y in the equation $y = mx + b$ and solve for b.

$y = mx + b$	This is the slope–intercept form.
$9 = 4(5) + b$	Substitute 4 for m, 5 for x, and 9 for y.
$9 = 20 + b$	Simplify.
$-11 = b$	Subtract 20 from both sides.

Because $m = 4$ and $b = -11$, the equation is $y = 4x - 11$.

Comment
When we are given a point and slope, we can determine the equation of the line by substituting into point–slope form or slope–intercept form.

Self Check 3 Use slope–intercept form to write the equation of the line with slope $\frac{7}{3}$ and passing through $(3, 1)$.

3. Graph Linear Equations Using the Slope and y-Intercept

It is easy to graph a linear equation when it is written in slope–intercept form. For example, to graph $y = \frac{4}{3}x - 2$ we note that $b = -2$ and that the y-intercept is $P(0, b) = P(0, -2)$. See Figure 2-31.

Because the slope is $\frac{\Delta y}{\Delta x} = \frac{4}{3}$, we can locate another point Q on the line by starting at point P and counting 3 units to the right and 4 units up. The change in x from point P to point Q is $\Delta x = 3$ and the corresponding change in y is $\Delta y = 4$. The line joining points P and Q is the graph of the equation.

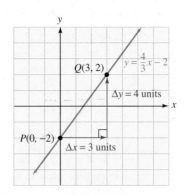

Figure 2-31

EXAMPLE 4 Find the slope and the y-intercept of the line with equation $3(y + 2) = 6x - 1$ and graph it.

Solution We will write the equation in the form $y = mx + b$ to find the slope m and the y-intercept $(0, b)$. Then we will use m and b to graph the line.

$3(y + 2) = 6x - 1$	
$3y + 6 = 6x - 1$	Remove parentheses.
$3y = 6x - 7$	Subtract 6 from both sides.
$y = 2x - \dfrac{7}{3}$	Divide both sides by 3.

The slope of the graph is 2, and the y-intercept is $\left(0, -\frac{7}{3}\right)$. We plot the y-intercept. Then we find a second point on the line by moving 1 unit to the right and 2 units up to the point $\left(1, -\frac{1}{3}\right)$. To get the graph, we draw a line through the two points, as shown in Figure 2-32.

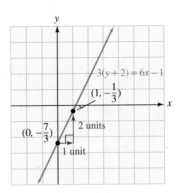

Figure 2-32

Self Check 4   Find the slope and the y-intercept of the line with equation $2(x - 3) = -3(y + 5)$. Then graph it.

$-\dfrac{2}{3}$

$2x - 6 = -3y - 15$

4.   Determine Whether Graphs of Linear Equations Are Parallel, Perpendicular, or Neither

EXAMPLE 5   Determine whether the lines represented by $4x + 8y = 10$ and $2x = 12 - 4y$ are parallel, perpendicular, or neither.

Solution   We will find the slope of each line and compare them. If the slopes are the same, the lines are parallel. If the product of the slopes is -1, the lines are perpendicular. Otherwise, they are neither parallel nor perpendicular.

We solve each equation for y and write each equation in slope–intercept form.

$$
\begin{aligned}
4x + 8y &= 10 \\
8y &= -4x + 10 \\
y &= \frac{-4x}{8} + \frac{10}{8} \\
y &= -\frac{1}{2}x + \frac{5}{4}
\end{aligned}
\qquad\qquad
\begin{aligned}
2x &= 12 - 4y \\
4y &= -2x + 12 \\
y &= \frac{-2x}{4} + \frac{12}{4} \\
y &= -\frac{1}{2}x + 3
\end{aligned}
$$

Since the values of b are different, the lines are distinct. Since each slope is $-\frac{1}{2}$, the lines are parallel.

Self Check 5   Are the lines represented by $y = 3x + 2$ and $6x - 2y = 5$ parallel, perpendicular, or neither?

EXAMPLE 6   Determine whether the lines represented by $4x + 8y = 10$ and $4x - 2y = 21$ are parallel, perpendicular, or neither.

Solution   We will find the slope of each line and compare them. If the slopes are the same, the lines are parallel. If the product of the slopes is -1, the lines are perpendicular. Otherwise, they are neither parallel nor perpendicular.

We solve each equation for y and write each equation in slope–intercept form.

$$4x + 8y = 10 \qquad\qquad 4x - 2y = 21$$
$$8y = -4x + 10 \qquad\qquad -2y = -4x + 21$$
$$y = \frac{-4x}{8} + \frac{10}{8} \qquad\qquad y = \frac{-4x}{-2} + \frac{21}{-2}$$
$$y = -\frac{1}{2}x + \frac{5}{4} \qquad\qquad y = 2x - \frac{21}{2}$$

Since the product of the slopes $\left(-\frac{1}{2} \text{ and } 2\right)$ is -1, the lines are perpendicular.

Self Check 6 Are the lines represented by $3x + 2y = 7$ and $y = \frac{2}{3}x + 3$ parallel, perpendicular, or neither?

5. Write Equations of Parallel and Perpendicular Lines

EXAMPLE 7 Write the equation of the line passing through $P(-2, 5)$ and parallel to the line $y = 8x - 3$.

Solution We will substitute the coordinates of $P(-2, 5)$ and the slope of the line parallel to $y = 8x - 3$ into point–slope form and simplify the results to write the equation of the parallel line.

The slope of the line given by $y = 8x - 3$ is 8, the coefficient of x. Since the graph of the desired equation is to be parallel to the graph of $y = 8x - 3$, its slope also must be 8.

We will substitute -2 for x_1, 5 for y_1, and 8 for m in the point–slope form and simplify.

$$
\begin{aligned}
y - y_1 &= m(x - x_1) \\
y - 5 &= 8[x - (-2)] && \text{Substitute 5 for } y_1, \text{ 8 for } m, \text{ and } -2 \text{ for } x_1. \\
y - 5 &= 8(x + 2) && -(-2) = 2 \\
y - 5 &= 8x + 16 && \text{Use the distributive property to remove parentheses.} \\
y &= 8x + 21 && \text{Add 5 to both sides.}
\end{aligned}
$$

The equation of the desired line is $y = 8x + 21$.

Self Check 7 Write the equation of the line passing through $Q(1, 2)$ and parallel to the line $y = 8x - 3$.

EXAMPLE 8 Write the equation of the line passing through $P(-2, 5)$ and perpendicular to the line $y = 8x - 3$.

Solution We will substitute the coordinates of $P(-2, 5)$ and the slope of the line perpendicular to $y = 8x - 3$ into point–slope form and simplify to write the equation of the perpendicular line.

Because the slope of the given line is 8, the slope of the desired perpendicular line must be $-\frac{1}{8}$.

We substitute -2 for x_1, 5 for y_1, and $-\frac{1}{8}$ for m into the point–slope form and simplify.

$$y - y_1 = m(x - x_1)$$

$$y - 5 = -\frac{1}{8}[x - (-2)] \quad \text{Substitute 5 for } y_1, -\frac{1}{8} \text{ for } m, \text{ and } -2 \text{ for } x_1.$$

$$y - 5 = -\frac{1}{8}(x + 2) \quad -(-2) = 2$$

$$y = -\frac{1}{8}x - \frac{1}{4} + 5 \quad \text{Remove parentheses and add 5 to both sides.}$$

$$y = -\frac{1}{8}x + \frac{19}{4} \quad -\frac{1}{4} + 5 = -\frac{1}{4} + \frac{20}{4} = \frac{19}{4}$$

The equation of the line is $y = -\frac{1}{8}x + \frac{19}{4}$.

Self Check 8   Write the equation of the line passing through $Q(1, 2)$ and perpendicular to $y = 8x - 3$.

6. Recognize and Use the Standard Form of the Equation of a Line

We have shown that the graph of any equation of the form $y = mx + b$ is a line with slope m and y-intercept $(0, b)$. In Section 2.1, we saw that the graph of any equation of the form $Ax + By = C$ (where A and B are not *both* zero) is also a line. We consider three possibilities.

* If $A \neq 0$ and $B \neq 0$, the equation $Ax + By = C$ can be written in slope–intercept form.

$$Ax + By = C$$

$$By = -Ax + C \quad \text{Subtract } Ax \text{ from both sides.}$$

$$y = -\frac{A}{B}x + \frac{C}{B} \quad \text{Divide both sides by } B.$$

This is the equation of a line with slope $-\frac{A}{B}$ and y-intercept $\left(0, \frac{C}{B}\right)$.

* If $A = 0$ and $B \neq 0$, the equation $Ax + By = C$ can be written in the form $y = \frac{C}{B}$. This is the equation of a horizontal line with y-intercept $\left(0, \frac{C}{B}\right)$.
* If $A \neq 0$ and $B = 0$, the equation $Ax + By = C$ can be written in the form $x = \frac{C}{A}$. This is the equation of a vertical line with x-intercept at $\left(\frac{C}{A}, 0\right)$.

Recall that $Ax + By = C$ is called the **standard form of the equation of a line.**

Standard Form of the Equation of a Line   If A, B, and C are real numbers and $B \neq 0$, the graph of

$$Ax + By = C$$

is a nonvertical line with slope of $-\frac{A}{B}$ and a y-intercept of $\left(0, \frac{C}{B}\right)$.

If $B = 0$, the graph is a vertical line with x-intercept of $\left(\frac{C}{A}, 0\right)$.

Comment

When writing equations in $Ax + By = C$ form, we usually clear the equation of fractions and make A positive. For example, $-x + \frac{5}{2}y = 2$ can be changed to $2x - 5y = -4$ by multiplying both sides by -2. We also divide out any common integer factors of A, B, and C. We would write $4x + 8y = 12$ as $x + 2y = 3$.

Comment

We also may write linear equations with a constant term of 0 on the right side of the equation. This form is referred to as *general form*. The line $2x - 5y + 3 = 0$ is written in general form.

EXAMPLE 9   Find the slope and the y-intercept of the graph of $3x - 2y = 5$.

Solution   The equation $3x - 2y = 5$ is in standard form, with $A = 3$, $B = -2$, and $C = 5$. We will use $-\frac{A}{B}$ to determine the slope of the line and $\left(0, \frac{C}{B}\right)$ to determine the y-intercept.

The slope of the graph is

$$m = -\frac{A}{B} = -\frac{3}{-2} = \frac{3}{2}$$

and the y-intercept is

$$\left(0, \frac{C}{B}\right) = \left(0, \frac{5}{-2}\right)$$

The slope is $\frac{3}{2}$ and the y-intercept is $\left(0, -\frac{5}{2}\right)$.

Self Check 9   Find the slope and the y-intercept of the graph of $3x - 4y = 12$.

We summarize the various forms of the equation of a line as follows.

Standard form:	$Ax + By = C$
	A and B cannot both be 0.
Slope–intercept form:	$y = mx + b$
	The slope is m, and the y-intercept is $(0, b)$.
Point–slope form:	$y - y_1 = m(x - x_1)$
	The slope is m, and the line passes through (x_1, y_1).
A horizontal line:	$y = b$
	The slope is 0, and the y-intercept is $(0, b)$.
A vertical line:	$x = a$
	There is no defined slope, and the x-intercept is $(a, 0)$.

7. Write the Equation of a Line that Models a Real-Life Problem

For tax purposes, many businesses use *straight-line depreciation* to find the declining value of aging equipment.

EXAMPLE 10   A business purchases a digital multimedia projector for \$1,970 and expects it to last for ten years. It then can be sold as scrap for a *salvage value* of \$270. If y is the value of the projector after x years of use, and y and x are related by the equation of a line,

a. Find the equation of the line.

b. Find the value of the projector after $2\frac{1}{2}$ years.

c. Find the economic meaning of the *y*-intercept of the line.

d. Find the economic meaning of the slope of the line.

Solution   a. We will find the slope and use the point–slope form to write the equation of the line. (See Figure 2-33.)

When the projector is new, its age *x* is 0, and its value *y* is $1,970. When the projector is 10 years old, $x = 10$ and $y = \$270$. Since the line passes through the points (0, 1,970) and (10, 270), the slope of the line is

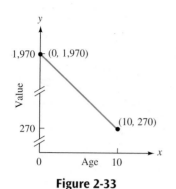

Figure 2-33

$$m = \frac{y_2 - y_1}{x_2 - x_1} \qquad \text{This is the slope formula.}$$

$$= \frac{270 - 1{,}970}{10 - 0} \qquad \text{Substitute 270 for } y_2, \text{ 1,970 for } y_1, \text{ 10 for } x_2, \text{ and 0 for } x_1.$$

$$= \frac{-1{,}700}{10}$$

$$= -170$$

To find the equation of the line, we substitute -170 for *m*, 0 for x_1, and 1,970 for y_1 in the point–slope form and simplify.

$$y - y_1 = m(x - x_1)$$
$$y - 1{,}970 = -170(x - 0)$$
(3) $$y = -170x + 1{,}970$$

The value *y* of the projector is related to its age *x* by the equation $y = -170x + 1{,}970$.

b. To find the value after $2\frac{1}{2}$ years, we will substitute 2.5 for *x* in Equation 3 and solve for *y*.

$$y = -170x + 1{,}970$$
$$= -170(\mathbf{2.5}) + 1{,}970 \qquad \text{Substitute 2.5 for } x.$$
$$= -425 + 1{,}970$$
$$= 1{,}545$$

In $2\frac{1}{2}$ years, the projector will be worth $1,545.

c. The *y*-intercept of the graph is (0, *b*), where *b* is the value of *y* when $x = 0$.

$$y = -170x + 1{,}970$$
$$y = -170(\mathbf{0}) + 1{,}970 \qquad \text{Substitute 0 for } x.$$
$$y = 1{,}970$$

The *y*-coordinate *b* of the *y*-intercept is the value of a 0-year-old projector, which is the projector's original cost, $1,970.

d. Each year, the value decreases by $170, because the slope of the line is -170. The slope of the depreciation line is called the *annual depreciation rate*.

8.   Use Linear Curve Fitting to Solve Problems

In statistics, the process of using one variable to predict another is called **regression.** For example, if we know a woman's height, we can make a good prediction about her weight, because taller women usually weigh more than shorter women.

Figure 2-34 shows the result of sampling ten women and finding their heights and weights. The graph of the ordered pairs (*h*, *w*) is called a **scattergram.**

Woman	Height (h) in inches	Weight (w) in pounds
1	60	100
2	61	105
3	62	120
4	62	130
5	63	135
6	64	120
7	64	125
8	65	155
9	67	155
10	69	160

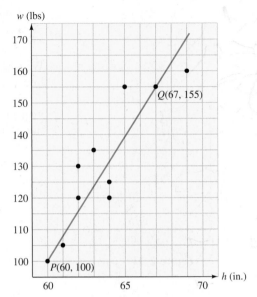

Figure 2-34

To write a **prediction equation** (sometimes called a **regression equation**), we must find the equation of the line that comes closer to all of the points in the scattergram than any other possible line. There are statistical methods to find this equation, but we can only approximate it here.

To write an approximation of the regression equation, we place a straightedge on the scattergram shown in Figure 2-34 and draw the line joining two points that seems to best fit all the points. In the figure, line PQ is drawn, where point P has coordinates of (60, 100) and point Q has coordinates of (67, 155).

Our approximation of the regression equation will be the equation of the line passing through points P and Q. To find the equation of this line, we first find its slope.

$$m = \frac{y_2 - y_1}{x_2 - x_1} \qquad \text{This is the slope formula.}$$

$$= \frac{155 - 100}{67 - 60} \qquad \text{Substitute 155 for } y_2 \text{, 100 for } y_1 \text{, 67 for } x_2 \text{, and 60 for } x_1.$$

$$= \frac{55}{7}$$

We then can use point–slope form to find the equation of the line.

$$y - y_1 = m(x - x_1) \qquad \text{This is the point–slope form.}$$

$$y - 100 = \frac{55}{7}(x - 60) \qquad \text{Choose (60, 100) for } (x_1, y_1).$$

$$y = \frac{55}{7}x - \frac{3,300}{7} + 100 \qquad \text{Remove parentheses and add 100 to both sides.}$$

$$\text{(4)} \qquad y = \frac{55}{7}x - \frac{2,600}{7} \qquad \text{Simplify.}$$

Our approximation of the regression equation is $y = \frac{55}{7}x - \frac{2,600}{7}$.

To predict the weight of a woman who is 66 inches tall, for example, we substitute 66 for x in Equation 4 and simplify.

$$y = \frac{55}{7}x - \frac{2{,}600}{7}$$

$$y = \frac{55}{7}(66) - \frac{2{,}600}{7}$$

$$y \approx 147.1428571$$

We would predict that a 66-inch-tall woman chosen at random will weigh about 147 pounds.

Self Check Answers

1. $y = -\frac{2}{3}x + \frac{7}{3}$ **2.** $y = -\frac{10}{13}x + \frac{2}{13}$ **3.** $y = \frac{7}{3}x - 6$

4. $-\frac{2}{3}, (0, -3)$ **5.** parallel **6.** perpendicular

7. $y = 8x - 6$ **8.** $y = -\frac{1}{8}x + \frac{17}{8}$

9. $\frac{3}{4}, (0, -3)$

2.3 Exercises

Vocabulary and Concepts *Fill in the blanks.*

1. The formula for the point–slope form of a line is
_____.

2. In the equation $y = mx + b$, __ is the slope of the graph of the line, and $(0, b)$ is the _____.

3. The equation $y = mx + b$ is called the _____ form of the equation of a line.

4. The standard form of the equation of a line is _____.

5. The slope of the graph of $Ax + By = C$ is ___.

6. The y-intercept of the graph of $Ax + By = C$ is _____.

Practice *Use point–slope form to write the equation of the line with the given properties. Write each equation in standard form.*

7. $m = 2$ passing through $P(2, 4)$

8. $m = -3$ passing through $P(3, 5)$

9. $m = 2$ passing through $P\left(-\frac{3}{2}, \frac{1}{2}\right)$

10. $m = -6$ passing through $P\left(\frac{1}{4}, -2\right)$

11. $m = \frac{2}{5}$ passing through $P(-1, 1)$

12. $m = -\frac{1}{5}$ passing through $P(-2, -3)$

13. $m = 0$ passing through $P(-6, -3)$

14. m is undefined passing through $P(-6, -3)$

15. $m = \pi$ passing through $P(\pi, 0)$

16. $m = \pi$ passing through $P(0, \pi)$

Use point–slope form to write the equation of each line. Write the equation in standard form.

17.

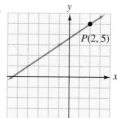

18.

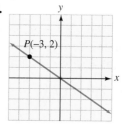

Use point–slope form to write the equation of the line passing through the two given points. Write each equation in slope–intercept form.

19. $P(0, 0), Q(4, 4)$ **20.** $P(-5, -5), Q(0, 0)$

21. $P(3, 4), Q(0, -3)$ **22.** $P(4, 0), Q(6, -8)$

Use point–slope form to write the equation of each line. Write each answer in slope–intercept form.

23.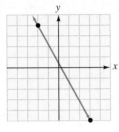

24.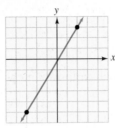

Use slope–intercept form to write the equation of the line with the given properties. Write each equation in slope–intercept form.

25. $m = 3; b = -2$

26. $m = -\dfrac{1}{3}; b = \dfrac{2}{3}$

27. $m = 5; b = -\dfrac{1}{5}$

28. $m = \sqrt{2}; b = \sqrt{2}$

29. $m = a; b = \dfrac{1}{a}$

30. $m = a; b = 2a$

31. $m = a; b = a$

32. $m = \dfrac{1}{a}; b = a$

Use slope–intercept form to write the equation of a line passing through the given point and having the given slope. Express the answer in standard form.

33. $P(0, 0); m = \dfrac{3}{2}$

34. $P(-3, -7); m = -\dfrac{2}{3}$

35. $P(-3, 5); m = -3$

36. $P(-5, 1); m = 1$

37. $P\left(0, \sqrt{2}\right); m = \sqrt{2}$

38. $P\left(-\sqrt{3}, 0\right); m = 2\sqrt{3}$

Write each equation in slope–intercept form to find the slope and the y-intercept. Then use the slope and y-intercept to draw the line.

39. $x - y = 1$

40. $x + y = 2$

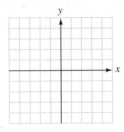

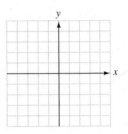

41. $x = \dfrac{3}{2}y - 3$

42. $x = -\dfrac{4}{5}y + 2$

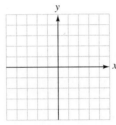

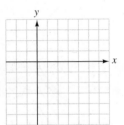

43. $3(y - 4) = -2(x - 3)$

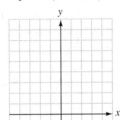

44. $-4(2x + 3) = 3(3y + 8)$

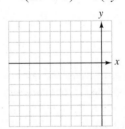

Find the slope and the y-intercept of the line determined by the given equation.

45. $3x - 2y = 8$

46. $-2x + 4y = 12$

47. $-2(x + 3y) = 5$

48. $5(2x - 3y) = 4$

49. $x = \dfrac{2y - 4}{7}$ **50.** $3x + 4 = -\dfrac{2(y - 3)}{5}$

Determine whether the graphs of each pair of equations are parallel, perpendicular, or neither.

51. $y = 3x + 4, y = 3x - 7$

52. $y = 4x - 13, y = \dfrac{1}{4}x + 13$

53. $x + y = 2, y = x + 5$

54. $x = y + 2, y = x + 3$

55. $y = 3x + 7, 2y = 6x - 9$

56. $2x + 3y = 9, 3x - 2y = 5$

57. $x = 3y + 4, y = -3x + 7$

58. $3x + 6y = 1, y = \dfrac{1}{2}x$

59. $y = 3, x = 4$

60. $y = -3, y = -7$

61. $x = \dfrac{y - 2}{3}, 3(y - 3) + x = 0$

62. $2y = 8, 3(2 + x) = 3(y + 2)$

Write the equation of the line that passes through the given point and is parallel to the given line. Write the answer in slope–intercept form.

63. $P(0, 0), y = 4x - 7$

64. $P(0, 0), x = -3y - 12$

65. $P(2, 5), 4x - y = 7$

66. $P(-6, 3), y + 3x = -12$

67. $P(4, -2), x = \dfrac{5}{4}y - 2$

68. $P(1, -5), x = -\dfrac{3}{4}y + 5$

Write the equation of the line that passes through the given point and is perpendicular to the given line. Write the answer in slope–intercept form.

69. $P(0, 0), y = 4x - 7$

70. $P(0, 0), x = -3y - 12$

71. $P(2, 5), 4x - y = 7$

72. $P(-6, 3), y + 3x = -12$

73. $P(4, -2), x = \dfrac{5}{4}y - 2$

74. $P(1, -5), x = -\dfrac{3}{4}y + 5$

Use the method of Example 9 to find the slope and the y-intercept of the graph of each equation.

75. $4x + 5y = 20$ **76.** $9x - 12y = 17$

77. $2x + 3y = 12$ **78.** $5x + 6y = 30$

79. Find the equation of the line perpendicular to the line $y = 3$ and passing through the midpoint of the segment joining $(2, 4)$ and $(-6, 10)$.

80. Find the equation of the line parallel to the line $y = -8$ and passing through the midpoint of the segment joining $(-4, 2)$ and $(-2, 8)$.

81. Find the equation of the line parallel to the line $x = 3$ and passing through the midpoint of the segment joining $(2, -4)$ and $(8, 12)$.

82. Find the equation of the line perpendicular to the line $x = 3$ and passing through the midpoint of the segment joining $(-2, 2)$ and $(4, -8)$.

Applications   *In Exercises 83–93, assume straight-line depreciation or straight-line appreciation.*

83. Depreciation   A Toyota Tundra truck was purchased for $24,300. Its salvage value at the end of its 7-year useful life is expected to be $1,900. Find the depreciation equation.

84. Depreciation   A small business purchases the laptop computer shown. It will be depreciated over a 4-year period, when its salvage value will be $300. Find the depreciation equation.

$2,700

85. Appreciation   A condominium in San Diego was purchased for $475,000. The owners expect the condominium to double in value in 10 years. Find the appreciation equation.

86. Appreciation   A house purchased for $112,000 is expected to double in value in 12 years. Find its appreciation equation.

87. Depreciation   Find the depreciation equation for the TV in the following want ad.

> *For Sale*: 3-year-old 54-inch TV, $1,900 new. Asking $1,190. Call 875-5555. Ask for Mike.

88. Depreciation   A Bose Wave Radio cost $555 when new and is expected to be worth $80 after 5 years. What will it be worth after 3 years?

89. Salvage value   A copier cost $1,050 when new and will be depreciated at the rate of $120 per year. If the useful life of the copier is 8 years, find its salvage value.

90. Rate of depreciation   A ski boat that cost $27,600 when new will have no salvage value after 12 years. Find its annual rate of depreciation.

91. Value of an antique   An antique table is expected to appreciate $40 each year. If the table will be worth $450 in 2 years, what will it be worth in 13 years?

92. Value of an antique   An antique clock is expected to be worth $350 after 2 years and $530 after 5 years. What will the clock be worth after 7 years?

93. Purchase price of real estate   A cottage that was purchased 3 years ago is now appraised at $47,700. If the property has been appreciating $3,500 per year, find its original purchase price.

94. Computer repair   A computer repair company charges a fixed amount, plus an hourly rate, for a service call. Use the information in the illustration to find the hourly rate.

AAA Computer Repair

Typical Charges	
2 hours	$ 70
4 hours	$105

95. Automobile repair   An auto repair shop charges an hourly rate, plus the cost of parts. If the cost of labor for a $1\frac{1}{2}$-hour radiator repair is $69, find the cost of labor for a 5-hour transmission overhaul.

96. Printer charges   A printer charges a fixed setup cost, plus $1 for every 100 copies. If 700 copies cost $52, how much will it cost to print 1,000 copies?

97. Predicting fires   A local fire department recognizes that city growth and the number of reported fires are related by a linear equation. City records show that 300 fires were reported in a year when the local population was 57,000 persons, and 325 fires were reported in a year when the population was 59,000 persons. How many fires can be expected in the year when the population reaches 100,000 persons?

98. Estimating the cost of rain gutter   A neighbor tells you that an installer of rain gutter charges $60, plus a dollar amount per foot. If the neighbor paid $435 for the installation of 250 feet of gutter, how much will it cost you to have 300 feet installed?

99. Converting temperatures   Water freezes at 32° Fahrenheit, or 0° Celsius. Water boils at 212° F, or 100° C. Find a formula for converting a temperature from degrees Fahrenheit to degrees Celsius.

100. Converting units   A speed of 1 mile per hour is equal to 88 feet per minute, and of course, 0 miles per hour is 0 feet per minute. Find an equation for converting a speed x, in miles per hour, to the corresponding speed y, in feet per minute.

101. Smoking   The percent y of 18- to 25-year-old smokers in the United States has been declining at a constant rate since 1974. If about 47% of this group smoked in 1974 and about 29% smoked in 1994, find a linear equation that models this decline. If this trend continues, estimate what percent will smoke in 2014.

102. Forensic science   Scientists believe there is a linear relationship between the height h (in centimeters) of a male and the length f (in centimeters) of his femur bone. Use the data in the table to find a linear equation that expresses the height h in terms of f. Round all constants to the nearest thousandth. How tall would you expect a man to be if his femur measures 50 cm? Round to the nearest centimeter.

Person	Length of femur (f)	Height (h)
A	62.5 cm	200 cm
B	40.2 cm	150 cm

103. Predicting stock prices The value of the stock of ABC Corporation has been increasing by the same fixed dollar amount each year. The pattern is expected to continue. Let 2008 be the base year corresponding to $x = 0$ with $x = 1, 2, 3, \ldots$ corresponding to later years. ABC stock was selling at $\$37\frac{1}{2}$ in 2008 and at $\$45$ in 2010. If y represents the price of ABC stock, find the equation $y = mx + b$ that relates x and y, and predict the price in the year 2012.

104. Estimating inventory Inventory of unsold goods showed a surplus of 375 units in January and 264 in April. Assume that the relationship between inventory and time is given by the equation of a line, and estimate the expected inventory in March. Because March lies between January and April, this estimation is called **interpolation.**

105. Oil depletion When a Petroland oil well was first brought on line, it produced 1,900 barrels of crude oil per day. In each later year, owners expect its daily production to drop by 70 barrels. Find the daily production after $3\frac{1}{2}$ years.

106. Waste management The corrosive waste in industrial sewage limits the useful life of the piping in a waste processing plant to 12 years. The piping system was originally worth $\$137,000$, and it will cost the company $\$33,000$ to remove it at the end of its 12-year useful life. Find the depreciation equation.

107. Crickets The table shows the approximate chirping rate at various temperatures for one type of cricket.

Temperature (°F)	Chirps per minute
50	20
60	80
70	115
80	150
100	250

a. Construct a scattergram as shown in the next column.

b. Assume a linear relationship and write a regression equation.

c. Estimate the chirping rate at a temperature of 90° F.

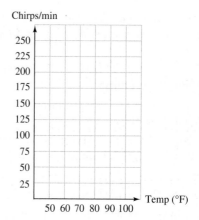

108. Fishing The table shows the lengths and weights of seven muskies captured by the Department of Natural Resources in Catfish Lake in Eagle River, Wisconsin.

Musky	Length (in.)	Weight (lb)
1	26	5
2	27	8
3	29	9
4	33	12
5	35	14
6	36	14
7	38	19

a. Construct a scattergram for the data.

b. Assume a linear relationship and write a regression equation.

c. Estimate the weight of a musky that is 32 inches long.

Discovery and Writing

109. Explain how to find the equation of a line passing through two given points.

110. In straight-line depreciation, explain why the slope of the line is called the *rate of depreciation.*

111. Prove that the equation of a line with x-intercept of $(a, 0)$ and y-intercept of $(0, b)$ can be written in the form

$$\frac{x}{a} + \frac{y}{b} = 1$$

112. Find the x- and y-intercepts of the line $bx + ay = ab$.

🔲 *Investigate the properties of slope and the y-intercept by experimenting with the following problems.*

113. Graph $y = mx + 2$ for several positive values of m. What do you notice?

114. Graph $y = mx + 2$ for several negative values of m. What do you notice?

115. Graph $y = 2x + b$ for several increasing positive values of b. What do you notice?

116. Graph $y = 2x + b$ for several decreasing negative values of b. What do you notice?

117. How will the graph of $y = \frac{1}{2}x + 5$ compare to the graph of $y = \frac{1}{2}x - 5$?

118. How will the graph of $y = \frac{1}{2}x - 5$ compare to the graph of $y = \frac{1}{2}x$?

Much of the hard work done in problems such as 107–108 can be done automatically by using a computer program, such as Excel.

119. The following table shows the length of a femur bone and the height for ten Caucasian males.

Length of femur bone (cm) x	Height (cm) y
49.3	180.8
47.4	176.7
47.1	176.1
48.5	176.5
45.2	170.6
47.8	176.8
49.4	178.6
49.6	179.6
50.9	185.2
47.8	176.2

a. Enter the data in an Excel Spreadsheet. Enter the femur values in Column A and the height values in Column B.

b. To plot the data, click and drag to highlight the two columns, select Insert on the menu bar, and then select Chart. When a window pops up, select XY (Scatter) as the chart type. When you click Finish, a **scattergram** will appear on the spreadsheet.

c. Observe that the points are approximately linear. To draw the regression line, select Chart on the menu bar and select Add Trendline. When the dialog box appears, be sure that the Linear Regression type is selected, and then click OK. The regression line will appear.

d. To find the equation of the regression line, estimate the coordinates of two points on the regression line. Then use the two points to write the equation of the line.

e. Use the equation of the regression line to predict the height of a male whose femur bone measures 46 cm.

120. The following table shows the personal income (per person, including the unemployed) and the personal outlays (per person, including the unemployed) for people living in the United States for 1994–2003.

Year	Per capita personal income ($) x	Per capita personal outlays ($) y
1994	5,842.5	4,902.4
1995	6,152.3	5,157.3
1996	6,520.6	5,460.0
1997	6,915.1	5,770.5
1998	7,423.0	6,119.1
1999	7,802.4	6,536.4
2000	8,429.7	7,025.6
2001	8,724.1	7,354.5
2002	8,878.9	7,668.5
2003	9,161.8	8,049.3

Source: Bureau of Economic Analysis

a. Use Excel to make a scattergram of the given data. See problem 119.

b. Use Excel to draw the regression line.

c. Write the equation of the regression line.

d. Interpret the slope of the regression line.

e. How much would you expect a person to spend if he/she had a personal income of $9,000?

Review *Simplify each expression.*

121. $x^7 x^3 x^{-5}$

122. $\dfrac{y^3 y^{-4}}{y^{-5}}$

123. $\left(\dfrac{81}{25}\right)^{-3/2}$

124. $\sqrt[3]{27x^7}$

125. $\sqrt{27} - 2\sqrt{12}$

126. $\dfrac{5}{\sqrt{5}}$

127. $\dfrac{5}{\sqrt{x} + 2}$

128. $\left(\sqrt{x} - 2\right)^2$

© PCL / Alamy

2.4 Graphs of Equations

Objectives

1. Find the *x*- and *y*-Intercepts of a Graph
2. Use Symmetry to Help Graph Equations
3. Identify the Center and Radius of a Circle
4. Write the Equation of a Circle
5. Find the General Form of the Equation of a Circle
6. Graph Circles Whose Equations Are Written in General Form
7. Solve Equations by Graphing

The London Eye opened in 2000 and is one of the world's tallest and most beautiful observation wheels.

It stands 443 feet high and was the vision of architects David Marks and Julia Barfield. Its circular design was used as a metaphor for the turning of the century. The Eye has been described as a breathtaking feat of design and engineering.

The graphs of many equations are curves and circles. In this section, we will plot several points (x, y) that satisfy such an equation and join them with a smooth curve. Usually the shape of the graph will become evident.

1. Find the *x*- and *y*-Intercepts of a Graph

In Figure 2-35(a), the **x-intercepts** of the graph are $(a, 0)$ and $(b, 0)$, the points where the graph intersects the *x*-axis. In Figure 2-35(b), the **y-intercept** is $(0, c)$, the point where the graph intersects the *y*-axis.

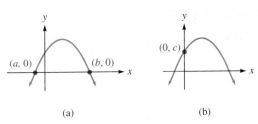

(a)   (b)

Figure 2-35

To graph the equation $y = x^2 - 4$, we first find the *x*- and *y*-intercepts. To find the *x*-intercepts, we let $y = 0$ and solve for *x*.

10. The standard form of the equation of a circle with center at (h, k) and radius r is

_____.

Practice *Find the x- and y-intercepts of each graph. Do not graph the equation.*

11. $y = x^2 - 4$

12. $y = x^2 - 9$

13. $y = 4x^2 - 2x$

14. $y = 2x - 4x^2$

15. $y = x^2 - 4x - 5$

16. $y = x^2 - 10x + 21$

17. $y = x^2 + x - 2$

18. $y = x^2 + 2x - 3$

19. $y = x^3 - 9x$

20. $y = x^3 + x$

21. $y = x^4 - 1$

22. $y = x^4 - 25x^2$

Graph each equation. Check your graph with a graphing calculator.

23. $y = x^2$

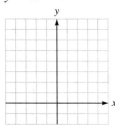

24. $y = -x^2$

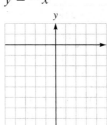

25. $y = -x^2 + 2$

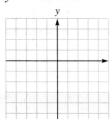

26. $y = x^2 - 1$

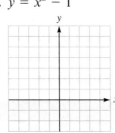

27. $y = x^2 - 4x$

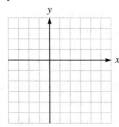

28. $y = x^2 + 2x$

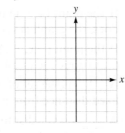

29. $y = \frac{1}{2}x^2 - 2x$

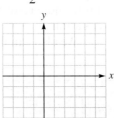

30. $y = \frac{1}{2}x^2 + 3$

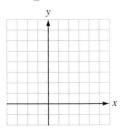

Find the symmetries, if any, of the graph of each equation. Do not graph the equation.

31. $y = x^2 + 2$

32. $y = 3x + 2$

33. $y^2 + 1 = x$

34. $y^2 + y = x$

35. $y^2 = x^2$

36. $y = 3x + 7$

37. $y = 3x^2 + 7$

38. $x^2 + y^2 = 1$

39. $y = 3x^3 + 7$

40. $y = 3x^3 + 7x$

41. $y^2 = 3x$

42. $y = 3x^4 + 7$

43. $y = |x|$

44. $y = |x + 1|$

45. $|y| = x$

46. $|y| = |x|$

Graph each equation. Be sure to find any intercepts and symmetries. Check your graph with a graphing calculator.

47. $y = x^2 + 4x$

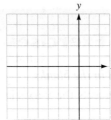

48. $y = x^2 - 6x$

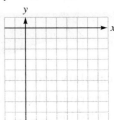

49. $y = x^3$

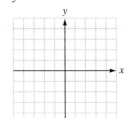

50. $y = x^3 + x$

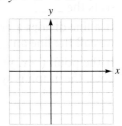

51. $y = |x - 2|$

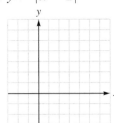

52. $y = |x| - 2$

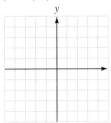

53. $y = 3 - |x|$

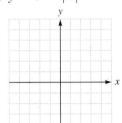

54. $y = 3|x|$

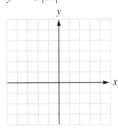

55. $y^2 = -x$

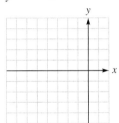

56. $y^2 = 4x$

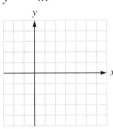

57. $y^2 = 9x$

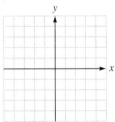

58. $y^2 = -4x$

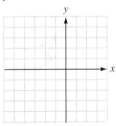

59. $y = \sqrt{x} - 1$

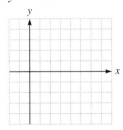

60. $y = 1 - \sqrt{x}$

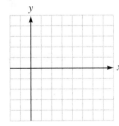

61. $xy = 4$

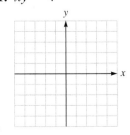

62. $xy = -9$

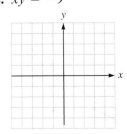

Identify the center and radius of each circle written in standard form.

63. $x^2 + y^2 = 100$

64. $x^2 + y^2 = 81$

65. $x^2 + (y - 5)^2 = 49$

66. $x^2 + (y + 3)^2 = 8$

67. $(x + 6)^2 + y^2 = \dfrac{1}{4}$

68. $(x - 5)^2 + y^2 = \dfrac{16}{25}$

69. $(x - 4)^2 + (y - 1)^2 = 9$

70. $(x + 11)^2 + (y + 7)^2 = 121$

71. $\left(x - \dfrac{1}{4}\right)^2 + (y + 2)^2 = 45$

72. $\left(x + \sqrt{5}\right)^2 + (y - 3)^2 = 1$

Write the equation in standard form of each circle with the given properties.

73. Center at the origin; $r = 5$

74. Center at the origin; $r = \sqrt{3}$

75. Center at $(0 - 6)$; $r = 6$

76. Center at $(0, 7)$; $r = 9$

77. Center at $(8, 0)$; $r = \dfrac{1}{5}$

78. Center at $(-10, 0)$; $r = \sqrt{11}$

79. Center at $(2, 12)$, $r = 13$

80. Center at $\left(\dfrac{2}{7}, -5\right)$; $r = 7$

Write the equation in general form of each circle with the given properties.

81. Center at the origin; $r = 1$

82. Center at the origin; $r = 4$

83. Center at $(6, 8)$; $r = 4$

84. Center at $(5, 3)$; $r = 2$

85. Center at $(3, -4)$; $r = \sqrt{2}$

86. Center at $(-9, 8)$; $r = 2\sqrt{3}$

87. Ends of diameter at $(3, -2)$ and $(3, 8)$

88. Ends of diameter at $(5, 9)$ and $(-5, -9)$

89. Center at $(-3, 4)$ and passing through the origin

90. Center at $(-2, 6)$ and passing through the origin

Discovery and Writing *The solution of the inequality $P(x) < 0$ consists of those numbers x for which the graph of $y = P(x)$ lies below the x-axis. To solve $P(x) < 0$, we graph $y = P(x)$ and trace to find numbers x that produce negative values of y. Solve each inequality.*

123. $x^2 + x - 6 < 0$ **124.** $x^2 - 3x - 10 > 0$

Review *Solve each equation.*

125. $3(x + 2) + x = 5x$
126. $12b + 6(3 - b) = b + 3$
127. $\dfrac{5(2 - x)}{3} - 1 = x + 5$

128. $\dfrac{r - 1}{3} = \dfrac{r + 2}{6} + 2$

129. **Mixing an alloy**   In 60 ounces of alloy for watch cases, there are 20 ounces of gold. How much copper must be added to the alloy so that a watch case weighing 4 ounces, made from the new alloy, will contain exactly 1 ounce of gold?

130. **Mixing coffee**   To make a mixture of 80 pounds of coffee worth $272, a grocer mixes coffee worth $3.25 a pound with coffee worth $3.85 a pound. How many pounds of cheaper coffee should the grocer use?

2.5 Proportion and Variation

Objectives

1. Solve Proportions
2. Use Direct Variation to Solve Problems
3. Use Inverse Variation to Solve Problems
4. Use Joint Variation to Solve Problems
5. Use Combined Variation to Solve Problems

Skydiving is an adventurous sport. Jumping out of an aircraft at a height of 14,000 feet and free falling delivers a rush of adrenaline that is an exhilarating experience and one that skydivers never forget.

The unique experience allows skydivers to fall approximately 1,300 feet every five seconds and reach speeds between 120 and 150 miles per hour.

For a free-falling object, we can calculate the distance fallen given the amount of time. In fact, the distance of the fall is proportional to the square of the time. In this section we will consider the ways that variables are related. These ways include: direct variation, inverse variation, and joint variation, and combinations of all three.

1. Solve Proportions

The quotient of two numbers is often called a **ratio.** For example, the fraction $\frac{3}{2}$ (or the expression 3:2) can be read as "the ratio of 3 to 2." Some examples are

$$\frac{3}{5}, \qquad 7:9, \qquad \frac{x + 1}{9}, \qquad \frac{a}{b}, \qquad \text{and} \qquad \frac{x^2 - 4}{x + 5}$$

An equation indicating that two ratios are equal is called a **proportion.** Some examples of proportions are

$$\frac{2}{3} = \frac{4}{6}, \qquad \frac{x}{y} = \frac{3}{5}, \qquad \text{and} \qquad \frac{x^2 + 8}{2(x + 3)} = \frac{17(x + 3)}{2}$$

In the proportion $\frac{a}{b} = \frac{c}{d}$, the numbers a and d are called the **extremes,** and the numbers b and c are called the **means.**

To develop an important property of proportions, we suppose that

$$\frac{a}{b} = \frac{c}{d}$$

and multiply both sides by bd to get

$$bd\left(\frac{a}{b}\right) = bd\left(\frac{c}{d}\right)$$

$$\frac{bda}{b} = \frac{bdc}{d}$$

$$da = bc$$

Thus, if $\frac{a}{b} = \frac{c}{d}$, then $ad = bc$. This proves the following statement.

Property of Proportions   In any proportion, the product of the extremes is equal to the product of the means.

We can use this property to solve proportions.

EXAMPLE 1   Solve the proportion: $\frac{x}{5} = \frac{2}{x+3}$.

Solution   We will use the property of proportions to solve the proportion.

$$\frac{x}{5} = \frac{2}{x+3}$$

$x(x+3) = 5 \cdot 2$	The product of the extremes equals the product of the means.
$x^2 + 3x = 10$	Remove parentheses and simplify.
$x^2 + 3x - 10 = 0$	Subtract 10 from both sides.
$(x-2)(x+5) = 0$	Factor the trinomial.
$x - 2 = 0$ or $x + 5 = 0$	Set each factor equal to 0.
$x = 2 \qquad\qquad x = -5$	

Thus, $x = 2$ or $x = -5$. Verify each solution.

Self Check 1   Solve: $\frac{2}{5} = \frac{3}{x-4}$.

EXAMPLE 2   Gasoline and oil for a Johnson outboard boat motor are to be mixed in a 50-to-1 ratio. How many ounces of oil should be mixed with 6 gallons of gasoline?

Solution   We will first express 6 gallons in terms of ounces.

6 gallons = $6 \cdot 128$ ounces = 768 ounces

Then we let x represent the number of ounces of oil needed, set up the proportion, and solve it.

$$\frac{50}{1} = \frac{768}{x}$$

$$50x = 768 \qquad \text{The product of the extremes equals the product of the means.}$$

$$x = \frac{768}{50} \qquad \text{Divide both sides by 50.}$$

$$x = 15.36$$

Approximately 15 ounces of oil should be added to 6 gallons of gasoline.

Self Check 2   How many ounces of oil should be mixed with 6 gallons of gas if the ratio is to be 40 parts of gas to 1 part of oil?

2.   Use Direct Variation to Solve Problems

Two variables are said to **vary directly** or be **directly proportional** if their ratio is a constant. The variables x and y vary directly when

$$\frac{y}{x} = k \qquad \text{or, equivalently,} \qquad y = kx \quad (k \text{ is a constant})$$

Direct Variation   The words "**y varies directly with x,**" or "**y is directly proportional to x,**" mean that $y = kx$ for some real-number constant k.

The number k is called the **constant of proportionality**.

EXAMPLE 3   Distance traveled in a given time varies directly with the speed. If a car travels 70 miles at 30 mph, how far will it travel in the same time at 45 mph?

Solution   We will use direct variation to solve the problem.

The phrase *distance varies directly with speed* translates into the formula $d = ks$, where d represents the distance traveled and s represents the speed. The constant of proportionality k can be found by substituting 70 for d and 30 for s in the equation $d = ks$.

$$d = ks$$

$$70 = k(30)$$

$$k = \frac{7}{3}$$

To evaluate the distance d traveled at 45 mph, we substitute $\frac{7}{3}$ for k and 45 for s into the formula $d = ks$.

$$d = \frac{7}{3}s$$

$$= \frac{7}{3}(45)$$

$$= 105$$

In the time it takes to go 70 miles at 30 mph, the car could travel 105 miles at 45 mph.

Self Check 3   How far will the car travel in the same time if its speed is 60 mph?

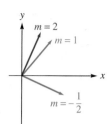

Figure 2-52

The statement *y varies directly with x* is equivalent to the equation $y = mx$, where *m* is the constant of proportionality. Because the equation is in the form $y = mx + b$ with $b = 0$, its graph is a line with slope *m* and *y*-intercept at 0. The graphs of $y = mx$ for several values of *m* are shown in Figure 2-52.

The graph of the relationship of direct variation is always a line that passes through the origin.

3.   Use Inverse Variation to Solve Problems

Two variables are said to **vary inversely** or be **inversely proportional** if their product is a constant.

$$xy = k \qquad \text{or, equivalently,} \qquad y = \frac{k}{x} \quad (k \text{ is a constant})$$

Inverse Variation   The words "**y varies inversely with x**," or "**y is inversely proportional to x**," mean that $y = \frac{k}{x}$ for some real-number constant *k*.

EXAMPLE 4   Intensity of illumination from a light source varies inversely with the square of the distance from the source. If the intensity of a light source is 100 lumens at a distance of 20 feet, find the intensity at 30 feet.

Solution   We will use inverse variation to solve the problem.

If *I* is the intensity and *d* is the distance from the light source, the phrase *intensity varies inversely with the square of the distance* translates into the formula

$$I = \frac{k}{d^2}$$

We can evaluate *k* by substituting 100 for *I* and 20 for *d* in the formula and solving for *k*.

$$I = \frac{k}{d^2}$$
$$100 = \frac{k}{20^2}$$
$$k = 40,000$$

To find the intensity at a distance of 30 feet, we substitute 40,000 for *k* and 30 for *d* in the formula.

$$I = \frac{k}{d^2}$$
$$I = \frac{40,000}{30^2}$$
$$= \frac{400}{9}$$

At 30 feet, the intensity of light would be $\frac{400}{9}$ lumens per square centimeter.

Self Check 4   Find the intensity at 50 feet.

The statement *y is inversely proportional to x* is equivalent to the equation $y = \frac{k}{x}$, where k is a constant. Figure 2-53 shows the graphs of $y = \frac{k}{x}$ $(x > 0)$ for three values of k. In each case, the equation determines one branch of a curve called a **hyperbola.** Verify these graphs with a graphing calculator.

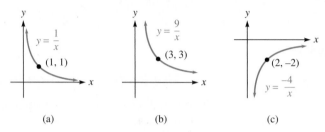

(a)   (b)   (c)

Figure 2-53

4.   Use Joint Variation to Solve Problems

Joint Variation   The words **"*y* varies jointly with *w* and *x*"** mean that $y = kwx$ for some real-number constant k.

EXAMPLE 5   Kinetic energy of an object varies jointly with its mass and the square of its velocity. A 25-gram mass moving at the rate of 30 centimeters per second has a kinetic energy of 11,250 dyne-centimeters. Find the kinetic energy of a 10-gram mass that is moving at 40 centimeters per second.

Solution   We will use joint variation to solve the problem.

If we let E, m, and v represent the kinetic energy, mass, and velocity, respectively, the phrase *energy varies jointly with the mass and the square of its velocity* translates into the formula

$$E = kmv^2$$

The constant k can be evaluated by substituting 11,250 for E, 25 for m, and 30 for v in the formula.

$$E = kmv^2$$
$$11{,}250 = k(25)(30)^2$$
$$11{,}250 = 22{,}500k$$
$$k = \frac{1}{2}$$

We now can substitute $\frac{1}{2}$ for k, 10 for m, and 40 for v in the formula and evaluate E.

$$E = kmv^2$$
$$= \frac{1}{2}(10)(40)^2$$
$$= 8{,}000$$

A 10-gram mass that is moving at 40 centimeters per second has a kinetic energy of 8,000 dyne-centimeters.

Self Check 5   Find the kinetic energy of a 25-gram mass that is moving at 100 centimeters per second.

5. Use Combined Variation to Solve Problems

The preceding terminology can be used in various combinations. In each of the statements shown in the table below, the formula on the left translates into the words on the right.

Formula	Words
$y = \dfrac{kx}{z}$	y varies directly with x and inversely with z.
$y = kx^2\sqrt[3]{z}$	y varies jointly with the square of x and the cube root of z.
$y = \dfrac{kx\sqrt{z}}{\sqrt[3]{t}}$	y varies jointly with x and the square root of z and inversely with the cube root of t.
$y = \dfrac{k}{xz}$	y varies inversely with the product of x and z.

EXAMPLE 6   The time it takes to build a highway varies directly with the length of the road, but inversely with the number of workers. If it takes 100 workers 4 weeks to build 2 miles of the George Washington Memorial Parkway, how long will it take 80 workers to build 10 miles of the parkway?

Solution   We will use combined variation to solve the problem.

We can let t represent the time in weeks, l represent the length in miles, and w represent the number of workers. Because time varies directly with the length of the parkway, but inversely with the number of workers, the relationship between these variables can be expressed by the equation

$$t = \frac{kl}{w}$$

We substitute 4 for t, 100 for w, and 2 for l to find k.

$$4 = \frac{k(2)}{100}$$

$400 = 2k$ **Multiply both sides by 100.**

$200 = k$ **Divide both sides by 2.**

We now substitute 80 for w, 10 for l, and 200 for k in the equation $t = \frac{kl}{w}$ and simplify.

$$t = \frac{kl}{w}$$

$$t = \frac{200(10)}{80}$$

$$= 25$$

It will take 25 weeks for 80 workers to build 10 miles of parkway.

Self Check 6   How long will it take 100 workers to build 20 miles of parkway?

Self Check Answers   **1.** $\frac{23}{2}$ **2.** 19.2 oz   **3.** 140 mi   **4.** 16 lumens per cm^2
5. 125,000 dyne-centimeters   **6.** 40 weeks

2.5   Exercises

Vocabulary and Concepts   *Fill in the blanks.*

1. A ratio is the _____ of two numbers.

2. A proportion is a statement that two _____ are equal.

3. In the proportion $\frac{a}{b} = \frac{c}{d}$, b and c are called the _____.

4. In the proportion $\frac{a}{b} = \frac{c}{d}$, a and d are called the _____.

5. In a proportion, the product of the _____ is equal to the product of the _____.

6. Direct variation translates into the equation _____.

7. The equation $y = \frac{k}{x}$ indicates _____ variation.

8. In the equation $y = kx$, k is called the _____ of proportionality.

9. The equation $y = kxz$ represents ____ variation.

10. In the equation $y = \frac{kx^2}{z}$, y varies directly with __ and inversely with __.

Practice   *Solve each proportion.*

11. $\dfrac{4}{x} = \dfrac{2}{7}$

12. $\dfrac{5}{2} = \dfrac{x}{6}$

13. $\dfrac{x}{2} = \dfrac{3}{x+1}$

14. $\dfrac{x+5}{6} = \dfrac{7}{8-x}$

Set up and solve a proportion to answer each question.

15. The ratio of women to men in a mathematics class is 3 to 5. How many women are in the class if there are 30 men?

16. The ratio of lime to sand in mortar is 3 to 7. How much lime must be mixed with 21 bags of sand to make mortar?

Find the constant of proportionality.

17. y is directly proportional to x. If $x = 30$, then $y = 15$.

18. z is directly proportional to t. If $t = 7$, then $z = 21$.

19. I is inversely proportional to R. If $R = 20$, then $I = 50$.

20. R is inversely proportional to the square of I. If $I = 25$, then $R = 100$.

21. E varies jointly with I and R. If $R = 25$ and $I = 5$, then $E = 125$.

22. z is directly proportional to the sum of x and y. If $x = 2$ and $y = 5$, then $z = 28$.

Solve each problem.

23. y is directly proportional to x. If $y = 15$ when $x = 4$, find y when $x = \frac{7}{5}$.

24. w is directly proportional to z. If $w = -6$ when $z = 2$, find w when $z = -3$.

25. w is inversely proportional to z. If $w = 10$ when $z = 3$, find w when $z = 5$.

26. y is inversely proportional to x. If $y = 100$ when $x = 2$, find y when $x = 50$.

27. P varies jointly with r and s. If $P = 16$ when $r = 5$ and $s = -8$, find P when $r = 2$ and $s = 10$.

28. m varies jointly with the square of n and the square root of q. If $m = 24$ when $n = 2$ and $q = 4$, find m when $n = 5$ and $q = 9$.

Determine whether the graph could represent direct variation, inverse variation, or neither.

29.
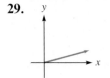

30.
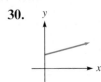

31.
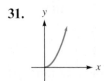

32.
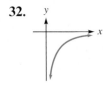

Applications   *Set up and solve the required proportion.*

33. **Caffeine**   Many convenience stores sell supersize 44-ounce soft drinks in refillable cups. For each of the products listed in the table on the next page, find the amount of caffeine contained in one of the large cups. Round to the nearest milligram.

Soft drink, 12 oz	Caffeine (mg)
Mountain Dew	55
Coca-Cola Classic	47
Pepsi	37

Based on data from the *Los Angeles Times*

34. Cellphones A country has 221 mobile cellular telephones per 250 inhabitants. If the country's population is about 280,000, how many mobile cellular telephones does the country have?

35. Wallpapering Read the instructions on the label of wallpaper adhesive. Estimate the amount of adhesive needed to paper 500 square feet of kitchen walls if a heavy wallpaper will be used.

COVERAGE: One-half gallon will hang approximately 4 single rolls (140 sq ft), depending on the weight of the wall covering and the condition of the wall.

36. Recommended dosages The recommended child's dose of the sedative hydroxine is 0.006 gram per kilogram of body mass. Find the dosage for a 30-kg child in milligrams.

37. Gas laws The volume of a gas varies directly with the temperature and inversely with the pressure. When the temperature of a certain gas is 330°, the pressure is 40 pounds per square inch and the volume is 20 cubic feet. Find the volume when the pressure increases 10 pounds per square inch and the temperature decreases to 300°.

38. Hooke's law The force f required to stretch a spring a distance d is directly proportional to d. A force of 5 newtons stretches a spring 0.2 meter. What force will stretch the spring 0.35 meter?

39. Free-falling objects The distance that an object will fall in t seconds varies directly with the square of t. An object falls 16 feet in 1 second. How long will it take the object to fall 144 feet?

40. Heat dissipation The power, in watts, dissipated as heat in a resistor varies jointly with the resistance, in ohms, and the square of the current, in amperes. A 10-ohm resistor carrying a current of 1 ampere dissipates 10 watts. How much power is dissipated in a 5-ohm resistor carrying a current of 3 amperes?

41. Heat dissipation The power, in watts, dissipated as heat in a resistor varies directly with the square of the voltage and inversely with the resistance. If 20 volts are placed across a 20-ohm resistor, it will dissipate 20 watts. What voltage across a 10-ohm resistor will dissipate 40 watts?

42. Period of a pendulum The time required for one complete swing of a pendulum is called the **period** of the pendulum. The period varies directly with the square of its length. If a 1-meter pendulum has a period of 1 second, find the length of a pendulum with a period of 2 seconds.

43. Frequency of vibration The **pitch,** or **frequency,** of a vibrating string varies directly with the square root of the tension. If a string vibrates at a frequency of 144 hertz due to a tension of 2 pounds, find the frequency when the tension is 18 pounds.

44. Kinetic energy The kinetic energy of an object varies jointly with its mass and the square of its velocity. What happens to the energy when the mass is doubled and the velocity is tripled?

45. Gravitational attraction The gravitational attraction between two massive objects varies jointly with their masses and inversely with the square of the distance between them. What happens to this force if each mass is tripled and the distance between them is doubled?

46. Gravitational attraction In Problem 45, what happens to the force if one mass is doubled and the other tripled and the distance between them is halved?

47. Plane geometry The area of an equilateral triangle varies directly with the square of the length of a side. Find the constant of proportionality.

48. Solid geometry The diagonal of a cube varies directly with the length of a side. Find the constant of proportionality.

Discovery and Writing

49. Explain the terms *extremes* and *means.*

50. Distinguish between a *ratio* and a *proportion.*

51. Explain the term *joint variation.*

52. Explain why $\frac{y}{x} = k$ indicates that y varies directly with x.

53. Explain why $xy = k$ indicates that y varies inversely with x.

54. As temperature increases on the Fahrenheit scale, it also increases on the Celsius scale. Is this direct variation? Explain.

Review   *Perform each operation and simplify.*

55. $\dfrac{1}{x + 2} + \dfrac{2}{x + 1}$

56. $\dfrac{x^2 - 1}{x + 1} \cdot \dfrac{x - 1}{x^2 - 2x + 1}$

57. $\dfrac{x^2 + 3x - 4}{x^2 - 5x + 4} \div \dfrac{x - 1}{x^2 - 3x - 4}$

58. $\dfrac{x + 2}{3x - 3} \div (2x + 4)$

59. $\dfrac{x^2 + 4 - (x + 2)^2}{4x^2}$

60. $\dfrac{\dfrac{1}{x} - \dfrac{1}{3}}{\dfrac{1}{x} - 1}$

CHAPTER REVIEW

2.1   The Rectangular Coordinate System

Definitions and Concepts	Examples
The rectangular coordinate system divides the plane into four quadrants.	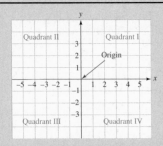

Definitions and Concepts	Examples
The **graph** of an equation in x and y is the set of all points (x, y) that satisfy the equation.	Use the x- and y-intercepts to graph the equation $6x + 4y = 24$.
The **y-intercept** of a line is the point $(0, b)$, where the line intersects the y-axis. To find b, substitute 0 for x in the equation of the line and solve for y.	**Find the y-intercept.** To find the y-intercept, we substitute 0 for x and solve for y.

$$6x + 4y = 24$$
$$6(0) + 4y = 24 \quad \text{Substitute 0 in for } x.$$
$$4y = 24 \quad \text{Simplify.}$$
$$y = 6 \quad \text{Divide both sides by 4.}$$

The y-intercept is the point $(0, 6)$.

The **x-intercept** of a line is the point $(a, 0)$, where the line intersects the x-axis. To find a, substitute 0 for y in the equation of the line and solve for x.

Find the x-intercept.   To find the x-intercept, we substitute 0 for y and solve for x.

$$6x + 4y = 24$$
$$6x + 4(0) = 24 \quad \text{Substitute 0 for } y.$$
$$6x = 24 \quad \text{Simplify.}$$
$$x = 4 \quad \text{Divide both sides by 6.}$$

The x-intercept is the point $(4, 0)$.

Find a third point as a check.   If we let $x = 2$, we will find that $y = 3$.

$$6x + 4y = 24$$
$$6(2) + 4y = 24 \quad \text{Substitute 2 for } x.$$
$$12 + 4y = 24 \quad \text{Simplify.}$$
$$4y = 12 \quad \text{Subtract 12 from both sides.}$$
$$y = 3 \quad \text{Divide both sides by 4.}$$

The point $(2, 3)$ satisfies the equation.

We plot each ordered pair and join them with a line to get the graph of the equation.

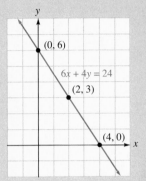

Equation of a vertical line through (a, b):

$x = a$

Equation of a horizontal line through (a, b):

$y = b$

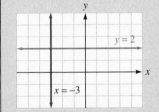

The distance formula: The distance d between points (x_1, y_1) and (x_2, y_2) is given by

$$d = \sqrt{(x_2 - x_1)^2 + (y_2 - y_1)^2}$$

Find the distance between $P(-5, 2)$ and $Q(-3, 4)$.

We can use the distance formula $d = \sqrt{(x_2 - x_1)^2 + (y_2 - y_1)^2}$. If we let $P(-5, 2) = P(x_1, y_1)$ and $Q(-3, 4) = Q(x_2, y_2)$, we can substitute -5 for x_1, 2 for y_1, -3 for x_2, and 4 for y_2 into the formula and simplify.

$$d(PQ) = \sqrt{(x_2 - x_1)^2 + (y_2 - y_1)^2}$$
$$d(PQ) = \sqrt{[-3 - (-5)]^2 + [4 - (2)]^2}$$
$$= \sqrt{(2)^2 + (2)^2}$$
$$= \sqrt{4 + 4} = \sqrt{8} = \sqrt{4 \cdot 2} = 2\sqrt{2}$$

The distance between the two points is $2\sqrt{2}$.

The midpoint formula: The midpoint of the line segment joining (x_1, y_1) and (x_2, y_2) is the point M

$$M\left(\frac{x_1 + x_2}{2}, \frac{y_1 + y_2}{2}\right)$$

To find the midpoint of the segment with endpoints at $(-4, 5)$ and $(6, 7)$, average the x-coordinates and average the y-coordinates:

The midpoint is $\left(\dfrac{-4 + 6}{2}, \dfrac{5 + 7}{2}\right) = \left(\dfrac{2}{2}, \dfrac{12}{2}\right) = (1, 6)$.

Exercises

Refer to the illustration and find the coordinates of each point.

1. A
2. B
3. C
4. D

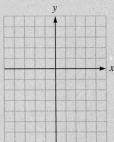

Graph each point. Indicate the quadrant in which the point lies, or the axis on which it lies.

5. $(-3, 5)$
6. $(5, -3)$
7. $(0, -7)$
8. $\left(-\frac{1}{2}, 0\right)$

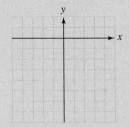

Solve each equation for y and graph the equation. Then check your graph with a graphing calculator.

9. $2x - y = 6$
10. $2x + 5y = -10$

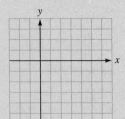

Use the x- and the y-intercepts to graph each equation.

11. $3x - 5y = 15$
12. $x + y = 7$

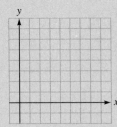

13. $x + y = -7$
14. $x - 5y = 5$

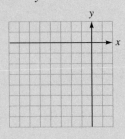
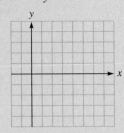

Graph each equation.

15. $y = 4$
16. $x = -2$

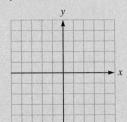
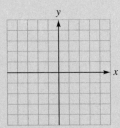

17. **Depreciation** A Ford Mustang purchased for $18,750 is expected to depreciate according to the formula $y = -2{,}200x + 18{,}750$. Find its value after 3 years.

18. **House appreciation** A house purchased for $250,000 is expected to appreciate according to the formula $y = 16{,}500x + 250{,}000$, where y is the value of the house after x years. Find the value of the house 5 years later.

Find the length of the segment PQ.

19. $P(-3, 7)$; $Q(3, -1)$
20. $P(-8, 6)$; $Q(-12, 10)$

21. $P\left(\sqrt{3}, 9\right)$; $Q\left(\sqrt{3}, 7\right)$
22. $P(a, -a)$; $Q(-a, a)$

Find the midpoint of the segment PQ.

23. $P(-3, 7)$; $Q(3, -1)$
24. $P(0, 5)$; $Q(-12, 10)$

25. $P\left(\sqrt{3}, 9\right)$; $Q\left(\sqrt{3}, 7\right)$
26. $P(a, -a)$; $Q(-a, a)$

2.2   The Slope of a Nonvertical Line

Definitions and Concepts	Examples
The slope of a nonvertical line passing through points $P(x_1, y_1)$ and $Q(x_2, y_2)$ is $$m = \frac{\text{change in } y}{\text{change in } x} = \frac{y_2 - y_1}{x_2 - x_1} \quad (x_2 \neq x_1)$$	Find the slope of the line passing through $P(-1, -3)$ and $Q(7, 9)$. We will substitute the points $P(-1, -3)$ and $Q(7, 9)$ into the slope formula $$m = \frac{\text{change in } y}{\text{change in } x} = \frac{y_2 - y_1}{x_2 - x_1}$$ to find the slope of the line. Let $P(x_1, y_1) = P(-1, -3)$ and $Q(x_2, y_2) = Q(7, 9)$. Then we substitute -1 for x_1, -3 for y_1, 7 for x_2, and 9 for y_2 to get $$m = \frac{\text{change in } y}{\text{change in } x}$$ $$m = \frac{y_2 - y_1}{x_2 - x_1}$$ $$= \frac{9 - (-3)}{7 - (-1)} = \frac{12}{8} = \frac{3}{2}$$ The slope of the line is $\frac{3}{2}$.
Slopes of horizontal and vertical lines The slope of a horizontal line (a line with an equation of the form $y = b$) is 0. The slope of a vertical line (a line with an equation of the form $x = a$) is not defined.	The slope of the graph of $y = 7$ is 0. The slope of the line $x = 6$ is not defined.
Slopes of parallel lines Nonvertical parallel lines have the same slope. **Slopes of perpendicular lines** The product of the slopes of two perpendicular lines is -1, provided neither line is vertical.	Determine whether the lines with the given slopes are parallel, perpendicular, or neither. $$m_1 = -6; m_2 = \frac{1}{6}$$ The product of the slopes $m_1 = -6$ and $m_2 = \frac{1}{6}$ is -1. The lines are perpendicular.

Exercises

Find the slope of the line PQ, if possible.

27. $P(3, -5)$; $Q(1, 7)$

28. $P(2, 7)$; $Q(-5, -7)$

29. $P(b, a)$; $Q(a, b)$

30. $P(a + b, b)$; $Q(b, b - a)$

Find two points on the line and find the slope of the line.

31. $y = 3x + 6$ **32.** $y = 5x - 6$

Determine whether the slope of each line is 0 or undefined.

33.   **34.**

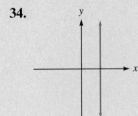

Determine whether the slope of each line is positive or negative.

35.   **36.**

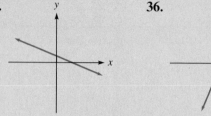

Determine whether the lines with the given slopes are parallel, perpendicular, or neither.

37. $m_1 = 5; m_2 = -\dfrac{1}{5}$ **38.** $m_1 = \dfrac{2}{7}; m_2 = \dfrac{7}{2}$

39. A line passes through $(-2, 5)$ and $(6, 10)$. A line parallel to it passes through $(2, 2)$ and $(10, y)$. Find y.

40. A line passes through $(-2, 5)$ and $(6, 10)$. A line perpendicular to it passes through $(-2, 5)$ and $(x, -3)$. Find x.

41. Rate of descent   If an airplane descends 3,000 feet in 15 minutes, what is the average rate of descent in feet per minute?

42. Rate of growth   A small business predicts sales according to a straight-line method. If sales were $50,000 in the first year and $147,500 in the third year, find the rate of growth in dollars per year (the slope of the line).

2.3	**Writing Equations of Lines**

Definitions and Concepts	**Examples**
Point–slope form: The equation of the line passing through $P(x_1, y_1)$ and with slope m is $y - y_1 = m(x - x_1)$	Write the equation of the line with slope $-\dfrac{4}{3}$ and passing through $P(3, -2)$. We will substitute $-\dfrac{4}{3}$ for m, 3 for x_1, and -2 for y_1 in the point–slope form $y - y_1 = m(x - x_1)$ and simplify. $$y - y_1 = m(x - x_1)$$ $$y - (-2) = -\frac{4}{3}(x - 3) \quad \text{Substitute } -\tfrac{4}{3} \text{ for } m, 3 \text{ for } x_1, \text{ and } -2 \text{ for } y_1.$$ $$y + 2 = -\frac{4}{3}x + 4 \quad \text{Remove parentheses.}$$ $$y = -\frac{4}{3}x + 2 \quad \text{Subtract 2 from both sides.}$$
Slope–intercept form: The equation of the line with slope m and y-intercept $(0, b)$ is $y = mx + b$.	The equation of the line $y = -\dfrac{5}{3}x + 4$ is written in slope–intercept form.
Standard form of the equation of a line: $Ax + By = C$	The above equation written in standard form $Ax + By = C$ is $5x + 3y = 12$.

Slope–intercept form can be used to find the slope and the y-intercept from the equation of a line.

Find the slope and the y-intercept of the line with equation $2x + 5y = -10$.

We will write the equation in the form $y = mx + b$ to find the slope m and the y-intercept $(0, b)$.

$$2x + 5y = -10$$
$$5y = -2x - 10 \quad \text{Subtract } 2x \text{ from both sides.}$$
$$y = -\frac{2}{5}x - 2 \quad \text{Divide both sides by 5.}$$

The slope of the graph is $-\frac{2}{5}$, and the y-intercept is $(0, -2)$.

Horizontal line: $y = b$
The slope is 0, and the y-intercept is $(0, b)$.

The equation of the horizontal line with slope 0 and y-intercept $(0, 7)$ is $y = 7$.

Vertical line: $x = a$
There is no defined slope, and the x-intercept is $(a, 0)$.

The equation of the vertical line with no defined slope and the x-intercept $(-6, 0)$ is $x = -6$.

Exercises

Use point–slope form to write the equation of each line. Write each equation in standard form.

43. The line passes through the origin and the point $(-5, 7)$.

44. The line passes through $(-2, 1)$ and has a slope of -4.

45. The line passes through $(2, -1)$ and has a slope of $-\frac{1}{5}$.

46. The line passes through $(7, -5)$ and $(4, 1)$.

Use slope–intercept form to write the equation of each line.

47. The line has a slope of $\frac{2}{3}$ and a y-intercept of 3.

48. The slope is $-\frac{3}{2}$ and the line passes through $(0, -5)$.

Use slope–intercept form to graph each equation.

49. $y = \frac{3}{5}x - 2$

50. $y = -\frac{4}{3}x + 3$

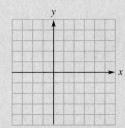
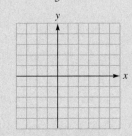

Find the slope and the y-intercept of the graph of each line.

51. $y = 2x$

52. $x = -2y$

53. $-2y = -3x + 10$

54. $2x = -4y - 8$

55. $5x + 2y = 7$

56. $3x - 4y = 14$

Write the equation of each line.

57. The line has a slope of 0 and passes through $(-5, 17)$.

58. The line has no defined slope and passes through $(-5, 17)$.

Write the equation of each line. Write the answer in slope–intercept form.

59. The line is parallel to $3x - 4y = 7$ and passes through $(2, 0)$.

60. The line passes through $(7, -2)$ and is parallel to the line segment joining $(2, 4)$ and $(4, -10)$.

61. The line passes through $(0, 5)$ and is perpendicular to the line $x + 3y = 4$.

62. The line passes through $(7, -2)$ and is perpendicular to the line segment joining $(2, 4)$ and $(4, -10)$.

Determine whether the graphs of each pair of equations are parallel, perpendicular, or neither.

63. $y = 3x + 8, 2y = 6x - 19$

64. $2x + 3y = 6, 3x - 2y = 15$

2.4   Graphs of Equations

Definitions and Concepts	Examples

Definitions and Concepts

To graph an equation:

1. Find the x- and y-intercepts
2. Find the symmetries of the graph
3. Plot some additional points, if necessary, and draw the graph.

Intercepts of a graph:
To find the x-intercepts, let $y = 0$ and solve for x.

To find the y-intercepts, let $x = 0$ and solve for y.

Test for x-axis symmetry
To test for x-axis symmetry, replace y with $-y$. If the resulting equation is equivalent to the original one, the graph is symmetric about the x-axis.

Test for y-axis symmetry
To test for y-axis symmetry, replace x with $-x$. If the resulting equation is equivalent to the original one, the graph is symmetric about the y-axis.

Examples

Graph: $y = x^3 - 4x$.

To graph $y = x^3 - 4x$, we will find the x- and y-intercepts, test for symmetries, plot points, and join the points with a smooth curve.

Step 1. Find the x- and y-intercepts. To find the x-intercepts, we let $y = 0$ and solve for x.

$$y = x^3 - 4x$$
$$0 = x^3 - 4x \quad \text{Substitute 0 for } y.$$
$$0 = x(x^2 - 4) \quad \text{Factor out } x.$$
$$0 = x(x + 2)(x - 2) \quad \text{Factor } x^2 - 4.$$
$$x = 0 \quad \text{or} \quad x + 2 = 0 \quad \text{or} \quad x - 2 = 0$$
$$x = -2 \qquad x = 2$$

The x-intercepts are $(0, 0)$, $(-2, 0)$, and $(2, 0)$.

To find the y-intercepts, we let $x = 0$ and solve for y.

$$y = x^3 - 4x$$
$$y = 0^3 - 4(0) \quad \text{Substitute 0 for } x.$$
$$y = 0$$

The y-intercept is $(0, 0)$.

Step 2. Test for symmetries. We test for symmetry about the x-axis by replacing y with $-y$.

(1) $\quad y = x^3 - 4x \quad$ This is the original equation.
$\quad -y = x^3 - 4x \quad$ Replace y with $-y$.
(2) $\quad y = -x^3 + 4x \quad$ Multiply both sides by -1.

Since Equations 1 and 2 are different, the graph is not symmetric about the x-axis.

To test for y-axis symmetry, we replace x with $-x$.

(1) $y = x^3 - 4x \quad$ This is the original equation.
$\quad y = (-x)^3 - 4(-x) \quad$ Replace x with $-x$.
(3) $y = -x^3 + 4x \quad$ Simplify.

Since Equations 1 and 3 are different, the graph is not symmetric about the y-axis.

Test for origin symmetry
To test for symmetry about the origin, replace x with $-x$ and y with $-y$. If the resulting equation is equivalent to the original one, the graph is symmetric about the origin.

To test for symmetry about the origin, we replace x with $-x$ and y with $-y$.

(1)	$y = x^3 - 4x$	This is the original equation.
	$-y = (-x)^3 - 4(-x)$	Replace x with $-x$ and y with $-y$.
	$-y = -x^3 + 4x$	Simplify.
(4)	$y = x^3 - 4x$	Multiply both sides by -1.

Since Equations 1 and 4 are the same, the graph is symmetric about the origin.

Plot some additional points and draw the graph.

Step 3. Graph the equation. To graph the equation, we plot the x- and y-intercepts and several other pairs (x, y) with positive values of x. We can use the property of symmetry about the origin to draw the graph for negative values of x.

$y = x^3 - 4x$

x	y	(x, y)
-2	0	$(-2, 0)$
0	0	$(0, 0)$
1	-3	$(1, -3)$
2	0	$(2, 0)$
3	15	$(3, 15)$

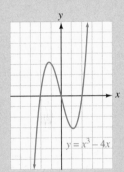

Circles: A **circle** is the set of all points in a plane that are a fixed distance from a point called its **center.** The fixed distance is the **radius of the circle.**

The standard equation of a circle with center (h, k):
The graph of any equation that can be written in the form

$$(x - h)^2 + (y - k)^2 = r^2$$

is a circle with radius r and center at point (h, k).

The standard equation of a circle with center $(0, 0)$:
The graph of any equation that can be written in the form

$$x^2 + y^2 = r^2$$

is a circle with radius r and center at the origin.

Find the center and radius of the circle with the equation $(x - 6)^2 + (y + 5)^2 = 4$.

Standard Form: $\qquad (x - h)^2 + (y - k)^2 = r^2$

Given Form: $\qquad (x - 6)^2 + [y - (-5)]^2 = 2^2$

We see that $h = 6$, $k = -5$, and $r = 2$. The center (h, k) of the circle is at $(6, -5)$ and the radius is 2.

The general form of the equation of a circle: The general form of the equation of a circle is

$$x^2 + y^2 + cx + dy + e = 0$$

where c, d, and e are real numbers.

To convert the general form of the equation of the circle $x^2 + y^2 + 4x - 2y - 20 = 0$ into standard form, we must find the coordinates of the center and the radius. To do so, we will complete the square on both x and y:

$$x^2 + y^2 + 4x - 2y = 20$$
$$x^2 + 4x + y^2 - 2y = 20$$
$$x^2 + 4x + 4 + y^2 - 2y + 1 = 20 + 4 + 1$$
$$(x + 2)^2 + (y - 1)^2 = 25$$

Exercises

Find the x- and y-intercepts of each graph. Do not graph the equation.

65. $y = 4x - 8x^2$

66. $y = x^2 - 10x - 24$

Find the symmetries, if any, of the graph of each equation. Do not graph the equation.

67. $y^2 = 8x$

68. $y = 3x^4 + 6$

69. $y = -2|x|$

70. $y = |x + 2|$

Graph each equation. Find all intercepts and symmetries.

71. $y = x^2 + 2$

72. $y = x^3 - 2$

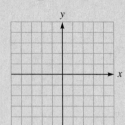

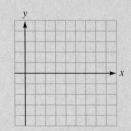

73. $y = \dfrac{1}{2}|x|$

74. $y = -\sqrt{x - 4}$

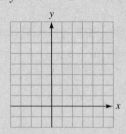

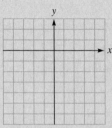

75. $y = \sqrt{x} + 2$

76. $y = |x + 1| + 2$

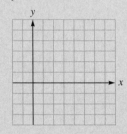

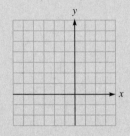

Use a graphing calculator to graph each equation.

77. $y = |x - 4| + 2$

78. $y = -\sqrt{x + 2} + 3$

79. $y = x + 2|x|$

80. $y^2 = x - 3$

Identify the center and radius of each circle written in standard form.

81. $x^2 + y^2 = 64$

82. $x^2 + (y - 6)^2 = 100$

83. $(x + 7)^2 + y^2 = \dfrac{1}{4}$

84. $(x - 5)^2 + (y + 1)^2 = 9$

Write the equation of each circle in standard form.

85. Center at the origin; $r = 7$

86. Center at $(3, 0)$; $r = \dfrac{1}{5}$

87. Center at $(-2, 12)$; $r = 5$

88. Center at $\left(\dfrac{2}{7}, 5\right)$; $r = 9$

Write the equation of each circle in standard form and general form.

89. Center at $(-3, 4)$; radius 12

90. Ends of diameter at $(-6, -3)$ and $(5, 8)$

Convert the general form of each circle given into standard form.

91. $x^2 + y^2 + 6x - 4y + 4 = 0$

92. $2x^2 + 2y^2 - 8x - 16y - 10 = 0$

Graph each equation.

93. $x^2 + y^2 - 16 = 0$ **94.** $x^2 + y^2 - 4x = 5$

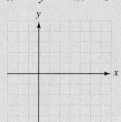

95. $x^2 + y^2 - 2y = 15$ **96.** $x^2 + y^2 - 4x + 2y = 4$

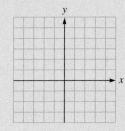
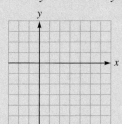

Use a graphing calculator to solve each equation. If an answer is not exact, round to the nearest hundredth.

97. $x^2 - 11 = 0$ **98.** $x^3 - x = 0$

99. $|x^2 - 2| - 1 = 0$ **100.** $x^2 - 3x = 5$

2.5   Proportion and Variation

Definitions and Concepts	Examples
An equation indicating that two ratios are equal is called a **proportion**. In the proportion $\frac{a}{b} = \frac{c}{d}$, the numbers a and d are called the **extremes,** and the numbers b and c are called the **means.**	In the proportion $\frac{3}{5} = \frac{9}{15}$, 3 and 15 are called the extremes, and 5 and 9 are called the means.
Property of Proportions: In any proportion, the product of the extremes is equal to the product of the means.	In the proportion above, the product of the means is equal to the product of the extremes. $5(9) = 3(15) = 45$

The property of proportions can be used to solve a proportion for a variable.

Solve the proportion $\frac{x}{6} = \frac{2}{x + 11}$ for x.

We will use the property of proportions to solve the proportion.

$$\frac{x}{6} = \frac{2}{x + 11}$$

$x(x + 11) = 6 \cdot 2$ The product of the extremes equals the product of the means.

$x^2 + 11x = 12$ Remove parentheses and simplify.

$x^2 + 11x - 12 = 0$ Subtract 12 from both sides.

$(x + 12)(x - 1) = 0$ Factor the trinomial.

$x + 12 = 0$ or $x - 1 = 0$ Set each factor equal to 0.

$x = -12$ $x = 1$

Thus, $x = -12$ or $x = 1$.

Direct Variation:
The words **"y varies directly with x,"** or **"y is directly proportional to x,"** mean that $y = kx$ for some real-number constant k. The number k is called the **constant of proportionality.**

In the direct variation formula, $y = 3x$, find y when $x = 5$.

$y = 3x = 3(5) = 15$

Inverse Variation:
The words **"y varies inversely with x,"** or **"y is inversely proportional to x,"** mean that $y = \frac{k}{x}$ for some real-number constant k.

In the inverse variation formula $y = \frac{k}{x}$, find the constant of variation if $y = 5$ when $x = 20$.

$$y = \frac{k}{x}$$

$5 = \dfrac{k}{20}$ Substitute 5 for y.
Substitute 20 for x.

$100 = k$

Joint Variation:
The words **"y varies jointly with w and x"** mean that $y = kwx$ for some real-number constant k.

Kinetic energy of an object varies jointly with its mass and the square of its velocity. A 50-gram mass moving at the rate of 20 centimeters per second has a kinetic energy of 40,000 dyne-centimeters. Find the kinetic energy of a 10-gram mass that is moving at 60 centimeters per second.

We will use joint variation to solve the problem. If we let E, m, and v represent the kinetic energy, mass, and velocity, respectively, the phrase *energy varies jointly with the mass and the square of its velocity* translates into the formula

$E = kmv^2$

The constant k can be evaluated by substituting 40,000 for E, 50 for m, and 20 for v in the formula.

$$E = kmv^2$$
$$\mathbf{40,000} = k(\mathbf{50})(\mathbf{20})^2$$
$$40,000 = 20,000k$$
$$k = 2$$

We can now substitute 2 for k, 10 for m, and 60 for v in the formula and evaluate E.

$$E = kmv^2$$
$$= 2(\mathbf{10})(\mathbf{60})^2$$
$$= 72,000$$

A 10-gram mass that is moving at 60 centimeters per second has a kinetic energy of 72,000 dyne-centimeters.

Exercises

Solve each proportion.

101. $\dfrac{x + 3}{10} = \dfrac{x - 1}{x}$

102. $\dfrac{x - 1}{2} = \dfrac{12}{x + 1}$

103. Hooke's law The force required to stretch a spring is directly proportional to the amount of stretch. If a 3-pound force stretches a spring 5 inches, what force would stretch the spring 3 inches?

104. Kinetic energy A moving body has a kinetic energy directly proportional to the square of its velocity. By what factor does the kinetic energy of an automobile increase if its speed increases from 30 mph to 50 mph?

105. Gas laws The volume of gas in a balloon varies directly with the temperature and inversely with the pressure. If the volume is 400 cubic centimeters when the temperature is 300 K and the pressure is 25 dynes per square centimeter, find the volume when the temperature is 200 K and the pressure is 20 dynes per square centimeter.

106. Geometry The area of a rectangle varies jointly with its length and width. Find the constant of proportionality.

107. Electrical resistance The resistance of a wire varies directly as the length of the wire and inversely as the square of its diameter. A 1,000-foot length of wire, 0.05 inches in diameter, has a resistance of 200 ohms. What would be the resistance of a 1,500-foot length of wire that is 0.08 inches in diameter?

108. Billing for services Angie's Painting and Decorating Service charges a fixed amount for accepting a wallpapering job and adds a fixed dollar amount for each roll hung. If the company bills a customer $177 to hang 11 rolls, and $294 to hang 20 rolls, find the cost to hang 27 rolls.

109. Paying for college Rolf must earn $5,040 for next semester's tuition. Assume he works x hours tutoring algebra at $14 per hour and y hours tutoring Spanish at $18 per hour, and makes his goal. Write an equation expressing the relationship between x and y, and graph the equation. If Rolf tutors algebra for 180 hours, how long must he tutor Spanish?

CHAPTER TEST

Indicate the quadrant in which the point lies, or the axis on which it lies.

1. $(-3, \pi)$
2. $(0, -8)$

Find the x- and y-intercepts and use them to graph the equation.

3. $x + 3y = 6$ 4. $2x - 5y = 10$

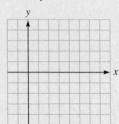

Graph each equation.

5. $2(x + y) = 3x + 5$ 6. $3x - 5y = 3(x - 5)$

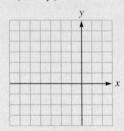
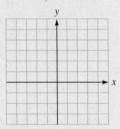

7. $\frac{1}{2}(x - 2y) = y - 1$ 8. $\frac{x + y - 5}{7} = 3x$

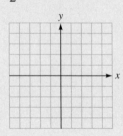
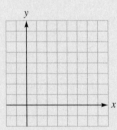

Find the distance between points P and Q.

9. $P(1, -1); Q(-3, 4)$ 10. $P(0, \pi); Q(-\pi, 0)$

Find the midpoint of the line segment PQ.

11. $P(3, -7); Q(-3, 7)$
12. $P\left(0, \sqrt{2}\right); Q\left(\sqrt{8}, \sqrt{18}\right)$

Find the slope of the line PQ.

13. $P(3, -9); Q(-5, 1)$
14. $P\left(\sqrt{3}, 3\right); Q\left(-\sqrt{12}, 0\right)$

Determine whether the two lines are parallel, perpendicular, or neither.

15. $y = 3x - 2; y = 2x - 3$
16. $2x - 3y = 5; 3x + 2y = 7$

Write the equation of the line in slope–intercept form with the given properties.

17. Passing through $(3, -5); m = 2$
18. $m = 3; b = \frac{1}{2}$
19. Parallel to $2x - y = 3; b = 5$
20. Perpendicular to $2x - y = 3; b = 5$
21. Passing through $\left(2, -\frac{3}{2}\right)$ and $\left(3, \frac{1}{2}\right)$
22. Parallel to the y-axis and passing through $(3, -4)$

Find the x- and y-intercepts of each graph.

23. $y = x^3 - 16x$
24. $y = |x - 4|$

Find the symmetries of each graph.

25. $y^2 = x - 1$ 26. $y = x^4 + 1$

Graph each equation. Find all intercepts and symmetries.

27. $y = x^2 - 9$ 28. $x = |y|$

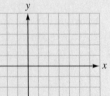

29. $y = 2\sqrt{x}$

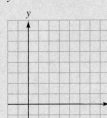

30. $x = y^3$

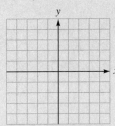

34. $x^2 - 4x + y^2 + 3 = 0$

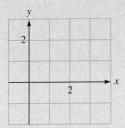

Write the equation of each circle in standard form.

31. Center at $(5, 7)$; radius of 8

32. Center at $(2, 4)$; passing through $(6, 8)$

Graph each equation.

33. $x^2 + y^2 = 9$

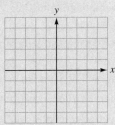

Write each statement as an equation.

35. y varies directly as the square of z.

36. w varies jointly with r and the square of s.

37. P varies directly with Q. $P = 7$ when $Q = 2$. Find P when $Q = 5$.

38. y is directly proportional to x and inversely proportional to the square of z, and $y = 16$ when $x = 3$ and $z = 2$. Find x when $y = 2$ and $z = 3$.

Use a graphing calculator to find the positive root of each equation.

39. $x^2 - 7 = 0$ **40.** $x^2 - 5x - 5 = 0$

Functions

In this chapter, we will discuss one of the most important concepts in mathematics—the concept of function.

Careers and Mathematics

Computer Programmer   Computer programmers write, test, and maintain the detailed instructions, called **programs,** that computers follow to perform their functions. With the help of other computer specialists, they determine which instructions to use to make computers do specific tasks. They are grouped into two broad types—applications programmers and systems programmers. Applications programmers write programs to handle a specific job, such as tracking inventory. Systems programmers write programs to maintain and control computer systems software. Computer programmers held about 435,000 jobs in 2006.

Education   Most programmers have a bachelor's degree, but a two-year degree or certificate may be adequate for some jobs. Some computer programmers hold a college degree in computer science, mathematics, or information systems, whereas others have taken special courses in computer programming to supplement their degree in a field such as accounting, finance, or another area of business. In 2006, more than 68 percent of computer programmers had a bachelor's degree or higher.

Job Outlook   Employment of computer programmers is expected to decline slowly, decreasing 4 percent through 2016. Median annual earnings of programmers were $65,510 in 2006.

For a sample application, see Exercise 66 in Section 3.3. For more information, see www.bls.gov/oco/ocos110.htm.

© Dana Hursey/Masterfile

3.1 **Functions and Function Notation**

Objectives

1. Understand the Concept of a Function
2. Find the Domain of a Function
3. Evaluate a Function
4. Evaluate the Difference Quotient for a Function
5. Graph a Function by Plotting Points
6. Use the Vertical Line Test to Identify Functions
7. Use Linear Functions to Model Applications

Correspondences between the elements of two sets are common occurrences in everyday life. For example,

- To every Motorola cell phone, there corresponds exactly one phone number.
- To every Honda Civic car, there corresponds exactly one vehicle identification number.
- To every case on the television show *Deal or No Deal,* there corresponds exactly one amount of money.
- To every item's barcode at a Target store, there corresponds exactly one price.

This table shows four of the most popular movies of all time and the year each movie was released.

Movie	Year Released
E. T., the Extra Terrestrial	1982
Titanic	1997
Shrek 2	2004
Pirates of the Caribbean: At World's End	2007

The information shown in the table sets up a correspondence between a movie and the year it was released. Note that for each of the movies, there corresponds exactly one year in which it was released. The phase *there corresponds exactly one* is extremely important in mathematics and we will use this idea to solve problems.

1. Understand the Concept of a Function

Correspondences in which exactly one quantity corresponds to (or depends on) another quantity according to some specific rule are called **functions.** Equations frequently are used in mathematics to represent functions. For example, the equation $y = x^2 - 1$ sets up a correspondence between two infinite sets of real numbers, x and y, according to the rule *square x and subtract 1.*

> **Equation:** $y = x^2 - 1$
>
> **Correspondence:**   Each real number x determines exactly one real number y.
>
> **Rule:**   Square x and subtract 1.

Since the value of y depends on the number x, we call y the **dependent variable** and x the **independent variable.**

The equation $y = x^2 - 1$ determines what **output value** y will result from each **input value** x. This idea of inputs and outputs is shown in Figure 3-1(a). In the equation $y = x^2 - 1$, if the input x is 2, the output y is

$$y = 2^2 - 1 = 3$$

This is illustrated in Figure 3-1(b).

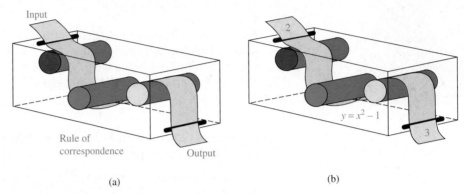

(a)   (b)

Figure 3-1

The equation $y = x^2 - 1$ also determines the table of ordered pairs and the graph shown in Figure 3-2. To see how the table determines the correspondence, we find an input in the x-column and read across to find the corresponding output in the y-column. If we select $x = 2$ as an input, we get $y = 3$ for the output.

To see how the graph of $y = x^2 - 1$ determines the correspondence, we draw a vertical and horizontal line through any point (say, point P) on the graph shown in Figure 3-2. Because these lines intersect the x-axis at 2 and the y-axis at 3, the point $P(2, 3)$ associates 3 on the y-axis with 2 on the x-axis.

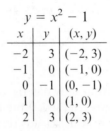
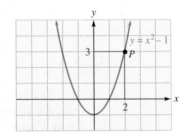

x	y	(x, y)
-2	3	$(-2, 3)$
-1	0	$(-1, 0)$
0	-1	$(0, -1)$
1	0	$(1, 0)$
2	3	$(2, 3)$

$y = x^2 - 1$

Figure 3-2

Comment

Correspondences can be set up by a verbal expression, by a table, by a set of ordered pairs, by an equation, and by a graph.

Any correspondence that assigns exactly one value of y to each number x is called a **function.** Assciated with every function are two sets of numbers called the **domain** and **range** of the function.

Function	A **function** f is a correspondence between a set of input values x and a set of output values y, where to each x-value there corresponds exactly one y-value.
Domain	The set of input values x is called the **domain** of the function.
Range	The set of output values y is called the **range** of the function.

EXAMPLE 1   Determine whether the following equations define y to be a function of x.
a. $7x + y = 5$ **b.** $y = |x| + 1$ **c.** $y^2 = x + 2$

Solution   In each case, we will solve each equation for y, if necessary, and determine whether each input value x determines exactly one output value y.

a. First, we solve $7x + y = 5$ for y to get

$$y = 5 - 7x$$

From this equation, we see that for each input value x there corresponds exactly one output value y. For example, if $x = 3$, then $y = 5 - 7(3) = -16$. The equation defines y to be a function of x.

b. Since each input number x that we substitute for x determines exactly one output y, the equation $y = |x| + 1$ defines y to be a function of x.

c. First, we solve $y^2 = x + 2$ for y using the square root property covered in Section 1.3 to get $y = \pm\sqrt{x + 2}$. Note that each input value x for which the function is defined (except -2) gives two outputs y. For example, if $x = 7$, then $y = \pm\sqrt{7 + 2} = \pm\sqrt{9} = \pm 3$. For this reason, the equation does not define y to be a function of x.

Comment

Some equations represent functions and some do not. If to some input x in an equation, there corresponds more than one output y, the equation will not represent a function.

Self Check 1   Determine whether each equation defines y to be a function of x.
a. $y = |x|$ **b.** $y = \sqrt{x}$ **c.** $y^2 = 2x$

2.   Find the Domain of a Function

The **domain** of a function is the set of all real numbers x for which the function is defined. Thus, to find the domain of a function, we must find the set of numbers that are permissible inputs for x. It is often helpful to determine any restrictions on the input values for x. For example, we consider the information in the following table.

Equation	Restriction	Domain
$y = x^3 - x^2 + x - 5$	There are no restrictions on x because any real number can be cubed, squared, and combined with constants.	all real numbers, $(-\infty, \infty)$
$y = \sqrt{x}$	The restriction that x cannot be a negative real number is placed on x because the square root of a negative number is an imaginary number.	$x \geq 0$, the interval $[0, \infty)$
$y = \dfrac{1}{x}$	The restriction that x cannot be 0 is placed on x, because division by 0 isn't defined.	$x \neq 0$, $(-\infty, 0) \cup (0, \infty)$

EXAMPLE 2   Find the domain of the function defined by each equation:

a. $y = 2x - 5$ **b.** $y = \sqrt{3x - 2}$ **c.** $y = \dfrac{3}{x + 2}$

Solution   We must determine what numbers are permissible inputs for x. This set of numbers is the domain.

a. Any real number that we input for x can be multiplied by 2 and then 5 can be subtracted from the result. Thus, the domain is the interval $(-\infty, \infty)$.

b. Since the radicand must be nonnegative, we have

$$3x - 2 \geq 0$$
$$3x \geq 2 \qquad \text{Add 2 to both sides.}$$
$$x \geq \frac{2}{3} \qquad \text{Divide both sides by 3.}$$

Thus, the domain is the interval $\left[\frac{2}{3}, \infty\right)$.

c. Since the fraction $\dfrac{3}{x + 2}$ is undefined when $x = -2$, -2 is not a permissible input value. Since all other values of x are permissible inputs, the domain is $(-\infty, -2) \cup (-2, \infty)$.

Self Check 2   Find the domain of each function: **a.** $y = |x| + 2$
b. $y = \sqrt[3]{x + 2}$

EXAMPLE 3   Find the domain of the function defined by the equation $y = \dfrac{1}{x^2 - 5x - 6}$.

Solution   We can factor the denominator to see what values of x will give 0's in the denominator. These values are not in the domain.

$$x^2 - 5x - 6 = 0$$
$$(x - 6)(x + 1) = 0$$
$$x - 6 = 0 \quad \text{or} \quad x + 1 = 0$$
$$x = 6 \qquad \qquad x = -1$$

The domain is $(-\infty, -1) \cup (-1, 6) \cup (6, \infty)$.

Self Check 3   Find the domain of the function defined by the equation $y = \dfrac{2}{x^2 - 16}$.

3.   Evaluate a Function

To indicate that y is a function of x, we often use **function notation** and write

$$y = f(x) \qquad \text{Read as "y is a function of x."}$$

The notation $y = f(x)$ provides a way of denoting the value of y (the *dependent variable*) that corresponds to some input number x (the *independent variable*). For example, if $y = f(x)$, the value of y that is determined when $x = 2$ is denoted by $f(2)$, read as "f of 2." If $f(x) = 5 - 7x$, we can evaluate $f(2)$ by substituting 2 for x.

$$f(x) = 5 - 7x$$
$$f(2) = 5 - 7(2) \quad \text{Substitute the input 2 for } x.$$
$$= -9$$

If $x = 2$, then $y = f(2) = -9$.

To evaluate $f(-5)$, we substitute -5 for x.

$$f(x) = 5 - 7x$$
$$f(-5) = 5 - 7(-5) \quad \text{Substitute the input } -5 \text{ for } x.$$
$$= 40$$

If $x = -5$, then $y = f(-5) = 40$.

Comment

To see why function notation is helpful, consider the following sentences. Note that the second sentence is much more concise.

1. In the function $y = 3x^2 + x - 4$, find the value of y when $x = -3$.
2. In $f(x) = 3x^2 + x - 4$, find $f(-3)$.

In this context, the notations y and $f(x)$ both represent the output of a function and can be used interchangeably, but function notation is more concise.

Sometimes functions are denoted by letters other than f. The notations $y = g(x)$ and $y = h(x)$ also denote functions involving the independent variable x.

EXAMPLE 4   Let $g(x) = 3x^2 + x - 4$. Find: **a.** $g(-3)$ **b.** $g(k)$ **c.** $g(-t^3)$ **d.** $g(k + 1)$

Solution   In each case, we will substitute the input value into the function for x and simplify.

a. $g(x) = 3x^2 + x - 4$
$$g(-3) = 3(-3)^2 + (-3) - 4$$
$$= 3(9) - 3 - 4$$
$$= 20$$

b. $g(x) = 3x^2 + x - 4$
$$g(k) = 3k^2 + k - 4$$
$$= 3k^2 + k - 4$$

c. $g(x) = 3x^2 + x - 4$
$$g(-t^3) = 3(-t^3)^2 + (-t^3) - 4$$
$$= 3t^6 - t^3 - 4$$

d. $g(x) = 3x^2 + x - 4$
$$g(k + 1) = 3(k + 1)^2 + (k + 1) - 4$$
$$= 3(k^2 + 2k + 1) + k + 1 - 4$$
$$= 3k^2 + 6k + 3 + k + 1 - 4$$
$$= 3k^2 + 7k$$

Self Check 4   Evaluate: **a.** $g(0)$ **b.** $g(2)$ **c.** $g(k - 1)$

4.   Evaluate the Difference Quotient for a Function

The fraction $\frac{f(x + h) - f(x)}{h}$ is called the **difference quotient** and is important in calculus. The difference quotient can be used to find quantities such as the velocity of a guided missile or the rate of change of a company's profit.

EXAMPLE 5   If $f(x) = x^2 - 2x - 5$, evaluate $\dfrac{f(x + h) - f(x)}{h}$.

Solution   We will evaluate the difference quotient in three steps. Find $f(x + h)$. Then subtract $f(x)$. Then divide by h.

Step 1:   Find $f(x + h)$.

$$f(x) = x^2 - 2x - 5$$
$$f(x + h) = (x + h)^2 - 2(x + h) - 5 \qquad \text{Substitute } x + h \text{ for } x.$$
$$= x^2 + 2xh + h^2 - 2x - 2h - 5$$

Step 2:   Find $f(x + h) - f(x)$. We can use the result from Step 1.

$$f(x + h) - f(x) = x^2 + 2xh + h^2 - 2x - 2h - 5 - (x^2 - 2x - 5) \qquad \text{Subtract } f(x).$$
$$= x^2 + 2xh + h^2 - 2x - 2h - 5 - x^2 + 2x + 5 \qquad \text{Remove parentheses.}$$
$$= 2xh + h^2 - 2h \qquad \text{Combine like terms.}$$

Step 3:   Find the difference quotient $\frac{f(x + h) - f(x)}{h}$. We can use the result from Step 2.

$$\frac{f(x + h) - f(x)}{h} = \frac{2xh + h^2 - 2h}{h} \qquad \text{Divide both sides by } h.$$
$$= \frac{h(2x + h - 2)}{h} \qquad \text{In the numerator, factor out } h.$$
$$= 2x + h - 2 \qquad \text{Divide out } h: \frac{h}{h} = 1.$$

Comment

After the completion of Step 2 in a polynomial function, when $f(x + h) - f(x)$ is simplified, each term always will include an h. Use that fact to check your work as you progress through the problem.

Self Check 5   If $f(x) = x^2 + 2$, evaluate $\frac{f(x + h) - f(x)}{h}$.

5.   Graph a Function by Plotting Points

If f is a function whose domain and range are sets of real numbers, its graph is the set of all points $(x, f(x))$ in the xy-plane that satisfy the equation $y = f(x)$. For example, the graph of the function $y = f(x) = -7x + 5$ is a line with slope -7 and y-intercept $(0, 5)$. (See Figure 3-3.) Since the graph of $f(x) = -7x + 5$ is a nonvertical line, $f(x)$ is called a **linear function.**

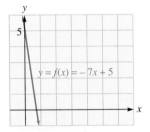

Figure 3-3

Graph of a Function   The **graph** of a function f in the xy-plane is the set of all points (x, y) where x is in the domain of f, y is in the range of f, and $y = f(x)$.

EXAMPLE 6   Graph the functions.   **a.** $f(x) = -2|x| + 3$ **b.** $f(x) = \sqrt{x - 2}$

Solution   In each case, we will make a table of solutions and plot the points given by the table. Then we will connect the points by drawing a smooth curve through them and obtain the graph of the function.

a. $f(x) = -2|x| + 3$

$f(x) = -2|x| + 3$

x	$f(x)$	$(x, f(x))$
-2	-1	$(-2, -1)$
-1	1	$(-1, -1)$
0	3	$(0, 3)$
1	1	$(1, 1)$
2	-1	$(2, -1)$

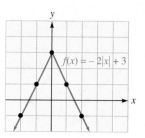

Figure 3-4

b. $f(x) = \sqrt{x - 2}$

$f(x) = \sqrt{x - 2}$

x	$f(x)$	$(x, f(x))$
2	0	$(2, 0)$
6	2	$(6, 2)$
11	3	$(11, 3)$
18	4	$(18, 4)$

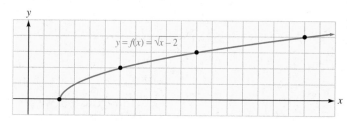

Figure 3-5

Self Check 6 Graph: $f(x) = |x + 3|$.

Both the **domain** and the **range** of a function can be identified by viewing the graph of the function. The inputs or x-values that correspond to points on the graph of the function can be identified on the x-axis and used to state the domain of the function. The outputs or $f(x)$-values that correspond to points on the graph of the function can be identified on the y-axis and used to state the range of the function. (See Figure 3-6.)

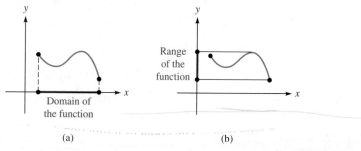

Figure 3-6

From Figure 3-4, we can see that the domain of the function $f(x) = -2|x| + 3$ is the set of all real numbers, and that the range is the set of real numbers that are less than or equal to 3.

From Figure 3-5, we can see that the domain of the function $f(x) = \sqrt{x - 2}$ is the set of all real numbers greater than or equal to 2, and that the range is the set of all real numbers greater than or equal to 0.

Accent on Technology **Calculator Graphs**

The calculator graphs of $f(x) = \sqrt{x - 2}$ and $f(x) = |x + 3|$ are shown in Figure 3-7. From Figure 3-7(a), we can see that the domain of $f(x) = \sqrt{x - 2}$ is $[2, \infty)$ and the range is $[0, \infty)$. From Figure 3-7(b), we can see that the domain of $f(x) = |x + 3|$ is $(-\infty, \infty)$ and the range is $[0, \infty)$.

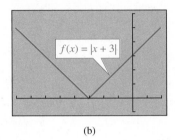

(a)

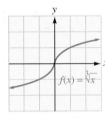

(b)

Figure 3-7

In Figure 3-8, we see the graphs of several basic functions.

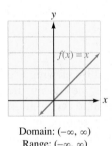

Domain: $(-\infty, \infty)$
Range: $(-\infty, \infty)$
Identity function
(a)

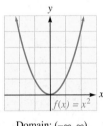

Domain: $(-\infty, \infty)$
Range: $[0, \infty)$
Squaring function
(b)

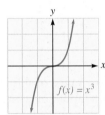

Domain: $(-\infty, \infty)$
Range: $(-\infty, \infty)$
Cubing function
(c)

Domain: $(-\infty, \infty)$
Range: $[0, \infty)$
Absolute value function
(d)

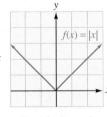

Domain: $[0, \infty)$
Range: $[0, \infty)$
Square root function
(e)

Domain: $(-\infty, \infty)$
Range: $(-\infty, \infty)$
Cube root function
(f)

Figure 3-8 Graphs of Several Basic Functions

6. Use the Vertical Line Test to Identify Functions

We can use a **vertical line test** to determine whether a graph represents a function.

Vertical Line Test   If every vertical line that intersects a graph does so exactly once, every number x determines exactly one value of y, and the graph represents a function.

If any vertical line intersects a graph more than once, more than one value of y corresponds to some numbers x, and the graph does not represent a function.

A graph that is a function is shown in Figure 3-9(a). A graph that is not a function is shown in Figure 3-9(b).

Comment
Some graphs represent functions
and some do not. Graphs that pass
the vertical line test are functions.

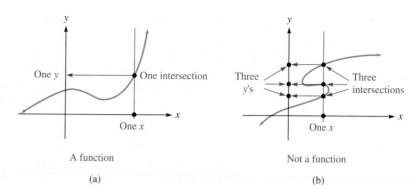

Figure 3-9

EXAMPLE 7   Determine which of the following graphs represent functions.

a.   **b.** 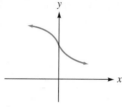   **c.**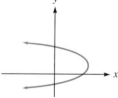

Solution   We will use the vertical line test by drawing several vertical lines through each
graph. If every vertical line that intersects the graph does so exactly once, the
graph represents a function. Otherwise, the graph does not represent a function.

a. This graph fails the vertical line
test, so it does not represent a
function.

b. This graph passes the vertical
line test, so it does represent a
function.

c. This graph fails the vertical line
test, so it does not represent a
function.

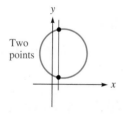   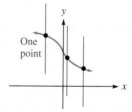   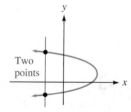

Self Check 7   Does the graph shown in Figure 3-2 represent a function?

Not all equations define functions. For example, the equation $x = |y|$ does
not define a function, because two values of y can correspond to one number x.
For example, if $x = 2$, then y can be either 2 or -2. The graph of the equation is
shown in Figure 3-10. Since the graph does not pass the vertical line test, it does
not represent a function.

$x = |y|$

x	y	(x, y)
2	-2	$(2, -2)$
1	-1	$(1, -1)$
0	0	$(0, 0)$
1	1	$(1, 1)$
2	2	$(2, 2)$

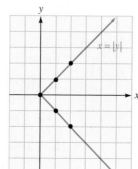

Figure 3-10

Correspondences between a set of input values x (called the *domain*) and a set of output values y (called the *range*), where to each x-value in the domain there corresponds one or more y-values in the range, are called **relations.** Although the graph in Figure 3-10 does not represent a function, it does represent a relation. We note that all functions are relations, but not all relations are functions.

Another way to visualize the definition of function is to consider the diagram shown in Figure 3-11(a). The function f that assigns the element y to the element x is represented by an arrow leaving x and pointing to y. The set of elements in **X** from which arrows originate is the domain of the function. The set of elements in **Y** to which arrows point is the range.

To constitute a function, each element of the domain must determine exactly one y-value in the range. However, the same value of y could correspond to several numbers x. In the function shown in Figure 3-11(b), the single value y corresponds to the three numbers x_1, x_2, and x_3 in the domain.

The correspondence shown in Figure 3-11(c) is not a function, because two values of y correspond to the same number x. However, it is still a relation.

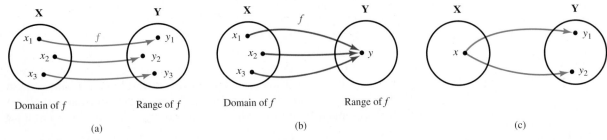

Figure 3-11

7.   Use Linear Functions to Model Applications

We have seen that the equation of a nonvertical line defines a linear function—an important function in mathematics and its applications.

Linear Functions   A **linear function** is a function determined by an equation of the form

$$f(x) = mx + b \qquad \text{or} \qquad y = mx + b$$

EXAMPLE 8   **Cost of a fraternity dance**   The cost associated with a fraternity dance is $300 for country club rental and $24 for each couple that attends.

a. Write the cost C of the dance in terms of the number of couples x attending.

b. Find the cost if 45 couples attend the dance.

Solution   Use the information stated in the problem to write a linear function that models the application.

a. The cost C for the dance is $24 per couple plus the $300 rental fee. If x couples attend, the cost is 24x$ plus 300. Therefore, the linear cost function is

$$C(x) = 24x + 300$$

b. To find the cost when 45 couples attend, we find $C(45)$.

$$C(x) = 24x + 300$$
$$C(\mathbf{45}) = 24(\mathbf{45}) + 300 \qquad \text{Substitute 45 for } x.$$

$$= 1{,}080 + 300$$
$$= 1{,}380$$

If 45 couples attend, the cost of the dance is $1,380.

Self Check 8   Find the cost when 60 couples attend.

EXAMPLE 9   **Heart rates**   The target heart rate at which a person should train to get an effective workout is a linear function of his age. For a 20-year-old starting an exercise program, the target heart rate should be 120 beats per minute. For a 40-year-old, it should be 108 beats per minute. Express the target heart rate R as a function of age A.

Solution   We will use the information stated in the problem to find the slope of the linear function and write a linear function that models the application.

Since the target heart rate (R) is given to be a linear function of age (A), there are constants m and b such that

$$R = mA + b$$

Since $R = 120$ when $A = 20$, the point $(A_1, R_1) = (20, 120)$ lies on the straight-line graph of this function. Since $R = 108$ when $A = 40$, the point $(A_2, R_2) = (40, 108)$ also lies on that line. The slope of the line is

$$
\begin{aligned}
m &= \frac{R_2 - R_1}{A_2 - A_1} \\
&= \frac{108 - 120}{40 - 20} \qquad \text{Substitute 108 for } R_2, \text{ 120 for } R_1, \text{ 40 for } A_2, \text{ and 20 for } A_1. \\
&= -\frac{12}{20} \\
&= -\frac{3}{5} \\
&= -0.6
\end{aligned}
$$

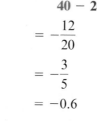

Srinivasa Ramanujan
(1887–1920)
Ramanujan was one of India's most prominent mathematicians. When in high school, he read *Synopsis of Elementary Results in Pure Mathematics* and from this book taught himself mathematics. He entered college but failed several times because he would only study mathematics. He went on to teach at Cambridge University in England.

Thus, $m = -0.6$. To determine b, we can substitute -0.6 for m and the coordinates of one point, say $P(20, 120)$, in the equation $R = mA + b$ and solve for b.

$$
\begin{aligned}
R &= mA + b \\
120 &= -0.6(20) + b \qquad \text{Substitute 120 for } R, \text{ 20 for } A, \text{ and } -0.6 \text{ for } m. \\
120 &= -12 + b \\
132 &= b \qquad\qquad\qquad \text{Add 12 to both sides.}
\end{aligned}
$$

If we substitute -0.6 for m and 132 for b in $R = mA + b$, we obtain $R = -0.6A + 132$.

Self Check 9   Find the target heart rate for a 30-year-old person.

Self Check Answers   **1. a.** a function   **b.** a function   **c.** not a function   **2. a.** $(-\infty, \infty)$
b. $(-\infty, \infty)$ **3.** $(-\infty, -4) \cup (-4, 4) \cup (4, \infty)$ **4. a.** -4 **b.** 10
c. $3k^2 - 5k - 2$ **5.** $2x + h$ **6.**   **7.** yes   **8.** $1,740
9. 114

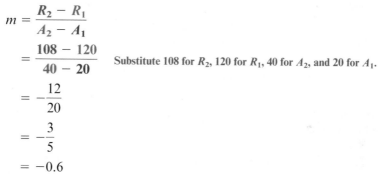

3.1   Exercises

Vocabulary and Concepts   *Fill in the blanks.*

1. A correspondence that assigns exactly one value of y to any number x is called a _____.

2. A correspondence that assigns one or more values of y to any number x is called a _____.

3. The set of input numbers x in a function is called the _____ of the function.

4. The set of all output values y in a function is called the _____ of the function.

5. The statement "y is a function of x" can be written as the equation _____.

6. The graph of a function $y = f(x)$ in the xy-plane is the set of all points _____ that satisfy the equation, where x is in the _____ of f and y is in the _____ of f.

7. In the function of Exercise 5, __ is called the independent variable.

8. In the function of Exercise 5, y is called the _____ variable.

9. If every _____ line that intersects a graph does so ___, the graph represents a function.

10. A function that can be written in the form $y = mx + b$ is called a _____ function.

Practice   *Assume that all variables represent real numbers. Determine whether each equation determines y to be a function of x.*

11. $y = x$

12. $y - 2x = 0$

13. $y^2 = x$

14. $|y| = x$

15. $y = x^2$

16. $y - 7 = 7$

17. $y^2 - 4x = 1$

18. $|x - 2| = y$

19. $|x| = |y|$

20. $x = 7$

21. $y = 7$

22. $|x + y| = 7$

Let the function f be defined by the equation $y = f(x)$, where x and $f(x)$ are real numbers. Find the domain of each function.

23. $f(x) = 3x + 5$

24. $f(x) = -5x + 2$

25. $f(x) = x^2$

26. $f(x) = x^3$

27. $f(x) = \dfrac{3}{x + 1}$

28. $f(x) = \dfrac{-7}{x + 3}$

29. $f(x) = \sqrt{x}$

30. $f(x) = \sqrt{x^2 - 1}$

31. $f(x) = \dfrac{x}{x + 3}$

32. $f(x) = \dfrac{x}{x - 3}$

33. $f(x) = \dfrac{x - 2}{x + 3}$

34. $f(x) = \dfrac{x + 2}{x - 1}$

35. $f(x) = \dfrac{x}{x^2 - 4}$

36. $f(x) = \dfrac{2x}{x^2 - 9}$

37. $f(x) = \dfrac{1}{x^2 - 4x - 5}$

38. $f(x) = \dfrac{x}{x^2 - 8x + 15}$

Let the function f be defined by $y = f(x)$, where x and $f(x)$ are real numbers. Find $f(2)$, $f(-3)$, $f(k)$, and $f(k^2 - 1)$.

39. $f(x) = 3x - 2$

40. $f(x) = 5x + 7$

41. $f(x) = \dfrac{1}{2}x + 3$

42. $f(x) = \dfrac{2}{3}x + 5$

43. $f(x) = x^2$

44. $f(x) = 3 - x^2$

45. $f(x) = |x^2 + 1|$

46. $f(x) = |x^2 + x + 4|$

47. $f(x) = \dfrac{2}{x + 4}$

48. $f(x) = \dfrac{3}{x - 5}$

49. $f(x) = \dfrac{1}{x^2 - 1}$

50. $f(x) = \dfrac{3}{x^2 + 3}$

51. $f(x) = \sqrt{x^2 + 1}$

52. $f(x) = \sqrt{x^2 - 1}$

Evaluate the difference quotient for each function f(x).

53. $f(x) = 3x + 1$

54. $f(x) = 5x - 1$

55. $f(x) = x^2 + 1$

56. $f(x) = x^2 - 3$

57. $f(x) = 4x^2 - 6$

58. $f(x) = 5x^2 + 3$

59. $f(x) = x^2 + 3x - 7$

60. $f(x) = x^2 - 5x + 1$

61. $f(x) = 2x^2 - 4x + 2$

62. $f(x) = 3x^2 + 2x - 3$

63. $f(x) = x^3$

64. $f(x) = \dfrac{1}{x}$

Graph each function. Use the graph to identify the domain and range of each function.

65. $f(x) = 2x + 3$

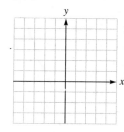

66. $f(x) = 3x + 2$

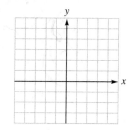

67. $f(x) = \dfrac{1}{2}x - 3$

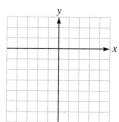

68. $f(x) = -\dfrac{3}{4}x + 4$

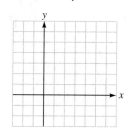

69. $2x = 3y - 3$

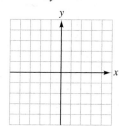

70. $3x = 2(y + 1)$

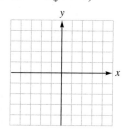

71. $f(x) = x^2 - 4$

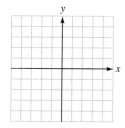

72. $f(x) = -x^2 + 3$

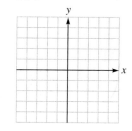

73. $f(x) = -x^3 + 2$

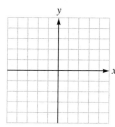

74. $f(x) = -x^3 + 1$

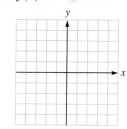

75. $f(x) = -\sqrt{x}$

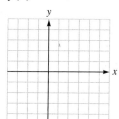

76. $f(x) = \sqrt{x + 1}$

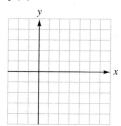

77. $f(x) = -\sqrt{x + 1}$

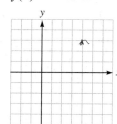

78. $f(x) = \sqrt{x} + 2$

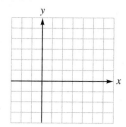

79. $f(x) = -|x|$

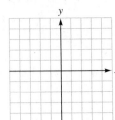

80. $f(x) = -|x| - 3$

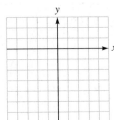

Use the vertical line test to determine whether each graph represents a function.

89.

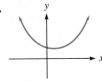

90.

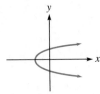

81. $f(x) = |x - 2|$

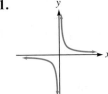

82. $f(x) = -|x - 2|$

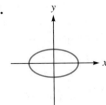

91.

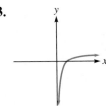

92.

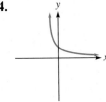

93.

94.

83. $f(x) = \sqrt{2x - 4}$

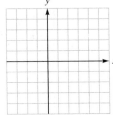

84. $f(x) = -\sqrt{2x - 4}$

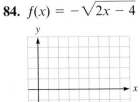

Use a graphing calculator to graph each function. Then determine the domain and range of the function.

95. $f(x) = \sqrt{2x - 5}$ **96.** $f(x) = |3x + 2|$

85. $f(x) = \left| \dfrac{1}{2}x + 3 \right|$ **86.** $f(x) = -\left| \dfrac{1}{2}x + 3 \right|$

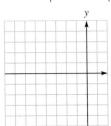
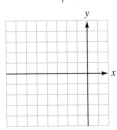

97. $f(x) = \sqrt[3]{5x - 1}$ **98.** $f(x) = -\sqrt[3]{3x + 2}$

Applications

99. Cost of t-shirts   A chapter of Phi Theta Kappa, an honors society for two-year college students, is purchasing t-shirts for each of its members. A local company has agreed to make the shirts for $8 each plus a graphic arts fee of $75.

  a. Write a linear function that describes the cost *C* for the shirts in terms of *x*, the number of t-shirts ordered.

  b. Find the total cost of 85 t-shirts.

Draw lines to indiate the domain and range of each function as intervals on the x- and y-axes.

87.

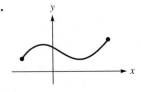

88.

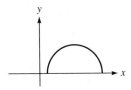

100. Service projects The Circle K Club is planning a service project for children at a local children's home. They plan to rent a *Dora the Explorer Moonwalk* for the event. The cost of the moonwalk will include a $60 delivery fee and $45 for each hour it is used. Express the total bill b in terms of the hours used h.

101. Cell phone plans A grandmother agrees to purchase a cell phone for emergency use only. AT&T now offers such a plan for $9.99 per month and $0.07 for each minute t the phone is used.

   a. Write a linear function that describes the monthly cost C in terms of the time in minutes t the phone is used.

   b. If the grandmother used her phone for 20 minutes during the first month, what was her bill?

102. Concessions A concessionaire at a football game pays a vendor $40 per game for selling hot dogs at $2.50 each.

   a. Write a linear function that describes the income I the vendor earns for the concessionaire during the game if the vendor sells h hot dogs.

   b. Find the income if the vendor sells 175 hot dogs.

103. Home construction In a proposal to prospective clients, a contractor listed the following costs:

| 1. Fees, permits, site preparation | $14,000 |
| 2. Construction, per square foot | $95 |

   a. Write a linear function the clients can use to determine the cost c of building a house having f square feet.

   b. Find the cost to build a 2,600-square-foot house.

104. Temperature conversion The Fahrenheit temperature reading F is a linear function of the Celsius reading C. If $C = 0$ when $F = 32$ and the readings are the same at $-40°$, express F as a function of C.

105. Cost of electricity The cost c of electricity in Eagle River is a linear function of x, the number of kilowatt-hours (kwh) used. If the cost of 100 kwh is $17 and the cost of 500 kwh is $57, find an equation that expresses c in terms of x.

106. Water billing The cost c of water is a linear function of n, the number of gallons used. If 1,000 gallons cost $4.70 and 9,000 gallons cost $14.30, express c as a function of n.

107. Coffee locations Suppose that in 2003 there were approximately 6,400 Starbucks locations. Suppose that in 2007 this number had grown to approximately 13,168. Write a linear function that represents the number of Starbucks n as a function of time t. Let $t = 0$ represent 2003.

108. Cliff divers The cliff divers of Acapulco amaze tourists with their diving skills. The velocity v of a diver is a function of the time t the diver has fallen. If $v = 2$ feet per second when $t = 0$ seconds and $v = 66$ feet per second when $t = 2$ seconds, express v as a function of time t.

Discovery and Writing *Find all values of x that will make f(x) = 0.*

109. $f(x) = 3x + 2$ **110.** $f(x) = -2x - 5$

111. Write a paragraph explaining how to find the domain of a function.

112. Write a paragraph explaining how to find the range of a function.

113. Explain why all functions are relations, but not all relations are functions.

114. Use a graphing calculator to graph the function $f(x) = \sqrt{x}$, and use TRACE and ZOOM to find $\sqrt{5}$ to 3 decimal places.

Review *Consider this set:*
$$\left\{-3, -1, 0, 0.5, \tfrac{3}{4}, 1, \pi, 7, 8\right\}.$$

115. Which numbers are natural numbers?

116. Which numbers are rational numbers?

117. Which numbers are prime numbers?

118. Which numbers are even numbers?

Write each set of numbers in interval notation.

119.

120.

Graph each union of two intervals.

121. $(-3,5) \cup [6,\infty)$ **122.** $(-\infty,0) \cup (0,\infty)$

3.2   Quadratic Functions

Objectives

1. Recognize the Characteristics of a Quadratic Function
2. Find the Vertex of a Parabola Whose Equation Is in Standard Form
3. Graph a Quadratic Function
4. Find the Vertex of a Parabola Whose Equation Is in General Form
5. Use a Quadratic Function to Solve Maximum and Minimum Problems

Ronald Martinez/Getty Images

Quadratic functions are important because we can use them to model many real-life problems. For example, the path of a basketball jump shot and the path of a guided missile can be modeled with quadratic functions. Businesses such as Coca Cola and Best Buy can use quadratic functions to help maximize the profit and revenue for the products they produce and sell.

1.   Recognize the Characteristics of a Quadratic Function

The linear function $f(x) = mx + b$ $(m \neq 0)$ is a first-degree polynomial function, because its right side is a first-degree polynomial in the variable x. A function defined by a polynomial of second-degree is called a **quadratic function.**

Quadratic Function   A **quadratic function** is a second-degree polynomial function in one variable of the form

$$f(x) = ax^2 + bx + c \quad \text{or} \quad y = ax^2 + bx + c$$

where a, b, and c are real numbers and $a \neq 0$.

Some examples of quadratic functions are:

$$f(x) = x^2 - 2x - 3 \text{ and } y = -2x^2 - 8x - 3$$

Quadratic functions can be graphed by plotting points. For example, to graph the function $f(x) = x^2 - 2x - 3$, we plot several points with coordinates that satisfy the equation. We then join them with a smooth curve to obtain the graph shown in Figure 3-12(a). A table of values and the graph of $f(x) = -2x^2 - 8x - 3$ are shown in Figure 3-12(b).

The graph of a quadratic function is called a **parabola,** a cup-shaped curve that opens either upward $\cup$ or downward $\cap$.

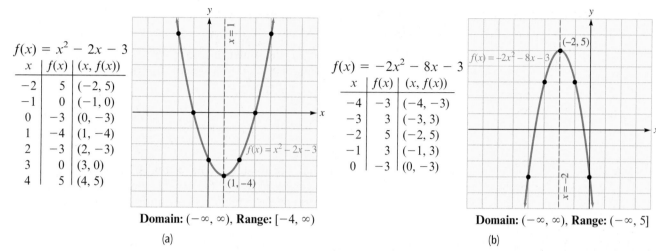

$f(x) = x^2 - 2x - 3$

x	$f(x)$	$(x, f(x))$
-2	5	$(-2, 5)$
-1	0	$(-1, 0)$
0	-3	$(0, -3)$
1	-4	$(1, -4)$
2	-3	$(2, -3)$
3	0	$(3, 0)$
4	5	$(4, 5)$

Domain: $(-\infty, \infty)$, **Range:** $[-4, \infty)$

(a)

$f(x) = -2x^2 - 8x - 3$

x	$f(x)$	$(x, f(x))$
-4	-3	$(-4, -3)$
-3	3	$(-3, 3)$
-2	5	$(-2, 5)$
-1	3	$(-1, 3)$
0	-3	$(0, -3)$

Domain: $(-\infty, \infty)$, **Range:** $(-\infty, 5]$

(b)

Figure 3-12

The graphs in Figure 3-12 suggest that the graph of a quadratic function has the following characteristics.

Characteristics of Quadratic Functions		
Characteristics	**Examples**	
Equation of a quadratic function $f(x) = ax^2 + bx + c$	$f(x) = x^2 - 2x - 3$	$f(x) = -2x^2 - 8x - 3$
If $a > 0$, the parabola **opens up.** If $a < 0$, the parabola **opens down.**	$a = 1$, opens up	$a = -2$, opens down
The **vertex** is the turning point of the parabola.	Vertex is $(1, -4)$.	Vertex is $(-2, 5)$.
The **minimum** or **maximum point** occurs at the vertex.	$(1, -4)$ is the minimum or lowest point on the graph.	$(-2, 5)$ is the maximum or highest point on the graph.
The **axis of symmetry** is the vertical line that intersects the parabola at the vertex. The parabola is symmetric about this vertical line.	The graph of $x = 1$ is the axis of symmetry.	The graph of $x = -2$ is the axis of symmetry.

Accent on Technology

Graphing Quadratic Functions

We can use a graphing calculator to graph quadratic functions. If we use window settings of $[-10, 10]$ for x and $[-10, 10]$ for y, the graph of $f(x) = x^2 - 2x - 3$ will look like Figure 3-13(a). The graph of $f(x) = -2x^2 - 8x - 3$ will look like Figure 3-13(b).

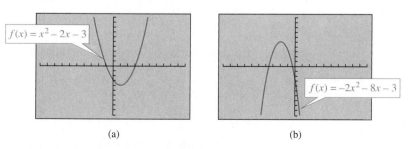

(a)   (b)

Figure 3-13

2. Find the Vertex of a Parabola Whose Equation Is in Standard Form

In Figure 3-12, we considered the function $f(x) = x^2 - 2x - 3$ with vertex $(1, -4)$. If we complete the square on the right side of the equation we will obtain

$$f(x) = (x^2 - 2x + 1) - 3 - 1 \qquad \text{One-half of } -2 \text{ is } -1, \text{ and } (-1)^2 \text{ is } 1. \text{ Add } 1 \text{ and subtract } 1 \text{ on the right side of the equation.}$$

$$f(x) = (x - 1)(x - 1) - 4 \qquad \text{Factor } x^2 - 2x + 1.$$
$$f(x) = (x - 1)^2 - 4$$

In this factored form, the coordinates of the vertex of the parabola can be read from the equation. The vertex of $f(x) = (x - 1)^2 - 4$ is $(1, -4)$. We call this factored form the **standard form** of the equation of a quadratic function.

Standard Form of the Equation of a Quadratic Function	The graph of the quadratic function $$y = f(x) = a(x - h)^2 + k \quad (a \neq 0)$$ is a parabola with vertex at (h, k). The parabola opens upward when $a > 0$ and downward when $a < 0$. The axis of symmetry of the parabola is the vertical line graph of the equation $x = h$.

EXAMPLE 1 Find the vertex of the graph of each quadratic function:
a. $f(x) = 2(x - 3)^2 + 5$ **b.** $f(x) = -3(x + 2)^2 - 4$

Solution In each case, the equation of the quadratic function is given in standard form. From the equations, we can identify h and k, because the vertex is the point with coordinates (h, k).

a. We identify the values of h and k.

Standard form: $f(x) = a(x - h)^2 + k$

Given function: $f(x) = 2(x - 3)^2 + 5$ $h = 3$ and $k = 5$.

Since $h = 3$ and $k = 5$, the vertex is the point with coordinates of $(3, 5)$.

b. We identify the values of h and k.

Standard form: $f(x) = a(x - h)^2 + k$
Given function: $f(x) = -3(x + 2)^2 - 4$
$$f(x) = -3[x - (-2)] + (-4) \qquad h = -2 \text{ and } k = -4.$$

Since $h = -2$ and $k = -4$, the vertex is the point with coordinates of $(-2, -4)$.

Self Check 1 Find the vertex of the graph of the quadratic function $f(x) = 2(x + 5)^2 - 4$.

3. Graph a Quadratic Function

The easiest way to graph a quadratic function is to follow these steps.

Graphing a Quadratic Function

To graph a quadratic function:

1. Determine whether the parabola opens upward or downward.
2. Find the vertex of the parabola.
3. Find the x-intercepts.
4. Find the y-intercept.
5. Identify one additional point on the graph.
6. Draw a smooth curve through the points found in Steps 2–5.

EXAMPLE 2

Graph the quadratic function $f(x) = 2(x + 1)^2 - 8$.

Solution

We first determine whether the parabola opens upward or downward. Then we will find the vertex and the x- and y-intercepts. Finally, we will find one additional point and draw a smooth curve through the plotted points.

Step 1: Determine whether the parabola opens upward or downward.

Standard form: $f(x) = a(x - h)^2 + k$
$$\downarrow$$
Given form: $f(x) = \mathbf{2}(x + 1)^2 - 8$

Since $a = 2$ and 2 is positive, the parabola opens upward.

Step 2: Find the vertex of the parabola.

Standard form: $f(x) = a(x - h)^2 + k$

Given form: $f(x) = 2(x + 1)^2 - 8$

$$f(x) = 2[x - (-1)]^2 + (-8)$$

Since $h = -1$ and $k = -8$ the vertex is the point with coordinates of $(-1, -8)$.

Step 3: Find the x-intercepts.

To find the x-intercepts, we substitute 0 for $f(x)$ and solve for x.

$$f(x) = 2(x + 1)^2 - 8$$

$0 = 2(x + 1)^2 - 8$	Substitute 0 for $f(x)$.
$8 = 2(x + 1)^2$	Add 8 to both sides of the equation.
$4 = (x + 1)^2$	Divide both sides by 2.
$x + 1 = \pm 2$	Write $(x + 1)^2$ on the left side and use the square root property.
$x = -1 \pm 2$	Subtract 1 from both sides.
$x = 1 \quad \text{or} \quad x = -3$	

The x-intercepts are the points with coordinates of $(1, 0)$ and $(-3, 0)$.

Step 4: Find the y-intercept.

To find the y-intercept, we substitute 0 in for x and solve for y.

$f(x) = 2(x + 1)^2 - 8$	
$y = 2(x + 1)^2 - 8$	Substitute y for $f(x)$.
$y = 2(0 + 1)^2 - 8$	Substitute 0 for x.
$y = 2(1)^2 - 8$	
$y = 2 - 8$	
$y = -6$	

The y-intercept is the point with coordinates of $(0, -6)$.

Step 5: Identify one additional point on the graph.
Because of symmetry, the point $(-2, -6)$ is on the graph.

Step 6: Draw a smooth curve through the points found in Steps 2–5.
We now can draw the graph of the function as shown in Figure 3-14.

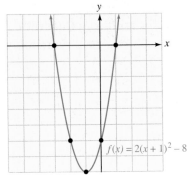

$f(x) = 2(x + 1)^2 - 8$

Figure 3-14

Self Check 2   Graph the function $f(x) = -(x - 2)^2 + 4$.

Comment
Sometimes the vertex of a quadratic function occurs at the origin, $(0, 0)$. In this case, both $h = 0$ and $k = 0$, and the equation of the parabola is of the form $f(x) = a(x - 0)^2 + 0$ or $f(x) = ax^2$ and $a \neq 0$.

4.   Find the Vertex of a Parabola Whose Equation Is in General Form

Comment
A quadratic function can have one, two, or no x-intercepts. However, it will always have one y-intercept.

To graph a quadratic function given in the form $f(x) = ax^2 + bx + c$ $(a \neq 0)$, called **general form,** we must find the coordinates of the vertex of the parabola. To do so, we can complete the square on $ax^2 + bx$ to change the equation into standard form $y = a(x - h)^2 + k$. As we have seen, we can read the coordinates (h, k) of the vertex from this form. Once the quadratic function is in standard form, we can graph the function following the steps given in Example 2.

EXAMPLE 3   Find the vertex of the parabola whose equation is $f(x) = -2x^2 + 12x - 16$.

Solution   We will complete the square on x, write the equation in standard form, and identify h and k, the coordinates of the vertex.
We begin by completing the square on $-2x^2 + 12x$.

$f(x) = -2x^2 + 12x - 16$	Identify a: $a = -2$.
$f(x) = -2(x^2 - 6x) - 16$	Factor $a = -2$ from $-2x^2 + 12x$.
$f(x) = -2(x^2 - 6x + 9 - 9) - 16$	One-half of -6 is -3 and $(-3)^2 = 9$. Add and subtract 9 within the parentheses.
$f(x) = -2(x^2 - 6x + 9) - 2(-9) - 16$	Distribute the multiplication by -2.
$f(x) = -2(x - 3)^2 + 18 - 16$	Factor $x^2 - 6x + 9$ and multiply.
$f(x) = -2(x - 3)^2 + 2$	Simplify.

The equation is now in standard form with $h = 3$ and $k = 2$. Therefore, the vertex is the point with coordinates $(h, k) = (3, 2)$.

Self Check 3   Find the vertex of the graph of $y = 4x^2 - 16x + 19$.

To find formulas for the coordinates of the vertex of a parabola defined by $y = ax^2 + bx + c$ $(a \neq 0)$, we can complete the square on x to write the equation in standard form $(y = a(x - h)^2 + k)$:

$$y = ax^2 + bx + c$$

$$y = a\left(x^2 + \frac{b}{a}x\right) + c \qquad \text{Factor } a \text{ from } ax^2 + bx.$$

$$y = a\left(x^2 + \frac{b}{a}x + \frac{b^2}{4a^2} - \frac{b^2}{4a^2}\right) + c \qquad \text{Add and subtract } \frac{b^2}{4a^2} \text{ within the parentheses.}$$

$$y = a\left(x^2 + \frac{b}{a}x + \frac{b^2}{4a^2}\right) - a\left(\frac{b^2}{4a^2}\right) + c \qquad \text{Distribute the multiplication of } a.$$

$$y = a\left(x + \frac{b}{2a}\right)^2 + c - \frac{b^2}{4a} \qquad \text{Factor } x^2 + \frac{b}{a}x + \frac{b^2}{4a^2} \text{ and simplify } a\left(\frac{b^2}{4a^2}\right).$$

$$y = a\left[x - \left(-\frac{b}{2a}\right)\right]^2 + c - \frac{b^2}{4a} \qquad -\left(-\frac{b}{2a}\right) = \frac{b}{2a}$$

If we compare the last equation to the form $y = a(x - h)^2 + k$, we see that $h = -\frac{b}{2a}$ and $k = c - \frac{b^2}{4a}$. This result gives the following fact.

Vertex of a Parabola   The graph of the function

$$y = f(x) = ax^2 + bx + c \quad (a \neq 0)$$

is a parabola with vertex at $\left(-\frac{b}{2a}, c - \frac{b^2}{4a}\right)$.

Comment

You don't need to memorize the formula for the y-coordinate of the vertex of a parabola. It is usually convenient to find the y-coordinate by substituting $-\frac{b}{2a}$ for x in the function and solving for y.

EXAMPLE 4   Graph the function: $y = f(x) = -2x^2 - 5x + 3$.

Solution   We begin by determining whether the parabola opens upward or downward. Then we find the vertex by using the formula $h = -\frac{b}{2a}$. Next we find the x- and y-intercepts and one additional point and then draw a smooth curve through the plotted points.

Step 1: Determine whether the parabola opens up or downward.
The equation has the form $y = ax^2 + bx + c$, where $a = -2, b = -5$, and $c = 3$. Since $a < 0$, the parabola opens downward.

Step 2: Find the vertex.
To find the x-coordinate of the vertex, we substitute the values of a and b into the formula $x = -\frac{b}{2a}$.

$$x = -\frac{b}{2a} = -\frac{-5}{2(-2)} = -\frac{5}{4}$$

The x-coordinate of the vertex is $-\frac{5}{4}$. To find the y-coordinate, we substitute $-\frac{5}{4}$ for x in the equation and solve for y.

$$y = -2x^2 - 5x + 3$$

$$y = -2\left(-\frac{5}{4}\right)^2 - 5\left(-\frac{5}{4}\right) + 3 \qquad \text{Substitute } -\frac{5}{4} \text{ for } x.$$

$$= -2\left(\frac{25}{16}\right) + \frac{25}{4} + 3$$

$$= -\frac{25}{8} + \frac{50}{8} + \frac{24}{8}$$

$$= \frac{49}{8}$$

Since the vertex is the point $\left(-\frac{5}{4}, \frac{49}{8}\right)$, we can plot it on the coordinate system in Figure 3-15(a) and draw the axis of symmetry.

Step 3: Find the *x*-intercepts.
To find the *x*-intercepts, we substitute 0 for *y* and solve for *x*.

$$y = -2x^2 - 5x + 3$$
$$0 = -2x^2 - 5x + 3 \qquad \text{Substitute 0 for } y.$$
$$0 = 2x^2 + 5x - 3 \qquad \text{Divide both sides by } -1 \text{ to make the leading}$$
$$\text{coefficient positive.}$$
$$0 = (2x - 1)(x + 3) \qquad \text{Factor the trinomial.}$$
$$2x - 1 = 0 \quad \text{or} \quad x + 3 = 0 \qquad \text{Set each factor equal to 0.}$$
$$x = \frac{1}{2} \qquad\qquad x = -3 \qquad \text{Solve each linear equation.}$$

The *x*-intercepts are $\left(\frac{1}{2}, 0\right)$ and $(-3, 0)$. We plot these intercepts as shown in Figure 3-15(a).

Step 4: Find the *y*-intercept.
To find the *y*-intercept, we let $x = 0$ and solve for *y*.

$$y = -2x^2 - 5x + 3$$
$$= -2(0)^2 - 5(0) + 3 \qquad \text{Substitute 0 for } x.$$
$$= 0 - 0 + 3$$
$$= 3$$

The *y*-intercept is $(0, 3)$. We plot the intercept as shown in Figure 3-15(a).

Step 5: Plot one additional point.
Because of symmetry, we know that the point $\left(-2\frac{1}{2}, 3\right)$ is on the graph. We plot this point on the coordinate system in Figure 3-15(a).

Step 6:   We can now draw the graph of the function, as shown in Figure 3-15(b).

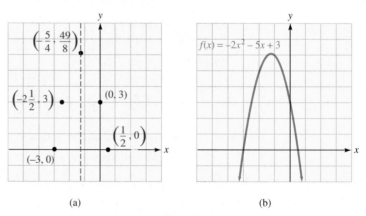

(a)   (b)

Figure 3-15

Comment
The y-intercept of a parabola written in the general form $f(x) = ax^2 + bx + c$ $(a \neq 0)$ is the point $(0, c)$. This is because y is always c when we substitute 0 for x.

Self Check 4   Graph the function: $f(x) = -3x^2 + 7x - 2$.

5.   Use a Quadratic Function to Solve Maximum and Minimum Problems

EXAMPLE 5   **Maximum area**   The Montana Dude Rancher's Association has 400 feet of fencing to enclose a rectangular corral. To save money and fencing, the association intends to use the bank of a river as one boundary of the corral, as in Figure 3-16. Find the dimensions that will enclose the largest area.

Solution   We will represent the fenced area with a quadratic function. Since its parabolic graph opens downward, the largest or maximum area will occur at the vertex. We can use the vertex formula to find the vertex.

Step 1: Represent area with a quadratic function.
Let x represent the width of the fenced area. Then $400 - 2x$ represents the length. Because the area A of a rectangle is the product of the length and the width, we have

$$A = (400 - 2x)x \qquad \text{or} \qquad A(x) = -2x^2 + 400x$$

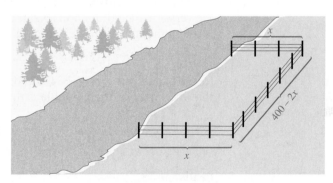

Figure 3-16

The graph of this area function is a parabola. Since the coefficient of x^2 is negative, the parabola opens downward and its vertex is its highest point. The A-coordinate of the vertex (x, A) represents the maximum area, and the x-coordinate represents the width of the corral that will give the maximum area.

Step 2: Find the vertex of the parabola.
We compare the equations

$$A(x) = -2x^2 + 400x \qquad \text{and} \qquad y = ax^2 + bx + c$$

to see that $a = -2$, $b = 400$, and $c = 0$. Using the vertex formula, the vertex of the parabola is the point with coordinates

$$\left(-\frac{b}{2a}, c - \frac{b^2}{4a}\right) = \left(-\frac{400}{2(-2)}, 0 - \frac{400^2}{4(-2)}\right) = (100, 20{,}000)$$

Comment

Note that we could have determined the y coordinate of 20,000 by finding $A(100)$.

$$A(x) = -2x^2 + 400x$$
$$A(100) = -2(100)^2 + 400(100) \quad \text{Substitute 100 for } x.$$
$$= -2(10,000) + 40,000$$
$$= -20,000 + 40,000$$
$$= 20,000$$

If the fence runs 100 feet out from the river, 200 feet parallel to the river, and 100 feet back to the river, it will enclose the largest possible area, which is 20,000 square feet.

Self Check 5   Find the largest area possible if the association has 1,200 feet of fencing available.

EXAMPLE 6   **Minimum cost**   A company that makes and sells webcams has found that the total weekly cost C of producing x webcams is given by the function $C(x) = 0.5x^2 - 210x + 26,250$. Find the production level that minimizes the weekly cost and find that weekly minimum cost.

Solution   The weekly cost function $C(x)$ is a quadratic function whose graph is a parabola that opens upward. The minimum value of $C(x)$ occurs at the vertex of the parabola. We will use the vertex formula to find the vertex of the parabola.

Since the coefficient of x^2 is 0.5 (a positive real number), the x-coordinate of the vertex is the production level that will minimize the cost, and the y-coordinate is that minimum cost. We compare the equations

$$C(x) = 0.5x^2 - 210x + 26,250 \qquad \text{and} \qquad y = ax^2 + bx + c$$

to see that $a = 0.5$, $b = -210$, and $c = 26,250$. Using the vertex formula, we see that the vertex of the parabola is the point with coordinates

$$\left(-\frac{b}{2a}, c - \frac{b^2}{4a}\right) = \left(-\frac{-210}{2(0.5)}, 26,250 - \frac{(-210)^2}{4(0.5)}\right) = (210, 4,200)$$

Comment

We can also determine the y-coordinate of 4,200 by finding $C(210)$.

$$C(x) = 0.5x^2 - 210x + 26,250$$
$$C(210) = 0.5(210)^2 - 210(210) + 26,250 \quad \text{Substitute 210 for } x.$$
$$= 0.5(44,100) - 44,100 + 26,250$$
$$= 22,050 - 17,850$$
$$= 4,200$$

If the company makes 210 webcams each week, it will minimize its production cost. The minimum weekly cost will be $4,200.

Self Check 6   A company that makes and sells baseball caps has found that the total monthly cost C of producing x caps is given by the function $C(x) = 0.2x^2 - 80x + 9,000$. Find the production level that will minimize the monthly cost and find the minimum cost.

Self Check Answers   **1.** $(-5, -4)$ **2.**

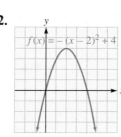

$f(x) = -(x - 2)^2 + 4$

3. $(2, 3)$ **4.**

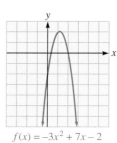

$f(x) = -3x^2 + 7x - 2$

5. 180,000 square feet   **6.** 200 caps at a minimum cost of $1,000

3.2 Exercises

Vocabulary and Concepts   *Fill in the blanks.*

1. A quadratic function is defined by the equation _____ $(a \neq 0)$.

2. The standard form for the equation of a parabola is _____ $(a \neq 0)$.

3. The vertex of the parabolic graph of the equation $y = 2(x - 3)^2 + 5$ will be at _____.

4. The vertical line that intersects the parabola at its vertex is the _____.

5. If the parabola opens _____, the vertex will be a minimum point.

6. If the parabola opens _____, the vertex will be a maximum point.

7. The x-coordinate of the vertex of the parabolic graph of $f(x) = ax^2 + bx + c$ is ____.

8. The y-coordinate of the vertex of the parabolic graph of $f(x) = ax^2 + bx + c$ is _____.

Practice   *Determine whether the graph of each quadratic function opens upward or downward. State whether a maximum or minimum point occurs at the vertex of the parabola.*

9. $f(x) = \dfrac{1}{2}x^2 + 3$ 10. $f(x) = 2x^2 - 3x$

11. $f(x) = -3(x + 1)^2 + 2$

12. $f(x) = -5(x - 1)^2 - 1$

13. $f(x) = -2x^2 + 5x - 1$

14. $f(x) = 2x^2 - 3x + 1$

Find the vertex of each parabola.

15. $y = x^2 - 1$

16. $y = -x^2 + 2$

17. $f(x) = (x - 3)^2 + 5$

18. $f(x) = -2(x - 3)^2 + 4$

19. $f(x) = -2(x + 6)^2 - 4$

20. $f(x) = \dfrac{1}{3}(x + 1)^2 - 5$

21. $f(x) = \dfrac{2}{3}(x - 3)^2$

22. $f(x) = 7(x + 2)^2 + 8$

23. $f(x) = x^2 - 4x + 4$ 24. $y = x^2 - 10x + 25$

25. $y = x^2 + 6x - 3$ 26. $y = -x^2 + 9x - 2$

27. $y = -2x^2 + 12x - 17$ 28. $y = 2x^2 + 16x + 33$

29. $y = 3x^2 - 4x + 5$ 30. $y = -4x^2 + 3x + 4$

31. $y = \dfrac{1}{2}x^2 + 4x - 3$ 32. $y = -\dfrac{2}{3}x^2 + 3x - 5$

Graph each quadratic function given in standard form.

33. $f(x) = x^2 - 4$ 34. $f(x) = x^2 + 1$

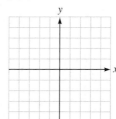

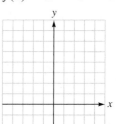

35. $f(x) = -3x^2 + 6$

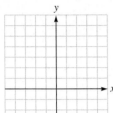

36. $f(x) = -4x^2 + 4$

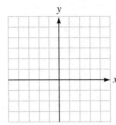

45. $f(x) = -3(x - 2)^2 + 6$ **46.** $f(x) = 2(x - 3)^2 - 4$

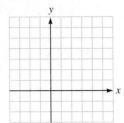
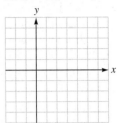

37. $f(x) = -\frac{1}{2}x^2 + 8$

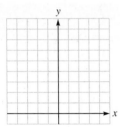

38. $f(x) = \frac{1}{2}x^2 - 2$

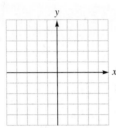

Graph each quadratic function.

47. $f(x) = x^2 + 2x$ **48.** $f(x) = x^2 - 6x$

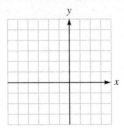
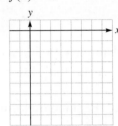

39. $f(x) = (x - 3)^2 - 1$ **40.** $f(x) = (x + 3)^2 - 1$

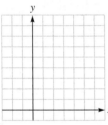
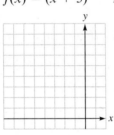

49. $f(x) = x^2 - 4x + 1$ **50.** $f(x) = x^2 - 6x - 7$

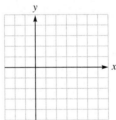
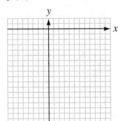

41. $f(x) = 2(x + 1)^2 - 2$ **42.** $f(x) = -\frac{3}{4}(x - 2)^2$

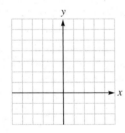
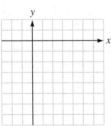

51. $f(x) = 2x^2 - 12x + 10$ **52.** $f(x) = -x^2 - 4x + 1$

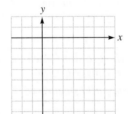
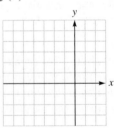

53. $f(x) = -3x^2 - 6x - 9$ **54.** $f(x) = -3x^2 - 3x + 18$

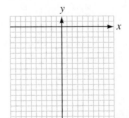
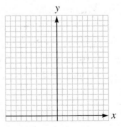

43. $f(x) = -(x + 4)^2 + 1$ **44.** $f(x) = -3(x - 4)^2 + 3$

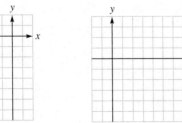

Applications

55. Police investigations   A police officer seals off the scene of an accident using a roll of yellow tape that is 300 feet long. What dimensions should be used to seal off the maximum rectangular area around the collision? Find the maximum area.

56. Maximizing area   The rectangular garden shown has a width of x and a perimeter of 100 feet. Find x such that the area of the rectangle is maximum.

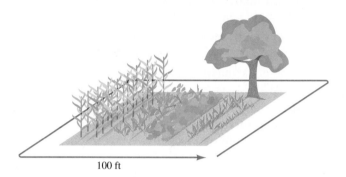

100 ft

57. Maximizing storage area   A farmer wants to partition a rectangular feed storage area in a corner of his barn, as shown in the illustration. The barn walls form two sides of the stall, and the farmer has 50 feet of partition for the remaining two sides. What dimensions will maximize the area?

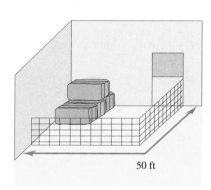

50 ft

58. Maximizing grazing area   A rancher wants to enclose a rectangular partitioned corral with 1,800 feet of fencing. (See the illustration.) What dimensions of the corral would enclose the largest possible area? Find the maximum area.

59. Sheet metal fabrication   A 24-inch-wide sheet of metal is to be bent into a rectangular trough with the cross section shown in the illustration. Find the dimensions that will maximize the amount of water the trough can hold. That is, find the dimensions that will maximize the cross-sectional area.

Depth

24 in.

Width

60. Landscape design   A gardener will use D feet of edging to border a rectangular plot of ground. Show that the maximum area will be enclosed if the rectangle is a square.

61. Architecture   A parabolic arch has an equation of $x^2 + 20y - 400 = 0$, where x is measured in feet. Find the maximum height of the arch.

62. Path of a guided missile   A guided missile is propelled from the origin of a coordinate system with the x-axis along the ground and the y-axis vertical. Its path, or **trajectory,** is given by the equation $y = 400x - 16x^2$. Find the object's maximum height.

63. Height of a basketball   The path of a basketball thrown from the free throw line can be modeled by the quadratic function

$$f(x) = -0.06x^2 + 1.5x + 6$$

where x is the horizontal distance (in feet) from the free throw line and $f(x)$ is the height (in feet) of the ball. Find the maximum height of the basketball.

64. Ballistics A child throws a ball up a hill that makes an angle of 45° with the horizontal. The ball lands 100 feet up the hill. Its trajectory is a parabola with equation $y = -x^2 + ax$ for some number a. Find a.

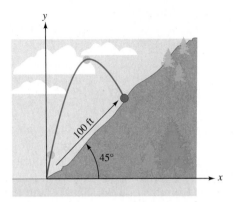

65. Maximizing height A ball is thrown straight up from the top of a building 144 ft tall with an initial velocity of 64 ft per second. The distance $s(t)$ (in feet) of the ball from the ground is given by $s(t) = 144 + 64t - 16t^2$. Find the maximum height attained by the ball.

66. Flat-panel television sets A wholesaler of appliances finds that she can sell $(2{,}400 - p)$ flat-panel television sets each week when the price is p dollars. What price will maximize revenue?

67. Digital cameras A company that produces and sells digital cameras has determined that the total weekly cost C of producing x digital cameras is given by the function

$$C(x) = 1.5x^2 - 144x + 5{,}856$$

Determine the production level that minimizes the weekly cost for producing the digital cameras and find that weekly minimum cost.

68. Finding mass transit fares The Municipal Transit Authority serves 150,000 commuters daily when the fare is $1.80. Market research has determined that every penny decrease in the fare will result in 1,000 new riders. What fare will maximize revenue?

69. Finding hotel rates A 300-room hotel is two-thirds filled when the nightly room rate is $90. Experience has shown that each $5 increase in cost results in 10 fewer occupied rooms. Find the nightly rate that will maximize income.

70. Selling concert tickets Tickets for a concert are cheaper when purchased in quantity. The first 100 tickets are priced at $10 each, but each additional block of 100 tickets purchased decreases the cost of each ticket by 50¢. How many blocks of tickets should be sold to maximize the revenue?

Use this information: At a time t seconds after an object is tossed vertically upward, it reaches a height s in feet given by the equation $s = 80t - 16t^2$.

71. In how many seconds does the object reach its maximum height?

72. In how many seconds does the object return to the point from which it was thrown?

73. What is the maximum height reached by the object?

74. Show that it takes the same amount of time for the object to reach its maximum height as it does to return from that height to the point from which it was thrown.

Use a graphing calculator to determine the coordinates of the vertex of each parabola. You will have to select appropriate viewing windows.

75. $y = 2x^2 + 9x - 56$ **76.** $y = 14x - \dfrac{x^2}{5}$

77. $y = (x - 7)(5x + 2)$ **78.** $y = -x(0.2 + 0.1x)$

Discovery and Writing   *Find all values of x that will make $f(x) = 0$.*

79. $f(x) = x^2 - 5x + 6$ **80.** $f(x) = 6x^2 + x - 2$

81. Find the dimensions of the largest rectangle that can be inscribed in the right triangle ABC shown in the illustration.

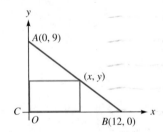

82. Point P lies in the first quadrant and on the line $x + y = 1$ in such a position that the area of triangle OPA is maximum. Find the coordinates of P. (See the illustration.)

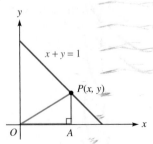

83. The sum of two numbers is 6, and the sum of the squares of those two numbers is as small as possible. What are the numbers?

84. What number most exceeds its square?

The maximum or minimum value of a quadratic function can be found automatically by using a computer program, such as Excel.

85. Find the minimum value of the function $f(x) = 2x^2 - 3x - 4$ by using the Solver in Excel. Give the value of x that minimizes the function as well as the minimum value of the function. See Problem 109 in Section 1.3.

86. Find the maximum value of the function $f(x) = -2x^2 + 3x + 4$ by using the Solver in Excel. Give the value of x that maximizes the function as well as the maximum value of the function. See Problem 109 in Section 1.3.

Review   *Find f(a) and f(−a).*

87. $f(x) = x^2 - 3x$

88. $f(x) = x^3 - 3x$

89. $f(x) = (5 - x)^2$

90. $f(x) = \dfrac{1}{x^2 - 4}$

91. $f(x) = 7$

92. $f(x) = -|x|$

3.3 Polynomial and Other Functions

Objectives

1. Understand the Characteristics of Polynomial Functions
2. Graph Polynomial Functions
3. Determine Whether a Function Is Even, Odd, or Neither
4. Identify the Intervals on Which a Function Is Increasing, Decreasing, or Constant
5. Graph Piecewise-Defined Functions
6. Evaluate and Graph the Greatest Integer Function

So far, we have discussed two types of polynomial functions—first-degree (or linear) functions, and second-degree (or quadratic) functions. In this section, we will discuss polynomial functions of higher degree.

Polynomial functions can be used to model the path of a roller coaster or to model the fluctuation of gasoline prices over the past few months. Goliath, a hypercoaster, opened in 2006 at Six Flags Over Georgia. It climbs to a height of 200 feet and reaches speeds of nearly 70 mph. It has more than 4,400 feet of steel track. Portions of Goliath's tracks can be modeled with a polynomial function.

Polynomial Functions   A **polynomial function in one variable (say, x)** is a function of the form

$$f(x) = a_n x^n + a_{n-1} x^{n-1} + \cdots + a_1 x + a_0$$

where $a_n, a_{n-1}, \ldots, a_1,$ and a_0 are real numbers and n is a whole number.

The **degree of a polynomial function** is the largest power of x that appears in the polynomial.

The table shows three basic polynomial functions.

Name	Function	Degree	Graph
Constant function	$f(x) = 5$	0	Horizontal line
Linear function	$f(x) = 2x - 7$	1	Nonvertical line
Quadratic function	$f(x) = -2x^2 + 4x - 5$	2	Parabola

Here are two examples of higher-degree polynomial functions, along with their graphs.

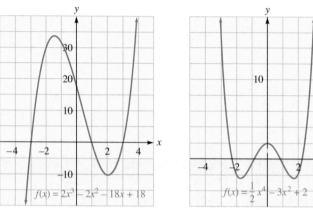

| **Figure 3-17** | **Figure 3-18** |

1.   Understand the Characteristics of Polynomial Functions

There are several basic characteristics common to all polynomial functions. We will list several of them.

1: The graphs of polynomial functions are smooth and continuous curves.
Like the graphs of linear and quadratic functions, the graphs of higher-degree polynomial functions are smooth, continuous curves. Because their graphs are smooth, they have no cusps or corners. Because they are continuous, their graphs have no breaks or holes. They always can be drawn without lifting the pencil from the paper. (See Figure 3-19.)

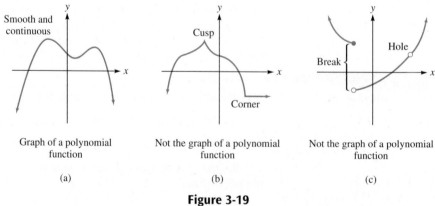

Figure 3-19

2: Many polynomial functions have graphs similar to the graphs of $f(x) = x$, $f(x) = x^2$, and $f(x) = x^3$.

Many polynomial functions are of the form $f(x) = x^n$ and several of their graphs are shown in Figure 3-20. Note that when n is even, the graph has the same general shape as $y = x^2$. When n is odd and greater than 1, the graph has the same general shape as $y = x^3$. However, the graphs are flatter at the origin and steeper as n becomes large.

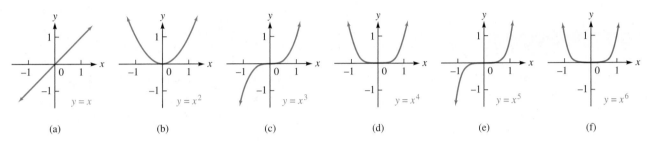

(a)   (b)   (c)   (d)   (e)   (f)

Figure 3-20

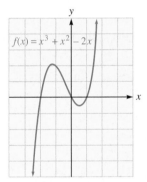

Figure 3-21

3: The end behavior of the graph of a polynomial function is similar to the graph of its term with highest degree.

The ends of the graph of any polynomial function will be similar to the graph of its term with the highest power of x, because when n becomes large, the other terms become relatively insignificant.

Consider the polynomial function $f(x) = x^3 + x^2 - 2x$. The end behavior of its graph will be similar to the ends of the graph of its leading term x^3. In Figure 3-20(c), we see that the graph of $y = x^3$ falls on the far left and rises on the far right. Therefore, the graph of $f(x)$ will also fall on the far left and rise on the far right. (See Figure 3-21.)

4: Polynomial functions can be symmetric about the y-axis or the origin.

In Section 2.4, we learned that a graph is symmetric about the y-axis if the graph of $y = f(x)$ has the same y-coordinate when the function is evaluated at x or at $-x$. Thus, a function is symmetric about the y-axis if $f(x) = f(-x)$ for all values of x that are in the domain of the function. (See Figure 3-22(a).) Also recall that a graph is symmetric about the origin if the point $(-x, -f(x))$ lies on the graph whenever $(x, f(x))$ does. In this case, $f(-x) = -f(x)$. (See Figure 3-22(b).)

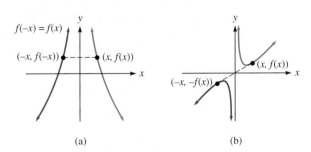

(a)   (b)

Figure 3-22

2.   Graph Polynomial Functions

To graph polynomial functions, we can use the following steps.

Graphing Polynomial Functions

1. Find any symmetries of the graph.
2. Find the x- and y-intercepts of the graph.
3. Determine where the graph is above and below the x-axis.
4. Plot a few points, if necessary, and draw the graph as a smooth, continuous curve.

EXAMPLE 1   Graph the function: $f(x) = x^3 - 4x$.

Solution   We will use the four steps stated above to graph the polynomial function.

Step 1: Find any symmetries of the graph.   To test for symmetry about the y-axis, we check to see whether $f(x) = f(-x)$. To test for symmetry about the origin, we check to see whether $f(-x) = -f(x)$.

$$f(x) = x^3 - 4x$$
$$f(-x) = (-x)^3 - 4(-x) \qquad \text{Substitute } -x \text{ for } x.$$
$$f(-x) = -x^3 + 4x \qquad \text{Simplify.}$$

Since $f(x) \neq f(-x)$, there is no symmetry about the y-axis. However, since $f(-x) = -f(x)$, there is symmetry about the origin.

Step 2: Find the x- and y-intercepts of the graph.   To find the x-intercepts, we let $f(x) = 0$ and solve for x.

$$x^3 - 4x = 0$$
$$x(x^2 - 4) = 0 \qquad \text{Factor out } x.$$
$$x(x + 2)(x - 2) = 0 \qquad \text{Factor } x^2 - 4.$$
$$x = 0 \quad \text{or} \quad x + 2 = 0 \quad \text{or} \quad x - 2 = 0 \qquad \text{Set each factor equal to 0.}$$
$$x = -2 \qquad\qquad x = 2$$

The x-intercepts are $(0, 0)$, $(-2, 0)$, and $(2, 0)$. If we let $x = 0$ and solve for $f(x)$, we see that the y-intercept is also $(0, 0)$.

Step 3: Determine where the graph is above or below the x-axis.   To determine where the graph is above or below the x-axis, we plot the solutions of $x^3 - 4x = 0$ (the x-intercepts) on a number line and establish the four intervals shown in Figure 3-23. We then test a number from each interval to determine the sign of $f(x)$. (For a review of this process, see Example 7 in Section 1.7.)

Sign of $f(x) = x^3 - 4x$	$-$	$+$	$-$	$+$
	$(-\infty, -2)$	$(-2, 0)$	$(0, 2)$	$(2, \infty)$
Test point	$f(-3) = -15$ -2	$f(-1) = 3$ 0	$f(1) = -3$ 2	$f(3) = 15$
Graph of $f(x)$	below the x-axis	above the x-axis	below the x-axis	above the x-axis

Figure 3-23

Step 4: Plot a few points and draw the graph as a smooth, continuous curve.   We now plot the intercepts and one additional point. In the previous step, we found that $f(1) = -3$. This will be the additional point we plot, $(1, -3)$. Making use of our knowledge of symmetry and where the graph is above and below the x-axis, we now draw the graph as shown in Figure 3-24(a). A calculator graph is shown in Figure 3-24(b).

$f(x) = x^3 - 4x$

x	$f(x)$	$(x, f(x))$
-2	0	$(-2, 0)$
0	0	$(0, 0)$
1	-3	$(1, -3)$
2	0	$(2, 0)$

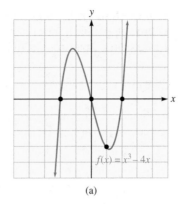

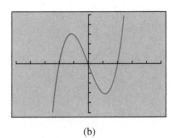

(a)　　　　　　　　　　　　　　　　(b)

Figure 3-24

Comment

Note that the far right and far left ends of the graph are similar to the ends of the graph of $f(x) = x^3$, which is the leading term of the function $f(x) = x^3 - 4x$ or the term with highest degree. On the far right, the graph rises and on the far left, the graph falls.

Self Check 1 Graph $f(x) = x^3 - 9x$.

The peak and valley of the graph shown in Figure 3-24 are called **turning points.** In calculus, such points are called **local minima** and **local maxima.** Although we cannot find these points without using calculus, we can approximate them by plotting points or by using the TRACE feature of a graphing calculator.

Note that Figure 3-24 shows the graph of a third-degree polynomial and the graph has 2 turning points. This suggests the following result from calculus that helps us understand the shape of many polynomial graphs.

Number of Turning Points If $f(x)$ is a polynomial function of nth degree, then the graph of $f(x)$ will have $n - 1$, or fewer, turning points.

EXAMPLE 2 Graph the function: $f(x) = x^4 - 5x^2 + 4$.

Solution We will use the four steps for graphing a polynomial function.

Step 1: Find any symmetries of the graph. Because x appears with only even exponents, $f(x) = f(-x)$, and the graph is symmetric about the y-axis. The graph is not symmetric about the origin.

Step 2: Find the x- and y-intercepts of the graph. To find the x-intercepts, we let $f(x) = 0$ and solve for x.

$$x^4 - 5x^2 + 4 = 0$$
$$(x^2 - 4)(x^2 - 1) = 0$$
$$(x + 2)(x - 2)(x + 1)(x - 1) = 0$$

$$x + 2 = 0 \quad \text{or} \quad x - 2 = 0 \quad \text{or} \quad x + 1 = 0 \quad \text{or} \quad x - 1 = 0$$
$$x = -2 \quad | \quad x = 2 \quad | \quad x = -1 \quad | \quad x = 1$$

The x-intercepts are $(-2, 0)$, $(2, 0)$, $(-1, 0)$, and $(1, 0)$. To find the y-intercept, we let $x = 0$ and see that the y-intercept is $(0, 4)$.

Step 3: Determine where the graph is above or below the x-axis. To determine where the graph is above or below the x-axis, we plot the x-coordinates of the x-intercepts on a number line and establish the intervals shown in Figure 3-25.

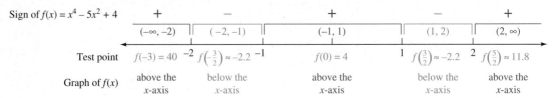

Sign of $f(x) = x^4 - 5x^2 + 4$	$+$	$-$	$+$	$-$	$+$
	$(-\infty, -2)$	$(-2, -1)$	$(-1, 1)$	$(1, 2)$	$(2, \infty)$
Test point	$f(-3) = 40$	$f\left(-\frac{3}{2}\right) \approx -2.2$	$f(0) = 4$	$f\left(\frac{3}{2}\right) \approx -2.2$	$f\left(\frac{5}{2}\right) \approx 11.8$
Graph of $f(x)$	above the x-axis	below the x-axis	above the x-axis	below the x-axis	above the x-axis

Figure 3-25

Step 4: Plot a few points and draw the graph as a smooth, continuous curve. We can now plot the intercepts and use our knowledge of symmetry and where the graph is above and below the x-axis to draw the graph, as in Figure 3-26(a). A calculator graph is shown in Figure 3-26(b).

$$f(x) = x^4 - 5x^2 + 4$$

x	$f(x)$	$(x, f(x))$
-2	0	$(-2, 0)$
-1	0	$(-1, 0)$
0	4	$(0, 4)$
1	0	$(1, 0)$
$\frac{3}{2}$	$-\frac{35}{16}$	$\left(\frac{3}{2}, -\frac{35}{16}\right)$
2	0	$(2, 0)$

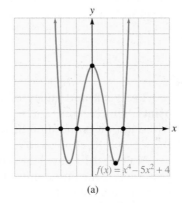

(a)

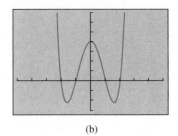

(b)

Figure 3-26

Comment

Note that the far right and far left ends of the graph are similar to the ends of the graph of $f(x) = x^4$, which is the leading term of the function $f(x) = x^4 - 5x^2 + 4$ or the term with highest degree. On the far right and far left, the graph rises. Also note that the graph has 3 turning points.

Self Check 2 Graph $f(x) = x^4 - 10x^2 + 9$.

3. Determine Whether a Function Is Even, Odd, or Neither

If $f(-x) = f(x)$ for all x in the domain of f, the graph of the function is symmetric about the y-axis, and the function is called an **even function.** If $f(-x) = -f(x)$ for all x in the domain of f, the function is symmetric about the origin, and the function is called an **odd function.**

Since the graph in Example 1 is symmetric about the origin, it represents an odd function. Since the graph in Example 2 is symmetric about the y-axis, it represents an even function. If $f(x)$ does not have either of these symmetries, it is neither even nor odd.

EXAMPLE 3 Determine whether each function is even, odd, or neither:
a. $f(x) = x^2$ **b.** $f(x) = x^3$

Solution To check whether the function is an even function, we find $f(-x)$ and see whether $f(-x) = f(x)$. To check whether the function is an odd function, we find $f(-x)$ and see whether $f(-x) = -f(x)$.

a. $f(-x) = (-x)^2 = x^2 = f(x)$
Since $f(-x) = f(x)$, the function $f(x) = x^2$ is an even function.

b. $f(-x) = (-x)^3 = -x^3 = -f(x)$
Since $f(-x) \neq f(x)$, the function $f(x) = x^3$ is not an even function. However, the function is an odd function, because $f(-x) = -f(x)$.

Self Check 3 Classify each function as even, odd, or neither: **a.** $f(x) = x^3 + x$
b. $f(x) = x^2 + 4$

4. Identify the Intervals on Which a Function Is Increasing, Decreasing, or Constant

If we trace the graph of a function from left to right and the values $f(x)$ increase as shown in Figure 3-27(a), we say that the function is **increasing on the interval** (a, b). If the values $f(x)$ decrease as in Figure 3-27(b), we say that the function is **decreasing on the interval** (a, b). If the values $f(x)$ remain unchanged as x increases, we say that the function is **constant on the interval** (a, b).

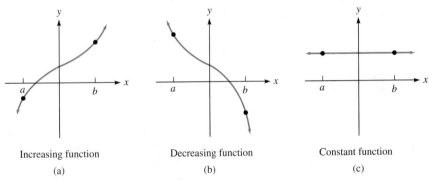

Increasing function
(a)

Decreasing function
(b)

Constant function
(c)

Figure 3-27

EXAMPLE 4 State the open intervals on which the function is increasing or decreasing.

Solution We trace the graph of the function from left to right and identify the open intervals on which $f(x)$-values increase and where they decrease.

We see from the graph shown in Figure 3-28 that the values of $f(x)$ increase on the open intervals $(-\infty, -2)$ and $(0, 2)$. They decrease on the open intervals $(-2, 0)$ and $(2, \infty)$. Hence, the function is increasing on $(-\infty, -2) \cup (0, 2)$ and is decreasing on $(-2, 0) \cup (2, \infty)$.

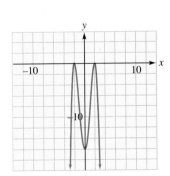

Figure 3-28

Intervals on which a function is increasing, decreasing, or constant are always written using open intervals. We also use *x*-values when writing open intervals for which the graph of a function is increasing, decreasing, or constant.

Self Check 4

Use the graph shown in Figure 3-28 to identify the open intervals, if any, on which the function is constant.

5. Graph Piecewise-Defined Functions

Some functions, called **piecewise-defined functions,** are defined by using different equations for different intervals in their domains. To illustrate, we will graph the piecewise-defined function *f* given by

$$f(x) = \begin{cases} -2 & \text{if } x \le 0 \\ x + 1 & \text{if } x > 0 \end{cases}$$

To evaluate this piecewise-defined function, we must determine which part of the function's definition to use. If $x \le 0$, we use the top part of the definition and the corresponding value of $f(x)$ is -2. Some examples are:

$$f(-2) = -2 \qquad f(-1) = -2 \qquad f(0) = -2$$

In the interval $(-\infty, 0]$, the function is constant and the graph is the horizontal line $y = -2$.

If $x > 0$, we use the bottom part of the definition to evaluate the function and the corresponding value of $f(x)$ is $x + 1$. Examples are:

$$f(1) = 1 + 1 = 2 \qquad f(2) = 2 + 1 = 3$$

In the interval $(0, \infty)$, the graph of the function is the line $f(x) = x + 1$. This is a linear function with slope $m = 1$ and *y*-intercept $(0, b) = (0, 1)$.

The graph of this piecewise-defined function appears in Figure 3-29.

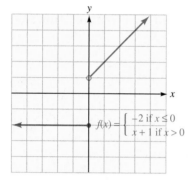

Figure 3-29

Comment

Note that when $x = 0$, we have $f(x) = -2$. For this reason, the point $(0, -2)$ is shown as a closed point and the point $(0, 1)$ is shown as an open point.

EXAMPLE 5

Graph the function: $f(x) = \begin{cases} -x & \text{if } x < 0 \\ x^2 & \text{if } 0 \le x \le 1. \\ 1 & \text{if } x > 1 \end{cases}$

Solution

This piecewise-defined function is defined in three parts. We will graph each part of the function. That is, we will graph

$f(x) = -x$ in the interval $(-\infty, 0)$,

$f(x) = x^2$ in the interval $[0, 1]$, and

$f(x) = 1$ in the interval $(1, \infty)$.

If $x < 0$, the value of $f(x)$ is determined by the equation $f(x) = -x$. We graph the line with slope $m = -1$ and the *y*-value of the *y*-intercept $b = 0$ in the interval $(-\infty, 0)$. In the interval $(-\infty, 0)$, the function is decreasing.

If $0 \le x \le 1$, the value of $f(x)$ is x^2. We graph the parabola in the interval $[0, 1]$. In this interval, the function is increasing.

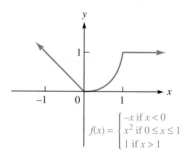

Figure 3-30

If $x > 1$, the value of $f(x)$ is 1. In the interval $(1, \infty)$, the function is constant and its graph is the same as the graph of $y = 1$.

The graph of the piecewise function appears in Figure 3-30.

Self Check 5   Graph: $f(x) = \begin{cases} 2x & \text{if } x \le 0 \\ x - 1 & \text{if } x > 0 \end{cases}$.

6. Evaluate and Graph the Greatest Integer Function

The **greatest integer function** is important in many business applications and in the field of computer science. This function is determined by the equation $f(x) = [x]$, where the value of $f(x)$ that corresponds to x is the greatest integer that is less than or equal to x. For example,

$$f(2.71) = [2.71] = 2 \qquad f(23.5) = [23.5] = 23$$
$$f(10) = [10] = 10 \qquad f(\pi) = [\pi] = 3$$
$$f(-2.5) = [-2.5] = -3$$

EXAMPLE 6   Graph: $f(x) = [x]$.

Solution   We will list several intervals and determine the corresponding values of the greatest integer function. Then we will use these values to graph the function.

$[0, 1)$ $f(x) = [x] = 0$ For numbers from 0 to 1 (not including 1), the greatest integer in the interval is 0.

$[1, 2)$ $f(x) = [x] = 1$ For numbers from 1 to 2 (not including 2), the greatest integer in the interval is 1.

$[2, 3)$ $f(x) = [x] = 2$ For numbers from 2 to 3 (not including 3), the greatest integer in the interval is 2.

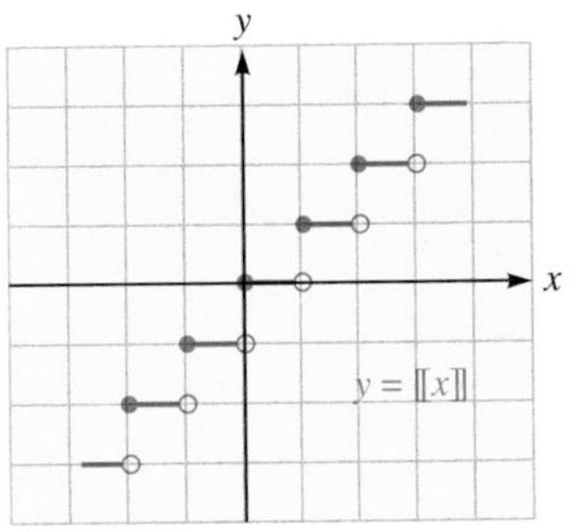

Within each interval, the values of y are constant, but they jump by 1 at integer values of x. The graph is shown in Figure 3-31. From the graph, we can see that the domain of the greatest integer function is the interval $(-\infty, \infty)$. The range is the set of integers.

Figure 3-31

Self Check 6   Find: **a.** $[7.61]$ **b.** $[-3.75]$.

Since the greatest integer function is made up of a series of horizontal line segments, it is an example of a group of functions called **step functions.**

EXAMPLE 7   To print business forms, a printing company charges customers $10 for the order, plus $20 for each box containing 200 forms. The printing company counts any portion of a box as a full box. Graph this step function.

Solution   To graph the step function, we will determine the cost for printing various amounts of boxes of forms. Then we will graph our result.

If we order the forms and then change our minds before the forms are printed, the cost will be $10. Thus, the ordered pair (0, 10) will be on the graph.

If we purchase up to one full box, the cost will be $10 for the order and $20 for the printing, for a total of $30. Thus, the ordered pair (1, 30) will be on the graph.

The cost for $1\frac{1}{2}$ boxes will be the same as the cost for 2 full boxes, or $50. Thus, the ordered pairs (1.5, 50) and (2, 50) are on the graph.

The complete graph is shown in Figure 3-32.

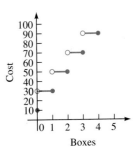

Figure 3-32

Self Check 7 Find the cost of $4\frac{1}{2}$ boxes.

Self Check Answers

1.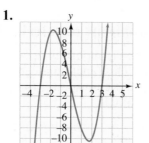

$f(x) = x^3 - 9x$

2.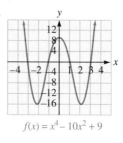

$f(x) = x^4 - 10x^2 + 9$

3. a. odd **b.** even

4. no intervals

5.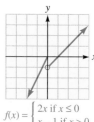

$f(x) = \begin{cases} 2x \text{ if } x \le 0 \\ x - 1 \text{ if } x > 0 \end{cases}$

6. a. 7 **b.** -4 **7.** $110

3.3 Exercises

Vocabulary and Concepts *Fill in the blanks.*

1. The degree of the function $y = f(x) = x^4 - 3$ is ___.

2. Peaks and valleys on a polynomial graph are called _____ points.

3. The graph of a nth degree polynomial function can have at most _____ turning points.

4. If the graph of a function is symmetric about the _____, it is called an even function.

5. If the graph of a function is symmetric about the origin, it is called an ____ function.

6. If the values of $f(x)$ get larger as x increases on an interval, we say that the function is _____ on the interval.

7. _____ functions are defined by different equations for different intervals in their domains.

8. If the values of $f(x)$ get smaller as x increases on an interval, we say that the function is _____ on the interval.

9. $[3.69] =$ __.

10. If the values of $f(x)$ do not change as x increases on an interval, we say that the function is _____ on the interval.

Practice *Graph each polynomial function.*

11. $f(x) = x^3 - 9x$

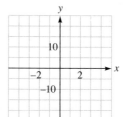

12. $f(x) = x^3 - 16x$

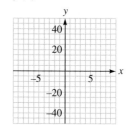

13. $f(x) = -x^3 - 4x^2$

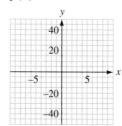

14. $f(x) = x^3 - x$

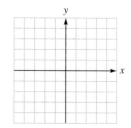

15. $f(x) = x^3 + x^2$

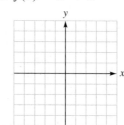

16. $f(x) = -x^3 + 1$

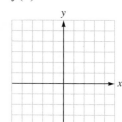

17. $f(x) = x^4 - 2x^2 + 1$ **18.** $f(x) = x^4 - 5x^2 + 4$

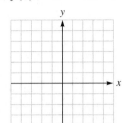   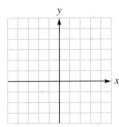

19. $f(x) = x^3 - x^2 - 4x + 4$

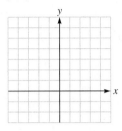

20. $f(x) = 4x^3 - 4x^2 - x + 1$

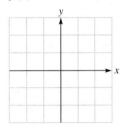

21. $f(x) = -x^4 + 5x^2 - 4$

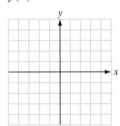

22. $f(x) = x(x - 3)(x - 2)(x + 1)$

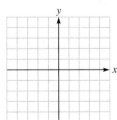

Determine whether each function is even or odd. If it is neither, so indicate.

23. $f(x) = x^4 + x^2$ **24.** $f(x) = x^3 - 2x$

25. $f(x) = x^3 + x^2$ **26.** $f(x) = x^6 - x^2$

27. $f(x) = x^5 + x^3$ **28.** $f(x) = x^3 - x^2$

29. $f(x) = 2x^3 - 3x$ **30.** $f(x) = 4x^2 - 5$

Determine where each function is increasing, decreasing, or constant.

31. 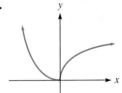   **32.**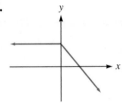

33. 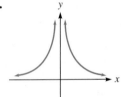   **34.**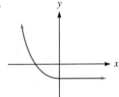

Determine the interval in which each function is decreasing, increasing, or is constant.

35. 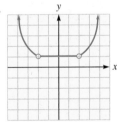   **36.**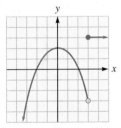

37. $f(x) = x^2 - 4x + 4$ **38.** $f(x) = 4 - x^2$

Evaluate each piecewise-defined function.

39. $f(x) = \begin{cases} 2x + 2 & \text{if } x < 0 \\ 3 & \text{if } x \geq 0 \end{cases}$

    a. $f(-2)$ **b.** $f(0)$

40. $f(x) = \begin{cases} x - 2 & \text{if } x < 1 \\ x^2 & \text{if } x \geq 1 \end{cases}$

    a. $f(1)$ **b.** $f(5)$

41. $f(x) = \begin{cases} 2 & \text{if } x < 0 \\ 2 - x & \text{if } 0 \le x < 2 \\ x + 1 & \text{if } x \ge 2 \end{cases}$

   a. $f(-1)$ **b.** $f(1)$ **c.** $f(2)$

42. $f(x) = \begin{cases} 2x & \text{if } x < 0 \\ 3 - x & \text{if } 0 \le x < 2 \\ |x| & \text{if } x \ge 2 \end{cases}$

   a. $f(-0.5)$ **b.** $f(0)$ **c.** $f(2)$

Graph each piecewise-defined function.

43. $f(x) = \begin{cases} x + 2 & \text{if } x < 0 \\ 2 & \text{if } x \ge 0 \end{cases}$

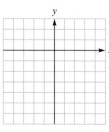

44. $f(x) = \begin{cases} 2x & \text{if } x < 0 \\ -2x & \text{if } x \ge 0 \end{cases}$

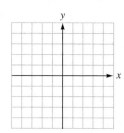

45. $f(x) = \begin{cases} -x & \text{if } x < 0 \\ x^2 & \text{if } x \ge 0 \end{cases}$

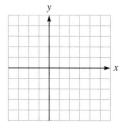

46. $f(x) = \begin{cases} |x| & \text{if } x < 0 \\ \sqrt{x} & \text{if } x \ge 0 \end{cases}$ **47.** $f(x) = \begin{cases} x & \text{if } x \le 0 \\ 2 & \text{if } x > 0 \end{cases}$

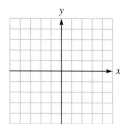

48. $f(x) = \begin{cases} -x & \text{if } x < 0 \\ \dfrac{1}{2}x & \text{if } x > 0 \end{cases}$

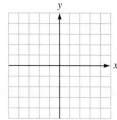

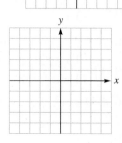

49. $f(x) = \begin{cases} 0 & \text{if } x < 0 \\ x^2 & \text{if } 0 \le x \le 2 \\ 4 - 2x & \text{if } x > 2 \end{cases}$

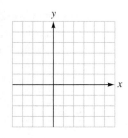

50. $f(x) = \begin{cases} 2 & \text{if } x < 0 \\ 2 - x & \text{if } 0 \le x < 2 \\ x & \text{if } x \ge 2 \end{cases}$

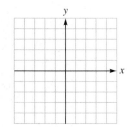

Evalute each function at the indicated x-values.

51. $f(x) = [\![x]\!]$ **a.** $f(3)$ **b.** $f(-4)$ **c.** $f(-2.3)$

52. $f(x) = [\![3x]\!]$ **a.** $f(4)$ **b.** $f(-2)$ **c.** $f(-1.2)$

53. $f(x) = [\![x + 3]\!]$ **a.** $f(-1)$ **b.** $f\left(\dfrac{2}{3}\right)$ **c.** $f(1.3)$

54. $f(x) = [\![4x]\!] - 1$ **a.** $f(-3)$ **b.** $f(0)$ **c.** $f(\pi)$

Graph each function.

55. $y = [\![2x]\!]$ **56.** $y = \left[\!\left[\dfrac{1}{3}x + 3\right]\!\right]$

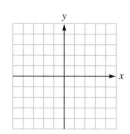

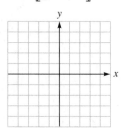

57. $y = [\![x]\!] - 1$ **58.** $y = [\![x + 2]\!]$

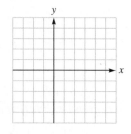

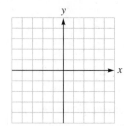

Applications

59. Grading scales A mathematics instructor assigns letter grades according to the following scale.

From	Up to but less than	Grade
60%	70%	D
70%	80%	C
80%	90%	B
90%	100% (including 100%)	A

Graph the ordered pairs (p, g), where p represents the percent and g represents the grade. Find the final semester grade of a student who has test scores of 67%, 73%, 84%, 87%, and 93%.

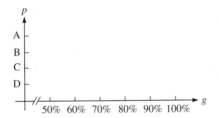

60. Calculating grades See Exercise 59 and find the final semester grade of a student who has test scores of 53%, 65%, 64%, 73%, 89%, and 82%.

61. Renting a Jeep A rental company charges $20 to rent a Jeep Wrangler for one day, plus $4 for every 100 miles (or portion of 100 miles) that it is driven. Graph the ordered pairs (m, c), where m represents the miles driven and c represents the cost. Find the cost if the car is driven 275 miles in one day.

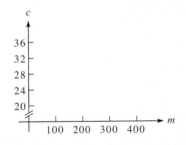

62. Riding in a taxi A taxicab company charges $3 for a trip up to 1 mile, and $2 for every extra mile (or portion of a mile). Graph the ordered pairs (m, c), where m represents the miles traveled and c represents the cost. Find the cost to ride $10\frac{1}{4}$ miles.

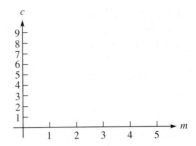

63. Computer communications An online information service charges for connect time at a rate of $12 per hour, computed for every minute or fraction of a minute. Graph the points (t, c), where c is the cost of t minutes of connect time. Find the cost of $7\frac{1}{2}$ minutes.

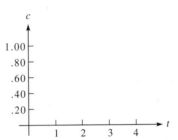

64. iPhone repair There is a charge of $30, plus $40 per hour (or fraction of an hour), to repair an iPhone. Graph the points (t, c), where t is the time it takes to do the job and c is the cost. If it takes 4 hours to repair an iPhone, how much did it cost?

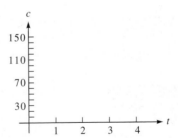

65. Rounding numbers Measurements are rarely exact; they often are *rounded* to an appropriate precision. Graph the points (x, y), where y is the result of rounding the number x to the nearest ten.

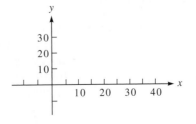

66. Signum function Computer programmers often use the following function, denoted by $y = \text{sgn } x$. Graph this function and find its domain and range.

$$y = \begin{cases} -1 & \text{if } x < 0 \\ 0 & \text{if } x = 0 \\ 1 & \text{if } x > 0 \end{cases}$$

67. Graph the function defined by $y = \dfrac{|x|}{x}$ and compare it to the graph in Exercise 66. Are the graphs the same?

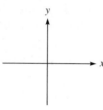

68. Graph: $y = x + |x|$.

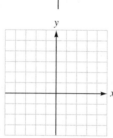

Discovery and Writing *Use a graphing calculator to explore the properties of graphs of polynomial functions. Write a paragraph summarizing your observations.*

69. Graph the function $y = x^2 + ax$ for several values of a. How does the graph change?

70. Graph the function $y = x^3 + ax$ for several values of a. How does the graph change?

71. Graph the function $y = (x - a)(x - b)$ for several values of a and b. What is the relationship between the x-intercepts and the equation?

72. Use the insight you gained in Exercise 71 to factor $x^3 - 3x^2 - 4x + 12$.

Review

73. If $f(x) = 3x + 2$, find $f(x + 1)$ and $f(x) + 1$.

74. If $f(x) = x^2$, find $f(x - 2)$ and $f(x) - 2$.

75. If $f(x) = \dfrac{3x + 1}{5}$, find $f(x - 3)$ and $f(x) - 3$.

76. If $f(x) = 8$, find $f(x + 8)$ and $f(x) + 8$.

77. Solve: $2x^2 - 3 = x$.

78. Solve: $4x^2 = 24x - 37$.

3.4 Translating and Stretching Graphs

Objectives

1. Use Vertical Translations to Graph Functions
2. Use Horizontal Translations to Graph Functions
3. Graph Functions Involving Two Translations
4. Use Reflections About the x- and y-Axes to Graph Functions
5. Use Vertical Stretching and Shrinking to Graph Functions
6. Use Horizontal Stretching and Shrinking to Graph Functions
7. Graph Functions Involving a Combination of Translations and Stretchings

We often can transform the graph of a function into the graph of another function by shifting the graph vertically or horizontally. Also, we can reflect a graph about the x- or y-axis, and stretch or shrink a graph horizontally or vertically to transform the graph of a function into the graph of another function. In this section, we will graph new functions from known ones using these methods.

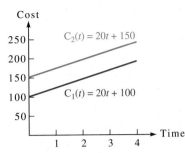

Figure 3-33

Consider a white water rafting trip on the Ocoee River in Tennessee. Suppose one company charges a group of students $20 for each hour on the river, plus $100 for a guide and equipment. The cost of the rafting trip can be represented by the function

$$C_1(t) = 20t + 100$$

where $C_1(t)$ represents the cost in dollars to raft t hours on the river.

If the company increases its charge for the guide and equipment to $150, the new cost function can be represented by

$$C_2(t) = 20t + 150$$

The graphs of the two cost functions are shown in Figure 3-33.

Note that if we shift the graph of $C_1(t)$ 50 units vertically upward, we obtain the graph of $C_2(t)$. This shift is called a *translation*.

As we continue our study of translations, it will be helpful to review the graphs of the basic functions that are shown in Figure 3-8 in Section 3.1. In this section, the graphs of $f(x) = x^2$, $f(x) = x^3$, $f(x) = |x|$, $f(x) = \sqrt{x}$, and $f(x) = \sqrt[3]{x}$ will be translated and stretched in various ways.

1. Use Vertical Translations to Graph Functions

Comment

The graphs of the functions shown in Figure 3-34 are exactly the graphs we would expect, based on our previous study of quadratic functions.

The graphs of functions can be identical except for their positions in the xy-plane. For example, Figure 3-34 shows the graph of $y = x^2 + k$ for three values of k. If $k = 0$, we have the graph of $y = x^2$. The graph of $y = x^2 + 2$ is identical to the graph of $y = x^2$, except that it is shifted 2 units upward. The graph of $y = x^2 - 3$ is identical to the graph of $y = x^2$, except that it is shifted 3 units downward. These shifts are called **vertical translations**.

$y = x^2$		
x	y	(x, y)
-2	4	$(-2, 4)$
-1	1	$(-1, 1)$
0	0	$(0, 0)$
1	1	$(1, 1)$
2	4	$(2, 4)$

$y = x^2 + 2$		
x	y	(x, y)
-2	6	$(-2, 6)$
-1	3	$(-1, 3)$
0	2	$(0, 2)$
1	3	$(1, 3)$
2	6	$(2, 6)$

$y = x^2 - 3$		
x	y	(x, y)
-2	1	$(-2, 1)$
-1	-2	$(-1, -2)$
0	-3	$(0, -3)$
1	-2	$(1, -2)$
2	1	$(2, 1)$

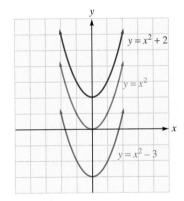

Figure 3-34

In general, we can make the following observations.

Vertical Translations

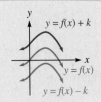

If f is a function and k is a positive number, then

- The graph of $y = f(x) + k$ is identical to the graph of $y = f(x)$, except that it is translated k units up.

- The graph of $y = f(x) - k$ is identical to the graph of $y = f(x)$, except that it is translated k units down.

EXAMPLE 1 Graph each function: **a.** $g(x) = |x| - 2$ **b.** $h(x) = |x| + 3$

Solution We will use vertical translations of $f(x) = |x|$ to graph each function.

a. The graph of $g(x) = |x| - 2$ is identical to the graph of $f(x) = |x|$, except that it is translated 2 units downward. It is translated downward because 2 is subtracted from $|x|$. The graph of $g(x)$ is shown in Figure 3-35(a).

b. The graph of $h(x) = |x| + 3$ is identical to the graph of $f(x) = |x|$, except that it is translated 3 units upward. It is translated upward because 3 is added to $|x|$. The graph of $h(x)$ is shown in Figure 3-35(b).

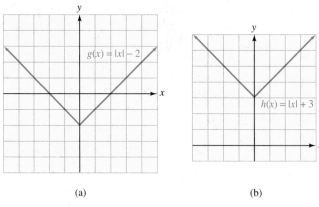

(a)   (b)

Figure 3-35

Self Check 1 *Fill in the blanks.* The graph of $g(x) = x^2 + 3$ is identical to the graph of $f(x) = x^2$, except that it is translated __ units _____.

The graph of $h(x) = x^2 - 4$ is identical to the graph of $f(x) = x^2$, except that it is translated __ units _____.

Comment

The graphs of the functions shown in Figure 3-36 are exactly the graphs we would expect, based on our previous knowledge of quadratic functions.

2. Use Horizontal Translations to Graph Functions

Figure 3-36 shows the graph of $y = (x + h)^2$ for three values of h. If $h = 0$, we have the graph of $y = x^2$. The graph of $y = (x - 2)^2$ is identical to the graph of $y = x^2$, except that it is shifted 2 units to the right. The graph of $y = (x + 3)^2$ is identical to the graph of $y = x^2$, except that it is shifted 3 units to the left. These shifts are called **horizontal translations**.

$y = x^2$			$y = (x - 2)^2$			$y = (x + 3)^2$		
x	y	(x, y)	x	y	(x, y)	x	y	(x, y)
-2	4	$(-2, 4)$	0	4	$(0, 4)$	-5	4	$(-5, 4)$
-1	1	$(-1, 1)$	1	1	$(1, 1)$	-4	1	$(-4, 1)$
0	0	$(0, 0)$	2	0	$(2, 0)$	-3	0	$(-3, 0)$
1	1	$(1, 1)$	3	1	$(3, 1)$	-2	1	$(-2, 1)$
2	4	$(2, 4)$	4	4	$(4, 4)$	-1	4	$(-1, 4)$

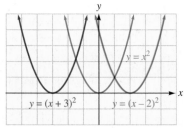

Figure 3-36

In general, we can make the following observations.

Horizontal Translations

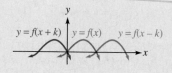

If f is a function and k is a positive number, then

- The graph of $y = f(x - k)$ is identical to the graph of $y = f(x)$, except that it is translated k units to the right.

- The graph of $y = f(x + k)$ is identical to the graph of $y = f(x)$, except that it is translated k units to the left.

EXAMPLE 2 Graph each function: **a.** $g(x) = |x - 4|$ **b.** $h(x) = |x + 2|$

Solution We will use horizontal translations of the graph of $f(x) = |x|$ to graph each function.

a. The graph of $g(x) = |x - 4|$ is identical to the graph of $f(x) = |x|$, except that it is translated 4 units to the right. It is translated 4 units to the right because within the absolute value symbols 4 is subtracted from x. The graph of $g(x)$ is shown in Figure 3-37(a).

b. The graph of $h(x) = |x + 2|$ is identical to the graph of $f(x) = |x|$, except that it is translated 2 units to the left. It is translated 2 units to the left because within the absolute value symbols 2 is added to x. The graph of $h(x)$ is shown in Figure 3-37(b).

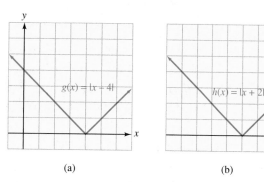

(a)   (b)

Figure 3-37

Comment

When we use horizontal translations to graph a function, it is easy to shift the function in the wrong direction. When we see a positive constant subtracted from x, we have the tendency to shift the graph to the left. This is **incorrect.**

When we see a positive constant added to x, we have the tendency to shift the graph right. This too is **incorrect.** We should avoid making these common errors.

Self Check 2 *Fill in the blanks.* The graph of $g(x) = (x - 3)^2$ is identical to the graph of $f(x) = x^2$, except that it is translated __ units to the ____. The graph of $h(x) = (x + 2)^2$ is identical to the graph of $f(x) = x^2$, except that it is translated __ units to the ____.

3. Graph Functions Involving Two Translations

Sometimes we can obtain a graph by using both a horizontal and a vertical translation.

EXAMPLE 3 Graph each function: **a.** $g(x) = (x - 5)^3 + 4$ **b.** $h(x) = (x + 2)^2 - 2$

Solution By inspection, we see that the function in part **a** involves two translations of $f(x) = x^3$ and the function in part **b** involves two translations of $f(x) = x^2$. We will perform the horizontal translation first, followed by a vertical translation to obtain the graph of each function.

a. The graph of $g(x) = (x - 5)^3 + 4$ is identical to the graph of $f(x) = x^3$, except that it is translated 5 units to the right and 4 units upward, as shown in Figure 3-38(a).

b. The graph of $h(x) = (x + 2)^2 - 2$ is identical to the graph of $f(x) = x^2$, except that it is translated 2 units to the left and 2 units downward, as shown in Figure 3-38(b).

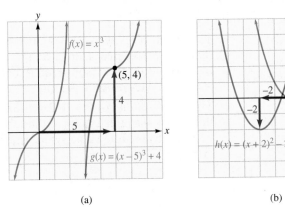

(a)   (b)

Figure 3-38

Self Check 3 *Fill in the blanks.* The graph of $g(x) = |x - 4| + 5$ is identical to the graph of $f(x) = |x|$, except that it is translated __ units to the _____ and __ units _____.

Accent on Technology

Calculator Graphs

We can use a graphing calculator to show the effects of vertical and horizontal translations by graphing $f(x) = x^2$ and $g(x) = (x - 3)^2 + 2$. The graph of $f(x) = x^2$ is a parabola opening upward, with vertex at the origin. The graph of $g(x) = (x - 3)^2 + 2$ should be that same parabola translated 3 units to the right and 2 units upward. Figure 3-39 shows the result of graphing these functions.

After tracing, we see that the vertex of the translated graph is the point (3, 2), as expected. (More zooms and traces will give more accurate results.)

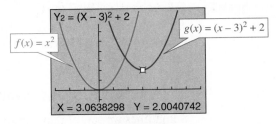

Figure 3-39

4. Use Reflections About the *x*- and *y*-Axes to Graph Functions

Figure 3-40(a) shows that the graph of $y = -\sqrt{x}$ is identical to the graph of $y = \sqrt{x}$, except that it is reflected about the *x*-axis. Figure 3-40(b) shows that the graph of $y = \sqrt{-x}$ is identical to the graph of $y = \sqrt{x}$, except that it is reflected about the *y*-axis.

$y = -\sqrt{x}$

x	y	(x, y)
0	0	$(0, 0)$
1	-1	$(1, -1)$
4	-2	$(4, -2)$

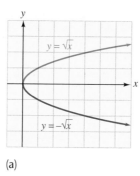

$y = \sqrt{-x}$

x	y	(x, y)
0	0	$(0, 0)$
-1	1	$(-1, 1)$
-4	2	$(-4, 2)$

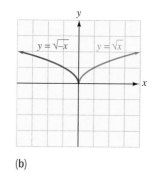

(a) (b)

Figure 3-40

In general, we can make the following observations.

Reflections

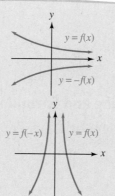

If *f* is a function, then

- The graph of $y = -f(x)$ is identical to the graph of $y = f(x)$, except that it is reflected about the *x*-axis.

- The graph of $y = f(-x)$ is identical to the graph of $y = f(x)$, except that it is reflected about the *y*-axis.

EXAMPLE 4   Graph each function: **a.** $g(x) = -|x + 1|$ **b.** $h(x) = |-x + 1|$

Solution   By inspection, we see that the function given in part **a** involves a reflection about the *x*-axis and the function given in part **b** involves a reflection about the *y*-axis. We will use reflections to draw the graph of each function.

Comment

It is often helpful to think of a reflection as a mirror image of the graph about the *x*- or *y*-axis.

a. The graph of $g(x) = -|x + 1|$ is identical to the graph of $f(x) = |x + 1|$, except that it is reflected about the *x*-axis. This is because $g(x) = -f(x)$. The graphs of both functions are shown in Figure 3-41(a).

b. The graph of $h(x) = |-x + 1|$ is identical to the graph of $f(x) = |x + 1|$, except that it is reflected about the *y*-axis. This is because $f(-x) = h(x)$. The graphs of both functions are shown in Figure 3-41(b).

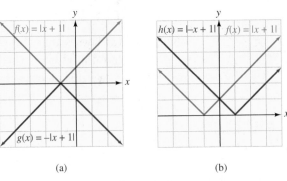

(a)   (b)

Figure 3-41

Self Check 4   *Fill in the blanks.* The graph of $g(x) = -\sqrt[3]{x}$ is identical to the graph of $f(x) = \sqrt[3]{x}$, except that it is reflected about the ___-axis. The graph of $h(x) = \sqrt{-x-4}$ is identical to the graph of $f(x) = \sqrt{x-4}$, except that it is reflected about the ___-axis.

5.   Use Vertical Stretching and Shrinking to Graph Functions

Figure 3-42 shows the graphs of $y = x^2$, $y = 3x^2$, and $y = \frac{1}{3}x^2$.

$y = x^2$

x	y	(x, y)
-2	4	$(-2, 4)$
-1	1	$(-1, 1)$
0	0	$(0, 0)$
1	1	$(1, 1)$
2	4	$(2, 4)$

$y = 3x^2$

x	y	(x, y)
-2	12	$(-2, 12)$
-1	3	$(-1, 3)$
0	0	$(0, 0)$
1	3	$(1, 3)$
2	12	$(2, 12)$

$y = \frac{1}{3}x^2$

x	y	(x, y)
-2	$\frac{4}{3}$	$\left(-2, \frac{4}{3}\right)$
-1	$\frac{1}{3}$	$\left(-1, \frac{1}{3}\right)$
0	0	$(0, 0)$
1	$\frac{1}{3}$	$\left(1, \frac{1}{3}\right)$
2	$\frac{4}{3}$	$\left(2, \frac{4}{3}\right)$

Figure 3-42

Because each value of $y = 3x^2$ is 3 times greater than the corresponding value of $y = x^2$, its graph is stretched vertically by a factor of 3. Because each value of $y = \frac{1}{3}x^2$ is $\frac{1}{3}$ times the corresponding value of $y = x^2$, its graph shrinks vertically by a factor of $\frac{1}{3}$.

In general, we can make the following observations.

Vertical Stretching and Shrinking

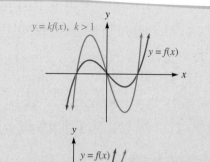

Figure 3-43

If f is a function and $k > 1$, then

- The graph of $y = kf(x)$ can be obtained by stretching the graph of $y = f(x)$ vertically by multiplying each value of $f(x)$ by k.

If f is a function and $0 < k < 1$, then

- The graph of $y = kf(x)$ can be obtained by shrinking the graph of $y = f(x)$ vertically by multiplying each value of $f(x)$ by k.

EXAMPLE 5   Graph each function: **a.** $g(x) = 2|x|$ **b.** $h(x) = \frac{1}{2}|x|$

Solution   We will vertically stretch or vertically shrink the basic function $f(x) = |x|$ to graph each of the given functions.

a. The graph of $g(x) = 2|x|$ is identical to the graph of $f(x) = |x|$, except that it is vertically stretched by a factor of 2. This is because each value of $|x|$ is multiplied by 2. The graphs of both functions are shown in Figure 3-44.

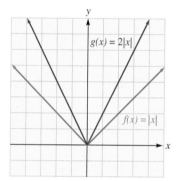

Figure 3-44

b. The graph of $g(x) = \frac{1}{2}|x|$ is identical to the graph of $f(x) = |x|$, except that it is vertically shrunk by a factor of $\frac{1}{2}$. This is because each value of $|x|$ is multiplied by $\frac{1}{2}$. The graphs of both functions are shown in Figure 3-45.

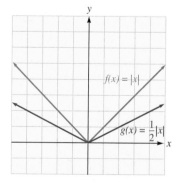

Figure 3-45

Comment
Note that vertically stretching the graph of a function narrows the graph of the function. Vertically shrinking the graph of a function widens the graph.

Self Check 5   *Fill in the blanks.* The graph of $g(x) = 5x^3$ is identical to the graph of $f(x) = x^3$, except that it is vertically _____ by a factor of __. The graph of $h(x) = \frac{1}{5}x^3$ is identical to the graph $f(x) = x^3$, except that it is vertically _____ by a factor of __.

6. Use Horizontal Stretching and Shrinking to Graph Functions

Functions also can be graphed by using horizontal stretchings and shrinkings.

Horizontal Shrinking and Stretching

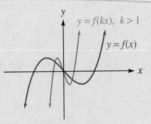

Figure 3-46

If f is a function and $k > 1$, then

- The graph of $y = f(kx)$ can be obtained by shrinking the graph of $y = f(x)$ horizontally by multiplying each x-value of $f(x)$ by $\frac{1}{k}$.

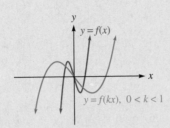

Figure 3-47

If f is a function and $0 < k < 1$, then

- The graph of $y = f(kx)$ can be obtained by stretching the graph of $y = f(x)$ horizontally by multiplying each x-value of $f(x)$ by $\frac{1}{k}$.

EXAMPLE 6 Graph $y = (3x)^2 - 1$ using the graph of the function $y = x^2 - 1$.

Solution Since $3 > 1$, the graph of $y = (3x)^2 - 1$ can be obtained by shrinking the graph of $y = x^2 - 1$ horizontally by dividing each x-coordinate of $y = x^2 - 1$ by 3.

First, we complete the table of solutions for $y = x^2 - 1$, shown in Figure 3-48. Next, we divide each x-value in the table by 3 to obtain the table of solutions for $y = (3x)^2 - 1$, also shown in the figure.

$y = x^2 - 1$

x	y	(x, y)
-2	3	$(-2, 3)$
-1	0	$(-1, 0)$
0	-1	$(0, -1)$
1	0	$(1, 0)$
2	3	$(2, 3)$

$y = (3x)^2 - 1$

x	y	(x, y)
$-\frac{2}{3}$	3	$\left(-\frac{2}{3}, 3\right)$
$-\frac{1}{3}$	0	$\left(-\frac{1}{3}, 0\right)$
0	-1	$(0, -1)$
$\frac{1}{3}$	0	$\left(\frac{1}{3}, 0\right)$
$\frac{2}{3}$	3	$\left(\frac{2}{3}, 3\right)$

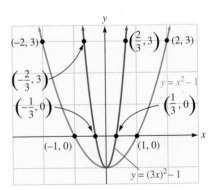

Figure 3-48

We now draw each graph as shown in Figure 3-48.

Self Check 6 *Fill in the blanks.* The graph of $g(x) = \left(\frac{1}{3}x\right)^2 - 1$ is identical to the graph of $f(x) = x^2 - 1$, except that it is horizontally _____ by a factor of __.

Accent on Technology

Calculator Graphs

We can use a graphing calculator to show the effect of a vertical stretching and reflection by graphing $g(x) = x^2 - 4$ and $h(x) = -2(x^2 - 4)$. The graph of $g(x) = x^2 - 4$ is a parabola opening up with vertex at $(0, -4)$. If that graph is stretched vertically by a factor of 2 and then reflected about the x-axis, the graph of $h(x) = -2(x^2 - 4)$ should be the result. That is what happens, as shown in Figure 3-49. Notice that the x-intercepts of the graphs are the same.

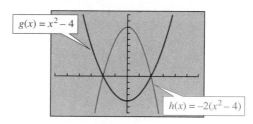

Figure 3-49

We can summarize the ideas in this section as follows.

Translations, Reflections, Stretchings, and Shrinkings

If f is a function and k represents a positive number then

The graph of	***can be obtained by graphing $y = f(x)$ and***
$y = f(x) + k$	translating the graph k units up.
$y = f(x) - k$	translating the graph k units down.
$y = f(x + k)$	translating the graph k units to the left.
$y = f(x - k)$	translating the graph k units to the right.
$y = -f(x)$	reflecting the graph about the x-axis.
$y = f(-x)$	reflecting the graph about the y-axis.
$y = kf(x) \quad k > 1$	stretching the graph vertically by multiplying each value of $f(x)$ by k.
$y = kf(x) \quad 0 < k < 1$	shrinking the graph vertically by multiplying each value $f(x)$ by k.
$y = f(kx) \quad k > 1$	shrinking the graph horizontally by multiplying each x-value of $f(x)$ by $\frac{1}{k}$.
$y = f(kx) \quad 0 < k < 1$	stretching the graph horizontally by multiplying each x-value of $f(x)$ by $\frac{1}{k}$.

EXAMPLE 7 Figure 3-50 shows the graph of $y = f(x)$. Use this graph and a translation to find the graph of: **a.** $y = f(x) + 2$ **b.** $y = f(x - 2)$ **c.** $y = 2f(x)$

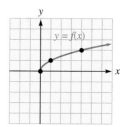

Figure 3-50

Solution We will use the summary of ideas in the section stated above to graph each function.

a. The graph of $y = f(x) + 2$ is identical to the graph of $y = f(x)$, except that it is translated 2 units up. (See Figure 3-51(a).)

b. The graph of $y = f(x - 2)$ is identical to the graph of $y = f(x)$, except that it is translated 2 units to the right. (See Figure 3-51(b).)

c. The graph of $y = 2f(x)$ is identical to the graph of $y = f(x)$, except that it is stretched vertically by multiplying each y-value of $f(x)$ by 2. (See Figure 3-51(c).)

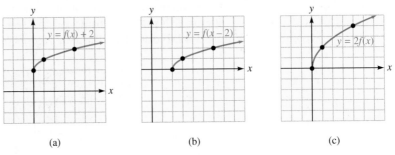

(a) (b) (c)

Figure 3-51

Self Check 7 Use Figure 3-50 and a reflection to find the graph of: **a.** $y = -f(x)$ **b.** $y = f(-x)$

7. Graph Functions Involving a Combination of Translations and Stretchings

To graph functions involving a combination of the translations and stretchings, we must apply each translation or stretching to the function. We usually will perform these translations and stretchings in the following order:

1. horizontal translation

2. stretching or shrinking

3. reflection

4. vertical translation

EXAMPLE 8 Use the function $f(x) = |x|$ to graph $g(x) = 3|x - 2| + 4$.

Solution We will graph $g(x) = 3|x - 2| + 4$ by applying three translations to the basic function $f(x) = |x|$:

  Step 1: Translate $f(x) = |x|$ horizontally 2 units to the right.

  Step 2: Vertically stretch the graph by a factor of 3.

  Step 3: Translate the graph vertically 4 units upward.

Step 1: Translate $f(x) = |x|$ horizontally 2 units to the right to obtain the graph of $y = |x - 2|$.

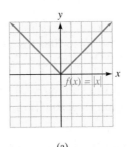 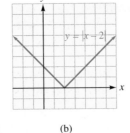

(a) (b)

Figure 3-52

Step 2:   Vertically stretch the graph of $y = |x - 2|$ by a factor of 3 to obtain the graph of $y = 3|x - 2|$.

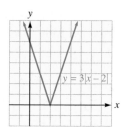

Figure 3-53

Step 3:   Translate the graph of $y = 3|x - 2|$ vertically 4 units upward to obtain the graph of $y = 3|x - 2| + 4$.

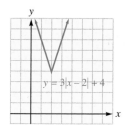

Figure 3-54

Self Check 8   Use the graph of the function $f(x) = x^3$ to graph $g(x) = \frac{1}{3}(x + 1)^3 - 2$.

Self Check Answers   **1.** 3, upward; 4, downward   **2.** 3, right; 2, left   **3.** 4, right; 5, upward
4. x; y **5.** stretched, 5; shrunk, $\frac{1}{5}$ **6.** stretched, 3
7. a.   **b.**   **8.**

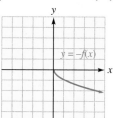 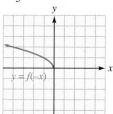 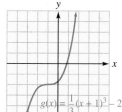

3.4 **Exercises**

Vocabulary and Concepts   *Fill in the blanks.*

1. The graph of $y = f(x) + 5$ is identical to the graph of $y = f(x)$, except that it is translated 5 units ___.

2. The graph of $y =$ _____ is identical to the graph of $y = f(x)$, except that it is translated 7 units down.

3. The graph of $y = f(x - 3)$ is identical to the graph of $y = f(x)$, except that it is translated 3 units _____.

4. The graph of $y = f(x + 2)$ is identical to the graph of $y = f(x)$, except that it is translated 2 units _____.

5. To draw the graph of $y = (x + 2)^2 - 3$, translate the graph of $y = x^2$ by __ units to the left and 3 units _____.

6. To draw the graph of $y = (x - 3)^3 + 1$, translate the graph of $y = x^3$ by 3 units to the _____ and 1 unit ___.

7. The graph of $y = f(-x)$ is a reflection of the graph of $y = f(x)$ about the _____.

8. The graph of _____ is a reflection of the graph of $y = f(x)$ about the x-axis.

9. The graph of $y = f(4x)$ shrinks the graph of $y = f(x)$ _____ by dividing each x-value of $f(x)$ by 4.

10. The graph of $y = 8f(x)$ stretches the graph of $y = f(x)$ _____ by a factor of 8.

Practice *The graph of each function is a translation of the graph of* $f(x) = x^2$. *Graph each function.*

11. $g(x) = x^2 - 2$

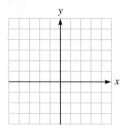

12. $g(x) = (x - 2)^2$

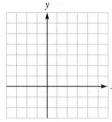

13. $g(x) = (x + 3)^2$

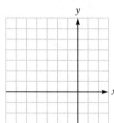

14. $g(x) = x^2 + 3$

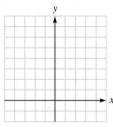

15. $h(x) = (x + 1)^2 + 2$

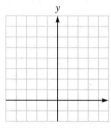

16. $h(x) = (x - 3)^2 - 1$

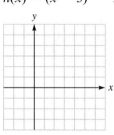

17. $h(x) = \left(x + \dfrac{1}{2}\right)^2 - \dfrac{1}{2}$

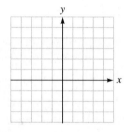

18. $h(x) = \left(x - \dfrac{3}{2}\right)^2 + \dfrac{5}{2}$

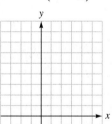

The graph of each function is a translation of the graph of $f(x) = x^3$. *Graph each function.*

19. $g(x) = x^3 + 1$

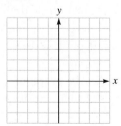

20. $g(x) = x^3 - 3$

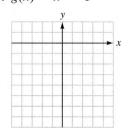

21. $g(x) = (x - 2)^3$

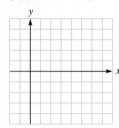

22. $g(x) = (x + 3)^3$

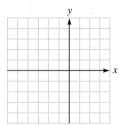

23. $h(x) = (x - 2)^3 - 3$

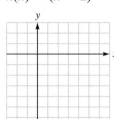

24. $h(x) = (x + 1)^3 + 4$

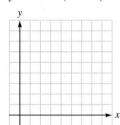

25. $y + 2 = x^3$

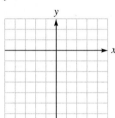

26. $y - 7 = (x - 5)^3$

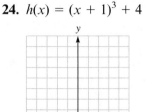

The graph of each function is a translation of the graph of $f(x) = |x|$. *Graph each function.*

27. $g(x) = |x| + 2$

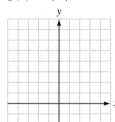

28. $g(x) = |x| - 2$

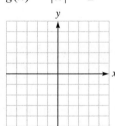

29. $g(x) = |x - 5|$

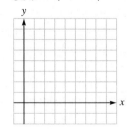

30. $g(x) = |x + 4|$

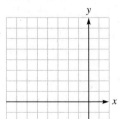

31. $f(x) = |x + 2| - 1$ **32.** $h(x) = |x - 3| + 3$

   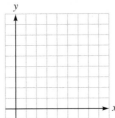

41. $g(x) = \sqrt[3]{x} - 2$ **42.** $g(x) = \sqrt[3]{x} + 5$

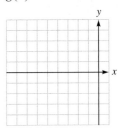   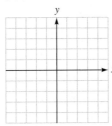

43. $h(x) = \sqrt[3]{x} + 1 - 1$ **44.** $h(x) = \sqrt[3]{x} - 1 - 1$

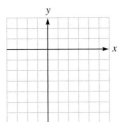   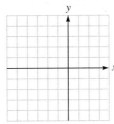

The graph of each function is a translation of the graph of $f(x) = \sqrt{x}$. Graph each function.

33. $g(x) = \sqrt{x} + 1$ **34.** $g(x) = \sqrt{x} - 3$

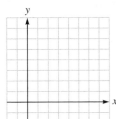   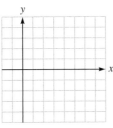

35. $g(x) = \sqrt{x + 2}$ **36.** $g(x) = \sqrt{x - 4}$

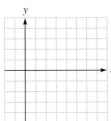   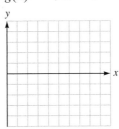

37. $h(x) = \sqrt{x - 2} - 1$ **38.** $h(x) = \sqrt{x + 2} + 3$

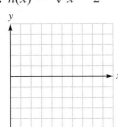   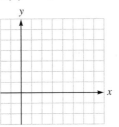

The graph of each function is a reflection of the graph of $y = x^2$, $y = x^3$, $y = |x|$, $y = \sqrt{x}$, or $y = \sqrt[3]{x}$. Graph each function.

45. $f(x) = -x^2$ **46.** $g(x) = (-x)^3$

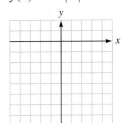   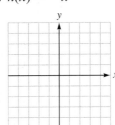

47. $h(x) = -x^3$ **48.** $f(x) = -|x|$

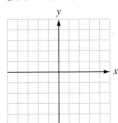   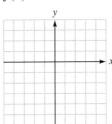

The graph of each function is a translation of the graph of $f(x) = \sqrt[3]{x}$. Graph each function.

39. $g(x) = \sqrt[3]{x} - 4$ **40.** $g(x) = \sqrt[3]{x} + 3$

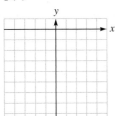   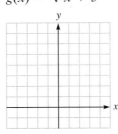

49. $f(x) = -\sqrt{x}$ **50.** $g(x) = \sqrt[3]{-x}$

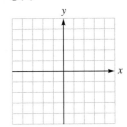   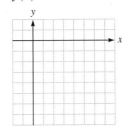

51. $f(x) = |-x|$

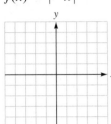

52. $g(x) = (-x)^2$

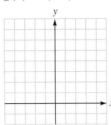

The graph of each function is a vertical stretching or shrinking of the graph of $y = x^2$, $y = x^3$, or $y = |x|$. Graph each function.

53. $f(x) = 2x^2$

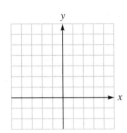

54. $g(x) = \dfrac{1}{2}x^2$

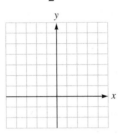

55. $h(x) = -3x^2$

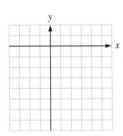

56. $f(x) = -\dfrac{1}{3}x^2$

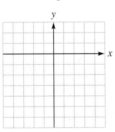

57. $f(x) = \dfrac{1}{2}x^3$

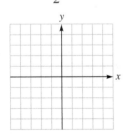

58. $g(x) = 2x^3$

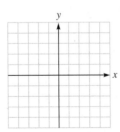

59. $h(x) = -3|x|$

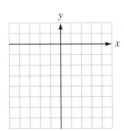

60. $f(x) = \dfrac{1}{3}|x|$

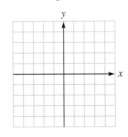

The graph of each function is a horizontal stretching or shrinking of the graph of $y = x^2$ or $y = x^3$. Graph each function.

61. $f(x) = \left(\dfrac{1}{2}x\right)^3$

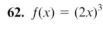

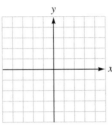

62. $f(x) = (2x)^3$

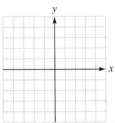

63. $f(x) = (2x)^2$

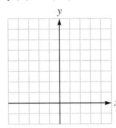

64. $f(x) = (-2x)^3$

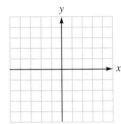

Graph each function using a combination of translations, stretchings, and shrinkings.

65. $g(x) = 3(x + 2)^2 - 1$

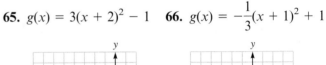

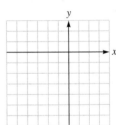

66. $g(x) = -\dfrac{1}{3}(x + 1)^2 + 1$

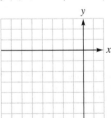

67. $h(x) = -2|x| + 3$

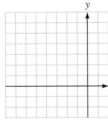

68. $f(x) = -2|x + 3|$

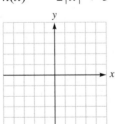

69. $f(x) = 2|x - 2| + 1$

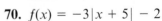

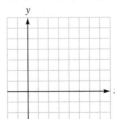

70. $f(x) = -3|x + 5| - 2$

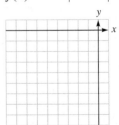

71. $f(x) = 2\sqrt{x} + 3$
$(x \geq 0)$

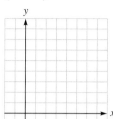

72. $g(x) = 2\sqrt{x + 3}$
$(x \geq -3)$

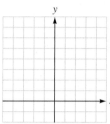

79. $y = f(x) + 1$

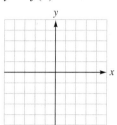

80. $y = f(x + 1)$

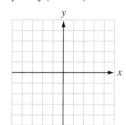

73. $h(x) = 2\sqrt{x - 2} + 1$
$(x \geq 2)$

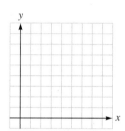

74. $h(x) = \dfrac{1}{2}\sqrt{x + 5} - 2$
$(x \geq -5)$

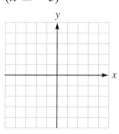

81. $y = 2f(x)$

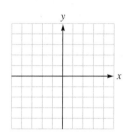

82. $y = f\left(\dfrac{x}{2}\right)$

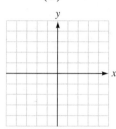

83. $y = f(x - 2) + 1$

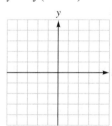

84. $y = -f(x)$

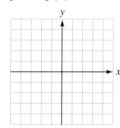

75. $g(x) = -2(x + 2)^3 - 1$

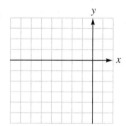

76. $g(x) = \dfrac{1}{3}(x + 1)^3 - 1$

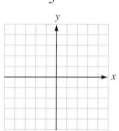

85. $y = 2f(-x)$

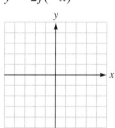

86. $y = f(x + 1) - 2$

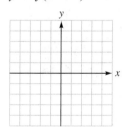

77. $f(x) = 2\sqrt[3]{x} + 4$

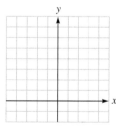

78. $f(x) = -2\sqrt[3]{x} + 1$

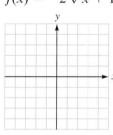

Use the following graph and a translation, stretching, or reflection to find the graph of each function.

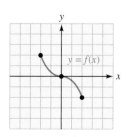
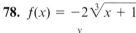

Discovery and Writing *Use a graphing calculator to perform each experiment. Write a brief paragraph describing your findings.*

87. Investigate the translations of the graph of a function by graphing the parabola $y = (x - k)^2 + k$ for several values of k. What do you observe about successive positions of the vertex?

88. Investigate the translations of the graph of a function by graphing the parabola $y = (x - k)^2 + k^2$ for several values of k. What do you observe about successive positions of the vertex?

89. Investigate the horizontal stretching of the graph of a function by graphing $y = \sqrt{ax}$ for several values of a. What do you observe?

90. Investigate the vertical stretching of the graph of a function by graphing $y = b\sqrt{x}$ for several values of b. What do you observe? Are these graphs different from the graphs in Exercise 89?

Write a paragraph using your own words.

91. Explain why the effect of vertically stretching a graph by a factor of -1 is to reflect the graph in the x-axis.

92. Explain why the effect of horizontally stretching a graph by a factor of -1 is to reflect the graph in the y-axis.

Review *Simplify each function.*

93. $\dfrac{x^2 + x - 6}{x^2 + 5x + 6}$

94. $\dfrac{2x^2 + 3x}{2x^2 + x - 3}$

Find the domain of each function.

95. $f(x) = \dfrac{x + 7}{x - 3}$

96. $f(x) = \dfrac{x^2 + 1}{x^2 + 3x + 2}$

Perform each division and write the answer in quotient $+ \frac{remainder}{divisor}$ form.

97. $\dfrac{x^2 + 3x}{x + 1}$

98. $\dfrac{x^2 + 3}{x + 1}$

3.5 Rational Functions

Objectives

1. Find the Domain of a Rational Function
2. Understand the Characteristics of Rational Functions and Their Graphs
3. Find Vertical Asymptotes of Rational Functions
4. Find Horizontal Asymptotes of Rational Functions
5. Identify Slant Asymptotes
6. Graph Rational Functions
7. Understand When a Graph Has a Hole or Missing Point
8. Solve Problems Modeled by Rational Functions

Nick Laham/Getty Images

We have discussed polynomial functions and now focus on another class of functions called **rational functions.** Rational functions are defined by rational expressions that are quotients of polynomials.

For example, consider the time t it takes a NASCAR driver to drive the 500 miles of the Daytona 500 race. The time t can be defined as a function of the average rate of the driver's speed r. That is

$$t = f(r) = \frac{500}{r}$$

If a driver averages a speed of 170 mph, we can evaluate $f(170)$ to determine the time in hours driven.

$$f(170) = \frac{500}{170} = \frac{50}{17} \quad \text{This is approximately 2.94 hours.}$$

We will discuss this type of function in this section.

Rational Functions   A **rational function** is a function defined by an equation of the form

$$y = \frac{P(x)}{Q(x)}$$

where $P(x)$ and $Q(x)$ are polynomials and $Q(x) \neq 0$.

1.   Find the Domain of a Rational Function

Because rational functions are quotients of polynomials and $Q(x)$ is the denominator of a fraction, $Q(x)$ cannot equal 0. Thus, the domain of a rational function must exclude all values of x for which $Q(x) = 0$.

Here are some examples of rational functions and their domains:

Function	Domain
$f(x) = \dfrac{3}{x - 2}$	$(-\infty, 2) \cup (2, \infty)$, x cannot equal 2
$f(x) = \dfrac{5x + 2}{x^2 - 4}$	$(-\infty, -2) \cup (-2, 2) \cup (2, \infty)$, x cannot equal 2 or -2
$f(x) = \dfrac{2x}{x^2 + 3}$	$(-\infty, \infty)$, x can equal any real number

EXAMPLE 1   Find the domain of $f(x) = \dfrac{3x + 2}{x^2 - 7x + 12}$.

Solution   We can factor the denominator to see what values of x will give 0's in the denominator. These values are not in the domain.

To find the numbers x that make the denominator 0, we set $x^2 - 7x - 12$ equal to 0 and solve for x.

$$x^2 - 7x + 12 = 0$$
$$(x - 4)(x - 3) = 0 \qquad \text{Factor } x^2 - 7x + 12.$$
$$x - 4 = 0 \quad \text{or} \quad x - 3 = 0 \qquad \text{Set each factor equal to 0.}$$
$$x = 4 \mid \qquad x = 3 \qquad \text{Solve each linear equation.}$$

Since 4 and 3 make the denominator 0, the domain is the set of all real numbers except $x = 4$ and $x = 3$. In interval notation, we have $(-\infty, 3) \cup (3, 4) \cup (4, \infty)$.

Self Check 1   Find the domain of $f(x) = \dfrac{2x - 3}{x^2 - x - 2}$.

Calculator Graphs

We can use graphing calculators to find domains and ranges of rational functions. If we use settings of $[-10, 10]$ for x and $[-10, 10]$ for y and graph $f(x) = \frac{2x + 1}{x - 1}$, we will obtain Figure 3-55.

From the graph, we can see that every real number x except 1 gives a value of y. Thus, the domain of the function is $(-\infty, 1) \cup (1, \infty)$. We also can see that y can be any value except 2. The range of the function is $(-\infty, 2) \cup (2, \infty)$.

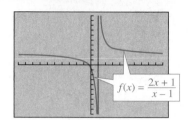

Figure 3-55

2. Understand the Characteristics of Rational Functions and Their Graphs

Consider the function $t = f(r) = \frac{500}{r}$ given at the beginning of the section. A graph of this function is shown in Figure 3-56.

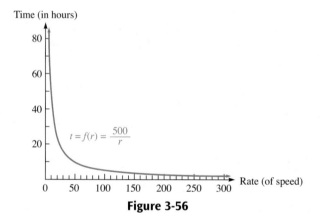

Figure 3-56

We see from the graph that as the rate of speed increases, the time it takes to complete the race decreases. In fact, if we drive at rocket speed, we will arrive in almost no time at all. We can express this by saying that "as the rate increases without bound (or approaches ∞), the time it takes to complete the race approaches 0 hours." When a graph approaches a line as shown in the figure, we call the line an **asymptote**. The horizontal line representing the rate axis shown in the graph is a **horizontal asymptote.**

We also see from the graph that as the rate of speed decreases, the time it takes to complete the race increases. In fact, if the car goes at turtle speed, it will take almost forever to finish the race. We can express this by saying that "as the rate gets slower and slower (or approaches 0 mph), the time approaches ∞." The vertical line representing the time axis shown on the graph is a **vertical asymptote.** A vertical asymptote is a vertical line that the graph approaches, but never touches.

Vertical Asymptote

The line $x = a$ is a **vertical asymptote** of the graph of a function $y = f(x)$ if $f(x)$ either increases or decreases without bound (approaches ∞ or $-\infty$) as x approaches a.

| **Horizontal Asymptote** | The line $y = b$ is a **horizontal asymptote** of the graph of a function $y = f(x)$ if $f(x)$ approaches b as x increases or decreases without bound (approaches ∞ or $-\infty$). |

Figure 3-57 shows typical vertical and horizontal asymptotes.

Vertical asymptote at x = a

$f(x)$ approaches ∞ as x approaches a from left.

$f(x)$ approaches $-\infty$ as x approaches a from right.

Horizontal asymptote at y = b

$f(x)$ approaches b as x approaches ∞ or as x approaches $-\infty$.

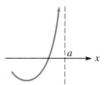
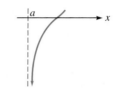
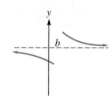

Figure 3-57

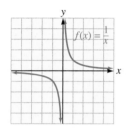

Figure 3-58

The graph of the rational function $f(x) = \frac{1}{x}$, called the **reciprocal function,** is shown in Figure 3-58. The domain of the function is $(-\infty, 0) \cup (0, \infty)$ and the range is $(-\infty, 0) \cup (0, \infty)$.

The reciprocal function has a vertical asymptote of $x = 0$ because:

• We see from the graph that as x approaches 0 from the right, y or $f(x)$ approaches ∞.

• We see from the graph that as x approaches 0 from the left, y or $f(x)$ approaches $-\infty$.

The reciprocal function has a horizontal asymptote of $y = 0$ because:

• We see from the graph that as x approaches ∞, y or $f(x)$ approaches 0.

• We see from the graph that as x approaches $-\infty$, y or $f(x)$ approaches 0.

3.   Find Vertical Asymptotes of Rational Functions

To find the vertical asymptotes of a rational function written in simplest form, we must find the values of x for which the denominator of the rational function is 0 and the function is undefined. For example, since the denominator of $f(x) = \frac{2x - 1}{x + 2}$ is 0 when $x = -2$, there are no corresponding values of y and the line $x = -2$ is a vertical asymptote. We note that when x approaches -2 from the right or from the left, $f(x)$ approaches $-\infty$ and ∞ respectively. A calculator graph of the function appears in Figure 3-59.

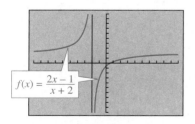

Figure 3-59

Comment

Graphing calculators do not graph asymptotes. When viewing a calculator graph of a rational function, we may see a line on the screen that usually is interpreted incorrectly as being a vertical asymptote. These lines occur because a graphing calculator graphs by connecting dots whose x-coordinates are close together. When two points straddle a vertical asymptote and their y-coordinates are far apart, the calculator draws a steep line that appears to be vertical.

Locating Vertical Asymptotes

To locate the vertical asymptotes of a rational function, $f(x) = \frac{P(x)}{Q(x)}$, we follow these steps:

Step 1: Factor $P(x)$ and $Q(x)$ and remove any common factors.

Step 2: Set the denominator equal to 0 and solve the equation.

If a is a solution of the equation found in Step 2, $x = a$ is a vertical asymptote.

EXAMPLE 2   Find the vertical asymptotes, if any, of each function:

a. $f(x) = \dfrac{2x}{x^2 - 16}$ **b.** $g(x) = \dfrac{x - 4}{x^2 - 16}$ **c.** $h(x) = \dfrac{5x}{x^2 + 16}$

Solution   We will locate the vertical asymptotes of the graph of each function by factoring its numerator and/or denominator and removing any common factors. Then we will set the resulting denominator equal to zero, solve the equation, and identify the vertical asymptotes.

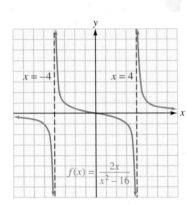

a. $f(x) = \dfrac{2x}{x^2 - 16}$

$f(x) = \dfrac{2x}{(x + 4)(x - 4)}$ Factor the denominator completely.

We then set the denominator equal to 0 and solve for x.

$$(x + 4)(x - 4) = 0$$
$$x + 4 = 0 \quad \text{or} \quad x - 4 = 0$$
$$x = -4 \quad | \quad x = 4$$

Figure 3-60

The vertical asymptotes are $x = 4$ and $x = -4$. (See Figure 3-60.)

b. $g(x) = \dfrac{x - 4}{x^2 - 16}$

$g(x) = \dfrac{x - 4}{(x + 4)(x - 4)}$ Factor the denominator completely.

$g(x) = \dfrac{1}{x + 4}$ Simplify: $\dfrac{x - 4}{x - 4} = 1$.

We set the denominator equal to 0 and solve for x.

$$x + 4 = 0$$
$$x = -4$$

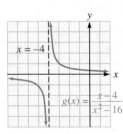

Figure 3-61

The vertical asymptote is $x = -4$. (See Figure 3-61.)

c. $h(x) = \dfrac{5x}{x^2 + 16}$

Comment

A rational function can have zero, one, or several vertical asymptotes.

Because the denominator is the sum of two squares without a common factor, it cannot be factored over the set of real numbers. Thus, no values of x will make the denominator equal to 0. Thus, the rational function has no vertical asymptotes. (See Figure 3-62.)

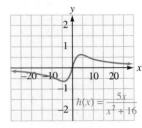

Figure 3-62

Self Check 2   Find the vertical asymptotes, if any, of each function.

     a. $g(x) = \dfrac{x + 5}{x^2 - 25}$ **b.** $h(x) = \dfrac{5}{x^2 + 25}$

4. Find Horizontal Asymptotes of Rational Functions

Comment

The graph of a rational function can cross horizontal and slant asymptotes but can never cross a vertical asymptote.

To find the horizontal asymptote of a rational function, we must find the value of y that the function approaches as x approaches ∞ or $-\infty$. To illustrate, we will consider three functions and some given function values.

First we consider a function where the degree of the numerator is less than the degree of the denominator.

$$f(x) = \frac{x + 2}{x^2 - 1} \qquad f(999) \approx 0.001 \qquad f(-999) \approx -0.001$$

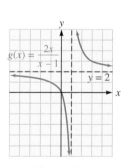

Figure 3-63

- The degree of the numerator is 1 and the degree of the denominator is 2. Since $1 < 2$, the degree of the numerator is less than the degree of the denominator.
- If f is evaluated at large x-values (in both the positive and negative directions), such as 999 and -999, we obtain y-values that are close to 0.
- **Conclusion:** The line $y = 0$ (the x-axis) is a horizontal asymptote. (See Figure 3-63.)

Secondly we consider a function where the degree of the numerator is equal to the degree of the denominator.

$$g(x) = \frac{2x}{x - 1} \qquad g(999) = 2.002 \qquad g(-999) = 1.998$$

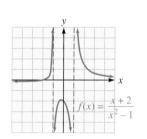

Figure 3-64

- The degree of the numerator is 1 and the degree of the denominator is 1. Thus, the degrees of the numerator and denominator are equal.
- The leading coefficient of the numerator is 2 and the leading coefficient of the denominator is 1. Note that $2 \div 1$ is 2.
- If g is evaluated at large x-values (in both the positive and negative directions), such as 999 and -999, we obtain y-values that are close to 2.
- **Conclusion:** The line $y = 2$ is a horizontal asymptote. (See Figure 3-64.)

Finally we consider a function where the degree of the numerator is greater than the degree of the denominator.

$$h(x) = \frac{x^3}{x - 1} \qquad h(999) \approx 999{,}001 \qquad h(-999) \approx 997{,}003$$

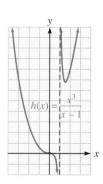

Figure 3-65

- The degree of the numerator is 3 and the degree of the denominator is 1. Thus, the degree of the numerator is greater than the degree of the denominator.
- If h is evaluated at large x-values (in both the positive and negative directions), such as 999 and -999, we obtain y-values that are extremely large and do not seem to approach a finite number.
- **Conclusion:** There is no horizontal asymptote. (See Figure 3-65.)

We summarize the conclusions made in the previous examples.

Locating Horizontal Asymptotes

To locate the horizontal asymptote of a rational function $f(x) = \frac{P(x)}{Q(x)}$, we consider three cases:

Case 1: If the degree of $P(x)$ *is less than* the degree of $Q(x)$, the line $y = 0$ is a horizontal asymptote.

Case 2: If the degree of $P(x)$ and $Q(x)$ are equal, the line $y = \frac{p}{q}$, where p and q are the leading coefficients of $P(x)$ and $Q(x)$, is a horizontal asymptote.

Case 3: If the degree of $P(x)$ is greater than the degree of $Q(x)$, there is no horizontal asymptote.

EXAMPLE 3   Find the horizontal asymptote, if any, of each function:

a. $f(x) = \dfrac{3x}{2x^2 - 1}$ **b.** $g(x) = \dfrac{3x^2 + 1}{x^2 - 2x + 1}$ **c.** $h(x) = \dfrac{x^3 + 2}{x - 5}$

Solution   In each part, we will locate the horizontal asymptote by comparing the degree of the numerator to the degree of the denominator. Then we will decide which case applies, and identify the horizontal asymptote.

a. Because the degree of the numerator is 1, the degree of the denominator is 2, and $1 < 2$, Case 1 applies. The horizontal asymptote is the line $y = 0$.

b. The degree of the numerator is 2 and the degree of the denominator is 2. Since the degrees are the same, Case 2 applies.

Because the leading coefficient of the numerator is 3 and the leading coefficient of the denominator is 1, we divide 3 by 1 to obtain the horizontal asymptote: $\frac{3}{1} = 3$. The horizontal asymptote is the line $y = 3$.

c. Because the degree of the numerator is 3, the degree of the denominator is 1, and $3 > 1$, Case 3 applies. There is no horizontal asymptote.

Self Check 3   Find the horizontal asymptotes, if any, of each function.

a. $g(x) = \dfrac{4x - 5}{5 - x}$ **b.** $h(x) = \dfrac{3x^2}{x^3 - 5}$

5.   Identify Slant Asymptotes

A third type of asymptote is called a **slant asymptote.** These asymptotes occur when the degree of the numerator of a rational function is one more than the degree of the denominator. As the name implies, it is a slanted line, neither vertical nor horizontal.

To illustrate a slant asymptote, we consider the graph of $f(x) = \dfrac{x^2}{x - 2}$ shown in Figure 3-66.

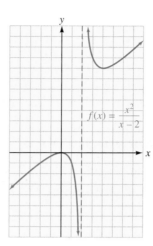

Figure 3-66

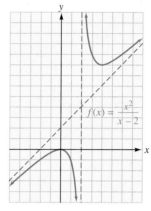

Figure 3-67

When x increases without bound to the right and to the left, the graph of the rational function approaches the slant asymptote shown in Figure 3-67. The equation of the slant asymptote is $y = x + 2$. To find this equation, we perform a long division, write the result in quotient $+ \frac{\text{remainder}}{\text{divisor}}$ form, and ignore the remainder.

$$
\begin{array}{r}
x + 2 \\
x - 2 \overline{\smash{)} x^2 + 0x + 0} \\
\underline{x^2 - 2x} \\
2x + 0 \\
\underline{2x - 4} \\
4
\end{array}
$$

Thus, $y = \frac{x^2}{x - 2} = x + 2 + \frac{4}{x - 2}$. Because the last fraction approaches 0 as x approaches ∞ and $-\infty$, the equation of the slant asymptote is $y = x + 2$.

Locating Slant
Asymptotes

If the degree of $P(x)$ is 1 greater than the degree of $Q(x)$ for the rational function $f(x) = \frac{P(x)}{Q(x)}$, there is a slant asymptote. To find it, divide $P(x)$ by $Q(x)$ and ignore the remainder.

EXAMPLE 4   Find the slant asymptote of $y = f(x) = \dfrac{3x^3 + 2x^2 + 2}{x^2 - 1}$.

Solution   To find the horizontal asymptote, we divide the numerator by the denominator, write the result in quotient $+ \frac{\text{remainder}}{\text{divisor}}$ form, and ignore the remainder.

$$
\begin{array}{r}
3x + 2 \\
x^2 - 1 \overline{\smash{)} 3x^3 + 2x^2 \qquad + 2} \\
\underline{3x^3 \qquad - 3x} \\
2x^2 + 3x + 2 \\
\underline{2x^2 \qquad - 2} \\
3x + 4
\end{array}
$$

Thus,

$$y = \frac{3x^3 + 2x^2 + 2}{x^2 - 1} = 3x + 2 + \frac{3x + 4}{x^2 - 1}$$

The last fraction approaches 0 as x approaches ∞ and $-\infty$. Thus, the graph of the rational function approaches the slant asymptote with equation of $y = 3x + 2$.

Self Check 4   Find the slant asymptote of $f(x) = \dfrac{2x^3 - 3x + 1}{x^2 - 4}$.

6.   Graph Rational Functions

We will use the following strategy to graph the rational function $f(x) = \frac{P(x)}{Q(x)}$, where $P(x)$ and $Q(x)$ are polynomials written in descending powers of x and $\frac{P(x)}{Q(x)}$ is in simplest form (no common factors).

Strategy for Graphing Rational Functions of the Form $\dfrac{P(x)}{Q(x)}$

Step 1:   Check for symmetry.

- If $f(-x) = f(x)$, the graph is symmetric about the y-axis.
- If $f(-x) = -f(x)$, the graph is symmetric about the origin.

Step 2:   Find the vertical asymptotes.

- If there are no common factors of $P(x)$ and $Q(x)$, the real roots of $Q(x) = 0$, if any, determine the vertical asymptotes of the graph.

Step 3:   Find the y- and x-intercepts, if any.

- $f(0)$ is the y-coordinate of the y-intercept of the graph.
- Set $P(x) = 0$ and solve to find the x-intercepts of the graph.

Step 4:   Find the horizontal asymptotes, if any.

- If the degree of $P(x)$ is less than the degree of $Q(x)$, the line $y = 0$ is a horizontal asymptote.
- If the degree of $P(x)$ is equal to the degree of $Q(x)$, the graph of $y = \frac{p}{q}$ is a horizontal asymptote, where p and q are the leading coefficients of $P(x)$ and $Q(x)$.

Step 5:   Find the slant asymptotes, if any.

- If the degree of $P(x)$ is 1 greater than the degree of $Q(x)$, there is a slant asymptote. To find it, divide $P(x)$ by $Q(x)$ and ignore the remainder.

Step 6:   Draw the graph.   Find additional points (if necessary) near the asymptotes.

EXAMPLE 5   Graph: $y = f(x) = \dfrac{x^2 - 4}{x^2 - 1}$.

Solution   We will use the steps outlined above to graph the rational function.

Step 1:   Symmetry   Because x is raised to even powers only, $f(-x) = f(x)$ and there is symmetry about the y-axis. There is no symmetry about the origin.

Step 2:   Vertical asymptotes   To find the vertical asymptotes, we factor the numerator and denominator of $f(x)$ and simplify, if possible.

$$f(x) = \frac{x^2 - 4}{x^2 - 1} = \frac{(x + 2)(x - 2)}{(x + 1)(x - 1)} \qquad \text{There are no common factors.}$$

We then set the denominator equal to 0 and solve for x.

$$(x + 1)(x - 1) = 0$$
$$x + 1 = 0 \quad \text{or} \quad x - 1 = 0$$
$$x = -1 \quad | \quad x = 1$$

There will be vertical asymptotes at $x = -1$ and $x = 1$.

Step 3: y- and x-intercepts   We can find the y-intercept by finding $f(0)$.

$$f(0) = \frac{0^2 - 4}{0^2 - 1} = \frac{-4}{-1} = 4$$

The y-intercept is $(0, 4)$.

We can find the x-intercepts by setting the numerator equal to 0 and solving for x:

$$x^2 - 4 = 0$$
$$(x + 2)(x - 2) = 0$$
$$x + 2 = 0 \quad \text{or} \quad x - 2 = 0$$
$$x = -2 \quad | \quad x = 2$$

The x-intercepts are $(2, 0)$ and $(-2, 0)$.

Step 4:   Horizontal asymptotes   Since the degrees of the numerator and denominator of the polynomials are the same, the line

$$y = \frac{1}{1} = 1 \qquad \begin{array}{l} \text{The leading coefficient of the numerator is 1.} \\ \text{The leading coefficient of the denominator is 1.} \end{array}$$

is a horizontal asymptote, whose equation is $y = 1$.

Step 5:   Slant asymptotes   Since the degree of the numerator is not 1 greater than the degree of the denominator, there are no slant asymptotes.

Step 6:   Graph   We plot the intercepts, draw the asymptotes, and make use of symmetry to graph the rational function. The graph is shown in Figure 3-68.

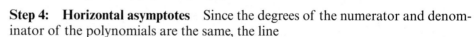

Figure 3-68

Comment

When using a graphing calculator to graph a rational function, make sure that the function is entered properly. To avoid making an error, place both the numerator and denominator in parentheses.

Self Check 5   Use a graphing calculator to graph the function in Example 5.

EXAMPLE 6   Graph: $y = f(x) = \dfrac{3x}{x - 2}$.

Solution   We will use the steps outlined earlier to graph the rational function.

Step 1:   Symmetry   We find $f(-x)$.

$$f(-x) = \frac{3(-x)}{(-x) - 2} = \frac{-3x}{-x - 2} = \frac{3x}{x + 2}$$

Because $f(-x) \neq f(x)$ and $f(-x) \neq -f(x)$, there is no symmetry about the y-axis or the origin.

Step 2:   Vertical asymptotes   We first note that $f(x)$ is in simplest form. We then set the denominator equal to 0 and solve for x. Since the solution is 2, there will be a vertical asymptote at $x = 2$.

Step 3: y- and x-intercepts   We can find the y-intercept by finding $f(0)$.

$$f(0) = \frac{3(0)}{(0) - 2} = \frac{0}{-2} = 0$$

The y-intercept is $(0, 0)$.

We can find the x-intercepts by setting the numerator equal to 0 and solving for x:

$$3x = 0$$
$$x = 0$$

The x-intercept is $(0, 0)$.

Step 4: Horizontal asymptotes Since the degrees of the numerator and denominator of the polynomials are the same, the line

$$y = \frac{3}{1} = 3$$

The leading coefficient of the numerator is 3.

The leading coefficient of the denominator is 1.

is a horizontal asymptote, whose equation is $y = 3$.

Step 5: Slant asymptotes Since the degree of the numerator is not 1 greater than the degree of the denominator, there are no slant asymptotes.

Step 6: Graph First, we plot the intercept $(0, 0)$ and draw the asymptotes. We then find additional points on our graph to see what happens when x is greater than 2. To do so, we choose 3, a value of x that is greater than 2, and evaluate $f(3)$.

$$f(3) = \frac{3(3)}{(3) - 2} = \frac{9}{1} = 9$$

Since $f(3) = 9$, the point $(3, 9)$ lies on our graph. We sketch the graph as shown in Figure 3-69.

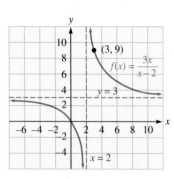

Figure 3-69

Self Check 6 Use a graphing calculator to graph the function in Example 6.

Everyday Connections

Genetic Variation and Human Evolution
Relationship between genetic and geographic distance.

The genetic distance of population pairs measured by F_{ST} is a function of geographic distance between the pairs.

The past decade of advances in molecular genetic technology has heralded a new era for all evolutionary studies, but especially the science of human evolution. Data on various kinds of DNA variation in human populations have accumulated rapidly. There is increasing recognition of the importance of this variation for medicine and developmental biology, and for understanding the history of our species.

1. Write the equation of the horizontal asymptote for the African population.

2. What is the genetic distance of an American population pair whose geographic distance from each other is 3,000 miles?

Source: http:www.nature.com/ng/journal/v33/n3s/full/ng1113.html

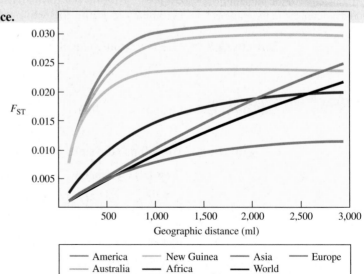

Only samples from indigenous people show high asymptotes. Asia and the world do not show asymptote within the range shown.

EXAMPLE 7 Graph the function: $y = \dfrac{1}{x(x-1)^2}$.

Solution We will use the steps outlined earlier to graph the rational function.

Step 1: Symmetry We find $f(-x)$.

$$f(x) = \frac{1}{x(x-1)^2}$$

$$f(-x) = \frac{1}{(-x)(-x-1)^2} \qquad \text{Replace } x \text{ with } -x.$$

$$= \frac{1}{(-x)[(-1)(x+1)]^2} \qquad \text{Factor out } -1.$$

$$= \frac{1}{(-x)(-1)^2(x+1)^2} \qquad \text{Square } -1 \text{ and } x+1.$$

$$= \frac{1}{(-x)(x+1)^2}$$

$$= \frac{-1}{x(x+1)^2} \qquad \text{Simplify.}$$

Because $f(-x) \neq f(x)$ and $f(-x) \neq -f(x)$, there is no symmetry about the y-axis or origin.

Step 2: Vertical asymptotes We set the denominator equal to 0 and solve for x. Since 0 and 1 make the denominator 0, the vertical asymptotes are $x = 0$ and $x = 1$.

Step 3: y- and x-intercepts Since x cannot be 0, the graph has no y-intercept. Since the numerator cannot be 0, y cannot be 0, and the graph has no x-intercepts.

Step 4: Horizontal asymptotes Since the degree of the numerator is 0 and the degree of the denominator is 1, and $0 < 1$, we conclude that the horizontal asymptote is the line $y = 0$.

Step 5: Slant asymptotes There are no slant asymptotes because the degree of the numerator isn't one greater than the degree of the denominator.

Step 6: Graph Because there are no intercepts to plot, we draw the asymptotes, and find a few additional points to plot. The table gives three points lying in different intervals separated by the asymptotes. We sketch the graph as shown in Figure 3-70.

$$y = \frac{1}{x(x-1)^2}$$

x	y	(x, y)
-1	$-\frac{1}{4}$	$\left(-1, -\frac{1}{4}\right)$
$\frac{1}{2}$	8	$\left(\frac{1}{2}, 8\right)$
2	$\frac{1}{2}$	$\left(2, \frac{1}{2}\right)$

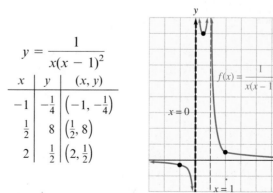

Figure 3-70

Self Check 7 Use a graphing calculator to graph the function in Example 7.

EXAMPLE 8 Graph: $y = f(x) = \dfrac{1}{x^2 + 1}$.

Solution We will use the steps outlined earlier to graph the rational function.

Step 1: Symmetry Because x is raised to only even powers and $f(-x) = f(x)$, there is symmetry about the y-axis. There is no symmetry about the origin.

Step 2: Vertical asymptotes Since no number x makes the denominator 0, the graph has no vertical asymptotes.

Step 3: y- and x-intercepts We can find the y-intercept by finding $f(0)$.

$$f(0) = \frac{1}{0^2 + 1} = \frac{1}{1} = 1$$

The y-intercept is $(0, 1)$.

Because the denominator and numerator are always positive, the fraction is always positive and the graph lies entirely above the x-axis. There are no x-intercepts. We also see that if we set the numerator equal to 0, there are no solutions for x.

Step 4: Horizontal asymptotes Since the degree of the numerator is less than the degree of the denominator, the line $y = 0$ is a horizontal asymptote.

Step 5: Slant asymptotes There are no slant asymptotes because the degree of the numerator isn't one greater than the degree of the denominator.

Step 6: Graph We first plot the y-intercept $(0, 1)$. Then we make use of the fact that the graph has y-axis symmetry and a horizontal asymptote of $y = 0$, and sketch the graph. The graph is shown in Figure 3-71.

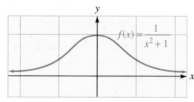

Figure 3-71

Self Check 8 Use a graphing calculator to graph the function in Example 8.

EXAMPLE 9 Graph: $y = f(x) = \dfrac{x^2 + x - 2}{x - 3}$.

Solution We first factor the numerator of the expression $y = \dfrac{(x - 1)(x + 2)}{x - 3}$ and use the steps outlined earlier to graph the rational function.

Step 1: Symmetry We find $f(-x)$.

$$f(x) = \frac{x^2 + x - 2}{x - 3}$$

$$f(-x) = \frac{(-x)^2 + (-x) - 2}{(-x) - 3} \qquad \text{Replace } x \text{ with } -x.$$

$$= \frac{x^2 - x - 2}{-x - 3} \qquad \text{Simplify.}$$

$$= \frac{x^2 - x - 2}{(-1)(x + 3)} \qquad \text{Factor } -1 \text{ out of the denominator.}$$

$$= \frac{-x^2 + x + 2}{x + 3} \qquad \text{Divide by } -1.$$

Because $f(-x) \neq f(x)$ and $f(x) \neq -f(x)$, there is no symmetry about the y-axis or origin.

Step 2:   Vertical asymptotes   We set the denominator equal to 0 and solve for x. Since 3 makes the denominator 0, the vertical asymptote is $x = 3$.

Step 3: y- and x-intercepts   The y-intercept is $\left(0, \frac{2}{3}\right)$ because $f(0) = \frac{2}{3}$. The x-intercepts are $(1, 0)$ and $(-2, 0)$ because the x-values of 1 and -2 make the numerator 0.

Step 4:   Horizontal asymptotes   Since the degree of the numerator is 2, the degree of the denominator is 1, and $2 > 1$, we conclude that there is no horizontal asymptote.

Step 5:   Slant asymptotes   Because the degree of the numerator is 1 greater than the degree of the denominator, this graph will have a slant asymptote. To find it, we perform a long division.

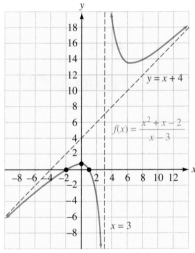

$$\begin{array}{r} x + 4 \\ x - 3 \overline{)\, x^2 + x - 2} \\ \underline{x^2 - 3x} \\ 4x - 2 \\ \underline{2x - 12} \\ 10 \end{array}$$

We write the function as

$$y = \frac{x^2 + x - 2}{x - 3} = x + 4 + \frac{10}{x - 3}$$

The fraction $\frac{10}{x - 3}$ approaches 0 as x approaches ∞ or $-\infty$, and the graph approaches a slant asymptote: the line $y = x + 4$.

Figure 3-72

Step 6:   Graph   The graph appears in Figure 3-72.

Self Check 9   Use a graphing calculator to graph the function in Example 9.

7.   Understand When a Graph Has a Hole or a Missing Point

We now consider a rational function $f(x) = \frac{P(x)}{Q(x)}$ where $P(x)$ and $Q(x)$ have a common factor. Graphs of such functions have gaps or missing points that are not the result of vertical asymptotes.

EXAMPLE 10   Find the domain of the function $f(x) = \dfrac{x^2 - x - 12}{x - 4}$ and graph it.

Solution   We find the values of x that make the denominator 0. These values of x will not be included in the domain. We then will write the rational function in simplest form and graph it using the methods of this section.

7. In the function $f(x) = \frac{P(x)}{Q(x)}$, if the degree of $P(x)$ and $Q(x)$ are _____, the horizontal asymptote is

$$y = \frac{\text{the leading coefficient of the numerator}}{\text{the leading coefficient of the denominator}}$$

8. In a rational function, if the degree of the numerator is 1 greater than the degree of the denominator, the graph will have a _____.

9. A graph can cross a _____ asymptote but can never cross a _____ asymptote.

10. The graph of $f(x) = \dfrac{x^2 - 4}{x + 2}$ will have a _____ point.

Find the equations of the vertical and horizontal asymptotes of each graph. Find the domain and range.

11.

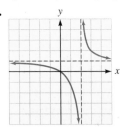

12.

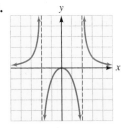

Practice *The time t it takes to travel 600 miles is a function of the mean rate of speed r:*

$$t = f(r) = \frac{600}{r}$$

Find t for the given values of r.

13. 30 mph   **14.** 40 mph

15. 50 mph   **16.** 60 mph

Suppose the cost (in dollars) of removing p% of the pollution in a river is given by the function

$$c = f(p) = \frac{50{,}000p}{100 - p} \quad (0 \le p < 100)$$

Find the cost of removing each percent of pollution.

17. 10%   **18.** 30%

19. 50%   **20.** 80%

Find the domain of each rational function. Do not graph the function.

21. $f(x) = \dfrac{x^2}{x - 2}$ **22.** $f(x) = \dfrac{x^3 - 3x^2 + 1}{x + 3}$

23. $f(x) = \dfrac{2x^2 + 7x - 2}{x^2 - 25}$ **24.** $f(x) = \dfrac{5x^2 + 1}{x^2 + 5}$

25. $f(x) = \dfrac{x - 1}{x^3 - x}$ **26.** $f(x) = \dfrac{x + 2}{2x^2 - 9x + 9}$

27. $f(x) = \dfrac{3x^2 + 5}{x^2 + 1}$ **28.** $f(x) = \dfrac{7x^2 - x + 2}{x^4 + 4}$

Find the vertical asymptotes, if any, of each rational function. Do not graph the function.

29. $f(x) = \dfrac{x}{x - 3}$ **30.** $f(x) = \dfrac{2x}{2x + 5}$

31. $f(x) = \dfrac{x + 2}{x^2 - 1}$ **32.** $f(x) = \dfrac{x - 4}{x^2 - 16}$

33. $f(x) = \dfrac{1}{x^2 - x - 6}$ **34.** $f(x) = \dfrac{x + 2}{2x^2 - 6x - 8}$

35. $f(x) = \dfrac{x^2}{x^2 + 5}$ **36.** $f(x) = \dfrac{x^3 - 3x^2 + 1}{2x^2 + 3}$

Find the horizontal asymptotes, if any, of each rational function. Do not graph the function.

37. $f(x) = \dfrac{2x - 1}{x}$ **38.** $f(x) = \dfrac{x^2 + 1}{3x^2 - 5}$

39. $f(x) = \dfrac{x^2 + x - 2}{2x^2 - 4}$ **40.** $f(x) = \dfrac{5x^2 + 1}{5 - x^2}$

41. $f(x) = \dfrac{x + 1}{x^3 - 4x}$ **42.** $f(x) = \dfrac{x}{2x^2 - x + 11}$

43. $f(x) = \dfrac{x^2}{x - 2}$ **44.** $f(x) = \dfrac{x^4 + 1}{x - 3}$

Find the slant asymptote, if any, of each rational function. Do not graph the function.

45. $f(x) = \dfrac{x^2 - 5x - 6}{x - 2}$

46. $f(x) = \dfrac{x^2 - 2x + 11}{x + 3}$

47. $f(x) = \dfrac{2x^2 - 5x + 1}{x - 4}$

48. $f(x) = \dfrac{5x^3 + 1}{x + 5}$

49. $f(x) = \dfrac{x^3 + 2x^2 - x - 1}{x^2 - 1}$

50. $f(x) = \dfrac{-x^3 + 3x^2 - x + 1}{x^2 + 1}$

Find all vertical, horizontal, and slant asymptotes, x- and y-intercepts, and symmetries, and then graph each function. Check your work with a graphing calculator.

51. $y = \dfrac{1}{x - 2}$

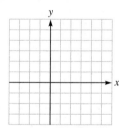

52. $y = \dfrac{3}{x + 3}$

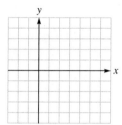

53. $y = \dfrac{x}{x - 1}$

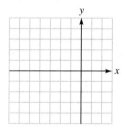

54. $y = \dfrac{x}{x + 2}$

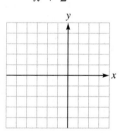

55. $f(x) = \dfrac{x + 1}{x + 2}$

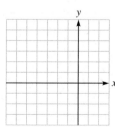

56. $f(x) = \dfrac{x - 1}{x - 2}$

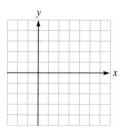

57. $f(x) = \dfrac{2x - 1}{x - 1}$

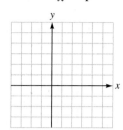

58. $f(x) = \dfrac{3x + 2}{x^2 - 4}$

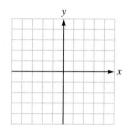

59. $g(x) = \dfrac{x^2 - 9}{x^2 - 4}$

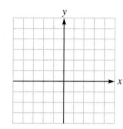

60. $g(x) = \dfrac{x^2 - 4}{x^2 - 9}$

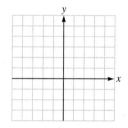

61. $g(x) = \dfrac{x^2 - x - 2}{x^2 - 4x + 3}$

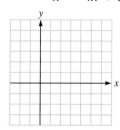

62. $g(x) = \dfrac{x^2 + 7x + 12}{x^2 - 7x + 12}$

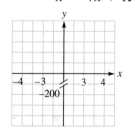

63. $y = \dfrac{x^2 + 2x - 3}{x^3 - 4x}$

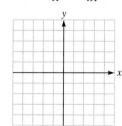

64. $y = \dfrac{3x^2 - 4x + 1}{2x^3 + 3x^2 + x}$

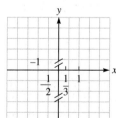

65. $y = \dfrac{x^2 - 9}{x^2}$

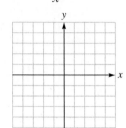

66. $y = \dfrac{3x^2 - 12}{x^2}$

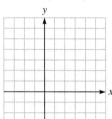

75. $h(x) = \dfrac{x^2 - 2x - 8}{x - 1}$

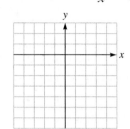

76. $h(x) = \dfrac{x^2 + x - 6}{x + 2}$

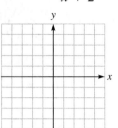

67. $f(x) = \dfrac{x}{(x + 3)^2}$

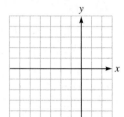

68. $f(x) = \dfrac{x}{(x - 1)^2}$

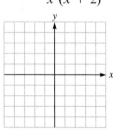

77. $f(x) = \dfrac{x^3 + x^2 + 6x}{x^2 - 1}$

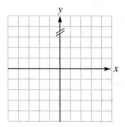

78. $f(x) = \dfrac{x^3 - 2x^2 + x}{x^2 - 4}$

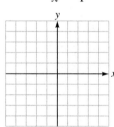

69. $f(x) = \dfrac{x + 1}{x^2(x - 2)}$

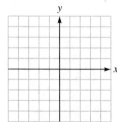

70. $f(x) = \dfrac{x - 1}{x^2(x + 2)^2}$

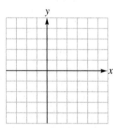

Graph each rational function. Note that the numerator and denominator of the fraction share a common factor.

79. $f(x) = \dfrac{x^2}{x}$

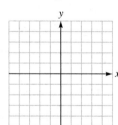

80. $f(x) = \dfrac{x^2 - 1}{x - 1}$

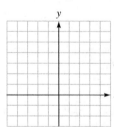

71. $y = \dfrac{x}{x^2 + 1}$

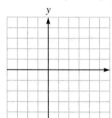

72. $y = \dfrac{x - 1}{x^2 + 2}$

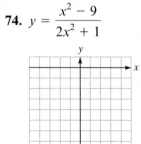

81. $f(x) = \dfrac{x^3 + x}{x}$

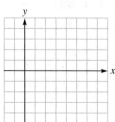

82. $f(x) = \dfrac{x^3 - x^2}{x - 1}$

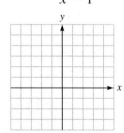

73. $y = \dfrac{3x^2}{x^2 + 1}$

74. $y = \dfrac{x^2 - 9}{2x^2 + 1}$

83. $f(x) = \dfrac{x^2 - 2x + 1}{x - 1}$

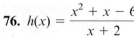

84. $f(x) = \dfrac{2x^2 + 3x - 2}{x + 2}$

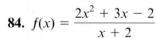

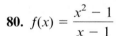

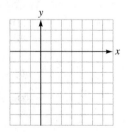

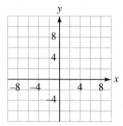

85. $f(x) = \dfrac{x^3 - 1}{x - 1}$ **86.** $f(x) = \dfrac{x^2 - x}{x^2}$

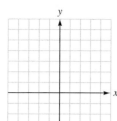   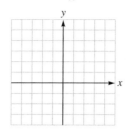

Applications

A service club wants to publish a directory of its members. Some investigation shows that the cost of typesetting and photography will be $700, and the cost of printing each directory will be $1.25.

87. Find a function that gives the total cost c of printing x directories.

88. Find the mean cost per directory if 500 directories are printed.

89. Find a function that gives the mean cost per directory $\bar{c}$ of printing x directories.

90. Find the mean cost per directory if 1,000 directories are printed.

91. Find the total cost of printing 500 directories.

92. Find the mean cost per directory if 2,000 directories are printed.

An electric company charges $7.50 per month plus 9¢ for each kilowatt hour (kwh) of electricity used.

93. Find a function that gives the total cost c of n kwh of electricity.

94. Find a function that gives the mean cost per kwh, $\bar{c}$, when using n kwh.

95. Find the total cost for using 775 kwh.

96. Find the mean cost per kwh when 775 kwh are used.

97. Find the mean cost per kwh when 1,000 kwh are used.

98. Find the mean cost per kwh when 1,200 kwh are used.

99. Utility costs An electric company charges $8.50 per month plus 9.5¢ for each kilowatt hour (kwh) of electricity used.

　a. Find a linear function that gives the total cost of n kwh of electricity.

　b. Find a rational function that gives the average cost per kwh when using n kwh.

　c. Find the average cost per kwh when 850 kwh are used.

100. Scheduling work crews The following rational function gives the number of days it would take two construction crews, working together, to frame a house that crew 1 (working alone) could complete in t days and crew 2 (working alone) could complete in $(t + 3)$ days.

$$f(t) = \frac{t^2 + 3t}{2t + 3}$$

　a. If crew 1 could frame a certain house in 21 days, how long would it take both crews working together?

　b. If crew 2 could frame a certain house in 25 days, how long would it take both crews working together?

Discovery and Writing

101. Can a rational function have two horizontal asymptotes? Explain.

102. Can a rational function have two slant asymptotes? Explain.

In Exercises 103–106, a, b, c, and d are nonzero constants.

103. Show that $y = 0$ is a horizontal asymptote of the graph of $y = \dfrac{ax + b}{cx^2 + d}$.

104. Show that $y = \dfrac{a}{c}x$ is a slant asymptote of the graph of $y = \dfrac{ax^3 + b}{cx^2 + d}$.

105. Show that $y = \dfrac{a}{c}$ is a horizontal asymptote of the graph of $y = \dfrac{ax^2 + b}{cx^2 + d}$.

106. Graph the rational function $y = \dfrac{x^3 + 1}{x}$ and explain why the curve is said to have a *parabolic asymptote*.

Use a graphing calculator to perform each experiment. Write a brief paragraph describing your findings.

107. Investigate the positioning of the vertical asymptotes of a rational function by graphing $y = \dfrac{x}{x - k}$ for several values of k. What do you observe?

108. Investigate the positioning of the vertical asymptotes of a rational function by graphing $y = \dfrac{x}{x^2 - k}$ for $k = 4, 1, -1,$ and 0. What do you observe?

109. Find the range of the rational function $y = \frac{kx^2}{x^2 + 1}$ for several values of k. What do you observe?

110. Investigate the positioning of the x-intercepts of a rational function by graphing $y = \frac{x^2 - k}{x}$ for $k = 1, -1$, and 0. What do you observe?

Review   *Perform each operation.*

111. $(2x^2 + 3x) + (x^2 - 2x)$

112. $(3x + 2) - (x^2 + 2)$

113. $(5x + 2)(2x + 5)$

114. $\dfrac{2x^2 + 3x + 1}{x + 1}$

115. If $f(x) = 3x + 2$, find $f(x + 1)$.

116. If $f(x) = x^2 + x$, find $f(2x + 1)$.

3.6   Operations on Functions

Objectives

1. Add, Subtract, Multiply and Divide Functions, Specifying Domains
2. Write Functions as Sums, Differences, Products, or Quotients of Other Functions
3. Evaluate Composite Functions
4. Determine Domains of Composite Functions
5. Write Functions as Compositions
6. Use Operations on Functions to Solve Problems

Functions can be combined by addition, subtraction, multiplication, and division. In this section, we will explore these operations on functions and give careful attention to their domains and ranges.

Suppose that the functions $R(x) = 140x$ and $C(x) = 120{,}000 + 40x$ model a company's yearly revenue and cost for producing and selling surfboards. By subtracting the functions, $R(x) - C(x)$, we would arrive at a new function represented by

$$
\begin{aligned}
(R - C)(x) &= 140x - (120{,}000 + 40x) \\
&= 140x - 120{,}000 - 40x \qquad \text{Remove parentheses.} \\
&= 100x - 120{,}000
\end{aligned}
$$

This function represents the profit made by the company when it sells x surfboards.

We now will discuss how to add, subtract, multiply, and divide functions.

1.   Add, Subtract, Multiply, and Divide Functions, Specifying Domains

With the following definitions, it is possible to perform arithmetic operations on algebraic functions.

Adding, Subtracting, Multiplying, and Dividing Functions

If the ranges of functions f and g are subsets of the real numbers, then

1. The **sum** of f and g, denoted as $f + g$, is defined by

$$(f + g)(x) = f(x) + g(x)$$

2. The **difference** of f and g, denoted as $f - g$, is defined by

$$(f - g)(x) = f(x) - g(x)$$

3. The **product** of f and g, denoted as $f \cdot g$, is defined by

$$(f \cdot g)(x) = f(x)g(x)$$

4. The **quotient** of f and g, denoted as f/g, is defined by

$$(f/g)(x) = \frac{f(x)}{g(x)} \quad (g(x) \neq 0)$$

The **domain** of each function, unless otherwise restricted, is the set of real numbers x that are in the domains of both f and g. In the case of the quotient f/g, there is the restriction that $g(x) \neq 0$.

EXAMPLE 1   Let $f(x) = 3x + 1$ and $g(x) = 2x - 3$. Find each function and its domain:
a. $f + g$ **b.** $f - g$

Solution   We will find the sum of f and g by using the definition

$$(f + g)(x) = f(x) + g(x)$$

We will find the difference of f and g by using the definition

$$(f - g)(x) = f(x) - g(x)$$

To find the domain of each result, we will consider the domains of both f and g.

a. $(f + g)(x) = f(x) + g(x)$
$$= (3x + 1) + (2x - 3)$$
$$= 5x - 2$$

Since the domain of both f and g is the set of real numbers, the domain of $f + g$ is the interval $(-\infty, \infty)$.

b. $(f - g)(x) = f(x) - g(x)$
$$= (3x + 1) - (2x - 3)$$
$$= 3x + 1 - 2x + 3 \qquad \text{Remove parentheses.}$$
$$= x + 4$$

Since the domain of both f and g is the set of real numbers, the domain of $f - g$ is the interval $(-\infty, \infty)$.

Self Check 1   Find $g - f$.

EXAMPLE 2   Let $f(x) = 3x + 1$ and $g(x) = 2x - 3$. Find each function and its domain:
a. $f \cdot g$ **b.** f/g

Solution   We will multiply f and g by using the definition

$$(f \cdot g)(x) = f(x)g(x)$$

We will divide f by g by using the definition

$$(f/g)(x) = \frac{f(x)}{g(x)} \quad (g(x) \neq 0)$$

We will find the domain of each result by considering the domains of both f and g.

a. $(f \cdot g)(x) = f(x) \cdot g(x)$
$$= (3x + 1)(2x - 3)$$
$$= 6x^2 - 7x - 3$$

Since the domain of both f and g is the set of real numbers, the domain of $f \cdot g$ is the interval $(-\infty, \infty)$.

b. $(f/g)(x) = \dfrac{f(x)}{g(x)} \quad (g(x) \neq 0)$

$$= \frac{3x + 1}{2x - 3} \quad (2x - 3 \neq 0)$$

Since $\frac{3}{2}$ will make $2x - 3$ equal to 0, the domain of f/g is the set of all real numbers except $\frac{3}{2}$. This is $\left(-\infty, \frac{3}{2}\right) \cup \left(\frac{3}{2}, \infty\right)$.

Self Check 2   Find g/f and its domain.

EXAMPLE 3   Let $f(x) = x^2 - 4$ and $g(x) = \sqrt{x}$. Find each function and its domain:
a. $f + g$ **b.** $f \cdot g$ **c.** f/g **d.** g/f

Solution   To determine each result, we use the definitions of the sum, product, and quotient of two functions. We will determine the domain of each result by considering the domains of both f and g.

First, we find the domains of f and g. Because 4 can be subtracted from any real number squared, the domain of f is the interval $(-\infty, \infty)$. Because $\sqrt{x}$ is to be a real number, the domain of g is the interval $[0, \infty)$.

a. $(f + g)(x) = f(x) + g(x)$
$$= x^2 - 4 + \sqrt{x}$$

The domain of $f + g$ consists of the numbers x that are in the domain of both f and g. This is $(-\infty, \infty) \cap [0, \infty)$, which is $[0, \infty)$. The domain of $f + g$ is $[0, \infty)$.

b. $(f \cdot g)(x) = f(x)g(x)$
$$= (x^2 - 4)\sqrt{x}$$
$$= x^2\sqrt{x} - 4\sqrt{x} \quad \text{Distribute the multiplication of } \sqrt{x}.$$

The domain of $f \cdot g$ consists of the numbers x that are in the domain of both f and g. The domain of $f \cdot g$ is $[0, \infty)$.

c. $(f/g)(x) = \dfrac{f(x)}{g(x)} \quad (g(x) \neq 0)$

$$= \frac{x^2 - 4}{\sqrt{x}}$$

The domain of f/g consists of the numbers x that are in the domain of both f and g, except 0 (because division by 0 is undefined). The domain of f/g is $(0, \infty)$.

To write $(f/g)(x)$ in a different form, we can rationalize the denominator of $\frac{x^2 - 4}{\sqrt{x}}$ and simplify.

$$
\begin{aligned}
(f/g)(x) &= \frac{x^2 - 4}{\sqrt{x}} \\
&= \frac{(x^2 - 4)\sqrt{x}}{\sqrt{x} \cdot \sqrt{x}} \qquad \text{Multiply numerator and denominator by } \sqrt{x}. \\
&= \frac{x^2\sqrt{x} - 4\sqrt{x}}{x}
\end{aligned}
$$

d. $(g/f)(x) = \dfrac{g(x)}{f(x)} \quad (f(x) \neq 0)$

$$
= \frac{\sqrt{x}}{x^2 - 4}
$$

The domain of g/f consists of the numbers x that are in $[0, \infty)$, the domain of both f and g, except 2 (because division by 0 is undefined). The domain of g/f is $[0, 2) \cup (2, \infty)$.

Self Check 3   Find $g - f$ and its domain.

EXAMPLE 4   Find $(f + g)(3)$ when $f(x) = x^2 + 1$ and $g(x) = 2x + 1$.

Solution   We will first find $(f + g)(x)$ and then find $(f + g)(3)$. To find $(f + g)(x)$, we proceed as follows:

$$
\begin{aligned}
(f + g)(x) &= f(x) + g(x) \\
&= x^2 + 1 + 2x + 1 \\
&= x^2 + 2x + 2
\end{aligned}
$$

To find $(f + g)(3)$, we proceed as follows:

$$
\begin{aligned}
(f + g)(x) &= x^2 + 2x + 2 \\
(f + g)(3) &= 3^2 + 2(3) + 2 \qquad \text{Substitute 3 for } x. \\
&= 9 + 6 + 2 \\
&= 17
\end{aligned}
$$

Comment
We also can find $(f + g)(3)$ by finding $f(3) + g(3)$. Note that $f(3) = 10$, $g(3) = 7$, and $f(3) + g(3) = 10 + 7 = 17$.

Self Check 4   Find $(f \cdot g)(-2)$.

2. Write Functions as Sums, Differences, Products, or Quotients of Other Functions

EXAMPLE 5   Let $h(x) = x^2 + 3x + 2$. Find two functions f and g such that
a. $f + g = h$ **b.** $f \cdot g = h$

Solution   For part **a**, we must find two functions f and g whose sum is h. For part **b**, we must find two functions f and g whose product is h.

a. There are many possibilities. One is $f(x) = x^2$ and $g(x) = 3x + 2$, for then

$$
\begin{aligned}
(f + g)(x) &= f(x) + g(x) \\
&= (x^2) + (3x + 2)
\end{aligned}
$$

$$= x^2 + 3x + 2$$
$$= h(x)$$

Another possibility is $f(x) = x^2 + 2x$ and $g(x) = x + 2$.

b. Again, there are many possibilities. One is suggested by factoring $x^2 + 3x + 2$.

$$x^2 + 3x + 2 = (x + 1)(x + 2)$$

If we let $f(x) = x + 1$ and $g(x) = x + 2$, then

$$(f \cdot g)(x) = f(x) \cdot g(x)$$
$$= (x + 1)(x + 2)$$
$$= x^2 + 3x + 2$$
$$= h(x)$$

Another possibility is $f(x) = 3$ and $g(x) = \frac{x^2}{3} + x + \frac{2}{3}$.

Self Check 5 Find two functions f and g such that $f - g = h$.

3. Evaluate Composite Functions

Often one quantity is a function of a second quantity that depends, in turn, on a third quantity. For example, the cost of a car trip is a function of the gasoline consumed. The amount of gasoline consumed, in turn, is a function of the number of miles driven. Such chains of dependence are analyzed mathematically as *composition of functions.*

Suppose that $y = f(x)$ and $y = g(x)$ define two functions. Any number x in the domain of g will produce a corresponding value $g(x)$ in the range of g. If $g(x)$ is in the domain of function f, then $g(x)$ can be substituted into f, and a corresponding value $f(g(x))$ will be determined. This two-step process defines a new function, called a **composite function,** denoted by $f \circ g$. (See Figure 3-75.)

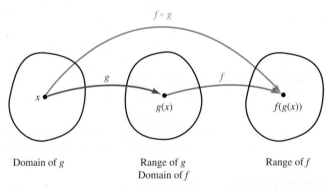

Figure 3-75

Composite Functions The **composite function** $f \circ g$ is defined by

$$(f \circ g)(x) = f(g(x))$$

The **domain** of $f \circ g$ consists of all those numbers in the domain of g for which $g(x)$ is in the domain of f.

To illustrate the previous definition, we consider the functions $f(x) = 5x + 1$ and $g(x) = 4x - 3$, and find $(f \circ g)(x)$ and $(g \circ f)(x)$.

$$
\begin{aligned}
(f \circ g)(x) &= f(g(x)) & (g \circ f)(x) &= g(f(x)) \\
&= f(4x - 3) & &= g(5x + 1) \\
&= 5(4x - 3) + 1 & &= 4(5x + 1) - 3 \\
&= 20x - 14 & &= 20x + 1
\end{aligned}
$$

Since we get different results, the composition of functions is not commutative.

We have seen that a function can be represented by a machine. If we put a number from the domain into the machine (the input), a number from the range comes out (the output). For example, if we put 2 into the machine shown in Figure 3-76(a), the number $f(2) = 5(2) - 2 = 8$ comes out. In general, if we put x into the machine shown in Figure 3-76(b), the value $f(x)$ comes out.

<del>**Comment**</del>

Note that in the previous example $(f \circ g)(x) \neq (g \circ f)(x)$.

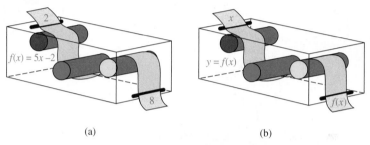

(a)   (b)

Figure 3-76

The function machines shown in Figure 3-77 illustrate the composition $f \circ g$. When we put a number x into the function g, the value $g(x)$ comes out. The value $g(x)$ then goes into function f, and $f(g(x))$ comes out.

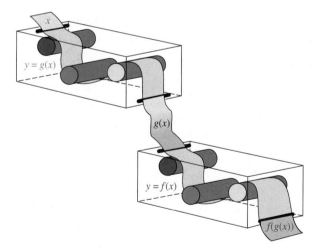

Figure 3-77

To further illustrate these ideas, we let $f(x) = 2x + 1$ and $g(x) = x - 4$.

- $(f \circ g)(9)$ means $f(g(9))$. In Figure 3-78(a), function g receives the number 9 and subtracts 4, and the number $g(9) = 5$ comes out. The 5 goes into the f function, which doubles it and adds 1. The final result, 11, is the output of the composite function $f \circ g$:

$$(f \circ g)(9) = f(g(9)) = f(5) = 2(5) + 1 = 11$$

- $(f \circ g)(x)$ means $f(g(x))$. In Figure 3-78(a), function g receives the number x and subtracts 4, and the number $x - 4$ comes out. The $x - 4$ goes into the f function, which doubles it and adds 1. The final result, $2x - 7$, is the output of the composite function $f \circ g$.

$$(f \circ g)(x) = f(g(x)) = f(x - 4) = 2(x - 4) + 1 = 2x - 7$$

- $(g \circ f)(-2)$ means $g(f(-2))$. In Figure 3-78(b), function f receives the number -2, doubles it and adds 1, and releases -3 into the g function. Function g subtracts 4 from -3 and releases a final output of -7. Thus,

$$(g \circ f)(-2) = g(f(-2)) = g(-3) = -3 - 4 = -7$$

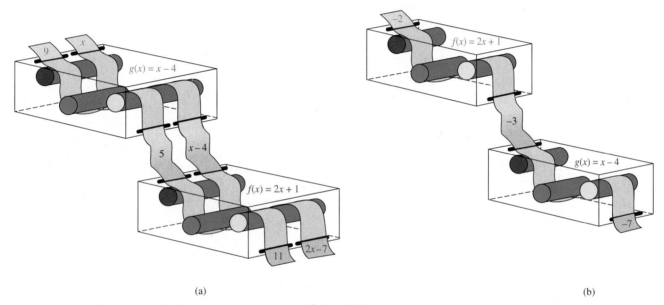

(a)   (b)

Figure 3-78

EXAMPLE 6   If $f(x) = 2x + 7$ and $g(x) = x^2 - 1$. Find
a. $(f \circ g)(x)$ **b.** $(g \circ f)(x)$

Solution   In part **a,** because $(f \circ g)(x)$ means $f(g(x))$, we will replace x in $f(x) = 2x + 7$ with $g(x)$. In part **b,** because $(g \circ f)(x)$ means $g(f(x))$ we will replace x in $g(x) = x^2 - 1$ with $f(x)$.

a. $(f \circ g)(x) = f(g(x))$

$\qquad\qquad\quad = f(x^2 - 1)$ Substitute $x^2 - 1$ for $g(x)$.

$\qquad\qquad\quad = 2(x^2 - 1) + 7$ Evaluate $f(x^2 - 1)$.

$\qquad\qquad\quad = 2x^2 - 2 + 7$ Remove parentheses.

$\qquad\qquad\quad = 2x^2 + 5$

b. $(g \circ f)(x) = g(f(x))$

$\qquad\qquad\quad = g(2x + 7)$ Substitute $2x + 7$ for $f(x)$.

$\qquad\qquad\quad = (2x + 7)^2 - 1$ Evaluate $g(2x + 7)$.

$\qquad\qquad\quad = 4x^2 + 28x + 49 - 1$ Square the binomial.

$\qquad\qquad\quad = 4x^2 + 28x + 48$

Self Check 6   If $h(x) = x + 1$, find $(f \circ h)(x)$.

EXAMPLE 7 If $f(x) = 3x - 2$ and $g(x) = 3x^2 + 6x - 5$, find $(f \circ g)(-2)$.

Solution Because $(f \circ g)(-2)$ means $f(g(-2))$, we first find $g(-2)$. We then find $f(g(-2))$.

$$g(x) = 3x^2 + 6x - 5$$
$$g(-2) = 3(-2)^2 + 6(-2) - 5$$
$$= 3(4) - 12 - 5$$
$$= -5$$

$$(f \circ g)(-2) = f(g(-2))$$
$$= f(-5)$$
$$= 3(-5) - 2$$
$$= -17$$

Self Check 7 Find $(g \circ f)(-1)$.

4. Determine Domains of Composite Functions

To be in the domain of the composite function $f \circ g$, a number x has to be in the domain of g, and the output of g must be in the domain of f. Thus, the domain of $f \circ g$ consists of those inputs x that are in the domain of g, and for which $g(x)$ is in the domain of f.

Strategy to Determine the Domain of $f \circ g$ To determine the domain of $(f \circ g)(x) = f(g(x))$, apply the following restrictions to the composition:

1. If x is not in the domain of g, it will not be in the domain of $f \circ g$.
2. Any x that has an output $g(x)$ that is not in the domain of f will not be in the domain of $f \circ g$.

EXAMPLE 8 Let $f(x) = \sqrt{x}$ and $g(x) = x - 3$. Find the domain of
a. $f \circ g$ **b.** $g \circ f$

Solution We first will find the domains of $f(x)$ and $g(x)$. Then we will find the domain of $f \circ g$ and $g \circ f$ by applying the restrictions stated above.

For $\sqrt{x}$ to be a real number, x must be a nonnegative real number. Thus, the domain of f is the interval $[0, \infty)$. Since any real number x can be an input into g, the domain of g is the interval $(-\infty, \infty)$.

a. The domain of $f \circ g$ is the set of real numbers x such that x is in the domain of g and $g(x)$ is in the domain of f. We have seen that all values of x are in the domain of g. However, $g(x)$ must be nonnegative, because $g(x)$ must be in the domain of f. So we must find the values of x such that $g(x)$ is greater than or equal to 0.

$$g(x) \geq 0 \quad \text{g(x) must be nonnegative.}$$
$$x - 3 \geq 0 \quad \text{Substitute } x - 3 \text{ for } g(x).$$
$$x \geq 3 \quad \text{Add 3 to both sides.}$$

Since $x \geq 3$, the domain of $f \circ g$ is the interval $[3, \infty)$.

b. The domain of $g \circ f$ is the set of real numbers x such that x is in the domain of f and $f(x)$ is in the domain of g. We have seen that only nonnegative values of x are in the domain of f. Because all values of $f(x)$ are in the domain of g, the domain of $g \circ f$ is the domain of f, which is the interval $[0, \infty)$.

Self Check 8   Find the domain of $f \circ f$.   ▬

EXAMPLE 9   Let $f(x) = \dfrac{x + 3}{x - 2}$ and $g(x) = \dfrac{1}{x}$.

  a. Find the domain of $f \circ g$.   **b.** Find $f \circ g$.

Solution   We first will find the domains of $f(x)$ and $g(x)$. Then, we will find the domain of $f \circ g$ by applying the restrictions stated above.

    For $\dfrac{1}{x}$ to be a real number, x cannot be 0. Thus, the domain of g is $(-\infty, 0) \cup (0, \infty)$.

    Since any real number except 2 can be an input into f, the domain of f is $(-\infty, 2) \cup (2, \infty)$.

a. The domain of $f \circ g$ is the set of real numbers x such that x is in the domain of g and $g(x)$ is in the domain of f. We have seen that all values of x but 0 are in the domain of g and that all values of $g(x)$ but 2 are in the domain of f. So we must exclude 0 from the domain of $f \circ g$ and all values of x where $g(x) = 2$. To find the excluded values, we proceed as follows:

$$g(x) = 2$$

$$\frac{1}{x} = 2 \qquad \text{Substitute } \frac{1}{x} \text{ for } g(x).$$

$$1 = 2x \qquad \text{Since } x \neq 0, \text{ we can multiply both sides by } x.$$

$$x = \frac{1}{2} \qquad \text{Divide both sides by 2.}$$

    The domain of $f \circ g$ is the set of all real numbers except 0 and $\frac{1}{2}$, which is $(-\infty, 0) \cup \left(0, \frac{1}{2}\right) \cup \left(\frac{1}{2}, \infty\right)$.

b. To find $f \circ g$, we proceed as follows.

$$(f \circ g)(x) = f(g(x))$$

$$= f\left(\frac{1}{x}\right) \qquad \text{Substitute } \frac{1}{x} \text{ for } g(x).$$

$$= \frac{\dfrac{1}{x} + 3}{\dfrac{1}{x} - 2} \qquad \text{Substitute } \frac{1}{x} \text{ for } x \text{ in } f.$$

$$= \frac{1 + 3x}{1 - 2x} \qquad \text{Multiply numerator and denominator by } x.$$

    Thus, $(f \circ g)(x) = \dfrac{1 + 3x}{1 - 2x}$.

Comment

Example 9 illustrates that the domain of the composite function $f \circ g$ cannot always be found by finding $f \circ g$ and analyzing it. In the example, we found that the domain is all real numbers, except for 0 and $\frac{1}{2}$. If we analyze only the form $f \circ g$, we would state the domain incorrectly as all real numbers except $\frac{1}{2}$.

Self Check 9   Let $f(x) = \dfrac{x}{x - 1}$ and $g(x) = \dfrac{1}{x}$.

  a. Find the domain of $f \circ g$.

  b. Find $f \circ g$.

5. Write Functions as Compositions

Comment

The result of a decomposition is not unique. There are several possibilities. In the example to the right, another possibility is

$f(x) = x^2$ and
$g(x) = (2x^2 - 6x + 3)^2$

When we have formed a composite function $f \circ g$, we obtained a function h. It is possible to reverse this composition process and begin with a function h and express it as a composition of two functions . This process is called **decomposition.**

For example, consider $h(x) = (2x^2 - 6x + 3)^4$. The function h takes $2x^2 - 6x + 3$ and raises it to the fourth power. To write the function h as a composition of functions f and g, we can let

$$f(x) = x^4 \quad \text{and} \quad g(x) = 2x^2 - 6x + 3$$

Then

$$(f \circ g)(x) = f(g(x))$$
$$= f(2x^2 - 6x + 3)$$
$$= (2x^2 - 6x + 3)^4$$
$$= h(x)$$

EXAMPLE 10 Let $h(x) = \sqrt{x + 1}$. Find functions f and g such that $f \circ g = h$.

Solution Because h takes the square root of the algebraic function $x + 1$, we let $f(x) = \sqrt{x}$ and $g(x) = x + 1$.

$$f(x) = \sqrt{x} \quad \text{and} \quad g(x) = x + 1$$

We can check the composition $f \circ g$ to see that it gives the original function h.

$$(f \circ g)(x) = f(g(x))$$
$$= f(x + 1)$$
$$= \sqrt{x + 1}$$
$$= h(x)$$

Self Check 10 Let $h(x) = \sqrt[3]{x^2 - 5}$. Find functions f and g such that $f \circ g = h$.

6. Use Operations on Functions to Solve Problems

EXAMPLE 11 A laboratory sample is removed from a cooler at a temperature of 15° F. Technicians then warm the sample at the rate of 3° F per hour. Express the sample's temperature in degrees Celsius as a function of the time t (in hours) since it was removed from the cooler.

Solution We first write a Fahrenheit temperature function that represents the warming of the sample. We then use composition of functions to write degrees Celsius as a function of degrees Fahrenheit.

The temperature of the sample is 15° F when $t = 0$. Because the sample warms at 3° F per hour, it warms $3t°$ after t hours. Thus, the Fahrenheit temperature after t hours is given by the function

$$F(t) = 3t + 15 \quad \text{\small $F(t)$ is the Fahrenheit temperature and t represents the time in hours.}$$

The Celsius temperature is a function of the Fahrenheit temperature $F(t)$, given by the formula

$$C(F(t)) = \frac{5}{9}(F(t) - 32)$$

To express the sample's Celsius temperature as a function of time, we find the composition function $C \circ F$.

$$
\begin{aligned}
(C \circ F)(t) &= C(F(t)) \\
&= C(3t + 15) && \text{Substitute for } F(t). \\
&= \frac{5}{9}[(3t + 15) - 32] && \text{Substitute } 3t + 15 \text{ for } F(t) \text{ in } C(F(t)). \\
&= \frac{5}{9}(3t - 17) && \text{Simplify.} \\
&= \frac{15}{9}t - \frac{85}{9} \\
&= \frac{5}{3}t - \frac{85}{9}
\end{aligned}
$$

Self Check 11   Find the $(C \circ F)(10)$.

Self Check Answers
1. $(g - f)(x) = -x - 4$ **2.** $(g/f)(x) = \frac{2x - 3}{3x + 1}, \left(-\infty, -\frac{1}{3}\right) \cup \left(-\frac{1}{3}, \infty\right)$
3. $(g - f)(x) = \sqrt{x} - x^2 + 4, [0, \infty)$ **4.** -15 **5.** One possibility is $f(x) = 2x^2$ and $g(x) = x^2 - 3x - 2$.   **6.** $(f \circ h)(x) = 2x + 9$ **7.** 40
8. $[0, \infty)$ **9.** $(-\infty, 0) \cup (0, 1) \cup (1, \infty); (f \circ g)(x) = \frac{1}{1 - x}$ **10.** One
possibility is $f(x) = \sqrt[3]{x}, g(x) = x^2 - 5$ **11.** approximately 7.2

3.6   **Exercises**

Vocabulary and Concepts   *Fill in the blanks.*
1. $(f + g)(x) = $ _____
2. $(f - g)(x) = $ _____
3. $(f \cdot g)(x) = $ _____
4. $(f/g)(x) = $ _____, where $g(x) \neq 0$
5. The domain of $f + g$ is the _____ of the domains of f and g.
6. $(f \circ g)(x) = $ _____
7. $(g \circ f)(x) = $ _____
8. To determine $(f \circ g)(-5)$, first find _____.
9. Composition of functions is not _____.
10. To be in the domain of the composite function $f \circ g$, a number x must be in the _____ of g, and the output of g must be in the _____ of f.

Practice   *Let $f(x) = 2x + 1$ and $g(x) = 3x - 2$. Find each function and its domain.*
11. $f + g$ **12.** $f - g$

13. $f \cdot g$ **14.** f/g

Let $f(x) = x^2 + x$ and $g(x) = x^2 - 1$. Find each function and its domain.
15. $f - g$ **16.** $f + g$

17. f/g **18.** $f \cdot g$

Let $f(x) = x^2 - 1$ and $g(x) = 3x - 2$. Find each value, if possible.
19. $(f + g)(2)$ **20.** $(f + g)(-3)$
21. $(f - g)(0)$ **22.** $(f - g)(-5)$
23. $(f \cdot g)(2)$ **24.** $(f \cdot g)(-1)$
25. $(f/g)\left(\frac{2}{3}\right)$ **26.** $(f/g)(t)$

Find two functions f and g such that h(x) can be expressed as the function indicated. Several answers are possible.

27. $h(x) = 3x^2 + 2x$; $f + g$

28. $h(x) = 3x^2$; $f \cdot g$

29. $h(x) = \dfrac{3x^2}{x^2 - 1}$; f/g

30. $h(x) = 5x + x^2$; $f - g$

31. $h(x) = x(3x^2 + 1)$; $f - g$

32. $h(x) = (3x - 2)(3x + 2)$; $f + g$

33. $h(x) = x^2 + 7x - 18$; $f \cdot g$

34. $h(x)$; f/g

Let f(x) = 2x − 5 and g(x) = 5x − 2. Find each value.

35. $(f \circ g)(2)$

36. $(g \circ f)(-3)$

37. $(f \circ f)\left(-\dfrac{1}{2}\right)$

38. $(g \circ g)\left(\dfrac{3}{5}\right)$

Let f(x) = 3x² − 2 and g(x) = 4x + 4. Find each value.

39. $(f \circ g)(-3)$

40. $(g \circ f)(3)$

41. $(f \circ f)\left(\sqrt{3}\right)$

42. $(g \circ g)(-4)$

Let f(x) = 3x and g(x) = x + 1. Determine the domain of each composite function and then find the composite function.

43. $f \circ g$

44. $g \circ f$

45. $f \circ f$

46. $g \circ g$

Let f(x) = x² and g(x) = 2x. Determine the domain of each composite function and then find the composite function.

47. $g \circ f$

48. $f \circ g$

49. $g \circ g$

50. $f \circ f$

Let f(x) = √x and g(x) = x + 1. Determine the domain of each composite function and then find the composite function.

51. $f \circ g$

52. $g \circ f$

53. $f \circ f$

54. $g \circ g$

Let f(x) = √(x + 1) and g(x) = x² − 1. Determine the domain of each composite function and then find the composite function.

55. $g \circ f$

56. $f \circ g$

57. $g \circ g$

58. $f \circ f$

Let f(x) = $\dfrac{1}{x - 1}$ and g(x) = $\dfrac{1}{x - 2}$. Determine the domain of each composite function and then find the composite function.

59. $f \circ g$

60. $g \circ f$

61. $f \circ f$

62. $g \circ g$

Find two functions f and g such that the composition f ∘ g = h expresses the given correspondence. Several answers are possible.

63. $h(x) = 3x - 2$

64. $h(x) = 7x - 5$

65. $h(x) = x^2 - 2$

66. $h(x) = x^3 - 3$

67. $h(x) = (x - 2)^2$

68. $h(x) = (x - 3)^3$

69. $h(x) = \sqrt{x + 2}$

70. $h(x) = \dfrac{1}{x - 5}$

71. $h(x) = \sqrt{x} + 2$

72. $h(x) = \dfrac{1}{x} - 5$

73. $h(x) = x$

74. $f(x) = 3$

Use the tables of values of f and g to find each composition value.

x	$f(x)$
2	4
4	9
6	13
8	17

x	$g(x)$
0	0
2	4
3	9
4	16

75. $(f \circ g)(2)$

76. $(g \circ f)(2)$

Applications

77. Picture tubes Refer to the television picture tube shown.

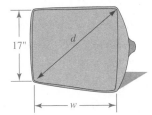

   a. Write a formula to find the area of the tube.

   b. Use the Pythagorean theorem to write a formula to find the width w of the tube.

   c. Write a formula to find the area of the tube as a function of the diagonal d.

78. Area of a square Write a formula for the area of a square in terms of its perimeter.

79. Perimeter of a square Write a formula for the perimeter of a square in terms of its area.

80. Ceramics When the temperature of a pot in a kiln is 1,200° F, an artist turns off the heat and leaves the pot to cool at a controlled rate of 81° F per hour. Express the temperature of the pot in degrees Celsius as a function of the time t (in hours) since the kiln was turned off.

Discovery and Writing

81. Let $f(x) = 3x$. Show that $(f + f)(x) = f(x + x)$.

82. Let $g(x) = x^2$. Show that $(g + g)(x) \neq g(x + x)$.

83. Let $f(x) = \dfrac{x - 1}{x + 1}$. Find $(f \circ f)(x)$.

84. Let $g(x) = \dfrac{x}{x - 1}$. Find $(g \circ g)(x)$.

Let $f(x) = x^2 - x$, $g(x) = x - 3$, and $h(x) = 3x$. Use a graphing calculator to graph both functions on the same axes. Write a brief paragraph summarizing your observations.

85. f and $f \circ g$

86. f and $g \circ f$

87. f and $f \circ h$

88. f and $h \circ f$

Review *Solve each equation for y.*

89. $x = 3y - 7$

90. $x = \dfrac{7}{y}$

91. $x = \dfrac{y}{y + 3}$

92. $x = \dfrac{y - 1}{y}$

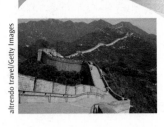

3.7 Inverse Functions

Objectives

  1. Understand the Definition of a One-to-One Function
  2. Determine Whether a Function Is One-to-One
  3. Verify Inverse Functions
  4. Find the Inverse of a One-to-One Function
  5. Understand the Relationship Between the Graphs of f and f^{-1}

In this section, we will discuss inverse functions. A function and its inverse do opposite things.

    Suppose we climb the Great Wall of China on a summer day when the temperature reaches a high of 35°C.

The linear function defined by $F = \frac{9}{5}C + 32$ gives a formula to convert degrees Celsius to degrees Fahrenheit. If we substitute a Celsius reading into the formula, a Fahrenheit reading comes out. For example, if we substitute 35 for C, we obtain a Fahrenheit reading of 95°:

$$F = \frac{9}{5}C + 32$$
$$= \frac{9}{5}(35) + 32$$
$$= 63 + 32$$
$$= 95$$

If we want to find a Celsius reading from a Fahrenheit reading, we need a formula into which we can substitute a Fahrenheit reading and have a Celsius reading come out. Such a formula is $C = \frac{5}{9}(F - 32)$, which takes the Fahrenheit reading of 95° and turns it back into a Celsius reading of 35°.

$$C = \frac{5}{9}(F - 32)$$
$$= \frac{5}{9}(95 - 32)$$
$$= \frac{5}{9}(63)$$
$$= 35$$

The functions defined by these two formulas do opposite things. The first turns 35°C into 95° Fahrenheit, and the second turns 95° Fahrenheit back into 35°C. Such functions are called *inverse functions*.

Some functions have inverses that are functions and some do not. To guarantee that the inverse of a function will also be a function, we must know that the function is *one-to-one*.

1. Understand the Definition of a One-to-One Function

In this section, we will find inverses of functions that are one-to-one. *One-to-one functions* are functions whose inverses are also functions.

We now examine what it means for a function to be one-to-one. Consider the following two functions:

Function 1: To each student, there corresponds exactly one eye color.

Function 2: To each student, there corresponds exactly one college identification number.

Function 1 is **not a one-to-one function** because two different students can have the same eye color.

Function 2 **is a one-to-one function** because two different students will always have two different ID numbers.

Recall that each element x in the domain of a function has a single output y. For some functions, different numbers x in the domain can have the same output. (See Figure 3-79(a).) For other functions, called **one-to-one functions,** different numbers x have different outputs. (See Figure 3-79(b).)

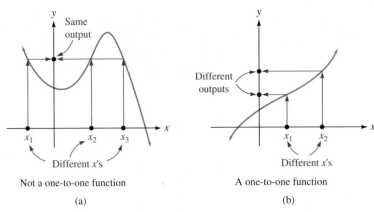

Figure 3-79

| One-to-One Functions | A function f from a set **X** to a set **Y** is called a **one-to-one function** if and only if different numbers in the domain of f have different outputs in the range of f. |

The previous definition implies that if x_1 and x_2 are two numbers in the domain of f and $x_1 \neq x_2$, then $f(x_1) \neq f(x_2)$.

2. Determine Whether a Function Is One-to-One

EXAMPLE 1   Determine whether each function is one-to-one.
a. $f(x) = x^4 + x^2$ **b.** $f(x) = x^3$

Solution   We will examine the functions and determine whether the definition of a one-to-one function applies. If different x-values always produce different y-values, the function is one-to-one.

a. The function $f(x) = x^4 + x^2$ is not one-to-one, because different numbers in the domain have the same output. For example, 2 and -2 have the same output: $f(2) = f(-2) = 20$.

b. The function $f(x) = x^3$ is one-to-one, because different numbers x produce different outputs $f(x)$. This is because different numbers have different cubes.

Self Check 1   Determine whether $f(x) = \sqrt{x}$ is one-to-one.

A **horizontal line test** can be used to determine whether the graph of a function represents a one-to-one function. If every horizontal line that intersects the graph of a function does so exactly once, the function passes the horizontal line test and is one-to-one. (See Figure 3-80(a).) If any horizontal line intersects the graph of a function more than once, the function fails the horizontal line test and is not one-to-one. (See Figure 3-80(b).)

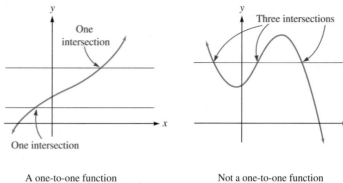

One intersection

Three intersections

One intersection

A one-to-one function

Not a one-to-one function

(a)

(b)

Figure 3-80

EXAMPLE 2   Use the horizontal line test to determine whether each graph represents a one-to-one function.

a.

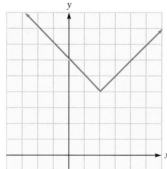

b.

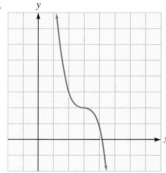

Solution   We will use the horizontal line test and draw many horizontal lines. If every horizontal line that intersects the graph does so exactly once, the function is one-to-one. If any horizontal line intersects the graph more than once, the function is not one-to-one.

a. Because the horizontal line drawn in Figure 3-81 intersects the graph in two places, the function fails the horizontal line test and is not a one-to-one function.

b. Several horizontal lines are drawn in Figure 3-82 and each one intersects the graph exactly once. We conclude that the graph passes the horizontal line test and represents a one-to-one function.

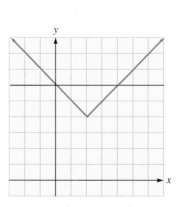

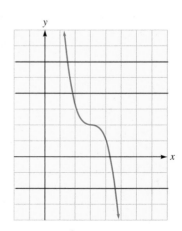

Figure 3-81

Figure 3-82

Determine whether the graph represents a one-to-one function.

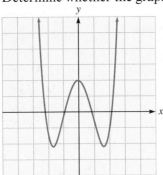

3. Verify Inverse Functions

Figure 3-83(a) illustrates a function f from set **X** to set **Y**. Since three arrows point to a single y, the function f is not one-to-one. If the arrows in Figure 3-83(a) were reversed, the diagram would not represent a function.

If the arrows of the one-to-one function f in Figure 3-83(b) were reversed, as in Figure 3-83(c), the diagram would represent a function. This function is called the **inverse of function f** and is denoted by the symbol f^{-1}.

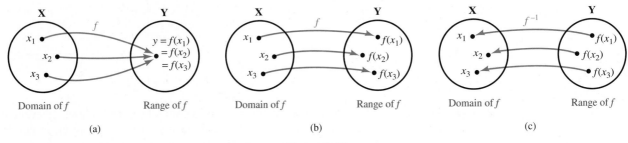

Figure 3-83

Consider the functions:

$$f(x) = 4x \qquad \text{and} \qquad g(x) = \frac{x}{4}$$

These functions are inverses of each other because function f multiplies any input x by 4, and function g will take the result and divide it by 4. The final result will be the original input x.

We can show that the composition of these functions (in either order) is x, called the *identity function.*

$$(f \circ g)(x) = f(g(x)) = f\left(\frac{x}{4}\right) = 4\left(\frac{x}{4}\right) = x$$

and

$$(g \circ f)(x) = g(f(x)) = g(4x) = \frac{4x}{4} = x$$

Since $g(x)$ is the inverse of $f(x)$, we can write $g(x)$ using inverse notation as $f^{-1}(x) = \frac{x}{4}$. Thus, $(f \circ f^{-1}) = x$ and $(f^{-1} \circ f)(x) = x$.

We now can define inverse functions.

Comment

The -1 in the notation for inverse function is not an exponent. Remember that

$$f^{-1}(x) \neq \frac{1}{f(x)}$$

Inverse Functions

If f and g are two one-to-one functions such that $(f \circ g)(x) = x$ for every x in the domain of g and $(g \circ f)(x) = x$ for every x in the domain of f, then f and g are **inverse functions.** Function g can be denoted as f^{-1}, and is called the **inverse function of f.**

We also can list two important properties of one-to-one functions.

Properties of a One-to-One Function

Property 1: If f is a one-to-one function, there is a one-to-one function $f^{-1}(x)$ such that

$$(f^{-1} \circ f)(x) = x \qquad \text{and} \qquad (f \circ f^{-1})(x) = x$$

Property 2: The domain of f is the range of f^{-1} and the range of f is the domain of f^{-1}.

Figure 3-84 shows a one-to-one function f and its inverse f^{-1}. To the number x in the domain of f, there corresponds an output $f(x)$ in the range of f. Since $f(x)$ is in the domain of f^{-1}, the output for $f(x)$ under the function f^{-1} is $f^{-1}(f(x)) = x$. Thus, $(f^{-1} \circ f)(x) = f^{-1}(f(x)) = x$.

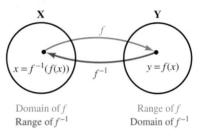

Domain of f Range of f
Range of f^{-1} Domain of f^{-1}

Figure 3-84

Comment

To show that one function is the inverse of another, we must show that their compositions are the identity function, x.

EXAMPLE 3

Verify that $f(x) = x^3$ and $g(x) = \sqrt[3]{x}$ are inverse functions.

Solution

To show that f and g are inverse functions, we must show that $f \circ g$ and $g \circ f$ are x, the identity function.

$$(f \circ g)(x) = f(g(x)) = f\left(\sqrt[3]{x}\right) = \left(\sqrt[3]{x}\right)^3 = x$$
$$(g \circ f)(x) = g(f(x)) = g(x^3) = \sqrt[3]{x^3} = x$$

Because g is the inverse of f, we can use inverse notation and write $g(x) = \sqrt[3]{x}$ as $f^{-1}(x) = \sqrt[3]{x}$. Because f is the inverse of g, we can use inverse notation and write $f(x) = x^3$ as $g^{-1}(x) = x^3$.

Self Check 3

If $x \geq 0$, are $f(x) = x^2$ and $g(x) = \sqrt{x}$ inverse functions?

4. Find the Inverse of a One-to-One Function

If f is the one-to-one function $y = f(x)$, then f^{-1} reverses the correspondence of f. That is, if $f(a) = b$, then $f^{-1}(b) = a$. To determine f^{-1}, we follow these steps.

Strategy for Finding $f^{-1}(x)$ from a Given Function $f(x)$

Step 1: Replace $f(x)$ with y.

Step 2: Interchange the variables x and y.

Step 3: Solve the resulting equation for y, if possible.

Step 4: Replace y with $f^{-1}(x)$.

Comment

After completing Step 3, if y does not represent a function of x, the process ends and f does not have an inverse.

Once $f^{-1}(x)$ is determined, it should be verified by showing that $(f \circ f^{-1})(x) = x$ and $(f^{-1} \circ f)(x) = x$.

EXAMPLE 4 Find the inverse of $f(x) = \dfrac{3}{2}x + 2$ and verify the result.

Solution We will use the strategy given above to find f^{-1}. We then will verify the result by showing that $(f \circ f^{-1})(x) = x$ and $(f^{-1} \circ f)(x) = x$.

To find f^{-1}, we use the following steps.

Step 1: Replace $f(x)$ with y.

$$f(x) = \frac{3}{2}x + 2$$

$$y = \frac{3}{2}x + 2$$

Step 2: Interchange the variables x and y.

$$x = \frac{3}{2}y + 2$$

Step 3: Solve the resulting equation for y.

$$x = \frac{3}{2}y + 2$$

$$2x = 3y + 4 \qquad \text{Multiply both sides by 2.}$$

$$2x - 4 = 3y \qquad \text{Subtract 4 from both sides.}$$

$$y = \frac{2x - 4}{3} \qquad \text{Divide both sides by 3.}$$

Step 4: Replace y with $f^{-1}(x)$.

$$y = \frac{2x - 4}{3}$$

$$f^{-1}(x) = \frac{2x - 4}{3}$$

The inverse of $f(x) = \frac{3}{2}x + 2$ is $f^{-1}(x) = \frac{2x - 4}{3}$.

To verify the result, we will use $f(x) = \frac{3}{2}x + 2$ and $f^{-1}(x) = \frac{2x - 4}{3}$ and show that $(f \circ f^{-1})(x) = x$ and $(f^{-1} \circ f)(x) = x$.

$$(f \circ f^{-1})(x) = f(f^{-1}(x))$$
$$= f\left(\frac{2x - 4}{3}\right)$$
$$= \frac{3}{2}\left(\frac{2x - 4}{3}\right) + 2$$
$$= x - 2 + 2$$
$$= x$$

$$(f^{-1} \circ f)(x) = f^{-1}(f(x))$$
$$= f^{-1}\left(\frac{3}{2}x + 2\right)$$
$$= \frac{2\left(\frac{3}{2}x + 2\right) - 4}{3}$$
$$= \frac{3x + 4 - 4}{3}$$
$$= x$$

Self Check 4   Find $f(2)$. Then find $f^{-1}(5)$. Explain the significance of the results.

5. Understand the Relationships Between the Graphs of f and f^{-1}

Because we interchange the positions of x and y to find the inverse of a function, the point (b, a) lies on the graph of $y = f^{-1}(x)$ whenever the point (a, b) lies on the graph of $y = f(x)$. Thus, the graph of a function and its inverse are reflections of each other about the line $y = x$.

EXAMPLE 5   Find the inverse of $f(x) = x^3 + 3$. Graph the function and its inverse on the same set of coordinate axes.

Solution   We will find the inverse of the function $f(x)$ using the strategy given in this section. We will use translations to graph both f and f^{-1}.
We first find f^{-1} and proceed as follows:

Step 1:   Replace $f(x)$ with y.

$$f(x) = x^3 + 3$$
$$y = x^3 + 3$$

Step 2:   Interchange the variables x and y.

$$x = y^3 + 3$$

Step 3:   Solve the resulting equation for y.

$$x - 3 = y^3$$
$$y = \sqrt[3]{x - 3}$$

Step 4:   Replace y with $f^{-1}(x)$

$$f^{-1}(x) = \sqrt[3]{x - 3}$$

We now graph f and f^{-1}. To graph $f(x) = x^3 + 3$, we translate the graph of $y = x^3$ vertically upward 3 units. To graph $f^{-1}(x) = \sqrt[3]{x - 3}$, we translate the graph of $y = \sqrt[3]{x}$ horizontally 3 units to the right. The graphs of f and f^{-1} are shown in Figure 3-85 in which the line $y = x$ is the axis of symmetry.

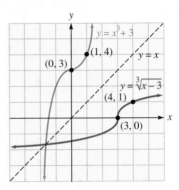

Figure 3-85

Comment

We also can graph f and f^{-1} by completing a table of solutions. The x and y columns of f can be reversed to obtain the table of solutions for f^{-1}.

Self Check 5 Find $f(2)$. Then find $f^{-1}(11)$. Explain the significance of the result.

In the next example, we will consider a function that is not one-to-one, but becomes so when we restrict its domain. By restricting the domain of the function and making it one-to-one, we are able to find its inverse and examine the function and its inverse graphically.

EXAMPLE 6 The function $y = f(x) = x^2 + 3$ is not one-to-one. However, it becomes one-to-one when we restrict its domain to the interval $(-\infty, 0]$. Under this restriction:
a. Find the inverse of f.   **b.** Graph each function and state each one's domain and range.

Solution We will find f^{-1} by using the four step strategy given in this section. We will then graph f and f^{-1} by using translations and then identify the domain and range from the graphs of each.

a. We first find f^{-1} and follow these steps:

Step 1: Replace $f(x)$ with y.

$f(x) = x^2 + 3$ $(x \le 0)$ The domain is restricted to $(-\infty, 0]$.
$y = x^2 + 3$

Step 2: Interchange the variables x and y.

$x = y^2 + 3$ $(y \le 0)$ Interchange x and y.

Step 3: Solve the resulting equation for y.

$x - 3 = y^2$ $(y \le 0)$

To solve this equation for y, we take the square root of both sides. Because $y \le 0$, we have

$-\sqrt{x - 3} = y$ $(y \le 0)$

Step 4: Replace y with $f^{-1}(x)$.

The inverse of f is defined by $f^{-1}(x) = -\sqrt{x - 3}$.

b. We graph the function $f(x) = x^2 + 3$, with domain $(-\infty, 0]$ by translating the graph of the parabola $y = x^2$, with domain $(-\infty, 0]$ vertically upward 3 units. From the graph, we see that the y coordinates are 3 and above and thus the range is the interval $[3, \infty)$. (See Figure 3-86.)

We graph the function $f^{-1}(x) = -\sqrt{x - 3}$ by translating the graph of $y = \sqrt{x}$ horizontally to the right 3 units and then reflecting the graph about the x-axis. It has domain $[3, \infty)$ and range $(-\infty, 0]$. (See Figure 3-86.) Note that the line of symmetry is shown and is $y = x$.

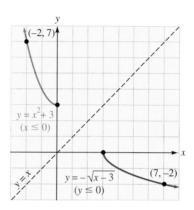

Figure 3-86

Domain of f and **range** of f^{-1}: $(-\infty, 0]$
Range of f and **domain** of f^{-1}: $[3, \infty)$

Self Check 6   Find the inverse of f when its domain is restricted to the interval $[0, \infty)$.

If a function is defined by the equation $y = f(x)$, we can often find the domain of f by inspection. Finding the range can be more difficult. One way to find the range of f is to find the domain of f^{-1}.

EXAMPLE 7   Find the domain and range of $f(x) = \dfrac{2}{x} + 3$. Find its range by finding the domain of $f^{-1}(x)$.

Solution   We will find the domain of $f(x) = \dfrac{2}{x} + 3$ by identifying the values of x that make the function undefined. We then will find $f^{-1}(x)$ and find its domain. The domain of $f^{-1}(x)$ will be the range of $f(x)$.

Because x cannot be 0, the domain of f is $(-\infty, 0) \cup (0, \infty)$. Next we find $f^{-1}(x)$.

Step 1:   Replace $f(x)$ with y.

$$f(x) = \frac{2}{x} + 3$$

$$y = \frac{2}{x} + 3$$

Step 2:   Interchange the variables x and y.

$$x = \frac{2}{y} + 3 \qquad \text{Interchange } x \text{ and } y.$$

Step 4: Find the y-intercept.
To find the y-intercept, we substitute 0 in for x and solve for y.

$$f(x) = 3(x + 2)^2 - 3$$
$$y = 3(x + 2)^2 - 3$$
$$y = 3(0 + 2)^2 - 3$$
$$y = 3(2)^2 - 3$$
$$y = 12 - 3$$
$$y = 9$$

The y-intercept is the point $(0, 9)$.

Step 5: Identify one additional point on the graph.
Because of symmetry, the point $(-4, 9)$ is on the graph.

Step 6: Draw a smooth curve through the points found in Steps 2–5.

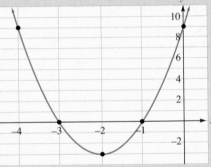

The axis of symmetry is $x = -2$ because $h = -2$.

If the graph of the parabola opens downward, then the vertex is the **maximum point** on the graph of the parabola.

If the graph of the parabola opens upward, then the vertex is the **minimum point** on the graph of the parabola.

Minimum cost   A company has found that the total monthly cost C of producing x air-hockey tables is given by $C(x) = 1.5x^2 - 270x + 28{,}665$. Find the production level that minimizes the monthly cost and find that monthly minimum cost.

The function $C(x)$ is a quadratic function whose graph is a parabola that opens upward. The minimum value of $C(x)$ occurs at the vertex of the parabola. We will use the vertex formula to find the vertex of the parabola.

We compare the equations

$$C(x) = 1.5x^2 - 270x + 28{,}665$$
and
$$y = ax^2 + bx + c$$

to see that $a = 1.5$, $b = -270$, and $c = 28{,}665$. Using the vertex formula, we see that the vertex of the parabola is the point with coordinates

$$\left(-\frac{b}{2a}, c - \frac{b^2}{4a}\right) = \left(-\frac{-270}{2(1.5)}, 28{,}665 - \frac{(-270)^2}{4(1.5)}\right)$$
$$= (90, 16{,}515)$$

If the company makes 90 air hockey tables each month, it will minimize its production cost. The minimum monthly cost will be \$16,515.

Exercises

Determine whether the graph of each quadratic function opens upward or downward. State whether a maximum or minimum point occurs at the vertex of the parabola.

21. $f(x) = \dfrac{1}{2}x^2 + 4$

22. $f(x) = -4(x + 1)^2 + 5$

Find the vertex of each parabola.

23. $f(x) = 2(x - 1)^2 + 6$ **24.** $y = -2(x + 4)^2 - 5$

25. $y = x^2 + 6x - 4$ **26.** $y = -4x^2 + 4x - 9$

Graph each quadratic function and find its vertex.

27. $f(x) = (x - 2)^2 - 3$ **28.** $f(x) = -(x - 4)^2 + 4$

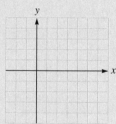 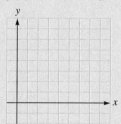

29. $y = x^2 - x$ **30.** $y = x - x^2$

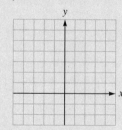 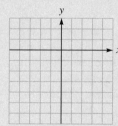

31. $y = x^2 - 3x - 4$ **32.** $y = 3x^2 - 8x - 3$

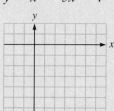 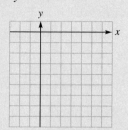

33. Architecture   A parabolic arch has an equation of $3x^2 + y - 300 = 0$. Find the maximum height of the arch.

34. Puzzle problem   The sum of two numbers is 1, and their product is as small as possible. Find the numbers.

35. Maximizing area   A rancher wants to enclose a rectangular corral with 1,400 feet of fencing. What are the dimensions of the corral that will maximize the area? Find the maximum area.

36. Digital cameras   A company that produces and sells digital cameras has determined that the total weekly cost C of producing x digital cameras is given by the function

$$C(x) = 1.5x^2 - 150x + 4{,}850$$

Determine the production level that minimizes the weekly cost for producing the digital cameras and find that weekly minimum cost.

3.3   Polynomial and Other Functions

Definitions and Concepts	Examples
A **polynomial function in one variable** (say, x) is a function of the form $$f(x) = a_n x^n + a_{n-1} x^{n-1} + \cdots + a_1 x + a_0$$ where $a_n, a_{n-1}, \ldots, a_1$, and a_0 are real numbers and n is a whole number. The **degree of a polynomial function** is the largest power of x that appears in the polynomial.	$f(x) = 3x^2 + 4x - 7$, degree of 2 $f(x) = -17x^4 + 3x^3 - 2x^2 + 13$, degree of 4

Graphing polynomial functions:

1. Find any symmetries of the graph.
2. Find the x- and y-intercepts of the graph.
3. Determine where the graph is above and below the x-axis.
4. Plot a few points, if necessary, and draw the graph as a smooth, continuous curve.

If $f(-x) = f(x)$ for all x in the domain of f, the graph of the function is symmetric about the y-axis, and the function is called an **even function.** If $f(-x) = -f(x)$ for all x in the domain of f, the function is symmetric about the origin, and the function is called an **odd function.**

If we trace the graph of a function from left to right and the values $f(x)$ increase, we say that the function is **increasing on the interval** (a, b). If the values $f(x)$ decrease, we say that the function is **decreasing on the interval** (a, b). If the values $f(x)$ remain unchanged as x increases we say that the function is **constant on the interval** (a, b).

Some functions, called **piecewise-defined functions,** are defined by using different equations for different intervals in their domains.

Graph the function $f(x) = -x^3 + 9x$ and determine whether it is even, odd, or neither.

Step 1: Find any symmetries of the graph.
To test for symmetry about the y-axis, we check to see whether $f(x) = f(-x)$. To test for symmetry about the origin, we check to see whether $f(x) = -f(x)$.

$$f(x) = -x^3 + 9x$$
$$f(-x) = -(-x)^3 + 9(-x) \quad \text{Substitute } -x \text{ for } x.$$
$$f(-x) = x^3 - 9x$$

Since $f(x) \neq f(-x)$, there is no symmetry about the y-axis. However, since $f(-x) = -f(x)$, there is symmetry about the origin.

Step 2: Find the x- and y-intercepts of the graph.
To find the x-intercepts, we let $f(x) = 0$ and solve for x.

$$-x^3 + 9x = 0$$
$$-x(x^2 - 9) = 0$$
$$-x(x + 3)(x - 3) = 0$$
$$-x = 0 \quad \text{or} \quad x + 3 = 0 \quad \text{or} \quad x - 3 = 0$$
$$x = 0 \qquad\qquad x = -3 \qquad\qquad x = 3$$

The x-intercepts are $(0, 0)$, $(-3, 0)$, and $(3, 0)$.

If we let $x = 0$ and solve for $f(x)$, we see that the y-intercept is also $(0, 0)$.

Step 3: Determine where the graph is above or below the x-axis.
To determine where the graph is above or below the x-axis, we plot the solutions of $-x^3 + 9x = 0$ (the x-intercepts) on a number line and establish the four intervals shown in the figure. We then test a number from each interval to determine the sign of $f(x)$.

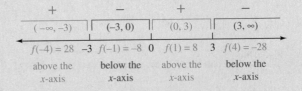

Step 4: Plot a few points and draw the graph as a smooth, continuous curve.
We now plot the intercepts and one additional point. In the previous step, we found that $f(1) = 8$. This will be the additional point we plot $(1, 8)$. Making use of our knowledge of symmetry and where the graph is above and below the x-axis, we now draw the graph as shown.

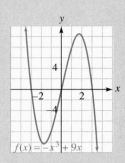

The function $f(x) = -x^3 + 9x$ is an odd function because the function is symmetric about the origin.

The **greatest integer function** is important in many business applications and in the field of computer science. This function is determined by the equation $f(x) = [x]$, where the value of $f(x)$ that corresponds to x is the greatest integer that is less than or equal to x.	Let $f(x) = [x - 3]$. Find $f(2.2)$ $f(x) = [x - 3]$ $f(2.2) = [2.2 - 3] = [-0.8] = -1$ -1 is the greatest integer less than or equal to -0.8.

Exercises

Graph each polynomial function and determine whether it is even, odd, or neither.

37. $y = x^3 - x$

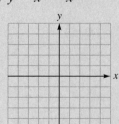

38. $y = x^2 - 4x$

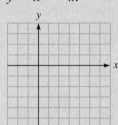

39. $y = x^3 - x^2$

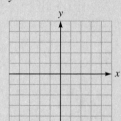

40. $y = 1 - x^4$

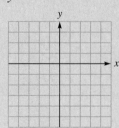

Evaluate each piecewise-defined function.

41. $f(x) = \begin{cases} x - 2 & \text{if } x < 3 \\ x^2 & \text{if } x \geq 3 \end{cases}$

    a. $f(-2)$ **b.** $f(3)$

42. $f(x) = \begin{cases} 2 & \text{if } x < 0 \\ 2 - x & \text{if } 0 \leq x < 2 \\ x + 1 & \text{if } x \geq 2 \end{cases}$

    a. $f\left(\frac{3}{2}\right)$ **b.** $f(2)$

Graph each piecewise-defined function and determine when it is increasing, decreasing, or constant.

43. $y = f(x) = \begin{cases} x + 5 & \text{if } x \leq 0 \\ 5 - x & \text{if } x > 0 \end{cases}$

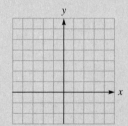

44. $y = f(x) = \begin{cases} x + 3 & \text{if } x \leq 0 \\ 3 & \text{if } x > 0 \end{cases}$

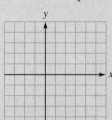

Evaluate each function at the indicated x-values.

45. $f(x) = [2x]$ Find $f(1.7)$.

46. $f(x) = [x - 5]$ Find $f(4.99)$.

Graph each function.

47. $f(x) = \lfloor x \rfloor + 2$

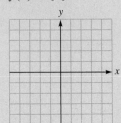

48. $f(x) = \lfloor x - 1 \rfloor$

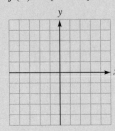

49. Renting a Jeep A rental company charges $20 to rent a Jeep Wrangler for one day, plus $8 for every 100 miles (or portion of 100 miles) that it is driven. Find the cost if the car is driven 295 miles in one day.

50. Riding in a taxi A taxicab company charges $4 for a trip up to 1 mile, and $2 for every extra mile (or portion of a mile). Find the cost to ride $11\frac{1}{2}$ miles.

| **3.4** | **Translating and Stretching Graphs** |

Definitions and Concepts

Vertical translations:

If $k > 0$, the graph of $\begin{cases} y = f(x) + k \\ y = f(x) - k \end{cases}$ is identical to the graph of $y = f(x)$, except that it is translated k units $\begin{cases} \text{up} \\ \text{down} \end{cases}$.

Horizontal translations:

If $k > 0$, the graph of $\begin{cases} y = f(x - k) \\ y = f(x + k) \end{cases}$ is identical to the graph of $y = f(x)$, except that it is translated k units to the $\begin{cases} \text{right} \\ \text{left} \end{cases}$.

Vertical stretchings:
If f is a function and $k > 1$, then

- The graph of $y = kf(x)$ can be obtained by stretching the graph of $y = f(x)$ vertically by multiplying each value of $f(x)$ by k.

If f is a function and $0 < k < 1$, then

- The graph of $y = kf(x)$ can be obtained by shrinking the graph of $y = f(x)$ vertically by multiplying each value of $f(x)$ by k.

Horizontal stretchings:
If f is a function and $k > 1$, then

- The graph of $y = f(kx)$ can be obtained by shrinking the graph of $y = f(x)$ horizontally by multiplying each x-value of $f(x)$ by $\frac{1}{k}$.

If f is a function and $0 < k < 1$,

- The graph of $y = f(kx)$ can be obtained by stretching the graph of $y = f(x)$ horizontally by multiplying each x-value of $f(x)$ by $\frac{1}{k}$.

Examples

The function $g(x) = \sqrt{x + 3} - 2$ is a translation of the graph of $f(x) = \sqrt{x}$. Graph both on one set of coordinate axes.

By inspection, we see that the function $g(x) = \sqrt{x + 3} - 2$ involves two translations of $f(x) = \sqrt{x}$. The graph of $g(x) = \sqrt{x + 3} - 2$ is identical to the graph of $f(x) = \sqrt{x}$ except it is translated 3 units to the left and 2 units downward as shown in the figure.

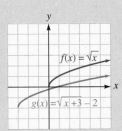

Graph $g(x) = -\frac{1}{3}|x|$.

The graph of $g(x) = -\frac{1}{3}|x|$ is identical to the graph of $f(x) = |x|$ except that it is vertically shrunk by a factor of $\frac{1}{3}$ and reflected about the x-axis. This is because each value of $|x|$ is multiplied by $-\frac{1}{3}$. The graphs of both functions are shown in the figure.

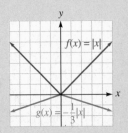

Reflections:

If f is a function, then

- The graph of $y = -f(x)$ is identical to the graph of $y = f(x)$ except that it is reflected about the x-axis.

- The graph of $y = f(-x)$ is identical to the graph of $y = f(x)$ except that it is reflected about the y-axis.

To graph functions involving a combination of the translations and stretchings, we must apply each translation or stretching to the function. We will apply these translations and stretchings in the following order:

1. horizontal translation

2. stretching or shrinking

3. reflection

4. vertical translation

Graph $g(x) = 2(x - 4)^3 + 1$.

We will graph $g(x) = 2(x - 4)^3 + 1$ by applying three translations to the basic function $f(x) = x^3$: translate $f(x) = x^3$ horizontally 4 units to the right, stretch the graph vertically by a factor of 2, and translate the graph vertically 1 unit upward. The graphs of both functions are shown in the figure.

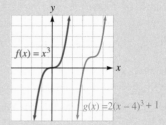

Exercises

Each function is a translation of a basic function. Graph both on one set of coordinate axes.

51. $g(x) = x^2 + 5$

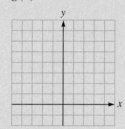

52. $g(x) = (x - 7)^3$

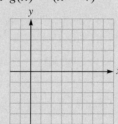

53. $g(x) = \sqrt{x + 2} + 3$

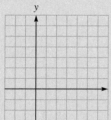

54. $g(x) = |x - 4| + 2$

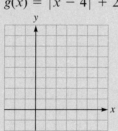

Each function is a stretching of $f(x) = x^3$. Graph both on one set of coordinate axes.

55. $g(x) = \dfrac{1}{3}x^3$

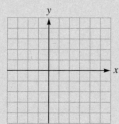

56. $g(x) = (-5x)^3$

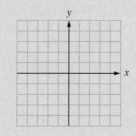

Graph each function using a combination of translations and stretchings.

57. $g(x) = -|x - 4| + 3$

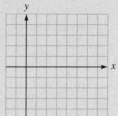

58. $g(x) = \dfrac{1}{4}|x - 4| + 1$

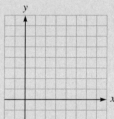

59. $g(x) = 3\sqrt{x + 3} + 2$

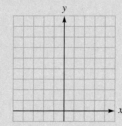

60. $g(x) = \dfrac{1}{3}(x + 3)^3 + 2$

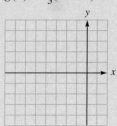

61. $f(x) = \sqrt{-x} + 3$

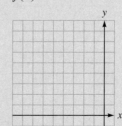

62. $g(x) = 2\sqrt[3]{x} - 5$

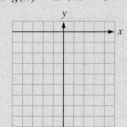

3.5 Rational Functions

Definitions and Concepts	Examples

Definitions and Concepts

A **rational function** is a function defined by an equation of the form $y = \frac{P(x)}{Q(x)}$ where $P(x)$ and $Q(x)$ are polynomials and $Q(x) \neq 0$.

Locating Vertical Asymptotes:
To locate the vertical asymptotes of a rational function, $f(x) = \frac{P(x)}{Q(x)}$, we follow these steps:

Step 1: Factor $P(x)$ and $Q(x)$ and remove any common factors.

Step 2: Set the denominator equal to 0 and solve the equation.

If a is a solution of the equation found in Step 2, $x = a$ is a vertical asymptote.

Locating Horizontal Asymptotes:
To locate the horizontal asymptote of a rational function $f(x) = \frac{P(x)}{Q(x)}$, we consider three cases:

Case 1: If the degree of $P(x)$ *is less than* the degree of $Q(x)$, the line $y = 0$ is a horizontal asymptote.

Case 2: If the degree of $P(x)$ and $Q(x)$ are equal, the line $y = \frac{p}{q}$, where p and q are the leading coefficients of $P(x)$ and $Q(x)$, is a horizontal asymptote.

Case 3: If the degree of $P(x)$ is greater than the degree of $Q(x)$, there is no horizontal asymptote.

A third type of asymptote is called a **slant asymptote.** These asymptotes occur when the degree of the numerator of a rational function is one more than the degree of the denominator. As the name implies, it is a slanted line, neither vertical nor horizontal.

Locating slant asymptotes:
If the degree of $P(x)$ is 1 greater than the degree of $Q(x)$ for the rational function $f(x) = \frac{P(x)}{Q(x)}$, there is a slant asymptote. To find it, divide $P(x)$ by $Q(x)$ and ignore the remainder.

Examples

$$f(x) = \frac{3}{x + 2} \qquad f(x) = \frac{3x + 4}{x^2 - 3x + 4}$$

Graph: $y = f(x) = \frac{4x}{x - 2}$.

We will use the steps outlined to graph the function.

Step 1: Symmetry
We find $f(-x)$.

$$f(-x) = \frac{4(-x)}{(-x) - 2} = \frac{-4x}{-x - 2} = \frac{4x}{x + 2}$$

Because $f(-x) \neq f(x)$ and $f(-x) \neq -f(x)$, there is no symmetry about the y-axis or the origin.

Step 2: Vertical asymptotes
We first note that $f(x)$ is in simplest form. We then set the denominator equal to 0 and solve for x. Since the solution is 2, there will be a vertical asymptote at $x = 2$.

Step 3: y- and x-intercepts
We can find the y-intercept by finding $f(0)$.

$$f(0) = \frac{4(0)}{(0) - 2} = \frac{0}{-2} = 0$$

The y-intercept is $(0, 0)$.

We can find the x-intercepts by setting the numerator equal to 0 and solving for x:

$$4x = 0$$
$$x = 0$$

The x-intercept is $(0, 0)$.

Step 4: Horizontal asymptotes
Since the degrees of the numerator and denominator of the polynomials are the same, the line

$$y = \frac{4}{1} = 4 \quad \text{The leading coefficient of the numerator is 4.}$$
The leading coefficient of the denominator is 1.

is a horizontal asymptote whose equation is $y = 4$.

Step 5: Slant asymptotes
Since the degree of the numerator is not 1 greater than the degree of the denominator, there are no slant asymptotes.

Steps to graph a rational function:
We will use the following steps to graph the rational function, $f(x) = \frac{P(x)}{Q(x)}$ where $\frac{P(x)}{Q(x)}$ is in simplest form (no common factors).

1. Check symmetries.
2. Look for vertical asymptotes.
3. Look for the y- and x-intercepts.
4. Look for horizontal asymptotes.
5. Look for slant asymptotes.
6. Graph the function.

Step 6: Graph
First, we plot the intercept $(0, 0)$ and draw the asymptotes. We then find one additional point on our graph to see what happens when x is greater than 2. To do so, we choose 3, a value of x that is greater than 2, and evaluate $f(3)$.

$$f(3) = \frac{4(3)}{(3) - 2} = \frac{12}{1} = 12$$

Since $f(3) = 12$, the point $(3, 12)$ lies on our graph. We sketch the graph as shown in the figure.

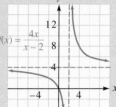

Exercises

Find the domain of each rational function.

63. $f(x) = \dfrac{3x^2 + x - 2}{x^2 - 25}$

64. $f(x) = \dfrac{2x^2 + 1}{x^2 + 7}$

Find the vertical asymptotes, if any, of each rational function.

65. $f(x) = \dfrac{x + 5}{x^2 - 1}$ **66.** $f(x) = \dfrac{x - 7}{x^2 - 49}$

67. $f(x) = \dfrac{x}{x^2 + x - 6}$ **68.** $f(x) = \dfrac{5x + 2}{2x^2 - 6x - 8}$

Find the horizontal asymptotes, if any, of each rational function.

69. $f(x) = \dfrac{2x^2 + x - 2}{4x^2 - 4}$ **70.** $f(x) = \dfrac{5x^2 + 4}{4 - x^2}$

71. $f(x) = \dfrac{x + 1}{x^3 - 4x}$ **72.** $f(x) = \dfrac{x^3}{2x^2 - x + 11}$

Find the slant asymptote, if any, for each rational function.

73. $f(x) = \dfrac{2x^2 - 5x + 1}{x - 4}$ **74.** $f(x) = \dfrac{5x^3 + 1}{x + 5}$

Graph each rational function.

75. $f(x) = \dfrac{2x}{x - 4}$ **76.** $f(x) = \dfrac{-4x}{x + 4}$

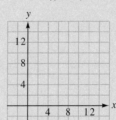

77. $f(x) = \dfrac{x}{(x - 1)^2}$ **78.** $f(x) = \dfrac{(x - 1)^2}{x}$

79. $f(x) = \dfrac{x^2 - x - 2}{x^2 + x - 2}$ **80.** $f(x) = \dfrac{x^3 + x}{x^2 - 4}$

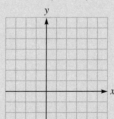
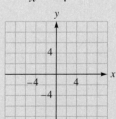

3.6   Operations on Functions

Definitions and Concepts	Examples
Adding, subtracting, multiplying, and dividing functions:	Let $f(x) = 3x + 5$ and $g(x) = 4x - 7$. Find each function and its domain: **a.** $f + g$ **b.** $f - g$ **c.** $f \cdot g$ **d.** f/g

If the ranges of functions f and g are subsets of the real numbers, then

1. The **sum of f and g,** denoted as $f + g$, is defined by

$$(f + g)(x) = f(x) + g(x)$$

a. $(f + g)(x) = f(x) + g(x)$
$$= (3x + 5) + (4x - 7)$$
$$= 7x - 2$$

Since the domain of both f and g is the set of real numbers, the domain of $f + g$ is the interval $(-\infty, \infty)$.

2. The **difference of f and g,** denoted as $f - g$, is defined by

$$(f - g)(x) = f(x) - g(x)$$

b. $(f - g)(x) = f(x) - g(x)$
$$= (3x + 5) - (4x - 7)$$
$$= 3x + 5 - 4x + 7$$
$$= -x + 12$$

Since the domain of both f and g is the set of real numbers, the domain of $f - g$ is the interval $(-\infty, \infty)$.

3. The **product of f and g,** denoted as $f \cdot g$, is defined by

$$(f \cdot g)(x) = f(x) \cdot g(x)$$

c. $(f \cdot g)(x) = f(x) \cdot g(x)$
$$= (3x + 5)(4x - 7)$$
$$= 12x^2 - x - 35$$

Since the domain of both f and g is the set of real numbers, the domain of $f \cdot g$ is the interval $(-\infty, \infty)$.

4. The **quotient of f and g,** denoted as f/g, is defined by

$$(f/g)(x) = \frac{f(x)}{g(x)} \quad g(x) \neq 0$$

The domain of each function, unless otherwise restricted, is the set of real numbers x that are in the domains of both f and g. In the case of the quotient f/g, there is the restriction that $g(x) \neq 0$.

d. $(f/g)(x) = \dfrac{f(x)}{g(x)}$
$$= \frac{3x + 5}{4x - 7} \quad (4x - 7 \neq 0)$$

Since $\frac{7}{4}$ will make $4x - 7$ equal to 0, the domain of f/g is the set of all real numbers except $\frac{7}{4}$. This is $\left(-\infty, \frac{7}{4}\right) \cup \left(\frac{7}{4}, \infty\right)$.

The **composite function** $f \circ g$ is defined by

$$(f \circ g)(x) = f(g(x))$$

The domain of $f \circ g$ consists of all those numbers in the domain of g for which $g(x)$ is in the domain of f.

If $f(x) = 2x + 7$ and $g(x) = x^2 + 1$, find $(f \circ g)(x)$ and its domain.

Because $(f \circ g)(x)$ means $f(g(x))$, we will replace x in $f(x) = 2x + 7$ with $g(x)$.

$$(f \circ g)(x) = f(g(x))$$
$$= f(x^2 + 1)$$
$$= 2(x^2 + 1) + 7$$
$$= 2x^2 + 9$$

The domain of $(f \circ g)(x)$ is the interval $(-\infty, \infty)$ because the domain of both f and g consists of all real numbers.

Exercises

Let $f(x) = x^2 - 1$ and $g(x) = 2x + 1$. Find each function and its domain.

81. $f + g$ **82.** $f \cdot g$

83. $f - g$ **84.** f/g

Let $f(x) = 2x^2 - 1$ and $g(x) = 2x - 1$. Find each value, if possible.

85. $(f + g)(-3)$ **86.** $(f - g)(-5)$

87. $(f \cdot g)(2)$ **88.** $(f/g)\left(\dfrac{1}{2}\right)$

Let $f(x) = x^2 - 1$ and $g(x) = 2x + 1$. Find each function and its domain.

89. $f \circ g$ **90.** $g \circ f$

Let $f(x) = x^2 - 5$ and $g(x) = 3x + 1$. Find each value.

91. $(f \circ g)(-2)$ **92.** $(g \circ f)(-2)$

Find two functions f and g such that the composition $f \circ g = h$ expresses the given correspondence. Several answers are possible.

93. $h(x) = (x - 5)^2$ **94.** $h(x) = (x + 6)^3$

3.7   Inverse Functions

Definitions and Concepts	Examples
A function f from a set **X** to a set **Y** is called a **one-to-one function** if and only if different numbers in the domain of f have different outputs in the range of f. A **horizontal line test** can be used to determine whether the graph of a function represents a one-to-one function. If every horizontal line that intersects the graph of a function does so exactly once, the function passes the horizontal line test and is one-to-one.	Determine whether the function $f(x) = x^4 - 2x^2$ is one-to-one. The function $f(x) = x^4 - 2x^2$ is not one-to-one, because different numbers in the domain have the same output. For example, 2 and -2 have the same output: $f(2) = f(-2) = 8$.
Inverse functions: If f and g are two one-to-one functions such that $(f \circ g)(x) = x$ for every x in the domain of g and $(g \circ f)(x) = x$ for every x in the domain of f, then f and g are inverse functions. Function g is denoted as f^{-1}, and is called the **inverse function of f**. **Properties of a one-to-one function:** **Property 1:** If f is a one-to-one function, there is a one-to-one function $f^{-1}(x)$ such that $\quad (f^{-1} \circ f)(x) = x \quad$ and $\quad (f \circ f^{-1})(x) = x$ **Property 2:** The domain of f is the range of f^{-1} and the range of f is the domain of f^{-1}.	Verify that $f(x) = x^5$ and $g(x) = \sqrt[5]{x}$ are inverse functions. To show that f and g are inverse functions, we must show that $f \circ g$ and $g \circ f$ are x, the identity function. $\quad (f \circ g)(x) = f(g(x)) = f\left(\sqrt[5]{x}\right) = \left(\sqrt[5]{x}\right)^5 = x$ $\quad (g \circ f)(x) = g(f(x)) = g(x^5) = \sqrt[5]{x^5} = x$ Because g is the inverse of f, we can use inverse notation and write $f(x) = x^5$ and $f^{-1}(x) = \sqrt[5]{x}$. Because f is the inverse of g, we can use inverse notation and write $g(x) = \sqrt[5]{x}$ and $g^{-1}(x) = x^5$.

Let $f(x) = 3x$ and $g(x) = x^2 + 2$. Find each function.

23. $f + g$

24. $g \circ f$

25. f/g

26. $f \circ g$

Assume that $f(x)$ is one-to-one. Find f^{-1}.

27. $f(x) = \dfrac{x + 1}{x - 1}$

28. $f(x) = x^3 - 3$

Find the range of f by finding the domain of f^{-1}.

29. $y = \dfrac{3}{x} - 2$

30. $y = \dfrac{3x - 1}{x - 3}$

CUMULATIVE REVIEW EXERCISES

Use the x- and y-intercepts to graph each equation.

1. $5x - 3y = 15$

2. $3x + 2y = 12$

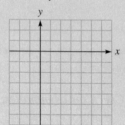

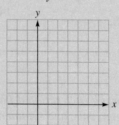

Find the length, the midpoint, and the slope of the line segment PQ.

3. $P\left(-2, \dfrac{7}{2}\right)$; $Q\left(3, -\dfrac{1}{2}\right)$

4. $P(3, 7)$; $Q(-7, 3)$

Write the equation of the line with the given properties. Give the answer in slope–intercept form.

5. The line passes through $(-3, 5)$ and $(3, -7)$.

6. The line passes through $\left(\dfrac{3}{2}, \dfrac{5}{2}\right)$ and has a slope of $\dfrac{7}{2}$.

7. The line is parallel to $3x - 5y = 7$ and passes through $(-5, 3)$.

8. The line is perpendicular to $x - 4y = 12$ and passes through the origin.

Graph each equation. Make use of intercepts and symmetries.

9. $x^2 = y - 2$

10. $y^2 = x - 2$

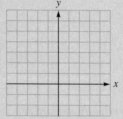

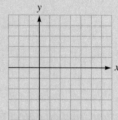

11. $x^2 + y^2 = 100$

12. $x^2 - 2x + y^2 = 8$

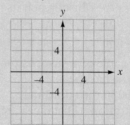

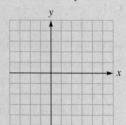

Solve each proportion.

13. $\dfrac{x - 2}{x} = \dfrac{x - 6}{5}$

14. $\dfrac{x + 2}{x - 6} = \dfrac{3x + 1}{2x - 11}$

15. Dental billing The billing schedule for dental X-rays specifies a fixed amount for the office visit plus a fixed amount for each X-ray exposure. If 2 X-rays cost \$37 and 4 cost \$54, find the cost of 5 exposures.

16. Automobile collisions   The energy dissipated in an automobile collision varies directly with the square of the speed. By what factor does the energy increase in a 50-mph collision compared with a 20-mph collision?

Determine whether each equation defines a function.

17. $y = 3x - 1$

18. $y = x^2 + 3$

19. $y = \dfrac{1}{x - 2}$

20. $y^2 = 4x$

Find the domain of each function.

21. $f(x) = x^2 + 5$

22. $f(x) = \dfrac{7}{x + 2}$

23. $y = -\sqrt{x - 2}$

24. $f(x) = \sqrt{x + 4}$

Find the vertex of the parabolic graph of each equation.

25. $y = x^2 + 5x - 6$

26. $f(x) = -x^2 + 5x + 6$

Graph each function.

27. $f(x) = x^2 - 4$

28. $f(x) = -x^2 + 4$

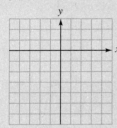
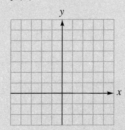

29. $f(x) = x^3 + x$

30. $f(x) = -x^4 + 2x^2 + 1$

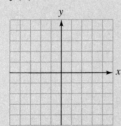
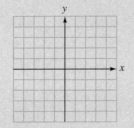

Graph each function. Show all asymptotes.

31. $f(x) = \dfrac{x}{x - 3}$

32. $f(x) = \dfrac{x^2 - 1}{x^2 - 9}$

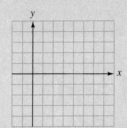
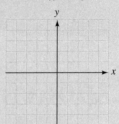

Let $f(x) = 3x - 4$ and $g(x) = x^2 + 1$. Find each function and its domain.

33. $(f + g)(x)$

34. $(f - g)(x)$

35. $(f \cdot g)(x)$

36. $(f/g)(x)$

Let $f(x) = 3x - 4$ and $g(x) = x^2 + 1$. Find each value.

37. $(f \circ g)(2)$

38. $(g \circ f)(2)$

39. $(f \circ g)(x)$
40. $(g \circ f)(x)$

Find the inverse of the function defined by each equation.

41. $y = 3x + 2$

42. $y = \dfrac{1}{x - 3}$

43. $y = x^2 + 5 \ (x \geq 0)$

44. $3x - y = 1$

Write each sentence as an equation.

45. y varies directly with the product of w and z.

46. y varies directly with x and inversely with the square of t.

Exponential and Logarithmic Functions

Careers and Mathematics

Electrical and Electronics Engineers   Although the terms electrical and electronics engineering often are used interchangeably in academia and industry, electrical engineers traditionally have focused on the generation and supply of power, whereas electronics engineers have worked on applications of electricity to control systems or signal processing.

Electrical engineers design, develop, test, and supervise the manufacture of electrical equipment. Some of this equipment includes electric motors; machinery controls, lighting, and wiring in buildings; automobiles; aircraft; radar and navigation systems; and power generation devices used by electric utilities.

Electronics engineers are responsible for a wide range of technologies, from portable music players to the global positioning system (GPS). Electronics engineers design, develop, test, and supervise the manufacture of electronic equipment such as broadcast and communications systems.

Electrical and electronics engineers held about 291,000 jobs in 2006.

Education   Electrical and electronics engineers typically enter the occupation with a bachelor's degree, but some basic research positions may require a graduate degree. Engineers offering their services directly to the public must be licensed.

Job Outlook   Employment of electrical engineers is expected to grow by 6 percent through 2016. Median annual earnings of electrical engineers were $75,930 in 2006. Employment of electronics engineers is expected to grow by 4 percent through 2016. Median annual earnings of electronic engineers were $81,050 in 2006.

For a sample application, see Example 1 in Section 4.4. For more information, see www.bls.gov/cos/ocos027.htm.

In this chapter, we will discuss exponential functions, which are often used in banking, ecology, and science. We also will discuss logarithmic functions, which are applied in chemistry, geology, and environmental science.

© SCPhotos/Alamy

4.1 Exponential Functions and Their Graphs

Objectives

1. Define and Use Irrational Exponents
2. Graph Exponential Functions
3. Solve Compound Interest Problems
4. Define *e* and Graph Base-*e* Exponential Functions
5. Use Translations to Graph Exponential Functions

Extreme water slides, called *plunge* or *plummet* slides, are fearsome water slides because of their heights. With nearly vertical drops, the slides are designed to allow riders to reach the greatest possible speeds. Summit Plummet at Blizzard Beach, a part of Walt Disney World Resort in Florida, stands 120 ft tall. On this slide, riders can achieve speeds up to 55 mph.

The shapes of extreme water slides can be modeled using *exponential functions,* the topic of this section.

Exponential functions are also important in business. Consider the graph shown in Figure 4-1. It shows the balance in a bank account in which $5,000 was invested in 1990 at 8%, compounded monthly. The graph shows that in the year 2015, the value of the account will be approximately $38,000, and in the year 2030, the value will be approximately $121,000.

The curve in Figure 4-1 is the graph of an exponential function. From the graph, we can see that the longer the money is kept on deposit, the more rapidly it will grow.

Sir Isaac Newton
(1642–1727)

Newton was an English scientist and mathematician. Because he was not a good farmer, he went to Cambridge University to become a preacher. When he had to leave Cambridge because of the plague, he made some of his most important discoveries. He is best known in mathematics for developing calculus and in physics for discovering the laws of motion. Newton probably contributed more to science and mathematics than anyone else in history.

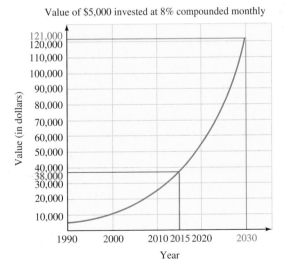

Value of $5,000 invested at 8% compounded monthly

Figure 4-1

Before we can discuss exponential functions, we must define irrational exponents.

1. Define and Use Irrational Exponents

We have discussed expressions of the form b^x, where x is a rational number.

$$5^2 \quad \text{means} \quad \text{"the square of 5."}$$

$$4^{1/3} \quad \text{means} \quad \text{"the cube root of 4."}$$

$$6^{-2/5} = \frac{1}{6^{2/5}} \quad \text{means} \quad \text{"the reciprocal of the fifth root of } 6^2\text{."}$$

To give meaning to b^x when x is an irrational number, we consider the expression

$$3^{\sqrt{2}} \quad \text{where } \sqrt{2} \text{ is the irrational number } 1.414213562\ldots$$

As a rational number r gets closer to the irrational number $\sqrt{2}$, the expression 3^r gets closer to some number p as shown in the table below. This number is defined to be $3^{\sqrt{2}}$.

3^r	Scientific calculator	Graphing calculator
$3^{1.4} \approx 4.655536722$	Press 3 y^x 1.4 $=$	3 $\wedge$ 1.4 ENTER
$3^{1.41} \approx 4.706965002$	Press 3 y^x 1.41 $=$	3 $\wedge$ 1.41 ENTER
$3^{1.414} \approx 4.727695035$	Press 3 y^x 1.414 $=$	3 $\wedge$ 1.414 ENTER
$3^{1.4142} \approx 4.72873393$	Press 3 y^x 1.4142 $=$	3 $\wedge$ 1.4142 ENTER

From the last two lines in the table, we see that the results agree to two decimal places: $3^{\sqrt{2}} \approx 4.72$. To find a better approximation of $3^{\sqrt{2}}$, we can use a calculator.

	Scientific calculator	Graphing calculator
$3^{\sqrt{2}} \approx 4.728804388$	Press 3 y^x 2 $\sqrt{}$ $=$	3 $\wedge$ $\sqrt{}$ 2 ENTER

If b is a positive number and x is a real number, the expression b^x always represents a positive number. It is also true that the familiar properties of exponents hold for irrational exponents.

EXAMPLE 1 Simplify each expression: **a.** $\left(3^{\sqrt{2}}\right)^{\sqrt{2}}$ **b.** $a^{\sqrt{8}} \cdot a^{\sqrt{2}}$

Solution We will use properties of exponents to simplify each expression.

a. $\left(3^{\sqrt{2}}\right)^{\sqrt{2}} = 3^{\sqrt{2}\sqrt{2}}$ **Keep the base and multiply the exponents.**

$\qquad\qquad\quad = 3^2 \qquad \sqrt{2}\sqrt{2} = \sqrt{4} = 2$

$\qquad\qquad\quad = 9$

b. $a^{\sqrt{8}} \cdot a^{\sqrt{2}} = a^{\sqrt{8}+\sqrt{2}}$ **Keep the base and add the exponents.**

$\qquad\qquad\quad = a^{2\sqrt{2}+\sqrt{2}} \qquad \sqrt{8} = \sqrt{4}\sqrt{2} = 2\sqrt{2}$

$\qquad\qquad\quad = a^{3\sqrt{2}} \qquad 2\sqrt{2} + \sqrt{2} = 3\sqrt{2}$

Self Check 1 Simplify: **a.** $\left(2^{\sqrt{3}}\right)^{\sqrt{12}}$ **b.** $x^{\sqrt{20}} \cdot x^{\sqrt{5}}$

2. Graph Exponential Functions

If $b > 0$ and $b \neq 1$, the equation $y = b^x$ defines a function, because for each input x, there is exactly one output y. Since x can be any real number, the domain of the function is the set of real numbers. Since the base b of the expression b^x is positive, y is always positive, and the range is the set of positive numbers. Since b^x is an exponential expression, the function is called an **exponential function.**

We make the restriction that $b > 0$ to exclude any imaginary numbers that might result from taking even roots of negative numbers. The restriction that $b \neq 1$ excludes the constant function $f(x) = 1^x$, in which $f(x) = 1$ for every real number x.

Exponential Functions

An **exponential function with base b** is defined by the equation

$$f(x) = b^x \qquad \text{or} \qquad y = b^x \quad (b > 0, b \neq 1, \text{ and } x \text{ is a real number})$$

The **domain of any exponential function** is the interval $(-\infty, \infty)$. The **range** is the interval $(0, \infty)$.

Since the domain and range of $f(x) = b^x$ are sets of real numbers, we can graph exponential functions. For example, to graph

$$f(x) = 2^x$$

we find several points $(x, f(x))$ whose coordinates satisfy the equation, plot the points, and join them with a smooth curve, as in Figure 4-2(a). To graph the function

$$f(x) = \left(\frac{1}{2}\right)^x$$

we find several points $(x, f(x))$ whose coordinates satisfy the equation, plot the points, and join them with a smooth curve, as shown in Figure 4-2(b).

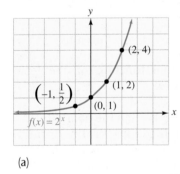

$f(x) = 2^x$		
x	$f(x)$	$(x, f(x))$
-1	$\frac{1}{2}$	$\left(-1, \frac{1}{2}\right)$
0	1	$(0, 1)$
1	2	$(1, 2)$
2	4	$(2, 4)$

(a)

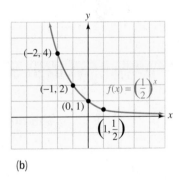

$f(x) = \left(\frac{1}{2}\right)^x$		
x	$f(x)$	$(x, f(x))$
-2	4	$(-2, 4)$
-1	2	$(-1, 2)$
0	1	$(0, 1)$
1	$\frac{1}{2}$	$\left(1, \frac{1}{2}\right)$

(b)

Figure 4-2

By looking at the graphs in Figure 4-2, we can see that the domain of each function is the interval $(-\infty, \infty)$ and that the range is the interval $(0, \infty)$.

EXAMPLE 2 Graph: $f(x) = 4^x$

Solution We will find several points (x, y) that satisfy the equation, plot the points, and join them with a smooth curve, as in Figure 4-3.

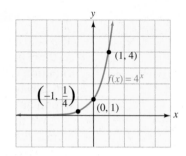

$$f(x) = 4^x$$

x	$f(x)$	$(x, f(x))$
-1	$\frac{1}{4}$	$\left(-1, \frac{1}{4}\right)$
0	1	$(0, 1)$
1	4	$(1, 4)$

Figure 4-3

Self Check 2   Graph: $f(x) = \left(\dfrac{1}{4}\right)^x$

The graph of $f(x) = 4^x$ in Example 2 has the following properties:

1. It passes through the point $(0, 1)$.
2. It passes through the point $(1, 4)$.
3. It approaches the x-axis. The x-axis is a horizontal asymptote.
4. The domain is the interval $(-\infty, \infty)$, and the range is the interval $(0, \infty)$.

This example illustrates the following properties of exponential functions.

Properties of Exponential Functions

The **domain of the exponential function** $f(x) = b^x$ is $(-\infty, \infty)$, the set of real numbers.

The **range** is $(0, \infty)$, the set of positive real numbers.

The graph has a y-intercept at $(0, 1)$.

The x-axis is an asymptote of the graph.

The graph of $f(x) = b^x$ passes through the point $(1, b)$.

EXAMPLE 3   The graph of an exponential function of the form $f(x) = b^x$ is shown in Figure 4-4. Find the value of b.

Solution   We note that the graph passes through the point $(0, 1)$, a property of exponential functions of this form. Since the graph also passes through the point $(2, 25)$, we can find the base b by substituting 2 for x and 25 for $f(2)$ in the equation $f(x) = b^x$ and solving for b.

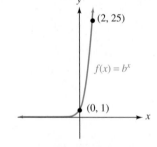

$$f(x) = b^x$$
$$f(2) = b^2$$
$$25 = b^2$$
$$5 = b \quad \textit{b \textbf{must be positive.}}$$

Figure 4-4

The base b is 5. Note that the points $(0, 1)$ and $(2, 25)$ satisfy the equation $f(x) = 5^x$.

Self Check 3   Can a graph passing through $(0, 2)$ and $\left(1, \frac{3}{2}\right)$ be the graph of $f(x) = b^x$?

Comment
Recall that $b^{-x} = \frac{1}{b^x} = \left(\frac{1}{b}\right)^x$. If $b > 1$, any function of the form $f(x) = b^{-x}$ models exponential decay, because $0 < \frac{1}{b} < 1$.

In Figure 4-2(a) (where $b = 2$ and $2 > 1$), the values of y increase as the values of x increase. Since the graph rises as we move to the right, the function is an increasing function. Such a function is said to model *exponential growth.*

In Figure 4-2(b) $\left(\text{where } b = \frac{1}{2} \text{ and } 0 < \frac{1}{2} < 1\right)$, the values of y decrease as the values of x increase. Since the graph drops as we move to the right, the function is a decreasing function. Such a function is said to model *exponential decay.*

Increasing and Decreasing Functions

If $b > 1$, then $f(x) = b^x$ is an **increasing function.** This function models **exponential growth.**

If $0 < b < 1$, then $f(x) = b^x$ is a **decreasing function.** This function models **exponential decay.**

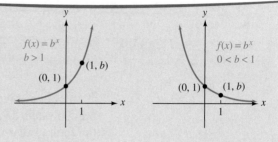

Increasing function   Decreasing function

An exponential function $f(x) = b^x$ is either increasing (for $b > 1$) or decreasing (for $0 < b < 1$). Because different real numbers x always determine different values of b^x, an exponential function is one-to-one.

An exponential function defined by

$$f(x) = b^x \quad \text{or} \quad y = b^x, \quad \text{where } b > 0 \text{ and } b \neq 1$$

is one-to-one. This implies that

1. If $b^r = b^s$, then $r = s$.
2. If $r \neq s$, then $b^r \neq b^s$.

3. Solve Compound Interest Problems

Banks pay **interest** for using their customers' money. Interest is calculated as a percent of the amount on deposit in an account and is paid annually (once a year), quarterly (four times per year), monthly, or daily. Interest left on deposit in a bank account also will earn interest. Such accounts are said to earn **compound interest.**

Compound Interest Formula

If P dollars are deposited in an account earning interest at an annual rate r, compounded k times each year, the amount A in the account after t years is given by

$$A = P\left(1 + \frac{r}{k}\right)^{kt}$$

EXAMPLE 4

The parents of a newborn child invest \$8,000 in a plan that earns 9% interest, compounded quarterly. If the money is left untouched, how much will the child have in the account in 55 years?

Solution

We will substitute 8,000 for P, 0.09 for r, and 55 for t in the formula for compound interest. Because quarterly compounding means four times per year, we will substitute 4 for k.

$$A = P\left(1 + \frac{r}{k}\right)^{kt}$$

$$A = 8{,}000\left(1 + \frac{0.09}{4}\right)^{4 \cdot 55}$$

$$= 8{,}000(1.0225)^{220}$$

$$\approx 1{,}069{,}103.266 \qquad \text{Use a calculator.}$$

In 55 years, the account will be worth \$1,069,103.27.

Self Check 4 Would \$20,000 invested at 7% interest, compounded monthly, have provided more income at age 55?

In financial calculations, the initial amount deposited is often called the **present value,** denoted by PV. The amount to which the account will grow is called the **future value,** denoted by FV. The interest rate for each compounding period is called the **periodic interest rate,** i, and the number of times interest is compounded is the **number of compounding periods,** n. Using these definitions, an alternate formula for compound interest is as follows.

$$FV = PV(1 + i)^n$$

To use this formula to solve Example 4, we proceed as follows:

$$FV = PV(1 + i)^n$$

$$FV = 8{,}000(1 + 0.0225)^{220} \qquad i = \frac{0.09}{4} = 0.0225 \text{ and } n = 4(55) = 220.$$

$$= 8{,}000(1.0225)^{220}$$

$$\approx 1{,}069{,}103.266 \qquad \text{Use a calculator.}$$

4. Define e and Graph Base-e Exponential Functions

In mathematical models of natural events, the number

$$e = 2.71828182845904\ldots$$

often appears as the base of an exponential function. We can introduce this number by considering the compound interest formula

$$A = P\left(1 + \frac{r}{k}\right)^{kt} \qquad \text{\footnotesize A is the amount, P is the initial deposit, r is the annual rate, k is the number of compoundings per year, and t is the time in years.}$$

and allowing k to become very large. To see what happens, we let $k = rx$, where x is another variable.

Leonhard Euler
(1707–1783)
Euler first used the letter i to represent $\sqrt{-1}$, the letter e for the base of natural logarithms, and the symbol Σ for summation. Euler was one of the most prolific mathematicians of all time, contributing to almost all areas of mathematics. Much of his work was accomplished after he became blind.

$$A = P\left(1 + \frac{r}{k}\right)^{kt}$$

$$A = P\left(1 + \frac{r}{rx}\right)^{rxt} \qquad \text{Substitute } rx \text{ for } k.$$

$$A = P\left(1 + \frac{1}{x}\right)^{rxt} \qquad \text{Simplify } \frac{r}{rx}.$$

$$A = P\left[\left(1 + \frac{1}{x}\right)^x\right]^{rt} \qquad \text{Remember that } (a^m)^n = a^{mn}.$$

Since all variables in this formula are positive, r is a constant rate, and $k = rx$, it follows that as k becomes large, so does x. What happens to the value of A as k becomes large will depend on the value of $\left(1 + \frac{1}{x}\right)^x$ as x becomes large. Some results calculated for increasing values of x appear in Table 4-1.

x	$\left(1 + \frac{1}{x}\right)^x$
1	2
10	2.5937425
100	2.7048138
1,000	2.7169239
1,000,000	2.7182805
1,000,000,000	2.7182818

Table 4-1

From the table, we can see that as x increases, the value of $\left(1 + \frac{1}{x}\right)^x$ approaches the value of e, and the formula

$$A = P\left[\left(1 + \frac{1}{x}\right)^x\right]^{rt}$$

becomes

$$A = Pe^{rt} \quad \text{Substitute } e \text{ for } \left(1 + \frac{1}{x}\right)^x.$$

When the amount invested grows exponentially according to the formula $A = Pe^{rt}$, we say that interest is **compounded continuously.**

Continuous Compound Interest Formula

If P dollars are deposited in an account earning interest at an annual rate r, compounded continuously, the amount A after t years is given by the formula

$$A = Pe^{rt}$$

EXAMPLE 5 If the parents of the newborn child in Example 4 had invested $8,000 at an annual rate of 9%, compounded continuously, how much would the child have in the account in 55 years?

Solution We will substitute $8,000 for P, 0.09 for r, and 55 for t in the continuous compound interest formula $A = Pe^{rt}$.

$$A = Pe^{rt}$$
$$A = 8,000e^{(0.09)(55)}$$
$$= 8,000e^{4.95}$$
$$\approx 1,129,399.711 \quad \text{Use a calculator.}$$

In 55 years, the balance will be $1,129,399.71, which is $60,296.44 more than the amount earned with quarterly compounding.

Self Check 5 Find the balance in 60 years.

To graph the exponential function $f(x) = e^x$, we plot several points and join them with a smooth curve (as in Figure 4-5(a)) or use a graphing calculator (as in Figure 4-5(b)).

$f(x) = e^x$

x	$f(x)$	$(x, f(x))$
-1	0.37	$(-1, 0.37)$
0	1	$(0, 1)$
1	2.72	$(1, 2.72)$
2	7.39	$(2, 7.39)$

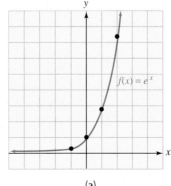

(a)

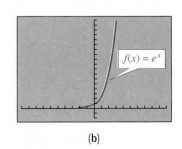

(b)

Figure 4-5

Comment
The graph of $f(x) = e^x$ is very important in mathematics and should be memorized.

5. Use Translations to Graph Exponential Functions

In Section 3.4, translations, reflections, and stretchings were applied to functions. These also may be applied to the graphs of exponential functions. A summary of these transformations, when $k > 0$, is shown in the table.

Equation	Translations, reflections, and stretchings of the graph of $f(x) = b^x$
$y = b^x + k$	Translates the graph of $f(x) = b^x$ upward k units
$y = b^x - k$	Translates the graph of $f(x) = b^x$ downward k units
$y = b^{x-k}$	Translates the graph of $f(x) = b^x$ to the right k units
$y = b^{x+k}$	Translates the graph of $f(x) = b^x$ to the left k units
$y = -b^x$	Reflects the graph of $f(x) = b^x$ about the x-axis
$y = b^{-x}$	Reflects the graph of $f(x) = b^x$ about the y-axis
$y = kb^x$	• Vertically stretches the graph of $f(x) = b^x$ if $k > 1$ • Vertically shrinks the graph of $f(x) = b^x$ if $0 < k < 1$
$y = b^{kx}$	• Horizontally stretches the graph of $f(x) = b^x$ if $0 < k < 1$ • Horizontally shrinks the graph of $f(x) = b^x$ if $k > 1$

EXAMPLE 6   On one set of axes, graph $f(x) = 2^x$ and $f(x) = 2^x + 3$.

Solution   The graph of $f(x) = 2^x + 3$ is identical to the graph of $f(x) = 2^x$, except that it is translated 3 units up. (See Figure 4-6.)

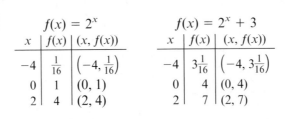

$$f(x) = 2^x$$

x	$f(x)$	$(x, f(x))$
-4	$\frac{1}{16}$	$\left(-4, \frac{1}{16}\right)$
0	1	$(0, 1)$
2	4	$(2, 4)$

$$f(x) = 2^x + 3$$

x	$f(x)$	$(x, f(x))$
-4	$3\frac{1}{16}$	$\left(-4, 3\frac{1}{16}\right)$
0	4	$(0, 4)$
2	7	$(2, 7)$

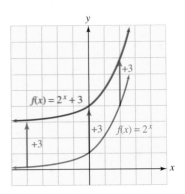

Figure 4-6

Self Check 6   On one set of axes, graph $f(x) = 2^x$ and $f(x) = 2^x - 2$.

EXAMPLE 7 On one set of axes, graph $f(x) = e^x$ and $f(x) = e^{x-3}$.

Solution The graph of $f(x) = e^{x-3}$ is identical to the graph of $f(x) = e^x$, except that it is translated 3 units to the right. (See Figure 4-7.)

$f(x) = e^x$

x	$f(x)$	$(x, f(x))$
-1	0.37	$(-1, 0.37)$
0	1	$(0, 1)$
1	2.72	$(1, 2.72)$
2	7.39	$(2, 7.39)$

$f(x) = e^{x-3}$

x	$f(x)$	$(x, f(x))$
2	0.37	$(2, 0.37)$
3	1	$(3, 1)$
4	2.72	$(4, 2.72)$
5	7.39	$(5, 7.39)$

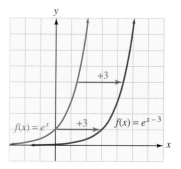

Figure 4-7

Self Check 7 On one set of axes, graph $f(x) = e^x$ and $f(x) = e^{x+2}$.

We can use a graphing calculator to graph exponential functions that are vertically or horizontally stretched or shrunk.

Accent on Technology **Graphing Exponential Functions**

To use a graphing calculator to graph the exponential function $f(x) = 2(3^{x/2})$, we enter the right side of the equation after the symbol $Y_1 =$. The display will show the equation

$$Y_1 = 2(3^{\wedge}(X/2))$$

If we use window settings of $[-10, 10]$ for x and $[-2, 18]$ for y and press GRAPH, we will obtain the graph shown in Figure 4-8(a).

To graph the exponential function $f(x) = 3e^{-x/2}$, we enter the right side of the equation after the symbol $Y_1 =$. The display will show the equation

$$Y_1 = 3(e^{\wedge}(-x/2))$$

If we use window settings of $[-10, 10]$ for x and $[-2, 18]$ for y and press GRAPH, we will obtain the graph shown in Figure 4-8(b).

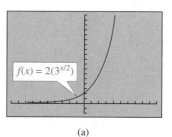

(a)

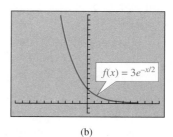

(b)

Figure 4-8

Self Check Answers   **1. a.** 64   **b.** $x^{3\sqrt{5}}$ **2.**

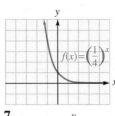

3. no
4. no
5. $1,771,251.33

6.

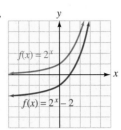

7.

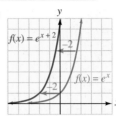

4.1 Exercises

Vocabulary and Concepts   *Fill in the blanks.*

1. If $b > 0$ and $b \neq 1$, $y = b^x$ represents an _____ function.

2. If $f(x) = b^x$ represents an increasing function, then $b > \underline{}$.

3. In interval notation, the domain of the exponential function $f(x) = b^x$ is _____.

4. The number b is called the ____ of the exponential function $y = b^x$.

5. The range of the exponential function $f(x) = b^x$ is _____.

6. The graphs of all exponential functions $y = b^x$ have the same __-intercept, the point _____.

7. If $b > 0$ and $b \neq 1$, the graph of $y = b^x$ approaches the x-axis, which is called an _____ of the curve.

8. If $f(x) = b^x$ represents a decreasing function, then $\underline{} < b < \underline{}$.

9. The graph of $y = b^x + 3$ is __ units above the graph of $y = b^x$.

10. The graph of an exponential function $y = b^x$ always passes through the points $(0, 1)$ and _____.

11. To two decimal places, the value of e is _____.

12. The continuous compound interest formula is $A = \underline{}$.

13. Since $e > 1$, the base-e exponential function is a (an) _____ function.

14. The graph of the exponential function $y = e^x$ passes through the points $(0, 1)$ and _____.

Practice   *Use a calculator to find each value to four decimal places.*

15. $4^{\sqrt{3}}$ **16.** $5^{\sqrt{2}}$

17. 7^{π} **18.** $3^{-\pi}$

Simplify each expression.

19. $5^{\sqrt{2}}5^{\sqrt{2}}$ **20.** $\left(5^{\sqrt{2}}\right)^{\sqrt{2}}$

21. $\left(a^{\sqrt{8}}\right)^{\sqrt{2}}$ **22.** $a^{\sqrt{12}}a^{\sqrt{3}}$

Find $f(0)$ and $f(2)$ for each of the given exponential functions.

23. $f(x) = 5^x$ **24.** $f(x) = 4^{-x}$

25. $f(x) = \left(\dfrac{1}{3}\right)^{-x}$ **26.** $f(x) = \left(\dfrac{1}{4}\right)^x$

Graph each exponential function.

27. $f(x) = 3^x$ **28.** $f(x) = 5^x$

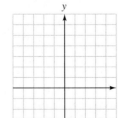

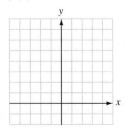

29. $f(x) = \left(\dfrac{1}{5}\right)^{x}$

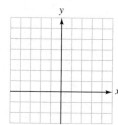

30. $f(x) = \left(\dfrac{1}{3}\right)^{x}$

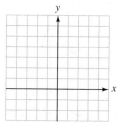

Determine whether the graph could represent an exponential function of the form $f(x) = b^{x}$.

39.

(0, 1)

40.

(1, 0)

31. $f(x) = \left(\dfrac{3}{4}\right)^{x}$

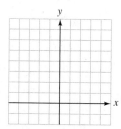

32. $f(x) = \left(\dfrac{4}{3}\right)^{x}$

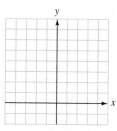

41.

(0, 2)

42.

(0, 1)

33. $f(x) = (1.5)^{x}$

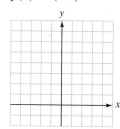

34. $f(x) = (0.3)^{x}$

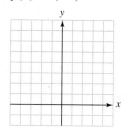

Find the value of b, if any, that would cause the graph of $y = b^{x}$ to look like the graph indicated.

43.

(0, 1) $\left(1, \dfrac{1}{2}\right)$

44.

(0, 1) (1, 7)

35. $f(x) = 3^{-x}$

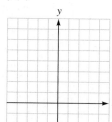

36. $f(x) = -5^{x}$

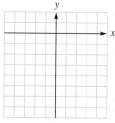

45.

(0, 2) (1, 5)

46.

(0, 1) (1, 3)

47.

(0, 1) (1, 2)

48.

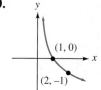

$\left(-1, \dfrac{1}{3}\right)$ (0, 1)

37. $f(x) = -\left(\dfrac{1}{5}\right)^{x}$

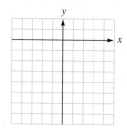

38. $f(x) = \left(\dfrac{1}{3}\right)^{-x}$

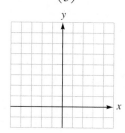

49.

(0, 1) $(2, e^{2})$

50.

(1, 0) (2, −1)

Graph each function using translations and reflections. Do not use a graphing calculator.

51. $f(x) = 3^x - 1$

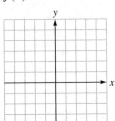

52. $f(x) = 2^x + 3$

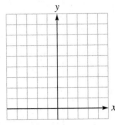

53. $f(x) = 2^x + 1$

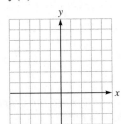

54. $f(x) = 4^x - 4$

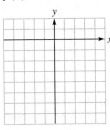

55. $f(x) = 3^{x-1}$

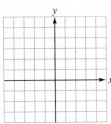

56. $f(x) = 2^{x+3}$

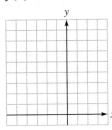

57. $f(x) = 3^{x+1}$

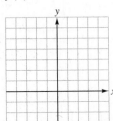

58. $f(x) = 2^{x-3}$

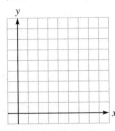

59. $f(x) = e^x - 4$

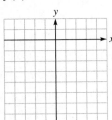

60. $f(x) = e^x + 2$

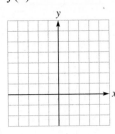

61. $f(x) = e^{x-2}$

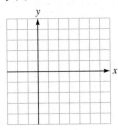

62. $f(x) = e^{x+3}$

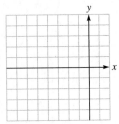

63. $f(x) = 2^{x+1} - 2$

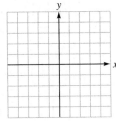

64. $f(x) = 3^{x-1} + 2$

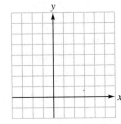

65. $y = 3^{x-2} + 1$

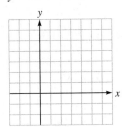

66. $y = 3^{x+2} - 1$

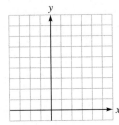

67. $f(x) = -3^x + 1$

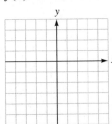

68. $f(x) = -2^x - 3$

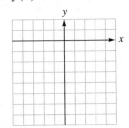

69. $f(x) = 2^{-x} - 3$

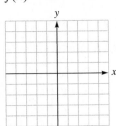

70. $f(x) = 4^{-x} + 4$

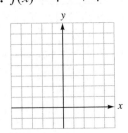

71. $f(x) = -e^x + 2$

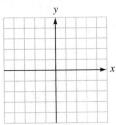

72. $f(x) = e^{-x} + 3$

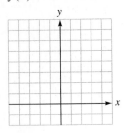

Use a graphing calculator to graph each function.

73. $f(x) = 5(2^x)$ **74.** $f(x) = 2(5^x)$

75. $f(x) = 3^{-x}$ **76.** $f(x) = 2^{-x}$

77. $f(x) = 2e^x$ **78.** $f(x) = 3e^{-x}$

79. $f(x) = 5e^{-0.5x}$ **80.** $f(x) = -3e^{2x}$

Applications   *In Exercises 81–84, assume that there are no deposits or withdrawals.*

81. Compound interest   An initial deposit of $10,000 earns 8% interest, compounded quarterly. How much will be in the account in 10 years?

82. Compound interest   An initial deposit of $1,000 earns 9% interest, compounded monthly. How much will be in the account in $4\frac{1}{2}$ years?

83. Comparing interest rates   How much more interest could $500 earn in 5 years, compounded semi-annually (two times a year), if the annual interest rate were $5\frac{1}{2}$% instead of 5%?

84. Comparing savings plans   Which institution in the ads provides the better investment?

> ### *Fidelity Savings & Loan*
> Earn 5.25%
> compounded monthly

> ### Union Trust
> Money Market Account
> paying 5.35%
> compounded annually

85. Compound interest   If $1 had been invested on July 4, 1776, at 5% interest, compounded annually, what would it be worth on July 4, 2076?

86. 360/365 method   Some financial institutions pay daily interest, compounded by the 360/365 method, using the following formula. Using this method, what will an initial investment of $1,000 be worth in 5 years, assuming a 7% annual interest rate?

$$A = A_0\left(1 + \frac{r}{360}\right)^{365t} \quad (t \text{ is in years}).$$

87. Carrying charges   A college student takes advantage of the ad shown and buys a bedroom set for $1,100. He plans to pay the $1,100 plus interest when his income tax refund comes in 8 months. At that time, what will he need to pay?

> ### BUY NOW,
> ### PAY LATER!
> Only $1\frac{3}{4}$% interest per month.

88. Credit card interest   A bank credit card charges interest at the rate of 21% per year, compounded monthly. If a senior in college charges $1,500 to pay for college expenses, and intends to pay it in one year, what will she have to pay?

89. Continuous compound interest   An initial investment of $5,000 earns 8.2% interest, compounded continuously. What will the investment be worth in 12 years?

90. Continuous compound interest   An initial investment of $2,000 earns 8% interest, compounded continuously. What will the investment be worth in 15 years?

91. Comparison of compounding methods   An initial deposit of $5,000 grows at an annual rate of 8.5% for 5 years. Compare the final balances resulting from continuous compounding and annual compounding.

92. Comparison of compounding methods   An initial deposit of $30,000 grows at an annual rate of 8% for 20 years. Compare the final balances resulting from continuous compounding and annual compounding.

93. Frequency of compounding $10,000 is invested in each of two accounts, both paying 6% annual interest. In the first account, interest compounds quarterly, and in the second account, interest compounds daily. Find the difference between the accounts after 20 years.

94. Determining an initial deposit An account now contains $11,180 and has been accumulating interest at a 7% annual rate, compounded continuously, for 7 years. Find the initial deposit.

95. Saving for college In 20 years, a father wants to accumulate $40,000 to pay for his daughter's college expenses. If he can get 6% interest, compounded quarterly, how much must he invest now to achieve his goal?

96. Saving for college In Problem 95, how much should he invest to achieve his goal if he can get 6% interest, compounded continuously?

97. Population of a city The population $P(t)$ of a small city can be approximated by the exponential function, $P(t) = 1,200e^{0.2t}$, where t represents time in years. What will be the population of the city in 12 years?

98. Amount of drug present The amount of a drug $A(t)$, in mg, present in the bloodstream t hours after being intravenously administered can be approximated by the exponential function, $A(t) = -1,000e^{-0.3t} + 1,250$. How much of the drug is present in the bloodstream after 14 hours?

Discovery and Writing

99. Financial planning To have P available in n years, A can be invested now in an account paying interest at an annual rate r, compounded annually. Show that

$$A = P(1 + r)^{-n}$$

100. If $2^{t+4} = k2^t$, find k.

101. If $5^{3t} = k^t$, find k.

102. a. If $e^{t+3} = ke^t$, find k.
   b. If $e^{3t} = k^t$, find k.

Review *Factor each expression completely.*

103. $x^2 + 9x^4$

104. $x^2 - 9x^4$

105. $x^2 + x - 12$

106. $x^3 + 27$

4.2 Applications of Exponential Functions

Objectives

1. Solve Radioactive Decay Problems
2. Solve Oceanography Problems
3. Solve Malthusian Population Growth Problems
4. Solve Epidemiology Problems

Flu kills an estimated 36,000 Americans each year and results in a much larger number of hospitalizations. The influenza virus replicates quickly and can infect a population rapidly. The most effective method of preventing the virus infection and its severe complications is a flu vaccination.

An event that changes with time, such as the spread of the influenza virus, can be modeled by an exponential function. In this section, we will see several important applications of these functions: radioactive decay, oceanography, population growth, and epidemiology.

A mathematical description of an observed event is called a **model** of that event. Many real-world occurrences change with time and can be modeled by exponential functions of the form

$$y = f(t) = ab^{kt} \quad \text{Remember that } ab^{kt} \text{ means } a(b^{kt}).$$

where a, b, and k are constants and t represents time. If f is an increasing function, we say that y *grows exponentially*. If f is a decreasing function, we say that y *decays exponentially*.

1. Solve Radioactive Decay Problems

The atomic structure of a radioactive material changes as the material emits radiation. Uranium, for example, changes (decays) into thorium, then into radium, and eventually into lead.

Experiments have determined the time it takes for one-half of a sample of a given radioactive element to decompose. That time is a constant, called the element's **half-life.** The amount present decays exponentially according to this formula.

Radioactive Decay Formula	The amount A of radioactive material present at time t is given by $$A = A_0 2^{-t/h}$$ where A_0 is the amount that was present initially (at $t = 0$) and h is the material's half-life.

EXAMPLE 1

The half-life of radium is approximately 1,600 years. How much of a 1-gram sample will remain after 1,000 years?

Solution

In this example, $A_0 = 1$, $h = 1,600$, and $t = 1,000$. We substitute these values into the formula for radioactive decay and simplify.

$$A = A_0 2^{-t/h}$$
$$A = 1 \cdot 2^{-1,000/1,600}$$
$$\approx 0.648419777 \quad \text{Use a calculator.}$$

After 1,000 years, approximately 0.65 gram of radium will remain.

Self Check 1

After 800 years, how much radium will remain?

2. Solve Oceanography Problems

Intensity of Light Formula	The intensity I of light (in lumens) at a distance x meters below the surface of a body of water decreases exponentially according to the formula $$I = I_0 k^x$$ where I_0 is the intensity of light above the water and k is a constant that depends on the clarity of the water.

Comment

A lumen is a unit of standard measurement that describes how much light is contained in a certain area. The lumen is part of the photometry group that measures different aspects of light.

EXAMPLE 2   At one location in the Atlantic Ocean, the intensity of light above water I_0 is 12 lumens and $k = 0.6$. Find the intensity of light at a depth of 5 meters.

Solution   We will substitute 12 for I_0, 0.6 for k, and 5 for x into the formula for light intensity and then simplify.

$$I = I_0 k^x$$
$$I = 12(0.6)^5$$
$$I = 0.93312$$

At a depth of 5 meters, the intensity of the light is slightly less than 1 lumen.

Self Check 2   Find the intensity at a depth of 10 meters.

3. Solve Malthusian Population Growth Problems

An equation based on the exponential function provides a model for **population growth.** One such model, called the **Malthusian model of population growth,** assumes a constant birth rate and a constant death rate. In this model, the population P grows exponentially according to the following formula.

Malthusian Model of Population Growth

If b is the annual birth rate, d is the annual death rate, t is the time (in years), P_0 is the initial population at $t = 0$, and P is the current population, then

$$P = P_0 e^{kt}$$

where $k = b - d$ is the **annual growth rate,** the difference between the annual birth rate and death rate.

EXAMPLE 3   The population of the United States is approximately 300 million people. Assuming that the annual birth rate is 19 per 1,000 and the annual death rate is 7 per 1,000, what does the Malthusian model predict the U.S. population will be in 50 years?

Solution   We can use the stated information to write the Malthusian model for U.S. population. We then will substitute into the model to predict the population in 50 years.

Since k is the difference between the birth and death rates, we have

$$k = b - d$$
$$k = \frac{19}{1,000} - \frac{7}{1,000} \qquad \text{Substitute } \frac{19}{1,000} \text{ for } b \text{ and } \frac{7}{1,000} \text{ for } d.$$
$$k = 0.019 - 0.007$$
$$= 0.012$$

We now can substitute 300,000,000 for P_0, 50 for t, and 0.012 for k in the formula for the Malthusian model of population growth and simplify.

$$P = P_0 e^{kt}$$
$$P = (300,000,000)e^{(0.012)(50)}$$
$$= (300,000,000)e^{0.6}$$
$$\approx 546,635,640.1 \qquad \text{Use a calculator.}$$

After 50 years, the U.S. population will exceed 546 million people.

Self Check 3   Find the population in 100 years.

The English economist Thomas Robert Malthus (1766–1834) pioneered in population study. He believed that poverty and starvation were unavoidable, because the human population tends to grow exponentially, whereas the food supply tends to grow linearly.

EXAMPLE 4   Suppose that a country with a population of 1,000 people is growing exponentially according to the formula

$$P = 1,000e^{0.02t}$$

where t is in years. Furthermore, assume that the food supply, measured in adequate food per day per person, is growing linearly according to the formula

$$y = 30.625x + 2,000$$

In how many years will the population outstrip the food supply?

Solution   We can use a graphing calculator with window settings of [0, 100] for x and [0, 10,000] for y. After graphing the functions as shown in Figure 4-9, we trace and zoom to find the point where the two graphs intersect. From the graph, we can see that the food supply will be adequate for about 71 years. At that time, the population of approximately 4,200 people will begin to have problems.

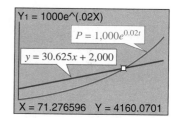

Figure 4-9

Self Check 4   In 80 years, what is the approximate number of people per day who will not have adequate food?

4. Solve Epidemiology Problems

Many infectious diseases, including some caused by viruses, spread most rapidly when they first infect a population, but then more slowly as the number of uninfected individuals decreases. These situations often are modeled by a function, called a *logistic function*.

Logistic Epidemiology Model   The size P of an infected population at any time t in years, is given by the logistic function

$$P = \frac{M}{1 + \left(\dfrac{M}{P_0} - 1\right)e^{-kt}}$$

where P_0 is the infected population size at $t = 0$, k is a constant determined by how contagious the virus is in a given environment, and M is the theoretical maximum size of the population P.

EXAMPLE 5   In a city with a population of 1,200,000, there are currently 1,000 cases of infection with the HIV virus. If the spread of the disease is projected by the formula

$$P = \frac{1,200,000}{1 + (1,200 - 1)e^{-0.4t}}$$

how many people will be infected in 3 years?

Solution   We can substitute 3 for t in the logistic formula and calculate P.

$$P = \frac{1,200,000}{1 + (1,200 - 1)e^{-0.4t}}$$

$$P = \frac{1,200,000}{1 + (1,199)e^{-0.4(3)}}$$

$$\approx 3,313.710094$$

In 3 years, approximately 3,300 people are expected to be infected.

Self Check 5   How many will be infected in 10 years?

Self Check Answers   **1.** about 0.71 g   **2.** 0.073 lumen   **3.** more than 996 million   **4.** about 503   **5.** about 52,000

4.2 Exercises

Vocabulary and Concepts   *Fill in the blanks.*

1. The Malthusian model assumes a constant _____ rate and a constant _____ rate.

2. The Malthusian prediction is pessimistic, because a _____ grows exponentially, but food supplies grow _____.

Applications   *Use a calculator to help solve each problem.*

3. **Tritium decay**   Tritium, a radioactive isotope of hydrogen, has a half-life of 12.4 years. Of an initial sample of 50 grams, how much will remain after 100 years?

4. **Chernobyl**   In April 1986, the world's worst nuclear power disaster occurred at Chernobyl in the former USSR. An explosion released about 1,000 kilograms of radioactive cesium-137 (^{137}Cs) into the atmosphere. If the half-life of ^{137}Cs is 30.17 years, how much will remain in the atmosphere in 100 years?

5. **Chernobyl**   Refer to Exercise 4. How much ^{137}Cs will remain in 200 years?

6. **Carbon-14 decay**   The half-life of radioactive carbon-14 is 5,700 years. How much of an initial sample will remain after 3,000 years?

7. **Plutonium decay**   One of the isotopes of plutonium, ^{237}Pu, decays with a half-life of 40 days. How much of an initial sample will remain after 60 days?

8. **Comparing radioactive decay**   One isotope of holmium, ^{162}Ho, has a half-life of 22 minutes. The half-life of a second isotope, ^{164}Ho, is 37 minutes. Starting with a sample containing equal amounts, find the ratio of the amounts of ^{162}Ho to ^{164}Ho after one hour.

9. **Drug absorption in smokers**   The biological half-life of the asthma medication theophylline is 4.5 hours for smokers. Find the amount of the drug retained in a smoker's system 12 hours after a dose of 1 unit is taken.

10. **Drug absorption in nonsmokers**   For a non-smoker, the biological half-life of theophylline is 8 hours. Find the amount of the drug retained in a nonsmoker's system 12 hours after taking a one-unit dose.

11. **Oceanography**   The intensity I of light (in lumens) at a distance x meters below the surface is given by $I = I_0 k^x$, where I_0 is the intensity at the surface and k depends on the clarity of the water. At one location in the Arctic Ocean, $I_0 = 8$ and $k = 0.5$. Find the intensity at a depth of 2 meters.

12. **Oceanography**   At one location in the Atlantic Ocean, $I_0 = 14$ and $k = 0.7$. Find the intensity of light at a depth of 12 meters. (See Exercise 11.)

13. **Oceanography**   At a depth of 3 meters at one location in the Pacific Ocean, the intensity I of light is 1 lumen and $k = 0.5$. Find the intensity I_0 of light at the surface.

14. **Oceanography**   At a depth of 2 meters at one location off the coast of Belize, the intensity I of light is 2 lumens and $k = 0.63$. Find the intensity I_0 of light at the surface.

15. **Bluegill population** A Wisconsin lake is stocked with 10,000 bluegill. The population is expected to grow exponentially according to the model $P = P_0 2^{t/2}$. How many bluegill will be in the lake in 5 years?

16. **Community growth** The population of Eagle River is growing exponentially according to the model $P = 375(1.3)^t$, where t is measured in years from the present date. Find the population in 3 years.

17. **Newton's law of cooling** Some hot water, initially at 100°C, is in a room with a temperature of 40°C. The temperature T of the water after t hours is given by $T = 40 + 60(0.75)^t$. Find the temperature in $3\frac{1}{2}$ hours.

18. **Bacterial cultures** A colony of 6 million bacteria is growing in a culture medium. The population P after t hours is given by the formula $P = (6 \times 10^6)(2.3)^t$. Find the population after 4 hours.

19. **Population growth** The growth of a town's population is modeled by $P = 173e^{0.03t}$. How large will the population be when $t = 20$?

20. **Population decline** The decline of a city's population is modeled by $P = 1.2 \times 10^6 e^{-0.008t}$. How large will the population be when $t = 30$?

21. **Epidemics** The spread of hoof and mouth disease through a herd of cattle can be modeled by the formula $P = P_0 e^{0.27t}$, where P is the size of the infected population, P_0 is the infected population size at $t = 0$, and t is in days. If a rancher does not act quickly to treat two cases, how many cattle will have the disease in one week?

22. **Alcohol absorption** In one individual, the percent of alcohol absorbed into the bloodstream after drinking two shots of whiskey is given by the following formula. Find the percent of alcohol absorbed into the blood after $\frac{1}{2}$ hour.

$$P = 0.3(1 - e^{-0.05t}) \quad \text{where } t \text{ is in minutes}$$

23. **World population growth** The population of the Earth is approximately 6 billion people and is growing at an annual rate of 1.9%. Assuming a Malthusian growth model, find the world population in 30 years.

24. **World population growth** See Exercise 23. Assuming a Malthusian growth model, find the world population in 40 years.

25. **World population growth** See Exercise 23. By what factor will the current population of the Earth increase in 50 years?

26. **World population growth** See Exercise 23. By what factor will the current population of the Earth increase in 100 years?

27. **Drug absorption** The percent P of the drug triazolam (a drug for treating insomnia) remaining in a person's bloodstream after t hours is given by $P = e^{-0.3t}$. What percent will remain in the bloodstream after 24 hours?

28. **Medicine** The concentration x of a certain drug in an organ after t minutes is given by $x = 0.08(1 - e^{-0.1t})$. Find the concentration of the drug in $\frac{1}{2}$ hour.

29. **Medicine** Refer to Exercise 28. Find the initial concentration of the drug (*Hint:* when $t = 0$).

30. **Spreading the news** Suppose the function

$$N = P(1 - e^{-0.1t})$$

is used to model the length of time t (in hours) it takes for N people living in a town with population P to hear a news flash. How many people in a town of 50,000 will hear the news between 1 and 2 hours after it happened?

31. **Spreading the news** How many people in the town described in Problem 30 will not have heard the news after 10 hours?

32. **Epidemics** Refer to Example 5. How many people will have the HIV virus in 5 years?

33. **Epidemics** Refer to Example 5. How many people will have HIV in 8 years?

34. **Epidemics** In a city with a population of 450,000, there are currently 1,000 cases of hepatitis. If the spread of the disease is projected by the following logistic function, how many people will contract the hepatitis virus after 6 years?

$$P = \frac{450,000}{1 + (450 - 1)e^{-0.2t}}$$

35. **Epidemics** In a Indonesian city with a population of 55,000, there are currently 100 cases of the avian bird flu. If the spread of the disease is projected by the following formula, how many people will contract the bird flu after 2 years?

$$P = \frac{55,000}{1 + (550 - 1)e^{-0.8t}}$$

36. Life expectancy The life expectancy of white females can be estimated by using the function $l = 78.5(1.001)^x$, where x is the current age. Find the life expectancy of a white female who is currently 50 years old. Give the answer to the nearest tenth.

37. Oceanography The width w (in millimeters) of successive growth spirals of the sea shell *Catapulus voluto,* shown in the illustration, is given by the function $w = 1.54e^{0.503n}$, where n is the spiral number. To the nearest tenth of a millimeter, find the width of the fifth spiral.

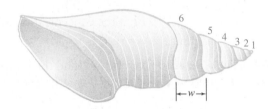

38. Skydiving Before the parachute opens, the velocity v (in meters per second) of a skydiver is given by $v = 50(1 - e^{-0.2t})$ where t is in seconds. Find the initial velocity.

39. Skydiving Refer to Exercise 38 and find the velocity after 20 seconds.

40. Free-falling objects After t seconds, a certain falling object has a velocity v given by $v = 50(1 - e^{-0.3t})$. Which is falling faster after 2 seconds, this object or the skydiver in Exercise 38?

41. Population growth In 1999, the male population of the United States was about 133 million, and the female population was about 139 million. Assuming a Malthusian growth model with a 1% annual growth rate, how many more females than males will there be in 20 years?

42. Population growth See Exercise 41. How many more females than males will there be in 50 years?

🖳 *Use a graphing calculator to solve each problem.*

43. In Example 4, suppose that better farming methods change the formula for food growth to $y = 31x + 2,000$. How long will the food supply be adequate?

44. In Example 4, suppose that a birth control program changed the formula for population growth to $P = 1,000e^{0.01t}$. How long will the food supply be adequate?

Discovery and Writing

45. The value of e can be calculated to any degree of accuracy by adding the first several terms of the following list.

$$1, 1, \frac{1}{2}, \frac{1}{2 \cdot 3}, \frac{1}{2 \cdot 3 \cdot 4}, \frac{1}{2 \cdot 3 \cdot 4 \cdot 5}, \cdots$$

The more terms that are added, the closer the sum will be to e. Add the first six numbers in the preceding list. To how many decimal places is the sum accurate?

46. 🖳 Graph the function defined by the equation $f(x) = \dfrac{e^x + e^{-x}}{2}$ from $x = -2$ to $x = 2$. The graph will look like a parabola, but it is not. The graph, called a **catenary,** is important in the design of power distribution networks, because it represents the shape of a uniform flexible cable whose ends are suspended from the same height. The function is called the **hyperbolic cosine function.**

47. 🖳 Graph the following logistic function, first discussed in Example 5. Use window settings of [0, 20] for x and [0, 1,500,000] for y.

$$P = \frac{1,200,000}{1 + (1,199)e^{-0.4t}}$$

48. 🖳 Use the trace capabilities of your graphing calculator to explore the logistic function of Example 5 and Exercise 47. As time passes, what value does P approach? How many years does it take for 20% of the population to become infected? For 80%?

Review *Find the value of x that makes each statement true.*

49. $2^3 = x$

50. $3^x = 9$

51. $x^3 = 27$

52. $3^{-2} = x$

53. $x^{-3} = \dfrac{1}{8}$

54. $3^x = \dfrac{1}{3}$

55. $9^{1/2} = x$

56. $x^{1/3} = 3$

4.3 Logarithmic Functions and Their Graphs

Objectives

1. Evaluate Logarithms
2. Evaluate Common Logarithms
3. Evaluate Natural Logarithms
4. Graph Logarithmic Functions
5. Use Translations to Graph Logarithmic Functions

Guests aboard the Royal Caribbean's cruise ship *Freedom of the Seas* can now "hang ten" while out to sea. The flowrider surf simulator allows riders to body board surf against a wave-like water flow of 34,000 gallons per minute.

It is important that water in the flowrider has the proper pH value. For example, if it is too acidic, the water will make our eyes and nose burn, and it will make our skin get dry and itchy. The pH of the water is one of the most important factors in pool water balance and should be tested frequently. To calculate pH, we need to understand *logarithms,* the topic of this section. As we continue through this chapter, we will see several real-life applications of logarithms.

Since exponential functions are one-to-one functions, each one has an inverse. For example, to find the inverse of the function $y = 3^x$, we interchange the positions of x and y to obtain $x = 3^y$. The graphs of these two functions are shown in Figure 4-10(a).

To find the inverse of the function $y = \left(\frac{1}{3}\right)^x$, we again interchange the positions of x and y to obtain $x = \left(\frac{1}{3}\right)^y$. The graphs of these two functions are shown in Figure 4-10(b).

Comment

The graphs of the functions are inverses and therefore symmetric about the line $y = x$.

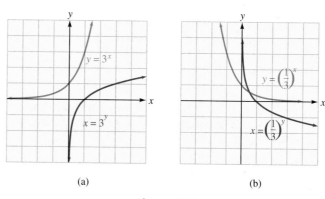

(a) (b)

Figure 4-10

In general, the inverse of the function $y = b^x$ is $x = b^y$. When $b > 1$, their graphs appear as shown in Figure 4-11(a) on the next page. When $0 < b < 1$, their graphs appear as shown in Figure 4-11(b).

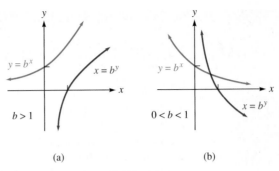

Figure 4-11

1. Evaluate Logarithms

Since an exponential function defined by $y = b^x$ is one-to-one, it has an inverse function that is defined by the equation $x = b^y$. To express this inverse function in the form $y = f^{-1}(x)$, we must solve the equation $x = b^y$ for y. To do this, we need the following definition.

Logarithmic Functions   If $b > 0$ and $b \neq 1$, the **logarithmic function with base b** is defined by

$$y = \log_b x \qquad \text{if and only if} \qquad x = b^y$$

The **domain of the logarithmic function** is the interval $(0, \infty)$. The **range** is the interval $(-\infty, \infty)$. The logarithmic function also is denoted as $f(x) = \log_b x$.

The range of the logarithmic function is the set of real numbers, because the value of y in the equation $x = b^y$ can be any real number. The domain is the set of positive numbers, because the value of x in the equation $x = b^y$ $(b > 0)$ is always positive.

Since the function $y = \log_b x$ is the inverse of the one-to-one exponential function $y = b^x$, the logarithmic function is also one-to-one.

The expression $x = b^y$ is said to be written in *exponential form*. The equivalent expression $y = \log_b x$ is said to be written in *logarithmic form*. To translate from one form to the other, it is helpful to keep track of the base and the exponent.

Comment

Since the domain of the logarithmic function is the set of positive numbers, the logarithm of 0 and the logarithm of a negative number are undefined in the set of real numbers.

Exponential form

$$x = b^y$$
↑↑
Base Exponent

Logarithmic form

$$y = \log_b x$$
↑   ↑
Exponent   Base

The definition of logarithm guarantees that any pair (x, y) that satisfies the equation $y = \log_b x$ also satisfies the equation $x = b^y$. Several examples are shown in the table.

$\log_b x = y$	because	$x = b^y$
$\log_5 25 = 2$	because	$25 = 5^2$
$\log_7 1 = 0$	because	$1 = 7^0$
$\log_{16} 4 = \dfrac{1}{2}$	because	$4 = \sqrt{16} = 16^{1/2}$
$\log_2 \dfrac{1}{8} = -3$	because	$\dfrac{1}{8} = 2^{-3}$

In each of these examples, the logarithm of a number is an exponent. In fact,

$\log_b x$ *is the exponent to which b is raised to get x.*

To express this as an equation, we write

$$b^{\log_b x} = x$$

EXAMPLE 1   Find y in each equation: **a.** $\log_2 8 = y$ **b.** $\log_5 1 = y$ **c.** $\log_7 \dfrac{1}{49} = y$

Solution   For each part, we can use the definition of the logarithm of a number to find y.

a. $\log_2 8 = y$ is equivalent to $8 = 2^y$. Since $8 = 2^3$, we have $2^y = 2^3$ and $y = 3$.

b. $\log_5 1 = y$ is equivalent to $1 = 5^y$. Since $1 = 5^0$, we have $5^y = 5^0$ and $y = 0$.

c. $\log_7 \dfrac{1}{49} = y$ is equivalent to $\dfrac{1}{49} = 7^y$. Since $\dfrac{1}{49} = 7^{-2}$, we have $7^y = 7^{-2}$ and $y = -2$.

Self Check 1   Find y in each equation: **a.** $\log_3 9 = y$ **b.** $\log_2 16 = y$

c. $\log_5 \dfrac{1}{25} = y$

EXAMPLE 2   Find a in each equation: **a.** $\log_a 32 = 5$ **b.** $\log_9 a = -\dfrac{1}{2}$ **c.** $\log_9 3 = a$

Solution   For each part, we will use the definition of the logarithm of a number to find a.

a. $\log_a 32 = 5$ is equivalent to $a^5 = 32$. Since $2^5 = 32$, we have $a^5 = 2^5$ and $a = 2$.

b. $\log_9 a = -\dfrac{1}{2}$ is equivalent to $9^{-1/2} = a$. Since $9^{-1/2} = \dfrac{1}{3}$, it follows that $a = \dfrac{1}{3}$.

c. $\log_9 3 = a$ is equivalent to $3 = 9^a$. Since $3 = 9^{1/2}$, we have $9^a = 9^{1/2}$ and $a = \dfrac{1}{2}$.

Self Check 2   Find d in each equation: **a.** $\log_4 \dfrac{1}{16} = d$ **b.** $\log_d 36 = 2$

c. $\log_8 d = -\dfrac{1}{3}$

2. Evaluate Common Logarithms

Many applications use base-10 logarithms (also called **common logarithms**). When the base b is not indicated in the notation $\log x$, we assume that $b = 10$:

$\log x$ means $\log_{10} x$

Because base-10 logarithms appear so often, you should become familiar with the following base-10 logarithms:

$\log_{10} \dfrac{1}{100} = -2$	because	$10^{-2} = \dfrac{1}{100}$
$\log_{10} \dfrac{1}{10} = -1$	because	$10^{-1} = \dfrac{1}{10}$
$\log_{10} 1 = 0$	because	$10^0 = 1$
$\log_{10} 10 = 1$	because	$10^1 = 10$
$\log_{10} 100 = 2$	because	$10^2 = 100$
$\log_{10} 1,000 = 3$	because	$10^3 = 1,000$

In general, we have

$$\log_{10} 10^x = x$$

Accent on Technology

Using Calculators to Find Logarithms

Before calculators, extensive tables were used to provide logarithms of numbers. Today, logarithms are easy to find with a calculator. For example, to find log 2.34 with a scientific calculator, we enter these numbers and press these keys:

2.34 **LOG**

The display will read .369215857 . To four decimal places, log 2.34 = 0.3692.

To use a graphing calculator, we enter these numbers and press these keys:

LOG 2.34 **ENTER**

The display will read

log 2.34
.3692158574

EXAMPLE 3   Find x in the equation $\log x = 0.7482$ to four decimal places.

Solution   We can use the definition of the common logarithm and a calculator to find x.

The equation $\log x = 0.7482$ is equivalent to $10^{0.7482} = x$. To find x with a calculator, we enter these numbers and press these keys:

Scientific calculator	*Graphing calculator*
10 y^x .7482 $=$	10 $\wedge$ .7482 **ENTER**

or

.7482 10^x	10^x .7482 **ENTER**

Either way, the result is 5.600154388. To four decimal places, $x = 5.6002$.

Self Check 3   Solve: $\log x = 1.87737$. Give the result to four decimal places.

3.   Evaluate Natural Logarithms

We have seen the importance of the number e in mathematical models of events in nature. Base-e logarithms are just as important. They are called **natural logarithms** or **Napierian logarithms** after John Napier (1550–1617). They usually are written as $\ln x$, rather than $\log_e x$:

$\ln x$ means $\log_e x$

Like all logarithmic functions, the domain of $f(x) = \ln x$ is the interval $(0, \infty)$, and the range is the interval $(-\infty, \infty)$.

To estimate the base-e logarithms of numbers, we can use a calculator.

Accent on Technology

Using Calculators to Find Logarithms

To use a calculator to estimate the value of $\ln 2.34$, we enter these numbers and press these keys:

Scientific calculator	*Graphing calculator*
2.34 **LN**	**LN** 2.34 **ENTER**

Either way, the result is .8501509294. To four decimal places, $\ln 2.34 = 0.8502$.

EXAMPLE 4   Use a calculator to find: **a.** ln 17.32   **b.** ln (log 0.05)

Solution   **a.** Enter these numbers and press these keys:

Scientific calculator	*Graphing calculator*
17.32 LN	LN 17.32 ENTER

Either way, the result is 2.851861903.

b. Enter these numbers and press these keys:

Scientific calculator	*Graphing calculator*
0.05 LOG LN	LN LOG 0.05) ENTER

Either way, we obtain an error, because log 0.05 is a negative number, and we cannot take the logarithm of a negative number.

Self Check 4   Find each value to four decimal places: **a.** $\ln \pi$
b. $\ln\left(\log \frac{1}{3}\right)$

EXAMPLE 5   Solve each equation and give the result to four decimal places:
a. $\ln x = 1.335$ **b.** $\ln x = \log 5.5$

Solution   We can write each equation in exponential form and use our calculator to find x.

a. The equation $\ln x = 1.335$ is equivalent to $e^{1.335} = x$. To use a calculator to find x, we enter these numbers and press these keys:

Scientific calculator	*Graphing calculator*
1.335 e^x	e^x 1.335 ENTER

Either way, the result is 3.799995946. To four decimal places, $x = 3.8000$.

b. The equation $\ln x = \log 5.5$ is equivalent to $e^{\log 5.5} = x$. To use a calculator to find x, we enter these numbers and press these keys:

Scientific calculator	*Graphing calculator*
5.5 LOG e^x	e^x LOG 5.5 ENTER

Either way, the result is 2.096695826. To four decimal places, $x = 2.0967$.

Self Check 5   Solve each equation and give each result to four decimal places:
a. $\ln x = 1.9344$ **b.** $\log x = \ln 3.2$

4.   Graph Logarithmic Functions

To graph the logarithmic function $y = f(x) = \log_2 x$, we calculate and plot several points with coordinates (x, y) that satisfy the equivalent equation $x = 2^y$. After joining these points with a smooth curve, we have the graph shown in Figure 4-12(a) on the next page.

To graph $y = f(x) = \log_{1/2} x$, we calculate and plot several points with coordinates (x, y) that satisfy the equation $x = \left(\frac{1}{2}\right)^y$. After joining these points with a smooth curve, we have the graph shown in Figure 4-12(b).

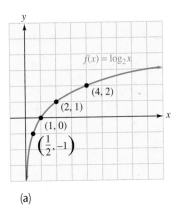

$f(x) = \log_2 x$

x	$f(x)$	$(x, f(x))$
$\frac{1}{4}$	-2	$\left(\frac{1}{4}, -2\right)$
$\frac{1}{2}$	-1	$\left(\frac{1}{2}, -1\right)$
1	0	$(1, 0)$
2	1	$(2, 1)$
4	2	$(4, 2)$
8	3	$(8, 3)$

(a)

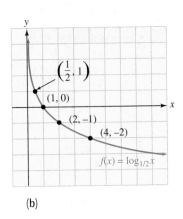

$f(x) = \log_{1/2} x$

x	$f(x)$	$(x, f(x))$
$\frac{1}{4}$	2	$\left(\frac{1}{4}, 2\right)$
$\frac{1}{2}$	1	$\left(\frac{1}{2}, 1\right)$
1	0	$(1, 0)$
2	-1	$(2, -1)$
4	-2	$(4, -2)$
8	-3	$(8, -3)$

(b)

Figure 4-12

A table of values and the graph of $f(x) = \log_{10} x$ appears in Figure 4-13.

$f(x) = \log x$

x	$f(x)$	$(x, f(x))$
$\frac{1}{100}$	-2	$\left(\frac{1}{100}, -2\right)$
$\frac{1}{10}$	-1	$\left(\frac{1}{10}, -1\right)$
1	0	$(1, 0)$
10	1	$(10, 1)$
100	2	$(100, 2)$

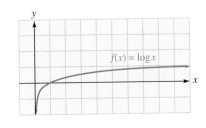

Figure 4-13

The graphs of all logarithmic functions are similar to those in Figure 4-14. If $b > 1$, the logarithmic function is an increasing function, as in Figure 4-14(a). If $0 < b < 1$, the logarithmic function is a decreasing function, as in Figure 4-14(b). As Figures 4-14(a) and (b) show, the graph of $f(x) = \log_b x$ has these properties:

> **Properties of the graph of $f(x) = \log_b x$**
>
> 1. It passes through the point $(1, 0)$.
> 2. It passes through the point $(b, 1)$.
> 3. The y-axis is an asymptote.
> 4. The domain is $(0, \infty)$, and the range is $(-\infty, \infty)$.

Figures 4-14(c) and (d) show that the exponential and logarithmic functions are inverses of each other and have symmetry about the line $y = x$.

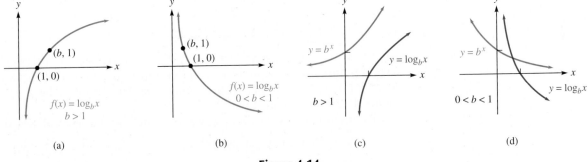

(a)   (b)   (c)   (d)

Figure 4-14

To graph $f(x) = \ln x$, we can plot points that satisfy the equation $x = e^y$ and join them with a smooth curve, as shown in Figure 4-15.

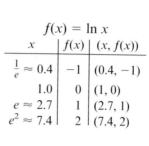

$f(x) = \ln x$

x	$f(x)$	$(x, f(x))$
$\frac{1}{e} \approx 0.4$	-1	$(0.4, -1)$
1.0	0	$(1, 0)$
$e \approx 2.7$	1	$(2.7, 1)$
$e^2 \approx 7.4$	2	$(7.4, 2)$

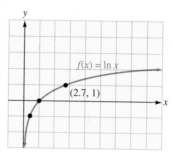

Figure 4-15

Comment

The function $y = \ln x$ is very important and its graph should be memorized. The functions $y = e^x$ and $y = \ln x$ are inverses.

From the definition of logarithm, we know that

$$y = \ln x \quad \text{if and only if} \quad x = e^y \quad \text{Remember that } \ln x = \log_e x.$$

Thus, $\ln e = 1$ because $e^1 = e$. This is an important property of the natural logarithm function. Two other important properties follow from the fact that $y = \ln x$ and $y = e^x$ are inverses:

$$\ln e^x = x \quad \text{and} \quad e^{\ln x} = x.$$

These two properties and examples are shown in the table:

Property	Examples
$\ln e^x = x$	$\ln e^8 = 8$ $\ln e^{-5x} = -5x$ $\ln e^{3x+2} = 3x + 2$
$e^{\ln x} = x$	$e^{\ln 4} = 4$ $e^{\ln(7x)} = 7x$ $e^{\ln(9x-2)} = 9x - 2$

5. Use Translations to Graph Logarithmic Functions

The graphs of many functions involving logarithms are translations of the basic logarithmic graphs. A summary of these are shown in the table for $k > 0$.

Equation	Translations, reflections, and stretchings of the graph of $f(x) = b^x$
$y = \log_b x + k$	Translates the graph of $y = \log_b x$ upward k units
$y = \log_b x - k$	Translates the graph of $y = \log_b x$ downward k units
$y = \log_b(x - k)$	Translates the graph of $y = \log_b x$ to the right k units
$y = \log_b(x + k)$	Translates the graph of $y = \log_b x$ to the left k units
$y = -\log_b x$	Reflects the graph of $y = \log_b x$ about the x-axis
$y = \log_b(-x)$	Reflects the graph of $f(x) = \log_b x$ about the y-axis
$y = k \log_b x$	• Vertically stretches the graph of $y = \log_b x$ if $k > 1$ • Vertically shrinks the graph of $y = \log_b x$ if $0 < k < 1$
$y = \log_b(kx)$	• Horizontally stretches the graph of $y = \log_b x$ if $0 < k < 1$ • Horizontally shrinks the graph of $y = \log_b x$ if $k > 1$

EXAMPLE 6   Graph: $f(x) = 3 + \log_2 x$

Solution   We can use translations to graph the function. The graph of $f(x) = 3 + \log_2 x$ is identical to the graph of $f(x) = \log_2 x$, except that it is translated 3 units upward. (See Figure 4-16.)

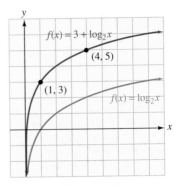

Figure 4-16

Self Check 6   Graph: $f(x) = \log_3 x - 2$

EXAMPLE 7   Graph: $f(x) = \log_{1/2}(x - 1)$

Solution   We can use translations to graph the function. The graph of $f(x) = \log_{1/2}(x - 1)$ is identical to the graph of $f(x) = \log_{1/2} x$, except that it is translated 1 unit to the right. (See Figure 4-17.)

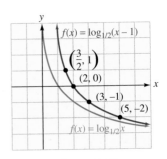

Figure 4-17

Self Check 7   Graph: $f(x) = \log_{1/3}(x + 2)$

Many graphs of logarithmic functions involve translations of the graph of $f(x) = \ln x$. For example, Figure 4-18 shows a calculator graph of the functions $f(x) = \ln x$, $f(x) = \ln x + 2$, and $f(x) = \ln x - 3$.

John Napier
(1550–1617)
Napier is famous for his work with natural logarithms. In fact, natural logarithms are often called Napierian logarithms. He also invented a device called Napier's rods, which did multiplications mechanically. This was a forerunner of modern-day computers.

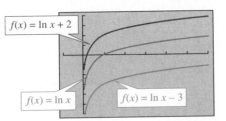

The graph of $f(x) = \ln x + 2$ is 2 units above the graph of $f(x) = \ln x$.

The graph of $f(x) = \ln x - 3$ is 3 units below the graph of $f(x) = \ln x$.

Figure 4-18

Figure 4-19 on the next page shows a calculator graph of the functions $f(x) = \ln x$, $f(x) = \ln(x - 2)$, and $f(x) = \ln(x + 3)$.

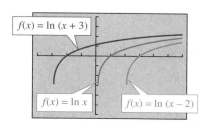

The graph of $f(x) = \ln(x + 3)$ is 3 units to the left of the graph of $f(x) = \ln x$.

The graph of $f(x) = \ln(x - 2)$ is 2 units to the right of the graph of $f(x) = \ln x$.

Figure 4-19

Accent on Technology

Using Calculators to Graph Logarithmic Functions

Graphing calculators can draw graphs of logarithmic functions. To use a calculator to graph $f(x) = -2 + \log_{10}\left(\frac{1}{2}x\right)$, we enter the right side of the equation after the symbol $Y_1 =$. The display will show the equation

$$Y_1 = -2 + \log(1/2 * x)$$

If we use window settings of $[-1, 5]$ for x and $[-5, 1]$ for y and press GRAPH , we will obtain the graph shown in Figure 4-20.

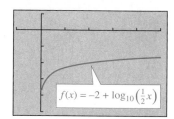

Figure 4-20

Self Check Answers

1. a. 2   **b.** 4   **c.** -2 **2. a.** -2 **b.** 6   **c.** $\frac{1}{2}$ **3.** 75.3998
4. a. 1.1447   **b.** no value   **5. a.** 6.9199   **b.** 14.5596
6. 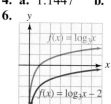   **7.**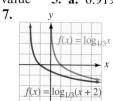

4.3 Exercises

Vocabulary and Concepts   *Fill in the blanks.*

1. The equation $y = \log_b x$ is equivalent to _____.

2. The domain of a logarithmic function is the interval _____.

3. The _____ of a logarithmic function is the interval $(-\infty, \infty)$.

4. $b^{\log_b x} = $ ___.

5. Because the exponential function is one-to-one, it has an _____ function.

6. The inverse of an exponential function is called a _____ function.

7. $\log_b x$ is the _____ to which b is raised to get x.

8. The y-axis is an _____ of the graph of $f(x) = \log_b x$.

9. The graph of $f(x) = \log_b x$ passes through the points _____ and _____.

10. $\log_{10} 10^x = $ ___.

11. $\ln x$ means _____.

12. The domain of the function $f(x) = \ln x$ is the interval _____.

13. The range of the function $f(x) = \ln x$ is the interval _____.

14. The graph of $f(x) = \ln x$ has the _____ as an asymptote.

15. In the expression log x, the base is understood to be ___.

16. In the expression ln x, the base is understood to be ___.

Practice *Write each equation in exponential form.*

17. $\log_3 81 = 4$ **18.** $\log_7 7 = 1$

19. $\log_{1/2} \dfrac{1}{8} = 3$ **20.** $\log_{1/5} 1 = 0$

21. $\log_4 \dfrac{1}{64} = -3$ **22.** $\log_6 \dfrac{1}{36} = -2$

23. $\log_\pi \pi = 1$ **24.** $\log_7 \dfrac{1}{49} = -2$

Write each equation in logarithmic form.

25. $8^2 = 64$ **26.** $10^3 = 1{,}000$

27. $4^{-2} = \dfrac{1}{16}$ **28.** $3^{-4} = \dfrac{1}{81}$

29. $\left(\dfrac{1}{2}\right)^{-5} = 32$ **30.** $\left(\dfrac{1}{3}\right)^{-3} = 27$

31. $x^y = z$ **32.** $m^n = p$

Find each value of x.

33. $\log_2 8 = x$ **34.** $\log_3 9 = x$

35. $\log_4 64 = x$ **36.** $\log_6 216 = x$

37. $\log_{1/2} \dfrac{1}{8} = x$ **38.** $\log_{1/3} \dfrac{1}{81} = x$

39. $\log_9 3 = x$ **40.** $\log_{125} 5 = x$

41. $\log_{1/2} 8 = x$ **42.** $\log_{1/2} 16 = x$

43. $\log_8 x = 2$ **44.** $\log_7 x = 0$

45. $\log_7 x = 1$ **46.** $\log_2 x = 8$

47. $\log_{25} x = \dfrac{1}{2}$ **48.** $\log_4 x = \dfrac{1}{2}$

49. $\log_5 x = -2$ **50.** $\log_3 x = -4$

51. $\log_{36} x = -\dfrac{1}{2}$ **52.** $\log_{27} x = -\dfrac{1}{3}$

53. $\log_x 5^3 = 3$ **54.** $\log_x 5 = 1$

55. $\log_x \dfrac{9}{4} = 2$ **56.** $\log_x \dfrac{\sqrt{3}}{3} = \dfrac{1}{2}$

57. $\log_x \dfrac{1}{64} = -3$ **58.** $\log_x \dfrac{1}{100} = -2$

59. $\log_x \dfrac{9}{4} = -2$ **60.** $\log_x \dfrac{\sqrt{3}}{3} = -\dfrac{1}{2}$

61. $2^{\log_2 5} = x$ **62.** $3^{\log_3 4} = x$

63. $x^{\log_4 6} = 6$ **64.** $x^{\log_3 8} = 8$

Use a calculator to find each value to four decimal places.

65. log 3.25 **66.** log 0.57

67. log 0.00467 **68.** log 375.876

69. ln 45.7 **70.** ln 0.005

71. $\ln \dfrac{2}{3}$ **72.** $\ln \dfrac{12}{7}$

73. ln 35.15 **74.** ln 0.675

75. ln 7.896 **76.** ln 0.00465

77. log (ln 1.7) **78.** ln (log 9.8)

79. ln (log 0.1) **80.** log (ln 0.01)

Use a calculator to find y to four decimal places, if possible.

81. log $y = 1.4023$ **82.** log $y = 0.926$

83. log $y = -3.71$ **84.** log $y = \log \pi$

85. ln $y = 1.4023$ **86.** ln $y = 2.6490$

87. ln $y = 4.24$ **88.** ln $y = 0.926$

89. ln $y = -3.71$ **90.** ln $y = -0.28$

91. log $y = \ln 8$ **92.** ln $y = \log 7$

Find each value without using a calculator.

93. log 10,000 **94.** log 1,000,000

95. log 0.001 **96.** $\log \dfrac{1}{100{,}000}$

97. $e^{(\ln 7)}$ **98.** $e^{(\ln 9)}$

99. $\ln(e^4)$ **100.** $\ln(e^{-6})$

Find the value of b, if any, that would cause the graph of y = log_b x to look like the graph shown.

101.

102.

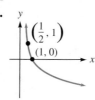

103.

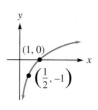

104.

Graph each function.

105. $f(x) = \log_3 x$

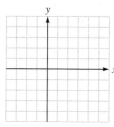

106. $f(x) = \log_4 x$

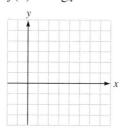

107. $f(x) = \log_{1/3} x$

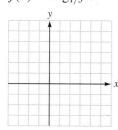

108. $f(x) = \log_{1/4} x$

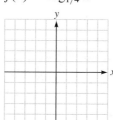

109. $f(x) = 2 + \log_2 x$

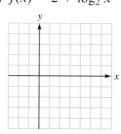

110. $f(x) = \log_2(x - 1)$

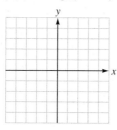

111. $f(x) = \log_3(x + 2)$

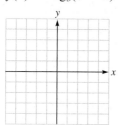

112. $f(x) = -3 + \log_3 x$

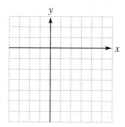

113. $f(x) = -3 + \ln x$

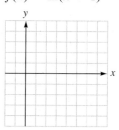

114. $f(x) = \ln(x + 1)$

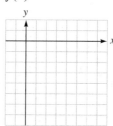

115. $f(x) = \ln(x - 4)$

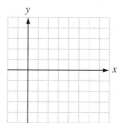

116. $f(x) = 2 + \ln x$

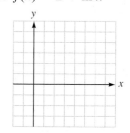

Use a graphing calculator to graph each function.

117. $f(x) = \log(3x)$

118. $f(x) = \log\left(\dfrac{x}{3}\right)$

119. $f(x) = \log(-x)$

120. $f(x) = -\log x$

121. $f(x) = \ln\left(\dfrac{1}{2}x\right)$

122. $f(x) = \ln x^2$

123. $f(x) = \ln(-x)$

124. $f(x) = \ln(3x)$

Discovery and Writing

125. Consider the following graphs. Which is larger, *a* or *b*, and why?

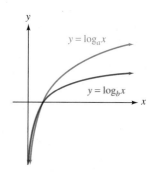

126. Consider the following graphs. Which is larger, a or b, and why?

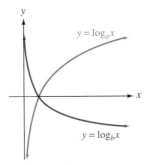

127. Pick two numbers and add their logarithms. Then find the logarithm of the product of those two numbers. What do you observe? Does it work for three numbers?

128. If $\log_a b = 7$, find $\log_b a$.

Review   *Find the vertex of each parabola.*

129. $y = x^2 + 7x + 3$ **130.** $y = 3x^2 - 8x - 1$

131. Fencing a pasture   A farmer will use 3,400 feet of fencing to enclose and divide the pasture shown in the illustration. What dimensions will enclose the greatest area?

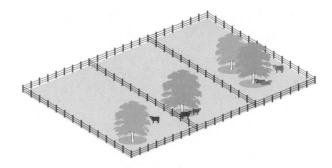

132. Selling appliances   When the price is p dollars, an appliance dealer can sell $(2,200 - p)$ refrigerators. What price will maximize his revenue?

133. Write the equation of a line passing through the origin and parallel to the line $y = 5x - 8$.

134. Write the equation of a line passing through $(3, 2)$ and perpendicular to the line $y = \frac{2}{3}x - 12$.

4.4   Applications of Logarithmic Functions

Objectives

1. Use Logarithms to Solve Electrical Engineering Problems
2. Use Logarithms to Solve Geology Problems
3. Use Logarithms to Solve Charging Battery Problems
4. Use Logarithms to Solve Population Growth Problems
5. Use Logarithms to Solve Isothermal Expansion Problems

Banyu Sakti/AFP/Getty Images

On December 24, 2004, the deadliest natural disaster in history occurred when an earthquake in the Indian Ocean triggered a series of tsunamis killing more than 225,000 people in eleven countries. Indonesia, Sri Lanka, Thailand, and India were hit the hardest. The devastation prompted worldwide aid and more than seven billion dollars were donated to help with relief efforts.

The earthquake was the second largest ever recorded and had a magnitude between 9.1 and 9.3 on the Richter scale. The intensity of an earthquake is based on a logarithmic function, the topic of this section.

In this section, we also will use logarithmic functions to solve problems in engineering, geology, social science, and physics.

1.   Use Logarithms to Solve Electrical Engineering Problems

Electronic engineers use common logarithms to measure the voltage gain of devices such as amplifiers or the length of a transmission line. The unit of gain, called the **decibel,** is defined by a logarithmic function.

Decibel Voltage Gain   If E_O is the output voltage of a device and E_I is the input voltage, the **decibel voltage gain** is given by

$$\text{db gain} = 20 \log \frac{E_O}{E_I}$$

EXAMPLE 1   Find the db gain of an amplifier if its input is 0.5 volt and its output is 40 volts.

Solution   We can find the decibel voltage gain by substituting the given values into the db gain formula.

$$\text{db voltage gain} = 20 \log \frac{E_O}{E_I}$$

$$\text{db voltage gain} = 20 \log \frac{40}{0.5} \qquad \text{Substitute 40 for } E_O \text{ and 0.5 for } E_I.$$

$$= 20 \log 80$$

$$\approx 38.06179974 \qquad \text{Use a calculator.}$$

To the nearest decibel, the db gain is 38 decibels.

Self Check 1   Find the db gain if the input is 0.7 volt.

2.   Use Logarithms to Solve Geology Problems

Seismologists measure the intensity of earthquakes on the **Richter scale,** which is based on a logarithmic function.

Richter Scale   If R is the intensity of an earthquake, A is the amplitude (measured in micrometers), and P is the period (the time of one oscillation of the Earth's surface, measured in seconds), then

$$R = \log \frac{A}{P}$$

EXAMPLE 2   Find the intensity of an earthquake with amplitude of 5,000 micrometers $\left(\frac{1}{2} \text{ centimeter}\right)$ and a period of 0.07 second.

Solution   We substitute 5,000 for A and 0.07 for P in the Richter scale formula and simplify.

$$R = \log \frac{A}{P}$$

$$R = \log \frac{5,000}{0.07}$$

$$\approx \log 71,428.57143 \qquad \text{Use a calculator.}$$

$$\approx 4.853871964$$

To the nearest tenth, the earthquake measures 4.9 on the Richter scale.

Self Check 2 Find the intensity of an aftershock with the same period but one-half of the amplitude.

3. Use Logarithms to Solve Charging Battery Problems

A battery charges at a rate that depends on how close it is to being fully charged—it charges fastest when it is most discharged. The formula that determines the time required to charge a battery to a certain level is based on a natural logarithmic function.

Charging Batteries   If M is the theoretical maximum charge that a battery can hold and k is a positive constant that depends on the battery and the charger, the length of time t (in minutes) required to charge the battery to a given level C is given by

$$t = -\frac{1}{k} \ln\left(1 - \frac{C}{M}\right)$$

EXAMPLE 3   How long will it take to bring a fully discharged battery to 90% of full charge? Assume that $k = 0.025$ and that time is measured in minutes.

Solution   90% of full charge means 90% of M. We can substitute $0.90M$ for C and 0.025 for k in the formula for charging batteries to find t.

$$t = -\frac{1}{k} \ln\left(1 - \frac{C}{M}\right)$$

$$t = -\frac{1}{0.025} \ln\left(1 - \frac{0.90M}{M}\right)$$

$$= -40 \ln(1 - 0.9)$$

$$= -40 \ln(0.1)$$

$$\approx 92.10340372 \qquad \text{Use a calculator.}$$

The battery will reach 90% charge in about 92 minutes.

Self Check 3 How long will it take this battery to reach 80% of full charge?

4. Use Logarithms to Solve Population Growth Problems

If a population grows exponentially at a certain annual rate, the time required for the population to double is called the **doubling time** and is given by the following formula. You will be asked to prove this formula in Exercise 104 in Section 4.6.

Population Doubling Time   If r is the annual growth rate and t is the time (in years) required for a population to double, then

$$t = \frac{\ln 2}{r}$$

EXAMPLE 4   The population of the Earth is growing at the approximate rate of 2% per year. If this rate continues, how long will it take the population to double?

Solution   Because the population is growing at the rate of 2% per year, we can substitute 0.02 for r in the formula for doubling time and simplify.

$$t = \frac{\ln 2}{r}$$

$$t = \frac{\ln 2}{0.02}$$

$$\approx 34.65735903$$

It will take about 35 years for the Earth's population to double.

Self Check 4   If the world population's annual growth rate could be reduced to 1.5% per year, what would be the doubling time?

5.   Use Logarithms to Solve Isothermal Expansion Problems

When energy is added to a gas, its temperature and volume could increase. In **isothermal expansion,** the temperature remains constant—only the volume changes. The energy required is calculated as follows.

Isothermal Expansion   If the temperature T is constant, the energy E required to increase the volume of 1 mole of gas from an initial volume V_i to a final volume V_f is given by

$$E = RT \ln\left(\frac{V_f}{V_i}\right)$$

E is measured in joules and T in Kelvins. R is the universal gas constant, which is 8.314 joules/mole/K.

EXAMPLE 5   Find the amount of energy that must be supplied to triple the volume of 1 mole of gas at a constant temperature of 300 K.

Solution   We substitute 8.314 for R and 300 for T in the formula. Since the final volume is to be three times the initial volume, we also substitute $3V_i$ for V_f.

$$E = RT \ln\left(\frac{V_f}{V_i}\right)$$

$$E = (8.314)(300) \ln\left(\frac{3V_i}{V_i}\right)$$

$$= 2,494.2 \ln 3$$

$$\approx 2,740.15877$$

Approximately 2,740 joules of energy must be added to triple the volume.

Self Check 5   What energy is required to double the volume?

Self Check Answers   **1.** about 35 decibels   **2.** about 4.6   **3.** about 64 min   **4.** about 46 years
5. 1,729 joules

4.4   Exercises

Vocabulary and Concepts   *Fill in the blanks.*

1. db gain = _____
2. The intensity of an earthquake is measured by the formula $R = $ ____.
3. The formula for charging batteries is _____.
4. If a population grows exponentially at a rate r, the time it will take for the population to double is given by the formula $t = $ ___.
5. The formula for isothermal expansion is _____.
6. The logarithm of a negative number is _____.

Applications   *Use a calculator to solve each problem.*

7. **Gain of an amplifier**   An amplifier produces an output of 17 volts when the input signal is 0.03 volt. Find the decibel voltage gain.
8. **Transmission lines**   A 4.9-volt input to a long transmission line decreases to 4.7 volts at the other end. Find the decibel voltage loss.
9. **Gain of an amplifier**   Find the db gain of an amplifier whose input voltage is 0.71 volt and whose output voltage is 20 volts.
10. **Gain of an amplifier**   Find the db gain of an amplifier whose output voltage is 2.8 volts and whose input voltage is 0.05 volt.
11. **db gain**   Find the db gain of the amplifier shown below.

12. **db gain**   Find the db gain of the amplifier shown below.

13. **Earthquakes**   An earthquake has an amplitude of 5,000 micrometers and a period of 0.2 second. Find its measure on the Richter scale.
14. **Earthquakes**   An earthquake has an amplitude of 8,000 micrometers and a period of 0.008 second. Find its measure on the Richter scale.
15. **Earthquakes**   An earthquake with a period of $\frac{1}{4}$ second has an amplitude of 2,500 micrometers. Find its measure on the Richter scale.
16. **Earthquakes**   An earthquake has a period of $\frac{1}{2}$ second and an amplitude of 5 cm. Find its measure on the Richter scale.   (*Hint:* 1 cm = 10,000 micrometers)
17. **Earthquakes**   An earthquake measuring between 3.5 and 5.4 on the Richter scale is often felt, but rarely causes damage. Suppose an earthquake in Northern California has an amplitude of 6,000 micrometers and a period of 0.3 second. Is it likely to cause damage?
18. **Earthquakes**   An earthquake measuring between 7 and 7.9 on the Richter scale is a major earthquake and can cause serious damage over larger areas. Suppose an earthquake in Chile has an amplitude of 198.5 cm and a period of 0.1 second. Would it cause serious damage over large areas?   (*Hint:* 1 cm = 10,000 micrometers)

19. **Battery charge**   If $k = 0.116$, how long will it take a battery to reach a 90% charge? Assume that the battery was fully discharged when it began charging.
20. **Battery charge**   If $k = 0.201$, how long will it take a battery to reach a 40% charge? Assume that the battery was fully discharged when it began charging.
21. **Population growth**   A town's population grows at the rate of 12% per year. If this growth rate remains constant, how long will it take the population to double?
22. **Fish population growth**   One thousand bass were stocked in Catfish Lake in Eagle River, Wisconsin, a lake with no bass population. If the population of bass is expected to grow at a rate of 25% per year, how long will it take the population to double?

23. **Population growth** A population growing at an annual rate r will triple in a time t given by the formula $t = \frac{\ln 3}{r}$. How long will it take the population of the town in Exercise 21 to triple?

24. **Fish population growth** How long would it take the fish population in Exercise 22 to triple?

25. **Isothermal expansion** One mole of gas expands isothermally to triple its volume. If the gas temperature is 400 K, what energy is absorbed?

26. **Isothermal expansion** One mole of gas expands isothermally to double its volume. If the gas temperature is 300K, what energy is absorbed?

If an investment is growing continuously for t years, its annual growth rate r is given by the following formula, where P is the current value and P_0 is the amount originally invested.

$$r = \frac{1}{t} \ln \frac{P}{P_0}$$

27. **Investing** An investment of $10,400 in America Online in 1992 was worth $10,400,000 in 1999. Find AOL's average annual growth rate during this period.

28. **Investing** A $5,000 investment in Dell Computer in 1995 was worth $237,000 in 1999. Find the average annual growth rate of the stock.

29. **Depreciation** In business, equipment often is depreciated using the double declining-balance method. In this method, a piece of equipment with a life expectancy of N years, costing $C, will depreciate to a value of $V in n years, where n is given by the following formula.

$$n = \frac{\log V - \log C}{\log\left(1 - \frac{2}{N}\right)}$$

If a computer that cost $37,000 has a life expectancy of 5 years and has depreciated to a value of $8,000, how old is it?

30. **Depreciation** A word processor worth $470 when new had a life expectancy of 12 years. If it is now worth $189, how old is it? (See Exercise 29.)

31. **Annuities** If $P is invested at the end of each year in an annuity earning interest at an annual rate r, the amount in the account will be $A after n years, where

$$n = \frac{\log\left[\dfrac{Ar}{P} + 1\right]}{\log(1 + r)}$$

If $1,000 is invested each year in an annuity earning 12% annual interest, when will the account be worth $20,000?

32. **Annuities** If $5,000 is invested each year in an annuity earning 8% annual interest, when will the account be worth $50,000? (See Exercise 31.)

33. **Breakdown voltage** The coaxial power cable shown has a central wire with radius $R_1 = 0.25$ centimeters. It is insulated from a surrounding shield with inside radius $R_2 = 2$ centimeters. The maximum voltage the cable can withstand is called the **breakdown voltage** V of the insulation. V is given by the formula

$$V = ER_1 \ln \frac{R_2}{R_1}$$

where E is the **dielectric strength** of the insulation. If $E = 400,000$ volts/centimeter, find V.

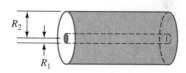

34. **Breakdown voltage** In Exercise 33, if the inside diameter of the shield were doubled, what voltage could the cable withstand?

35. Suppose you graph the function $f(x) = \ln x$ on a coordinate grid with a unit distance of 1 centimeter on the x- and y-axes. How far out must you go on the x-axis so that $f(x) = 12$? Give your result to the nearest mile.

36. Suppose you graph the function $f(x) = \log x$ on a coordinate grid with a unit distance of 1 centimeter on the x- and y-axes. How far out must you go on the x-axis so that $f(x) = 12$? Give the result to the nearest mile. Why is this result so much larger than the result in Exercise 35?

Discovery and Writing

37. One form of the logistic function is given by the following equation. Explain how you would find the y-intercept of its graph.

$$y = \frac{1}{1 + e^{-2x}}$$

38. Graph the function $y = \ln|x|$. Explain why the graph looks the way it does.

Review *Write the equation of the required line.*

39. Having a slope of 7 and a y-intercept of 3

40. Parallel to the line $3x + 2y = 9$ and passing through the point $(-3, 5)$

41. A vertical line passing through $(2, 3)$

42. A horizontal line passing through $(2, 3)$

Simplify each expression.

43. $\dfrac{2(x + 2) - 1}{4x^2 - 9}$

44. $\dfrac{x + 1}{x} + \dfrac{x - 1}{x + 1}$

45. $\dfrac{x^2 + 3x + 2}{3x + 9} \cdot \dfrac{x + 3}{x^2 - 4}$

46. $\dfrac{1 + \dfrac{y}{x}}{\dfrac{y}{x} - 1}$

Claudio Pozo/AFP/Getty Images

4.5 Properties of Logarithms

Objectives

1. Use Properties of Logarithms to Simplify Expressions
2. Use the Change-of-Base Formula
3. Use Logarithms to Solve pH Problems
4. Use Logarithms to Solve Problems in Electronics
5. Use Logarithms to Solve Physiology Problems

Rolling Stone magazine lists the rock band U2 at number 22 in their list of the top 100 artists of all time. The Irish band is noted for lead singer Bono's vocals and their anthem-like sound.

Attending a U2 concert or cranking-up a car stereo and playing U2's song *With or Without You* is an enjoyable activity, and music is an important part of our pop culture. Since many students prefer their music loud, the loudness of sound and its intensity are interesting concepts. In this section, we will see that loudness of sound and the intensity of sound are related by a formula involving the natural logarithmic function.

1. Use Properties of Logarithms to Simplify Expressions

Since logarithms are exponents, the properties of exponents have counterparts in the theory of logarithms. We begin with four basic properties.

Properties of Logarithms If b is a positive number and $b \neq 1$, then

1. $\log_b 1 = 0$ **2.** $\log_b b = 1$

3. $\log_b b^x = x$ **4.** $b^{\log_b x} = x$ $(x > 0)$

Properties 1 through 4 follow directly from the definition of logarithm.

1. $\log_b 1 = 0$, because $b^0 = 1$.
2. $\log_b b = 1$, because $b^1 = b$.
3. $\log_b b^x = x$, because $b^x = b^x$.
4. $b^{\log_b x} = x$, because $\log_b x$ is the exponent to which b is raised to get x.

Properties 3 and 4 also indicate that the composition of the exponential and logarithmic functions (in both directions) is the identity function. This is expected, because the exponential and logarithmic functions with the same base are inverse functions.

EXAMPLE 1 Simplify each expression: **a.** $\log_3 1$ **b.** $\log_4 4$ **c.** $\log_7 7^3$ **d.** $b^{\log_b 3}$

Solution We can simplify each expression using the properties of logarithms.

  a. By Property 1, $\log_3 1 = \mathbf{0}$, because $3^0 = 1$.
  b. By Property 2, $\log_4 4 = \mathbf{1}$, because $4^1 = 4$.
  c. By Property 3, $\log_7 7^3 = \mathbf{3}$, because $7^3 = 7^3$.
  d. By Property 4, $b^{\log_b 3} = \mathbf{3}$, because $\log_b 3$ is the power to which b is raised to get 3.

Self Check 1 Simplify: **a.** $\log_4 1$ **b.** $\log_3 3$ **c.** $\log_2 2^4$ **d.** $5^{\log_5 2}$

The four properties also hold for natural logarithms.

1. $\ln 1 = \mathbf{0}$, because $e^0 = 1$.
2. $\ln e = \mathbf{1}$, because $e^1 = e$.
3. $\ln e^x = x$, because $e^x = e^x$.
4. $e^{\ln x} = x$, because $\ln x$ is the exponent to which e is raised to get x.

The next two properties state that

The logarithm of a product is the sum of the logarithms.
The logarithm of a quotient is the difference of the logarithms.

The Logarithm of a Product and a Difference

If M, N, and b are positive numbers and $b \neq 1$, then

5. $\log_b MN = \log_b M + \log_b N$ **6.** $\log_b \dfrac{M}{N} = \log_b M - \log_b N$

Property 5 is known as the **product rule.** Property 6 is known as the **quotient rule.**

Proof To prove Property 5, we let $x = \log_b M$ and $y = \log_b N$ and use the definition of logarithm to write each equation in exponential form.

  $M = b^x$ and $N = b^y$

Then $MN = b^x b^y$ and a property of exponents gives

  $MN = b^{x+y}$ $b^x b^y = b^{x+y}$; keep the base and add the exponents.

We write this exponential equation in logarithmic form as

  $\log_b MN = x + y$

Substituting the values of x and y completes the proof.

$$\log_b MN = \log_b M + \log_b N$$

The proof of Property 6 is similar. You will be asked to do it in an exercise.

Comment

By Property 5 of logarithms, the logarithm of a *product* is equal to the *sum* of the logarithms. The logarithm of a sum or a difference usually does not simplify. In general,

$$\log_b(M + N) = \log_b M + \log_b N \quad \text{and} \quad \log_b(M - N) = \log_b M - \log_b N$$

By Property 6, the logarithm of a *quotient* is equal to the *difference* of the logarithms. The logarithm of a quotient is not the quotient of the logarithms:

$$\log_b \frac{M}{N} = \frac{\log_b M}{\log_b N}$$

Properties 5 and 6 also hold for natural logarithms.

5. $\ln MN = \ln M + \ln N$ Product rule

6. $\ln \dfrac{M}{N} = \ln M - \ln N$ Quotient rule

Accent on Technology

We can use a calculator to illustrate Property 5 of logarithms by showing that

$$\ln[(3.7)(15.9)] = \ln 3.7 + \ln 15.9$$

We calculate the left and right sides of the equation separately and compare the results. To use a calculator to find $\ln[(3.7)(15.9)]$, we enter these numbers and press these keys:

Scientific calculator
3.7 ☒ 15.9 ☐ LN

Graphing calculator
LN 3.7 ☒ 15.9) ENTER

The display will read 4.074651929.

To find $\ln 3.7 + \ln 15.9$, we enter these numbers and press these keys:

Scientific calculator
3.7 LN ☐ 15.9 LN ☐

Graphing calculator
LN 3.7) ☐ LN 15.9) ENTER

The display will read 4.074651929. Since the left and right sides are equal, the equation is true.

Two more properties state that

The logarithm of a power is the power times the logarithm.

If the logarithms of two numbers are equal, the numbers are equal.

More Properties of Logarithms

If M and b are positive numbers and $b \neq 1$, then

7. $\log_b M^p = p \log_b M$ **8.** If $\log_b x = \log_b y$, then $x = y$.

Property 7 is known as the **power rule.**

Proof   To prove Property 7, we let $x = \log_b M$, write the expression in exponential form, and raise both sides to the pth power:

$$M = b^x$$

$$(M)^p = (b^x)^p \quad \text{Raise both sides to the } p\text{th power.}$$

$$M^p = b^{px} \quad \text{Keep the base and multiply the exponents.}$$

Using the definition of logarithms gives

$$\log_b M^p = px$$

Substituting the value for x completes the proof.

$$\log_b M^p = p \log_b M$$

Property 8 follows from the fact that the logarithmic function is a one-to-one function. Property 8 will be important in the next section when we solve logarithmic equations.

Properties 7 and 8 also hold for natural logarithms.

7. $\ln M^p = p \ln M$ **Power rule**

8. If $\ln x = \ln y$, then $x = y$.

We can use the properties of logarithms to write a logarithm as the sum or difference of several logarithms.

EXAMPLE 2   Assume that x, y, and z are positive numbers. Write each expression in terms of the logarithms of x, y, and z: **a.** $\log_b xyz$ **b.** $\ln \frac{x}{yz}$

Solution   We can use the properties of logarithms to write each logarithm as the sum or difference of several logarithms.

a. $\log_b xyz = \log_b(xy)z$

$$= \log_b(xy) + \log_b z \qquad \text{The log of a product is the sum of the logs.}$$

$$= \log_b x + \log_b y + \log_b z \qquad \text{The log of a product is the sum of the logs.}$$

b. $\ln \dfrac{x}{yz} = \ln x - \ln(yz)$ The ln of a quotient is the difference of the natural logs.

$$= \ln x - (\ln y + \ln z) \qquad \text{The ln of a product is the sum of the natural logs.}$$

$$= \ln x - \ln y - \ln z \qquad \text{Remove parentheses.}$$

Self Check 2   Write the expression in terms of the logarithms of x, y, and z: $\log_b \frac{xy}{z}$

EXAMPLE 3   Assume that x, y, z, and b are positive numbers and $b \neq 1$. Write each expression in terms of the logarithms of x, y, and z: **a.** $\log_b(x^3y^2z)$ **b.** $\ln \dfrac{y^2\sqrt{z}}{x}$

Solution   We can use the properties of logarithms to write each logarithm as the sum or difference of several logarithms.

a. $\log_b(x^3y^2z) = \log_b x^3 + \log_b y^2 + \log_b z$ The log of a product is the sum of the logs.

$$= 3 \log_b x + 2 \log_b y + \log_b z \qquad \text{The log of a power is the power times the log.}$$

b. $\ln \dfrac{y^2 \sqrt{z}}{x} = \ln\left(y^2 \sqrt{z}\right) - \ln x$ The ln of a quotient is the difference of the natural logs.

$$= \ln y^2 + \ln z^{1/2} - \ln x \qquad \text{The ln of a product is the sum of the natural logs: } \sqrt{z} = z^{1/2}.$$

$$= 2 \ln y + \frac{1}{2} \ln z - \ln x \qquad \text{The ln of a power is the power times the ln.}$$

Self Check 3   Write the expression in terms of the logarithms of x, y, and z: $\log_b \sqrt[3]{\dfrac{x^2 y}{z}}$.

We can use the properties of logarithms to combine several logarithms into one logarithm.

EXAMPLE 4   Assume that x, y, z, and b are positive numbers and $b \neq 1$. Write each expression as one logarithm: **a.** $2 \log_b x + \frac{1}{3} \log_b y$ **b.** $\frac{1}{2} \log_b(x - 2) - \log_b y + 3 \log_b z$

Solution   We can use the properties of logarithms to combine several logarithms into one logarithm.

a. $2 \log_b x + \dfrac{1}{3} \log_b y = \log_b x^2 + \log_b y^{1/3}$ A power times a log is the log of the power.

$$= \log_b(x)^2 \sqrt[3]{y}$$

$$= \log_b(x^2 y^{1/3}) \qquad \text{The sum of two logs is the log of the product.}$$

b. $\dfrac{1}{2} \log_b(x - 2) - \log_b y + 3 \log_b z$

$$= \log_b(x - 2)^{1/2} - \log_b y + \log_b z^3 \qquad \text{A power times a log is the log of the power.}$$

$$= \log_b \frac{(x - 2)^{1/2}}{y} + \log_b z^3 \qquad \text{The difference of two logs is the log of the quotient.}$$

$$= \log_b \frac{z^3 \sqrt{x - 2}}{y} \qquad \text{The sum of two logs is the log of the product.}$$

Self Check 4   Write as one logarithm: $2 \ln x + \frac{1}{2} \ln y - 3 \ln(x - y)$

We summarize the eight properties of logarithms and natural logarithms as follows.

Properties of Logarithms   If b, M, and N are positive numbers and $b \neq 1$, then

1. $\log_b 1 = 0$ **1.** $\ln 1 = 0$

2. $\log_b b = 1$ **2.** $\ln e = 1$

3. $\log_b b^x = x$ **3.** $\ln e^x = x$

4. $b^{\log_b x} = x$ **4.** $e^{\ln x} = x$

5. $\log_b MN = \log_b M + \log_b N$ **5.** $\ln MN = \ln M + \ln N$

6. $\log_b \dfrac{M}{N} = \log_b M - \log_b N$ **6.** $\ln \dfrac{M}{N} = \ln M - \ln N$

7. $\log_b M^p = p \log_b M$ **7.** $\ln M^p = p \ln M$

8. If $\log_b x = \log_b y$, then $x = y$.   **8.** If $\ln x = \ln y$, then $x = y$.

EXAMPLE 5   Given that $\log_{10} 2 \approx 0.3010$ and $\log_{10} 3 \approx 0.4771$, find approximations for:
a. $\log_{10} 18$ **b.** $\log_{10} 2.5$

Solution   We will use the properties of logarithms to write each expression in terms of known logarithms. Then we can substitute the values of the known logarithms and simplify.

a. $\log_{10} 18 = \log_{10}(2 \cdot 3^2)$

$$= \log_{10} 2 + \log_{10} 3^2 \qquad \text{The log of a product is the sum of the logs.}$$

$$= \log_{10} 2 + 2\log_{10} 3 \qquad \text{The log of a power is the power times the log.}$$

$$\approx 0.3010 + 2(0.4771)$$

$$\approx 1.2552$$

b. $\log_{10} 2.5 = \log_{10}\left(\dfrac{5}{2}\right)$

$$= \log_{10} 5 - \log_{10} 2 \qquad \text{The log of a quotient is the difference of the logs.}$$

$$= \log_{10} \dfrac{10}{2} - \log_{10} 2 \qquad \text{Write 5 as } \dfrac{10}{2}.$$

$$= \log_{10} 10 - \log_{10} 2 - \log_{10} 2 \qquad \text{The log of a quotient is the difference of the logs.}$$

$$= 1 - 2\log_{10} 2 \qquad \log_{10} 10 = 1$$

$$\approx 1 - 2(0.3010)$$

$$\approx 0.3980$$

Comment
In Example 5, it is important to note that we can use logarithmic values that are not stated in the problem. For example, $\log 10 = 1$, $\log 100 = 2$, and $\log 1,000 = 3$.

Self Check 5   Use the information given in Example 5 to find an approximation for $\log_{10} 0.75$.

2.   Use the Change-of-Base Formula

We have seen how to use a calculator to find base-10 and base-e logarithms. To use a calculator to find logarithms with different bases, such as $\log_7 63$, we can divide the base-10 (or base-e) logarithm of 63 by the base-10 (or base-e) logarithm of 7.

$$\log_7 63 = \frac{\log 63}{\log 7} \qquad\qquad \log_7 63 = \frac{\ln 63}{\ln 7}$$

$$\approx 2.129150068 \qquad\qquad\qquad \approx 2.129150068$$

To check the result, we verify that $7^{2.129150068} \approx 63$. This example suggests that if we know the base-a logarithm of a number, we can find its logarithm to some other base b.

Change-of-Base Formula   If a, b, and x are positive numbers and $a \neq 1$ and $b \neq 1$, then

$$\log_b x = \frac{\log_a x}{\log_a b}$$

To prove this formula, we begin with the equation $\log_b x = y$.

$$y = \log_b x$$

$$x = b^y \qquad \text{Change the equation from logarithmic to exponential form.}$$

$$\log_a x = \log_a b^y \qquad \text{Take the base-}a\text{ logarithm of both sides.}$$

$$\log_a x = y \log_a b \qquad \text{The log of a power is the power times the log.}$$

$$y = \frac{\log_a x}{\log_a b} \qquad \text{Divide both sides by } \log_a b.$$

$$\log_b x = \frac{\log_a x}{\log_a b} \qquad \text{Refer to the first equation and substitute } \log_b x \text{ for } y.$$

If we know logarithms to base a (for example, $a = 10$), we can find the logarithm of x to a new base b by dividing the base-a logarithm of x by the base-a logarithm of b.

Comment

$\frac{\log_a x}{\log_a b}$ means that one logarithm is to be divided by the other. They are not to be subtracted.

EXAMPLE 6   Use the change-of-base formula to find $\log_3 5$.

Solution   We can substitute 3 for b, 10 for a, and 5 for x into the change-of-base formula and simplify.

$$\log_b x = \frac{\log_a x}{\log_a b}$$

$$\log_3 5 = \frac{\log_{10} 5}{\log_{10} 3} \qquad \text{Divide the base-10 logarithm of 5 by the base-10 logarithm of 3.}$$

$$\approx 1.464973521$$

To four decimal places, $\log_3 5 = 1.4650$.

Self Check 6   Find $\log_5 3$ to four decimal places.

We can use properties of logarithms to solve many problems.

3.   Use Logarithms to Solve pH Problems

The more acidic a chemical solution, the greater the concentration of hydrogen ions. Chemists measure this concentration indirectly by the **pH scale,** or the **hydrogen ion index.**

pH of a Solution   If [H$^+$] is the hydrogen ion concentration in gram-ions per liter, then

$$pH = -\log[H^+]$$

Since pure water has approximately 10^{-7} gram-ions per liter, its pH is

$$pH = -\log[H^+]$$
$$pH = -\log 10^{-7}$$
$$= -(-7)\log 10 \qquad \text{The log of a power is the power times the log.}$$
$$= -(-7) \cdot 1 \qquad \text{Use Property 2 of logarithms: } \log_b b = 1.$$
$$= 7$$

EXAMPLE 7   Seawater has a pH of approximately 8.5. Find its hydrogen ion concentration.

Solution   We can substitute 8.5 for pH and solve the equation $pH = -\log[H^+]$ for [H$^+$].

$$8.5 = -\log[H^+]$$
$$-8.5 = \log[H^+]$$
$$[H^+] = 10^{-8.5} \qquad \text{Change the equation from logarithmic form to exponential form.}$$

We then can use a calculator to find that $[H^+] \approx 3.2 \times 10^{-9}$ gram-ions per liter.

Self Check 7 The pH of a solution is 5.7. Find the hydrogen ion concentration.

4.   Use Logarithms to Solve Problems in Electronics

Recall that if E_O is the output voltage of a device and E_I is the input voltage, the decibel voltage gain is given by

(1) $$\text{db gain} = 20 \log \frac{E_O}{E_I}$$

If input and output are measured in watts instead of volts, a different formula is needed.

EXAMPLE 8   Show that an alternate formula for db voltage gain is

$$\text{db gain} = 10 \log \frac{P_O}{P_I}$$

where P_I is the power input and P_O is the power output.

Solution   Power is directly proportional to the square of the voltage. So for some constant k,

$$P_I = k(E_I)^2 \qquad \text{and} \qquad P_O = k(E_O)^2$$

and

$$\frac{P_O}{P_I} = \frac{k(E_O)^2}{k(E_I)^2} = \left(\frac{E_O}{E_I}\right)^2$$

We raise both sides to the $\frac{1}{2}$ power to get

$$\frac{E_O}{E_I} = \left(\frac{P_O}{P_I}\right)^{1/2}$$

which we substitute into formula (1) for db gain.

$$\begin{aligned}
\text{db gain} &= 20 \log \frac{E_O}{E_I} \\
&= 20 \log\left(\frac{P_O}{P_I}\right)^{1/2} \\
&= 20 \cdot \frac{1}{2} \log \frac{P_O}{P_I} \qquad \text{The log of a power is the power times the log.} \\
\text{db gain} &= 10 \log \frac{P_O}{P_I} \qquad \text{Simplify.}
\end{aligned}$$

Self Check 8   Find the db gain of a device to the nearest hundredth when $P_O = 30$ watts and $P_I = 2$ watts

5.   Use Logarithms to Solve Physiology Problems

In physiology, experiments suggest that the relationship between the loudness and the intensity of sound is a logarithmic one known as the Weber–Fechner law.

Weber–Fechner Law If *L* is the apparent loudness of a sound and *I* is the intensity, then

$$L = k \ln I$$

EXAMPLE 9 What increase in the intensity of a sound is necessary to cause a doubling of the apparent loudness?

Solution We use the formula $L = k \ln I$. To double the apparent loudness, we multiply both sides of the equation by 2 and use Property 7 of logarithms.

$$L = k \ln I$$
$$2L = 2k \ln I$$
$$= k \ln I^2$$

To double the apparent loudness, we must square the intensity.

Self Check 9 What increase is necessary to triple the apparent loudness?

Self Check Answers **1. a.** 0 **b.** 1 **c.** 4 **d.** 2 **2.** $\log_b x + \log_b y - \log_b z$
3. $\frac{1}{3}(2\log_b x + \log_b y - \log_b z)$ **4.** $\ln \frac{x^2\sqrt{y}}{(x-y)^3}$ **5.** -0.1249 **6.** 0.6826
7. 2×10^{-6} **8.** 11.76 **9.** Cube the intensity.

4.5 Exercises

Vocabulary and Concepts *Fill in the blanks.*

1. $\log_b 1 = $ ___
2. $\log_b b = $ ___
3. $\log_b MN = \log_b$ ___ $+ \log_b$ ___
4. $b^{\log_b x} = $ ___
5. If $\log_b x = \log_b y$, then ___ $=$ ___.
6. $\log_b \frac{M}{N} = \log_b M$ ___ $\log_b N$
7. $\log_b x^p = p \cdot \log_b$ ___
8. $\log_b b^x = $ ___
9. $\log_b(A + B)$ ___ $\log_b A + \log_b B$
10. $\log_b A + \log_b B$ ___ $\log_b AB$

Simplify each expression.

11. $\log_4 1 = $ ___
12. $\log_4 4 = $ ___
13. $\log_4 4^7 = $ ___
14. $4^{\log_4 8} = $ ___
15. $5^{\log_5 10} = $ ___
16. $\log_5 5^2 = $ ___
17. $\log_5 5 = $ ___
18. $\log_5 1 = $ ___

Practice *Use a calculator to verify each equation.*

19. $\log[(3.7)(2.9)] = \log 3.7 + \log 2.9$
20. $\ln \frac{9.3}{2.1} = \ln 9.3 - \ln 2.1$
21. $\ln(3.7)^3 = 3 \ln 3.7$
22. $\log\sqrt{14.1} = \frac{1}{2}\log 14.1$
23. $\log 3.2 = \frac{\ln 3.2}{\ln 10}$
24. $\ln 9.7 = \frac{\log 9.7}{\log e}$

Assume that x, y, z, and b are positive numbers. Use the properties of logarithms to write each expression in terms of the logarithms of x, y, and z.

25. $\log_b 2xy$
26. $\log_b 3xz$
27. $\log_b \frac{2x}{y}$
28. $\log_b \frac{x}{yz}$

29. $\log_b x^2 y^3$

30. $\log_b x^3 y^2 z$

31. $\log_b (xy)^{1/3}$

32. $\log_b x^{1/2} y^3$

33. $\log_b x \sqrt{z}$

34. $\log_b \sqrt{xy}$

35. $\log_b \dfrac{\sqrt[3]{x}}{\sqrt[3]{yz}}$

36. $\log_b \sqrt[4]{\dfrac{x^3 y^2}{z^4}}$

37. $\ln x^7 y^8$

38. $\ln \dfrac{4x}{y}$

39. $\ln \dfrac{x}{y^4 z}$

40. $\ln x\sqrt{y}$

Assume that x, y, and z are positive numbers. Use the properties of logarithms to write each expression as the logarithm of one quantity.

41. $\log_b (x + 1) - \log_b x$

42. $\log_b x + \log_b (x + 2) - \log_b 8$

43. $2 \log_b x + \dfrac{1}{3} \log_b y$

44. $-2 \log_b x - 3 \log_b y + \log_b z$

45. $-3 \log_b x - 2 \log_b y + \dfrac{1}{2} \log_b z$

46. $3 \log_b (x + 1) - 2 \log_b (x + 2) + \log_b x$

47. $\log_b \left(\dfrac{x}{z} + x \right) - \log_b \left(\dfrac{y}{z} + y \right)$

48. $\log_b (xy + y^2) - \log_b (xz + yz) + \log_b z$

49. $\ln x + \ln(x + 5) - \ln 9$

50. $5 \ln x + \dfrac{1}{5} \ln y$

51. $-6 \ln x - 2 \ln y + \ln z$

52. $-2 \ln x - 3 \ln y + \dfrac{1}{3} \ln z$

Determine whether each statement is true or false.

53. $\log_b ab = \log_b a + 1$

54. $\log_b \dfrac{1}{a} = -\log_b a$

55. $\log_b 0 = 1$

56. $\log_b 2 = \log_2 b$

57. $\log_b (x + y) \neq \log_b x + \log_b y$

58. $\log_b xy = (\log_b x)(\log_b y)$

59. If $\log_a b = c$, then $\log_b a = c$.

60. If $\log_a b = c$, then $\log_b a = \dfrac{1}{c}$.

61. $\log_7 7^7 = 7$

62. $7^{\log_7 7} = 7$

63. $\log_b(-x) = -\log_b x$

64. If $\log_b a = c$, then $\log_b a^p = pc$.

65. $\dfrac{\log_b A}{\log_b B} = \log_b A - \log_b B$

66. $\log_b(A - B) = \dfrac{\log_b A}{\log_b B}$

67. $\log_b \dfrac{1}{5} = -\log_b 5$

68. $3 \log_b \sqrt[3]{a} = \log_b a$

69. $\dfrac{1}{3} \log_b a^3 = \log_b a$

70. $\log_{4/3} y = -\log_{3/4} y$

71. $\log_b y + \log_{1/b} y = 0$

72. $\log_{10} 10^3 = 3(10^{\log_{10} 3})$

73. $\ln xy = (\ln x)(\ln y)$

74. $\dfrac{\ln A}{\ln B} = \ln A - \ln B$

75. $\dfrac{1}{5} \ln a^5 = \ln a$

76. $\ln y = -\ln \dfrac{1}{y}$

Assume that $\log_{10} 4 = 0.6021$, $\log_{10} 7 = 0.8451$, and $\log_{10} 9 = 0.9542$. Use these values and the properties of logarithms to find each value. Do not use a calculator.

77. $\log_{10} 28$

78. $\log_{10} \dfrac{7}{4}$

79. $\log_{10} 2.25$

80. $\log_{10} 36$

81. $\log_{10} \dfrac{63}{4}$

82. $\log_{10} \dfrac{4}{63}$

83. $\log_{10} 252$

84. $\log_{10} 49$

85. $\log_{10} 112$

86. $\log_{10} 324$

87. $\log_{10} \dfrac{144}{49}$

88. $\log_{10} \dfrac{324}{63}$

Use a calculator and the change-of-base formula to find each logarithm.

89. $\log_3 7$

90. $\log_7 3$

91. $\log_\pi 3$

92. $\log_3 \pi$

93. $\log_3 8$

94. $\log_5 10$

95. $\log_{\sqrt{2}} \sqrt{5}$

96. $\log_\pi e$

Applications

97. pH of water slide The water in the Abyss, a water slide at the Atlantis Resort in the Bahamas, has a hydrogen ion concentration of 6.3×10^{-8} gram-ions per liter. Find the pH.

98. pH of swimming pool The ideal pH for a swimming pool is 7.2, the same pH as our eyes. The swimming pool at the local YMCA has a hydrogen ion concentration of 1.6×10^{-7} gram-ions per liter. Find the pH of the pool. Is this ideal?

99. pH of a solution Find the pH of a solution with a hydrogen ion concentration of 1.7×10^{-5} gram-ions per liter.

100. pH of calcium hydroxide Find the hydrogen ion concentration of a saturated solution of calcium hydroxide whose pH is 13.2.

101. pH of apples The pH of apples can range from 2.9 to 3.3. Find the range in the hydrogen ion concentration.

102. pH of sour pickles The hydrogen ion concentration of sour pickles is 6.31×10^{-4}. Find the pH.

103. db gain An amplifier produces a 40-watt output with a $\frac{1}{2}$-watt input. Find the db gain.

104. db loss Losses in a long telephone line reduce a 12-watt input signal to an output of 3 watts. Find the db gain. (Because it is a loss, the "gain" will be negative.)

105. Weber–Fechner law What increase in intensity is necessary to quadruple the loudness?

106. Weber–Fechner law What decrease in intensity is necessary to make a sound half as loud?

107. Isothermal expansion If a certain amount E of energy is added to one mole of a gas, it expands from an initial volume of 1 liter to a final volume V without changing its temperature according to the formula

$$E = 8{,}300 \ln V$$

Find the volume if twice that energy is added to the gas.

108. Richter scale By what factor must the amplitude of an earthquake change to increase its severity by 1 point on the Richter scale? Assume that the period remains constant. The Richter scale is given by

$$R = \log \frac{A}{P}$$

where A is the amplitude and P the period of the tremor.

Discovery and Writing

109. Simplify: $\quad 3^{4 \log_3 2} + 5^{\frac{1}{2} \log_5 25}$

110. Find the value of $a - b$:

$$5 \log x + \frac{1}{3} \log y - \frac{1}{2} \log x - \frac{5}{6} \log y = \log(x^a y^b)$$

111. Prove Property 6 of logarithms:

$$\log_b \frac{M}{N} = \log_b M - \log_b N$$

112. Show that $-\log_b x = \log_{1/b} x$.

113. Show that $e^{x \ln a} = a^x$.

114. Show that $e^{\ln x} = x$.

115. Show that $\ln(e^x) = x$.

116. If $\log_b 3x = 1 + \log_b x$, find b.

117. Explain why $\ln(\log 0.9)$ is undefined.

118. Explain why $\log_b(\ln 1)$ is undefined.

In Exercises 119–120, A and B are both negative. Thus, AB and $\frac{A}{B}$ are positive, and $\log AB$ and $\log \frac{A}{B}$ are defined.

119. Is it still true that $\log AB = \log A + \log B$? Explain.

120. Is it still true that $\log \frac{A}{B} = \log A - \log B$? Explain.

Review *Determine whether each equation defines a function.*

121. $y = 3x - 1$

122. $y = \dfrac{x + 3}{x - 1}$

123. $y^2 = 4x$

124. $y = 4x^2$

Find the domain of each function.

125. $f(x) = x^2 - 4$

126. $f(x) = \dfrac{1}{x^2 - 4}$

127. $f(x) = \sqrt{x^2 + 4}$

128. $f(x) = \sqrt{x^2 - 4}$

4.6 Exponential and Logarithmic Equations

Objectives

1. Use Like Bases to Solve Exponential Equations
2. Use Logarithms to Solve Exponential Equations
3. Solve Logarithmic Equations
4. Solve Carbon-14 Dating Problems
5. Solve Population Growth Problems

Rob Fiocca/Jupiter Images

Coffee is one of the most popular beverages worldwide. Suppose we visit a coffee shop and order a white chocolate mocha. After blending smooth white chocolate with rich espresso and steamed milk and topping it with whipped cream, the coffee is served to us at a temperature of 180°F.

Suppose the exponential function $T = 70 + 110e^{-0.2t}$ models the temperature T of the mocha after t minutes. If we are interested in determining how long it will take for the temperature of the coffee to reach 80°F, we would substitute 80 in for T and solve the resulting equation $80 = 70 + 110e^{-0.2t}$ for t. This equation is called an *exponential equation* because t occurs as an exponent.

In this section, we will learn to solve exponential equations. We also will learn to solve *logarithmic equations*.

An **exponential equation** is an equation with a variable in one of its exponents. Some examples of exponential equations are

$$3^x = 5 \qquad e^{2x} = 7 \qquad 6^{x-3} = 2^x \qquad 3^{2x+1} - 10(3^x) + 3 = 0$$

A **logarithmic equation** is an equation with logarithmic expressions that contain a variable. Some examples of logarithmic equations are

$$\log 2x = 25 \qquad \ln x - \ln(x - 12) = 24 \qquad \log x = \log \frac{1}{x} + 4$$

1. Use Like Bases to Solve Exponential Equations

Some exponential equations can be solved by using like bases.

Property of Exponents If $b^x = b^y$, then $x = y$.
That is, equal quantities with like bases have equal exponents.

EXAMPLE 1   Solve: $4^{x+3} = 8^{2x}$

Solution   We will use like bases to solve the exponential equation.

$$4^{x+3} = 8^{2x}$$
$$(2^2)^{x+3} = (2^3)^{2x} \qquad \text{Write 4 as } 2^2 \text{ and 8 as } 2^3.$$
$$2^{2(x+3)} = 2^{6x} \qquad \text{Multiply exponents.}$$
$$2(x + 3) = 6x \qquad \text{Equal quantities with like bases have equal exponents.}$$
$$2x + 6 = 6x \qquad \text{Use the distributive property.}$$
$$-4x = -6 \qquad \text{Subtract 6 and } 6x \text{ from both sides.}$$
$$x = \frac{3}{2} \qquad \text{Divide both sides by } -4 \text{ and simplify.}$$

Self Check 1   Solve: $3^{3x-5} = 81$.

EXAMPLE 2   Solve: $2^{x^2+2x} = \dfrac{1}{2}$

Solution   Since $\frac{1}{2} = 2^{-1}$, we can write the equation in the form $2^{x^2+2x} = 2^{-1}$ and use like bases to solve the equation.

Because equal quantities with like bases have equal exponents, we have

$$x^2 + 2x = -1$$
$$x^2 + 2x + 1 = 0 \qquad \text{Add 1 to both sides.}$$
$$(x + 1)(x + 1) = 0 \qquad \text{Factor the trinomial.}$$
$$x + 1 = 0 \quad \text{or} \quad x + 1 = 0 \qquad \text{Set each factor equal to 0.}$$
$$x = -1 \quad \Big| \quad x = -1$$

Verify that -1 satisfies the equation.

Self Check 2   Solve: $3^{x^2+2x} = 27$

EXAMPLE 3   Solve: $e^{6x^2} = e^{-x+1}$

Solution   We can use like bases to solve the exponential equation.

Because equal quantities with like bases have equal exponents, we have

$$6x^2 = -x + 1$$
$$6x^2 + x - 1 = 0 \qquad \text{Add } x - 1 \text{ to both sides.}$$
$$(3x - 1)(2x + 1) = 0 \qquad \text{Factor the trinomial.}$$
$$3x - 1 = 0 \quad \text{or} \quad 2x + 1 = 0 \qquad \text{Set each factor equal to 0.}$$
$$x = \frac{1}{3} \quad \Big| \quad x = -\frac{1}{2}$$

Verify that $\frac{1}{3}$ and $-\frac{1}{2}$ satisfy the equation.

Self Check 3   Solve: $e^{4x^2} = e^{24}$

2. Use Logarithms to Solve Exponential Equations

Some exponential equations can be solved using logarithms.

EXAMPLE 4   Solve the exponential equation: $3^x = 5$

Solution   Since logarithms of equal numbers are equal, we can take the common logarithm of each side of the equation. We then can use the power rule and move the variable x from its position as an exponent to a position as a coefficient and solve the equation.

$$3^x = 5$$
$$\log 3^x = \log 5 \qquad \text{Take the common logarithm of each side.}$$
$$x \log 3 = \log 5 \qquad \text{The log of a power is the power times the log.}$$
$$(1) \qquad x = \frac{\log 5}{\log 3} \qquad \text{Divide both sides by log 3.}$$
$$\approx 1.464973521 \qquad \text{Use a calculator.}$$

To four decimal places, $x = 1.4650$.

Self Check 4   Solve $5^x = 3$ to four decimal places.

Comment

A careless reading of Equation 1 leads to a common error. The right side of the equation calls for a division, not a subtraction.

$$\frac{\log 5}{\log 3} \qquad \text{means} \qquad (\log 5) \div (\log 3)$$

It is the expression $\log \frac{5}{3}$ that means $\log 5 - \log 3$.

EXAMPLE 5   Solve the exponential equation: $6^{x-3} = 2^x$

Solution   We will use logarithms to solve the exponential equation.

$$6^{x-3} = 2^x$$
$$\log 6^{x-3} = \log 2^x \qquad \text{Take the common logarithm of each side.}$$
$$(x - 3)\log 6 = x \log 2 \qquad \text{The log of a power is the power times the log.}$$
$$x \log 6 - 3 \log 6 = x \log 2 \qquad \text{Use the distributive property.}$$
$$x \log 6 - x \log 2 = 3 \log 6 \qquad \text{Add 3 log 6 and subtract } x \text{ log 2 from both sides.}$$
$$x(\log 6 - \log 2) = 3 \log 6 \qquad \text{Factor out } x \text{ on the left side.}$$
$$x = \frac{3 \log 6}{\log 6 - \log 2} \qquad \text{Divide both sides by log 6} - \text{log 2.}$$
$$x \approx 4.892789261 \qquad \text{Use a calculator.}$$

To four decimal places, $x = 4.8928$.

Comment

We can use common logarithms or natural logarithms to solve many exponential equations. In Examples 4 and 5, we took the common logarithm of both sides of an equation to solve the equation. We could just as well have taken the natural logarithm of both sides and obtained the same decimal result.

Self Check 5   Solve: $5^{x+3} = 3^x$

In the next example, we will take the natural logarithm of both sides of an exponential equation. However, before we do, it is a good idea to review the following properties of natural logarithms.

Natural Logarithm Properties	
$\ln 1 = 0$ $\ln e = 1$ $\ln e^x = x$	
Product Rule:	$\ln MN = \ln M + \ln N$
Quotient Rule:	$\ln \dfrac{M}{N} = \ln M - \ln N$
Power Rule:	$\ln M^p = p \ln M$

EXAMPLE 6   Use natural logarithms to solve: **a.** $e^x = 7$ **b.** $4^{x+3} = 8^{2x}$

Solution   For each problem, we will take the natural logarithm of both sides, simplify using the natural logarithms properties, and solve for x.

a. $e^x = 7$

$\ln e^x = \ln 7$ Take the natural logarithm of both sides.

$x = \ln 7$ Substitute x for $\ln e^x$: $\ln(e^x) = x$.

$x \approx 1.95$ Use a calculator and round to two decimal places.

Comment

As we see in Example 6a, if an exponential equation has a term with a base of e, then taking the natural logarithm of both sides is a good choice.

b. $4^{x+3} = 8^{2x}$

$\ln 4^{x+3} = \ln 8^{2x}$ Take the natural logarithm of both sides.

$(x + 3) \ln 4 = (2x) \ln 8$ The log of a power is the power times the log.

$x \ln 4 + 3 \ln 4 = 2x \ln 8$ Use the distributive property on the left side.

$x \ln 4 - 2x \ln 8 = -3 \ln 4$ Subtract $2x \ln 8$ and $3 \ln 4$ from both sides.

$x(\ln 4 - 2 \ln 8) = -3 \ln 4$ Factor out x on the left side.

$x = \dfrac{-3 \ln 4}{\ln 4 - 2 \ln 8}$ Divide both sides by $\ln 4 - 2 \ln 8$.

$x = 1.5$ Use a calculator.

Comment

Note that the exponential equations in Examples 1 and 6b are identical and the answers are the same. As we see, sometimes we can use like bases or logarithms to solve the same equation.

Self Check 6   Use natural logarithms to solve: **a.** $3e^x = 12$ **b.** $8^{x+1} = 4^{2x}$

3. Solve Logarithmic Equations

We can use the following property to change some logarithmic equations into algebraic equations that we can solve.

Property of Logarithms   If $\log_b x = \log_b y$, then $x = y$.
That is, logarithms of equal numbers are equal.

EXAMPLE 7   Solve: $\log_5(3x + 2) = \log_5(2x - 3)$

Solution   We will use the property of logarithms stated above to solve the equation. Then we will check our solutions.

$\log_5(3x + 2) = \log_5(2x - 3)$

$3x + 2 = 2x - 3$ If the logs of two numbers are equal, the numbers are equal.

$x = -5$ Subtract $2x$ and 2 from both sides.

Comment
Example 7 illustrates that we must check the solutions of a logarithmic equation.

Check: $\log_5(3x + 2) = \log_5(2x - 3)$

$$\log_5[3(-5) + 2] \overset{?}{=} \log_5[2(-5) - 3]$$

$$\log_5(-13) \overset{?}{=} \log_5(-13)$$

Since the logarithm of a negative number does not exist, -5 is extraneous and must be discarded. The equation has no roots.

Self Check 7   Solve: $\log_3(5x + 2) = \log_3(6x + 1)$.

EXAMPLE 8   Solve: $\log x + \log(x - 3) = 1$

Solution   We first will combine the two logarithms into a single logarithm. Next, we will use the definition of logarithm to write the equation in exponential form. Finally, we will solve the equation and check the solutions.

$$\log x + \log(x - 3) = 1$$

$$\log x(x - 3) = 1 \qquad \text{The sum of two logs is the log of a product.}$$

$$x(x - 3) = 10^1 \qquad \text{Use the definition of logarithms to change the equation to exponential form.}$$

$$x^2 - 3x - 10 = 0 \qquad \text{Remove parentheses and subtract 10 from both sides.}$$

$$(x + 2)(x - 5) = 0 \qquad \text{Factor the trinomial.}$$

$$x + 2 = 0 \quad \text{or} \quad x - 5 = 0$$

$$x = -2 \qquad \qquad x = 5$$

Check: The number -2 is not a solution, because it does not satisfy the equation (a negative number does not have a logarithm). We check the remaining number, 5.

$$\log x + \log(x - 3) = 1$$

$$\log 5 + \log(5 - 3) \overset{?}{=} 1 \qquad \text{Substitute 5 for } x.$$

$$\log 5 + \log 2 \overset{?}{=} 1$$

$$\log 10 \overset{?}{=} 1 \qquad \text{The sum of two logs is the log of a product.}$$

$$1 = 1 \qquad \log_b b = 1$$

Since 5 does check, it is a root.

Self Check 8   Solve: $\log x + \log(x - 15) = 2$

EXAMPLE 9   Solve: **a.** $\ln x = 5$ **b.** $\dfrac{\ln(5x - 6)}{\ln x} = 2$

Solution   For part **a,** we will write the natural logarithm as $\log_e x = 5$ and use the definition of logarithm to solve the equation. For part **b,** we will multiply both sides by $\ln x$ and then use properties of logarithms to write an algebraic equation. We then will solve the equation. We must verify the solutions to both parts.

a. $\ln x = 5$

$$\log_e x = 5 \qquad \text{Rewrite using the definition of natural logarithm.}$$

$$x = e^5 \qquad \text{Write in exponential form.}$$

Verify that e^5 satisfies the equation.

b. We multiply both sides of $\dfrac{\ln(5x - 6)}{\ln x} = 2$ by $\ln x$ to get

$$\ln(5x - 6) = 2 \ln x$$

and apply the power rule of logarithms to get

$$\ln(5x - 6) = \ln x^2$$

By the property of logarithms stated earlier, $5x - 6 = x^2$, because they have equal logarithms. So

$$x^2 = 5x - 6$$
$$x^2 - 5x + 6 = 0$$
$$(x - 3)(x - 2) = 0$$
$$x - 3 = 0 \quad \text{or} \quad x - 2 = 0$$
$$x = 3 \quad | \quad x = 2$$

Verify that both 2 and 3 satisfy the equation.

Self Check 9   Solve: **a.** $6 \ln x = 12$ **b.** $\dfrac{\ln(8x - 15)}{\ln x} = 2$.

Accent on Technology

We can use a graphing calculator to solve logarithmic equations. For example, to solve $\log x + \log(x - 3) = 1$, we subtract 1 from both sides to get

$$\log x + \log(x - 3) - 1 = 0$$

and then graph the corresponding function

$$y = \log x + \log(x - 3) - 1$$

with settings of $[0, 10]$ for x and $[-2, 2]$ for y, to obtain the graph in Figure 4-21.

Since the root of the equation is the x-intercept, we can find the root by tracing to find the value of the x-intercept. The root is $x = 5$. (See Figure 4-22.)

We also can find the x-intercept using the ZERO command in the CALC menu.

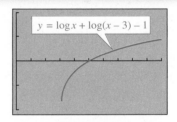

Figure 4-21

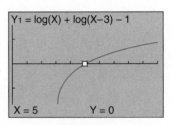

Figure 4-22

---

4.   Solve Carbon-14 Dating Problems

When a living organism dies, the oxygen/carbon dioxide cycle common to all living things ceases; then carbon-14, a radioactive isotope with a half-life of 5,700 years, is no longer absorbed. By measuring the amount of carbon-14 present in ancient objects, archaeologists can estimate the object's age.

The amount A of radioactive material present at time t is given by the model

$$A = A_0 \, 2^{-t/h}$$

where A_0 is the amount present initially and h is the half-life of the material.

EXAMPLE 10   An archeologist finds a wooden statue in the tomb of all ancient Egyptian ruler. If the statue contains two-thirds of its original carbon-14 content, how old is it?

Solution   To find the time t when $A = \frac{2}{3} A_0$, we substitute $\frac{2A_0}{3}$ for A and 5,700 for h in the radioactive decay formula and solve for t:

$$A = A_0\, 2^{-t/h}$$

$$\frac{2A_0}{3} = A_0\, 2^{-t/5,700}$$

$$1 = \frac{3}{2}(2^{-t/5,700})$$ Divide both sides by A_0 and multiply both sides by $\frac{3}{2}$

$$\log 1 = \log \frac{3}{2}(2^{-t/5,700})$$ Take the common logarithm of each side.

$$0 = \log \frac{3}{2} + \log 2^{-t/5,700}$$ The log of a product is the sum of the logs.

$$-\log \frac{3}{2} = -\frac{t}{5,700} \log 2$$ Subtract $\log \frac{3}{2}$ from both sides and use the power rule of logarithms.

$$5,700 \left(\frac{\log \frac{3}{2}}{\log 2} \right) = t$$ Multiply both sides by $-\dfrac{5,700}{\log 2}$.

$$t \approx 3,334.286254$$ Use a calculator.

The wooden statue is approximately 3,300 years old.

Self Check 10   How old is an artifact that has 60% of its original carbon-14 content?

Everyday Connections

Carbon-14 Dating
Turin Shroud Older Than Thought

January 25, 2005—The Shroud of Turin, the piece of linen long believed to have been wrapped around Jesus's body after the crucifixion, is much older than the date suggested by radiocarbon tests, according to new microchemical research.

Published in an issue of *Thermochimica Acta,* a chemistry peer-reviewed scientific journal, the study dismisses the results of the 1988 carbon-14 dating.

Victor Boswell/Getty Images

At that time, three reputable laboratories in Oxford, Zurich, and Tucson, Ariz., concluded that the cloth on which the smudged outline of the body of a man is indelibly impressed, was a medieval fake dating from 1260 to 1390, and not the burial cloth wrapped around the body of Christ.

Suppose a fossil sample contains 12% of the amount of carbon-14 that was present originally. Given that the half-life of carbon-14 is 5,700 years, answer the following questions.

1. Approximate the age of the fossil sample.

2. Approximate the amount of carbon-14 remaining when the fossil sample is 40,000 years old.

Source: http://dsc.discovery.com/news/briefs/20050124/shroud.html

5. Solve Population Growth Problems

When there is sufficient food and space, populations of living organisms tend to increase exponentially according to the Malthusian growth model

$$P = P_0 e^{kt}$$

where P_0 is the initial population at $t = 0$ and k depends on the rate of growth.

EXAMPLE 11   Streptococcus bacteria in a laboratory culture increased from an initial population of 500 to 1,500 in 3 hours. Find the time it will take for the population to reach 10,000.

Solution   We will apply the following two steps to solve the problem:

Step 1.   Substitute 1,500 for P, 500 for P_0, and 3 for t into the Malthusian growth model and find k.

Step 2.   Substitute 10,000 for P, 500 for P_0, and the value of k found in Step 1 into the model and use logarithms to solve for t.

1. Substitute 1,500 for P, 500 for P_0, and 3 for t into the Malthusian growth model and find k.

$$P = P_0 e^{kt}$$

$1,500 = 500(e^{k3})$ Substitute 1,500 for P, 500 for P_0, and 3 for t.

$3 = e^{3k}$ Divide both sides by 500.

$3k = \ln 3$ Change the equation from exponential to logarithmic form.

$k = \dfrac{\ln 3}{3}$ Divide both sides by 3.

2. To find out when the population will reach 10,000, we substitute 10,000 for P, 500 for P_0, and $\frac{\ln 3}{3}$ for k in the equation $P = P_0 e^{kt}$, and solve for t:

$$P = P_0 e^{kt}$$

$10,000 = 500 e^{\left(\frac{\ln 3}{3}\right)t}$

$20 = e^{\left(\frac{\ln 3}{3}\right)t}$ Divide both sides by 500.

$\ln(20) = \ln[e^{\left(\frac{\ln 3}{3}\right)t}]$ Take the natural log of both sides.

$\ln(20) = \dfrac{\ln 3}{3}t$ Simplify the right side using the natural log property $\ln(e^x) = x$.

$t = \dfrac{3 \ln 20}{\ln 3}$ Multiply both sides by $\dfrac{3}{\ln 3}$.

≈ 8.180499084 Use a calculator.

The culture will reach 10,000 bacteria in a little more than 8 hours.

Self Check 11   If the population increases from 1,000 to 3,000 in 3 hours, how long will it take to reach 20,000?

Self Check Answers   **1.** 3   **2.** 1, −3   **3.** $\pm\sqrt{6}$ **4.** $x = 0.6826$ **5.** −9.4520
6. a. $\ln 4 \approx 1.3863$ **b.** 3   **7.** 1   **8.** 20; −5 is extraneous   **9. a.** e^2
b. 3, 5   **10.** about 4,200 years   **11.** about 8 hr

4.6 Exercises

Vocabulary and Concepts *Fill in the blanks.*

1. An equation with a variable in its exponent is called a(n) _____ equation.
2. An equation with a logarithmic expression that contains a variable is a(n) _____ equation.
3. The formula for carbon dating is $A =$ _____.
4. The formula for population growth is $P =$ _____.

Practice *Solve each exponential equation using like bases.*

5. $4^{x+2} = 8^x$
6. $27^{x+1} = 3^{2x+1}$
7. $3^{x-1} = 9^{2x}$
8. $5^{2x+1} = 125^x$
9. $2^{x^2-2x} = 8$
10. $5^{x^2-3x} = 625$
11. $3^{x^2+4x} = \dfrac{1}{81}$
12. $7^{x^2+3x} = \dfrac{1}{49}$
13. $e^{-x+6} = e^x$
14. $e^{2x+1} = e^{3x-11}$
15. $e^{x^2-1} = e^{24}$
16. $e^{x^2+7x} = \dfrac{1}{e^{12}}$

Solve each equation using logarithms. If an answer is not exact, give the answer to four decimal places.

17. $4^x = 5$
18. $7^x = 12$
19. $13^{x-1} = 2$
20. $5^{x+1} = 3$
21. $2^{x+1} = 3^x$
22. $5^{x-3} = 3^{2x}$
23. $2^x = 3^x$
24. $3^{2x} = 4^x$
25. $7^{x^2} = 10$
26. $8^{x^2} = 11$
27. $8^{x^2} = 9^x$
28. $5^{x^2} = 2^{5x}$
29. $e^x = 10$
30. $8e^x = 16$
31. $4e^{2x} = 24$
32. $2e^{5x} = 18$

Solve each equation. If an answer is not exact, give the answer to four decimal places.

33. $4^{x+2} - 4^x = 15$ (*Hint:* $4^{x+2} = 4^x 4^2$.)
34. $3^{x+3} + 3^x = 84$ (*Hint:* $3^{x+3} = 3^x 3^3$.)
35. $2(3^x) = 6^{2x}$
36. $2(3^{x+1}) = 3(2^{x-1})$
37. $2^{2x} - 10(2^x) + 16 = 0$ (*Hint:* Let $y = 2^x$.)
38. $3^{2x} - 10(3^x) + 9 = 0$ (*Hint:* Let $y = 3^x$.)
39. $2^{2x+1} - 2^x = 1$ (*Hint:* $2^{a+b} = 2^a 2^b$.)
40. $3^{2x+1} - 10(3^x) + 3 = 0$ (*Hint:* $3^{a+b} = 3^a 3^b$.)

Solve each equation using the definition of logarithm or the definition of natural logarithm.

41. $\log x^2 = 2$
42. $\log x^3 = 3$
43. $\log \dfrac{4x + 1}{2x + 9} = 0$
44. $\log \dfrac{5x + 2}{2(x + 7)} = 0$
45. $\ln x = 6$
46. $\ln x = 3$
47. $\ln(2x - 7) = 4$
48. $\ln(3x - 5) = 7$

Solve each equation.

49. $\log(2x - 3) = \log(x + 4)$
50. $\log(3x + 5) - \log(2x + 6) = 0$
51. $\log x + \log(x - 48) = 2$
52. $\log x + \log(x + 9) = 1$
53. $\log x + \log(x - 15) = 2$
54. $\log x + \log(x + 21) = 2$
55. $\log(x + 90) = 3 - \log x$
56. $\log(x - 3) - \log 6 = 2$
57. $\log(5,000) - \log(x - 2) = 3$
58. $\log(2x - 3) - \log(x - 1) = 0$
59. $\log_7 x + \log_7(x - 5) = \log_7 6$
60. $\ln x + \ln(x - 2) = \ln 120$
61. $\ln 15 - \ln(x - 2) = \ln x$
62. $\ln 10 - \ln(x - 3) = \ln x$
63. $\log_6 8 - \log_6 x = \log_6(x - 2)$
64. $\log(x - 6) - \log(x - 2) = \log \dfrac{5}{x}$
65. $\log(x - 1) - \log 6 = \log(x - 2) - \log x$
66. $\log x^2 = (\log x)^2$
67. $\log(\log x) = 1$
68. $\log_3(\log_3 x) = 1$
69. $\dfrac{\log(3x - 4)}{\log x} = 2$
70. $\dfrac{\ln(8x - 7)}{\ln x} = 2$
71. $\dfrac{\ln(5x + 6)}{2} = \ln x$
72. $\dfrac{1}{2}\log(4x + 5) = \log x$
73. $\log_3 x = \log_3\left(\dfrac{1}{x}\right) + 4$
74. $\log_5(7 + x) + \log_5(8 - x) - \log_5 2 = 2$

75. $2 \log_2 x = 3 + \log_2(x - 2)$

76. $2 \log_3 x - \log_3(x - 4) = 2 + \log_3 2$

77. $\ln(7y + 1) = 2 \ln(y + 3) - \ln 2$

78. $2 \log(y + 2) = \log(y + 2) - \log 12$

Use a graphing calculator to solve each equation. If an answer is not exact, give the result to the nearest hundredth.

79. $\log x + \log(x - 15) = 2$

80. $\log x + \log(x + 3) = 1$

81. $2^{x+1} = 7$

82. $\ln(2x + 5) - \ln 3 = \ln(x - 1)$

Applications *Use a calculator to help solve each problem.*

83. **Tritium decay** The half-life of tritium is 12.4 years. How long will it take for 25% of a sample of tritium to decompose?

84. **Radioactive decay** In 2 years, 20% of a radioactive element decays. Find its half-life.

85. **Thorium decay** An isotope of thorium, ^{227}Th, has a half-life of 18.4 days. How long will it take 80% of the sample to decompose?

86. **Lead decay** An isotope of lead, ^{201}Pb, has a half-life of 8.4 hours. How many hours ago was there 30% more of the substance?

87. **Carbon-14 dating** A cloth fragment is found in an ancient tomb. It contains 70% of the carbon-14 that it is assumed to have had initially. How old is the cloth?

88. **Carbon-14 dating** Only 25% of the carbon-14 in a wooden bowl remains. How old is the bowl?

89. **Compound interest** If $500 is deposited in an account paying 8.5% annual interest, compounded semiannually, how long will it take for the account to increase to $800?

90. **Continuous compound interest** In Exercise 89, how long will it take if the interest is compounded continuously?

91. **Compound interest** If $1,300 is deposited in a savings account paying 9% interest, compounded quarterly, how long will it take the account to increase to $2,100?

92. **Compound interest** A sum of $5,000 deposited in an account grows to $7,000 in 5 years. Assuming annual compounding, what interest rate is being paid?

93. **Rule of seventy** A rule of thumb for finding how long it takes an investment to double is called the **rule of seventy.** To apply the rule, divide 70 by the interest rate (expressed as a percent). At 5%, it takes $\frac{70}{5} = 14$ years to double the investment. At 7%, it takes $\frac{70}{7} = 10$ years. Explain why this formula works.

94. **Bacterial growth** A staphylococcus bacterial culture grows according to the formula $P = P_0 a^t$. If it takes 5 days for the culture to triple in size, how long will it take to double in size?

95. **Oceanography** The intensity I of a light a distance x meters beneath the surface of a lake decreases exponentially. If the light intensity at 6 meters is 70% of the intensity at the surface, at what depth will the intensity be 20%?

96. **Rodent control** The rodent population in a city is currently estimated at 30,000. If it is expected to double every 5 years, when will the population reach 1 million?

97. **Temperature of coffee** Refer to the section opener and find the time it takes for the white chocolate mocha to reach a temperature of 80°F.

98. **Time of death** The exponential function $T(t) = 17e^{-0.0626t} + 20$ models the temperature T in °C of a person's body t hours after death. If a dead body is discovered at 8:30 AM and the body's temperature is 30°C, what was the person's approximate time of death?

99. **Newton's law of cooling** Water whose temperature is at 100°C is left to cool in a room where the temperature is 60°C. After 3 minutes, the water temperature is 90°. If the water temperature T is a function of time t given by $T = 60 + 40e^{kt}$, find k.

100. **Newton's law of cooling** Refer to Exercise 99 and find the time for the water temperature to reach 70°C.

101. **Newton's law of cooling** A block of steel, initially at 0°C, is placed in an oven heated to 300°C. After 5 minutes, the temperature of the steel is 100°C. If the steel temperature T is a function of time t given by $T = 300 - 300e^{kt}$, find the value of k.

102. **Newton's law of cooling** Refer to Exercise 101 and find the time for the steel temperature to reach 200°C.

Discovery and Writing

103. Explain why it is necessary to check the solutions of a logarithmic equation.

104. Use the population growth formula to show that the doubling time for population growth is given by

$$t = \frac{\ln 2}{r}$$

105. Use the population growth formula to show that the tripling time for population growth is given by

$$t = \frac{\ln 3}{r}$$

106. Can you solve $x = \log x$ algebraically? Can you find an approximate solution?

Find x.

107. $\log_2(\log_5(\log_7 x)) = 2$

108. $\log_8\left[16\sqrt[3]{4{,}096}\right]^{\frac{1}{6}} = x$

Review   *Find the inverse of the function defined by each equation.*

109. $y = 3x + 2$

110. $y = \dfrac{1}{x-3}$

Let $f(x) = 5x - 1$ and $g(x) = x^2$. Find each value.

111. $(f \circ g)(2)$

112. $(g \circ f)(2)$

113. $(f \circ g)(x)$

114. $(g \circ f)(x)$

CHAPTER REVIEW

4.1   Exponential Functions and Their Graphs

Definitions and Concepts	Examples
An **exponential function** with base b is defined by the equation $$y = f(x) = b^x \quad (b > 0, b \neq 1)$$	Graph: $f(x) = 6^x$ We will find several points (x, y) that satisfy the equation, plot the points, and join them with a smooth curve, as shown in the figure. $f(x) = 6^x$ $\begin{array}{c\|c\|c} x & f(x) & (x, f(x)) \\ \hline -1 & \frac{1}{6} & \left(-1, \frac{1}{6}\right) \\ 0 & 1 & (0, 1) \\ 1 & 6 & (1, 6) \end{array}$ 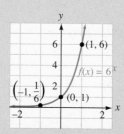
Compound interest formula: If P dollars are deposited in an account earning interest at an annual rate r, compounded k times each year, the amount A in the account after t years is given by $$A = P\left(1 + \frac{r}{k}\right)^{kt}$$	The grandparents of a newborn child invest \$10,000 in an educational savings plan that earns 8% interest, compounded quarterly. If the money is left untouched, how much will the child have in the account in 18 years?

We will substitute 10,000 for P, 0.08 for r, and 18 for t in the formula for compound interest. Because quarterly compounding means four times per year, we will substitute 4 for k.

$$A = P\left(1 + \frac{r}{k}\right)^{kt}$$

$$A = 10{,}000\left(1 + \frac{0.08}{4}\right)^{4 \cdot 18}$$

$$= 10{,}000(1.02)^{72}$$

$$\approx 41{,}611.40 \qquad \text{Use a calculator and round to two decimals.}$$

In 18 years, the account will be worth \$41,611.40.

The number $e \approx 2.718281828$.

The graph of $f(x) = e^x$ is:

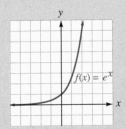

Continuous compound interest formula:
If P dollars are deposited in an account earning interest at an annual rate r, compounded continuously, the amount A after t years is given by the formula

$$A = Pe^{rt}$$

If the grandparents of the newborn child in the previous example had invested \$10,000 at an annual rate of 8%, compounded continuously, how much would the child have in the account in 18 years?

We will substitute \$10,000 for P, 0.08 for r, and 18 for t in the continuous compound interest formula $A = Pe^{rt}$.

$$A = Pe^{rt}$$

$$A = 10{,}000e^{(0.08)(18)}$$

$$= 10{,}000e^{1.44}$$

$$\approx 42{,}206.96 \qquad \text{Use a calculator and round to two decimal places.}$$

In 18 years, the balance will be \$42,206.96.

Exercises

Use properties of exponents to simplify.

1. $5^{\sqrt{2}} \cdot 5^{\sqrt{2}}$ **2.** $\left(2^{\sqrt{5}}\right)^{\sqrt{2}}$

Graph the function defined by each equation.

3. $f(x) = 3^x$ **4.** $f(x) = \left(\frac{1}{3}\right)^x$

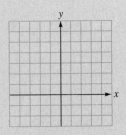 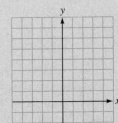

5. The graph of $f(x) = 7^x$ will pass through the points $(0, p)$ and $(1, q)$. Find p and q.

6. Give the domain and range of the function $f(x) = b^x$, with $b > 0$ and $b \neq 1$.

Use a translation to help graph each function.

7. $f(x) = \left(\frac{1}{2}\right)^x - 2$ **8.** $f(x) = \left(\frac{1}{2}\right)^{x+2}$

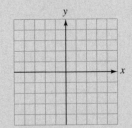 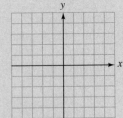

Graph each function.

9. $f(x) = -5^x$

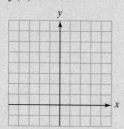

10. $f(x) = -5^x + 4$

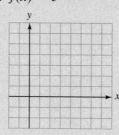

13. How much will \$10,500 become if it earns 9% per year for 60 years, compounded quarterly?

14. If \$10,500 accumulates interest at an annual rate of 9%, compounded continuously, how much will be in the account in 60 years?

11. $f(x) = e^x + 1$

12. $f(x) = e^{x-3}$

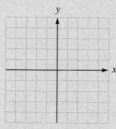

4.2   Applications of Exponential Functions

Definitions and Concepts	Examples
Radioactive decay formula: The amount A of radioactive material present at time t is given by $$A = A_0 2^{-t/h}$$ where A_0 is the amount that was present initially (at $t = 0$) and h is the material's half-life.	The half-life of radium is approximately 1,600 years. To find how much of a 1-gram sample will remain after 500 years, we will substitute $A_0 = 1$, $h = 1,600$, and $t = 500$ into the formula for radioactive decay and simplify. $A = A_0 2^{-t/h}$ $A = 1 \cdot 2^{-500/1,600}$ ≈ 0.81 **Use a calculator. Round to two decimal places.** After 500 years, approximately 0.81 gram of radium will remain.
Intensity of light formula: The intensity I of light (in lumens) at a distance x meters below the surface of a body of water decreases exponentially according to the formula $$I = I_0 k^x$$ where I_0 is the intensity of light above the water and k is a constant that depends on the clarity of the water.	At one location in the Atlantic Ocean, the intensity of light above water I_0 is 10 lumens and $k = 0.4$. To find the intensity of light at a depth of 3 meters, we will substitute 10 for I_0, 0.4 for k, and 3 for x into the formula for light intensity and simplify. $I = I_0 k^x$ $I = 10(0.4)^3$ $I = 0.64$ At a depth of 3 meters, the intensity of the light is 0.64 lumen.

Malthusian model of population growth:
If b is the annual birth rate, d is the annual death rate, t is the time (in years), P_0 is the initial population at $t = 0$, and P is the current population, then

$$P = P_0 e^{kt}$$

where $k = b - d$ is the **annual growth rate,** the difference between the annual birth rate and death rate.

The population of the United States is approximately 300 million people. Assuming that the annual birth rate is 19 per 1,000 and the annual death rate is 7 per 1,000, what does the Malthusian model predict the U.S. population will be in 30 years?

We can use the stated information to write the Malthusian model for U.S. population. We then will substitute into the model to predict the population in 50 years.

Since k is the difference between the birth and death rates, we have

$$k = b - d$$
$$k = \frac{19}{1,000} - \frac{7}{1,000} \quad \text{Substitute } \tfrac{19}{1,000} \text{ for } b \text{ and } \tfrac{7}{1,000} \text{ for } d.$$
$$k = 0.019 - 0.007$$
$$= 0.012$$

We can substitute $300,000,000$ for P_0, 30 for t, and 0.012 for k in the formula for the Malthusian model of population growth and simplify.

$$P = P_0 e^{kt}$$
$$P = (300,000,000)e^{(0.012)(30)}$$
$$= (300,000,000)e^{0.36}$$
$$\approx 429,998,824.4 \qquad \text{Use a calculator.}$$

After 30 years, the U.S. population will be approximately 430 million people.

Exercises

15. The half-life of a radioactive material is about 34.2 years. How much of the material is left after 20 years?

16. Find the intensity of light at a depth of 12 meters if $I_0 = 14$ and $k = 0.7$.

17. The population of the United States is approximately 300,000,000 people. Find the population in 50 years if $k = 0.015$.

18. In a city with a population of 450,000, there are currently 1,000 cases of hepatitis. If the spread of the disease is projected by the following logistic function, how many people will contract the hepatitis virus after 5 years?

$$P = \frac{450,000}{1 + (450 - 1)e^{-0.2t}}$$

4.3 Logarithmic Functions and Their Graphs

Definitions and Concepts	**Examples**
Logarithmic functions: If $b > 0$ and $b \neq 1$, the **logarithmic function with base** b is defined by $y = \log_b x$ if and only if $x = b^y$. The **domain of the logarithmic function** is the interval $(0, \infty)$. The **range** is the interval $(-\infty, \infty)$.	$\log_5 125 = 3 \qquad$ because $\qquad 125 = 5^3$ $\log_8 1 = 0 \qquad$ because $\qquad 1 = 8^0$ $\log_9 3 = \dfrac{1}{2} \qquad$ because $\qquad 3 = 9^{1/2} = \sqrt{9}$ $\log_2 \dfrac{1}{16} = -4 \qquad$ because $\qquad \dfrac{1}{16} = 2^{-4}$ $\log 10,000 = 4 \qquad$ because $\qquad 10,000 = 10^4$

Base-10 logarithms are called **common logarithms.** The notation log x represents $\log_{10} x$.

To find x in the equation $\log_8 \frac{1}{64} = x$, we note that $\log_8 \frac{1}{64} = x$ is equivalent to $\frac{1}{64} = 8^x$. Since $\frac{1}{64} = 8^{-2}$, we have $8^x = 8^{-2}$ and $x = -2$.

To graph $f(x) = \log_5 x$, we will find several points (x, y) that satisfy the equation, plot the points, and join them with a smooth curve, as shown in the figure.

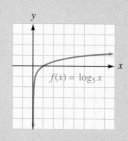

$f(x) = \log_5 x$		
x	$f(x)$	$(x, f(x))$
$\frac{1}{25}$	-2	$\left(\frac{1}{25}, -2\right)$
$\frac{1}{5}$	-1	$\left(\frac{1}{5}, -1\right)$
1	0	$(1, 0)$
5	1	$(5, 1)$
25	2	$(25, 2)$

Base-e logarithms are called **natural logarithms.** The notation ln x means $\log_e x$

The graph of $f(x) = \ln x$ is:

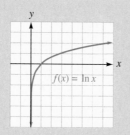

To graph $f(x) = \ln(x + 4)$, we will translate the graph of $f(x) = \ln x$ four units to the left as shown in the figure.

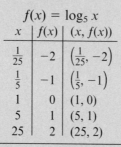

$\ln e^{-5x} = -5x$ $\qquad$ $e^{\ln(7x)} = 7x$

Since the functions $y = \ln x$ and $y = e^x$ are inverses

$\ln e^x = x$ $\qquad$ and $\qquad$ $e^{\ln x} = x$

Exercises

19. Give the domain and range of the logarithmic function $f(x) = \log_3 x$.

20. Give the domain and range of the natural logarithm function, $f(x) = \ln x$.

Find each value.

21. $\log_3 9$

22. $\log_9 \frac{1}{3}$

23. $\log_x 1$

24. $\log_5 0.04$

25. $\log_a \sqrt{a}$

26. $\log_a \sqrt[3]{a}$

Find x.

27. $\log_2 x = 5$

28. $\log_{\sqrt{3}} x = 4$

29. $\log_{\sqrt{2}} x = 6$

30. $\log_{0.1} 10 = x$

31. $\log_x 2 = -\frac{1}{3}$

32. $\log_x 32 = 5$

33. $\log_{0.25} x = -1$

34. $\log_{0.125} x = -\frac{1}{3}$

35. $\log_{\sqrt{2}} 32 = x$

36. $\log_{\sqrt{5}} x = -4$

37. $\log_{\sqrt{3}} 9\sqrt{3} = x$

38. $\log_{\sqrt{5}} 5\sqrt{5} = x$

Graph each function.

39. $f(x) = \log(x - 2)$

40. $f(x) = 3 + \log x$

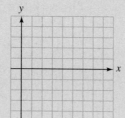

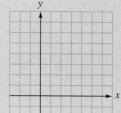

Graph each pair of equations on one set of coordinate axes.

41. $y = 4^x$ and $y = \log_4 x$

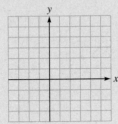

42. $y = \left(\dfrac{1}{3}\right)^x$ and $y = \log_{1/3} x$

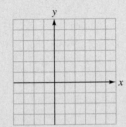

Use a calculator to find each value to four decimal places.

43. $\ln 452$

44. $\ln(\log 7.85)$

Use a calculator to solve each equation. Round each answer to four decimal places.

45. $\ln x = 2.336$

46. $\ln x = \log 8.8$

Graph each function.

47. $y = f(x) = 1 + \ln x$

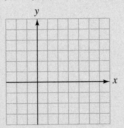

48. $y = f(x) = \ln(x + 1)$

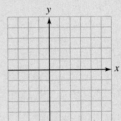

Simplify each expression.

49. $\ln(e^{12})$

50. $e^{\ln 14x}$

4.4 Applications of Logarithmic Functions

Definitions and Concepts	**Examples**
Decibel voltage gain: If E_O is the output voltage of a device and E_I is the input voltage, the **decibel voltage gain** is given by $$\text{db gain} = 20 \log \frac{E_O}{E_I}$$	To find the db gain of an amplifier with an input of 0.4 volt and an output of 50 volts, we can substitute the given values into the db gain formula. $$\text{db gain} = 20 \log \frac{E_O}{E_I}$$ $\text{db gain} = 20 \log \dfrac{50}{0.4}$ Substitute 50 for E_O and 0.4 for E_I. $= 20 \log 125$ ≈ 41.93820026 Use a calculator. To the nearest decibel, the db gain is 42 decibels.
Richter scale: If R is the intensity of an earthquake, A is the amplitude (measured in micrometers), and P is the period (the time of one oscillation of the Earth's surface, measured in seconds), then $$R = \log \frac{A}{P}$$	To find the intensity of an earthquake with amplitude of 5,000 micrometers $\left(\frac{1}{2}\text{ cm}\right)$ and a period of 0.08 second, we substitute 5,000 for A and 0.08 for P in the Richter scale formula and simplify. $R = \log \dfrac{A}{P}$ $R = \log \dfrac{5{,}000}{0.08}$ $= \log 62{,}500$ ≈ 4.795880017 Use a calculator. To the nearest tenth, the earthquake measures 4.8 on the Richter scale.

Charging batteries:
If M is the theoretical maximum charge that a battery can hold and k is a positive constant that depends on the battery and the charger, the length of time (in minutes) required to charge the battery to a given level C is given by

$$t = -\frac{1}{k} \ln\left(1 - \frac{C}{M}\right)$$

To find how long will it take to bring a fully discharged battery to 80% of full charge, we will assume that $k = 0.02$ and that time is measured in minutes.

Since 80% of full charge means 80% of M, we can substitute $0.80M$ for C and 0.02 for k in the formula for charging batteries and find t.

$$t = -\frac{1}{k} \ln\left(1 - \frac{C}{M}\right)$$

$$t = -\frac{1}{0.02} \ln\left(1 - \frac{0.80M}{M}\right)$$

$$= -50 \ln(1 - 0.8)$$

$$= -50 \ln(0.2)$$

$$\approx 80.47189562 \qquad \text{Use a calculator.}$$

The battery will reach 80% charge in about 80 minutes.

Population doubling time:
If r is the annual growth rate and t is the time (in years) required for a population to double, then

$$t = \frac{\ln 2}{r}$$

The population of the Earth is growing at the approximate rate of 2.5% per year. If this rate continues, how long will it take the population to double?

Because the population is growing at the rate of 2.5% per year, we can substitute 0.025 for r in the formula for doubling time and simplify.

$$t = \frac{\ln 2}{r}$$

$$t = \frac{\ln 2}{0.025}$$

$$\approx 27.72588722$$

It will take about 28 years for the Earth's population to double.

Isothermal expansion:
If the temperature T is constant, the energy E required to increase the volume of 1 mole of gas from an initial volume V_i to a final volume V_f is given by

$$E = RT \ln\left(\frac{V_f}{V_i}\right)$$

E is measured in joules and T in Kelvins. R is the universal gas constant, which is 8.314 joules/mole/K.

To find the amount of energy that must be supplied to double the volume of 1 mole of gas at a constant temperature of 300 K, we substitute 8.314 for R and 300 for T in the formula. Since the final volume is to be two times the initial volume, we also substitute $2V_i$ for V_f.

$$E = RT \ln\left(\frac{V_f}{V_i}\right)$$

$$E = (8.314)(300) \ln\left(\frac{2V_i}{V_i}\right)$$

$$= 2{,}494.2 \ln 2$$

$$\approx 1728.847698$$

Approximately 1,729 joules of energy must be added to double the volume.

Exercises

51. An amplifier has an output of 18 volts when the input is 0.04 volt. Find the db gain.

52. An earthquake had a period of 0.3 second and an amplitude of 7,500 micrometers. Find its measure on the Richter scale.

53. How long will it take a dead battery to reach an 80% charge? (Assume $k = 0.17$.)

54. How long will it take the population of the United States to double if the growth rate is 3% per year?

55. Find the amount of energy that must be supplied to double the volume of 1 mole of gas at a constant temperature of 350K. (*Hint:* $R = 8.314$.)

4.5 Properties of Logarithms

Definitions and Concepts	Examples
Properties of logarithms: If b is a positive number and $b \neq 1$,	
1. $\log_b 1 = 0$	By property 1, $\log_9 1 = \mathbf{0}$, because $9^0 = 1$.
2. $\log_b b = 1$	By property 2, $\log_{11} 11 = \mathbf{1}$, because $11^1 = 11$.
3. $\log_b b^x = x$	By property 3, $\log_4 4^3 = \mathbf{3}$, because $4^3 = 4^3$.
4. $b^{\log_b x} = x$	By property 4, $6^{\log_6 3} = 3$, because $\log_6 3$ is the power to which 6 is raised to get 3.
5. $\log_b MN = \log_b M + \log_b N$ (product rule)	By property 5, $\log(2 \cdot 3) = \log 2 + \log 3$.
6. $\log_b \dfrac{M}{N} = \log_b M - \log_b N$ (quotient rule)	By property 6, $\log \dfrac{15}{5} = \log 15 - \log 5$.
7. $\log_b M^p = p \log_b M$ (power rule)	By property 7, $\log 3^2 = 2 \log 3$.
8. If $\log_b x = \log_b y$, then $x = y$.	By property 8, if $\log 8 = \log y$, then $8 = y$.
The properties of logarithms also hold for natural logarithms.	
Properties of natural logarithms:	
1. $\ln 1 = 0$	By property 1, $\ln 1 = \mathbf{0}$, because $e^0 = 1$.
2. $\ln e = 1$	By property 2, $\ln e = \mathbf{1}$, because $e^1 = e$.
3. $\ln e^x = x$	By property 3, $\ln(e^{13}) = 13$.
4. $e^{\ln x} = x$	By property 4, $e^{\ln 23} = 23$.
5. $\ln MN = \ln M + \ln N$ (product rule)	By property 5, $\ln(2 \cdot 3) = \ln 2 + \ln 3$.
6. $\ln \dfrac{M}{N} = \ln M - \ln N$ (quotient rule)	By property 6, $\ln \dfrac{15}{5} = \ln 15 - \ln 5$.
7. $\ln M^p = p \ln M$ (power rule)	By property 7, $\ln 3^2 = 2 \ln 3$.
8. If $\ln x = \ln y$, then $x = y$.	By property 8, if $\ln 8 = \ln y$, then $8 = y$.
	Write the expression $\log_2 \dfrac{x^2}{y}$ in terms of logarithms of x and y. $$\log_2 \frac{x^2}{y} = \log_2 x^2 - \log_2 y \quad \text{Use the quotient rule.}$$ $$= 2 \log_2 x - \log_2 y \quad \text{Use the power rule.}$$

Write $5 \ln x + \frac{1}{2} \ln y$ as a single natural logarithm.

$$5 \ln x + \frac{1}{2} \ln y = \ln x^5 + \ln y^{1/2} \quad \text{Use the power rule.}$$

$$= \ln(x^5 y^{1/2}) \quad \text{Use the product rule.}$$

$$= \ln\left(x^5 \sqrt{y}\right)$$

Change-of-base formula: $$\log_b y = \frac{\log_a y}{\log_a b}$$	Use the change-of-base formula to find $\log_3 10$. We will substitute 3 for b and 10 for y in the change-of-base formula. $$\log_3 10 = \frac{\log_{10} 10}{\log_{10} 3} = \frac{1}{0.4771212547} = 2.095903274$$

pH of a solution:
If $[\text{H}^+]$ is the hydrogen ion concentration in gram-ions per liter, then

$$\text{pH} = -\log[\text{H}^+]$$

Lemon juice in a bottle has a pH of approximately 4. To find its hydrogen ion concentration, we can substitute 4 for pH in the pH formula and solve it for $[\text{H}^+]$.

$$4 = -\log[\text{H}^+]$$
$$-4 = \log[\text{H}^+]$$
$$[\text{H}^+] = 10^{-4} \quad \text{Change the equation from logarithmic form to exponential form.}$$

We then can use a calculator to find that $[\text{H}^+] = 1 \times 10^{-4}$ gram-ions per liter.

Weber–Fechner law:
If L is the apparent loudness of a sound and I is the intensity, then

$$L = k \ln I$$

To find what increase in the intensity of a sound is necessary to cause a quadrupling of the apparent loudness, we can use the formula $L = k \ln I$.

To quadruple the apparent loudness, we multiply both sides of the equation by 4 and use Property 7 of natural logarithms.

$$L = k \ln I$$
$$4L = 4k \ln I$$
$$= k \ln I^4$$

To quadruple the apparent loudness, we must raise the intensity to the fourth power.

Exercises

Simplify each expression.

56. $\log_7 1$ **57.** $\log_7 7$

58. $\log_7 7^3$ **59.** $7^{\log_7 4}$

60. $\ln e^4$ **61.** $\ln 1$

62. $10^{\log_{10} 7}$ **63.** $e^{\ln 3}$

64. $\log_b b^4$ **65.** $\ln e^9$

Write each expression in terms of the logarithms of x, y, and z.

66. $\log_b \dfrac{x^2 y^3}{z^4}$ **67.** $\log_8 \sqrt{\dfrac{x}{yz^2}}$

68. $\ln \dfrac{x^4}{y^5 z^6}$ **69.** $\ln \sqrt[3]{xyz}$

Write each expression as the logarithm of one quantity.

70. $3 \log_b x - 5 \log_b y + 7 \log_b z$

71. $\dfrac{1}{2}(\log_b x + 3 \log_b y) - 7 \log_b z$

72. $4 \ln x - 5 \ln y - 6 \ln z$

73. $\dfrac{1}{2} \ln x + 3 \ln y - \dfrac{1}{3} \ln z$

Assume that $\log a = 0.6$, $\log b = 0.36$, *and* $\log c = 2.4$.
Find the value of each expression.

74. $\log abc$

75. $\log a^2 b$

76. $\log \dfrac{ac}{b}$

77. $\log \dfrac{a^2}{c^3 b^2}$

78. To four decimal places, find $\log_5 17$.

79. pH of grapefruit The pH of grapefruit juice is about 3.1. Find its hydrogen ion concentration.

80. Find the decrease in loudness if the intensity is cut in half.

4.6 Exponential and Logarithmic Equations

Definitions and Concepts	Examples
Property of exponents: If $b^x = b^y$ then $x = y$. That is, equal quantities with like bases have equal exponents.	Solve: $5^{x^2 - 2x} = \dfrac{1}{5}$ Since $\dfrac{1}{5} = 5^{-1}$, we can write the equation in the form $5^{x^2-2x} = 5^{-1}$ and use like bases to solve the equation. Because equal quantities with like bases have equal exponents, we have $x^2 - 2x = -1$ $x^2 - 2x + 1 = 0$ Add 1 to both sides. $(x-1)(x-1) = 0$ Factor the trinomial. $x - 1 = 0$ or $x - 1 = 0$ Set each factor equal to 0. $x = 1$ $x = 1$
	Solve the exponential equation: $3^x = 15$ Since logarithms of equal numbers are equal, we can take the common logarithm of each side of the equation. We then can use the power rule and move the variable x from its position as an exponent to a position as a coefficient and solve the equation. $3^x = 5$ $\log 3^x = \log 15$ Take the common logarithm of each side. $x \log 3 = \log 15$ Use the power rule. $x = \dfrac{\log 15}{\log 3}$ Divide both sides by log 3. ≈ 2.464973521 Use a calculator. To four decimal places, $x = 2.4650$.
	Solve: $e^x = 19$ We will take the natural logarithm of both sides, simplify using the natural logarithms properties, and solve for x. $e^x = 19$ $\ln e^x = \ln 19$ Take the natural logarithm of both sides. $x = \ln 19$ Use the property: $\ln(e^x) = x$. $x \approx 2.94$ Use a calculator and round to two decimal places.

Solve: $\log_4(5x + 3) = \log_4(-5x + 23)$

We will use the property "if $\log_b x = \log_b y$, then $x = y$" to solve the equation.

$\log_4(5x + 3) = \log_4(-5x + 23)$

$5x + 3 = -5x + 23$	Use the log property stated above.
$10x = 20$	Add $5x$ and subtract 3 from both sides.
$x = 2$	Divide both sides by 10.

Check: $\log_4(5x + 3) = \log_4(-5x + 23)$

$$\log_4[5(2) + 3] \stackrel{?}{=} \log_4[-5(2) + 23]$$

$$\log_4(13) = \log_4(13)$$

Since $x = 2$ checks, it is the solution.

Solve: $\log x + \log(x - 9) = 1$

$\log x + \log(x - 9) = 1$

$\log x(x - 9) = 1$	Use the product rule.
$x(x - 9) = 10^1$	Use the definition of logarithm and write in exponential form.
$x^2 - 9x - 10 = 0$	Remove parentheses and subtract 10 from both sides.
$(x - 10)(x + 1) = 0$	Factor the trinomial.
$x - 10 = 0 \quad$ or $\quad x + 1 = 0$	
$x = 10 \quad\quad\quad x = -1$	

Check: The number -1 is not a solution, because it does not satisfy the equation. (A negative number does not have a logarithm.) We check the remaining number, 10.

$\log x + \log(x - 9) = 1$	
$\log 10 + \log(10 - 9) \stackrel{?}{=} 1$	Substitute 10 for x.
$\log 10 + \log 1 \stackrel{?}{=} 1$	
$1 + 0 \stackrel{?}{=} 1$	$\log_{10} 10 = 1, \log_{10} 1 = 0$
$1 = 1$	

Since 10 does check, it is a root.

Carbon dating:

The amount A of radioactive material present at time t is given by the model

$$A = A_0 \, 2^{-t/h}$$

where A_0 is the amount present initially and h is the half-life of the material.

To find the age of an artifact that contains three-fourths of its original carbon-14 content, we substitute $\frac{3}{4} A_0$ for A and 5,700 for h in the formula for carbon dating and solve for t.

$$A = A_0 \, 2^{-t/5,700}$$

$$\frac{3}{4} A_0 = A_0 \, 2^{-t/5,700}$$

$\dfrac{3}{4} = 2^{-t/5,700}$	Divide both sides by A_0.
$\log \dfrac{3}{4} = \log 2^{-t/5,00}$	Take the base-10 log of both sides.

$$-0.1249387366 = -\frac{t}{5{,}700}\log 2$$ Use property 7 of logarithms.

$$t \approx 2{,}365$$ Solve for t.

The artifact is about 2,300 years old.

Exercises

Solve each equation for x.

81. $5^{x+2} = 625$

82. $2^{x^2+4x} = \dfrac{1}{8}$

83. $e^x = e^{-6x+14}$

84. $e^{2x^2} = e^{18}$

85. $3^x = 7$

86. $2^x = 3^{x-1}$

87. $2e^x = 16$

88. $-5e^x = -35$

Solve each equation for x.

89. $\log_7(-7x + 2) = \log_7(3x + 32)$

90. $\ln(x + 3) = \ln(-5x + 51)$

91. $\log x + \log(29 - x) = 2$

92. $\log_2 x + \log_2(x - 2) = 3$

93. $\log_2(x + 2) + \log_2(x - 1) = 2$

94. $\dfrac{\log(7x - 12)}{\log x} = 2$

95. $\ln x + \ln(x - 5) = \ln 6$

96. $\log 3 - \log(x - 1) = -1$

97. $e^{x \ln 2} = 9$

98. $\ln x = \ln(x - 1)$

99. $\ln x = \ln(x - 1) + 1$

100. $\ln x = \log_{10} x$ (*Hint:* Use the change-of-base formula.)

101. Carbon-14 dating A wooden statue found in Egypt has a carbon-14 content that is two-thirds of that found in living wood. If the half-life of carbon-14 is 5,700 years, how old is the statue?

CHAPTER TEST

Graph each function.

1. $f(x) = 2^x + 1$

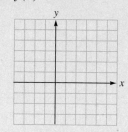

2. $f(x) = e^{x-2}$

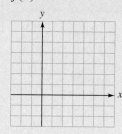

Solve each problem.

3. A radioactive material decays according to the formula $A = A_0(2)^{-t}$. How much of a 3-gram sample will be left in 6 years?

4. An initial deposit of \$1,000 earns 6% interest, compounded twice a year. How much will be in the account in one year?

5. An account contains \$2,000 and has been earning 8% interest, compounded continuously. How much will be in the account in 10 years?

Find each value.

6. $\log_7 343$

7. $\log_3 \dfrac{1}{27}$

8. $\log_{10} 10^{12} + 10^{\log_{10} 5}$

9. $\log_{3/2} \dfrac{9}{4}$

10. $\log_{2/3} \dfrac{27}{8}$

Graph each function.

11. $f(x) = \log(x - 1)$

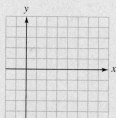

12. $f(x) = 2 + \ln x$

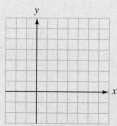

Write each expression in terms of the logarithms of a, b, and c.

13. $\log a^2 b c^3$

14. $\ln\sqrt{\dfrac{a}{b^2 c}}$

Write each expression as a logarithm of a single quantity.

15. $\dfrac{1}{2}\log(a + 2) + \log b - 2 \log c$

16. $\dfrac{1}{3}(\ln a - 2 \ln b) - \ln c$

Assume that $\log 2 = 0.3010$ *and* $\log 3 = 0.4771$. *Find each value. Do not use a calculator.*

17. $\log 24$

18. $\log \dfrac{8}{3}$

Use the change-of-base formula to find each logarithm. Do not attempt to simplify the answer.

19. $\log_7 3$

20. $\log_\pi e$

Determine whether each statement is true or false.

21. $\log_a ab = 1 + \log_a b$

22. $\dfrac{\log a}{\log b} = \log a - \log b$

23. Find the pH of a solution with a hydrogen ion concentration of 3.7×10^{-7}.
(*Hint:* pH $= \log[H^+]$.)

24. Find the db gain of an amplifier when $E_O = 60$ volts and $E_I = 0.3$ volt.
(*Hint:* db gain $= 20 \log(E_O/E_I)$.)

Solve each equation.

25. $3^{x^2 - 2x} = 27$

26. $3^{x-1} = 100^x$

27. $5e^x = 45$

28. $\ln(5x + 2) = \ln(2x + 5)$

29. $\log x + \log(x - 9) = 1$

30. $\log_6 18 - \log_6(x - 3) = \log_6 3$

5

Solving Polynomial Equations

A goal throughout all of algebra has been to solve equations. In this chapter, we will develop methods to solve polynomial equations of any degree.

Careers and Mathematics

Agricultural Engineer   Agricultural engineers apply knowledge of engineering technology and biological science to agriculture and the efficient use of biological resources. They are also referred to as biological engineers. They design agricultural machinery, equipment, sensors, processes, and structures, such as those used for crop storage. Some engineers specialize in areas such as power systems and machinery design; structures and environment engineering; and food and bioprocess engineering. They develop ways to conserve soil and water and to improve the processing of agricultural products.

Agricultural engineers held about 3,100 jobs in 2006.

Education   Agricultural engineers typically enter the occupation with a bachelor's degree in agricultural engineering, but some basic research positions may require a graduate degree. Engineers offering their services directly to the public must be licensed. Continuing education to keep current with rapidly changing technology is important for engineers.

Job Outlook   Employment of agricultural engineers is expected to grow by 9 percent through 2016. Employment growth should result from the need to increase crop yields to feed an expanding population and produce crops used as renewable energy sources. Median annual earnings of agricultural engineers were $66,030 in 2006.

For a sample application, see Example 3 in Section 5.3. For more information, see www.bls.gov/oco/ocos027.htm.

452

© Steve Bloom Images/Alamy

5.1 The Remainder and Factor Theorems; Synthetic Division

Objectives

1. Understand the Definition of a Zero of a Polynomial
2. Use the Remainder Theorem
3. Use the Factor Theorem
4. Use Synthetic Division to Divide Polynomials
5. Use Synthetic Division to Evaluate Polynomials
6. Use Synthetic Division to Solve Polynomial Equations

The cheetah is the fastest land animal and is known for its speed and stealth. Cheetahs can run at speeds between 70 mph and 75 mph. They have the ability to accelerate from 0 mph to 68 mph in three seconds. This acceleration is greater than most sports cars.

Although we will never run as fast as a cheetah, we do live in a society that wants convenience and speed. We demand fast service, fast food, fast communication, fast technology, and fast vehicles.

In this section, we will learn about synthetic division, a fast way to divide certain polynomials. Synthetic division can help us find the roots of many polynomial equations.

1. Understand the Definition of a Zero of a Polynomial

We have seen that many polynomial equations can be solved by factoring. For example, to solve the third-degree polynomial equation $x^3 - 3x^2 + 2x = 0$, we proceed as follows:

$$x^3 - 3x^2 + 2x = 0$$
$$x(x^2 - 3x + 2) = 0 \qquad \text{Factor out } x.$$
$$x(x - 1)(x - 2) = 0 \qquad \text{Factor } x^2 - 3x + 2.$$
$$x = 0 \quad \text{or} \quad x - 1 = 0 \quad \text{or} \quad x - 2 = 0 \qquad \text{Set each factor equal to 0.}$$
$$x = 1 \qquad\qquad x = 2$$

The solution set of this equation is $\{0, 1, 2\}$.

In this chapter, we will solve higher-degree polynomial equations. In general, an nth-degree polynomial equation is defined as follows:

Polynomial Equation   A **polynomial equation** is an equation that can be written in the form $P(x) = 0$, where

$$P(x) = a_n x^n + a_{n-1}x^{n-1} + a_{n-2}x^{n-2} + \cdots + a_1 x + a_0$$

where n is a natural number and the polynomial is of degree n.

Any number that makes $P(x) = 0$ when substituted for x is called a *zero of the polynomial*.

Zero of a Polynomial Function   A **zero of the polynomial** $P(x)$ is any number r for which $P(r) = 0$.

The real numbers $0, 1,$ and 2 are zeros of the polynomial function $P(x) = x^3 - 3x^2 + 2x$ because

$$P(0) = 0^3 - 3(0)^2 + 2(0) \qquad P(1) = 1^3 - 3(1)^2 + 2(1) \qquad P(2) = 2^3 - 3(2)^2 + 2(2)$$
$$= 0 - 0 + 0 \qquad\qquad = 1 - 3 + 2 \qquad\qquad = 8 - 12 + 4$$
$$= 0 \qquad\qquad\qquad = 0 \qquad\qquad\qquad = 0$$

As we have shown, the roots $0, 1,$ and 2 of the polynomial equation $x^3 - 3x^2 + 2x = 0$ are also zeros of the polynomial function $P(x) = x^3 - 3x^2 + 2x$. In general, the roots of the polynomial equation $P(x) = 0$ are the zeros of the polynomial $P(x)$.

2. Use the Remainder Theorem

There is a relationship between a zero r of a polynomial $P(x)$ and the results of a long division of $P(x)$ by $x - r$. We will examine this relationship in this section.

EXAMPLE 1 Let $P(x) = 3x^3 - 5x^2 + 3x - 10$.
a. Find $P(1)$. **b.** Divide $P(x)$ by $x - 1$.

Solution **a.** To find $P(1)$, we will substitute 1 for x in the polynomial.

$$P(1) = 3(1)^3 - 5(1)^2 + 3(1) - 10$$
$$= 3 - 5 + 3 - 10$$
$$= -9$$

b. To divide $P(x)$ by $x - 1$, we proceed as follows:

$$
\require{enclose}
\begin{array}{r}
3x^2 - 2x + 1 \\
x - 1 \enclose{longdiv}{3x^3 - 5x^2 + 3x - 10} \\
\underline{3x^3 - 3x^2} \\
-2x^2 + 3x \\
\underline{-2x^2 + 2x} \\
+\; x - 10 \\
\underline{x - 1} \\
-\; 9
\end{array}
$$

Note that the remainder is equal to $P(1)$.

Self Check 1 Let $P(x) = 2x^2 - 3x + 5$. Find $P(2)$ and divide $P(x)$ by $x - 2$. What do you notice about the results?

EXAMPLE 2 Let $P(x) = 3x^3 - 5x^2 + 3x - 10$.
a. Find $P(-2)$. **b.** Divide $P(x)$ by $x + 2$.

Solution **a.** To find $P(-2)$, we will substitute -2 for x in the polynomial.

$$P(-2) = 3(-2)^3 - 5(-2)^2 + 3(-2) - 10$$
$$= 3(-8) - 5(4) + 3(-2) - 10$$
$$= -24 - 20 - 6 - 10$$
$$= -60$$

b. To divide $P(x)$ by $x + 2$, we proceed as follows:

$$
\begin{array}{r}
3x^2 - 11x + 25 \\
x + 2\overline{)\,3x^3 - 5x^2 + 3x - 10} \\
\underline{3x^3 + 6x^2} \\
-11x^2 + 3x \\
\underline{-11x^2 - 22x} \\
25x - 10 \\
\underline{25x + 50} \\
\mathbf{-60}
\end{array}
$$

Note that the remainder is equal to $P(-2)$.

Self Check 2   Let $P(x) = 2x^2 - 3x + 5$. Find $P(-3)$ and divide $P(x)$ by $x + 3$. What do you notice about the results?

When $P(x)$ was divided by $x - 1$ in Example 1, the remainder was $P(1) = -9$. When $P(x)$ was divided by $x + 2$, or $x - (-2)$, in Example 2, the remainder was $P(-2) = -60$. These results are not coincidental. The **remainder theorem** states that a division of any polynomial $P(x)$ by $x - r$ gives $P(r)$ as the remainder.

The Remainder Theorem   If $P(x)$ is a polynomial, r is any number, and $P(x)$ is divided by $x - r$, the remainder is $P(r)$.

Proof   To divide $P(x)$ by $x - r$, we must find a quotient $Q(x)$ and a remainder $R(x)$ such that

Dividend $=$ divisor $\cdot$ quotient $+$ remainder
$$P(x) = (x - r) \cdot Q(x) + R(x)$$

Since the degree of the remainder $R(x)$ must be less than the degree of the divisor $x - r$, and the degree of $x - r$ is 1, $R(x)$ must be a constant R.

In the equation

$$P(x) = (x - r)Q(x) + R$$

the polynomial on the left side is the same as the polynomial on the right side, and the values that they assume for any number x are equal. If we replace x with r, we have

$$
\begin{aligned}
P(r) &= (r - r)Q(r) + R \\
&= (0)Q(r) + R \\
&= R
\end{aligned}
$$

Thus, $P(r) = R$.

EXAMPLE 3   Use the remainder theorem to find the remainder that will occur when $P(x) = 2x^4 - 10x^3 + 17x^2 - 14x - 3$ is divided by $x - 3$.

Solution   By the remainder theorem, the remainder will be $P(3)$.

$$
\begin{aligned}
P(x) &= 2x^4 - 10x^3 + 17x^2 - 14x - 3 \\
P(3) &= 2(3)^4 - 10(3)^3 + 17(3)^2 - 14(3) - 3 \qquad \text{Substitute 3 for } x. \\
&= 0
\end{aligned}
$$

The remainder will be 0. Although this calculation is tedious, it is easy to do with a calculator.

Self Check 3   Find the remainder when $P(x)$ is divided by $x - 2$.

3.   Use the Factor Theorem

If $R = P(r) = 0$ in the equation $P(x) = (x - r)Q(x) + R$, then $P(x)$ factors as $(x - r)Q(x)$. This fact can help us factor polynomials.

The Factor Theorem   If $P(x)$ is a polynomial and r is any number, then

If $P(r) = 0$, then $x - r$ is a factor of $P(x)$.

If $x - r$ is a factor of $P(x)$, then $P(r) = 0$.

Proof   **Part 1:**   First, we assume that $P(r) = 0$ and prove that $x - r$ is a factor of $P(x)$. If $P(r) = 0$, then $R = 0$, and the equation $P(x) = (x - r)Q(x) + R$ becomes

$$P(x) = (x - r)Q(x) + 0$$
$$P(x) = (x - r)Q(x)$$

Therefore, $x - r$ divides $P(x)$ exactly, and $x - r$ is a factor of $P(x)$.

Part 2:   Conversely, we assume that $x - r$ is a factor of $P(x)$ and prove that $P(r) = 0$. Because, by assumption, $x - r$ is a factor of $P(x)$, $x - r$ divides $P(x)$ exactly, and the division has a remainder of 0. By the remainder theorem, this remainder is $P(r)$. Hence, $P(r) = 0$.

We can state the factor theorem in a slightly different way.

Alternate Form of the Factor Theorem   If r is a zero of the polynomial $P(x)$, then $x - r$ is a factor of $P(x)$.

If $x - r$ is a factor of $P(x)$, then r is a zero of the polynomial.

It is important to note that for each factor of a polynomial, there corresponds a zero. Several possible factors of a polynomial are shown with their corresponding zeros.

Factor	Zero
$x - 5$	5
$x + 2$	-2
$x - \frac{1}{2}$	$\frac{1}{2}$

EXAMPLE 4   Determine whether $x + 2$ is a factor of $P(x) = x^4 - 7x^2 - 6x$.

Solution   By the factor theorem, $x + 2$, or $x - (-2)$, is a factor of $P(x)$ if $P(-2) = 0$. So we find the value of $P(-2)$.

$$P(x) = x^4 - 7x^2 - 6x$$
$$P(-2) = (-2)^4 - 7(-2)^2 - 6(-2) \quad \text{Substitute } -2 \text{ for } x.$$
$$= 16 - 28 + 12$$
$$= 0$$

Since $P(-2) = 0$, we know that $x - (-2)$, or $x + 2$, is a factor of $P(x)$.

Comment

In Example 4, $x + 2$ is a factor of the polynomial $P(x) = x^4 - 7x^2 - 6x$. By the alternate form of the factor theorem, we know that -2 is a zero of the polynomial function.

Self Check 4   Determine whether $x + 3$ is a factor of $P(x)$.

4. Use Synthetic Division to Divide Polynomials

Synthetic division is an easy way to divide higher-degree polynomials by binomials of the form $x - r$, and it is much faster than long division. To see how it works, we consider the following long division. On the left is a complete division. On the right is a modified version in which the variables have been removed.

$$
\begin{array}{r}
2x^2 + 10x + 27 \\
x - 3\overline{)2x^3 + 4x^2 - 3x + 10} \\
\underline{2x^3 - 6x^2} \\
10x^2 - 3x \\
\underline{10x^2 - 30x} \\
27x + 10 \\
\underline{27x - 81} \\
\text{(remainder) } 91
\end{array}
\qquad
\begin{array}{r}
2 + 10 + 27 \\
1 - 3\overline{)2 + 4 - 3 + 10} \\
\underline{2 - 6} \\
10 - 3 \\
\underline{10 - 30} \\
27 + 10 \\
\underline{27 - 81} \\
\text{(remainder) } 91
\end{array}
$$

We can shorten the work even more by omitting the numbers printed in color.

$$
\begin{array}{r}
2 + 10 + 27 \\
-3\overline{)2 + 4 - 3 + 10} \\
\underline{- 6} \\
10 - \\
\underline{- 30} \\
27 \\
\underline{- 81} \\
\text{(remainder) } 91
\end{array}
$$

We then can compress the work vertically to get

$$
\begin{array}{r}
2 + 10 + 27 \\
-3\overline{)2 + 4 - 3 + 10} \\
\underline{- 6 - 30 - 81} \\
10 \quad 27 \quad 91
\end{array}
$$

If we write the 2 in the quotient on the bottom line, the bottom line gives both the coefficients of the quotient and the remainder. The top line now can be eliminated, and the division appears as

$$
\begin{array}{r}
\underline{-3|}\ \ 2 + 4 - 3 + 10 \\
\underline{- 6 - 30 - 81} \\
2 \quad 10 \quad 27 \quad 91
\end{array}
$$

The bottom line was obtained by subtracting the middle line from the top line. If we replace the -3 in the divisor with $+3$, the signs of each number in the middle line will be reversed in the division process. Then the bottom line can be obtained by addition, and we have the final form of the synthetic division.

$$
\begin{array}{r}
\underline{+3|}\ \ 2 + 4 - 3 + 10 \qquad \text{\footnotesize The coefficients of the dividend.}\\
\underline{+ 6 + 30 + 81} \\
2 \quad 10 \quad 27 \quad 91 \qquad \text{\footnotesize The coefficients of the quotient and the remainder.}
\end{array}
$$

Thus,

$$
\frac{2x^3 + 4x^2 - 3x + 10}{x - 3} = 2x^2 + 10x + 27 + \frac{91}{x - 3}
$$

EXAMPLE 5   Use synthetic division to divide $10x + 3x^4 - 8x^3 + 3$ by $x - 2$.

Solution   We first write the terms of $P(x)$ in descending powers of x:

$$3x^4 - 8x^3 + 10x + 3$$

We then write the coefficients of the dividend, with its terms in descending powers of x, and the 2 from the divisor in the following form:

$$\underline{2|}\quad 3\quad -8\quad \mathbf{0}\quad 10\quad 3 \qquad \text{Write 0 for the coefficient of the missing } x^2 \text{ term.}$$

Then we follow these steps:

$$\underline{2|}\quad 3\quad -8\quad 0\quad 10\quad 3 \qquad \text{Bring down the 3.}$$
$$\quad \downarrow$$
$$\quad 3$$

$$\underline{2|}\quad 3\quad -8\quad 0\quad 10\quad 3 \qquad \text{Multiply 2 and 3 together to get 6, and add 6 and } -8$$
$$\qquad 6 \qquad\qquad\qquad \text{to get } -2.$$
$$\quad 3\quad -2$$

$$\underline{2|}\quad 3\quad -8\quad 0\quad 10\quad 3 \qquad \text{Multiply 2 and } -2 \text{ together to get } -4, \text{ and add } -4$$
$$\qquad 6\quad -4 \qquad\qquad \text{and 0 to get } -4.$$
$$\quad 3\quad -2\quad -4$$

$$\underline{2|}\quad 3\quad -8\quad 0\quad 10\quad 3 \qquad \text{Multiply 2 and } -4 \text{ together to get } -8, \text{ and add } -8$$
$$\qquad 6\quad -4\quad \mathbf{-8} \qquad\quad \text{and 10 to get 2.}$$
$$\quad 3\quad -2\quad -4\quad 2$$

$$\underline{2|}\quad 3\quad -8\quad 0\quad 10\quad 3 \qquad \text{Multiply 2 and 2 together to get 4, and add 4 and 3}$$
$$\qquad 6\quad -4\quad -8\quad \mathbf{4} \qquad \text{to get 7.}$$
$$\quad 3\quad -2\quad -4\quad 2\quad 7 \qquad \text{The coefficients of the quotient and the remainder.}$$

Thus,

$$\frac{10x + 3x^4 - 8x^3 + 3}{x - 2} = 3x^3 - 2x^2 - 4x + 2 + \frac{7}{x - 2}$$

Self Check 5   Divide $2x^3 - 4x^2 + 5x - 7$ by $x - 3$.

5.   Use Synthetic Division to Evaluate Polynomials

EXAMPLE 6   Use synthetic division to find $P(-2)$ when $P(x) = 5x^3 + 3x^2 - 21x - 1$.

Solution   Because of the remainder theorem, $P(-2)$ is the remainder when $P(x)$ is divided by $x - (-2)$. Earlier in the section, we would have found the remainder by using long division. We now can simplify the work and find the remainder by using synthetic division.

$$\underline{-2|}\quad 5\quad 3\quad -21\quad -1$$
$$\qquad -10 \qquad\qquad\qquad -2(5) = -10;\; 3 + (-10) = -7$$
$$\quad 5\quad -7$$

$$\begin{array}{r|rrrr} -2 & 5 & 3 & -21 & -1 \\ & & -10 & 14 & \\ \hline & 5 & -7 & -7 & \end{array} \qquad -2(-7) = 14; \; -21 + 14 = -7$$

$$\begin{array}{r|rrrr} -2 & 5 & 3 & -21 & -1 \\ & & -10 & 14 & 14 \\ \hline & 5 & -7 & -7 & 13 \end{array} \qquad -2(-7) = 14; \; -1 + 14 = 13$$

Because the remainder is 13, $P(-2) = 13$.

Self Check 6   Find $P(3)$.

EXAMPLE 7   If $P(x) = x^3 - x^2 + x - 1$, find $P(i)$, where $i = \sqrt{-1}$.

Solution   We will use synthetic division and find the remainder.

$$\begin{array}{r|rrrr} i & 1 & -1 & +1 & -1 \\ & & i & -1-i & +1 \\ \hline & 1 & i-1 & -i & 0 \end{array}$$

Since the remainder is 0, $P(i) = 0$ and i is a zero of $P(x)$.

Comment
Example 7 illustrates that synthetic division can be used with complex numbers.

Self Check 7   Find $P(-i)$.

6. Use Synthetic Division to Solve Polynomial Equations

EXAMPLE 8   Let $P(x) = 3x^3 - 5x^2 + 3x - 10$. Completely solve the polynomial equation $P(x) = 0$ given that 2 is one solution.

Solution   Since 2 is a solution of the equation $P(x) = 0$, we know that 2 is a zero of $P(x)$. We will use synthetic division and divide $P(x)$ by $x - 2$, obtaining a remainder of 0. Then we will use the result of the synthetic division to help factor the polynomial and solve the equation.

1. We use synthetic division to divide $P(x)$ by $x - 2$.

$$\begin{array}{r|rrrr} 2 & 3 & -5 & 3 & -10 \\ & & 6 & 2 & 10 \\ \hline & 3 & 1 & 5 & 0 \end{array}$$

2. We then write the quotient and factor it.

$$3x^3 - 5x^2 + 3x - 10 = (x - 2)(3x^2 + x + 5)$$

3. Finally, we solve the polynomial equation $P(x) = 0$.

$$3x^3 - 5x^2 + 3x - 10 = 0$$
$$(x - 2)(3x^2 + x + 5) = 0$$

To solve for x, we set each factor equal to 0 and apply the quadratic formula to the equation $3x^2 + x + 5 = 0$.

$$x - 2 = 0 \qquad \text{or} \qquad 3x^2 + x + 5 = 0$$
$$x = 2 \qquad\qquad\qquad x = \frac{-1 \pm \sqrt{1^2 - 4(3)(5)}}{2(3)}$$
$$x = \frac{-1 \pm i\sqrt{59}}{6}$$

The solution set is $\left\{ 2, -\dfrac{1}{6} + \dfrac{\sqrt{59}}{6}i, -\dfrac{1}{6} - \dfrac{\sqrt{59}}{6}i \right\}$.

Self Check 8   Solve: $x^3 - 1 = 0$

EXAMPLE 9   Find a third-degree polynomial $P(x)$ with three zeros of 3, 3, and -5.

Solution   We will use the factor theorem and write the three factors that correspond to the three zeros of 3, 3, and -5. We will then multiply the resulting binomials.

If 3, 3, and -5 are the three zeros of $P(x)$, then $x - 3$, $x - 3$, and $x - (-5)$ are the three factors of $P(x)$.

$$P(x) = (x - 3)(x - 3)(x + 5)$$
$$P(x) = (x^2 - 6x + 9)(x + 5) \qquad \text{Multiply } x - 3 \text{ and } x - 3.$$
$$P(x) = x^3 - x^2 - 21x + 45 \qquad \text{Multiply using the distributive property.}$$

The polynomial $P(x) = x^3 - x^2 - 21x + 45$ has zeros of 3, 3, and -5.
Because 3 occurs twice as a zero, we say that 3 is a **zero of multiplicity 2.**

Self Check 9   Find a polynomial $P(x)$ with zeros of -2, 2, and 3.

Self Check Answers   **1.** $P(2)$ is the remainder.   **2.** $P(-3)$ is the remainder.   **3.** -11 **4.** no
5. $2x^2 + 2x + 11 + \dfrac{26}{x - 3}$ **6.** 98   **7.** 0   **8.** $1, \dfrac{-1 \pm i\sqrt{3}}{2}$
9. $P(x) = x^3 - 3x^2 - 4x + 12$

5.1   Exercises

Vocabulary and Concepts   *Fill in the blanks.*

1. The variables in a polynomial have _____-number exponents.
2. A zero of $P(x)$ is any number r for which _____.
3. The remainder theorem holds when r is ___ number.
4. If $P(x)$ is a polynomial and $P(x)$ is divided by _____, the remainder will be $P(r)$.
5. If $P(x)$ is a polynomial, then $P(r) = 0$ if and only if $x - r$ is a _____ of $P(x)$.
6. A shortcut method for dividing a polynomial by a binomial of the form $x - r$ is called _____ division.

Practice   *Use long division to perform each division.*

7. $\dfrac{4x^3 - 2x^2 - x + 1}{x - 1}$

8. $\dfrac{2x^3 + 3x^2 - 5x + 1}{x + 3}$

9. $\dfrac{2x^4 + x^3 + 2x^2 + 15x - 5}{x + 2}$

10. $\dfrac{x^4 + 6x^3 - 2x^2 + x - 1}{x - 1}$

Find each value by substituting the given value of x into the polynomial and simplifying. Then find the value by doing a long division and finding the remainder.

11. $P(x) = 3x^3 - 2x^2 - 5x - 7$; $P(2)$
12. $P(x) = 5x^3 + 4x^2 + x - 1$; $P(-2)$
13. $P(x) = 7x^4 + 2x^3 + 5x^2 - 1$; $P(-1)$
14. $P(x) = 2x^4 - 2x^3 + 5x^2 - 1$; $P(2)$
15. $P(x) = 2x^5 + x^4 - x^3 - 2x + 3$; $P(1)$
16. $P(x) = 3x^5 + x^4 - 3x^2 + 5x + 7$; $P(-2)$

Use the remainder theorem to find the remainder when $P(x) = 3x^4 + 5x^3 - 4x^2 - 2x + 1$ is divided by each binomial. Use a calculator on Exercises 21–24.

17. $x + 2$ **18.** $x - 1$

19. $x - 2$ **20.** $x + 1$

21. $x - 12$ **22.** $x + 15$

23. $x + 3.25$ **24.** $x - 7.12$

Use the factor theorem to determine whether each statement is true. If the statement is not true, so indicate.

25. $x - 1$ is a factor of $P(x) = x^7 - 1$.

26. $x - 2$ is a factor of
$P(x) = x^3 - x^2 + 2x - 8$.

27. $x - 1$ is a factor of $P(x) = 3x^5 + 4x^2 - 7$.

28. $x + 1$ is a factor of $P(x) = 3x^5 + 4x^2 - 7$.

29. $x + 3$ is a factor of $P(x) = 2x^3 - 2x^2 + 1$.

30. $x - 3$ is a factor of
$P(x) = 3x^5 - 3x^4 + 5x^2 - 13x - 6$.

31. $x - 1$ is a factor of
$P(x) = x^{1,984} - x^{1,776} + x^{1,492} - x^{1,066}$.

32. $x + 1$ is a factor of
$P(x) = x^{1,984} + x^{1,776} - x^{1,492} - x^{1,066}$.

Use synthetic division to express the polynomial $P(x) = 3x^3 - 2x^2 - 6x - 4$ in the form (divisor)(quotient) + remainder for each divisor.

33. $x - 1$

34. $x - 2$

35. $x - 3$

36. $x - 4$

37. $x + 1$

38. $x + 2$

39. $x + 3$

40. $x + 4$

Use synthetic division to perform each division.

41. $\dfrac{x^3 + x^2 + x - 3}{x - 1}$

42. $\dfrac{x^3 - x^2 - 5x + 6}{x - 2}$

43. $\dfrac{7x^3 - 3x^2 - 5x + 1}{x + 1}$

44. $\dfrac{2x^3 + 4x^2 - 3x + 8}{x - 3}$

45. $\dfrac{4x^4 - 3x^3 - x + 5}{x - 3}$

46. $\dfrac{x^4 + 5x^3 - 2x^2 + x - 1}{x + 1}$

47. $\dfrac{3x^5 - 768x}{x - 4}$

48. $\dfrac{x^5 - 4x^2 + 4x + 4}{x + 3}$

Let $P(x) = 5x^3 + 2x^2 - x + 1$. Use synthetic division to find each value.

49. $P(2)$ **50.** $P(-2)$

51. $P(-5)$ **52.** $P(3)$

Let $P(x) = 2x^4 - x^2 + 2$. Use synthetic division to find each value.

53. $P\left(\dfrac{1}{2}\right)$ **54.** $P\left(\dfrac{1}{3}\right)$

55. $P(i)$ **56.** $P(-i)$

Let $P(x) = x^4 - 8x^3 + 8x + 14x^2 - 15$. Write the terms of $P(x)$ in descending powers of x and use synthetic division to find each value.

57. $P(1)$ **58.** $P(0)$

59. $P(-3)$ **60.** $P(-1)$

Let $P(x) = 8 - 8x^2 + x^5 - x^3$. Write the terms of $P(x)$ in descending powers of x and use synthetic division to find each value.

61. $P(i)$ **62.** $P(-i)$

63. $P(-2i)$ **64.** $P(2i)$

A partial solution set is given for each equation. Find the complete solution set.

65. $x^3 + 3x^2 - 13x - 15 = 0$; $\{-1\}$

66. $x^3 + 6x^2 + 5x - 12 = 0$; $\{1\}$

67. $2x^3 + x^2 - 18x - 9 = 0$; $\left\{-\dfrac{1}{2}\right\}$

68. $2x^3 - 3x^2 - 11x + 6 = 0$; $\left\{\dfrac{1}{2}\right\}$

69. $x^4 - 2x^3 - 2x^2 + 6x - 3 = 0$; $\{1, 1\}$

70. $x^5 + 4x^4 + 4x^3 - x^2 - 4x - 4 = 0$; $\{1, -2, -2\}$

71. $x^4 - 5x^3 + 7x^2 - 5x + 6 = 0$; $\{2, 3\}$

72. $x^4 + 2x^3 - 3x^2 - 4x + 4 = 0$; $\{1, -2\}$

Find a polynomial with the given zeros.

73. $4, 5$

74. $-3, 5$

75. $1, 1, 1$

76. $1, 0, -1$

77. $2, 4, 5$

78. $7, 6, 3$

79. $1, -1, \sqrt{2}, -\sqrt{2}$

80. $0, 0, 0, \sqrt{3}, -\sqrt{3}$

81. $\sqrt{2}, i, -i$

82. $i, i, 2$

83. $0, 1 + i, 1 - i$

84. $i, 2 + i, 2 - i$

Discovery and Writing

85. If 0 is a zero of
$P(x) = a_n x^n + a_{n-1} x^{n-1} + \cdots + a_1 x + a_0,$
find a_0.

86. If 0 occurs twice as a zero of
$P(x) = a_n x^n + a_{n-1} x^{n-1} + \cdots + a_1 x + a_0,$
find a_1.

87. If $P(2) = 0$ and $P(-2) = 0$, explain why $x^2 - 4$ is a factor of $P(x)$.

88. If $P(x) = x^4 - 3x^3 + kx^2 + 4x - 1$ and $P(2) = 11$, find k.

Review *Find the quadrant in which each point lies.*

89. $P(3, -2)$

90. $Q(-2, -5)$

91. $R(8, \pi)$

92. $S(-9, 9)$

Find the distance between each pair of points.

93. $A(3, -3), B(-5, 3)$

94. $C(-8, 2), D(2, -22)$

Find the slope of the line passing through each pair of points.

95. $E(3, 5), F(-5, -3)$

96. $G\left(3, \dfrac{3}{5}\right), H\left(-\dfrac{3}{5}, 1\right)$

5.2 Descartes' Rule of Signs and Bounds on Roots

Shaun Botterill/Getty Images

Objectives

1. Understand the Fundamental Theorem of Algebra

2. Use the Conjugate Pairs Theorem

3. Use Descartes' Rule of Signs

4. Find Integer Bounds on Roots

Englishman David Beckham currently plays major league soccer for the Los Angeles Galaxy. Beckham is one of the highest paid players in the world, and he has been runner-up for the *International Federation of Association Football's* player of the year award twice.

In soccer, it is important for players to know the boundaries of the playing field. They need to know when it is advantageous to kick the ball out of bounds and when it is advantageous to keep the ball in bounds.

Knowing where the boundaries are also can be helpful in mathematics. Establishing integer bounds on roots can be an important aid in locating and finding the roots of polynomial equations. In this section, we will learn how to find integer bounds for the roots.

The remainder theorem and synthetic division provide a way of verifying that a particular number is a root of a polynomial equation, but they do not provide the roots. We need some guidelines that indicate how many roots to expect, what kind of roots to expect, and where they are located. This section develops several theorems that provide such guidelines.

1. Understand the Fundamental Theorem of Algebra

Before attempting to find the roots of a polynomial equation, it would be useful to know whether any roots exist. This question was answered by Carl Friedrich Gauss (1777–1855) when he proved the **fundamental theorem of algebra.**

The Fundamental Theorem of Algebra

If $P(x)$ is a polynomial with positive degree, then $P(x)$ has at least one zero.

The fundamental theorem of algebra guarantees that polynomials such as

$$2x^3 - 2x^2 + 7x - 3 \quad \text{and} \quad 32.75x^{1,984} + ix^3 - (2 + i)x - 5$$

all have zeros. Since all polynomials with positive degree have zeros, their corresponding polynomial equations all have roots. They are the zeros of the polynomial.

The next theorem will help us show that every nth-degree polynomial equation has exactly n roots.

The Polynomial Factorization Theorem

If $n > 0$ and $P(x)$ is an nth-degree polynomial, then $P(x)$ has exactly n linear factors:

$$P(x) = a_n(x - r_1)(x - r_2)(x - r_3) \cdot \cdots \cdot (x - r_n)$$

where $r_1, r_2, r_3, \ldots, r_n$ are numbers and a_n is the leading coefficient of $P(x)$.

Proof Let $P(x)$ be a polynomial of degree n ($n > 0$). Because of the fundamental theorem of algebra, we know that $P(x)$ has a zero r_1 and that the equation $P(x) = 0$ has r_1 for a root. By the factor theorem, we know that $x - r_1$ is a factor of $P(x)$. Thus,

$$P(x) = (x - r_1)Q_1(x)$$

If the leading coefficient of the nth-degree polynomial $P(x)$ is a_n, then $Q_1(x)$ is a polynomial of degree $n - 1$ whose leading coefficient is also a_n.

By the fundamental theorem of algebra, we know that $Q_1(x)$ also has a zero, r_2. By the factor theorem, $x - r_2$ is a factor of $Q_1(x)$, and

$$P(x) = (x - r_1)(x - r_2)Q_2(x)$$

where $Q_2(x)$ is a polynomial of degree $n - 2$ with leading coefficient a_n.

This process can continue only to n factors of the form $x - r_i$ until the final quotient $Q_n(x)$ is a polynomial of degree $n - n$, or degree 0. Thus, the polynomial $P(x)$ factors completely as

$$(1) \quad P(x) = a_n(x - r_1)(x - r_2)(x - r_3) \cdot \cdots \cdot (x - r_n)$$

We can use the polynomial factorization theorem and make the following conclusions:

1. If we substitute any one of the numbers $r_1, r_2, r_3, \ldots, r_n$ for x in Equation 1, $P(x)$ will equal 0. Thus, each value of r is a zero of $P(x)$ and a root of the equation $P(x) = 0$.

2. There can be no other roots, because no single factor in Equation 1 is 0 for any value of x not included in the list $r_1, r_2, r_3, \ldots, r_n$.

3. The values of r in the previous list need not be distinct. Any number r_i that occurs k times as a root of a polynomial equation is called a **root of multiplicity k.**

The following theorem summarizes the previous discussion.

> **Theorem**   If multiple roots are counted individually, the polynomial equation $P(x) = 0$ with degree n $(n > 0)$ has exactly n roots among the complex numbers.

The previous theorems are illustrated by these examples.

Polynomial Function	Properties
$P(x) = x - 6$	**Degree:** First **Number of linear factors:** 1 **Linear factor:** $x - 6$ **Number of zeros:** 1 **Zero:** 6 **Multiplicity of each zero:** 6 has a multiplicity of one.
$\begin{aligned} P(x) &= x^2 - 8x + 16 \\ &= (x - 4)(x - 4) \end{aligned}$	**Degree:** Second **Number of linear factors:** 2 **Linear factors:** $x - 4$, $x - 4$ **Number of zeros:** 2 **Zeros:** 4 and 4 **Multiplicity of each zero:** 4 has a multiplicity of two.

$\begin{aligned} P(x) &= x^3 + 100x \\ &= x(x^2 + 100) \\ &= x(x - 10i)(x + 10i) \end{aligned}$	**Degree:** Third
	Number of linear factors: 3
	Linear factors: x, $x - 10i$, and $x + 10i$
	Number of zeros: 3
	Zeros: 0, $10i$, and $-10i$
	Multiplicity of each zero:
	    0 has a multiplicity of one.
	    $10i$ has a multiplicity of one.
	    $-10i$ has a multiplicity of one.

$\begin{aligned} P(x) &= x^4 - 7x^2 \\ &= x^2(x^2 - 7) \\ &= x \cdot x\left(x + \sqrt{7}\right)\left(x - \sqrt{7}\right) \end{aligned}$	**Degree:** Fourth
	Number of linear factors: 4
	Linear factors: x, x, $x + \sqrt{7}$, $x - \sqrt{7}$
	Number of zeros: 4
	Zeros: 0, 0, $\sqrt{7}$, and $-\sqrt{7}$
	Multiplicity of each zero:
	    0 has a multiplicity of two.
	    $\sqrt{7}$ has a multiplicity of one.
	    $-\sqrt{7}$ has a multiplicity of one.

Comment
Remember that every nth degree polynomial has exactly n linear factors and exactly n zeros. The n zeros are also the roots of the polynomial equation $P(x) = 0$.

2.   Use the Conjugate Pairs Theorem

Recall that the complex numbers $a + bi$ and $a - bi$ are called **complex conjugates** of each other. The next theorem points out that *complex roots of polynomial equations with real coefficients occur in complex conjugate pairs.*

> **The Conjugate Pairs Theorem**   If a polynomial equation $P(x) = 0$ with real-number coefficients has a complex root $a + bi$ with $b \neq 0$, then its conjugate $a - bi$ is also a root.

EXAMPLE 1 Find a second-degree polynomial equation with real coefficients that has a root of $2 + i$.

Solution We can use the conjugate pairs theorem to determine the second root. Then we will use the two roots to write the linear factorization and equation of the polynomial.

1. **Determine the roots.** Because $2 + i$ is a root, its complex conjugate $2 - i$ is also a root. The solution set is $\{2 + i, 2 - i\}$.

2. **Write the linear factorization of the polynomial and equation.** Because $2 + i$ is a root, $x - (2 + i)$ is a factor. Because $2 - i$ is a root, $x - (2 - i)$ is a factor. The polynomial equation is

$$[x - (2 + i)][x - (2 - i)] = 0$$
$$(x - 2 - i)(x - 2 + i) = 0$$
$$x^2 - 4x + 5 = 0 \quad \text{Multiply.}$$

The equation $x^2 - 4x + 5 = 0$ will have roots of $2 + i$ and $2 - i$.

Self Check 1 Find a second-degree polynomial equation with real coefficients that has a root of $1 - i$.

EXAMPLE 2 Find a fourth-degree polynomial equation with real coefficients and i as a root of multiplicity 2.

Solution We can use the conjugate pairs theorem to determine the other two roots. Then we will use the four roots to write the linear factorization and equation of the polynomial.

1. **Determine the roots.** Because i is a root twice and a fourth-degree polynomial equation has four roots, we must find the other two roots. According to the conjugate pairs theorem, the missing roots are the conjugates of the given roots. Thus, the complete solution set is

$$\{i, i, -i, -i\}$$

2. **Write the linear factorization of the polynomial equation.** Because i is a root of multiplicity 2, the factor $x - i$ occurs twice. Because $-i$ is a root of multiplicity 2, the factor $x - (-i)$ occurs twice. The equation is

$$(x - i)(x - i)[x - (-i)][x - (-i)] = 0$$
$$(x - i)(x + i)(x - i)(x + i) = 0$$
$$(x^2 + 1)(x^2 + 1) = 0$$
$$x^4 + 2x^2 + 1 = 0$$

Comment
The conjugate pairs theorem applies only to polynomials with real-number coefficients. In Examples 1 and 2, the theorem does apply and the results were polynomials with real-number coefficients.

Self Check 2 Find a fourth-degree polynomial equation with real coefficients and $-i$ as a root of multiplicity 2.

EXAMPLE 3 Find a quadratic equation with a double root of i.

Solution We will use the two given roots of i to write the linear factorization of the polynomial and quadratic equation.

Comment
Since the resulting polynomial equation in Example 3 does not have real-number coefficients, the conjugate pairs theorem does not apply.

Because i is a root twice, the linear factorization of the polynomial and the quadratic equation are

$$(x - i)(x - i) = 0$$
$$x^2 - 2ix - 1 = 0 \quad \text{Multiply.}$$

In this equation, the coefficient of x is $-2i$, which is not a real number. Therefore, it is not surprising that the roots are not in complex conjugate pairs.

Self Check 3   Find a quadratic equation with a double root of $-i$.

3. Use Descartes' Rule of Signs

René Descartes (1596–1650) is credited with a theorem known as **Descartes' rule of signs,** which enables us to estimate the number of positive, negative, and nonreal roots of a polynomial equation.

If a polynomial is written in descending powers of x and we scan it from left to right, we say that a variation in sign occurs whenever successive terms have opposite signs. For example, the polynomial

$$P(x) = \overbrace{3x^5 - 2x^4}^{+\text{ to }-} \overbrace{- 5x^3 + x^2}^{-\text{ to }+} \overbrace{- x - 9}^{+\text{ to }-}$$

has three variations in sign, and the polynomial

$$P(-x) = 3(-x)^5 - 2(-x)^4 - 5(-x)^3 + (-x)^2 - (-x) - 9$$

$$= \overbrace{-3x^5 - 2x^4}^{-\text{ to }+} + 5x^3 + \overbrace{x^2 + x - 9}^{+\text{ to }-}$$

has two variations in sign.

Descartes' Rule of Signs   If $P(x)$ is a polynomial with real coefficients, the number of positive roots of $P(x) = 0$ is either equal to the number of variations in sign of $P(x)$ or less than that by an even number.

The number of negative roots of $P(x) = 0$ is either equal to the number of variations in sign of $P(-x)$ or less than that by an even number.

EXAMPLE 4   Find the number of possible positive, negative, and nonreal roots of $P(x) = 3x^3 - 2x^2 + x - 5 = 0$.

Solution   We can use Descartes' rule of signs to determine the number of possible positive, negative, and nonreal roots.

1. **Determine the number of possible positive roots.**   Since there are three variations of sign in $P(x) = 3x^3 - 2x^2 + x - 5 = 0$, there can be either 3 positive roots or only 1. (1 is less than 3 by the even number 2.)

2. **Determine the number of possible negative roots.**   Because

$$P(-x) = 3(-x)^3 - 2(-x)^2 + (-x) - 5$$
$$= -3x^3 - 2x^2 - x - 5$$

has no variations in sign, there are 0 negative roots. Furthermore, 0 is not a root, because the terms of the polynomial do not have a common factor of x.

3. **Determine the number of nonreal roots.**   If there are 3 positive roots, then all of the roots are accounted for. If there is 1 positive root, the 2 remaining roots must be nonreal complex numbers. The following table shows these possibilities.

Comment
To obtain $P(-x)$ quickly, we can simply multiply each odd degree term of $P(x)$ by negative one.

Number of positive roots	Number of negative roots	Number of nonreal roots
3	0	0
1	0	2

The number of nonreal complex roots is the number needed to bring the total number of roots up to 3.

Self Check 4   Find the possibilities for the roots of $5x^3 + 2x^2 - x + 3 = 0$.

EXAMPLE 5   Find the number of possible positive, negative, and nonreal roots of $P(x) = 5x^5 - 3x^3 - 2x^2 + x - 1 = 0$.

Solution   We can used Descartes' rule of signs to determine the number of possible positive, negative, and nonreal roots.

1. **Determine the number of possible positive roots.**   Since there are three variations of sign in $P(x)$, there are either 3 or 1 positive roots.

2. **Determine the number of possible negative roots.**   Because $P(-x) = -5x^5 + 3x^3 - 2x^2 - x - 1 = 0$ has two variations in sign, there are 2 or 0 negative roots.

3. **Determine the number of nonreal roots.**   The possibilities are as follows:

Number of positive roots	Number of negative roots	Number of nonreal roots
1	0	4
3	0	2
1	2	2
3	2	0

In each case, the number of nonreal complex roots is an even number. This is expected, because this polynomial has real coefficients, and its nonreal complex roots will occur in conjugate pairs.

Self Check 5   Find the possibilities for the roots of $5x^5 - 2x^2 - x - 1 = 0$.

4.   Find Integer Bounds on Roots

A final theorem provides a way to find **bounds** on the roots of a polynomial equation, enabling us to look for roots where they can be found.

Upper and Lower Bounds on Roots

Let $P(x)$ be a polynomial with real coefficients and a positive leading coefficient.

1. If $P(x)$ is synthetically divided by a positive number c and each term in the last row of the division is nonnegative, then no number greater than c can be a root of $P(x) = 0$. (c is called an **upper bound** of the real roots.)

2. If $P(x)$ is synthetically divided by a negative number d and the signs in the last row alternate,* no value less than d can be a root of $P(x) = 0$. (d is called a **lower bound** of the real roots.)

*If 0 appears in the third row, that 0 can be assigned either a + or a − sign to help the signs alternate.

EXAMPLE 6   Establish integer bounds for the roots of $2x^3 + 3x^2 - 5x - 7 = 0$.

Solution   We will use the upper and lower bounds rules stated above to establish the integer bounds for the roots.

1. We will perform several synthetic divisions by positive integers, looking for nonnegative values in the last row.
   Trying 1 first gives

$$
\begin{array}{r|rrrr}
1 & 2 & 3 & -5 & -7 \\
  &   & 2 & 5 & 0 \\
\hline
  & +2 & +5 & 0 & -7
\end{array}
$$

Because one of the signs in the last row is negative, we cannot claim that 1 is an upper bound. We now try 2.

$$
\begin{array}{r|rrrr}
2 & 2 & 3 & -5 & -7 \\
  &   & 4 & 14 & 18 \\
\hline
  & +2 & +7 & +9 & +11
\end{array}
$$

Because the last row is entirely nonnegative, we can claim that 2 is an upper bound. That is, no number greater than 2 can be a root of the equation.

2. We perform several synthetic divisions by negative integers, looking for alternating signs in the last row.
   We begin with −3.

$$
\begin{array}{r|rrrr}
-3 & 2 & 3 & -5 & -7 \\
   &   & -6 & 9 & -12 \\
\hline
   & +2 & -3 & +4 & -19
\end{array}
$$

Since the signs in the last row alternate, −3 is a lower bound. That is, no number less than −3 can be a root. To see whether there is a greater lower bound, we try −2.

$$
\begin{array}{r|rrrr}
-2 & 2 & 3 & -5 & -7 \\
   &   & -4 & 2 & 6 \\
\hline
   & +2 & -1 & -3 & -1
\end{array}
$$

Since the signs in the last row do not alternate, we cannot claim that −2 is a lower bound.

Since −3 is a lower bound and 2 is an upper bound, then all of the real roots must be in the interval $(-3, 2)$.

Self Check 6   Establish integer bounds for the roots of $2x^3 + 3x^2 - 11x - 7 = 0$.

It is important to understand what the theorem on the bounds of roots says and what it doesn't say. The following two explanations will help clarify that for us.

1. If we divide synthetically by a positive number c and the last row of the synthetic division is entirely nonnegative, the theorem guarantees that c is an upper bound of the roots. However, if the last row contains some negative values, c could still be an upper bound.

2. If we divide by a negative number d and the signs in the last row alternate, the theorem guarantees that d is a lower bound of the roots. However, if the signs in the last row do not alternate, d could still be a lower bound. This is illustrated in Example 6. It can be shown that the smallest negative root of the equation is approximately -1.81. Thus, -2 is a lower bound for the roots of the equation. However, when we checked -2, the last row of the synthetic division did not have alternating signs. Unfortunately, the theorem does not always determine the best bounds for the roots of the equation.

Self Check Answers   **1.** $x^2 - 2x + 2 = 0$ **2.** $x^4 + 2x^2 + 1 = 0$ **3.** $x^2 + 2ix - 1 = 0$
4.

Num. of pos. roots	Num. of neg. roots	Num. of non-real roots
2	1	0
0	1	2

5.

Num. of pos. roots	Num. of neg. roots	Num. of non-real roots
1	2	2
1	0	4

6. Roots are in $(-4, 3)$.

5.2   Exercises

Vocabulary and Concepts   *Fill in the blanks.*

1. If $P(x)$ is a polynomial with positive degree, then $P(x)$ has at least one ____.

2. The statement in Exercise 1 is called the

   _____.

3. The _____ of $a + bi$ is $a - bi$.

4. The polynomial $6x^4 + 5x^3 - 2x^2 + 3$ has __ variations in sign.

5. The polynomial $(-x)^3 - (-x)^2 - 4$ has __ variations in sign.

6. The equation $7x^4 + 5x^3 - 2x + 1 = 0$ can have at most __ positive roots.

7. The equation $7x^4 + 5x^3 - 2x + 1 = 0$ can have at most __ negative roots.

8. Complex roots occur in complex _____ pairs. (Assume that the equation has real coefficients.)

9. If no number less than d can be a root of $P(x) = 0$, then d is called a(n) _____.

10. If no number greater than c can be a root of $P(x) = 0$, then c is called a(n) _____.

Practice   *Determine how many roots each equation has.*

11. $x^{10} = 1$ 12. $x^{40} = 1$

13. $3x^4 - 4x^2 - 2x = -7$

14. $-32x^{111} - x^5 = 1$

15. One root of $x(3x^4 - 2) = 12x$ is 0. How many other roots are there?

16. Two roots of $3x^2(x^7 - 14x + 3) = 0$ are 0. How many other roots are there?

Determine how many linear factors and zeros each polynomial function has.

17. $P(x) = x^4 - 81$ 18. $P(x) = x^{40} + x^{39}$

19. $P(x) = 4x^5 + 8x^3$ **20.** $P(x) = x^3 + 144x$

46. $-7x^5 - 6x^4 + 3x^3 - 2x^2 + 7x - 4 = 0$

Write a second-degree polynomial equation with real coefficients and the given root.

21. $2i$

22. $-3i$

23. $3 - i$

24. $4 + 2i$

Find integer bounds for the roots of each equation.

47. $x^2 - 2x - 4 = 0$ **48.** $9x^2 - 6x - 1 = 0$

49. $18x^2 - 6x - 1 = 0$ **50.** $2x^2 - 10x - 9 = 0$

Write a third-degree polynomial equation with real coefficients and the given roots.

25. $3, -i$

26. $1, i$

27. $2, 2 + i$

28. $-2, 3 - i$

51. $6x^3 - 13x^2 - 110x = 0$

52. $12x^3 + 20x^2 - x - 6 = 0$

53. $x^5 + x^4 - 8x^3 - 8x^2 + 15x + 15 = 0$

54. $3x^4 - 5x^3 - 9x^2 + 15x = 0$

55. $3x^5 - 11x^4 - 2x^3 + 38x^2 - 21x - 15 = 0$

56. $3x^6 - 4x^5 - 21x^4 + 4x^3 + 8x^2 + 8x + 32 = 0$

Write a fourth-degree polynomial equation with real coefficients and the given roots.

29. $3, 2, i$

30. $1, 2, 1 + i$

31. $i, 1 - i$

32. $i, 2 - i$

Discovery and Writing

57. Explain why the fundamental theorem of algebra guarantees that every polynomial equation of positive degree has at least one root.

58. Explain why the fundamental theorem of algebra and the factor theorem guarantee that an nth-degree polynomial equation has n roots.

Use Descartes' rule of signs to find the number of possible positive, negative, and nonreal roots of each equation.

33. $3x^3 + 5x^2 - 4x + 3 = 0$

59. Prove that any odd-degree polynomial equation with real coefficients must have at least one real root.

34. $3x^3 - 5x^2 - 4x - 3 = 0$

60. If a, b, c, and d are positive numbers, prove that $ax^4 + bx^2 + cx - d = 0$ has exactly two nonreal roots.

35. $2x^3 + 7x^2 + 5x + 5 = 0$

36. $-2x^3 - 7x^2 - 5x - 4 = 0$

Review   *Assume that k represents a positive real number, and complete each sentence.*

37. $8x^4 = -5$

38. $-3x^3 = -5$

39. $x^4 + 8x^2 - 5x - 10 = 0$

61. The graph of $y = f(x - k)$ looks like the graph of $y = f(x)$, except that it has been translated _____.

62. The graph of $y = f(x) - k$ looks like the graph of $y = f(x)$, except that it has been translated _____.

40. $5x^7 + 3x^6 - 2x^5 + 3x^4 + 9x^3 + x^2 + 1 = 0$

63. The graph of $y = f(-x)$ looks like the graph of $y = f(x)$, except that it has been _____.

41. $-x^{10} - x^8 - x^6 - x^4 - x^2 - 1 = 0$

64. The graph of $y = -f(x)$ looks like the graph of $y = f(x)$, except that it has been _____.

42. $x^{10} + x^8 + x^6 + x^4 + x^2 + 1 = 0$

43. $x^9 + x^7 + x^5 + x^3 + x = 0$ (Is 0 a root?)

65. If $k > 1$, the graph of $y = kf(x)$ looks like the graph of $y = f(x)$, except that it has been _____.

44. $-x^9 - x^7 - x^5 - x^3 - x = 0$ (Is 0 a root?)

66. If $0 < k < 1$, the graph of $y = f(kx)$ looks like the graph of $y = f(x)$, except that it has been _____.

45. $-2x^4 - 3x^2 + 2x + 3 = 0$

Every package matters

fedex.com

5.3   Roots of Polynomial Equations

Objectives

1. Find Possible Rational Roots of Polynomial Equations
2. Find Rational Roots of Polynomial Equations
3. Find Real and Nonreal Roots of Polynomial Equations
4. Solve Problems

FedEx is a company that offers reliable shipping services across town and across the globe. It has a wide range of envelopes and boxes available to accommodate the shipping needs of most individuals and companies.

The box shown above has the following characteristics:

- The length of the box is 6 inches more than its height.
- The width of the box is 3 inches more than its height.
- The volume of the box is 2,080 cubic inches.

To find the dimensions of the box, we can let h represent the height (in inches). Then $h + 6$ will represent the length and $h + 3$ will represent the width. Since the volume of the box is given to be 2,080 cubic inches, we have

$$V = l \cdot w \cdot h$$
$$2{,}080 = (h + 6)(h + 3)h \qquad \text{Substitute.}$$
$$2{,}080 = h^3 + 9h^2 + 18h \qquad \text{Multiply.}$$
$$0 = h^3 + 9h^2 + 18h - 2{,}080 \qquad \text{Subtract 2,080 from both sides.}$$

To find the height h, we must find the roots of the polynomial equation. One root of the equation is $x = 10$, because 10 satisfies the equation:

$$(10)^3 + 9(10)^2 + 18(10) - 2{,}080 = 0$$

Therefore, the height of the box is $h = 10$ inches, the length is $h + 6 = 16$ inches, and the width is $h + 3 = 13$ inches.

In this section, we will develop a strategy for finding roots of higher-degree polynomial equations.

1.   Find Possible Rational Roots of Polynomial Equations

Recall that a rational number is any number that can be written in the form $\frac{p}{q}$, where p and q are integers and $q \neq 0$. The following theorem enables us to list the possible rational roots of such equations.

Rational Root Theorem   Let the polynomial equation

$$P(x) = a_n x^n + a_{n-1} x^{n-1} + a_{n-2} x^{n-2} + \cdots + a_1 x + a_0 = 0$$

have integer coefficients. If the rational number $\frac{p}{q}$ (written in lowest terms) is a root of $P(x) = 0$, then p is a factor of the constant a_0, and q is a factor of the leading coefficient a_n.

Proof Let $\frac{p}{q}$ (written in lowest terms) be a rational root of $P(x) = 0$. Then the equation is satisfied by $\frac{p}{q}$:

$$(1) \quad a_n\left(\frac{p}{q}\right)^n + a_{n-1}\left(\frac{p}{q}\right)^{n-1} + a_{n-2}\left(\frac{p}{q}\right)^{n-2} + \cdots + a_1\left(\frac{p}{q}\right) + a_0 = 0$$

We can clear Equation 1 of fractions by multiplying both sides by q^n.

$$(2) \quad a_n p^n + a_{n-1}p^{n-1}q + a_{n-2}p^{n-2}q^2 + \cdots + a_1 pq^{n-1} + a_0 q^n = 0$$

We can factor p from all but the last term and subtract $a_0 q^n$ from both sides to get

$$p(a_n p^{n-1} + a_{n-1}p^{n-2}q + a_{n-2}p^{n-3}q^2 + \cdots + a_1 q^{n-1}) = -a_0 q^n$$

Since p is a factor of the left side, it is also a factor of the right side. So p is a factor of $-a_0 q^n$, but because $\frac{p}{q}$ is written in lowest terms, p cannot be a factor of q^n. Therefore, p is a factor of a_0.

We can factor q from all but the first term of Equation 2 and subtract $a_n p^n$ from both sides to get

$$q(a_{n-1}p^{n-1} + a_{n-2}p^{n-2}q + a_{n-3}p^{n-3}q^2 + \cdots + a_0 q^{n-1}) = -a_n p^n$$

Since q is a factor of the left side, it is also a factor of the right side. Because q is not a factor of p^n, it must be a factor of a_n. ∎

Everyday Connections

Gasoline Prices

"In the next 20 years, oil production from conventional reservoirs may begin to decline, creating a gap between supply and demand." Stephen Holditch, President, Society of Petroleum Engineers

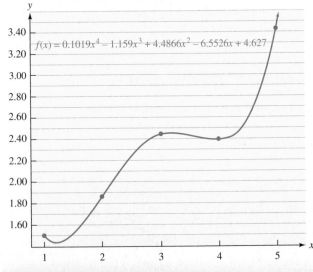

U.S. City Average Price-per-Gallon (dollars)

$f(x) = 0.1019x^4 - 1.159x^3 + 4.4866x^2 - 6.5526x + 4.627$

The polynomial function shown in the graph models the average January U.S. fuel oil price in dollars per gallon during the time period 2004–2008, where x equals the number of years since 2003. Use the graph of the function to answer the following questions.

1. How many times was the average January price per gallon $2.40 between 2004 and 2008?

2. What was the average January price per gallon in June 2007?

Source: http://data.bls.gov/cgi-bin/surveymost

To illustrate the rational root theorem, we consider the equation

$$\frac{1}{2}x^4 + \frac{2}{3}x^3 + 3x^2 - \frac{3}{2}x + 3 = 0$$

Because the theorem requires integer coefficients, we multiply both sides of the equation by 6 to clear it of fractions.

$$3x^4 + 4x^3 + 18x^2 - 9x + 18 = 0$$

By the previous theorem, the only possible numerators for the rational roots of the equation are the factors of the constant term 18:

$$\pm 1, \quad \pm 2, \quad \pm 3, \quad \pm 6, \quad \pm 9, \quad \text{and} \quad \pm 18$$

The only possible denominators are the factors of the leading coefficient 3:

$$\pm 1 \quad \text{and} \quad \pm 3$$

Comment

When we list possible rational roots for a polynomial equation, the list will not contain radicals or complex numbers.

We can form a list of all possible rational solutions by listing the combinations of possible numerators and denominators:

$$\pm\frac{1}{1}, \pm\frac{2}{1}, \pm\frac{3}{1}, \pm\frac{6}{1}, \pm\frac{9}{1}, \pm\frac{18}{1}, \pm\frac{1}{3}, \pm\frac{2}{3}, \pm\frac{3}{3}, \pm\frac{6}{3}, \pm\frac{9}{3}, \pm\frac{18}{3}$$

Since several of these possibilities are duplicates, we can condense the list to get

Possible rational roots

$$\pm 1, \quad \pm 2, \quad \pm 3, \quad \pm 6, \quad \pm 9, \quad \pm 18, \quad \pm\frac{1}{3}, \quad \pm\frac{2}{3}$$

2. Find Rational Roots of Polynomial Equations

To find the rational roots of a polynomial equation, we will use the following steps.

Finding Rational Roots
1. Use Descartes' rule of signs to determine the number of possible positive, negative, and nonreal roots.
2. Use the rational root theorem to list possible rational roots.
3. Use synthetic division to find a root.
4. If there are more roots, repeat the previous steps.

EXAMPLE 1 Find the solution set of $P(x) = 2x^3 + 3x^2 - 8x + 3 = 0$.

Solution We will use the steps outlined above to solve the polynomial equation. Since the equation is of third degree, it has 3 roots.

1. To determine the number of possible positive, negative, and nonreal roots, we will use Descartes' rule of signs. We find that there are two possible combinations of positive, negative, and nonreal roots. They are as follows:

Number of positive roots	Number of negative roots	Number of nonreal roots
2	1	0
0	1	2

2. We then find the possible rational roots that have the form $\frac{\text{factor of the constant 3}}{\text{factor of the leading coefficient 2}}$. They are

$$\pm\frac{3}{1}, \quad \pm\frac{1}{1}, \quad \pm\frac{3}{2}, \quad \pm\frac{1}{2}$$

or, written in order of increasing size,

$$-3, \quad -\frac{3}{2}, \quad -1, \quad -\frac{1}{2}, \quad \frac{1}{2}, \quad 1, \quad \frac{3}{2}, \quad 3$$

3. We then use synthetic division and check each possibility to see whether it is a root. We can start with $\frac{3}{2}$.

$$\begin{array}{r|rrrr} \frac{3}{2} & 2 & 3 & -8 & 3 \\ & & 3 & 9 & \frac{3}{2} \\ \hline & 2 & 6 & 1 & \frac{9}{2} \end{array}$$

Since the remainder is not 0, $\frac{3}{2}$ is not a root and we can cross it off the list. Since every number in the last row of the synthetic division is positive, $\frac{3}{2}$ is an upper bound. So 3 cannot be a root either and we can cross it off the list.

$$-3, \quad -\frac{3}{2}, \quad -1, \quad -\frac{1}{2}, \quad \frac{1}{2}, \quad 1, \quad \cancel{\frac{3}{2}}, \quad \cancel{3}$$

We now try $\frac{1}{2}$:

$$\begin{array}{r|rrrr} \frac{1}{2} & 2 & 3 & -8 & 3 \\ & & 1 & 2 & -3 \\ \hline & 2 & 4 & -6 & 0 \end{array}$$ **This row represents the quotient $2x^2 + 4x - 6$.**

Since the remainder is 0, $\frac{1}{2}$ is a root and the binomial $x - \frac{1}{2}$ is a factor of $P(x)$.

4. The remaining roots must be supplied by the remaining factor, which is the quotient $2x^2 + 4x - 6$. We can find the other roots by solving the equation $2x^2 + 4x - 6 = 0$, called the **depressed equation.**

$$\begin{array}{ll} 2x^2 + 4x - 6 = 0 & \\ x^2 + 2x - 3 = 0 & \text{Divide both sides by 2.} \\ (x - 1)(x + 3) = 0 & \text{Factor } x^2 + 2x - 3. \\ x - 1 = 0 \quad \text{or} \quad x + 3 = 0 & \\ x = 1 \quad \mid \quad x = -3 & \end{array}$$

Comment
Example 1 illustrates that the upper and lower bounds theorem is helpful when the list of possible rational roots is long.

The solution set of the equation is $\left\{\frac{1}{2}, 1, -3\right\}$. Note that two positive roots and one negative root is a predicted possibility.

Self Check 1 Find the solution set of $3x^3 - 10x^2 + 9x - 2 = 0$.

Accent on Technology **Confirming Roots of an Equation**

We can confirm that the roots found in Example 1 are correct by graphing the function $P(x) = 2x^3 + 3x^2 - 8x + 3$ and locating the resulting x-intercepts of the graph. If we use a graphing window of $x = [-4, 4]$ and $y = [-20, 20]$ and graph the function, we will obtain the graph shown in Figure 5-1. From the graph, we can see that the x-intercepts are at $x = -3$, $x = \frac{1}{2}$, and $x = 1$. These are the roots of the equation.

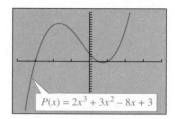

$$P(x) = 2x^3 + 3x^2 - 8x + 3$$

Figure 5-1

3.   Find Real and Nonreal Roots of Polynomial Equations

EXAMPLE 2   Find the solution set of:

$$P(x) = x^7 - 2x^6 - 5x^5 + 6x^4 - x^3 + 2x^2 + 5x - 6 = 0$$

Solution   We will use the steps outlined above to find the roots of the polynomial equation. Because the equation is of seventh degree, it has 7 roots.

1. To determine the number of possible positive, negative, and nonreal roots, we will use Descartes' rule of signs. We find that there are several possible combinations of positive, negative, and nonreal roots. They are as follows:

Number of positive roots	Number of negative roots	Number of nonreal roots
5	2	0
3	2	2
1	2	4
5	0	2
3	0	4
1	0	6

2. We then write the list of all possible rational roots that have the form $\frac{\text{factor of the constant 6}}{\text{factor of the leading coefficient 1}}$. They are as follows:

$$-6, \quad -3, \quad -2, \quad -1, \quad 1, \quad 2, \quad 3, \quad 6$$

3. We then use synthetic division and check each possibility to see whether it is a root. We can start with -3.

$$
\begin{array}{r|rrrrrrrr}
-3 & 1 & -2 & -5 & 6 & -1 & 2 & 5 & -6 \\
   &   & -3 & 15 & -30 & 72 & -213 & 633 & -1{,}914 \\
\hline
   & 1 & -5 & 10 & -24 & 71 & -211 & 638 & -1{,}920
\end{array}
$$

Since the last number in the synthetic division is not 0, -3 is not a root and can be crossed off the list. Since the signs in the last row alternate, -3 is a lower bound, and we can cross off -6 as well.

$$-\cancel{6}, \quad -\cancel{3}, \quad -2, \quad -1, \quad 1, \quad 2, \quad 3, \quad 6$$

We now try -2:

$$
\begin{array}{r|rrrrrrr}
-2 & 1 & -2 & -5 & 6 & -1 & 2 & 5 & -6 \\
   &   & -2 & 8 & -6 & 0 & 2 & -8 & 6 \\
\hline
   & 1 & -4 & 3 & 0 & -1 & 4 & -3 & 0
\end{array}
$$

Since the remainder is 0, -2 is a root.

4. Because the root -2 is negative, we can revise the chart of possibilities to eliminate the possibility that there are 0 negative roots.

Number of positive roots	Number of negative roots	Number of nonreal roots
5	2	0
3	2	2
1	2	4

The remaining roots must be supplied by the remaining factor, which is the quotient in the synthetic division shown above: $x^6 - 4x^5 + 3x^4 - x^2 + 4x - 3 = 0$. We can find the other roots by repeating Steps 1–3 and solving this equation, called the **depressed equation.**

Because the constant term of this depressed equation is different from the constant term of the original equation, we can cross off other possible rational roots. For example, the number -2 cannot be a root a second time, because it is not a factor of -3. The numbers 2 and 6 are no longer possible roots, because neither is a factor of -3.

The list of possible roots is now

$$-\cancel{6}, \quad -3, \quad -\cancel{2}, \quad -1, \quad 1, \quad \cancel{2}, \quad 3, \quad \cancel{6}$$

Since we know that there is one more negative root, we will synthetically divide the coefficients of the depressed equation by -1.

$$
\begin{array}{r|rrrrrr}
-1 & 1 & -4 & 3 & 0 & -1 & 4 & -3 \\
   &   & -1 & 5 & -8 & 8 & -7 & 3 \\
\hline
   & 1 & -5 & 8 & -8 & 7 & -3 & 0
\end{array}
$$

Since the remainder is 0, -1 is a root, and the solution set so far is $\{-2, -1, \ldots\}$. The number -1 cannot be a root again, because we have found both negative roots. When we cross off -1, we have only two possibilities left.

$$-\cancel{6}, \quad -3, \quad -\cancel{2}, \quad -\cancel{1}, \quad 1, \quad \cancel{2}, \quad 3, \quad \cancel{6}$$

The depressed equation is now $x^5 - 5x^4 + 8x^3 - 8x^2 + 7x - 3 = 0$.

We can synthetically divide the coefficients of this equation by 1 to get

$$
\begin{array}{r|rrrrrr}
1 & 1 & -5 & 8 & -8 & 7 & -3 \\
  &   & 1 & -4 & 4 & -4 & 3 \\
\hline
  & 1 & -4 & 4 & -4 & 3 & 0
\end{array}
$$

The depressed equation is now $x^4 - 4x^3 + 4x^2 - 4x + 3 = 0.$

Since the remainder is 0, 1 joins the solution set $\{-2, -1, 1, \ldots\}$. To see whether 1 is a root a second time, we synthetically divide the coefficients of the new depressed equation by 1.

$$
\begin{array}{r|rrrrr}
1 & 1 & -4 & 4 & -4 & 3 \\
  &   & 1 & -3 & 1 & 3 \\
\hline
  & 1 & -3 & 1 & -3 & 0
\end{array}
$$

The depressed equation is now $x^3 - 3x^2 + x - 3 = 0.$

Again, 1 is a root, and the solution set is now $\{-2, -1, 1, 1, \dots\}$.

To see whether 1 is a root a third time, we synthetically divide the coefficients of the new depressed equation by 1.

$$\begin{array}{r|rrrr} 1 & 1 & -3 & 1 & -3 \\ & & 1 & -2 & -1 \\ \hline & 1 & -2 & -1 & -4 \end{array}$$

Since the remainder is not 0, the number 1 is not a root for a third time, and we can cross 1 off the list of possibilities, leaving only 3.

$$-\cancel{6}, \quad -\cancel{3}, \quad -\cancel{2}, \quad -\cancel{1}, \quad \cancel{1}, \quad \cancel{2}, \quad 3, \quad \cancel{6}$$

To see whether 3 is a root, we synthetically divide the coefficients of $x^3 - 3x^2 + x - 3 = 0$ by 3.

$$\begin{array}{r|rrrr} 3 & 1 & -3 & 1 & -3 \\ & & 3 & 0 & 3 \\ \hline & 1 & 0 & 1 & 0 \end{array}$$

Since the remainder is $0, 3$ joins the solution set, which is now $\{-2, -1, 1, 1, 3, \dots\}$.

The depressed equation is now $x^2 + 1 = 0$, which can be solved as a quadratic equation.

$$x^2 + 1 = 0$$
$$x^2 = -1$$
$$x = i \quad \text{or} \quad x = -i$$

The complete solution set is $\{-2, -1, 1, 1, 3, i, -i\}$. The solution set contains 3 positive roots, 2 negative roots, and 2 nonreal roots that are complex conjugates. This combination was one of the predicted possibilities.

Comment

Example 2 illustrates that finding the seven roots of a seventh-degree polynomial equation can be a lengthy process. In such cases, the four step process for finding rational roots must be repeated.

Self Check 2   Find the solution set of $x^5 - x^4 + 3x^3 - 3x^2 - 4x + 4 = 0$.

Accent on Technology

Confirming Roots of an Equation

We can confirm that the real roots found in Example 2 are correct by graphing the function

$$P(x) = x^7 - 2x^6 - 5x^5 + 6x^4 - x^3 + 2x^2 + 5x - 6$$

and locating the resulting x-intercepts of the graph. If we use a graphing window of $x = [-4, 4]$ and $y = [-130, 50]$ and graph the function, we will obtain the graph shown in Figure 5-2. From the graph, we can see that the x-intercepts are at $x = -2$, $x = -1$, $x = 1$, and $x = 3$. We cannot detect the complex roots from the graph.

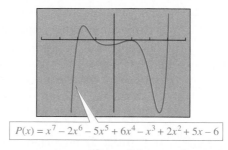

$P(x) = x^7 - 2x^6 - 5x^5 + 6x^4 - x^3 + 2x^2 + 5x - 6$

Figure 5-2

4. Solve Problems

EXAMPLE 3   To protect cranberry crops from the damage of early freezes, growers flood the cranberry bogs. Three irrigation sources, used together, can flood a cranberry bog in one day. If the sources are used one at a time, the second source requires one day longer to flood the bog than the first, and the third requires four days longer than the first. If the bog must be flooded before a freeze that is predicted in three days, can the water in the last two sources be diverted to other bogs?

Solution   We will use the given information to write an equation that models the situation. We will then solve the equation.

We can let x represent the number of days it would take the first irrigation source to flood the bog. Then $x + 1$ and $x + 4$ represent the number of days it would take the second and third sources, respectively, to flood the bog.

Because the first source, alone, requires x days to flood the bog, that source could fill $\frac{1}{x}$ of the bog in one day. In one day's time, the remaining sources could flood $\frac{1}{x + 1}$ and $\frac{1}{x + 4}$ of the bog. This gives the equation

The part of the bog the first source can flood in one day	plus	the part the second source can flood in one day	plus	the part the third source can flood in one day	equals	one bog.
$\frac{1}{x}$	$+$	$\frac{1}{x + 1}$	$+$	$\frac{1}{x + 4}$	$=$	1

We multiply both sides of the equation by $x(x + 1)(x + 4)$ to clear it of fractions and then simplify to get

$$x(x + 1)(x + 4)\left(\frac{1}{x} + \frac{1}{x + 1} + \frac{1}{x + 4}\right) = 1 \cdot x(x + 1)(x + 4)$$

$$(x + 1)(x + 4) + x(x + 4) + x(x + 1) = x(x + 1)(x + 4)$$

$$x^2 + 5x + 4 + x^2 + 4x + x^2 + x = x^3 + 5x^2 + 4x$$

$$0 = x^3 + 2x^2 - 6x - 4$$

To solve the equation $x^3 + 2x^2 - 6x - 4 = 0$, we first list its possible rational roots, which are the factors of the constant term, -4.

$$-4, \quad -2, \quad -1, \quad 1, \quad 2, \quad \text{and} \quad 4$$

One solution of this equation is $x = 2$, because when we synthetically divide by 2, the remainder is 0:

$$\underline{2|} \quad 1 \quad 2 \quad -6 \quad -4$$
$$ \quad \quad 2 \quad 8 \quad 4$$
$$\overline{ \quad 1 \quad 4 \quad 2 \quad 0} \quad \text{The depressed equation is } x^2 + 4x + 2 = 0.$$

We can find the remaining solutions by using the quadratic formula to solve the depressed equation. The two solutions are $-2 + \sqrt{2}$ and $-2 - \sqrt{2}$. Since both of these numbers are negative and the time it takes to flood the bog cannot be negative, these roots must be discarded. The only meaningful solution is 2.

Since the first source, alone, can flood the bog in two days, and it is three days until the freeze, the other two water sources can be diverted to flood other bogs.

Self Check 3   If a freeze is predicted in one day, can the water in the last two sources be diverted?

Self Check Answers   1. $\left\{1, 2, \frac{1}{3}\right\}$ 2. $\{-1, 1, 1, 2i, -2i\}$ 3. no

5.3   Exercises

Vocabulary and Concepts   *Fill in the blanks.*

1. The rational roots of the equation
$3x^3 + 4x - 7 = 0$ will have the form $\frac{p}{q}$, where p is
a factor of ___ and q is a factor of 3.

2. The rational roots of the equation
$5x^3 + 3x^2 - 4 = 0$ will have the form $\frac{p}{q}$, where p
is a factor of -4 and q is a factor of ___.

3. Consider the synthetic division of
$5x^3 - 7x^2 - 3x - 63 = 0$.

$$
\begin{array}{r|rrrr}
3 & 5 & -7 & -3 & -63 \\
  &   & 15 & 24 & 63 \\
\hline
  & 5 & 8 & 21 & 0
\end{array}
$$

Since the remainder is 0, 3 is a ___ of the
equation.

4. In Problem 3, the depressed equation is
_____.

Practice   *Use the rational root theorem to list all possible rational roots of the polynomial equation.*

5. $x^3 + 10x^2 + 5x - 12 = 0$

6. $-x^3 + 3x^2 - 4x - 8 = 0$

7. $2x^4 - x^3 + 10x^2 + 5x - 6 = 0$

8. $3x^4 - x^3 + 7x^2 - 5x - 8 = 0$

9. $4x^5 - x^4 - x^3 + x^2 + 5x - 10 = 0$

10. $6x^4 - 2x^3 + x^2 - x + 3 = 0$

Find all rational roots of each equation.

11. $x^3 - 5x^2 - x + 5 = 0$
12. $x^3 + 7x^2 - x - 7 = 0$
13. $x^3 - 2x^2 - x + 2 = 0$
14. $x^3 + x^2 - 4x - 4 = 0$
15. $x^3 - x^2 - 4x + 4 = 0$

16. $x^3 + 2x^2 - x - 2 = 0$
17. $x^3 - 2x^2 - 9x + 18 = 0$
18. $x^3 + 3x^2 - 4x - 12 = 0$
19. $2x^3 - x^2 - 2x + 1 = 0$
20. $3x^3 + x^2 - 3x - 1 = 0$
21. $3x^3 + 5x^2 + x - 1 = 0$
22. $2x^3 - 3x^2 + 1 = 0$
23. $x^4 - 10x^3 + 35x^2 - 50x + 24 = 0$
24. $x^4 + 4x^3 + 6x^2 + 4x + 1 = 0$
25. $x^4 + 3x^3 - 13x^2 - 9x + 30 = 0$
26. $x^4 - 8x^3 + 14x^2 + 8x - 15 = 0$
27. $x^5 + 3x^4 - 5x^3 - 15x^2 + 4x + 12 = 0$

28. $x^5 - 3x^4 - 5x^3 + 15x^2 + 4x - 12 = 0$

29. $x^7 - 12x^5 + 48x^3 - 64x = 0$

30. $x^7 + 7x^6 + 21x^5 + 35x^4 + 35x^3 + 21x^2$
$+ 7x + 1 = 0$

31. $3x^3 - 2x^2 + 12x - 8 = 0$
32. $4x^4 - 8x^3 - x^2 + 8x - 3 = 0$
33. $3x^4 - 14x^3 + 11x^2 + 16x - 12 = 0$
34. $2x^4 - x^3 - 2x^2 - 4x - 40 = 0$
35. $12x^4 + 20x^3 - 41x^2 + 20x - 3 = 0$
36. $4x^5 - 12x^4 + 15x^3 - 45x^2 - 4x + 12 = 0$

37. $6x^5 - 7x^4 - 48x^3 + 81x^2 - 4x - 12 = 0$

38. $36x^4 - x^2 + 2x - 1 = 0$
39. $30x^3 - 47x^2 - 9x + 18 = 0$
40. $20x^3 - 53x^2 - 27x + 18 = 0$
41. $15x^3 - 61x^2 - 2x + 24 = 0$
42. $12x^4 + x^3 + 42x^2 + 4x - 24 = 0$
43. $20x^3 - 44x^2 + 9x + 18 = 0$
44. $24x^3 - 82x^2 + 89x - 30 = 0$

Find all roots of each equation.

45. $x^3 - 3x^2 - 2x + 6 = 0$

46. $x^3 + 3x^2 - 3x - 9 = 0$

47. $2x^3 - x^2 + 2x - 1 = 0$

48. $3x^3 + x^2 + 3x + 1 = 0$

49. $x^4 - 2x^3 - 8x^2 + 8x + 16 = 0$

50. $x^4 - 2x^3 - 2x^2 + 2x + 1 = 0$

51. $2x^4 + x^3 + 17x^2 + 9x - 9 = 0$

52. $2x^4 - 4x^3 + 2x^2 + 4x - 4 = 0$

53. $x^5 - 3x^4 + 28x^3 - 76x^2 + 75x - 25 = 0$

54. $x^5 + 3x^4 - 2x^3 - 14x^2 - 15x - 5 = 0$

*In Exercises 55–58, $1 + i$ is a root of each equation.
Find the other roots.*

55. $x^3 - 5x^2 + 8x - 6 = 0$

56. $x^3 - 2x + 4 = 0$

57. $x^4 - 2x^3 - 7x^2 + 18x - 18 = 0$

58. $x^4 - 2x^3 - 2x^2 + 8x - 8 = 0$

Solve each equation.

59. $x^3 - \dfrac{4}{3}x^2 - \dfrac{13}{3}x - 2 = 0$

60. $x^3 - \dfrac{19}{6}x^2 + \dfrac{1}{6}x + 1 = 0$

61. $x^{-5} - 8x^{-4} + 25x^{-3} - 38x^{-2} + 28x^{-1} - 8 = 0$

62. $1 - x^{-1} - x^{-2} - 2x^{-3} = 0$

Applications

63. Parallel resistance If three resistors with resistances of R_1, R_2, and R_3 are wired in parallel, their combined resistance R is given by the following formula.

$$\frac{1}{R} = \frac{1}{R_1} + \frac{1}{R_2} + \frac{1}{R_3}$$

The design of a voltmeter requires that the resistance R_2 be 10 ohms greater than the resistance R_1, that the resistance R_3 be 50 ohms greater than R_1, and that their combined resistance be 6 ohms. Find the value of each resistance.

64. Fabricating sheet metal The open tray shown in the illustration is to be manufactured from a 12-by-14-inch rectangular sheet of metal by cutting squares from each corner and folding up the sides. If the volume of the tray is to be 160 cubic inches and x is to be an integer, what size squares should be cut from each corner?

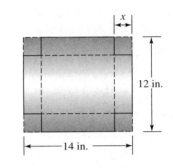

65. Packaging The length of a FedEx 25 kg box is 7 inches more than its height. The width of the box is 4 inches more than its height. If the volume of the box is 4,420 cubic inches, find the height of the box.

66. Pop cans A Mountain Dew aluminum can is approximately the shape of a cylinder. If the height of the can is 9 centimeters more than its radius and the volume of the can is approximately 108π cubic centimeters, find the radius of the can. The formula for the volume of a cylinder is $V = \pi r^2 h$.

Discovery and Writing

67. If n is an even integer and c is a positive constant, show that $x^n + c = 0$ has no real roots.

68. If n is an even positive integer and c is a positive constant, show that $x^n - c = 0$ has two real roots.

69. Precalculus A rectangle is inscribed in the parabola $y = 16 - x^2$, as shown in the illustration. Find the point (x, y) if the area of the rectangle is 42 square units.

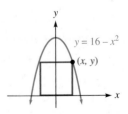

70. Precalculus One corner of the rectangle shown is at the origin, and the opposite corner (x, y) lies in the first quadrant on the curve $y = x^3 - 2x^2$. Find the point (x, y) if the area of the rectangle is 27 square units.

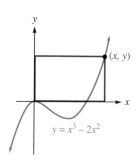

73. $\sqrt{18a^2b} + a\sqrt{50b}$

74. $\left(\sqrt{3} + \sqrt{5b}\right)\left(\sqrt{12} - \sqrt{45b}\right)$

75. $\dfrac{2}{\sqrt{3} - 1}$

76. $\dfrac{\sqrt{11} + \sqrt{x}}{\sqrt{11} - \sqrt{x}}$

Review *Simplify each radical expression. Assume that all variables represent positive numbers.*

71. $\sqrt{72a^3b^5c}$

72. $\dfrac{5a}{\sqrt{5a}}$

5.4

Approximating Irrational Roots of Polynomial Equations

Objectives

1. Use the Intermediate Value Theorem
2. Use the Bisection Method
3. Approximate Real Roots with a Graphing Calculator

Sometimes we find ourselves in a car wondering how to get somewhere. Fortunately, many cars now are equipped with a navigation system that can give us directions to a specific address or point of interest.

A navigation system uses satellite signals to determine the exact location of our car. Once the system knows the location of the car, it can guide us to our destination.

When solving more-complicated equations, there are several theorems that act as a navigation system does and can tell us the locations of the roots of the equation. Once we know their locations, it is much easier to find them.

Before developing these theorems, we will review some facts about solving equations.

• All first-degree equations are easy to solve by using simple algebraic methods.

• All quadratic equations can be solved by using the quadratic formula.

• There are formulas for solving third- and fourth-degree polynomial equations. However, they are complicated.

• No formulas exist for solving polynomial equations of degree 5 or greater. This fact was proved by the Norwegian mathematician Niels Henrik Abel (1802–1829) and (for equations of degree greater than 5) by the French mathematician Evariste Galois (1811–1832).

To solve a higher-degree polynomial equation with integer coefficients, we can use the methods of the previous section to find its *rational roots*. To find any irrational or complex roots, the final depressed equation would have to be a first- or second-degree equation.

In this section, we will discuss ways of approximating *irrational roots* of higher-degree polynomial equations that may not have any rational roots.

EXAMPLE 1 Show that $\sqrt{3}$ is a root of $x^4 - 9 = 0$ and that it is irrational.

Solution We know that $\sqrt{3}$ is a root of the equation because

$$x^4 - 9 = 0$$
$$\left(\sqrt{3}\right)^4 - 9 \stackrel{?}{=} 0 \qquad \text{Substitute } \sqrt{3} \text{ for } x.$$
$$9 - 9 \stackrel{?}{=} 0 \qquad \left(\sqrt{3}\right)^4 = 9$$
$$0 = 0$$

If there are any rational roots of this equation, they must have the form $\frac{\text{factor of the constant } 9}{\text{factor of the leading coefficient } 1}$. Therefore, the only possible rational roots of the equation would be

$$\pm\frac{1}{1} \qquad \text{and} \qquad \pm\frac{9}{1}$$

Since none of the numbers $1, -1, 9,$ or -9 satisfies the equation, no rational number can be a root. Thus, the root of $\sqrt{3}$ must be irrational.

Self Check 1 Show that $-\sqrt{5}$ is a root of $x^4 - 25 = 0$ and that it is irrational.

Comment

Roots of polynomial equations can be rational numbers, irrational numbers, or complex numbers. In this section, we will restrict ourselves to roots that are real. That is, rational or irrational numbers.

1. Use the Intermediate Value Theorem

The following theorem, called the **intermediate value theorem,** leads to a way of locating an interval that contains a root.

The Intermediate Value Theorem Let $P(x)$ be a polynomial with real coefficients. If $P(a) \neq P(b)$ for $a < b$, then $P(x)$ takes on all values between $P(a)$ and $P(b)$ on the closed interval $[a, b]$.

Justification This theorem becomes clear when we consider the graph of the polynomial $y = P(x)$, shown in Figure 5-3. We have seen that graphs of polynomials are *continuous* curves, a technical term that means, roughly, that they can be drawn without lifting the pencil from the paper. If $P(a) \neq P(b)$, the continuous curve joining the points $A(a, P(a))$ and $B(b, P(b))$ must take on all values between $P(a)$ and $P(b)$ in the interval $[a, b]$, because the curve has no gaps in it.

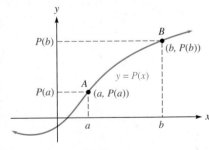

Figure 5-3

The next theorem follows from the intermediate value theorem.

| **The Location Theorem** | Let $P(x)$ be a polynomial with real coefficients. If $P(a)$ and $P(b)$ have opposite signs, there is at least one number r in the interval (a, b) for which $P(r) = 0$. |

Proof   See Figure 5-4. By the intermediate value theorem, $P(x)$ takes on all values between $P(a)$ and $P(b)$. Since $P(a)$ and $P(b)$ have opposite signs, the number 0 lies between them. Thus, there is a number r between a and b for which $P(r) = 0$. This number r is a zero of $P(x)$, and a root of the equation $P(x) = 0$.

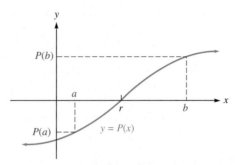

Figure 5-4

EXAMPLE 2   Show that $2x^3 - x^2 - 8x + 4 = 0$ has at least one real root between 0 and 1.

Solution   We will evaluate the polynomial at $x = 0$ and $x = 1$. If the resulting values have opposite signs, we know that the equation has a root that lies between 0 and 1.
   First, we let $P(x) = 2x^3 - x^2 - 8x + 4$ and evaluate $P(0)$ and $P(1)$ as follows:

$$P(0) = 2(0)^3 - (0)^2 - 8(0) + 4 = 4$$
$$P(1) = 2(1)^3 - (1)^2 - 8(1) + 4 = -3$$

Because $P(0)$ and $P(1)$ have opposite signs, we know by the location theorem that there is at least one real root between 0 and 1, as shown in Figure 5-5.

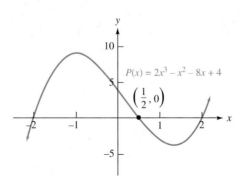

Figure 5-5

Self Check 2   Show that $2x^3 - 9x^2 + 7x + 6 = 0$ has at least one real root between -1 and 0.

2. Use the Bisection Method

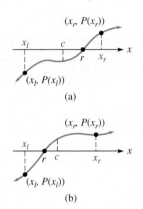

(a)

(b)

Figure 5-6

The location theorem provides a method (called the **bisection method**) for finding the roots of $P(x) = 0$ to any desired degree of accuracy.

Suppose we find, by trial and error, that the numbers x_l and x_r (for left and right) straddle a root—that is, $x_l < x_r$ and $P(x_l)$ and $P(x_r)$ have opposite signs. (See Figure 5-6.) Further suppose that $P(x_l) < 0$ and $P(x_r) > 0$. We can compute the number c that is halfway between x_l and x_r and then compute $P(c)$. If $P(c) = 0$, we've found a root. However, if $P(c)$ is not 0, we proceed in one of two ways:

1. If $P(c) < 0$, the root r lies between c and x_r, as shown in Figure 5-6(a). In this case, we let c become a new x_l and repeat the procedure.

2. If $P(c) > 0$, the root r lies between x_l and c, as shown in Figure 5-6(b). In this case, we let c become a new x_r and repeat the procedure.

At any stage in this procedure, the root is contained between the current values of x_l and x_r. If the original bounds were 1 unit apart, after 10 repetitions of this procedure, the root would be between bounds that were 2^{-10} units apart. This means that the zero of $P(x)$ would be within 0.0001 units of either x_l or x_r.

After 20 repetitions, the root would be between bounds that were 2^{-20} units apart. This means that the zero of $P(x)$ would be within 0.000001 of either x_l or x_r.

EXAMPLE 3   Use the bisection method to find the positive root of $x^2 - 2 = 0$ to the nearest tenth.

Solution   We will use the location theorem and the bisection method to find the positive root of the equation to the nearest tenth.

Since $P(1) = -1$ and $P(2) = 2$ have opposite signs, there is a root between 1 and 2. We can let $x_l = 1$ and $x_r = 2$ and compute the midpoint c:

$$c = \frac{x_l + x_r}{2} = \frac{1 + 2}{2} = 1.5$$

Because $P(c) = P(1.5) = 0.25$ is a positive number, we let c become a new x_r and calculate a new midpoint, which we will call c_1.

$$c_1 = \frac{x_l + x_r}{2} = \frac{1 + 1.5}{2} = 1.25$$

Because $P(c_1) = P(1.25) = -0.4375$ is a negative number, we let c_1 become a new x_l and calculate a new midpoint, which we will call c_2.

$$c_2 = \frac{x_l + x_r}{2} = \frac{1.25 + 1.5}{2} = 1.375$$

Because $P(c_2) = P(1.375) = -0.109375$ is a negative number, we let c_2 become a new x_l and calculate a new midpoint, which we will call c_3.

$$c_3 = \frac{x_l + x_r}{2} = \frac{1.375 + 1.5}{2} = 1.4375$$

Because $P(c_3) = P(1.4375) = 0.066406$ is a positive number, we let c_3 become a new x_r and calculate a new midpoint, which we will call c_4.

$$c_4 = \frac{x_l + x_r}{2} = \frac{1.375 + 1.4375}{2} = 1.40625$$

From here on, the first two digits of the midpoints will remain 1.4. The root r of the equation (to the nearest tenth) is 1.4.

Self Check 3   Find the negative root of $x^2 - 2 = 0$ to the nearest tenth.

3. Approximate Real Roots with a Graphing Calculator

We can approximate the real roots of an equation either by using the TRACE and ZOOM capabilities of a graphing calculator or by using the ZERO feature.

Accent on Technology

Solving Equations

To use the TRACE and ZOOM capabilities of a graphing calculator to solve the equation $x^4 - 6x^2 + 9 = 0$, we graph the function $y = x^4 - 6x^2 + 9$ and trace and zoom to find the x-intercepts. If we use window settings of $[-6, 6]$ for x and $[-2, 10]$ for y and graph the function, we will obtain Figure 5-7(a). If we trace and move the cursor near the positive x-intercept, we will obtain Figure 5-7(b). If we zoom in and trace, we will obtain Figure 5-7(c). To get better results, we can zoom in and trace again to obtain Figure 5-7(d), which shows that the x-coordinate of the x-intercept is approximately 1.731383. After zooming in and tracing a few more times, we will see that to three decimal places, $x = 1.732$. By symmetry, we know that the negative solution is -1.732. Note that we cannot find complex roots from the graph.

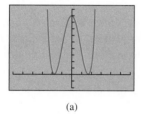
(a)

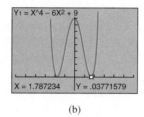
(b)

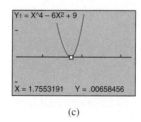
(c)

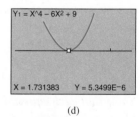
(d)

Figure 5-7

To use the ZERO feature, we first graph the equation to obtain Figure 5-8(a). We then select ZERO under the CALC menu to obtain Figure 5-8(b). To find the positive x-intercept, we guess a left bound of 0 and press ENTER , guess a right bound of 4 and press ENTER , and press ENTER again to obtain Figure 5-8(c), which shows that the x-coordinate of the x-intercept is approximately 1.732051.

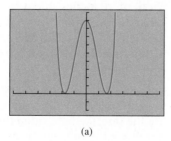
(a)

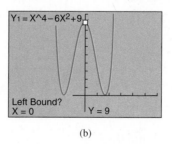
(b)

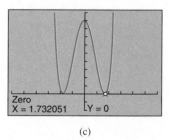
(c)

Figure 5-8

The graph shown in Figure 5-8(a) illustrates a situation in which the bisection method does not work. Since the graph doesn't cross the x-axis, the values of the functions on opposite sides of the intercepts do not have opposite signs. In this case, the bisection method would be useless.

5.4   Exercises

Vocabulary and Concepts   *Fill in the blanks.*

1. If $P(x)$ is a polynomial with real coefficients and $P(a) \neq P(b)$ for $a < b$, then $P(x)$ takes on all values between _____ in the interval $[a, b]$.

2. If $P(x)$ has real coefficients and $P(a)$ and $P(b)$ have opposite signs, there is at least one number r in (a, b) for which _____.

3. In the bisection method, we find two numbers x_l and x_r that straddle a root. We then compute a new guess for the root (a number c) by finding the average of _____.

4. A _____ curve has no gaps in it.

Practice   *Show that each equation has irrational roots.*

5. $x^4 - 49 = 0$

6. $x^4 - 64 = 0$

Show that each equation has at least one real root between the specified numbers.

7. $2x^2 + x - 3 = 0$; -2 and -1

8. $2x^3 + 17x^2 + 31x - 20 = 0$; -1 and 2

9. $3x^3 - 11x^2 - 14x = 0$; 4 and 5

10. $2x^3 - 3x^2 + 2x - 3 = 0$; 1 and 2

11. $x^4 - 8x^2 + 15 = 0$; 1 and 2

12. $x^4 - 8x^2 + 15 = 0$; 2 and 3

13. $30x^3 + 10 = 61x^2 + 39x$; 2 and 3

14. $30x^3 + 10 = 61x^2 + 39x$; -1 and 0

15. $30x^3 + 10 = 61x^2 + 39x$; 0 and 1

16. $5x^3 - 9x^2 - 4x + 9 = 0$; -1 and 0

Use the bisection method to find the following values to the nearest tenth.

17. The positive root of $x^2 - 3 = 0$.

18. The negative root of $x^2 - 3 = 0$.

19. The negative root of $x^2 - 5 = 0$.

20. The positive root of $x^2 - 5 = 0$.

21. The positive root of $x^3 - x^2 - 2 = 0$.

22. The negative root of $x^3 - x + 2 = 0$.

23. The negative root of $3x^4 + 3x^3 - x^2 - 4x - 4 = 0$.

24. The positive root of $x^5 + x^4 - 4x^3 - 4x^2 - 5x - 5 = 0$.

Use a graphing calculator to find the distinct real solutions of each equation to the nearest tenth. Which roots, if any, would the bisection method fail to find?

25. $x^2 - 5 = 0$

26. $x^2 - 10x + 25 = 0$

27. $x^3 - 5x^2 + 8x - 4 = 0$

28. $x^3 - 5x^2 - 2x + 10 = 0$

Applications

29. **Containers**   A box has a length of 16 inches, a width of 10 inches, and a height of between 4 inches and 8 inches. Can it have a volume of 1,000 in.^{3}?   (*Hint:* Model the volume of the box with a function of h and use the intermediate value theorem.)

30. **Lollipops**   If a candy company makes spherically shaped lollipops with radii between 1 and 4 cm, use the intermediate value theorem to determine whether a lollipop can be made with a volume of 200 cubic centimeters.   (*Hint:* The volume formula for a sphere is $V = \frac{4}{3}\pi r^3$.)

31. **Precalculus**   Use the bisection method or a graphing calculator to find the coordinates of the two points on the graph of $y = x^3$ that lie 1 unit from the origin. (See the illustration.) Give the result to the nearest hundredth.

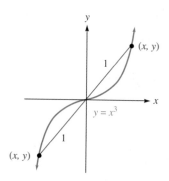

$y = x^3$

32. Building crates The width of the shipping crate shown is to be 2 feet greater than its height and the length is to be 5 feet greater than its width, and its volume is to be 170 cubic feet. Use the bisection method or a graphing calculator to find the height of the crate to the nearest tenth of a foot.

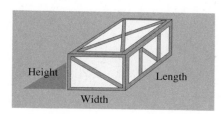

Discovery and Writing

33. Explain when the bisection method will not work.

34. Describe how you could tell from a graph that a polynomial equation has no real roots.

Review *Give the equations of all of the asymptotes of each rational function. Don't graph the functions.*

35. $y = \dfrac{5x - 3}{x^2 - 4}$

36. $y = \dfrac{5x^2 - 3}{x^2 - 25}$

37. $y = \dfrac{x^2}{x - 2}$

38. $y = \dfrac{x + 1}{x^2 + 1}$

CHAPTER REVIEW

5.1 The Remainder and Factor Theorems; Synthetic Division

Definitions and Concepts	Examples
A **polynomial equation** is an equation that can be written in the form $P(x) = 0$, where $$P(x) = a_n x^n + a_{n-1}x^{n-1} + a_{n-2}x^{n-2} + \cdots + a_1 x + a_0$$ where n is a natural number and the polynomial is of degree n.	$3x - 10 = 0$; degree: 1 $7x^2 + 3x - 1 = 0$; degree: 2 $4x^3 + 7x^2 + 3x - 1 = 0$; degree: 3 $x^4 + 7x^2 - 4x^3 - 1 = 0$; degree: 4
Zero of a polynomial function: A **zero of the polynomial** $P(x)$ is any number r for which $P(r) = 0$.	The real number 1 is a zero of the polynomial function $P(x) = x^3 - 4x^2 + 3x$ because $$P(1) = 1^3 - 4(1)^2 + 3(1)$$ $$= 1 - 4 + 3$$ $$= 0$$

The remainder theorem:
If $P(x)$ is a polynomial, r is any number, and $P(x)$ is divided by $x - r$, the remainder is $P(r)$.

Let $P(x) = 3x^3 - 5x^2 + 3x - 10$ and use the remainder theorem and long division to find $P(2)$.

We will divide $P(x)$ by $x - 2$ as follows:

$$
\begin{array}{r}
3x^2 + x + 5 \\
x - 2 \overline{\smash{)}3x^3 - 5x^2 + 3x - 10} \\
\underline{3x^3 - 6x^2} \\
x^2 + 3x \\
\underline{x^2 - 2x} \\
5x - 10 \\
\underline{5x - 10} \\
0
\end{array}
$$

The remainder is 0. $P(2) = 0$.

The factor theorem:
If $P(x)$ is a polynomial and r is any number, then if $P(r) = 0$, $x - r$ is a factor of $P(x)$. If $x - r$ is a factor of $P(x)$, then $P(r) = 0$.

Alternate form of the factor theorem:
If r is a zero of the polynomial $P(x)$, then $x - r$ is a factor of $P(x)$. If $x - r$ is a factor of $P(x)$, then r is a zero of the polynomial.

We see in the previous example that $P(2) = 0$. By the factor theorem, we know that $x - 2$ is a factor of the polynomial $P(x) = 3x^3 - 5x^2 + 3x - 10$. We also know that 2 is a zero of the polynomial.

Synthetic division is a fast way to divide higher-degree polynomials by binomials of the form $x - r$.

Divide $P(x) = 5x^3 + 3x^2 - 21x - 1$ by $x - 2$ using synthetic division.

We first write the coefficients of the dividend, with its terms in descending powers of x, and the 2 from the divisor in the following form:

$$\underline{2|}\quad 5 \quad 3 \quad -21 \quad -1$$

$$
\begin{array}{c|rrrr}
\underline{2|} & 5 & 3 & -21 & -1 \\
\hline
& 5 &
\end{array}
\qquad \text{Bring down the 5.}
$$

$$
\begin{array}{c|rrrr}
\underline{2|} & 5 & 3 & -21 & -1 \\
& & 10 & & \\
\hline
& 5 & 13 &
\end{array}
\qquad 2(5) = 10; \ 3 + 10 = 13
$$

$$
\begin{array}{c|rrrr}
\underline{2|} & 5 & 3 & -21 & -1 \\
& & 10 & 26 & \\
\hline
& 5 & 13 & 5 &
\end{array}
\qquad 2(13) = 26; \ -21 + 26 = 5
$$

$$
\begin{array}{c|rrrr}
\underline{2|} & 5 & 3 & -21 & -1 \\
& & 10 & 26 & 10 \\
\hline
& 5 & 13 & 5 & 9
\end{array}
\qquad 2(5) = 10; \ -1 + 10 = 9
$$

The quotient is $5x^2 + 13x + 5$. The remainder is 9. Thus,

$$\frac{5x^3 + 3x^2 - 21x - 1}{x - 2} = 5x^2 + 13x + 5 + \frac{9}{x - 2}$$

Because the remainder is 9, we know by the remainder theorem that $P(2) = 9$.

Because $P(2) \neq 0$, we know by the factor theorem, that $x - 2$ is not a factor of $5x^3 + 3x^2 - 21x - 1$.

Synthetic division can be used to solve polynomial equations.

Let $P(x) = x^3 + x^2 - 3x - 3$. Completely solve the polynomial equation $P(x) = 0$ given that -1 is a root.

1. We use synthetic division to divide $P(x)$ by $x + 1$.

$$\begin{array}{r|rrrr} -1 & 1 & 1 & -3 & -3 \\ & & -1 & 0 & 3 \\ \hline & 1 & 0 & -3 & 0 \end{array}$$

2. We then write the quotient and factor it.

$$x^3 + x^2 - 3x - 3 = (x + 1)(x^2 - 3)$$

3. Finally, we solve the polynomial equation $P(x) = 0$.

$$x^3 + x^2 - 3x - 3 = 0$$
$$(x + 1)(x^2 - 3) = 0$$

We then set each factor equal to 0 and solve for x.

$$\begin{array}{ccc} x + 1 = 0 & \text{or} & x^2 - 3 = 0 \\ x = -1 & & x^2 = 3 \\ & & x = \pm\sqrt{3} \end{array}$$

The solution set is $\{-1, \sqrt{3}, -\sqrt{3}\}$.

Exercises

Let $P(x) = 4x^4 + 2x^3 - 3x^2 - 2$. Find the remainder when $P(x)$ is divided by each binomial.

1. $x - 1$ **2.** $x - 2$

3. $x + 3$ **4.** $x + 2$

Use the factor theorem to determine whether each statement is true.

5. $x - 2$ is a factor of $x^3 + 4x^2 - 2x + 4$.

6. $x + 3$ is a factor of $2x^4 + 10x^3 + 4x^2 + 7x + 21$.

7. $x - 5$ is a factor of $x^5 - 3{,}125$.

8. $x - 6$ is a factor of $x^5 - 6x^4 - 4x + 24$.

Use synthetic division to perform each division.

9. $3x^4 + 2x^2 + 3x + 7; x - 3$

10. $2x^4 - 3x^2 + 3x - 1; x - 2$

11. $5x^5 - 4x^4 + 3x^3 - 2x^2 + x - 1; x + 2$

12. $4x^5 + 2x^4 - x^3 + 3x^2 + 2x + 1; x + 1$

Let $P(x) = 5x^3 + 2x^2 - x + 1$. Use synthetic division to find each value.

13. $P(3)$ **14.** $P(-3)$

15. $P\left(\dfrac{1}{2}\right)$ **16.** $P(i)$

A partial solution set is given for each equation. Find the complete solution set.

17. $2x^3 - 3x^2 - 11x + 6 = 0$; $\{3\}$

18. $x^4 + 4x^3 - x^2 - 20x - 20 = 0$; $\{-2, -2\}$

Find the polynomial of lowest degree with integer coefficients and the given zeros.

19. $-1, 2,$ and $\dfrac{3}{2}$

20. $1, -3,$ and $\dfrac{1}{2}$

21. $2, -5, i,$ and $-i$

22. $-3, 2, i,$ and $-i$

5.2   Descartes' Rule of Signs and Bounds on Roots

Definitions and Concepts	Examples
The fundamental theorem of algebra: If $P(x)$ is a polynomial with positive degree, then $P(x)$ has at least one zero.	The fundamental theorem of algebra guarantees that polynomials such as $3x^3 - 5x^2 + 7x - 2$ and $4x^5 + ix^3 - (2 + i)x - 15$ have zeros.
The polynomial factorization theorem: If $n > 0$ and $P(x)$ is an nth-degree polynomial, then $P(x)$ has exactly n linear factors: $\quad P(x) = a_n(x - r_1)(x - r_2)(x - r_3) \cdot \,\cdots\, \cdot (x - r_n)$ where $r_1, r_2, r_3, \ldots, r_n$ are numbers and a_n is the leading coefficient of $P(x)$. If multiple roots are counted individually, the polynomial equation $P(x) = 0$ with degree n $\ (n > 0)$ has exactly n roots among the complex numbers.	Consider the following polynomial of degree $n = 3$. $\begin{aligned} P(x) &= x^3 + 144x \\ &= x(x^2 + 144) \\ &= x(x - 12i)(x + 12i) \end{aligned}$ The polynomial factorization theorem guarantees that we have exactly 3 linear factors. They are $\quad x, \qquad x - 12i, \qquad$ and $\qquad x + 12i$ The polynomial equation $P(x) = x^3 + 144x = 0$ has exactly 3 roots. They are $\quad 0, \qquad 12i, \qquad$ and $\qquad -12i$ Each root occurs once and has a multiplicity of one.
The conjugate pairs theorem: If a polynomial equation $P(x) = 0$ with real-number coefficients has a complex root $a + bi$ with $b \neq 0$, then its conjugate $a - bi$ is also a root.	The conjugate pairs theorem applies to the $P(x) = x^3 + 144x = 0$ because the polynomial has real coefficients. The complex roots $10i$ and $-10i$ that occur are conjugate pairs.
Descartes' rule of signs: If $P(x)$ is a polynomial with real coefficients, the number of positive roots of $P(x) = 0$ is either equal to the number of variations in sign of $P(x)$ or less than that by an even number. The number of negative roots of $P(x) = 0$ is either equal to the number of variations in sign of $P(-x)$ or less than that by an even number.	Use Descartes' rule of signs to find the number of possible positive, negative, and nonreal roots of $P(x) = 5x^3 - 7x^2 + x - 6 = 0$. **1. Find the number of possible positive roots.** Since there are three variations of sign in $P(x) = 5x^3 - 7x^2 + x - 6 = 0$, there can be either 3 positive roots or only 1. (1 is less than 3 by the even number 2.) **2. Find the number of possible negative roots.** Because $\begin{aligned} P(-x) &= 5(-x)^3 - 7(-x)^2 + (-x) - 6 \\ &= -5x^3 - 7x^2 - x - 6 \end{aligned}$ has no variations in sign, there are 0 negative roots. Furthermore, 0 is not a root, because the terms of the polynomial do not have a common factor of x.

3. **Find the number of nonreal roots.**

   If there are 3 positive roots, then all of the roots are accounted for. If there is 1 positive root, the 2 remaining roots must be nonreal complex numbers. The following chart shows these possibilities.

Number of positive roots	Number of negative roots	Number of nonreal roots
3	0	0
1	0	2

The number of nonreal complex roots is the number needed to bring the total number of roots up to 3.

Upper and lower bounds on roots:

Let $P(x)$ be a polynomial with real coefficients and a positive leading coefficient.

1. If $P(x)$ is synthetically divided by a positive number c and each term in the last row of the division is nonnegative, then no number greater than c can be a root of $P(x) = 0$. (c is called an **upper bound** of the real roots.)

2. If $P(x)$ is synthetically divided by a negative number d and the signs in the last row alternate,* no value less than d can be a root of $P(x) = 0$. (d is called a **lower bound** of the real roots.)

*If 0 appears in the third row, that 0 can be assigned either a + or a − sign to help the signs alternate.

Use the upper and lower bounds rules to establish bounds for the roots of $2x^3 + 2x^2 - 8x - 8 = 0$.

1. We will perform several synthetic divisions by positive integers, looking for nonnegative values in the last row.

   Trying 1 first gives

$$
\begin{array}{r|rrrr}
1 & 2 & 2 & -8 & -8 \\
  &   & 2 & 4 & -4 \\
\hline
  & +2 & +4 & -4 & -12
\end{array}
$$

   Because one of the signs in the last row is negative, we cannot claim that 1 is an upper bound. We now try 2.

$$
\begin{array}{r|rrrr}
2 & 2 & 2 & -8 & -8 \\
  &   & 4 & 12 & 8 \\
\hline
  & +2 & +6 & +4 & 0
\end{array}
$$

   Because the last row is entirely nonnegative, we can claim that 2 is an upper bound. That is, no number greater than 2 can be a root of the equation. In fact, 2 is a root because the remainder is 0.

2. We perform several synthetic divisions by negative integers, looking for alternating signs in the last row. We begin with −3.

$$
\begin{array}{r|rrrr}
-3 & 2 & 2 & -8 & -8 \\
   &   & -6 & 12 & -12 \\
\hline
   & +2 & -4 & +4 & -20
\end{array}
$$

   Since the signs in the last row alternate, −3 is a lower bound. That is, no number less than −3 can be a root. All of the real roots must be in the interval $(-3, 2)$.

Exercises

How many roots does each equation have?

23. $3x^6 - 4x^5 + 3x + 2 = 0$

24. $2x^6 - 5x^4 + 5x^3 - 4x^2 + x - 12 = 0$

25. $3x^{65} - 4x^{50} + 3x^{17} + 2x = 0$

26. $x^{1,984} - 12 = 0$

Determine how many linear factors and zeros each polynomial function has.

27. $P(x) = x^4 - 16$

28. $P(x) = x^{40} + x^{30}$

29. $P(x) = 4x^5 + 2x^3$

30. $P(x) = x^3 - 64x$

Find another root of a polynomial equation with real coefficients if the given quantity is one root.

31. $2 + i$ **32.** $-i$

Write a third-degree polynomial equation with real coefficients and the given roots.

33. $4, -i$ **34.** $-5, i$

Find the number of possible positive, negative, and nonreal roots for each equation. Do not attempt to solve the equation.

35. $3x^4 + 2x^3 - 4x + 2 = 0$

36. $2x^4 - 3x^3 + 5x^2 + x - 5 = 0$

37. $4x^5 + 3x^4 + 2x^3 + x^2 + x = 7$

38. $3x^7 - 4x^5 + 3x^3 + x - 4 = 0$

39. $x^4 + x^2 + 24{,}567 = 0$

40. $-x^7 - 5 = 0$

Find integer bounds for the roots of each equation.

41. $5x^3 - 4x^2 - 2x + 4 = 0$

42. $x^4 + 3x^3 - 5x^2 - 9x + 1 = 0$

5.3   Roots of Polynomial Equations

Definitions and Concepts

Rational root theorem:

Let the polynomial equation

$$P(x) = a_n x^n + a_{n-1} x^{n-1} + a_{n-2} x^{n-2} + \cdots$$
$$+ a_1 x + a_0 = 0$$

have integer coefficients. If the rational number $\frac{p}{q}$ (written in lowest terms) is a root of $P(x) = 0$, then p is a factor of the constant a_0, and q is a factor of the leading coefficient a_n.

Examples

Consider $2x^3 - 7x^2 - 17x + 10 = 0$.

By the rational root theorem, the only possible numerators for the rational roots of the equation are the factors of the constant term 10:

$$\pm 1, \quad \pm 2, \quad \pm 5, \quad \text{and} \quad \pm 10$$

The only possible denominators are the factors of the leading coefficient 2:

$$\pm 1 \quad \text{and} \quad \pm 2$$

We can form a list of all possible rational solutions by listing the combinations of possible numerators and denominators:

$$\pm \frac{1}{1}, \quad \pm \frac{2}{1}, \quad \pm \frac{5}{1}, \quad \pm \frac{10}{1}, \quad \pm \frac{1}{2}, \quad \pm \frac{2}{2}, \quad \pm \frac{5}{2}, \quad \pm \frac{10}{2}$$

Since several of these possibilities are duplicates, we can condense the list to get

Possible rational roots

$$\pm 1, \quad \pm 2, \quad \pm 5, \quad \pm 10, \quad \pm \frac{1}{2}, \quad \text{and} \quad \pm \frac{5}{2}$$

Finding rational roots:

1. Use Descartes' rule of signs to determine the number of possible positive, negative, and nonreal roots.

2. Use the rational root theorem to list possible rational roots.

3. Use synthetic division to find a root.

4. If there are more roots, repeat the previous steps.

(Once the depressed equation is quadratic, synthetic division isn't required. It can be solved as a quadratic equation.)

Find the solution set of the polynomial equation $P(x) = 2x^3 - 7x^2 - 17x + 10 = 0$.

1. To determine the number of possible positive, negative, and nonreal roots, we will use Descartes' rule of signs.

   Since there are two variations in sign of $P(x)$, the number of positive roots is either 2 or 0.
   Next, we consider $P(-x)$.

   $$P(-x) = -2x^3 - 7x^2 + 17x + 10$$

   Since there is one variation in sign of $P(-x)$, the number of negative roots is 1.
   The following table shows these possibilities:

Number of positive roots	Number of negative roots	Number of nonreal roots
2	1	0
0	1	2

2. List the possible rational roots.
   The possible rational roots were found in the previous example. They are:

   Possible rational roots

   $$\pm 1, \quad \pm 2, \quad \pm 5, \quad \pm 10, \quad \pm\frac{1}{2}, \quad \text{and} \quad \pm\frac{5}{2}$$

3. Use synthetic division to find a root. We can start with -2.

   $$\begin{array}{r|rrrr} -2 & 2 & -7 & -17 & 10 \\ & & -4 & 22 & -10 \\ \hline & 2 & -11 & 5 & 0 \end{array}$$

   Since the remainder is 0, -2 is a root and the binomial $x + 2$ is a factor of $P(x)$.

4. The remaining roots must be supplied by the remaining factor, which is the quotient $2x^2 - 11x + 5$. We can find the other roots by solving the depressed equation $2x^2 - 11x + 5 = 0$.

   $2x^2 - 11x + 5 = 0$

   $(2x - 1)(x - 5) = 0$ Factor.

   $2x - 1 = 0$ or $x - 5 = 0$

   $\qquad 2x = 1 \qquad\qquad x = 5$

   $\qquad x = \frac{1}{2}$

The solution set of the equation is $\left\{-2, \frac{1}{2}, 5\right\}$.

   Note that two positive roots and one negative root is a predicted possibility.

Exercises

Use the rational root theorem to list all possible rational roots of the polynomial equation.

43. $2x^4 + x^3 - 3x^2 - 5x - 6 = 0$

44. $4x^5 - 2x^4 + 3x^3 - 5x - 10 = 0$

Find all rational roots of each equation.

45. $x^3 - 10x^2 + 29x - 20 = 0$

46. $x^3 - 8x^2 - x + 8 = 0$

47. $2x^3 + 17x^2 + 41x + 30 = 0$

48. $3x^3 + 2x^2 + 2x - 1 = 0$

49. $4x^4 - 25x^2 + 36 = 0$

50. $2x^4 - 11x^3 - 6x^2 + 64x + 32 = 0$

Find all roots of each equation.

51. $3x^3 - x^2 + 48x - 16 = 0$

52. $x^4 - 2x^3 - 9x^2 + 8x + 20 = 0$

| **5.4** | **Approximating Irrational Roots of Polynomial Equations** |

Definitions and Concepts	**Examples**
The intermediate value theorem: Let $P(x)$ be a polynomial with real coefficients. If $P(a) \neq P(b)$ for $a < b$, then $P(x)$ takes on all values between $P(a)$ and $P(b)$ on the closed interval $[a, b]$.	Consider the polynomial $P(x) = 4x^3 + 2x^2 - 12x + 3$ and note that $$P(0) = 4(0)^3 + 2(0)^2 - 12(0) + 3 = 3$$ $$P(1) = 4(1)^3 + 2(1)^2 - 12(1) + 3 = -3$$ Because $P(0) \neq P(1)$ and $0 < 1$, the intermediate value theorem guarantees that $P(x)$ takes on all values between 3 and -3 on the closed interval $[0, 1]$.
The location theorem: Let $P(x)$ be a polynomial with real coefficients. If $P(a)$ and $P(b)$ have opposite signs, there is at least one number r in the interval (a, b) for which $P(r) = 0$.	To show that the polynomial equation, $4x^3 + 2x^2 - 12x + 3 = 0$ has at least one real root between 0 and 1, we see from above that $P(0) = 3$ and $P(1) = -3$ have opposite signs. Therefore, we know by the location theorem that there is at least one real root between 0 and 1, as shown in the figure. 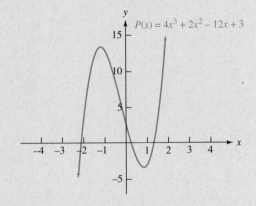

Bisection method:

The bisection method is a method for finding the roots of $P(x) = 0$ to any desired degree of accuracy.

Suppose we find, by trial and error, that the numbers x_l and x_r (for left and right) straddle a root—that is, $x_l < x_r$ and $P(x_l)$ and $P(x_r)$ have opposite signs. Further suppose that $P(x_l) < 0$ and $P(x_r) > 0$. We can compute the number c that is halfway between x_l and x_r and then compute $P(c)$. If $P(c) = 0$, we've found a root. However, if $P(c)$ is not 0, we proceed in one of two ways:

1. If $P(c) < 0$, the root r lies between c and x_r. In this case, we let c become a new x_l and repeat the procedure.

2. If $P(c) > 0$, the root r lies between x_l and c. In this case, we let c become a new x_r and repeat the procedure.

Use the bisection method to find the positive root of $x^2 - 5 = 0$ to the nearest tenth.

We let $P(x) = x^2 - 5$. Since $P(2) = -1$ and $P(3) = 4$ have opposite signs, there is a root between 2 and 3. We can let $x_l = 2$ and $x_r = 3$ and compute the midpoint c:

$$c = \frac{x_l + x_r}{2} = \frac{2 + 3}{2} = 2.5$$

Because $P(c) = P(2.5) = 1.25$ is a positive number, we let c become a new x_r and calculate a new midpoint, which we will call c_1.

$$c_1 = \frac{x_l + x_r}{2} = \frac{2 + 2.5}{2} = 2.25$$

Because $P(c_1) = P(2.25) = 0.0625$ is a positive number, we let c_1 become a new x_r and calculate a new midpoint, which we will call c_2.

$$c_2 = \frac{x_l + x_r}{2} = \frac{2 + 2.25}{2} = 2.125$$

Because $P(c_2) = P(2.125) = -0.484375$ is a negative number, we let c_2 become a new x_l and calculate a new midpoint, which we will call c_3.

$$c_3 = \frac{x_l + x_r}{2} = \frac{2.125 + 2.25}{2} = 2.1875$$

Because $P(c_3) = P(2.1875) = -0.21484375$ is a positive number, we let c_3 become a new x_l and calculate a new midpoint, which we will call c_4.

$$c_4 = \frac{x_l + x_r}{2} = \frac{2.1875 + 2.25}{2} = 2.21875$$

Because $P(c_4) = P(2.21875) = -0.07714843$ is a negative number, we calculate a new midpoint.

From here on, the first two digits of the midpoints will remain 2.2. The root r of the equation (to the nearest tenth) is 2.2.

Graphing calculator usage:

To approximate the roots of a polynomial equation using a graphing calculator, use the ZOOM and TRACE capabilities of the calculator or the ZERO feature of the calculator.

See the Accent on Technology, *Solving Equations,* in Section 5.4.

Exercises

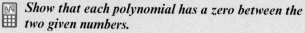 *Show that each polynomial has a zero between the two given numbers.*

53. $5x^3 + 37x^2 + 59x + 18 = 0$; -1 and 0

54. $6x^3 - x^2 - 10x - 3 = 0$; 1 and 2

Use the bisection method to find the positive root of each equation to the nearest tenth.

55. $x^3 - 2x^2 - 9x - 2 = 0$

56. $6x^2 - 13x - 5 = 0$

Use a graphing calculator to find the positive root of each equation to the nearest hundredth.

57. $6x^2 - 7x - 5 = 0$

58. $3x^2 + x - 2 = 0$

59. Designing solar collectors The space available for the installation of three solar collecting panels requires that their lengths differ by the amounts shown in the illustration, and that the total of their widths be 15 meters. To be equally effective, each panel must measure exactly 60 square meters. Find the dimensions of each panel.

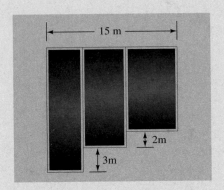

60. Designing a storage tank The design specifications for the cylindrical storage tank shown require that its height be 3 feet greater than the radius of its circular base and that the volume of the tank be 19,000 cubic feet. Use the bisection method or a graphing calculator to find the radius of the tank to the nearest hundredth of a foot.

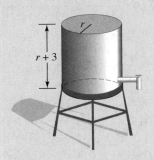

CHAPTER TEST

1. Is $x = -2$ a zero of $P(x) = x^2 + 5x + 6$?

Use long division and the remainder theorem to find each value.

2. $P(x) = 3x^3 - 9x - 5$; $P(2)$

3. $P(x) = x^5 + 2$; $P(-2)$

4. Use the factor theorem to determine whether $x - 3$ is a factor of $2x^4 - 10x^3 + 4x^2 + 7x + 21$.

Use synthetic division to express the polynomial $P(x) = 2x^3 - 3x^2 - 4x - 1$ in the form (divisor)(quotient) + remainder for each divisor.

5. $x - 2$

6. $x + 1$

Use synthetic division to perform each division.

7. $\dfrac{2x^2 - 7x - 15}{x - 5}$

8. $\dfrac{3x^3 + 7x^2 + 2x}{x + 2}$

Let $P(x) = 3x^3 - 2x^2 + 4$. Use synthetic division to find each value.

9. $P(1)$

10. $P(-2)$

11. $P\left(-\dfrac{1}{3}\right)$

12. $P(i)$

Find a polynomial with the given zeros.

13. $5, -1, 0$

14. $i, -i, \sqrt{3}, -\sqrt{3}$

Write a third-degree polynomial equation with real coefficients and the given roots.

15. $2, i$

16. $1, 2 + i$

17. How many linear factors and zeros does $P(x) = 3x^3 + 2x^2 - 4x + 1$ have?

18. If $3 - 2i$ is a root of $P(x) = 0$ where $P(x)$ has real-number coefficients, find another root.

Use Descartes' rule of signs to find the number of possible positive, negative, and nonreal roots of each equation.

19. $3x^5 - 2x^4 + 2x^2 - x - 3 = 0$

20. $2x^3 - 5x^2 - 2x - 1 = 0$

Find integer bounds for the roots of each equation.

21. $x^5 - x^4 - 5x^3 + 5x^2 + 4x - 5 = 0$

22. $2x^3 - 11x^2 + 10x + 3 = 0$

23. Use the rational root theorem to list all possible rational roots of $x^3 + 4x^2 + 3x + 8 = 0$.

24. Find all roots of the equation $2x^3 + 3x^2 - 11x - 6 = 0$.

25. Find all roots of the equation $x^3 - 2x^2 + x - 2 = 0$.

26. Does the polynomial $P(x) = 3x^3 + 2x^2 - 4x + 4$ have a zero between the values $x = 1$ and $x = 2$?

Use the bisection method to find the positive root of the equation to the nearest tenth.

27. $x^2 - 11 = 0$

CUMULATIVE REVIEW EXERCISES

Graph the function defined by each equation.

1. $f(x) = 3^x - 2$

2. $f(x) = 2e^x$

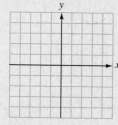

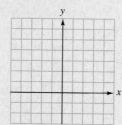

3. $f(x) = \log_3 x$

4. $f(x) = \ln(x - 2)$

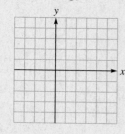

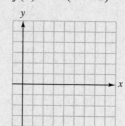

Find each value.

5. $\log_2 64$

6. $\log_{1/2} 8$

7. $\ln e^3$

8. $3 \log_3 2$

Write each expression in terms of the logarithms of a, b, and c.

9. $\log abc$

10. $\log \dfrac{a^2 b}{c}$

11. $\log \sqrt{\dfrac{ab}{c^3}}$

12. $\ln \dfrac{\sqrt{ab^2}}{c}$

Write each expression as the logarithm of a single quantity.

13. $3 \log a - 3 \log b$

14. $\dfrac{1}{2} \log a + 3 \log b - \dfrac{2}{3} \log c$

Solve each equation.

15. $3^{x+1} = 8$

16. $3^{x-1} = 3^{2x}$

17. $\log x + \log 2 = 3$

18. $\log(x + 1) + \log(x - 1) = 1$

Let $P(x) = 4x^3 + 3x + 2$. *Use synthetic division to find each value.*

19. $P(1)$ **20.** $P(-2)$

21. $P\left(\dfrac{1}{2}\right)$ **22.** $P(i)$

Determine whether each binomial is a factor of $P(x) = x^3 + 2x^2 - x - 2$. *Use synthetic division.*

23. $x + 1$ **24.** $x - 2$

25. $x - 1$ **26.** $x + 2$

Determine how many roots each equation has.

27. $x^{12} - 4x^8 + 2x^4 + 12 = 0$

28. $x^{2,000} - 1 = 0$

Determine the number of possible positive, negative, and nonreal roots of each equation.

29. $x^4 + 2x^3 - 3x^2 + x + 2 = 0$

30. $x^4 - 3x^3 - 2x^2 - 3x - 5 = 0$

Solve each equation.

31. $x^3 + x^2 - 9x - 9 = 0$

32. $x^3 - 2x^2 - x + 2 = 0$

Linear Systems

Careers and Mathematics

Applied Mathematician Applied mathematicians use mathematical theories and techniques, such as modeling and computational methods, to solve practical problems in business, government, engineering, and the physical, life, and social sciences. For example, they may analyze the most efficient way to schedule airline routes, the effects and safety of new drugs, the aerodynamics of an experimental automobile, or the cost-effectiveness of manufacturing processes. Some mathematicians, called **cryptanalysts,** analyze and decipher encryption systems designed to transmit military, political, financial, or law enforcement-related information in code.

Applied mathematicians held about 3,000 jobs in 2006. Many work for federal or state governments. In the private sector, mathematicians work for software publishers, insurance companies, and in aerospace or pharmaceutical manufacturing. In addition, many professionals with backgrounds in applied mathematics were among the 54,000 full-time mathematical science faculty in colleges and universities in 2006.

Education A Ph.D. degree in mathematics is usually the minimum educational requirement for prospective mathematicians, except in the federal government, where entry-level job candidates usually must have at least a bachelor's degree with a major in mathematics or 24 semester hours of mathematics courses.

Job Outlook Employment of mathematicians is expected to increase by 10 percent through 2016. More workers with a knowledge of mathematics will be required in the future because of advancements in technology. Median annual earnings of applied mathematicians were \$86,930 in 2006.

For sample applications, see Examples in Section 6.8. For more information, see www.bls.gov/oco/ocos.043.htm.

In this chapter, we learn to solve systems of equations—sets of several equations, most with more than one variable. One method of solution uses matrices, an important tool in mathematics and its applications.

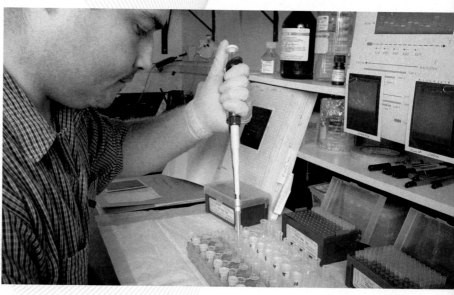

© CDC/PHIL/Corbis

6.1 Systems of Linear Equations

Objectives

1. Solve Systems Using the Graphing Method
2. Solve Systems Using the Substitution Method
3. Solve Systems Using the Addition Method
4. Solve Systems with Infinitely Many Solutions
5. Solve Inconsistent Systems
6. Solve Systems Involving Three Equations in Three Variables
7. Solve Problems Involving Systems of Equations

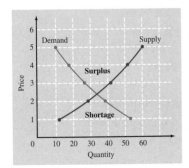

In economics, we often talk about the law of supply and demand.

- As the illustration shows, few units of a product will be in demand when the price is high. However, more units will be in demand when the price is low.
- As the illustration also shows, few units of product will be supplied when the price is low. However, more product will be supplied when the price is high.

In a free economy, the intersection of the graph of the supply function and the demand function will be the market price of the product.

To see that this is true, suppose that the price of a product is $3. At this price, less than 30 units of product would be in demand, but more than 40 units would be supplied. Since supply is greater than demand, the price would come down.

At a price of $2, a little less than 40 units would be in demand, but less than 30 units would be supplied. Since demand is greater than supply, the price would go up.

Only when the graphs cross would the demand and supply be in equilibrium.

In this section, we will begin to discuss how to find the common solution of two equations that occur simultaneously.

Equations with two variables have infinitely many solutions. For example, the following tables give a few of the solutions of $x + y = 5$ and $x - y = 1$.

$$x + y = 5 \qquad x - y = 1$$

x	y		x	y
1	4		8	7
2	3		3	2
3	2		1	0
5	0		0	-1

Only the pair $x = 3$ and $y = 2$ satisfies both equations. The pair of equations

$$\begin{cases} x + y = 5 \\ x - y = 1 \end{cases}$$

is called a **system of equations,** and the solution $x = 3$, $y = 2$ is called its **simultaneous solution,** or just its **solution.** The process of finding the solution of a system of equations is called **solving the system.**

The graph of an equation in two variables displays the equation's infinitely many solutions. For example, the graph of one of the lines in Figure 6-1 represents the infinitely many solutions of the equation $5x - 2y = 1$. The other line in the figure is the graph of the infinitely many solutions of $2x + 3y = 8$.

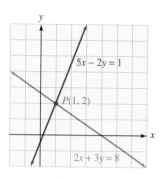

Figure 6-1

Because only point $P(1, 2)$ lies on both lines, only its coordinates satisfy both equations. Thus, the simultaneous solution of the system of equations

$$\begin{cases} 5x - 2y = 1 \\ 2x + 3y = 8 \end{cases}$$

is the pair of numbers $x = 1$ and $y = 2$, or simply the ordered pair $(1, 2)$.

This discussion suggests a graphical method of solving systems of equations in two variables.

1. Solve Systems Using the Graphing Method

We can use the following steps to solve a system of two equations in two variables.

Strategy for Using the Graphing Method

1. On one coordinate grid, graph each equation.
2. Find the coordinates of the point or points where all of the graphs intersect. These coordinates give the solutions of the system.
3. If the graphs have no point in common, the system has no solution.

EXAMPLE 1 Use the graphing method to solve each system:

a. $\begin{cases} 3x + y = 1 \\ -x + 2y = 9 \end{cases}$ **b.** $\begin{cases} 2x - 3y = 4 \\ 4x = -4 + 6y \end{cases}$ **c.** $\begin{cases} y = 4 - x \\ 2x + 2y = 8 \end{cases}$

Solution **a.** The graphs of the equations are the lines shown in Figure 6-2(a) on the next page. The solution of this system is given by the coordinates of the point $(-1, 4)$, where the lines intersect. By checking both values in both equations, we can verify that the solution is $x = -1$ and $y = 4$.

b. The graphs of the equations are the parallel lines shown in Figure 6-2(b). Since parallel lines do not intersect, the system has no solutions.

c. The graphs of the equations are the lines shown in Figure 6-2(c). Since the lines are the same, they have infinitely many points in common, and the system has infinitely many solutions. All ordered pairs whose coordinates satisfy one of the equations satisfy the other also.

To find some solutions, we substitute numbers for x in the first equation and solve for y. If $x = 3$, for example, then $y = 1$. One solution is the pair $(3, 1)$. Other solutions are $(0, 4)$ and $(5, -1)$.

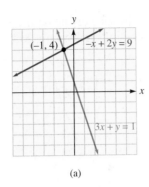

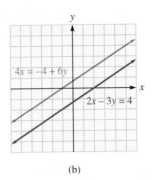

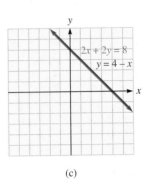

(a)   (b)   (c)

Figure 6-2

Self Check 1   Solve: $\begin{cases} y = 2x \\ x + y = 3 \end{cases}$.

Example 1 illustrates three possibilities that can occur when we solve systems of equations. If a system of equations has at least one solution, as in parts **a** and **c** of Example 1, the system is called **consistent.** If it has no solutions, as in part **b,** it is called **inconsistent.**

If a system of two equations in two variables has exactly one solution as in part **a,** or no solution as in part **b,** the equations in the system are called **independent.** If a system of linear equations has infinitely many solutions, as in part **c,** the equations of the system are called **dependent.**

There are three possibilities that can occur when two equations, each with two variables, are graphed.

Possible figure	If the	Then
	lines are distinct and intersect,	the equations are independent, and the system is consistent. The system has one solution.
	lines are distinct and parallel,	the equations are independent, and the system is inconsistent. The system has no solutions.
	lines coincide,	the equations are dependent, and the system is consistent. The system has infinitely many solutions.

Accent on Technology   **Solving Systems with a Graphing Calculator**

To use a graphing calculator to solve the system $\begin{cases} x - 4y = 7 \\ x + 2y = 4 \end{cases}$, we must solve both equations for y. We do so to get the equivalent system $\begin{cases} y = \dfrac{x - 7}{4} \\ y = \dfrac{4 - x}{2} \end{cases}$. Depending on the viewing window, the two graphs will be similar to those in Figure 6-3(a). We zoom in on the intersection, as in Figure 6-3(b), and trace to find the solution: $x = 5$, $y = -\frac{1}{2}$.

We also can find the intersection of the two lines by using the INTERSECT feature found on most graphing calculators. On a TI-84 Plus calculator, INTERSECT is found in the CALC menu. To learn how to use this feature, please consult the owner's manual. After graphing the lines and using INTERSECT, we obtain a graph similar to Figure 6-3(c). The display shows the coordinates of the point of intersection.

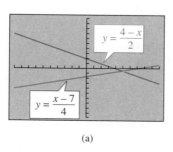
(a)

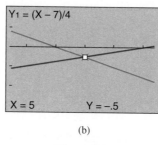
(b)

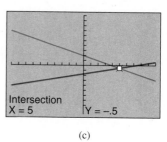
(c)

Figure 6-3

We now consider two algebraic methods to find solutions of systems containing any number of variables.

2.   Solve Systems Using the Substitution Method

We use the following steps to solve a system of two equations in two variables by substitution.

Strategy for Using the Substitution Method

1. Solve one equation for a variable—say, y.
2. Substitute the expression obtained for y for every y in the second equation.
3. Solve the equation that results.
4. Substitute the solution found in Step 3 into the equation found in Step 1 and solve for y.

EXAMPLE 2   Use the substitution method to solve $\begin{cases} 3x + y = 1 \\ -x + 2y = 9 \end{cases}$.

Solution   We can solve the first equation for y and then substitute that result for y in the second equation.

$$\begin{cases} 3x + y = 1 \rightarrow y = 1 - 3x \\ -x + 2y = 9 \end{cases}$$

The substitution gives one linear equation with one variable, which we can solve for x:

$$-x + 2(1 - 3x) = 9$$
$$-x + 2 - 6x = 9$$
$$-7x = 7 \qquad \text{Combine like terms and subtract 2 from both sides.}$$
$$x = -1 \qquad \text{Divide both sides by } -7.$$

Comment

We could solve either equation for x or y. However, it is easier to solve for a variable with a coefficient of 1 or −1. Then fractions will be avoided.

To find y, we substitute -1 for x in the equation $y = 1 - 3x$ and simplify:

$$y = 1 - 3x$$
$$= 1 - 3(-1)$$
$$= 1 + 3$$
$$= 4$$

The solution is the ordered pair $(-1, 4)$.

Self Check 2 Solve: $\begin{cases} 2x + 3y = 1 \\ x - y = 3 \end{cases}$.

3. Solve Systems Using the Addition Method

As with substitution, the addition method combines the equations of a system to eliminate terms involving one of the variables.

Strategy for Using the Addition Method

1. Write the equations of the system in general form so that terms with the same variable are aligned vertically.
2. Multiply all the terms of one or both of the equations by constants chosen to make the coefficients of x (or y) differ only in sign.
3. Add the equations. Solve the equation that results, if possible.
4. Substitute the value obtained in Step 3 into either of the original equations, and solve for the remaining variable.
5. The results obtained in Steps 3 and 4 are the solution of the system.

EXAMPLE 3 Use the addition method to solve $\begin{cases} 3x + 2y = 8 \\ 2x - 5y = 18 \end{cases}$.

Solution To eliminate y, we multiply the terms of the first equation by 5 and the terms of the second equation by 2, to obtain an equivalent system in which the coefficients of y differ only in sign. Then we add the equations to eliminate y.

$$\begin{cases} 3x + 2y = 8 \\ 2x - 5y = 18 \end{cases} \begin{array}{l} \rightarrow \ 15x + 10y = 40 \quad \text{Multiply by 5.} \\ \rightarrow \ \underline{\ 4x - 10y = 36} \quad \text{Multiply by 2.} \\ 19x = 76 \quad \text{Add the equations.} \\ x = 4 \quad \text{Divide both sides by 19.} \end{array}$$

We now substitute 4 for x into either of the original equations and solve for y. If we use the first equation,

$$3x + 2y = 8$$
$$3(4) + 2y = 8 \qquad \text{Substitute 4 for } x.$$
$$12 + 2y = 8 \qquad \text{Simplify.}$$
$$2y = -4 \qquad \text{Subtract 12 from both sides.}$$
$$y = -2 \qquad \text{Divide both sides by 2.}$$

The solution is $(4, -2)$.

Comment
To make the coefficients of y differ only in sign and be the smallest numbers possible, we multiply each equation by a number that makes the new coefficients of y the least common multiple of the original coefficients.

Self Check 3 Solve: $\begin{cases} 2x + 5y = 8 \\ 2x - 4y = -10 \end{cases}$.

4. Solve Systems with Infinitely Many Solutions

EXAMPLE 4   Use the addition method to solve $\begin{cases} x + 2y = 3 \\ 2x + 4y = 6 \end{cases}$.

Solution   To eliminate the variable y, we can multiply both sides of the first equation by -2 and add the result to the second equation to get

$$
\begin{array}{r}
-2x - 4y = -6 \\
2x + 4y = 6 \\
\hline
0 = 0
\end{array}
$$

Although the result $0 = 0$ is true, it does not give the value of y.

Since the second equation in the original system is twice the first, the equations are equivalent. If we were to solve this system by graphing, the two lines would coincide. The (x, y) coordinates of points on that one line form the infinite set of solutions to the given system. The system is consistent, but the equations are dependent.

To find a general solution, we can solve either equation of the system for y. If we solve the first equation for y, we obtain

$$
\begin{aligned}
x + 2y &= 3 \\
2y &= 3 - x \\
y &= \frac{3 - x}{2}
\end{aligned}
$$

All ordered pairs that are solutions will have the form $\left(x, \frac{3-x}{2}\right)$. For example:

- If $x = 3$, then $y = 0$, and $(3, 0)$ is a solution.
- If $x = 0$, then $y = \frac{3}{2}$, and $\left(0, \frac{3}{2}\right)$ is a solution.
- If $x = -7$, then $y = 5$, and $(-7, 5)$ is a solution.

Self Check 4   Solve: $\begin{cases} 3x - y = 2 \\ 6x - 2y = 4 \end{cases}$.

5. Solve Inconsistent Systems

EXAMPLE 5   Solve: $\begin{cases} x + y = 3 \\ x + y = 2 \end{cases}$.

Solution   We can multiply both sides of the second equation by -1 and add the results to the first equation to get

$$
\begin{array}{r}
x + y = 3 \\
-x - y = -2 \\
\hline
0 = 1
\end{array}
$$

Because $0 \neq 1$, the system has no solutions. If we graph each equation in this system, the graphs will be parallel lines. This system is inconsistent, and the equations are independent.

Self Check 5   Solve: $\begin{cases} 3x + 2y = 2 \\ 3x + 2y = 3 \end{cases}$.

EXAMPLE 6 Solve: $\begin{cases} \dfrac{x + 2y}{4} + \dfrac{x - y}{5} = \dfrac{6}{5} \\ \dfrac{x + y}{7} - \dfrac{x - y}{3} = -\dfrac{12}{7} \end{cases}$

Solution To clear the first equation of fractions, we multiply both sides by the lowest common denominator, 20. To clear the second equation of fractions, we multiply both sides of the second equation by the LCD, which is 21.

$$\begin{cases} 20\left(\dfrac{x + 2y}{4} + \dfrac{x - y}{5}\right) = 20\left(\dfrac{6}{5}\right) \\ 21\left(\dfrac{x + y}{7} - \dfrac{x - y}{3}\right) = 21\left(-\dfrac{12}{7}\right) \end{cases}$$

$$\begin{cases} 5(x + 2y) + 4(x - y) = 24 \\ 3(x + y) - 7(x - y) = -36 \end{cases}$$

$$\begin{cases} 9x + 6y = 24 \\ -4x + 10y = -36 \end{cases} \quad \text{Remove parentheses and combine like terms.}$$

$$\begin{cases} 3x + 2y = 8 \quad \text{Divide both sides by 3.} \\ 2x - 5y = 18 \quad \text{Divide both sides by } -2. \end{cases}$$

In Example 3, we saw that the solution to this final system is $(4, -2)$.

Self Check 6 Solve: $\begin{cases} \dfrac{x + y}{3} + \dfrac{x - y}{4} = \dfrac{3}{4} \\ \dfrac{x - y}{3} - \dfrac{2x - y}{2} = -\dfrac{1}{3} \end{cases}$

6. Solve Systems Involving Three Equations in Three Variables

To solve three equations in three variables, we use addition to eliminate one variable. This will produce a system of two equations in two variables, which we can solve using the methods previously discussed.

EXAMPLE 7 Solve the system:

(1) $\begin{cases} x + 2y + z = 8 \\ 2x + y - z = 1 \\ x + y - 2z = -3 \end{cases}$
(2)
(3)

Solution We can add Equations 1 and 2 to eliminate z

(1) $\quad x + 2y + z = 8$
(2) $\quad \underline{2x + \ y - z = 1}$
(4) $\quad 3x + 3y \quad\quad = 9$

and divide both sides of Equation 4 by 3 to get

(5) $\quad x + y = 3$

We now choose a different pair of equations—say, Equations 1 and 3—and eliminate z again. If we multiply both sides of Equation 1 by 2 and add the result to Equation 3, we get

$$
\begin{array}{rl}
& 2x + 4y + 2z = 16 \\
(3) & \underline{x + y - 2z = -3} \\
(6) & 3x + 5y \phantom{{}- 2z} = 13
\end{array}
$$

Equations 5 and 6 form the system $\begin{cases} x + y = 3 \\ 3x + 5y = 13 \end{cases}$, which we can solve by substitution.

$$
\begin{array}{rl}
(5) & x + y = 3 \rightarrow y = \boxed{3 - x} \\
(6) & 3x + 5y = 13
\end{array}
$$

$$
\begin{array}{ll}
3x + 5(3 - x) = 13 & \text{Substitute } 3 - x \text{ for } y. \\
3x + 15 - 5x = 13 & \text{Remove parentheses.} \\
-2x = -2 & \text{Combine like terms and subtract 15 from both sides.} \\
x = 1 & \text{Divide both sides by } -2.
\end{array}
$$

To find y, we substitute 1 for x in the equation $y = 3 - x$.

$$
\begin{aligned}
y &= 3 - 1 \\
y &= 2
\end{aligned}
$$

To find z, we substitute 1 for x and 2 for y in any one of the original equations that includes z, and we find that $z = 3$. The solution is the ordered triple

$$(x, y, z) = (1, 2, 3)$$

Because there is one solution, the system is consistent, and its equations are independent.

Self Check 7 Solve the system:

$$
\begin{cases}
x - y + 2z = 2 \\
2x + y + z = 4 \\
-x - 2y + 3z = 0
\end{cases}
$$

EXAMPLE 8 Solve the system:

$$
\begin{array}{rl}
(1) & \\
(2) & \begin{cases} x + 2y + z = 8 \\ 2x + y - z = 1 \\ x - y - 2z = -7 \end{cases} \\
(3) &
\end{array}
$$

Solution We can add Equations 1 and 2 to eliminate z.

$$
\begin{array}{rl}
& x + 2y + z = 8 \\
& \underline{2x + y - z = 1} \\
(4) & 3x + 3y \phantom{{}- z} = 9
\end{array}
$$

We now can multiply Equation 1 by 2 and add it to Equation 3 to eliminate z again.

$$
\begin{array}{rl}
& 2x + 4y + 2z = 16 \\
& \underline{x - y - 2z = -7} \\
(5) & 3x + 3y \phantom{{}+ 2z} = 9
\end{array}
$$

Since Equations 4 and 5 are the same, the system is consistent, but the equations are dependent. There will be infinitely many solutions. To find a general solution, we can solve either Equation 4 or 5 for y to get

(6) $\quad y = 3 - x$

We can find the value of z in terms of x by substituting the right side of Equation 6 into any of the first three equations—say Equation 1.

$$x + 2y + z = 8$$
$$x + 2(3 - x) + z = 8 \qquad \text{Substitute } 3 - x \text{ for } y.$$
$$x + 6 - 2x + z = 8 \qquad \text{Use the distributive property to remove parentheses.}$$
$$-x + 6 + z = 8 \qquad \text{Combine terms.}$$
$$z = x + 2 \qquad \text{Solve for } z.$$

A general solution to this system is $(x, y, z) = (x, 3 - x, x + 2)$. To find some specific solutions, we can substitute numbers for x and compute y and z. For example,

If $x = 1$, then $y = 2$ and $z = 3$. One possible solution is $(1, 2, 3)$.
If $x = 0$, then $y = 3$ and $z = 2$. Another possible solution is $(0, 3, 2)$.

Self Check 8   Solve the system:

$$\begin{cases} x + y + z = 3 \\ x - y + 3z = 4 \\ 2x + 2y + 2z = 5 \end{cases}$$

7. Solve Problems Involving Systems of Equations

EXAMPLE 9   An airplane flies 600 miles with the wind for 2 hours and returns against the wind in 3 hours. Find the speed of the wind and the air speed of the plane.

Solution   If a represents the air speed and w represents wind speed, the ground speed of the plane with the wind is the combined speed $a + w$. On the return trip, against the wind, the ground speed is $a - w$. The information in this problem is organized in Figure 6-4 and can be used to give a system of two equations in the variables a and w.

	d	$=$	r	$\cdot$	t
Outbound trip	600		$a + w$		2
Return trip	600		$a - w$		3

Figure 6-4

Since $d = rt$, we have $\begin{cases} 600 = 2(a + w) \\ 600 = 3(a - w) \end{cases}$, which can be written as

(7) $\quad \begin{cases} 300 = a + w \\ 200 = a - w \end{cases}$

We can add these equations to get

$$500 = 2a$$
$$a = 250 \qquad \text{Divide both sides by 2.}$$

To find w, we substitute 250 for a into either of the previous equations (we'll use Equation 7) and solve for w:

(7) $300 = a + w$

$300 = \mathbf{250} + w$

$w = 50$ Subtract 250 from both sides.

The air speed of the plane is 250 mph. With a 50-mph tailwind, the ground speed is $250 + 50$, or 300 mph. At 300 mph, the 600-mile trip will take 2 hours.

With a 50-mph headwind, the ground speed is $250 - 50$, or 200 mph. At 200 mph, the 600-mile trip will take 3 hours. The answers check.

Self Check Answers   **1.** $(1, 2)$ **2.** $(2, -1)$ **3.** $(-1, 2)$ **4.** ordered pairs of the form $(x, 3x - 2)$
5. The system is inconsistent; no solutions.   **6.** $(1, 2)$ **7.** $(1, 1, 1)$
8. no solution; inconsistent system

6.1 Exercises

Vocabulary and Concepts   *Fill in the blanks.*

1. A set of several equations with several variables is called a _____ of equations.

2. Any set of numbers that satisfies each equation of a system is called a _____ of the system.

3. If a system of equations has a solution, the system is _____.

4. If a system of equations has no solution, the system is _____.

5. If a system of equations has only one solution, the equations of the system are _____.

6. If a system of equations has infinitely many solutions, the equations of the system are _____.

7. The system $\begin{cases} x + y = 5 \\ x - y = 1 \end{cases}$ is _____ (consistent, inconsistent).

8. The system $\begin{cases} x + y = 5 \\ x + y = 1 \end{cases}$ is _____ (consistent, inconsistent).

9. The equations of the system $\begin{cases} x + y = 5 \\ 2x + 2y = 10 \end{cases}$ are _____ (dependent, independent).

10. The equations of the system $\begin{cases} x + y = 5 \\ x - y = 1 \end{cases}$ are _____ (dependent, independent).

11. The pair $(1, 3)$ __ (is, is not) a solution of the system $\begin{cases} x + 2y = 7 \\ 2x - y = -1 \end{cases}$.

12. The pair $(1, 3)$ __ (is, is not) a solution of the system $\begin{cases} 3x + y = 6 \\ x - 3y = -8 \end{cases}$.

Practice   *Solve each system of equations by graphing.*

13. $\begin{cases} y = -3x + 5 \\ x - 2y = -3 \end{cases}$ **14.** $\begin{cases} x - 2y = -3 \\ 3x + y = -9 \end{cases}$

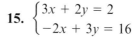   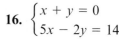

15. $\begin{cases} 3x + 2y = 2 \\ -2x + 3y = 16 \end{cases}$ **16.** $\begin{cases} x + y = 0 \\ 5x - 2y = 14 \end{cases}$

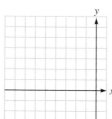   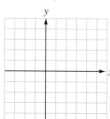

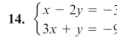 *Use a graphing calculator to approximate the solutions of each system. Give answers to the nearest tenth.*

17. $\begin{cases} y = -5.7x + 7.8 \\ y = 37.2 - 19.1x \end{cases}$ **18.** $\begin{cases} y = 3.4x - 1 \\ y = -7.1x + 3.1 \end{cases}$

19. $\begin{cases} y = \dfrac{5.5 - 2.7x}{3.5} \\ 5.3x - 9.2y = 6.0 \end{cases}$ **20.** $\begin{cases} 29x + 17y = 7 \\ -17x + 23y = 19 \end{cases}$

41. $\begin{cases} x + 2(x - y) = 2 \\ 3(y - x) - y = 5 \end{cases}$ **42.** $\begin{cases} 3x = 4(2 - y) \\ 3(x - 2) + 4y = 0 \end{cases}$

Solve each system by substitution, if possible.

21. $\begin{cases} y = x - 1 \\ y = 2x \end{cases}$ **22.** $\begin{cases} y = 2x - 1 \\ x + y = 5 \end{cases}$

23. $\begin{cases} 2x + 3y = 0 \\ y = 3x - 11 \end{cases}$ **24.** $\begin{cases} 2x + y = 3 \\ y = 5x - 11 \end{cases}$

25. $\begin{cases} 4x + 3y = 3 \\ 2x - 6y = -1 \end{cases}$ **26.** $\begin{cases} 4x + 5y = 4 \\ 8x - 15y = 3 \end{cases}$

27. $\begin{cases} x + 3y = 1 \\ 2x + 6y = 3 \end{cases}$ **28.** $\begin{cases} x - 3y = 14 \\ 3(x - 12) = 9y \end{cases}$

29. $\begin{cases} y = 3x - 6 \\ x = \dfrac{1}{3}y + 2 \end{cases}$ **30.** $\begin{cases} 3x - y = 12 \\ y = 3x - 12 \end{cases}$

Solve each system by the addition method, if possible.

31. $\begin{cases} 5x - 3y = 12 \\ 2x - 3y = 3 \end{cases}$ **32.** $\begin{cases} 2x + 3y = 8 \\ -5x + y = -3 \end{cases}$

33. $\begin{cases} x - 7y = -11 \\ 8x + 2y = 28 \end{cases}$ **34.** $\begin{cases} 3x + 9y = 9 \\ -x + 5y = -3 \end{cases}$

35. $\begin{cases} 3(x - y) = y - 9 \\ 5(x + y) = -15 \end{cases}$ **36.** $\begin{cases} 2(x + y) = y + 1 \\ 3(x + 1) = y - 3 \end{cases}$

37. $\begin{cases} 2 = \dfrac{1}{x + y} \\ 2 = \dfrac{3}{x - y} \end{cases}$ **38.** $\begin{cases} \dfrac{1}{x + y} = 12 \\ \dfrac{3x}{y} = -4 \end{cases}$

39. $\begin{cases} y + 2x = 5 \\ 0.5y = 2.5 - x \end{cases}$

40. $\begin{cases} -0.3x + 0.1y = -0.1 \\ 6x - 2y = 2 \end{cases}$

43. $\begin{cases} x + \dfrac{y}{3} = \dfrac{5}{3} \\ \dfrac{x + y}{3} = 3 - x \end{cases}$ **44.** $\begin{cases} 3x - y = 0.25 \\ x + \dfrac{3}{2}y = 2.375 \end{cases}$

45. $\begin{cases} \dfrac{3}{2}x + \dfrac{1}{3}y = 2 \\ \dfrac{2}{3}x + \dfrac{1}{9}y = 1 \end{cases}$ **46.** $\begin{cases} \dfrac{x + y}{2} + \dfrac{x - y}{5} = 2 \\ x = \dfrac{y}{2} + 1 \end{cases}$

47. $\begin{cases} \dfrac{x - y}{5} + \dfrac{x + y}{2} = 6 \\ \dfrac{x - y}{2} - \dfrac{x + y}{4} = 3 \end{cases}$ **48.** $\begin{cases} \dfrac{x - 2}{5} + \dfrac{y + 3}{2} = 5 \\ \dfrac{x + 3}{2} + \dfrac{y - 2}{3} = 6 \end{cases}$

Solve each system, if possible.

49. $\begin{cases} x + y + z = 3 \\ 2x + y + z = 4 \\ 3x + y - z = 5 \end{cases}$ **50.** $\begin{cases} x - y - z = 0 \\ x + y - z = 0 \\ x - y + z = 2 \end{cases}$

51. $\begin{cases} x - y + z = 0 \\ x + y + 2z = -1 \\ -x - y + z = 0 \end{cases}$ **52.** $\begin{cases} 2x + y - z = 7 \\ x - y + z = 2 \\ x + y - 3z = 2 \end{cases}$

53. $\begin{cases} 2x + y = 4 \\ x - z = 2 \\ y + z = 1 \end{cases}$ **54.** $\begin{cases} 3x + y + z = 0 \\ 2x - y + z = 0 \\ 2x + y + z = 0 \end{cases}$

55. $\begin{cases} x + y + z = 6 \\ 2x + y + 3z = 17 \\ x + y + 2z = 11 \end{cases}$ **56.** $\begin{cases} x + y + z = 3 \\ 2x + y + z = 6 \\ x + 2y + 3z = 2 \end{cases}$

57. $\begin{cases} x + y + z = 3 \\ x + z = 2 \\ 2x + 2y + 2z = 3 \end{cases}$ **58.** $\begin{cases} x + y + z = 3 \\ x + z = 2 \\ 2x + y + 2z = 5 \end{cases}$

59. $\begin{cases} x + 2y - z = 2 \\ 2x - y = -1 \\ 3x + y + z = 1 \end{cases}$ **60.** $\begin{cases} x + y = 2 \\ y + z = 2 \\ 3x + 3y = 2 \end{cases}$

61. $\begin{cases} 3x + 4y + 2z = 4 \\ 6x - 2y + z = 4 \\ 3x - 8y - 6z = -3 \end{cases}$ **62.** $\begin{cases} x + y = 2 \\ y + z = 2 \\ x - z = 0 \end{cases}$

63. $\begin{cases} 2x - y - z = 0 \\ x - 2y - z = -1 \\ x - y - 2z = -1 \end{cases}$ **64.** $\begin{cases} x + 3y - z = 5 \\ 3x - y + z = 2 \\ 2x + y = 1 \end{cases}$

65. $\begin{cases} (x + y) + (y + z) + (z + x) = 6 \\ (x - y) + (y - z) + (z - x) = 0 \\ x + y + 2z = 4 \end{cases}$

66. $\begin{cases} (x + y) + (y + z) = 1 \\ (x + z) + (x + z) = 3 \\ (x - y) - (x - z) = -1 \end{cases}$

Applications   *Use systems of equations to solve each problem.*

67. Planning for harvest   A farmer raises corn and soybeans on 350 acres of land. Because of expected prices at harvest time, he thinks it would be wise to plant 100 more acres of corn than of soybeans. How many acres of each does he plant?

68. Club memberships   There is an initiation fee to join the Pine River Country Club, as well as monthly dues. The total cost after 7 months' membership will be \$3,025, and after $1\frac{1}{2}$ years, \$3,850. Find both the initiation fee and the monthly dues.

69. Framing pictures   A rectangular picture frame has a perimeter of 1,900 centimeters and a width that is 250 centimeters less than its length. Find the area of the picture.

70. Boating   A Mississippi riverboat can travel 30 kilometers downstream in 3 hours and can make the return trip in 5 hours. Find the speed of the boat in still water.

71. Making an alloy   A metallurgist wants to make 60 grams of an alloy that is to be 34% copper. She has samples that are 9% copper and 84% copper. How many grams of each must she use?

72. Archimedes' law of the lever   The two weights shown will be in balance if the product of one weight and its distance from the fulcrum is equal to the product of the other weight and its distance from the fulcrum. Two weights are in balance when one is 2 meters and the other 3 meters from the fulcrum. If the fulcrum remained in the same spot and the weights were interchanged, the closer weight would need to be increased by 5 pounds to maintain balance. Find the weights.

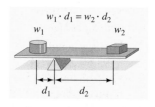

73. Lifting weights   A 112-pound force can lift the 448-pound load shown. If the fulcrum is moved 1 additional foot away from the load, a 192-pound force is required. Find the length of the lever.

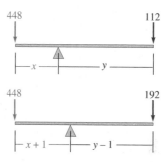

74. Writing test questions   For a test question, a mathematics teacher wants to find two constants a and b such that the test item "Simplify $a(x + 2y) - b(2x - y)$" will have an answer of $-3x + 9y$. What constants a and b should the teacher use?

75. Break-even point   Rollowheel, Inc. can manufacture a pair of in-line skates for \$43.53. Daily fixed costs of manufacturing in-line skates amount to \$742.72. A pair of in-line skates can be sold for \$89.95. Find equations expressing the expenses E and the revenue R as functions of x, the number of pairs manufactured and sold. At what production level will expenses equal revenues?

76. Choosing salary options   For its sales staff, a company offers two salary options. One is \$326 per week plus a commission of $3\frac{1}{2}$% of sales. The other is \$200 per week plus $4\frac{1}{4}$% of sales. Find equations that express incomes as functions of sales, and find the weekly sales level that produces equal salaries.

Use systems of three equations in three variables to solve each problem.

77. Work schedules A college student earns $198.50 per week working three part-time jobs. Half of his 30-hour work week is spent cooking hamburgers at a fast-food chain, earning $5.70 per hour. In addition, the student earns $6.30 per hour working at a gas station and $10 per hour doing janitorial work. How many hours per week does the student work at each job?

78. Investment income A woman invested a $22,000 rollover IRA account in three banks paying 5%, 6%, and 7% annual interest. She invested $2,000 more at 6% than at 5%. The total annual interest she earned was $1,370. How much did she invest at each rate?

79. Age distribution Approximately 3 million people live in Costa Rica. 2.61 million are less than 50 years old, and 1.95 million are older than 14. How many people are in each of the categories 0–14 years, 15–49 years, and 50 years and older?

80. Designing arches The engineer designing a parabolic arch knows that its equation has the form $y = ax^2 + bx + c$. Use the information in the illustration to find a, b, and c. Assume that the distances are given in feet.   (*Hint:* The coordinates of points on the parabola satisfy its equation.)

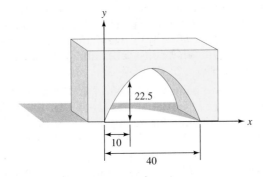

81. Geometry The sum of the angles of a triangle is 180°. In a certain triangle, the largest angle is 20° greater than the sum of the other two and is 10° greater than 3 times the smallest. How large is each angle?

82. Ballistics The path of a thrown object is a parabola with the equation $y = ax^2 + bx + c$. Use the information in the illustration to find a, b, and c. (Distances are in feet.)

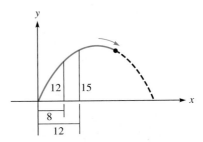

Discovery and Writing

83. 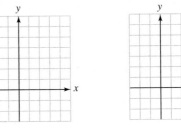 Use a graphing calculator to attempt to find the solution of the system $\begin{cases} x - 8y = -51 \\ 3x - 25y = -160 \end{cases}$.

84. Solve the system of Exercise 83 algebraically. Which method is easier, and why?

85. Use a graphing calculator to attempt to find the solution of the system $\begin{cases} 17x - 23y = -76 \\ 29x + 19y = -278 \end{cases}$.

86. Solve the system of Exercise 85 algebraically. Which method is easier, and why?

87. Invent a system of two equations with the solution $x = 1$, $y = 2$ that is difficult to solve with a graphing calculator (as in Exercise 85).

88. Invent a system of two equations with the solution $x = 1$, $y = 2$ that is easier to solve graphically than algebraically.

Review   *Graph each function.*

89. $f(x) = 3^x$

90. $f(x) = \log_2 x$

Solve for x:

91. $3 = \log_2 x$

92. $x = \log_5 \dfrac{1}{25}$

Use the properties of logarithms to write each expression in terms of the logarithms of x, y, and z.

93. $\log \dfrac{x}{y^2 z}$

94. $\log x(y^z)$

Use the properties of logarithms to write each expression as a single logarithm.

95. $\log x + 3 \log y - \dfrac{1}{2} \log z$

96. $\dfrac{1}{2}(\log A - 3 \log B)$

6.2 Gaussian Elimination and Matrix Methods

Objectives

1. Use Row Operations to Write Matrices in Row Echelon Form
2. Solve an Inconsistent System
3. Solve a System of Dependent Equations
4. Solve Systems Using Gauss–Jordan Elimination

Since the invention of modern-day electronics, dot matrix screens with rectangular arrays of dots (or pixels) have become important. For example, we all are familiar with signs like a football scoreboard, a sign in front of a bank that gives the date and temperature, or a sign that shows which lanes are open on the toll road.

In this section, we will show how rectangular arrays of numbers can be used to solve systems of equations. We begin by introducing a method for solving systems called **Gaussian elimination,** after the German mathematician Carl Friedrich Gauss (1777–1855). In this method, we transform a system of equations into an equivalent system that can be solved by a process called **back substitution.**

Using Gaussian elimination, we solve systems of equations by working only with the coefficients of the variables. From the system of equations

$$\begin{cases} x + 2y + z = 8 \\ 2x + y - z = 1 \\ x + y - 2z = -3 \end{cases}$$

for example, we can form three rectangular arrays of numbers. Each is called a **matrix.** The first is the **coefficient matrix,** which contains the coefficients of the variables of the system. The second matrix contains the constants from the right side of the system. We can write the third matrix, called the **augmented matrix** or the **system matrix,** by joining the two. The coefficient matrix appears to the left of the dashed line, and the matrix of constants appears to the right.

Coefficient matrix   **Constants**   **System matrix**

$$\begin{bmatrix} 1 & 2 & 1 \\ 2 & 1 & -1 \\ 1 & 1 & -2 \end{bmatrix} \qquad \begin{bmatrix} 8 \\ 1 \\ -3 \end{bmatrix} \qquad \left[\begin{array}{ccc:c} 1 & 2 & 1 & 8 \\ 2 & 1 & -1 & 1 \\ 1 & 1 & -2 & -3 \end{array}\right]$$

Each row of the system matrix represents one equation of the system.

- The first row represents the equation $x + 2y + z = 8$.
- The second row represents the equation $2x + y - z = 1$.
- The third row represents the equation $x + y - 2z = -3$.

Because the system matrix has three rows and four columns, it is called a 3×4 matrix (read as "3 by 4"). The coefficient matrix is 3×3 and the constants form a 3×1 matrix.

To illustrate Gaussian elimination, we solve this system by the addition method of the previous section. In a second column, we keep track of the changes to the system matrix.

EXAMPLE 1 Use Gaussian elimination to solve

$$
\begin{array}{l}
(1) \\
(2) \\
(3)
\end{array}
\begin{cases}
x + 2y + z = 8 \\
2x + y - z = 1 \\
x + y - 2z = -3
\end{cases}
$$

Solution We multiply each term in Equation 1 by -2 and add the result to Equation 2 to obtain Equation 4. Then we multiply each term of Equation 1 by -1 and add the result to Equation 3 to obtain Equation 5. This gives the equivalent system

$$
\begin{array}{l}
(1) \\
(4) \\
(5)
\end{array}
\begin{cases}
x + 2y + z = 8 \\
 -3y - 3z = -15 \\
 -y - 3z = -11
\end{cases}
\qquad
\left[\begin{array}{ccc|c}
1 & 2 & 1 & 8 \\
0 & -3 & -3 & -15 \\
0 & -1 & -3 & -11
\end{array}\right]
$$

Now we divide both sides of Equation 4 by -3 to obtain Equation 6

$$
\begin{array}{l}
(1) \\
(6) \\
(5)
\end{array}
\begin{cases}
x + 2y + z = 8 \\
 y + z = 5 \\
 -y - 3z = -11
\end{cases}
\qquad
\left[\begin{array}{ccc|c}
1 & 2 & 1 & 8 \\
0 & 1 & 1 & 5 \\
0 & -1 & -3 & -11
\end{array}\right]
$$

and add Equation 6 to Equation 5 to obtain Equation 7.

$$
\begin{array}{l}
(1) \\
(6) \\
(7)
\end{array}
\begin{cases}
x + 2y + z = 8 \\
 y + z = 5 \\
 -2z = -6
\end{cases}
\qquad
\left[\begin{array}{ccc|c}
1 & 2 & 1 & 8 \\
0 & 1 & 1 & 5 \\
0 & 0 & -2 & -6
\end{array}\right]
$$

Finally, we divide both sides of Equation 7 by -2 to obtain the system

$$
\begin{array}{l}
(1) \\
(6)
\end{array}
\begin{cases}
x + 2y + z = 8 \\
 y + z = 5 \\
 z = 3
\end{cases}
\qquad
\left[\begin{array}{ccc|c}
1 & 2 & 1 & 8 \\
0 & 1 & 1 & 5 \\
0 & 0 & 1 & 3
\end{array}\right]
$$

Carl Friedrich Gauss
(1777–1855)
Many people consider Gauss to be the greatest mathematician of all time. He made contributions in the areas of number theory, solutions of equations, the geometry of curved surfaces, and statistics. For his efforts, he earned the title "Prince of the Mathematicians."

The system now can be solved by back substitution. Because $z = 3$, we can substitute 3 for z in Equation 6 and solve for y:

$$
\begin{array}{l}
(6)
\end{array}
\begin{aligned}
y + z &= 5 \\
y + 3 &= 5 \\
y &= 2
\end{aligned}
$$

We now can substitute 2 for y and 3 for z in Equation 1 and solve for x:

$$
\begin{array}{l}
(1)
\end{array}
\begin{aligned}
x + 2y + z &= 8 \\
x + 2(2) + 3 &= 8 \\
x + 7 &= 8 \\
x &= 1
\end{aligned}
$$

The solution is the ordered triple $(1, 2, 3)$.

Self Check 1 Solve: $\begin{cases} x - 3y = 1 \\ 2x + y = 9 \end{cases}$.

1. Use Row Operations to Write Matrices in Row Echelon Form

In Example 1, we performed operations on the equations as well as on the corresponding matrices. These matrix operations are called **elementary row operations.**

Elementary Row Operations

If any of the following operations are performed on the rows of a system matrix, the matrix of an equivalent system results:

Type 1 row operation: Two rows of a matrix can be interchanged.

Type 2 row operation: The elements of a row of a matrix can be multiplied by a nonzero constant.

Type 3 row operation: Any row can be changed by adding a multiple of another row to it.

- A type 1 row operation is equivalent to writing the equations of a system in a different order.
- A type 2 row operation is equivalent to multiplying both sides of an equation by a nonzero constant.
- A type 3 row operation is equivalent to adding a multiple of one equation to another.

Any matrix that can be obtained from another matrix by a sequence of elementary row operations is **row equivalent** to the original matrix. This means that row equivalent matrices represent equivalent systems of equations.

The process of Example 1 changed a system matrix into a special row equivalent form, called **row echelon form.**

Row Echelon Form of a Matrix

A matrix is in **row echelon form** if it has these three properties:

1. The first nonzero entry in each row (called the **lead entry**) is 1.

2. Lead entries appear farther to the right as we move down the rows of the matrix.

3. Any rows containing only 0's are at the bottom of the matrix.

Examples of Matrices in Row Echelon Form The matrices

$$\begin{bmatrix} 1 & 3 & 0 & 7 \\ 0 & 0 & 1 & 8 \\ 0 & 0 & 0 & 0 \end{bmatrix}, \quad \begin{bmatrix} 0 & 1 & 2 \\ 0 & 0 & 1 \end{bmatrix}, \quad \text{and} \quad \begin{bmatrix} 1 & 1 \\ 0 & 0 \\ 0 & 0 \\ 0 & 0 \end{bmatrix}$$

are in row echelon form, because the first nonzero entry in each row is 1, the lead entries appear farther to the right as we move down the rows, and any rows that consist entirely of 0's are at the bottom of the matrix.

Examples of Matrices Not in Row Echelon Form The next three matrices are not in row echelon form.

$$\begin{bmatrix} 1 & 3 & 0 & 7 \\ 0 & 0 & 1 & 8 \\ 0 & 0 & 1 & 0 \end{bmatrix}$$

The lead entry in the last row is not to the right of the lead entry of the middle row.

$$\begin{bmatrix} 0 & 3 & 0 \\ 0 & 0 & 1 \end{bmatrix}$$ The lead entry of the first row is not 1.

$$\begin{bmatrix} 0 & 0 \\ 1 & 0 \\ 0 & 1 \end{bmatrix}$$ The row of 0's is not last.

EXAMPLE 2   Solve the following system by transforming its system matrix into row echelon form and back substituting.

$$\begin{cases} x + 2y + 3z = 4 \\ 2x - y - 2z = 0 \\ x - 3y - 3z = -2 \end{cases}$$

Solution   This system is represented by the system matrix

$$\begin{bmatrix} 1 & 2 & 3 & \vdots & 4 \\ 2 & -1 & -2 & \vdots & 0 \\ 1 & -3 & -3 & \vdots & -2 \end{bmatrix}$$

We will reduce the system matrix to row echelon form. We can use the 1 in the upper left corner to zero out the rest of the first column. To do so, we multiply the first row by -2 and add the result to the second row. We indicate this operation with the notation $(-2)R1 + R2$

$$\begin{bmatrix} 1 & 2 & 3 & \vdots & 4 \\ 2 & -1 & -2 & \vdots & 0 \\ 1 & -3 & -3 & \vdots & -2 \end{bmatrix} (-2)R1 + R2 \rightarrow \begin{bmatrix} 1 & 2 & 3 & \vdots & 4 \\ 0 & -5 & -8 & \vdots & -8 \\ 1 & -3 & -3 & \vdots & -2 \end{bmatrix}$$

Next, we multiply row one by -1 and add the result to row three to get a new row three. This fills the rest of the first column with 0's.

$$(-1)R1 + R3 \rightarrow \begin{bmatrix} 1 & 2 & 3 & \vdots & 4 \\ 0 & -5 & -8 & \vdots & -8 \\ 0 & -5 & -6 & \vdots & -6 \end{bmatrix}$$

Comment
When reducing a matrix to row echelon form, we obtain 0's below each lead entry in a column by adding multiples of the row containing the lead entry.

We multiply row two by -1 and add the result to row three to get

$$(-1)R2 + R3 \rightarrow \begin{bmatrix} 1 & 2 & 3 & \vdots & 4 \\ 0 & -5 & -8 & \vdots & -8 \\ 0 & 0 & 2 & \vdots & 2 \end{bmatrix}$$

Finally, we multiply row two by $-\frac{1}{5}$ and row three by $\frac{1}{2}$.

$$\begin{matrix} \left(-\frac{1}{5}\right)R2 \rightarrow \\ \left(\frac{1}{2}\right)R3 \rightarrow \end{matrix} \begin{bmatrix} 1 & 2 & 3 & \vdots & 4 \\ 0 & 1 & \frac{8}{5} & \vdots & \frac{8}{5} \\ 0 & 0 & 1 & \vdots & 1 \end{bmatrix}$$

The final matrix, now in row echelon form, represents the equivalent system

$$\begin{cases} x + 2y + 3z = 4 \\ y + \frac{8}{5}z = \frac{8}{5} \\ z = 1 \end{cases}$$

We now can use back substitution to solve the system. To find y, we substitute 1 for z in the second equation.

$$y + \frac{8}{5}z = \frac{8}{5}$$

$$y + \frac{8}{5}(1) = \frac{8}{5}$$

$$y = 0$$

To solve for x, we substitute 0 for y and 1 for z in the first equation.

$$x + 2y + 3z = 4$$
$$x + 2(0) + 3(1) = 4$$
$$x + 3 = 4$$
$$x = 1$$

The solution of the original system is the ordered triple $(x, y, z) = (1, 0, 1)$.

Self Check 2   Solve: $\begin{cases} x + y + 2z = 5 \\ y - 2z = 0 \\ x - z = 0 \end{cases}$.

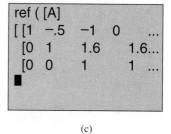

Accent on Technology

Reducing a Matrix to Row Echelon Form

We can use a TI-84 Plus graphing calculator to reduce a matrix to row echelon form. For example, to reduce the matrix of Example 2

$$\begin{bmatrix} 1 & 2 & 3 & 4 \\ 2 & -1 & -2 & 0 \\ 1 & -3 & -3 & -2 \end{bmatrix}$$

to row echelon form, we press $\boxed{\text{2nd}}$ $\boxed{\text{MATRIX}}$, select EDIT, and enter the size and entries of the matrix, as shown in Figure 6-5(a). We then press $\boxed{\text{2nd}}$ $\boxed{\text{QUIT}}$ to clear the screen. After pressing $\boxed{\text{2nd}}$ $\boxed{\text{MATRIX}}$ again, we select MATH, scroll down to A: ref (, and press $\boxed{\text{ENTER}}$. Finally, we press $\boxed{\text{2nd}}$ $\boxed{\text{MATRIX}}$ and select 1 : [A] by pressing $\boxed{\text{ENTER}}$ to obtain Figure 6-5(b). Pressing $\boxed{\text{ENTER}}$ one last time gives the ref (row echelon form) of the matrix, shown in Figure 6-5(c).

```
MATRIX[A]  3 ×4
[ 1      2       3    –
[ 2     -1      -2    –
[ 1     -3      -3    –



3,1 = 1
```
(a)

```
ref ( [A]




```
(b)

```
ref ( [A]
[ [1   -.5   -1    0    ...
  [0    1    1.6   1.6...
  [0    0    1     1   ...

```
(c)

Figure 6-5

Comment

The row echelon form shown in Figure 6-5(c) is not the same as the row echelon form found in Example 2. However, if we multiply row two of the matrix shown in Figure 6-5(c) by $\frac{5}{2}$ and add the results to row one, we obtain the equivalent row echelon form found in Example 2.

$$\begin{bmatrix} 1 & -.5 & -1 & 0 \\ 0 & 1 & 1.6 & 1.6 \\ 0 & 0 & 1 & 1 \end{bmatrix} \left(\tfrac{5}{2}\right)R2 + R1 \rightarrow \begin{bmatrix} 1 & 2 & 3 & 4 \\ 0 & 1 & \tfrac{8}{5} & \tfrac{8}{5} \\ 0 & 0 & 1 & 1 \end{bmatrix}$$

If we back substitute into the equations represented by either row echelon form, we will obtain the same values for x, y, and z.

Since the systems of Examples 1 and 2 have solutions, each system is consistent. The next two examples illustrate a system that is inconsistent and a system whose equations are dependent.

2. Solve an Inconsistent System

EXAMPLE 3   Use matrix methods to solve $\begin{cases} x + y + z = 3 \\ 2x - y + z = 2. \\ 3y + z = 1 \end{cases}$

Solution   We form the system matrix and use row operations to write it in row echelon form. We use the 1 in the top left position to zero out the rest of the first column.

$$\begin{bmatrix} 1 & 1 & 1 & \vdots & 3 \\ 2 & -1 & 1 & \vdots & 2 \\ 0 & 3 & 1 & \vdots & 1 \end{bmatrix} (-2)R1 + R2 \rightarrow \begin{bmatrix} 1 & 1 & 1 & \vdots & 3 \\ 0 & -3 & -1 & \vdots & -4 \\ 0 & 3 & 1 & \vdots & 1 \end{bmatrix}$$

$$(1)R2 + R3 \rightarrow \begin{bmatrix} 1 & 1 & 1 & \vdots & 3 \\ 0 & -3 & -1 & \vdots & -4 \\ 0 & 0 & 0 & \vdots & -3 \end{bmatrix}$$

Since the last row of the matrix represents the equation

$$0x + 0y + 0z = -3$$

and no values of x, y, and z could make $0 = -3$, there is no point in continuing. The given system has no solution and is inconsistent.

Self Check 3   Solve: $\begin{cases} x + y + z = 5 \\ y - z = 1 \\ x + 2y = 3 \end{cases}$.

3. Solve a System of Dependent Equations

EXAMPLE 4   Use matrices to solve $\begin{cases} x + 2y + z = 8 \\ 2x + y - z = 1 \\ x - y - 2z = -7 \end{cases}$.

Solution   We can set up the system matrix and use row operations to reduce it to row echelon form:

$$\begin{bmatrix} 1 & 2 & 1 & \vdots & 8 \\ 2 & 1 & -1 & \vdots & 1 \\ 1 & -1 & -2 & \vdots & -7 \end{bmatrix} \begin{matrix} (-2)R1 + R2 \rightarrow \\ (-1)R1 + R3 \rightarrow \end{matrix} \begin{bmatrix} 1 & 2 & 1 & \vdots & 8 \\ 0 & -3 & -3 & \vdots & -15 \\ 0 & -3 & -3 & \vdots & -15 \end{bmatrix}$$

$$\begin{pmatrix} -\frac{1}{3} \end{pmatrix}R2 \rightarrow \begin{pmatrix} -\frac{1}{3} \end{pmatrix}R3 \rightarrow \left[\begin{array}{ccc|c} 1 & 2 & 1 & 8 \\ 0 & 1 & 1 & 5 \\ 0 & 1 & 1 & 5 \end{array}\right]$$

$$(-1)R2 + R3 \rightarrow \left[\begin{array}{ccc|c} 1 & 2 & 1 & 8 \\ 0 & 1 & 1 & 5 \\ 0 & 0 & 0 & 0 \end{array}\right]$$

The final matrix is in echelon form and represents the system

(1) $\begin{cases} x + 2y + z = 8 \\ \end{cases}$
(2) $\begin{cases} y + z = 5 \\ \end{cases}$
(3) $\begin{cases} 0x + 0y + 0z = 0 \end{cases}$

Since all coefficients in Equation 3 are 0, it can be ignored. To solve this system by back substitution, we solve Equation 2 for y in terms of z:

$$y = 5 - z$$

and then substitute $5 - z$ for y in Equation 1 and solve for x in terms of z.

(1)
$$\begin{aligned} x + 2y + z &= 8 \\ x + 2(5 - z) + z &= 8 \\ x + 10 - 2z + z &= 8 \qquad \text{Remove parentheses.} \\ x + 10 - z &= 8 \qquad \text{Combine like terms.} \\ x &= -2 + z \qquad \text{Solve for } x. \end{aligned}$$

The solution of this system is $(x, y, z) = (-2 + z, 5 - z, z)$.

There are infinitely many solutions to this system. We can choose any real number for z, but once it is chosen, x and y are determined. For example,

- If $z = 3$, then $x = 1$ and $y = 2$. One possible solution of this system is $x = 1$, $y = 2$, and $z = 3$.
- If $z = 2$, then $x = 0$ and $y = 3$. Another solution is $x = 0$, $y = 3$, and $z = 2$.

Because this system has solutions, it is consistent. Because there are infinitely many solutions, the equations are dependent.

Self Check 4   Solve: $\begin{cases} x + 2y + z = 2 \\ x - z = 2 \\ 2x + 2y = 4 \end{cases}$.

4.   Solve Systems Using Gauss–Jordan Elimination

In Gaussian elimination, we perform row operations until the system matrix is in row echelon form. Then we convert the matrix back into equation form and solve by back substitution. A modification of this method, called **Gauss–Jordan elimination,** uses row operations to produce a matrix in *reduced* **row echelon** form. With this method, we can obtain the solution of the system from that matrix directly, and back substitution is not needed.

Reduced Row Echelon Form of a Matrix

A matrix is in **reduced row echelon form** if

1. It is in row echelon form.
2. Entries above each lead entry are also zero.

Examples of Matrices in Reduced Row Echelon Form The matrices

$$\begin{bmatrix} 1 & 0 & 0 \\ 0 & 1 & 0 \end{bmatrix}, \quad \begin{bmatrix} 1 & 0 & 0 & 7 \\ 0 & 1 & 0 & 8 \\ 0 & 0 & 1 & 3 \end{bmatrix}$$

are in reduced row echelon form, because each matrix is in row echelon form and the entries above each lead entry are 0's.

Examples of Matrices Not in Reduced Row Echelon Form

$$\begin{bmatrix} 1 & 3 & 0 & 7 \\ 0 & 0 & 1 & 8 \\ 0 & 0 & 1 & 0 \end{bmatrix} \quad \text{The matrix is not in row echelon form.}$$

$$\begin{bmatrix} 1 & 3 & 5 \\ 0 & 1 & 1 \end{bmatrix} \quad \text{The number above the lead entry of 1 is not 0.}$$

In the next example, we solve a system by Gauss–Jordan elimination.

EXAMPLE 5 Use Gauss–Jordan elimination to solve $\begin{cases} w + 2x + 3y + z = 4 \\ x + 4y - z = 0 \\ w - x - y + 2z = 2 \end{cases}$.

Solution We row reduce the system matrix as follows. We first use the 1 in the upper left corner to zero out the rest of the first column:

$$\begin{bmatrix} 1 & 2 & 3 & 1 & \vdots & 4 \\ 0 & 1 & 4 & -1 & \vdots & 0 \\ 1 & -1 & -1 & 2 & \vdots & 2 \end{bmatrix} \quad (-1)R1 + R3 \rightarrow \begin{bmatrix} 1 & 2 & 3 & 1 & \vdots & 4 \\ 0 & 1 & 4 & -1 & \vdots & 0 \\ 0 & -3 & -4 & 1 & \vdots & -2 \end{bmatrix}$$

The lead entry in the second row is already 1. We use it to zero out the rest of the second column.

$$\begin{matrix} (-2)R2 + R1 \rightarrow \\ \\ (3)R2 + R3 \rightarrow \end{matrix} \begin{bmatrix} 1 & 0 & -5 & 3 & \vdots & 4 \\ 0 & 1 & 4 & -1 & \vdots & 0 \\ 0 & 0 & 8 & -2 & \vdots & -2 \end{bmatrix}$$

To make the lead entry in the third row equal to 1, we multiply the third row by $\frac{1}{8}$ and use it to zero out the rest of the third column.

$$\left(\tfrac{1}{8}\right)R3 \rightarrow \begin{bmatrix} 1 & 0 & -5 & 3 & \vdots & 4 \\ 0 & 1 & 4 & -1 & \vdots & 0 \\ 0 & 0 & 1 & -\frac{1}{4} & \vdots & -\frac{1}{4} \end{bmatrix}$$

$$\begin{matrix} (5)R3 + R1 \rightarrow \\ (-4)R3 + R2 \rightarrow \\ \\ \end{matrix} \begin{bmatrix} 1 & 0 & 0 & \frac{7}{4} & \vdots & \frac{11}{4} \\ 0 & 1 & 0 & 0 & \vdots & 1 \\ 0 & 0 & 1 & -\frac{1}{4} & \vdots & -\frac{1}{4} \end{bmatrix}$$

The final matrix is in reduced row echelon form. Note that each lead entry is 1, and each is alone in its column. The matrix represents the system of equations

$$\begin{cases} w + \dfrac{7}{4}z = \dfrac{11}{4} \\ x = 1 \\ y - \dfrac{1}{4}z = -\dfrac{1}{4} \end{cases} \quad \text{or} \quad \begin{cases} w = \dfrac{11}{4} - \dfrac{7}{4}z \\ x = 1 \\ y = -\dfrac{1}{4} + \dfrac{1}{4}z \end{cases}$$

A general solution is $(w, x, y, z) = \left(\frac{11}{4} - \frac{7}{4}z, 1, -\frac{1}{4} + \frac{1}{4}z, z\right)$. To find some specific solutions, we choose values for z, and the corresponding values of w, x, and y will be determined. For example, if $z = 1$, then $w = 1$, $x = 1$, and $y = 0$. Thus, $(w, x, y, z) = (1, 1, 0, 1)$ is a solution. If $z = -1$, another solution is $(w, x, y, z) = \left(\frac{9}{2}, 1, -\frac{1}{2}, -1\right)$.

Self Check 5 Use Gauss–Jordan elimination to solve the system:

$$\begin{cases} w + x + y - z = 2 \\ 2w + 2x + 2y - 2z = 3 \\ w + x - y + 3z = 4 \\ w + 3x + 2y - 2z = -1 \end{cases}$$

To find the row reduced form of a matrix with a TI-84 Plus calculator, enter the matrix and use the rref (row reduced echelon form) feature.

In Example 5, there were more variables than equations. In the next example, there are more equations than variables.

EXAMPLE 6 Use Gauss–Jordan elimination to solve $\begin{cases} 2x + y = 4 \\ x - 3y = 9 \\ x + 4y = -5 \end{cases}$.

Solution To get a 1 in the top left corner, we could multiply the first row by $\frac{1}{2}$, but that would introduce fractions. Instead, we can exchange the first two rows and proceed as follows:

$$\begin{bmatrix} 2 & 1 & 4 \\ 1 & -3 & 9 \\ 1 & 4 & -5 \end{bmatrix} \begin{array}{c} R1 \leftrightarrow R2 \end{array} \begin{bmatrix} 1 & -3 & 9 \\ 2 & 1 & 4 \\ 1 & 4 & -5 \end{bmatrix}$$

$$\begin{array}{c} (-2)R1 + R2 \rightarrow \\ (-1)R1 + R3 \rightarrow \end{array} \begin{bmatrix} 1 & -3 & 9 \\ 0 & 7 & -14 \\ 0 & 7 & -14 \end{bmatrix}$$

$$\begin{array}{c} \left(\frac{1}{7}\right)R2 \rightarrow \\ \left(\frac{1}{7}\right)R3 \rightarrow \end{array} \begin{bmatrix} 1 & -3 & 9 \\ 0 & 1 & -2 \\ 0 & 1 & -2 \end{bmatrix}$$

$$\begin{array}{c} (3)R2 + R1 \rightarrow \\ \\ (-1)R2 + R3 \rightarrow \end{array} \begin{bmatrix} 1 & 0 & 3 \\ 0 & 1 & -2 \\ 0 & 0 & 0 \end{bmatrix}$$

This final matrix is in reduced echelon form. It represents the system

$$\begin{cases} x = 3 \\ y = -2 \end{cases}$$

Verify that $x = 3$ and $y = -2$ satisfy the equations of the original system.

Self Check 6 Solve: $\begin{cases} x + y = 5 \\ x - y = 1 \\ x + 2y = 7 \end{cases}$.

6.2 Exercises

Vocabulary and Concepts *Fill in the blanks.*

1. A rectangular array of numbers is called a _____.

2. A 3×5 matrix has __ rows and __ columns.

3. The matrix containing the coefficients of the variables is called the _____ matrix.

4. The coefficient matrix joined to the column of constants is called the _____ matrix or the _____ matrix.

5. Each row of a system matrix represents one _____.

6. The rows of the system matrix are changed using elementary _____.

7. If one system matrix is changed to another using row operations, the matrices are _____.

8. If two system matrices are row equivalent, then the systems have the _____.

9. In a type 1 row operation, two rows of a matrix can be _____.

10. In a type 2 row operation, one entire row can be _____ by a nonzero constant.

11. In a type 3 row operation, any row can be changed by _____ to it any _____ of another row.

12. The first nonzero entry in a row is called that row's _____.

Practice *Use Gaussian elimination to solve each system.*

13. $\begin{cases} x + y = 5 \\ x - 2y = -4 \end{cases}$

14. $\begin{cases} x + 3y = 8 \\ 2x - 5y = 5 \end{cases}$

15. $\begin{cases} x - y = 1 \\ 2x - y = 8 \end{cases}$

16. $\begin{cases} x - 5y = 4 \\ 2x + 3y = 21 \end{cases}$

17. $\begin{cases} x + 2y - z = 2 \\ x - 3y + 2z = 1 \\ x + y - 3z = -6 \end{cases}$

18. $\begin{cases} x + 5y - z = 2 \\ x + 2y + z = 3 \\ x + y + z = 2 \end{cases}$

19. $\begin{cases} x - y - z = -3 \\ 5x + y = 6 \\ y + z = 4 \end{cases}$

20. $\begin{cases} x + y = 1 \\ x + z = 3 \\ y + z = 2 \end{cases}$

Determine whether each matrix is in row echelon form, reduced row echelon form, or neither.

21. $\begin{bmatrix} 1 & 3 & 0 & 5 \\ 0 & 1 & 2 & 7 \\ 0 & 0 & 1 & 0 \end{bmatrix}$

22. $\begin{bmatrix} 1 & 3 & 0 & 5 \\ 0 & 1 & 2 & 7 \\ 0 & 0 & 0 & 0 \end{bmatrix}$

23. $\begin{bmatrix} 1 & 0 & 1 \\ 0 & 1 & 5 \\ 0 & 0 & 0 \\ 0 & 0 & 0 \end{bmatrix}$

24. $\begin{bmatrix} 1 & 0 & 1 \\ 0 & 1 & 5 \\ 0 & 0 & 1 \\ 0 & 0 & 0 \end{bmatrix}$

Write each system as a matrix and solve it by Gaussian elimination.

25. $\begin{cases} 2x + y = 3 \\ x - 3y = 5 \end{cases}$

26. $\begin{cases} x + 2y = -1 \\ 3x - 5y = 19 \end{cases}$

27. $\begin{cases} x - 7y = -2 \\ 5x - 2y = -10 \end{cases}$

28. $\begin{cases} 3x - y = 3 \\ 2x + y = -3 \end{cases}$

29. $\begin{cases} 2x - y = 5 \\ x + 3y = 6 \end{cases}$

30. $\begin{cases} 3x - 5y = -25 \\ 2x + y = 5 \end{cases}$

31. $\begin{cases} x - 2y = 3 \\ -2x + 4y = 6 \end{cases}$

32. $\begin{cases} 2(2y - x) = 6 \\ 4y = 2(x + 3) \end{cases}$

33. $\begin{cases} 2x - y = 7 \\ -x + \dfrac{1}{3}y = -\dfrac{7}{3} \end{cases}$

34. $\begin{cases} 45x - 6y = 60 \\ 30x + 15y = 63.75 \end{cases}$

35. $\begin{cases} x - y + z = 3 \\ 2x - y + z = 4 \\ x + 2y - z = -1 \end{cases}$

36. $\begin{cases} 2x + y - z = 1 \\ x + y - z = 0 \\ 3x + y + 2z = 2 \end{cases}$

37. $\begin{cases} x + y - z = -1 \\ 3x + y = 4 \\ y - 2z = -4 \end{cases}$ **38.** $\begin{cases} 3x + y = 7 \\ x - z = 0 \\ y - 2z = -8 \end{cases}$

39. $\begin{cases} x - y + z = 2 \\ 2x + y + z = 5 \\ 3x - 4z = -5 \end{cases}$ **40.** $\begin{cases} x + z = -1 \\ 3x + y = 2 \\ 2x + y + 5z = 3 \end{cases}$

41. $\begin{cases} x + y + 2z = 4 \\ -x - y - 3z = -5 \\ 2x + y + z = 2 \end{cases}$ **42.** $\begin{cases} x + y - z = 5 \\ x + y + z = 2 \\ 3x + 3y - z = 12 \end{cases}$

43. $\begin{cases} -x + 3y + 2z = -10 \\ 3x - 2y - 2z = 7 \\ -2x + y - z = -10 \end{cases}$ **44.** $\begin{cases} 2x - y + z = 6 \\ 3x + y - z = 2 \\ -x + 3y - 3z = 8 \end{cases}$

***Write each system as a matrix and solve it by
Gauss–Jordan elimination. If a system has infinitely
many solutions, show a general solution.***

45. $\begin{cases} x - 2y = 7 \\ y = 3 \end{cases}$ **46.** $\begin{cases} x - 2y = 7 \\ y = 8 \end{cases}$

47. $\begin{cases} x + 2y - z = 3 \\ y + 3z = 1 \\ z = -2 \end{cases}$ **48.** $\begin{cases} x - 3y + 2z = -1 \\ y - 2z = 3 \\ z = 5 \end{cases}$

49. $\begin{cases} x - y = 7 \\ x + y = 13 \end{cases}$ **50.** $\begin{cases} x + 2y = 7 \\ 2x - y = -1 \end{cases}$

51. $\begin{cases} x - \dfrac{1}{2}y = 0 \\ x + 2y = 0 \end{cases}$ **52.** $\begin{cases} x - y = 5 \\ -x + \dfrac{1}{5}y = -9 \end{cases}$

53. $\begin{cases} x + y + 2z = 0 \\ x + y + z = 2 \\ x + z = 1 \end{cases}$ **54.** $\begin{cases} x + 2y = -3 \\ x + 4y = -2 \\ 2x + z = -8 \end{cases}$

55. $\begin{cases} 2x + y - 2z = 1 \\ -x + y - 3z = 0 \\ 4x + 3y = 4 \end{cases}$ **56.** $\begin{cases} 3x + y = 3 \\ 3x + y - z = 2 \\ 6x + z = 5 \end{cases}$

57. $\begin{cases} 2x - 2y + 3z + t = 2 \\ x + y + z + t = 5 \\ -x + 2y - 3z + 2t = 2 \\ x + y + 2z - t = 4 \end{cases}$ **58.** $\begin{cases} x + y + 2z + t = 1 \\ x + 2y + z + t = 2 \\ 2x + y + z + t = 4 \\ x + y + z + 2t = 3 \end{cases}$

59. $\begin{cases} x + y + t = 4 \\ x + z + t = 2 \\ 2x + 2y + z + 2t = 8 \\ x - y + z - t = -2 \end{cases}$ **60.** $\begin{cases} x - y + 2z + t = 3 \\ 3x - 2y - z - t = 4 \\ 2x + y + 2z - t = 10 \\ x + 2y + z - 3t = 8 \end{cases}$

61. $\begin{cases} \dfrac{1}{3}x + \dfrac{3}{4}y - \dfrac{2}{3}z = -2 \\ x + \dfrac{1}{2}y + \dfrac{1}{3}z = 1 \\ \dfrac{1}{6}x - \dfrac{1}{8}y - z = 0 \end{cases}$ **62.** $\begin{cases} \dfrac{1}{4}x + y + 3z = 1 \\ \dfrac{1}{2}x - 4y + 6z = -1 \\ \dfrac{1}{3}x - 2y - 2z = -1 \end{cases}$

63. $\begin{cases} \dfrac{1}{2}x + \dfrac{1}{4}y - z = 2 \\ \dfrac{2}{3}x + \dfrac{1}{4}y + \dfrac{1}{2}z = \dfrac{3}{2} \\ \dfrac{2}{3}x + z = -\dfrac{1}{3} \end{cases}$ **64.** $\begin{cases} \dfrac{5}{7}x - \dfrac{1}{3}y + z = 0 \\ \dfrac{2}{7}x + y + \dfrac{1}{8}z = 9 \\ 6x + 4y - \dfrac{27}{4}z = 20 \end{cases}$

65. $\begin{cases} 3x - 6y + 9z = 18 \\ 2x - 4y + 3z = 12 \\ x - 2y + 3z = 6 \end{cases}$ **66.** $\begin{cases} x + 2y - z = 7 \\ 2x - y + z = 2 \\ 3x - 4y + 3z = -3 \end{cases}$

***Each system contains a different number of equations
than variables. Solve each system using Gauss–Jordan
elimination. If a system has infinitely many solutions,
show a general solution.***

67. $\begin{cases} x + y = -2 \\ 3x - y = 6 \\ 2x + 2y = -4 \\ x - y = 4 \end{cases}$ **68.** $\begin{cases} x - y = -3 \\ 2x + y = -3 \\ 3x - y = -7 \\ 4x + y = -7 \end{cases}$

69. $\begin{cases} x + 2y + z = 4 \\ 3x - y - z = 2 \end{cases}$ **70.** $\begin{cases} x + 2y - 3z = 5 \\ 5x + y - z = -11 \end{cases}$

71. $\begin{cases} w + x = 1 \\ w + y = 0 \\ x + z = 0 \end{cases}$ **72.** $\begin{cases} w + x - y + z = 2 \\ 2w - x - 2y + z = 0 \\ w - 2x - y + z = -1 \end{cases}$

73. $\begin{cases} x + y = 3 \\ 2x + y = 1 \\ 3x + 2y = 2 \end{cases}$ **74.** $\begin{cases} x + 2y + z = 4 \\ x - y + z = 1 \\ 2x + y + 2z = 2 \\ 3x + 3z = 6 \end{cases}$

Applications *Use matrix methods to solve each problem.*

75. Flight range The speed of an airplane with a tailwind is 300 miles per hour, and with a headwind 220 miles per hour. On a day with no wind, how far could the plane travel on a 5-hour fuel supply?

76. Resource allocation 120,000 gallons of fuel are to be divided between two airlines. Triple A Airways requires twice as much as UnityAir. How much fuel should be allocated to Triple A?

77. Library shelving To use space effectively, librarians like to fill shelves completely. One 35-inch shelf can hold 3 dictionaries, 5 atlases, and 1 thesaurus; or 6 dictionaries and 2 thesauruses; or 2 dictionaries, 4 atlases, and 3 thesauruses. How wide is one copy of each book?

78. Copying machine productivity When both copying machines A and B are working, secretaries can make 100 copies in one minute. In one minute's time, copiers A and C together produce 140 copies, and all three working together produce 180 copies. How many copies per minute can each machine produce separately?

79. Nutritional planning One ounce of each of three foods has the vitamin and mineral content shown in the table. How many ounces of each must be used to provide exactly 22 milligrams (mg) of niacin, 12 mg of zinc, and 20 mg of vitamin C?

Food	Niacin	Zinc	Vitamin C
A	1 mg	1 mg	2 mg
B	2 mg	1 mg	1 mg
C	2 mg	1 mg	2 mg

80. Chainsaw sculpting A wood sculptor carves three types of statues with a chainsaw. The number of hours required for carving, sanding, and painting a totem pole, a bear, and a deer are shown in the table. How many of each should be produced to use all available labor hours?

	Totem pole	Bear	Deer	Time available
Carving	2 hr	2 hr	1 hr	14 hr
Sanding	1 hr	2 hr	2 hr	15 hr
Painting	3 hr	2 hr	2 hr	21 hr

Discovery and Writing

81. Explain the difference between the row echelon form and the reduced row echelon form of a matrix.

82. If the upper-left corner entry of a matrix is zero, what row operation might you do first?

83. What characteristics of a row reduced matrix would let you conclude that the system is inconsistent?

84. Explain the differences between Gaussian elimination and Gauss–Jordan elimination.

Use matrix methods to solve each system.

85. $\begin{cases} x^2 + y^2 + z^2 = 14 \\ 2x^2 + 3y^2 - 2z^2 = -7 \\ x^2 - 5y^2 + z^2 = 8 \end{cases}$
(*Hint:* Solve first as a system in x^2, y^2, and z^2.)

86. $\begin{cases} 5\sqrt{x} + 2\sqrt{x} + \sqrt{z} = 22 \\ \sqrt{x} + \sqrt{y} - \sqrt{z} = 5 \\ 3\sqrt{x} - 2\sqrt{y} - 3\sqrt{z} = 10 \end{cases}$

Review *Fill in the blank.*

87. The slope–intercept form of the equation of a line is _____.

88. The point–slope form of the equation of a line is

_____.

89. The slopes of parallel lines are _____.

90. The slopes of _____ lines are negative reciprocals.

Write the equation of a line with the given properties.

91. The line has a slope of 2 and a *y*-intercept of 7.

92. The line has a slope of -3 and passes through the point $(2, -3)$.

93. The line is vertical and passes through $(2, -3)$.

94. The line is horizontal and passes through $(2, -3)$.

6.3 Matrix Algebra

Objectives

1. Add and Subtract Matrices
2. Multiply a Matrix by a Constant
3. Multiply Matrices
4. Solve Problems Using Matrices
5. Recognize the Identity Matrix

With certain restrictions, we can add, subtract, and multiply matrices. In fact, many problems can be solved using the arithmetic of matrices.

To illustrate the addition of matrices, suppose there are 108 police officers employed at two different locations:

Downtown Station		
	Male	Female
Day shift	21	18
Night shift	12	6

Suburban Station		
	Male	Female
Day shift	14	12
Night shift	15	10

The employment information about the police officers is contained in two matrices.

$$D = \begin{bmatrix} 21 & 18 \\ 12 & 6 \end{bmatrix} \quad \text{and} \quad S = \begin{bmatrix} 14 & 12 \\ 15 & 10 \end{bmatrix}$$

The entry 21 in matrix *D* indicates that 21 male officers work the day shift at the downtown station. The entry 10 in matrix *S* indicates that 10 female officers work the night shift at the suburban station.

To find the city-wide totals, we can add the corresponding entries of matrices *D* and *S*:

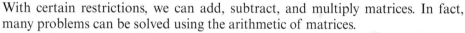

$$D + S = \begin{bmatrix} 21 & 18 \\ 12 & 6 \end{bmatrix} + \begin{bmatrix} 14 & 12 \\ 15 & 10 \end{bmatrix} = \begin{bmatrix} 35 & 30 \\ 27 & 16 \end{bmatrix}$$

We interpret the total to mean:

	Male	Female
Day shift	35	30
Night shift	27	16

To illustrate how to multiply a matrix by a number, suppose that one-third of the officers at the downtown station retire. The downtown staff then would consist of $\frac{2}{3}D$ officers. We can compute $\frac{2}{3}D$ by multiplying each entry of matrix D by $\frac{2}{3}$.

$$\frac{2}{3}D = \frac{2}{3}\begin{bmatrix} 21 & 18 \\ 12 & 6 \end{bmatrix} = \begin{bmatrix} 14 & 12 \\ 8 & 4 \end{bmatrix}$$

After retirements, the downtown staff will be

	Male	Female
Day shift	14	12
Night shift	8	4

These examples illustrate two calculations used in the algebra of matrices, which is the topic of this section. We begin by giving a formal definition of a matrix and defining when two matrices are equal.

Matrices An *m × n matrix* is a rectangular array of *mn* numbers arranged in *m* rows and *n* columns. We say that the matrix is of **size** (or **order**) $m \times n$.

Matrices often are denoted by letters such as A, B, and C. To denote the entries in an $m \times n$ matrix A, we use double-subscript notation: The entry in the first row, third column is a_{13}, and the entry in the ith row, jth column is a_{ij}. We can use any of the following notations to denote the $m \times n$ matrix A:

$$A, \qquad [a_{ij}], \qquad \left.\begin{bmatrix} a_{11} & a_{12} & a_{13} & \cdots & a_{1n} \\ a_{21} & a_{22} & a_{23} & \cdots & a_{2n} \\ \vdots & \vdots & \vdots & \ddots & \vdots \\ a_{m1} & a_{m2} & a_{m3} & \cdots & a_{mn} \end{bmatrix}\right\} m \text{ rows}$$

$$\underbrace{\phantom{a_{11} \quad a_{12} \quad a_{13} \quad \cdots \quad a_{1n}}}_{n \text{ columns}}$$

Two matrices are equal if they are the same size, with the same entries in corresponding positions.

Equality of Matrices If $A = [a_{ij}]$ and $B = [b_{ij}]$ are both $m \times n$ matrices, then

$$A = B$$

provided that each entry a_{ij} in matrix A is equal to the corresponding entry b_{ij} in matrix B.

The following matrices are equal, because they are the same size and corresponding entries are equal.

$$\begin{bmatrix} \sqrt{9} & 0.5 \\ 1 & 4 \end{bmatrix} = \begin{bmatrix} 3 & \frac{1}{2} \\ 1 & 2^2 \end{bmatrix}$$

The following matrices are not equal, because they are not the same size.

$$\begin{bmatrix} 1 & 2 & 3 \\ 1 & 2 & 3 \end{bmatrix} \neq \begin{bmatrix} 1 & 2 & 3 \\ 1 & 2 & 3 \\ 1 & 2 & 3 \end{bmatrix}$$

The first matrix is 2×3, and the second is 3×3.

1. Add and Subtract Matrices

We can add matrices of the same size by adding the entries in corresponding positions.

Sum of Two Matrices   Let A and B be two $m \times n$ matrices. The sum, $A + B$, is the $m \times n$ matrix C found by adding the corresponding entries of matrices A and B:

$$A + B = C$$

where each entry c_{ij} in C is equal to the sum of a_{ij} in A and b_{ij} in B.

EXAMPLE 1   Add: $\begin{bmatrix} 2 & 1 & 3 \\ 1 & -1 & 0 \end{bmatrix} + \begin{bmatrix} 1 & -1 & 2 \\ -1 & 1 & 5 \end{bmatrix}$.

Solution   Since each matrix is 2×3, we can find their sum by adding their corresponding entries.

$$\begin{bmatrix} 2 & 1 & 3 \\ 1 & -1 & 0 \end{bmatrix} + \begin{bmatrix} 1 & -1 & 2 \\ -1 & 1 & 5 \end{bmatrix} = \begin{bmatrix} 2+1 & 1-1 & 3+2 \\ 1-1 & -1+1 & 0+5 \end{bmatrix}$$

$$= \begin{bmatrix} 3 & 0 & 5 \\ 0 & 0 & 5 \end{bmatrix}$$

Self Check 1   Add: $\begin{bmatrix} 3 & -5 \\ 2 & 0 \\ -6 & 5 \end{bmatrix} + \begin{bmatrix} -3 & 4 \\ -1 & -3 \\ 7 & 0 \end{bmatrix}$.

Comment
Matrices that are not the same size cannot be added.

In arithmetic, 0 is the **additive identity,** because $a + 0 = 0 + a = a$ for any real number a. In matrix algebra, the matrix $\mathbf{0} = \begin{bmatrix} 0 & 0 \\ 0 & 0 \end{bmatrix}$ is called an **additive identity matrix,** because $A + \mathbf{0} = \mathbf{0} + A$. For example,

$$\begin{bmatrix} 1 & 2 \\ 3 & 4 \end{bmatrix} + \begin{bmatrix} 0 & 0 \\ 0 & 0 \end{bmatrix} = \begin{bmatrix} 0 & 0 \\ 0 & 0 \end{bmatrix} + \begin{bmatrix} 1 & 2 \\ 3 & 4 \end{bmatrix} = \begin{bmatrix} 1 & 2 \\ 3 & 4 \end{bmatrix}$$

The Additive Identity Matrix   Let A be any $m \times n$ matrix. There is an $m \times n$ matrix $\mathbf{0}$, called the **zero matrix** or the **additive identity matrix,** for which

$$A + \mathbf{0} = \mathbf{0} + A = A$$

The matrix $\mathbf{0}$ consists of m rows and n columns of 0's.

Every matrix also has an additive inverse.

The Additive Inverse of a Matrix   Any $m \times n$ matrix A has an **additive inverse,** an $m \times n$ matrix $-A$ with the property that the sum of A and $-A$ is the zero matrix:

$$A + (-A) = (-A) + A = \mathbf{0}$$

The entries of $-A$ are the negatives of the corresponding entries of A.

The additive inverse of $A = \begin{bmatrix} 1 & -3 & 2 \\ 0 & 1 & -5 \end{bmatrix}$ is the matrix

$$-A = \begin{bmatrix} -1 & 3 & -2 \\ 0 & -1 & 5 \end{bmatrix}$$ Each entry of $-A$ is the negative of the corresponding entry of A.

because their sum is the zero matrix:

$$A + (-A) = \begin{bmatrix} 1 & -3 & 2 \\ 0 & 1 & -5 \end{bmatrix} + \begin{bmatrix} -1 & 3 & -2 \\ 0 & -1 & 5 \end{bmatrix}$$

$$= \begin{bmatrix} 1-1 & -3+3 & 2-2 \\ 0+0 & 1-1 & -5+5 \end{bmatrix}$$

$$= \begin{bmatrix} 0 & 0 & 0 \\ 0 & 0 & 0 \end{bmatrix}$$

Subtraction of matrices is similar to the subtraction of real numbers.

Difference of Two Matrices   If A and B are $m \times n$ matrices, their difference, $A - B$, is the sum of A and the additive inverse of B:

$$A - B = A + (-B)$$

For example,

$$\begin{bmatrix} 3 & 7 \\ -4 & 0 \end{bmatrix} - \begin{bmatrix} -1 & 4 \\ -5 & 1 \end{bmatrix} = \begin{bmatrix} 3 & 7 \\ -4 & 0 \end{bmatrix} + \begin{bmatrix} 1 & -4 \\ 5 & -1 \end{bmatrix} = \begin{bmatrix} 4 & 3 \\ 1 & -1 \end{bmatrix}$$

2. Multiply a Matrix by a Constant

As we have seen in the police example at the beginning of the section, we can multiply a matrix by a constant by multiplying each of its entries by that constant. For example, if $A = \begin{bmatrix} 1 & 2 \\ 3 & 4 \end{bmatrix}$, then

$$5A = 5\begin{bmatrix} 1 & -2 \\ 3 & 4 \end{bmatrix} = \begin{bmatrix} 5 \cdot 1 & 5(-2) \\ 5 \cdot 3 & 5 \cdot 4 \end{bmatrix} = \begin{bmatrix} 5 & -10 \\ 15 & 20 \end{bmatrix}$$

$$-A = -1A = -1\begin{bmatrix} 1 & -2 \\ 3 & 4 \end{bmatrix} = \begin{bmatrix} -1 \cdot 1 & -1(-2) \\ -1 \cdot 3 & -1 \cdot 4 \end{bmatrix} = \begin{bmatrix} -1 & 2 \\ -3 & -4 \end{bmatrix}$$

The second example above illustrates that we can find the additive inverse of a matrix by multiplying the matrix by -1.

If A is a matrix, the real number k in the product kA is called a **scalar.**

Multiplying a Matrix by a Scalar   If A and B are two $m \times n$ matrices and k is a scalar, then $kA = B$ where each entry b_{ij} in B is equal to k times the corresponding entry a_{ij} in A.

EXAMPLE 2   Let $\begin{bmatrix} 5 & y \\ 15 & z \end{bmatrix} = 5\begin{bmatrix} x & 3 \\ 3 & y \end{bmatrix}$. Find y and z.

Solution   We simplify the right side of the expression by multiplying each entry of the matrix by 5.

$$\begin{bmatrix} 5 & y \\ 15 & z \end{bmatrix} = \begin{bmatrix} 5x & 15 \\ 15 & 5y \end{bmatrix}$$

Because the matrices are equal, their corresponding entries are equal. So $y = 15$, and $z = 5y$. We conclude that $y = 15$ and $z = 5 \cdot 15 = 75$.

Self Check 2   Find x.

Jack Hollingsworth/Getty Images

3. Multiply Matrices

To introduce multiplication of matrices, we consider the grading policy at a school where a student's final grade is based on a test score average and a homework average.

The grades for three students are shown below, along with how the test scores and the homework scores are weighted.

	Test score Average	Homework Average
Ann	83	95
Tonya	72	80
Carlos	85	92

	Weighting	
	System 1	System 2
Test average	0.5	0.3
Homework average	0.5	0.7

To calculate the students' final grades, we must use information from both tables. For example, if their teacher used weighting system 1, their final grades would be

Ann's grade = 83(0.5) + 95(0.5) = 41.5 + 47.5 = 89
Tonya's grade = 72(0.5) + 80(0.5) = 36 + 40 = 76
Carlos' grade = 85(0.5) + 92(0.5) = 42.5 + 46 = 88.5

If we calculate their final grades using weighting system 2, we get

Ann's grade = 83(0.3) + 95(0.7) = 24.9 + 66.5 = 91.4
Tonya's grade = 72(0.3) + 80(0.7) = 21.6 + 56 = 77.6
Carlos' grade = 85(0.3) + 92(0.7) = 25.5 + 64.4 = 89.9

The result of each grade calculation is the sum of the products of the elements in one row of the first matrix and one column of the second matrix. Each calculation is a part of the product of the matrices

$$A = \begin{bmatrix} 83 & 95 \\ 72 & 80 \\ 85 & 92 \end{bmatrix} \quad \text{and} \quad B = \begin{bmatrix} 0.5 & 0.3 \\ 0.5 & 0.7 \end{bmatrix}$$

Arthur Cayley
(1821–1895)
Cayley taught mathematics at Cambridge University. When he refused to take religious vows, he was fired and became a lawyer. After 14 years, he returned to mathematics and to Cambridge. Cayley was a major force in developing the theory of matrices.

To further illustrate how to find the product of two matrices, we will find the product of the 3×2 matrix A and the 2×2 matrix B. This product will exist because the number of columns in A is equal to the number of rows in B. The result will be a 3×2 matrix C.

$$AB = \begin{bmatrix} 83 & 95 \\ 72 & 80 \\ 85 & 92 \end{bmatrix} \begin{bmatrix} 0.5 & 0.3 \\ 0.5 & 0.7 \end{bmatrix} = \begin{bmatrix} ? & ? & ? \\ ? & ? & ? \\ ? & ? & ? \end{bmatrix} = C$$

Each entry of matrix C will be the result of multiplying the entries in one row of A and the entries in one column of B and adding the results. As shown below, the first-row, first-column entry of matrix C is the sum of the products of entries of the first row of A and the first column of B. The second-row, second-column entry of matrix C is the sum of the products of entries of the second row of A and the second column of B.

$$AB = \begin{bmatrix} 83 & 95 \\ 72 & 80 \\ 85 & 92 \end{bmatrix} \begin{bmatrix} 0.5 & 0.3 \\ 0.5 & 0.7 \end{bmatrix} = \begin{bmatrix} 83(0.5) + 95(0.5) & ? \\ ? & 72(0.3) + 80(0.7) \\ ? & ? \end{bmatrix} = \begin{bmatrix} 89 & ? \\ ? & 77.6 \\ ? & ? \end{bmatrix} = C$$

As shown below, the first-row, second-column entry of matrix C is the sum of the products of entries of the first row of A and the second column of B. The second-row, first-column entry of matrix C is the sum of the products of entries of the second row of A and the first column of B.

$$AB = \begin{bmatrix} 83 & 95 \\ 72 & 80 \\ 85 & 92 \end{bmatrix} \begin{bmatrix} 0.5 & 0.3 \\ 0.5 & 0.7 \end{bmatrix} = \begin{bmatrix} 83(0.5) + 95(0.5) & 83(0.3) + 95(0.7) \\ 72(0.5) + 80(0.5) & 72(0.3) + 80(0.7) \\ ? & ? \end{bmatrix} = \begin{bmatrix} 89 & 91.4 \\ 76 & 77.6 \\ ? & ? \end{bmatrix} = C$$

As shown below, the third-row, first-column entry of matrix C is the sum of the products of entries of the third row of A and the first column of B. The third-row, second-column entry of matrix C is the sum of the products of entries of the third row of A and the second column of B.

$$AB = \begin{bmatrix} 83 & 95 \\ 72 & 80 \\ 85 & 92 \end{bmatrix} \begin{bmatrix} 0.5 & 0.3 \\ 0.5 & 0.7 \end{bmatrix} = \begin{bmatrix} 83(0.5) + 95(0.5) & 83(0.3) + 95(0.7) \\ 72(0.5) + 80(0.5) & 72(0.3) + 80(0.7) \\ 85(0.5) + 92(0.5) & 85(0.3) + 92(0.7) \end{bmatrix} = \begin{bmatrix} 89 & 91.4 \\ 76 & 77.6 \\ 88.5 & 89.9 \end{bmatrix} = C$$

The resulting 3×2 matrix C is the product of matrices A and B, and gives the final grades of each student with respect to each weighting system.

For the product AB of two matrices to exist, the number of columns of A must equal the number of rows of B. If the product exists, it will have as many rows as A and as many columns as B:

	Weighting	
	System 1	System 2
Ann	89	91.4
Tonya	76	77.6
Carlos	88.5	89.9

$$\begin{matrix} A & \cdot & B & = & C \\ m \times n & & n \times p & & m \times p \end{matrix}$$

These must agree.

The product is of size $m \times p$.

We now formally define the product of two matrices.

Product of Two Matrices Let A be an $m \times n$ matrix and B be an $n \times p$ matrix. The product, AB, is the $m \times p$ matrix C

$$AB = C$$

where each entry c_{ij} in C is the sum of the products of the corresponding entries in the ith row of A and the jth column of B, where $i = 1, 2, 3, \ldots, m$ and $j = 1, 2, 3, \ldots, p$.

EXAMPLE 3 Find $C = AB$ if $A = \begin{bmatrix} 1 & 2 & 3 \\ -2 & 1 & -1 \end{bmatrix}$ and $B = \begin{bmatrix} 1 & 5 \\ -2 & 4 \\ 1 & -3 \end{bmatrix}$.

Solution Because matrix A is 2×3 and B is 3×2, the number of columns of A is the same as the number of rows of B. Therefore, the product C exists and it will be a 2×2 matrix. To find entry c_{11} of C, we find the sum of the products of the entries in the first row of A and the first column of B:

$$c_{11} = 1 \cdot 1 + 2 \cdot (-2) + 4 \cdot (1) = 1$$

$$\begin{bmatrix} 1 & 2 & 4 \\ -2 & 1 & -1 \end{bmatrix} \begin{bmatrix} 1 & 5 \\ -2 & 4 \\ 1 & -3 \end{bmatrix} = \begin{bmatrix} 1 & ? \\ ? & ? \end{bmatrix}$$

To find entry c_{12}, we move across the first row of A and down the second column of B:

$$c_{12} = 1 \cdot 5 + 2 \cdot 4 + 4 \cdot (-3) = 1$$

$$\begin{bmatrix} 1 & 2 & 4 \\ -2 & 1 & -1 \end{bmatrix} \begin{bmatrix} 1 & 5 \\ -2 & 4 \\ 1 & -3 \end{bmatrix} = \begin{bmatrix} 1 & 1 \\ ? & ? \end{bmatrix}$$

Comment

Remember that when you multiply matrices of the same size, do not multiply their corresponding entries.

To find entry c_{21}, we move across the second row of A and down the first column of B:

$$c_{21} = (-2) \cdot 1 + 1 \cdot (-2) + (-1) \cdot 1 = -5$$

$$\begin{bmatrix} 1 & 2 & 4 \\ -2 & 1 & -1 \end{bmatrix} \begin{bmatrix} 1 & 5 \\ -2 & 4 \\ 1 & -3 \end{bmatrix} = \begin{bmatrix} 1 & 1 \\ -5 & ? \end{bmatrix}$$

To find entry c_{22}, we move across the second row of A and down the second column of B:

$$c_{22} = (-2) \cdot 5 + 1 \cdot 4 + (-1) \cdot (-3) = -3$$

$$\begin{bmatrix} 1 & 2 & 4 \\ -2 & 1 & -1 \end{bmatrix} \begin{bmatrix} 1 & 5 \\ -2 & 4 \\ 1 & -3 \end{bmatrix} = \begin{bmatrix} 1 & 1 \\ -5 & -3 \end{bmatrix}$$

Self Check 3   Find $D = EF$ if $E = \begin{bmatrix} 1 & -2 \\ 2 & 0 \end{bmatrix}$ and $F = \begin{bmatrix} 2 & 1 \\ 3 & 5 \end{bmatrix}$.

EXAMPLE 4   Find the product: $\begin{bmatrix} 1 & -1 & 2 \\ 1 & 3 & 0 \\ 0 & 1 & 1 \end{bmatrix} \begin{bmatrix} 2 & 1 \\ 1 & 3 \\ 0 & 1 \end{bmatrix}$.

Solution   Because the matrices are 3×3 and 3×2, the number of columns of the first matrix is the same as the number of rows of the second matrix. The product will be a 3×2 matrix.

$$\begin{bmatrix} 1 & -1 & 2 \\ 1 & 3 & 0 \\ 0 & 1 & 1 \end{bmatrix} \begin{bmatrix} 2 & 1 \\ 1 & 3 \\ 0 & 1 \end{bmatrix} = \begin{bmatrix} 1 \cdot 2 + (-1) \cdot 1 + 2 \cdot 0 & 1 \cdot 1 + (-1) \cdot 3 + 2 \cdot 1 \\ 1 \cdot 2 + 3 \cdot 1 + 0 \cdot 0 & 1 \cdot 1 + 3 \cdot 3 + 0 \cdot 1 \\ 0 \cdot 2 + 1 \cdot 1 + 1 \cdot 0 & 0 \cdot 1 + 1 \cdot 3 + 1 \cdot 1 \end{bmatrix}$$

$$= \begin{bmatrix} 1 & 0 \\ 5 & 10 \\ 1 & 4 \end{bmatrix}$$

Self Check 4   Find the product: $\begin{bmatrix} 1 & -2 \\ 3 & 1 \\ -4 & 0 \end{bmatrix} \begin{bmatrix} 4 & -3 \\ 1 & -1 \end{bmatrix}$.

EXAMPLE 5 Find each product: **a.** $[1 \quad 2 \quad 3] \begin{bmatrix} 4 \\ 5 \\ 6 \end{bmatrix}$ **b.** $\begin{bmatrix} 1 \\ 2 \\ 3 \end{bmatrix} [4 \quad 5 \quad 6]$

Solution
a. Since the first matrix is 1×3 and the second matrix is 3×1, the product is a 1×1 matrix:

$$[1 \quad 2 \quad 3] \begin{bmatrix} 4 \\ 5 \\ 6 \end{bmatrix} = [1 \cdot 4 + 2 \cdot 5 + 3 \cdot 6] = [32]$$

b. Since the first matrix is 3×1 and the second matrix is 1×3, the product is a 3×3 matrix:

$$\begin{bmatrix} 1 \\ 2 \\ 3 \end{bmatrix} [4 \quad 5 \quad 6] = \begin{bmatrix} 1 \cdot 4 & 1 \cdot 5 & 1 \cdot 6 \\ 2 \cdot 4 & 2 \cdot 5 & 2 \cdot 6 \\ 3 \cdot 4 & 3 \cdot 5 & 3 \cdot 6 \end{bmatrix} = \begin{bmatrix} 4 & 5 & 6 \\ 8 & 10 & 12 \\ 12 & 15 & 18 \end{bmatrix}$$

Self Check 5 Find each product:

a. $[1 \quad 3 \quad 5] \begin{bmatrix} 1 \\ 0 \\ 1 \end{bmatrix}$ **b.** $\begin{bmatrix} 1 \\ 2 \end{bmatrix} [3 \quad 4]$

Comment
There are cases where the product of two matrices is commutative, $AB = BA$, but, in general, the multiplication of matrices is not commutative.

The multiplication of matrices is not commutative. To show this, we compute AB and BA, where $A = \begin{bmatrix} 1 & 1 \\ 0 & 0 \end{bmatrix}$ and $B = \begin{bmatrix} 0 & 1 \\ 0 & 1 \end{bmatrix}$.

$$AB = \begin{bmatrix} 1 & 1 \\ 0 & 0 \end{bmatrix} \begin{bmatrix} 0 & 1 \\ 0 & 1 \end{bmatrix} = \begin{bmatrix} 0 & 2 \\ 0 & 0 \end{bmatrix}$$

$$BA = \begin{bmatrix} 0 & 1 \\ 0 & 1 \end{bmatrix} \begin{bmatrix} 1 & 1 \\ 0 & 0 \end{bmatrix} = \begin{bmatrix} 0 & 0 \\ 0 & 0 \end{bmatrix}$$

Since the products are not equal, matrix multiplication is not commutative.

Accent on Technology   **Add and Multiply Matrices**

Several models of graphing calculators are able to do matrix arithmetic. For example, to find the sum and the product of the matrices

$$A = \begin{bmatrix} 2 & 3.7 \\ -2.1 & 3 \end{bmatrix} \quad \text{and} \quad B = \begin{bmatrix} 2 & -1 \\ 0 & 0.3 \end{bmatrix}$$

on a TI-84 Plus model, we press MATRIX , select EDIT, and enter the size and entries of the matrix A, as shown in Figure 6-6(a). Similarly, we enter matrix B as in Figure 6-6(b). Figure 6-6(c) displays the result $A + B$ as well as the product AB.

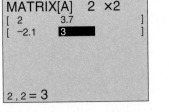
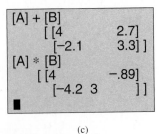

(a)   (b)   (c)

Figure 6-6

4. Solve Problems Using Matrices

EXAMPLE 6 Suppose that supplies must be purchased for the police officers discussed at the beginning of the section. The quantities and prices of each item required for each shift are as follows:

	Quantities				Unit Prices (in $)
	Uniforms	Badges	Whistles		
Day shift	17	13	19	Uniforms	47
Night shift	14	24	27	Badges	7
				Whistles	5

Find the cost of supplies of each shift.

Solution We can write the quantities Q and prices P in matrix form and multiply them to get a cost matrix.

$$C = QP$$

$$= \begin{bmatrix} 17 & 13 & 19 \\ 14 & 24 & 27 \end{bmatrix} \begin{bmatrix} 47 \\ 7 \\ 5 \end{bmatrix}$$

$$= \begin{bmatrix} 17 \cdot 47 + 13 \cdot 7 + 19 \cdot 5 \\ 14 \cdot 47 + 24 \cdot 7 + 27 \cdot 5 \end{bmatrix} \quad \begin{array}{l} \text{(17 uniforms)(\$47) + (13 badges)(\$7) + (19 whistles)(\$5)} \\ \text{(14 uniforms)(\$47) + (24 badges)(\$7) + (27 whistles)(\$5)} \end{array}$$

$$= \begin{bmatrix} 985 \\ 961 \end{bmatrix}$$

It will cost \$985 to buy supplies for the day shift and \$961 to buy supplies for the night shift.

Self Check 6 If the unit price (in $) for uniforms increases to \$50, what will it cost to buy supplies for each shift?

5. Recognize the Identity Matrix

The number 1 is called the *identity for multiplication* because multiplying a number by 1 does not change the number: $a \cdot 1 = 1 \cdot a = a$. There is a **multiplicative identity matrix** with a similar property.

The Identity Matrix Let A be an $n \times n$ matrix. There is an $n \times n$ **identity matrix** I for which

$$AI = IA = A$$

It is the matrix I consisting of 1's on its main diagonal and 0's elsewhere.

$$I = \begin{bmatrix} 1 & 0 & 0 & \cdots & 0 \\ 0 & 1 & 0 & \cdots & 0 \\ 0 & 0 & 1 & \cdots & 0 \\ \vdots & \vdots & \vdots & \ddots & \vdots \\ 0 & 0 & 0 & \cdots & 1 \end{bmatrix}$$

We illustrate the previous definition for the 3×3 identity matrix.

$$\begin{bmatrix} 1 & 0 & 0 \\ 0 & 1 & 0 \\ 0 & 0 & 1 \end{bmatrix}\begin{bmatrix} 1 & 2 & 3 \\ 4 & 5 & 6 \\ 7 & 8 & 9 \end{bmatrix} = \begin{bmatrix} 1\cdot 1 + 0\cdot 4 + 0\cdot 7 & 1\cdot 2 + 0\cdot 5 + 0\cdot 8 & 1\cdot 3 + 0\cdot 6 + 0\cdot 9 \\ 0\cdot 1 + 1\cdot 4 + 0\cdot 7 & 0\cdot 2 + 1\cdot 5 + 0\cdot 8 & 0\cdot 3 + 1\cdot 6 + 0\cdot 9 \\ 0\cdot 1 + 0\cdot 4 + 1\cdot 7 & 0\cdot 2 + 0\cdot 5 + 1\cdot 8 & 0\cdot 3 + 0\cdot 6 + 1\cdot 9 \end{bmatrix}$$

Comment

An identity matrix is always a square matrix—a matrix that has the same number of rows and columns.

$$= \begin{bmatrix} 1 & 2 & 3 \\ 4 & 5 & 6 \\ 7 & 8 & 9 \end{bmatrix}$$

$$\begin{bmatrix} 1 & 2 & 3 \\ 4 & 5 & 6 \\ 7 & 8 & 9 \end{bmatrix}\begin{bmatrix} 1 & 0 & 0 \\ 0 & 1 & 0 \\ 0 & 0 & 1 \end{bmatrix} = \begin{bmatrix} 1 & 2 & 3 \\ 4 & 5 & 6 \\ 7 & 8 & 9 \end{bmatrix}$$

The following properties of real numbers carry over to matrices. In the exercises, you will be asked to illustrate many of these properties.

Properties of Matrices Let A, B, and C be matrices and a and b be scalars.

Commutative property of addition:	$A + B = B + A$
Associative property of addition:	$A + (B + C) = (A + B) + C$
Associative properties of scalar multiplication:	$\begin{cases} a(bA) = (ab)A \\ a(AB) = (aA)B \end{cases}$
Distributive properties of scalar multiplication:	$(a + b)A = aA + bA$ $a(A + B) = aA + aB$
Associative property of multiplication:	$A(BC) = (AB)C$
Distributive properties of matrix multiplication:	$\begin{cases} A(B + C) = AB + AC \\ (A + B)C = AC + BC \end{cases}$

EXAMPLE 7 Verify the distributive property $A(B + C) = AB + AC$ using the matrices

$$A = \begin{bmatrix} 2 & 1 \\ 3 & 1 \end{bmatrix}, \quad B = \begin{bmatrix} 3 & -4 \\ 1 & -1 \end{bmatrix}, \quad \text{and} \quad C = \begin{bmatrix} -1 & 3 \\ 0 & 1 \end{bmatrix}$$

Solution We do the operations on the left side and the right side separately and compare the results.

$$\begin{bmatrix} 2 & 1 \\ 3 & 1 \end{bmatrix}\left(\begin{bmatrix} 3 & -4 \\ 1 & -1 \end{bmatrix} + \begin{bmatrix} -1 & 3 \\ 0 & 1 \end{bmatrix}\right) = \begin{bmatrix} 2 & 1 \\ 3 & 1 \end{bmatrix}\left(\begin{bmatrix} 2 & -1 \\ 1 & 0 \end{bmatrix}\right) \qquad \text{Do the addition within the parentheses.}$$

$$= \begin{bmatrix} 5 & -2 \\ 7 & -3 \end{bmatrix} \qquad \text{Multiply.}$$

$$\begin{bmatrix} 2 & 1 \\ 3 & 1 \end{bmatrix}\begin{bmatrix} 3 & -4 \\ 1 & -1 \end{bmatrix} + \begin{bmatrix} 2 & 1 \\ 3 & 1 \end{bmatrix}\begin{bmatrix} -1 & 3 \\ 0 & 1 \end{bmatrix} = \begin{bmatrix} 7 & -9 \\ 10 & -13 \end{bmatrix} + \begin{bmatrix} -2 & 7 \\ -3 & 10 \end{bmatrix} \qquad \text{Do the multiplications.}$$

$$= \begin{bmatrix} 5 & -2 \\ 7 & -3 \end{bmatrix} \qquad \text{Add.}$$

Because the left and right sides agree, this example illustrates the distributive property.

Self Check Answers

1. $\begin{bmatrix} 0 & -1 \\ 1 & -3 \\ 1 & 5 \end{bmatrix}$ **2.** 1 **3.** $\begin{bmatrix} -4 & -9 \\ 4 & 2 \end{bmatrix}$ **4.** $\begin{bmatrix} 2 & -1 \\ 13 & -10 \\ -16 & 12 \end{bmatrix}$

5. a. $[6]$ **b.** $\begin{bmatrix} 3 & 4 \\ 6 & 8 \end{bmatrix}$ **6.** day shift: \$1,036; nightshift: \$1,003

6.3 Exercises

Vocabulary and Concepts *Fill in the blanks.*

1. In a matrix A, the symbol a_{ij} is the entry in row ___ and column ___.

2. For matrices A and B to be equal, they must be the same ____, and corresponding entries must be ____.

3. To find the sum of matrices A and B, we add the _____ entries.

4. To multiply a matrix by a scalar, we multiply _____ by that scalar.

5. The product of a 3×2 matrix A and a 2×4 matrix B will exist because the number of _____ of A is equal to the number of ____ of B.

6. The product C of the matrices in Exercise 5 will be a _____ matrix.

7. Among 2×2 matrices, $\begin{bmatrix} 0 & 0 \\ 0 & 0 \end{bmatrix}$ is the _____ matrix.

8. Among 2×2 matrices, $\begin{bmatrix} 1 & 0 \\ 0 & 1 \end{bmatrix}$ is the _____ matrix.

Practice *Find values of x and y, if any, that will make the matrices equal.*

9. $\begin{bmatrix} x & y \\ 1 & 3 \end{bmatrix} = \begin{bmatrix} 2 & 5 \\ 1 & 3 \end{bmatrix}$

10. $\begin{bmatrix} x & 5 \\ 3 & y \end{bmatrix} = \begin{bmatrix} 0 & 5 \\ 3 & 2 \end{bmatrix}$

11. $\begin{bmatrix} x + y & 3 + x \\ -2 & 5y \end{bmatrix} = \begin{bmatrix} 3 & 4 \\ -2 & 10 \end{bmatrix}$

12. $\begin{bmatrix} x + y & x - y \\ 2x & 3y \end{bmatrix} = \begin{bmatrix} -x & x - 2 \\ -y & 8 - y \end{bmatrix}$

Find A + B.

13. $A = \begin{bmatrix} 2 & 1 & -1 \\ -3 & 2 & 5 \end{bmatrix}, B = \begin{bmatrix} -3 & 1 & 2 \\ -3 & -2 & -5 \end{bmatrix}$

14. $A = \begin{bmatrix} 3 & 2 & 1 \\ -2 & 3 & -3 \\ -4 & -2 & -1 \end{bmatrix}, B = \begin{bmatrix} -2 & 6 & -2 \\ 5 & 7 & -1 \\ -4 & -6 & 7 \end{bmatrix}$

Find A − B.

15. $A = \begin{bmatrix} -3 & 2 & -2 \\ -1 & 4 & -5 \end{bmatrix}, B = \begin{bmatrix} 3 & -3 & -2 \\ -2 & 5 & -5 \end{bmatrix}$

16. $A = \begin{bmatrix} 2 & 2 & 0 \\ -2 & 8 & 1 \\ 3 & -3 & -8 \end{bmatrix}, B = \begin{bmatrix} -4 & 3 & 7 \\ -1 & 2 & 0 \\ 1 & 4 & -1 \end{bmatrix}$

Find the additive inverse of each matrix.

17. $A = \begin{bmatrix} 5 & -2 & 7 \\ -5 & 0 & 3 \\ -2 & 3 & -5 \end{bmatrix}$

18. $A = \begin{bmatrix} 3 & -\frac{2}{3} & -5 & \frac{1}{2} \end{bmatrix}$

Find 5A.

19. $A = \begin{bmatrix} 3 & -3 \\ 0 & -2 \end{bmatrix}$ **20.** $A = \begin{bmatrix} 3 & \frac{3}{5} \\ 0 & -1 \end{bmatrix}$

21. $A = \begin{bmatrix} 5 & 15 & -2 \\ -2 & -5 & 1 \end{bmatrix}$

22. $A = \begin{bmatrix} -3 & 1 & 2 \\ -8 & -2 & -5 \end{bmatrix}$

Find $5A + 3B$.

23. $A = \begin{bmatrix} 3 & 1 & -2 \\ -4 & 3 & -2 \end{bmatrix}, B = \begin{bmatrix} 1 & -2 & 2 \\ -5 & -5 & 3 \end{bmatrix}$

24. $A = \begin{bmatrix} 2 & -5 \\ -5 & 2 \end{bmatrix}, B = \begin{bmatrix} 5 & -2 \\ 2 & -5 \end{bmatrix}$

Find each product, if possible.

25. $\begin{bmatrix} 2 & 3 \\ 3 & -2 \end{bmatrix}\begin{bmatrix} 1 & 2 \\ 0 & -2 \end{bmatrix}$

26. $\begin{bmatrix} -2 & 3 \\ 3 & -2 \end{bmatrix}\begin{bmatrix} 2 & 4 \\ -5 & 7 \end{bmatrix}$

27. $\begin{bmatrix} -4 & -2 \\ 21 & 0 \end{bmatrix}\begin{bmatrix} -5 & 6 \\ 21 & -1 \end{bmatrix}$

28. $\begin{bmatrix} -5 & 4 \\ 4 & -5 \end{bmatrix}\begin{bmatrix} 6 & -2 \\ 1 & 3 \end{bmatrix}$

29. $\begin{bmatrix} 2 & 1 & 3 \\ 1 & 2 & -1 \\ 0 & 1 & 0 \end{bmatrix}\begin{bmatrix} 1 & 2 & 3 \\ 2 & -2 & 1 \\ 0 & 0 & 1 \end{bmatrix}$

30. $\begin{bmatrix} 2 & 1 & 1 \\ 1 & 1 & 2 \\ 1 & -2 & -1 \end{bmatrix}\begin{bmatrix} 1 & 2 & 3 \\ 1 & 2 & -3 \\ -1 & -1 & 3 \end{bmatrix}$

31. $\begin{bmatrix} 1 & -2 & -3 \\ 2 & 0 & 1 \end{bmatrix}\begin{bmatrix} 4 \\ -5 \\ -6 \end{bmatrix}$

32. $\begin{bmatrix} 1 \\ -2 \\ -3 \end{bmatrix}[4 \quad -5 \quad -6]$

33. $[1 \quad 2 \quad 3]\begin{bmatrix} 4 & 5 & 6 \\ 7 & 8 & 9 \end{bmatrix}$

34. $\begin{bmatrix} 2 & 5 \\ -1 & 7 \end{bmatrix}\begin{bmatrix} 3 & 5 & -8 \\ -2 & 7 & 5 \\ 3 & -6 & 2 \end{bmatrix}$

35. $\begin{bmatrix} 2 & 3 & 4 \\ 1 & 2 & 3 \\ -2 & 2 & 2 \end{bmatrix}\begin{bmatrix} -1 \\ 2 \\ 3 \end{bmatrix}$

36. $\begin{bmatrix} 2 & 5 \\ -3 & 1 \\ 0 & -2 \\ 1 & -5 \end{bmatrix}\begin{bmatrix} 3 & -2 & 4 \\ -2 & -3 & 1 \end{bmatrix}$

37. $\begin{bmatrix} 1 & 2 & 3 \\ 4 & 5 & 6 \\ 7 & 8 & 9 \end{bmatrix}\begin{bmatrix} 1 & 2 \\ 3 & 4 \end{bmatrix}$

38. $\begin{bmatrix} 1 & 4 & 0 & 0 \\ -4 & 1 & 0 & -2 \\ 0 & 0 & 1 & 0 \\ 0 & 2 & 0 & 1 \end{bmatrix}\begin{bmatrix} 1 \\ 2 \\ -2 \\ -1 \end{bmatrix}$

Let $A = \begin{bmatrix} 2.3 & -1.7 & 3.1 \\ -2 & 3.5 & 1 \\ -8 & 4.7 & 9.1 \end{bmatrix}, B = \begin{bmatrix} -2.5 \\ 5.2 \\ -7 \end{bmatrix},$

and $C = \begin{bmatrix} -5.8 \\ 2.9 \\ 4.1 \end{bmatrix}$. *Use a graphing calculator to find each result.*

39. AB

40. $B + C$

41. A^2

42. $AB + C$

Let $A = \begin{bmatrix} 2 & 3 \\ 1 & 3 \end{bmatrix}, B = \begin{bmatrix} 2 & 1 & -5 \\ 1 & 1 & 2 \end{bmatrix},$

$C = \begin{bmatrix} -2 & -1 & 6 \\ 0 & -1 & -1 \end{bmatrix}, D = \begin{bmatrix} 1 & 2 \\ 1 & 3 \end{bmatrix},$ *and*

$E = \begin{bmatrix} 1 & -2 \\ 2 & 3 \end{bmatrix}$. *Verify each property by doing the operations on each side of the equation and comparing the results.*

43. Distributive property: $A(B + C) = AB + AC$
44. Associative property of scalar multiplication: $5(6A) = (5 \cdot 6)A$
45. Associative property of scalar multiplication: $3(AB) = (3A)B$
46. Associative property of multiplication: $A(DE) = (AD)E$

Let $A = \begin{bmatrix} 1 & 3 \\ 2 & 5 \end{bmatrix}, B = \begin{bmatrix} -1 \\ 3 \end{bmatrix},$ *and* $C = [3 \quad 2]$.
Perform the operations, if possible.

47. $A - BC$

48. $AB + B$

49. $CB - AB$

50. CAB

51. ABC

52. $CA + C$

53. A^2B **54.** $(BC)^2$

Applications *Use a graphing calculator to help solve each problem.*

55. Sporting goods Two suppliers manufactured footballs, baseballs, and basketballs in the quantities and costs given in the tables. Find matrices Q and C that represent the quantities and costs, find the product QC, and interpret the result.

	Quantities		
	Footballs	Baseballs	Basketballs
Supplier 1	200	300	100
Supplier 2	100	200	200

Unit costs	(in $)
Footballs	5
Baseballs	2
Basketballs	4

56. Retailing Three ice cream stores sold cones, sundaes, and milkshakes in the quantities and prices given in the tables. Find matrices Q and P that represent the quantities and prices, find the product QP, and interpret the results.

	Quantities		
	Cones	Sundaes	Shakes
Store 1	75	75	32
Store 2	80	69	27
Store 3	62	40	30

Unit price	
Cones	$1.50
Sundaes	$1.75
Shakes	$3.00

57. Beverage sales Beverages were sold to parents and children at a school basketball game in the quantities and prices given in the tables. Find matrices Q and P that represent the quantities and prices, find the product QP, and interpret the result.

	Quantities		
	Coffee	Milk	Cola
Adult males	217	23	319
Adult females	347	24	340
Children	3	97	750

	Price
Coffee	$0.75
Milk	$1.00
Cola	$1.25

58. Production costs Each of four factories manufactures three products in the daily quantities and unit costs given in the tables. Find a suitable matrix product to represent production costs.

	Production quantities		
Factory	Product A	Product B	Product C
Ashtabula	19	23	27
Boston	17	21	22
Chicago	21	18	20
Denver	27	25	22

	Unit production costs	
	Day shift	Night shift
Product A	$1.20	$1.35
Product B	$ .75	$ .85
Product C	$3.50	$3.70

59. Connectivity matrix An entry of 1 in the **connectivity matrix** A on the next page indicates that the person associated with that row knows the address of the person associated with that column. For example, the 1 in Bill's row and Al's column indicates that Bill can write to Al. The 0 in Bill's row and Carl's column indicates that Bill cannot write to Carl. However, Bill could ask Al to forward his letter to Carl. The matrix A^2 indicates the number of ways that one person can write to another with a letter that is forwarded exactly once. Find A^2.

$$\begin{array}{c}\begin{array}{ccc}\textbf{Al} & \textbf{Bill} & \textbf{Carl}\end{array}\\ \begin{array}{c}\textbf{Al}\\ \textbf{Bill}\\ \textbf{Carl}\end{array}\begin{bmatrix} 0 & 1 & 1\\ 1 & 0 & 0\\ 0 & 1 & 0 \end{bmatrix} = A\end{array}$$

60. Communication routing Refer to Exercise 59. Find and interpret the matrix $A + A^2$. Can everyone receive a letter from everyone else with at most one forwarding?

Discovery and Writing

61. Routing telephone calls A long-distance telephone carrier has established several direct microwave links among four cities. In the following connectivity matrix, entries a_{ij} and a_{ji} indicate the number of direct links between cities i and j. For example, cities 2 and 4 are not linked directly but could be connected through city 3. Find and interpret matrix A^2.

$$A = \begin{bmatrix} 0 & 2 & 1 & 0\\ 2 & 0 & 1 & 0\\ 1 & 1 & 0 & 2\\ 0 & 0 & 2 & 0 \end{bmatrix}$$

62. Communication on one-way channels Three communication centers are linked as indicated in the illustration, with communication only in the direction of the arrows. Thus, location 1 can send a message directly to location 2 along two paths, but location 2 can return a message directly on only one path. Entry c_{ij} of matrix C indicates the number of channels from i to j. Find and interpret C^2.

$$C = \begin{bmatrix} 0 & 2 & 2\\ 1 & 0 & 1\\ 1 & 0 & 0 \end{bmatrix}$$

63. If A and B are 2×2 matrices, is $(AB)^2$ equal to A^2B^2? Support your answer.

64. Let a, b, and c be real numbers. If $ab = ac$ and $a \neq 0$, then $b = c$. Find 2×2 matrices A, B, and C, where $A \neq \mathbf{0}$ to show that such a law does not hold for all matrices.

65. Another property of the real numbers is that if $ab = 0$, then either $a = 0$ or $b = 0$. To show that this property is not true for matrices, find two nonzero 2×2 matrices A and B, such that $AB = \mathbf{0}$.

66. Find 2×2 matrices to show that $(A + B)(A - B) \neq A^2 - B^2$.

Review *Perform the operations and simplify.*

67. $(3x + 2)(2x - 3) - (2 - x)$

68. $\dfrac{x^2 + 3x - 4}{2x + 5 - (x + 1)}$

69. $\dfrac{1 + \dfrac{1}{x}}{1 - \dfrac{1}{x}}$

70. $\dfrac{1 - x^{-1}}{1 + x^{-1}}$

71. Solve the formula for a.

$$s = \frac{n(a + l)}{2}$$

72. Solve the formula for x_1.

$$y - y_1 = m(x - x_1)$$

6.4 Matrix Inversion

Objectives

1. Find the Inverse of a Square Matrix Using Row Operations
2. Solve a System of Equations by Matrix Inversion
3. Solve Problems Using Matrix Inversion

Since the beginning of civilized society, governments and businesses have been interested in communicating in secret. This required the development of secret codes. Today, credit card information, bank account information, and even many e-mail messages are *encrypted.* Encryption is a process of converting ordinary text into unreadable text called *cipher text.*

Codes and code breaking have a long history in warfare. A very important event that helped the United States win WWII was the capture of German U-boat 110, which carried an Enigma encryption machine like the one shown here. By breaking the code, the British and United States navies knew the movements of German submarines.

Some of the most sophisticated codes we have today involve matrices. In Exercises 39 and 40, we must find the inverse of a matrix to decode a message.

Inverse of a Matrix   If A and B are $n \times n$ matrices, I is the $n \times n$ identity matrix, and

$$AB = BA = I$$

then A and B are called **multiplicative inverses.** Matrix A is the **inverse** of B, and B is the **inverse** of A.

It can be shown that if a matrix A has an inverse, it has only one inverse. The inverse of A is written as A^{-1}.

$$AA^{-1} = A^{-1}A = I$$

EXAMPLE 1   Show that A and B are inverses.

$$A = \begin{bmatrix} 1 & 1 & 0 \\ 4 & 3 & 0 \\ 2 & 1 & -1 \end{bmatrix} \qquad B = \begin{bmatrix} -3 & 1 & 0 \\ 4 & -1 & 0 \\ -2 & 1 & -1 \end{bmatrix}$$

Solution   We must show that both AB and BA are equal to I.

$$AB = \begin{bmatrix} 1 & 1 & 0 \\ 4 & 3 & 0 \\ 2 & 1 & -1 \end{bmatrix}\begin{bmatrix} -3 & 1 & 0 \\ 4 & -1 & 0 \\ -2 & 1 & -1 \end{bmatrix}$$

$$= \begin{bmatrix} -3 + 4 + 0 & 1 - 1 + 0 & 0 + 0 + 0 \\ -12 + 12 + 0 & 4 - 3 + 0 & 0 + 0 + 0 \\ -6 + 4 + 2 & 2 - 1 - 1 & 0 + 0 + 1 \end{bmatrix}$$

$$= \begin{bmatrix} 1 & 0 & 0 \\ 0 & 1 & 0 \\ 0 & 0 & 1 \end{bmatrix}$$

$$BA = \begin{bmatrix} -3 & 1 & 0 \\ 4 & -1 & 0 \\ -2 & 1 & -1 \end{bmatrix} \begin{bmatrix} 1 & 1 & 0 \\ 4 & 3 & 0 \\ 2 & 1 & -1 \end{bmatrix} = \begin{bmatrix} 1 & 0 & 0 \\ 0 & 1 & 0 \\ 0 & 0 & 1 \end{bmatrix}$$

Self Check 1 Are C and D inverses? $C = \begin{bmatrix} 2 & 5 \\ 1 & 3 \end{bmatrix}$ $D = \begin{bmatrix} 3 & -5 \\ -1 & 2 \end{bmatrix}$

1. Find the Inverse of a Square Matrix Using Row Operations

A matrix that has an inverse is called a **nonsingular matrix** and is said to be **invertible.** If it does not have an inverse, it is a **singular matrix** and is not invertible. The following method provides a way to find the inverse of an invertible matrix.

Finding the Inverse of a Matrix

If a sequence of row operations performed on the $n \times n$ matrix A reduces A to the $n \times n$ identity matrix I, then those same row operations, performed in the same order on I, will transform I into A^{-1}.

If no sequence of row operations will reduce A to I, then A is not invertible.

Comment

The row operations we use to change the matrix into the identity are the same row operations we use in Gauss–Jordan elimination.

To use this method to find the inverse of an invertible matrix, we perform row operations on matrix A to change it to the identity matrix I. At the same time, we perform the same row operations on I. This changes I into A^{-1}.

A notation for this process uses an n-row-by-$2n$-column matrix, with matrix A as the left half and matrix I as the right half. If A is invertible, the proper row operations performed on $[A \mid I]$ will transform it into $[I \mid A^{-1}]$.

EXAMPLE 2 Find the inverse of matrix A if $A = \begin{bmatrix} 2 & -4 \\ 4 & -7 \end{bmatrix}$.

Solution We can set up a 2×4 matrix with A on the left and I on the right of the dashed line:

$$[A \mid I] = \begin{bmatrix} 2 & -4 & 1 & 0 \\ 4 & -7 & 0 & 1 \end{bmatrix}$$

We perform row operations on the entire matrix to transform the left half into I:

$$\begin{bmatrix} 2 & -4 & 1 & 0 \\ 4 & -7 & 0 & 1 \end{bmatrix} \begin{matrix} \left(\frac{1}{2}\right)R1 \rightarrow \\ (-2)R1 + R2 \rightarrow \end{matrix} \begin{bmatrix} 1 & -2 & \frac{1}{2} & 0 \\ 0 & 1 & -2 & 1 \end{bmatrix}$$

$$(2)R2 + R1 \rightarrow \begin{bmatrix} 1 & 0 & -\frac{7}{2} & 2 \\ 0 & 1 & -2 & 1 \end{bmatrix}$$

Since matrix A has been transformed into I, the right side of the previous matrix is A^{-1}. We can verify this by finding AA^{-1} and $A^{-1}A$ and showing that each product is I:

$$AA^{-1} = \begin{bmatrix} 2 & -4 \\ 4 & -7 \end{bmatrix} \begin{bmatrix} -\frac{7}{2} & 2 \\ -2 & 1 \end{bmatrix} = \begin{bmatrix} 1 & 0 \\ 0 & 1 \end{bmatrix}$$

$$A^{-1}A = \begin{bmatrix} -\frac{7}{2} & 2 \\ -2 & 1 \end{bmatrix} \begin{bmatrix} 2 & -4 \\ 4 & -7 \end{bmatrix} = \begin{bmatrix} 1 & 0 \\ 0 & 1 \end{bmatrix}$$

Self Check 2 Find the inverse of $A = \begin{bmatrix} 3 & 2 \\ 4 & 3 \end{bmatrix}$.

EXAMPLE 3 Find the inverse of matrix A if $A = \begin{bmatrix} 1 & 1 & 0 \\ 1 & 2 & 1 \\ 2 & 3 & 2 \end{bmatrix}$.

Solution We set up a 3×6 matrix with A on the left and I on the right of the dashed line.

$$[A \mid I] = \begin{bmatrix} 1 & 1 & 0 & 1 & 0 & 0 \\ 1 & 2 & 1 & 0 & 1 & 0 \\ 2 & 3 & 2 & 0 & 0 & 1 \end{bmatrix}$$

We then perform row operations on the matrix to transform the left half into I.

$$\begin{bmatrix} 1 & 1 & 0 & 1 & 0 & 0 \\ 1 & 2 & 1 & 0 & 1 & 0 \\ 2 & 3 & 2 & 0 & 0 & 1 \end{bmatrix} \begin{array}{c} \\ (-1)R1 + R2 \to \\ (-2)R1 + R3 \to \end{array} \begin{bmatrix} 1 & 1 & 0 & 1 & 0 & 0 \\ 0 & 1 & 1 & -1 & 1 & 0 \\ 0 & 1 & 2 & -2 & 0 & 1 \end{bmatrix}$$

$$\begin{array}{c} (-1)R2 + R1 \to \\ \\ (-1)R2 + R3 \to \end{array} \begin{bmatrix} 1 & 0 & -1 & 2 & -1 & 0 \\ 0 & 1 & 1 & -1 & 1 & 0 \\ 0 & 0 & 1 & -1 & -1 & 1 \end{bmatrix}$$

$$\begin{array}{c} R3 + R1 \to \\ (-1)R3 + R2 \to \\ \\ \end{array} \begin{bmatrix} 1 & 0 & 0 & 1 & -2 & 1 \\ 0 & 1 & 0 & 0 & 2 & -1 \\ 0 & 0 & 1 & -1 & -1 & 1 \end{bmatrix}$$

Since the left half of the matrix has been transformed into the identity matrix, the right half has become A^{-1}, and

$$A^{-1} = \begin{bmatrix} 1 & -2 & 1 \\ 0 & 2 & -1 \\ -1 & -1 & 1 \end{bmatrix}$$

Self Check 3 Find the inverse of $B = \begin{bmatrix} 1 & 1 & 1 \\ 2 & 1 & 4 \\ 2 & 2 & 3 \end{bmatrix}$.

EXAMPLE 4 If possible, find the inverse of $A = \begin{bmatrix} 1 & 2 \\ 2 & 4 \end{bmatrix}$.

Solution We form the 2×4 matrix

$$[A \mid I] = \begin{bmatrix} 1 & 2 & 1 & 0 \\ 2 & 4 & 0 & 1 \end{bmatrix}$$

and begin to transform the left side of the matrix into the identity matrix I:

$$\begin{bmatrix} 1 & 2 & 1 & 0 \\ 2 & 4 & 0 & 1 \end{bmatrix} (-2)R1 + R2 \to \begin{bmatrix} 1 & 2 & 1 & 0 \\ 0 & 0 & -2 & 1 \end{bmatrix}$$

In obtaining the second-row, first-column position of A, the entire second row of A is zeroed out. Since we cannot transform matrix A to the identity, A is not invertible.

Self Check 4   If possible, find the inverse of $B = \begin{bmatrix} 1 & -2 \\ -3 & 6 \end{bmatrix}$.

The previous examples suggest the following facts:

- Matrices that are not square matrices do not have inverses.
- Some square matrices have inverses and others do not.

Accent on Technology

Finding the Inverse of a Matrix

We can use a graphing calculator to find the inverse of a matrix. For example, to find the inverse of

$$A = \begin{bmatrix} 2 & 2 & 3 \\ 1 & 2 & 3 \\ 1 & 0 & 1 \end{bmatrix}$$

using a TI-84 Plus calculator, we press MATRIX, select EDIT, and enter matrix A as shown in Figure 6-7(a). We exit entry mode by pressing 2nd QUIT. Then we display A^{-1} by pressing MATRIX 1 x^{-1} ENTER. The display appears in Figure 6-7(b). To verify that $AA^{-1} = I$, we press CLEAR MATRIX 1 × MATRIX 1 x^{-1} ENTER to obtain the result shown in Figure 6-7(c).

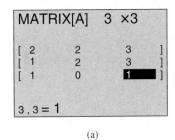

(a)

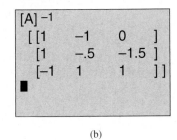

(b)

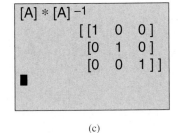

(c)

Figure 6-7

2.   Solve a System of Equations by Matrix Inversion

If we multiply the matrices on the left side of the equation

$$\begin{bmatrix} 1 & 1 & 0 \\ 1 & 2 & 1 \\ 2 & 3 & 2 \end{bmatrix} \begin{bmatrix} x \\ y \\ z \end{bmatrix} = \begin{bmatrix} 20 \\ 30 \\ 55 \end{bmatrix}$$

and set the corresponding entries equal, we get the following system of equations:

$$\begin{cases} x + y = 20 \\ x + 2y + z = 30 \\ 2x + 3y + 2z = 55 \end{cases}$$

A system of equations can always be written as a matrix equation $AX = B$, where A is the coefficient matrix of the system, X is a column matrix of variables, and B is the column matrix of constants. If matrix A is invertible, the matrix equation $AX = B$ is easy to solve.

EXAMPLE 5   Solve: $\begin{bmatrix} 1 & 1 & 0 \\ 1 & 2 & 1 \\ 2 & 3 & 2 \end{bmatrix}\begin{bmatrix} x \\ y \\ z \end{bmatrix} = \begin{bmatrix} 20 \\ 30 \\ 55 \end{bmatrix}$.

Solution   The 3×3 matrix on the left is the matrix whose inverse was found in Example 3. We multiply both sides of the equation on the left by this inverse to obtain an equivalent system of equations.

$$\begin{bmatrix} 1 & -2 & 1 \\ 0 & 2 & -1 \\ -1 & -1 & 1 \end{bmatrix}\begin{bmatrix} 1 & 1 & 0 \\ 1 & 2 & 1 \\ 2 & 3 & 2 \end{bmatrix}\begin{bmatrix} x \\ y \\ z \end{bmatrix} = \begin{bmatrix} 1 & -2 & 1 \\ 0 & 2 & -1 \\ -1 & -1 & 1 \end{bmatrix}\begin{bmatrix} 20 \\ 30 \\ 55 \end{bmatrix}$$

$$\begin{bmatrix} 1 & 0 & 0 \\ 0 & 1 & 0 \\ 0 & 0 & 1 \end{bmatrix}\begin{bmatrix} x \\ y \\ z \end{bmatrix} = \begin{bmatrix} 15 \\ 5 \\ 5 \end{bmatrix} \qquad \text{Multiply the matrices. On the left, remember that } A^{-1}A = I.$$

$$\begin{bmatrix} x \\ y \\ z \end{bmatrix} = \begin{bmatrix} 15 \\ 5 \\ 5 \end{bmatrix} \qquad IX = X$$

The solution of this system can be read directly from the matrix on the right side. Verify that the values $x = 15$, $y = 5$, $z = 5$ satisfy the original equations.

Self Check 5   Solve: $\begin{bmatrix} 2 & -4 \\ 4 & -7 \end{bmatrix}\begin{bmatrix} x \\ y \end{bmatrix} = \begin{bmatrix} -4 \\ 2 \end{bmatrix}$. (The inverse was found in Example 2.)

Example 5 suggests the following result.

Solving Systems of Equations   If A is invertible, the solution of the matrix equation $AX = B$ is
$$X = A^{-1}B$$

Accent on Technology   **Solving Equations**

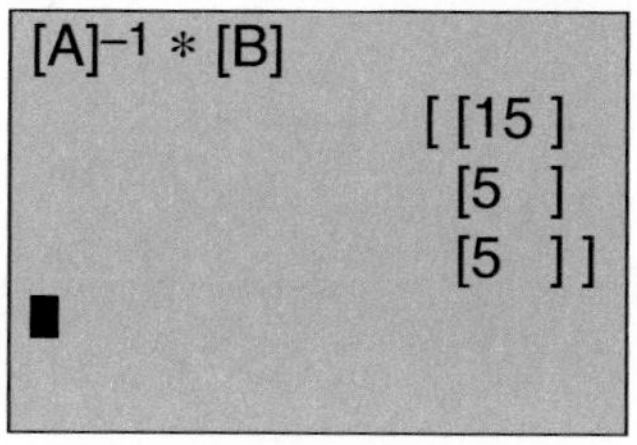

To use a TI-84 Plus graphing calculator to solve the system of Example 5, we enter
the matrix $\begin{bmatrix} 1 & 1 & 0 \\ 1 & 2 & 1 \\ 2 & 3 & 2 \end{bmatrix}$ into the calculator as matrix A and the matrix $\begin{bmatrix} 20 \\ 30 \\ 55 \end{bmatrix}$ as
matrix B and press 2nd QUIT. We then press MATRIX 1 x^{-1} $\times$ MATRIX 2 ENTER to get Figure 6-8. The solution is $x = 15$, $y = 5$, and $z = 5$. Always check the results in the original equation.

```
[A]−1 * [B]
          [ [15 ]
            [5  ]
            [5  ] ]
■
```

Figure 6-8

This method is especially useful for finding solutions of several systems of equations that differ from each other only in the column matrix B. If the coefficient matrix A remains unchanged from one system of equations to the next, then

A^{-1} needs to be found only once. The solution of each system is found by a single matrix multiplication, $A^{-1}B$.

3. Solve Problems Using Matrix Inversion

EXAMPLE 6 A company that manufactures medical equipment spends time on paperwork, manufacture, and testing for each of three versions of a circuit board. The times spent on each and the total time available are given in the following tables.

Hours Required per Unit			
	Product A	Product B	Product C
Paperwork	1	1	0
Manufacture	1	2	1
Testing	2	3	2

Hours Available	
Paperwork	20
Manufacture	30
Testing	55

Solution We can let x, y, and z represent the number of units of products A, B, and C to be manufactured, respectively. We then can set up the following system of equations:

Paperwork:	$x + y = 20$	One hour is needed for every A, and one hour for every B.
Manufacture:	$x + 2y + z = 30$	One hour is needed for every A and C, and two hours for every B.
Testing:	$2x + 3y + 2z = 55$	Two hours are needed for every A and C, and three hours for every B.

In matrix form, the system becomes

$$\begin{bmatrix} 1 & 1 & 0 \\ 1 & 2 & 1 \\ 2 & 3 & 2 \end{bmatrix} \begin{bmatrix} x \\ y \\ z \end{bmatrix} = \begin{bmatrix} 20 \\ 30 \\ 55 \end{bmatrix}$$

We solved this equation in Example 5 to get $x = 15$, $y = 5$, and $z = 5$. To use all of the available time, the company should manufacture 15 units of product A and 5 units each of products B and C.

Self Check Answers   1. yes   2. $\begin{bmatrix} 3 & -2 \\ -4 & 3 \end{bmatrix}$ 3. $\begin{bmatrix} 5 & 1 & -3 \\ -2 & -1 & 2 \\ -2 & 0 & 1 \end{bmatrix}$ 4. not possible

5. $\begin{bmatrix} x \\ y \end{bmatrix} = \begin{bmatrix} 18 \\ 10 \end{bmatrix}$

6.4 Exercises

Vocabulary and Concepts *Fill in the blanks.*

1. Matrices A and B are multiplicative inverses if _____.

2. A nonsingular matrix ___ (is, is not) invertible.

3. If A is invertible, elementary row operations can change $[A \mid I]$ into _____.

4. If A is invertible, the solution of $AX = B$ is _____.

Practice   *Find the inverse of each matrix, if possible.*

5. $\begin{bmatrix} 3 & -4 \\ -2 & 3 \end{bmatrix}$

6. $\begin{bmatrix} 2 & 3 \\ 3 & 5 \end{bmatrix}$

7. $\begin{bmatrix} 3 & 7 \\ 2 & 5 \end{bmatrix}$

8. $\begin{bmatrix} 1 & -2 \\ 2 & -5 \end{bmatrix}$

9. $\begin{bmatrix} 1 & 0 & 3 \\ -1 & 1 & 3 \\ -2 & 1 & 1 \end{bmatrix}$

10. $\begin{bmatrix} 2 & 1 & -1 \\ 2 & 2 & -1 \\ -1 & -1 & 1 \end{bmatrix}$

11. $\begin{bmatrix} 3 & 2 & 1 \\ 1 & 1 & -1 \\ 4 & 3 & 1 \end{bmatrix}$

12. $\begin{bmatrix} -2 & 1 & -3 \\ 2 & 3 & 0 \\ 1 & 0 & 1 \end{bmatrix}$

13. $\begin{bmatrix} 1 & 3 & 5 \\ 0 & 1 & 6 \\ 1 & 4 & 11 \end{bmatrix}$

14. $\begin{bmatrix} 1 & 2 & 3 \\ 4 & 5 & 6 \\ 7 & 8 & 9 \end{bmatrix}$

15. $\begin{bmatrix} 1 & 2 & 3 \\ 0 & 1 & 2 \\ 0 & 0 & 1 \end{bmatrix}$

16. $\begin{bmatrix} 1 & 2 & 3 \\ 0 & 1 & 1 \\ 0 & -1 & 0 \end{bmatrix}$

17. $\begin{bmatrix} 1 & 6 & 4 \\ 1 & -2 & -5 \\ 2 & 4 & -1 \end{bmatrix}$

18. $\begin{bmatrix} 1 & 1 & 1 \\ 1 & 0 & -1 \\ 1 & 2 & 3 \end{bmatrix}$

Use a graphing calculator to find the inverse of each matrix.

19. $\begin{bmatrix} 1 & 2 & 3 & 4 \\ 0 & 1 & 2 & 3 \\ 0 & 0 & 1 & 2 \\ 0 & 0 & 0 & 1 \end{bmatrix}$

20. $\begin{bmatrix} 1 & 0 & 0 & 0 \\ 1 & 1 & 0 & 0 \\ 1 & 1 & 1 & 0 \\ 1 & 2 & 2 & 1 \end{bmatrix}$

21. $\begin{bmatrix} 1 & 1 & -1 \\ 0.5 & 1 & 0.5 \\ 1 & 1 & -1.5 \end{bmatrix}$

22. $\begin{bmatrix} -2 & -1 & 1 \\ 0.5 & -1.5 & -0.5 \\ 0 & 1 & 0.5 \end{bmatrix}$

23. $\begin{bmatrix} 3 & 3 & -3 & 2 \\ 1 & -4 & 3 & -5 \\ 3 & 0 & -2 & -1 \\ -1 & 5 & -3 & 6 \end{bmatrix}$

24. $\begin{bmatrix} 1 & 0 & 0 & 0 \\ 2 & 1 & 0 & 0 \\ 3 & 2 & 1 & 0 \\ 4 & 3 & 2 & 1 \end{bmatrix}$

Use matrix inversion to solve each system of equations. Note that several systems have the same coefficient matrix.

25. $\begin{cases} 3x - 4y = 1 \\ -2x + 3y = 5 \end{cases}$

26. $\begin{cases} 3x - 4y = -1 \\ -2x + 3y = 3 \end{cases}$

27. $\begin{cases} 3x - 4y = 0 \\ -2x + 3y = 0 \end{cases}$

28. $\begin{cases} 3x - 4y = -3 \\ -2x + 3y = -2 \end{cases}$

29. $\begin{cases} 2x + y - z = 2 \\ 2x + 2y - z = 4 \\ -x - y + z = -1 \end{cases}$

30. $\begin{cases} 2x + y - z = 3 \\ 2x + 2y - z = -1 \\ -x - y + z = 4 \end{cases}$

31. $\begin{cases} -2x + y - 3z = 2 \\ 2x + 3y = -3 \\ x + z = 5 \end{cases}$

32. $\begin{cases} -2x + y - 3z = 5 \\ 2x + 3y = 1 \\ x + z = -2 \end{cases}$

Use a graphing calculator to solve each system of equations. Use matrix inversion.

33. $\begin{cases} 5x + 3y = 13 \\ -7x + 5y = -9 \end{cases}$

34. $\begin{cases} 8x - 3y = 7 \\ -3x + 2y = 0 \end{cases}$

35. $\begin{cases} 5x + 2y + 3z = 12 \\ 2x + 5z = 7 \\ 3x + z = 4 \end{cases}$

36. $\begin{cases} 3x + 2y - z = 0 \\ 5x - 2y = 5 \\ 3x + y + z = 6 \end{cases}$

Applications

37. **Manufacturing and testing**   The numbers of hours required to manufacture and test each of two models of heart monitor are given in the first table, and the numbers of hours available each week for manufacturing and testing are given in the second table.

	Hours Required per Unit	
	Model A	**Model B**
Manufacturing	23	27
Testing	21	22

Hours Available	
Manufacturing	127
Testing	108

How many of each model can be manufactured each week?

38. Making clothes A clothing manufacturer makes coats, shirts, and slacks. The time required for cutting, sewing, and packaging each item is shown in the table. How many of each should be made to use all available labor hours?

	Coats	**Shirts**	**Slacks**
Cutting	20 min	15 min	10 min
Sewing	60 min	30 min	24 min
Packaging	5 min	12 min	6 min

Time available	
Cutting	115 hr
Sewing	280 hr
Packaging	65 hr

39. Cryptography The letters of a message, called **plain text**, are assigned values 1–26 (for a–z) and are written in groups of 2 as 2×1 matrices. To write the message in **cipher text**, each 2×1 matrix B is multiplied by a matrix A, where

$$A = \begin{bmatrix} 1 & 1 \\ 2 & 3 \end{bmatrix}$$

Find the plain text if the cipher text of one message is

$$AB = \begin{bmatrix} 17 \\ 43 \end{bmatrix}$$

40. Cryptography The letters of a message, called **plain text,** are assigned values 1–26 (for a–z) and are written in groups of 3 as 3×1 matrices. To write the message in **cipher text**, each 3×1 matrix Y is multiplied by matrix A, where

$$A = \begin{bmatrix} 1 & 1 & 0 \\ 2 & 3 & 3 \\ 1 & 1 & 1 \end{bmatrix}$$

Find the plain text if the cipher text of one message is

$$AY = \begin{bmatrix} 30 \\ 122 \\ 49 \end{bmatrix}$$

Discovery and Writing *Use examples chosen from 2 × 2 matrices to support each answer.*

41. Does $(AB)^{-1} = A^{-1}B^{-1}$?

42. Does $(AB)^{-1} = B^{-1}A^{-1}$?

Let $A = \begin{bmatrix} -1 & -1 \\ 1 & 1 \end{bmatrix}$.

43. Show that $A^2 = \mathbf{0}$.

44. Show that the inverse of $I - A$ is $I + A$.

Let $A = \begin{bmatrix} 3 & 0 & 0 \\ -2 & -1 & -2 \\ 3 & 6 & 3 \end{bmatrix}$ *and* $X = \begin{bmatrix} x \\ y \\ z \end{bmatrix}$. *Solve each equation. Each solution is called an* **eigenvector** *of the matrix A.*

45. $(A - 2I)X = 0$ **46.** $(A - 3I)X = 0$

47. Suppose that A, B, and C are $n \times n$ matrices and A is invertible. If $AB = AC$, prove that $B = C$.

48. Prove that $\begin{bmatrix} a & b \\ c & d \end{bmatrix}$ has an inverse if and only if $ad - bc \neq 0$. (*Hint:* Try to find the inverse and see what happens.)

49. Suppose that B is any matrix for which $B^2 = \mathbf{0}$. Show that $I - B$ is invertible by showing that the inverse of $I - B$ is $I + B$.

50. Suppose that C is any matrix for which $C^3 = \mathbf{0}$. Show that $I - C$ is invertible by showing that the inverse of $I - C$ is $I + C + C^2$.

Review *Find the domain of each function.*

Review *Find the domain of each function.*

51. $y = \dfrac{3x - 5}{x^2 - 4}$

52. $y = \dfrac{3x - 5}{x^2 + 4}$

53. $y = \dfrac{3x - 5}{\sqrt{x^2 + 4}}$

54. $y = \dfrac{3x - 5}{\sqrt{x^2 - 4}}$

Find the range of each function.

55. $y = x^2$

56. $y = x^3$

57. $y = \log x$

58. $y = 2^x$

6.5 Determinants

Objectives

1. Evaluate Determinants of Higher-Order Matrices
2. Understand and Use Properties of Determinants
3. Use Determinants to Solve Systems of Equations
4. Write Equations of Lines
5. Find Areas of Triangles

There is a function, called the **determinant function,** that associates a number with every square matrix. The domain of this function is the set of all square matrices, and the range is the set of numbers. Historically, determinants were considered before matrices. They first appeared in the 3rd century BC when they were considered in a Chinese textbook called *The Nine Chapters on the Mathematical Art.* Gerolamo Cardano, an Italian mathematician, considered two-by-two determinants toward the end of the 16th century. However, it was the great Swiss mathematician, Gabriel Cramer, who made them popular.

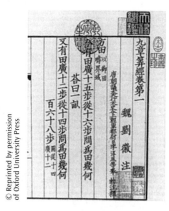

In this section, we will use determinants to solve systems of linear equations, find the area of a triangle, and find the equation of a line.

The determinant function is written as $\det(A)$ or as $|A|$.

Comment

Do not confuse the notation $|A|$ with absolute value symbols.

Determinants If a, b, c, and d are numbers, the determinant of $A = \begin{bmatrix} a & b \\ c & d \end{bmatrix}$ is

$$\det(A) = \begin{vmatrix} a & b \\ c & d \end{vmatrix} = ad - bc$$

EXAMPLE 1

a. $\begin{vmatrix} 1 & 2 \\ 3 & 4 \end{vmatrix} = 1 \cdot 4 - 2 \cdot 3$

$= 4 - 6$

$= -2$

b. $\begin{vmatrix} -2 & 3 \\ -\pi & \frac{1}{2} \end{vmatrix} = (-2)\left(\dfrac{1}{2}\right) - (3)(-\pi)$

$= -1 + 3\pi$

Self Check 1 Evaluate: $\begin{vmatrix} 3 & -2 \\ 5 & -4 \end{vmatrix}$.

1. Evaluate Determinants of Higher-Order Matrices

To evaluate determinants of higher-order matrices, we must define the **minor** and the **cofactor** of an entry in a matrix.

Minor and Cofactor of a Matrix

Let $A = [a_{ij}]$ be a square matrix of size (or order) $n \geq 2$.

1. The **minor** of a_{ij}, denoted as M_{ij}, is the determinant of the $n - 1 \times n - 1$ matrix formed by deleting the ith row and the jth column of A.

2. The **cofactor** of a_{ij}, denoted as C_{ij}, is $\begin{cases} M_{ij} \text{ when } i + j \text{ is even} \\ -M_{ij} \text{ when } i + j \text{ is odd} \end{cases}$.

EXAMPLE 2 In $A = \begin{bmatrix} 1 & 2 & 3 \\ 4 & 5 & 6 \\ 7 & 8 & 9 \end{bmatrix}$, find the minor and cofactor of **a.** a_{31} **b.** a_{12}

Solution **a.** The minor M_{31} is the minor of $a_{31} = 7$ appearing in row 3, column 1. It is found by deleting row 3 and column 1:

$$M_{31} = \begin{vmatrix} 1 & 2 & 3 \\ 4 & 5 & 6 \\ 7 & 8 & 9 \end{vmatrix} = \begin{vmatrix} 2 & 3 \\ 5 & 6 \end{vmatrix} = -3$$

Because $i + j$ is even $(3 + 1 = 4)$, the cofactor of the minor M_{31} is M_{31}:

$$C_{31} = M_{31} = \begin{vmatrix} 2 & 3 \\ 5 & 6 \end{vmatrix} = 2 \cdot 6 - 3 \cdot 5 = 12 - 15 = -3$$

b. The minor M_{12} is the minor of $a_{12} = 2$ appearing in row 1, column 2. It is found by deleting row 1 and column 2.

$$M_{12} = \begin{vmatrix} 1 & 2 & 3 \\ 4 & 5 & 6 \\ 7 & 8 & 9 \end{vmatrix} = \begin{vmatrix} 4 & 6 \\ 7 & 9 \end{vmatrix} = -6$$

Because $i + j$ is odd $(1 + 2 = 3)$, the cofactor of the minor M_{12} is $-M_{12}$:

$$C_{12} = -M_{12} = -\begin{vmatrix} 4 & 6 \\ 7 & 9 \end{vmatrix} = -(4 \cdot 9 - 6 \cdot 7) = -(36 - 42) = -(-6) = 6$$

Self Check 2 Find the cofactor of a_{23}.

We are now ready to evaluate determinants of higher-order matrices.

Value of a Determinant

If A is a square matrix of order $n \geq 2$, $|A|$ is the sum of the products of the elements in any row (or column) and the cofactors of those elements.

In Example 3, we evaluate a 3×3 determinant by expanding the determinant in three ways: along two different rows and along a column. This method is called **expanding a determinant by cofactors.**

EXAMPLE 3   Evaluate $\begin{vmatrix} 1 & 2 & -3 \\ -1 & 0 & 1 \\ -2 & 2 & 1 \end{vmatrix}$ by expanding by cofactors along the designated row or column.

Expanding on row 1:

$$\begin{vmatrix} 1 & 2 & -3 \\ -1 & 0 & 1 \\ -2 & 2 & 1 \end{vmatrix} = a_{11}C_{11} + a_{12}C_{12} + a_{13}C_{13}$$

$$= 1\begin{vmatrix} 0 & 1 \\ 2 & 1 \end{vmatrix} + 2\left(-\begin{vmatrix} -1 & 1 \\ -2 & 1 \end{vmatrix}\right) + (-3)\begin{vmatrix} -1 & 0 \\ -2 & 2 \end{vmatrix}$$

$$= 1(-2) + 2(-1) - 3(-2)$$

$$= -2 - 2 + 6$$

$$= 2$$

Gabriel Cramer
(1704–1752)
Cramer, a Swiss mathematician, made contributions to the fields of geometry, analysis, the study of algebraic curves, the history of mathematicians, and determinants. Although other mathematicians had previously worked with determinants, it was the work of Cramer that popularized them.

Expanding on row 3:

$$\begin{vmatrix} 1 & 2 & -3 \\ -1 & 0 & 1 \\ -2 & 2 & 1 \end{vmatrix} = a_{31}C_{31} + a_{32}C_{32} + a_{33}C_{33}$$

$$= -2\begin{vmatrix} 2 & -3 \\ 0 & 1 \end{vmatrix} + 2\left(-\begin{vmatrix} 1 & -3 \\ -1 & 1 \end{vmatrix}\right) + 1\begin{vmatrix} 1 & 2 \\ -1 & 0 \end{vmatrix}$$

$$= -2(2) + 2(+2) + 1(2)$$

$$= -4 + 4 + 2$$

$$= 2$$

Comment

Note that we don't need to evaluate the determinant $\begin{vmatrix} 1 & -3 \\ -2 & 1 \end{vmatrix}$ because it has a coefficient of 0. When we expand by cofactors along a row or column containing some 0s, the work is easier because we have fewer determinants to evaluate.

Expanding on column 2:

$$\begin{vmatrix} 1 & 2 & -3 \\ -1 & 0 & 1 \\ -2 & 2 & 1 \end{vmatrix} = a_{12}C_{12} + a_{22}C_{22} + a_{32}C_{32}$$

$$= 2\left(-\begin{vmatrix} -1 & 1 \\ -2 & 1 \end{vmatrix}\right) + 0\begin{vmatrix} 1 & -3 \\ -2 & 1 \end{vmatrix} + 2\left(-\begin{vmatrix} 1 & -3 \\ -1 & 1 \end{vmatrix}\right)$$

$$= 2(-1) + 0(-5) + 2(+2)$$

$$= -2 + 4$$

$$= 2$$

In each case, the result is 2.

Self Check 3   Evaluate $\begin{vmatrix} 1 & 0 & 1 \\ 2 & -1 & 0 \\ 3 & 1 & -1 \end{vmatrix}$ by expanding on its first row and second column.

Everyday Connections

Lewis Carroll

One of the more amusing historical anecdotes concerning matrices and determinants involves the English mathematician Charles Dodgson, also known as Lewis Carroll. The anecdote describes how England's Queen Victoria so enjoyed reading Carroll's book, *Alice in Wonderland,* that she requested a copy of his next publication. To her great surprise, she received an autographed copy of a mathematics text titled, *An Elementary Treatise on Determinants.* The story was repeated as fact so often that Carroll finally included an explicit disclaimer in his book, *Symbolic Logic,* insisting the incident never actually occurred.

Compute the determinant for each matrix.

a. $\begin{vmatrix} 3 & 4 \\ 2 & 1 \end{vmatrix}$

b. $\begin{vmatrix} 3 & 4 & 2 \\ 1 & -1 & 5 \\ 1 & 2 & -2 \end{vmatrix}$

Source: http://mathworld.wolfram.com/Determinant.html

EXAMPLE 4   Evaluate: $\begin{vmatrix} 0 & 0 & 2 & 0 \\ 1 & 2 & 17 & -3 \\ -1 & 0 & 28 & 1 \\ -2 & 2 & -37 & 1 \end{vmatrix}$.

Solution   Because row 1 contains three 0's, we expand the determinant along row 1. Then only one cofactor needs to be evaluated.

$$\begin{vmatrix} \mathbf{0} & \mathbf{0} & \mathbf{2} & \mathbf{0} \\ 1 & 2 & 17 & -3 \\ -1 & 0 & 28 & 1 \\ -2 & 2 & -37 & 1 \end{vmatrix} = 0|?| - 0|?| + 2\begin{vmatrix} 1 & 2 & -3 \\ -1 & 0 & 1 \\ -2 & 2 & 1 \end{vmatrix} - 0|?|$$

$$= 2(2) \quad \text{See Example 3.}$$

$$= 4$$

Self Check 4   Evaluate: $\begin{vmatrix} 0 & 1 & 0 \\ 2 & 11 & -2 \\ 1 & 13 & 1 \end{vmatrix}$.

Example 4 suggests the following theorem.

Zero Row or Column Theorem   If every entry in a row or column of a square matrix A is 0, then $|A| = 0$.

Accent on Technology

Evaluating Determinants

Many graphing calculators are able to evaluate determinants. Simply enter a square matrix (say, A) and press det(and A. For example, to evaluate the determinant

$$\begin{vmatrix} 2 & 3 & 0.5 & 6 \\ 1 & 4 & -2 & -3 \\ 3 & 4 & -3 & -2 \\ -0.7 & 6 & 2 & 1 \end{vmatrix}$$

with a TI-84 Plus calculator, enter the matrix as shown in Figure 6-9(a). Note that you have to scroll left and right to see the entire screen. Then press 2nd QUIT MATRIX , select MATH, and press 1. Then press MATRIX , 1, and ENTER to obtain Figure 6-9(b). The value of the determinant is 99.3.

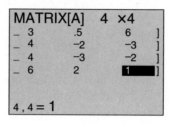
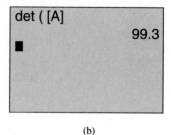

(a)   (b)

Figure 6-9

2. Understand and Use Properties of Determinants

We have seen that there are row operations for transforming matrices. There are similar row and column operations for transforming determinants.

Row and Column Operations for Determinants

Let A be a square matrix and k be a real number.

1. If a matrix B is obtained from matrix A by interchanging two rows (or columns), then $|B| = -|A|$.
2. If B is obtained from A by multiplying every element in a row (or column) of A by k, then $|B| = k|A|$.
3. If B is obtained from A by adding k times any row (or column) of A to another row (or column) of A, then $|B| = |A|$.

We will illustrate each row operation by showing that the operation holds true for the 2×2 determinant

$$(1) \qquad |A| = \begin{vmatrix} 3 & 4 \\ 1 & 2 \end{vmatrix} = 3(2) - 4(1) = 6 - 4 = 2$$

Interchanging two rows: Let $|B|$ be the determinant obtained from $|A|$ by interchanging its two rows. Then

$$|B| = \begin{vmatrix} 1 & 2 \\ 3 & 4 \end{vmatrix} = 1(4) - 2(3) = 4 - 6 = -2$$

Since $-2 = -1(2)$, we have $|B| = -|A|$.

In general, if $|A|$ and $|B|$ are as follows, we have

$$|A| = \begin{vmatrix} a & b \\ c & d \end{vmatrix} = ad - bc$$

$$|B| = \begin{vmatrix} c & d \\ a & b \end{vmatrix} = cb - da = -(ad - bc) = -|A|$$

Multiplying every element in a row by k: Let $|B|$ be the determinant obtained from $|A|$ in Equation 1 by multiplying its second row by 3. Then

$$|B| = \begin{vmatrix} 3 & 4 \\ 3(1) & 3(2) \end{vmatrix} = \begin{vmatrix} 3 & 4 \\ 3 & 6 \end{vmatrix} = 3(6) - 4(3) = 18 - 12 = 6$$

Since $6 = 3(2)$, we have $|B| = 3|A|$.

In general, if $|A|$ and $|B|$ are as follows, we have

(2) $\quad |A| = \begin{vmatrix} a & b \\ c & d \end{vmatrix} = ad - bc$

$$|B| = \begin{vmatrix} a & b \\ kc & kd \end{vmatrix} = akd - bkc = k(ad - bc) = k|A|$$

Adding k times any row of A to another row of A: Let $|B|$ be the determinant obtained from $|A|$ in Equation 1 by adding 3 times its first row to its second row. Then

$$|B| = \begin{vmatrix} 3 & 4 \\ 3(3) + 1 & 3(4) + 2 \end{vmatrix} = 3(14) - 4(10) = 42 - 40 = 2$$

Since $2 = 2$, we have $|A| = |B|$.

In general, if $|B|$ is obtained from $|A|$ in Equation 2 by adding k times its first row to its second row. Then

$$|B| = \begin{vmatrix} a & b \\ ka + c & kb + d \end{vmatrix} = a(kb + d) - b(ka + c) = akb + ad - bka - bc = ad - bc = |A|$$

Evaluating higher-order determinants can be very time consuming. However, as we have seen, if we use row and column operations to introduce as many 0's as possible in a row (or column) and then expand the determinant by cofactors along that row (or column), the work is much easier. In the next example, we will use several row and column operations to introduce 0's into the determinant.

EXAMPLE 5 Use row a column operations to help evaluate $\begin{vmatrix} 10 & 20 & -10 & 20 \\ 2 & 1 & 1 & 1 \\ 1 & 2 & -3 & 2 \\ 2 & -1 & -1 & 1 \end{vmatrix}$.

Solution To get smaller numbers in the first row, we can use a type 2 row operation and multiply each entry in the first row by $\frac{1}{10}$. However, we then must multiply the resulting determinant by 10 to retain its original value.

(1) $\quad \begin{vmatrix} 10 & 20 & -10 & 20 \\ 2 & 1 & 1 & 1 \\ 1 & 2 & -3 & 2 \\ 2 & -1 & -1 & 1 \end{vmatrix} \begin{matrix} \left(\frac{1}{10}\right)R1 \rightarrow \\ \\ = 10 \end{matrix} \begin{vmatrix} 1 & 2 & -1 & 2 \\ 2 & 1 & 1 & 1 \\ 1 & 2 & -3 & 2 \\ 2 & -1 & -1 & 1 \end{vmatrix}$

To get three 0's in the first row of the second determinant in Equation 1, we perform a type 3 row operation and expand the new determinant along its first row.

$$10\begin{vmatrix} 1 & 2 & -1 & 2 \\ 2 & 1 & 1 & 1 \\ 1 & 2 & -3 & 2 \\ 2 & -1 & -1 & 1 \end{vmatrix} \xrightarrow{(-1)R3 + R1} = 10\begin{vmatrix} 0 & 0 & 2 & 0 \\ 2 & 1 & 1 & 1 \\ 1 & 2 & -3 & 2 \\ 2 & -1 & -1 & 1 \end{vmatrix}$$

(2)
$$= 10(2)\begin{vmatrix} 2 & 1 & 1 \\ 1 & 2 & 2 \\ 2 & -1 & 1 \end{vmatrix}$$

To introduce 0's into the 3×3 determinant in Equation 2, we perform a column operation on the 3×3 determinant and expand the result on its first column.

$$20\begin{vmatrix} 2 & 1 & 1 \\ 1 & 2 & 2 \\ 2 & -1 & 1 \end{vmatrix} \overset{\underset{\displaystyle\downarrow}{(-2)C3 + C1}}{=} 20\begin{vmatrix} 0 & 1 & 1 \\ -3 & 2 & 2 \\ 0 & -1 & 1 \end{vmatrix}$$

$$= 20\left[-(-3)\begin{vmatrix} 1 & 1 \\ -1 & 1 \end{vmatrix} \right]$$

$$= 20(3)[1 - (-1)]$$

$$= 60(2)$$

$$= 120$$

Self Check 5   Evaluate: $\begin{vmatrix} 15 & 20 & 5 & 10 \\ 1 & 2 & 2 & 3 \\ 0 & 3 & -1 & 1 \\ 1 & 0 & 0 & -1 \end{vmatrix}$.

We will consider one final theorem.

Theorem   If A is a square matrix with two identical rows (or columns), then $|A| = 0$.

Proof   If the square matrix A has two identical rows (or columns), we can apply a type 3 row (or column) operation to zero out one of those rows (or columns). Since the matrix would then have an all-zero row (or column), its determinant would be 0, by the zero row or column theorem.

3. Use Determinants to Solve Systems of Equations

We can solve the system $\begin{cases} ax + by = e \\ cx + dy = f \end{cases}$ by multiplying the first equation by d, multiplying the second equation by $-b$, and adding to get

$$\begin{aligned} adx + bdy &= ed \\ \underline{-bcx - bdy} &= \underline{-bf} \\ adx - bcx &= ed - bf \end{aligned}$$

If $ad \neq bc$, we can solve the resulting equation for x:

$$adx - bcx = ed - bf$$

$$(ad - bc)x = ed - bf \qquad \text{Factor out } x.$$

$$(3) \qquad x = \frac{ed - bf}{ad - bc} \qquad \text{Divide both sides by } ad - bc.$$

If $ad \neq bc$, we also can solve the system for y to get

$$(4) \qquad y = \frac{af - ec}{ad - bc}$$

We can write the values of x and y in Equations 3 and 4 using determinants.

$$x = \frac{\begin{vmatrix} e & b \\ f & d \end{vmatrix}}{\begin{vmatrix} a & b \\ c & d \end{vmatrix}} = \frac{ed - bf}{ad - bc} \qquad y = \frac{\begin{vmatrix} a & e \\ c & f \end{vmatrix}}{\begin{vmatrix} a & b \\ c & d \end{vmatrix}} = \frac{af - ec}{ad - bc}$$

If we compare these formulas with the original system,

$$\begin{cases} ax + by = e \\ cx + dy = f \end{cases}$$

we see that the denominators are the determinant of the coefficient matrix:

$$\text{Denominator determinant} = \begin{vmatrix} a & b \\ c & d \end{vmatrix}$$

To find the numerator determinant for x, we replace the a and c in the first column of the denominator determinant with the constants e and f.

To find the numerator determinant for y, we replace the b and d in the second column of the denominator determinant with the constants e and f.

$$x = \frac{\begin{vmatrix} e & b \\ f & d \end{vmatrix}}{\begin{vmatrix} a & b \\ c & d \end{vmatrix}} \qquad y = \frac{\begin{vmatrix} a & e \\ c & f \end{vmatrix}}{\begin{vmatrix} a & b \\ c & d \end{vmatrix}}$$

This method of using determinants to solve systems of equations is called **Cramer's rule.**

Cramer's Rule for Two Equations in Two Variables

If the system $\begin{cases} ax + by = e \\ cx + dy = f \end{cases}$ has a single solution, it is given by

$$x = \frac{D_x}{D} \qquad \text{and} \qquad y = \frac{D_y}{D}$$

where $D = \begin{vmatrix} a & b \\ c & d \end{vmatrix}$, $D_x = \begin{vmatrix} e & b \\ f & d \end{vmatrix}$, and $D_y = \begin{vmatrix} a & e \\ c & f \end{vmatrix}$.

If D, D_x, and D_y are all 0, the system is consistent, but the equations are dependent. If $D = 0$ and $D_x \neq 0$ or $D_y \neq 0$, the system is inconsistent.

EXAMPLE 6 Use Cramer's rule to solve $\begin{cases} 3x + 2y = 7 \\ -x + 5y = 9 \end{cases}$.

Solution
$$x = \frac{\begin{vmatrix} 7 & 2 \\ 9 & 5 \end{vmatrix}}{\begin{vmatrix} 3 & 2 \\ -1 & 5 \end{vmatrix}} = \frac{7 \cdot 5 - 2 \cdot 9}{3 \cdot 5 - 2(-1)} = \frac{35 - 18}{15 + 2} = \frac{17}{17} = 1$$

$$y = \frac{\begin{vmatrix} 3 & 7 \\ -1 & 9 \end{vmatrix}}{\begin{vmatrix} 3 & 2 \\ -1 & 5 \end{vmatrix}} = \frac{3 \cdot 9 - 7(-1)}{3 \cdot 5 - 2(-1)} = \frac{27 + 7}{15 + 2} = \frac{34}{17} = 2$$

Verify that the ordered pair $(1, 2)$ satisfies both of the equations in the system.

Self Check 6 Solve: $\begin{cases} 2x + 5y = 9 \\ 3x + 7y = 13 \end{cases}$.

We can use Cramer's rule to solve systems of n equations in n variables where each equation has the form

$$a_1x_1 + a_2x_2 + \cdots + a_nx_n = c$$

To do so, we let D be the determinant of the coefficient matrix of the system and let D_{x_i} be the determinant formed by replacing the ith column of D by the column of constants from the right of the equal signs. If $D \neq 0$, Cramer's rule provides the following solution:

$$x_1 = \frac{D_{x_1}}{D}, x_2 = \frac{D_{x_2}}{D}, \ldots, x_n = \frac{D_{x_n}}{D}$$

EXAMPLE 7 Use Cramer's rule to solve the system $\begin{cases} 2x - y + 2z = 3 \\ x - y + z = 2 \\ x + y + 2z = 3 \end{cases}$.

Solution Each of the values x, y, and z is the quotient of two 3×3 determinants. The denominator of each quotient is the determinant consisting of the nine coefficients of the variables. The numerators for x, y, and z are modified copies of this denominator determinant. We substitute the column of constants for the coefficients of the variable for which we are solving.

$$\begin{cases} 2x - y + 2z = 3 \\ x - y + z = 2 \\ x + y + 2z = 3 \end{cases}$$

$$x = \frac{\begin{vmatrix} 3 & -1 & 2 \\ 2 & -1 & 1 \\ 3 & 1 & 2 \end{vmatrix}}{\begin{vmatrix} 2 & -1 & 2 \\ 1 & -1 & 1 \\ 1 & 1 & 2 \end{vmatrix}} = \frac{3\begin{vmatrix} -1 & 1 \\ 1 & 2 \end{vmatrix} - (-1)\begin{vmatrix} 2 & 1 \\ 3 & 2 \end{vmatrix} + 2\begin{vmatrix} 2 & -1 \\ 3 & 1 \end{vmatrix}}{2\begin{vmatrix} -1 & 1 \\ 1 & 2 \end{vmatrix} - (-1)\begin{vmatrix} 1 & 1 \\ 1 & 2 \end{vmatrix} + 2\begin{vmatrix} 1 & -1 \\ 1 & 1 \end{vmatrix}} = \frac{2}{-1} = -2$$

$$y = \frac{\begin{vmatrix} 2 & 3 & 2 \\ 1 & 2 & 1 \\ 1 & 3 & 2 \end{vmatrix}}{\begin{vmatrix} 2 & -1 & 2 \\ 1 & -1 & 1 \\ 1 & 1 & 2 \end{vmatrix}} = \frac{2\begin{vmatrix} 2 & 1 \\ 3 & 2 \end{vmatrix} - 3\begin{vmatrix} 1 & 1 \\ 1 & 2 \end{vmatrix} + 2\begin{vmatrix} 1 & 2 \\ 1 & 3 \end{vmatrix}}{-1} = \frac{1}{-1} = -1$$

$$z = \frac{\begin{vmatrix} 2 & -1 & 3 \\ 1 & -1 & 2 \\ 1 & 1 & 3 \end{vmatrix}}{\begin{vmatrix} 2 & -1 & 2 \\ 1 & -1 & 1 \\ 1 & 1 & 2 \end{vmatrix}} = \frac{2\begin{vmatrix} -1 & 2 \\ 1 & 3 \end{vmatrix} - (-1)\begin{vmatrix} 1 & 2 \\ 1 & 3 \end{vmatrix} + 3\begin{vmatrix} 1 & -1 \\ 1 & 1 \end{vmatrix}}{-1} = \frac{-3}{-1} = 3$$

Verify that the ordered triple $(-2, -1, 3)$ satisfies each equation in the system.

4. Write Equations of Lines

If we are given the coordinates of two points in the xy-plane, we can use determinants to write the equation of the line passing through those points.

Two-Point Form of the Equation of a Line

The equation of the line passing through points $P(x_1, y_1)$ and $Q(x_2, y_2)$ is given by

$$\begin{vmatrix} x & y & 1 \\ x_1 & y_1 & 1 \\ x_2 & y_2 & 1 \end{vmatrix} = 0$$

EXAMPLE 8 Write the equation of the line passing through $P(-2, 3)$ and $Q(4, -5)$.

Solution We set up the equation $\begin{vmatrix} x & y & 1 \\ -2 & 3 & 1 \\ 4 & -5 & 1 \end{vmatrix} = 0$ and expand along the first row to get

$$[3(1) - 1(-5)]x - [-2(1) - 1(4)]y + [(-2)(-5) - 3(4)]1 = 0$$
$$8x + 6y - 2 = 0$$
$$4x + 3y = 1 \quad \text{Add 2 to both sides and divide both sides by 2.}$$

The equation of the line is $4x + 3y = 1$.

Self Check 8 Find the equation of the line passing through $(1, 3)$ and $(3, 5)$.

5. Find Areas of Triangles

Area of a Triangle If points $P(x_1, y_1)$, $Q(x_2, y_2)$, and $R(x_3, y_3)$ are the vertices of a triangle, then the area of the triangle is given by

$$A = \pm\frac{1}{2}\begin{vmatrix} x_1 & y_1 & 1 \\ x_2 & y_2 & 1 \\ x_2 & y_3 & 1 \end{vmatrix} \quad \text{Pick either + or − to make the area positive.}$$

EXAMPLE 9 Find the area of the triangle shown in Figure 6-10.

Solution We set up the equation

$$A = \pm\frac{1}{2}\begin{vmatrix} 0 & 0 & 1 \\ 5 & 0 & 1 \\ 5 & 12 & 1 \end{vmatrix}$$

and expand the determinant along the first row to get

$$= \pm\frac{1}{2}\left(1\begin{vmatrix} 5 & 0 \\ 5 & 12 \end{vmatrix}\right)$$

$$= \pm\frac{1}{2}(60 - 0)$$

$$= 30$$

The area of the triangle is 30 square units.

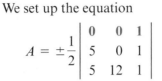

Figure 6-10

Self Check 9 Find the area of the triangle with vertices at $(1, 2)$, $(2, 3)$, and $(-1, 4)$.

Self Check Answers **1.** -2 **2.** 6 **3.** 6, 6 **4.** -4 **5.** 30 **6.** $(2, 1)$ **8.** $x - y = -2$
9. 2 sq. units

6.5 Exercises

Vocabulary and Concepts *Fill in the blanks.*

1. The determinant of a square matrix A is written as ____ or ____.

2. $\begin{vmatrix} a & b \\ c & d \end{vmatrix} =$ _____

3. If every entry in one row or one column of A is zero, then $|A| =$ __.

4. If a matrix B is obtained from matrix A by adding one row to another, then $|B| =$ ___.

5. If two columns of A are identical, then $|A| =$ __.

6. In Cramer's rule, the denominator is the determinant of the _____.

Practice *Evaluate each determinant.*

7. $\begin{vmatrix} 2 & 1 \\ -2 & 3 \end{vmatrix}$

8. $\begin{vmatrix} -3 & -6 \\ 2 & -5 \end{vmatrix}$

9. $\begin{vmatrix} 2 & -3 \\ -3 & 5 \end{vmatrix}$

10. $\begin{vmatrix} 5 & 8 \\ -6 & -2 \end{vmatrix}$

In Exercises 11–18, $A = \begin{vmatrix} 1 & -2 & 3 \\ 4 & 5 & -6 \\ -7 & 8 & 9 \end{vmatrix}$. *Find each minor or cofactor.*

11. M_{21} **12.** M_{13}

13. M_{33} **14.** M_{32}

15. C_{21} **16.** C_{13}

17. C_{33} **18.** C_{32}

Evaluate each determinant by expanding by cofactors.

19. $\begin{vmatrix} 2 & -3 & 5 \\ -2 & 1 & 3 \\ 1 & 3 & -2 \end{vmatrix}$ **20.** $\begin{vmatrix} 1 & 3 & 1 \\ -2 & 5 & 3 \\ 3 & -2 & -2 \end{vmatrix}$

21. $\begin{vmatrix} 1 & -1 & 2 \\ 2 & 1 & 3 \\ 1 & 1 & -1 \end{vmatrix}$ **22.** $\begin{vmatrix} 1 & 3 & 1 \\ 2 & 1 & -1 \\ 2 & -1 & 1 \end{vmatrix}$

23. $\begin{vmatrix} 2 & 1 & -1 \\ 1 & 3 & 5 \\ 2 & -5 & 3 \end{vmatrix}$

24. $\begin{vmatrix} 3 & 1 & -2 \\ -3 & 2 & 1 \\ 1 & 3 & 0 \end{vmatrix}$

25. $\begin{vmatrix} 0 & 1 & -3 \\ -3 & 5 & 2 \\ 2 & -5 & 3 \end{vmatrix}$

26. $\begin{vmatrix} 1 & -7 & -2 \\ -2 & 0 & 3 \\ -1 & 7 & 1 \end{vmatrix}$

27. $\begin{vmatrix} 0 & 0 & 1 & 0 \\ -2 & 1 & 0 & 1 \\ 1 & 0 & 1 & 2 \\ 2 & 0 & 1 & 2 \end{vmatrix}$

28. $\begin{vmatrix} 1 & 0 & -2 & 1 \\ 0 & 1 & 0 & 1 \\ 0 & 3 & -1 & 2 \\ 0 & -1 & 0 & 1 \end{vmatrix}$

29. $\begin{vmatrix} 1 & 2 & 1 & 3 \\ -2 & 1 & -3 & 1 \\ -1 & 0 & 1 & -2 \\ 2 & -1 & -1 & 3 \end{vmatrix}$

30. $\begin{vmatrix} -1 & 3 & -2 & 5 \\ 2 & 1 & 0 & 1 \\ 1 & 3 & -2 & 5 \\ 2 & -1 & 0 & -1 \end{vmatrix}$

Determine whether each statement is true. Do not evaluate the determinants.

31. $\begin{vmatrix} 1 & 3 & -4 \\ -2 & 1 & 3 \\ 1 & 3 & 2 \end{vmatrix} = - \begin{vmatrix} -2 & 1 & 3 \\ 1 & 3 & -4 \\ 1 & 3 & 2 \end{vmatrix}$

32. $\begin{vmatrix} 4 & 6 & 8 \\ 10 & 5 & 15 \\ 20 & 5 & 10 \end{vmatrix} = \begin{vmatrix} 2 & 3 & 4 \\ 10 & 5 & 15 \\ 20 & 5 & 10 \end{vmatrix}$

33. $\begin{vmatrix} -2 & -3 & -4 \\ 5 & -1 & 2 \\ 1 & 2 & 3 \end{vmatrix} = - \begin{vmatrix} 2 & 3 & 4 \\ -5 & 1 & -2 \\ 1 & 2 & 3 \end{vmatrix}$

34. $\begin{vmatrix} 1 & 2 & 3 \\ 4 & 5 & 6 \\ 7 & 8 & 9 \end{vmatrix} = \begin{vmatrix} 5 & 7 & 9 \\ 4 & 5 & 6 \\ 7 & 8 & 9 \end{vmatrix}$

If $\begin{vmatrix} a & b & c \\ d & e & f \\ g & h & i \end{vmatrix} = 3$, *find the value of each determinant.*

35. $\begin{vmatrix} d & e & f \\ a & b & c \\ -g & -h & -i \end{vmatrix}$

36. $\begin{vmatrix} 5a & 5b & 5c \\ -d & -e & -f \\ 3g & 3h & 3i \end{vmatrix}$

37. $\begin{vmatrix} a+g & b+h & c+i \\ d & e & f \\ g & h & i \end{vmatrix}$

38. $\begin{vmatrix} g & h & i \\ a & b & c \\ d & e & f \end{vmatrix}$

Use Cramer's rule to find the solution of each system, if possible.

39. $\begin{cases} 3x + 2y = 7 \\ 2x - 3y = -4 \end{cases}$

40. $\begin{cases} x - 5y = -6 \\ 3x + 2y = -1 \end{cases}$

41. $\begin{cases} x - y = 3 \\ 3x - 7y = 9 \end{cases}$

42. $\begin{cases} 2x - y = -6 \\ x + y = 0 \end{cases}$

43. $\begin{cases} x + 2y + z = 2 \\ x - y + z = 2 \\ x + y + 3z = 4 \end{cases}$

44. $\begin{cases} x + 2y - z = -1 \\ 2x + y - z = 1 \\ x - 3y - 5z = 17 \end{cases}$

45. $\begin{cases} 2x - y + z = 5 \\ 3x - 3y + 2z = 10 \\ x + 3y + z = 0 \end{cases}$

46. $\begin{cases} x - y - z = 2 \\ x + y + z = 2 \\ -x - y + z = -4 \end{cases}$

47. $\begin{cases} \dfrac{x}{2} + \dfrac{y}{3} + \dfrac{z}{2} = 11 \\ \dfrac{x}{3} + y - \dfrac{z}{6} = 6 \\ \dfrac{x}{2} + \dfrac{y}{6} + z = 16 \end{cases}$

48. $\begin{cases} \dfrac{x}{2} + \dfrac{y}{5} + \dfrac{z}{3} = 17 \\ \dfrac{x}{5} + \dfrac{y}{2} + \dfrac{z}{5} = 32 \\ x + \dfrac{y}{3} + \dfrac{z}{2} = 30 \end{cases}$

49. $\begin{cases} 2p - q + 3r - s = 0 \\ p + q - s = -1 \\ 3p - r = 2 \\ p - 2q + 3s = 7 \end{cases}$

50. $\begin{cases} a + b + c + d = 8 \\ a + b + c + 2d = 7 \\ a + b + 2c + 3d = 3 \\ a + 2b + 3c + 4d = 4 \end{cases}$

Use determinants to write the equation of the line that passes through the given points.

51. $P(0, 0)$, $Q(4, 6)$ **52.** $P(2, 3)$, $Q(6, 8)$

53. $P(-2, 3)$, $Q(5, -3)$ **54.** $P(1, -2)$, $Q(-4, 3)$

Use determinants to find the area of each triangle with vertices at the given points.

55. $P(0, 0)$, $Q(12, 0)$, $R(12, 5)$

56. $P(0, 0)$, $Q(0, 5)$, $R(12, 5)$

57. $P(2, 3)$, $Q(10, 8)$, $R(0, 20)$

58. $P(1, 1)$, $Q(6, 6)$, $R(2, 10)$

In Exercises 59–61, illustrate each column operation by showing that it is true for the determinant $\begin{vmatrix} a & b \\ c & d \end{vmatrix}$.

59. Interchanging two columns

60. Multiplying each element in a column by k

61. Adding k times any column to another column

62. Use the method of addition to solve
$$\begin{cases} ax + by = e \\ cx + dy = f \end{cases}$$ for y, and thereby show that
$$y = \frac{af - ec}{ad - bc}.$$

Expand the determinants and solve for x.

63. $\begin{vmatrix} 3 & x \\ 1 & 2 \end{vmatrix} = \begin{vmatrix} 2 & -1 \\ x & -5 \end{vmatrix}$ **64.** $\begin{vmatrix} 4 & x^2 \\ 1 & -1 \end{vmatrix} = \begin{vmatrix} x & 4 \\ 2 & 3 \end{vmatrix}$

65. $\begin{vmatrix} 3 & x & 1 \\ x & 0 & -2 \\ 4 & 0 & 1 \end{vmatrix} = \begin{vmatrix} 2 & x \\ x & 4 \end{vmatrix}$

66. $\begin{vmatrix} x & -1 & 2 \\ -2 & x & 3 \\ 4 & -3 & -1 \end{vmatrix} = \begin{vmatrix} 2 & 2 \\ 5 & x \end{vmatrix}$

Applications

67. Investing A student wants to average a 6.6% return by investing $20,000 in the three stocks listed in the table. Because HiTech is a high-risk investment, he wants to invest three times as much in SaveTel and OilCo combined as he invests in HiTech. How much should he invest in each stock?

Stock	Rate of return
HiTech	10%
SaveTel	5%
OilCo	6%

68. Ice skating The illustration shows three circles traced out by a figure skater during her performance. If the centers of the circles are the given distances apart, find the radius of each circle.

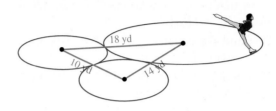

Discovery and Writing *Evaluate each determinant. What do you discover?*

69. $\begin{vmatrix} 1 & 3 & 4 \\ 0 & 5 & 2 \\ 0 & 0 & 2 \end{vmatrix}$ **70.** $\begin{vmatrix} 2 & 1 & -2 \\ 0 & 3 & 4 \\ 0 & 0 & -1 \end{vmatrix}$

71. $\begin{vmatrix} 1 & 2 & 4 & 3 \\ 0 & 2 & 2 & 1 \\ 0 & 0 & 3 & 2 \\ 0 & 0 & 0 & 4 \end{vmatrix}$ **72.** $\begin{vmatrix} 2 & 1 & -2 & 1 \\ 0 & 2 & 2 & -1 \\ 0 & 0 & 3 & 1 \\ 0 & 0 & 0 & 2 \end{vmatrix}$

73. Another way to evaluate a 3×3 determinant is to copy its first two columns to the right of the determinant as shown. Then find the product of the numbers on each red diagonal and find their sum. Then find the product of the numbers on each blue diagonal and find their sum. Then subtract the sum of the products on the blue diagonals from the sum of the products on the red diagonals. Find the value of the determinant.

74. Use the method of Exercise 73 to evaluate the determinant $\begin{vmatrix} 0 & 1 & -3 \\ -3 & 5 & 2 \\ 2 & -5 & 3 \end{vmatrix}$.

75. Use an example chosen from 2×2 matrices to show that the determinant of the product of two matrices is the product of the determinants of those two matrices.

76. Find an example among 2×2 matrices to show that the determinant of a sum of two matrices is not equal to the sum of the determinants of those matrices.

77. A determinant is a function that associates a number with every square matrix. Give the domain and the range of that function.

78. Use an example chosen from 2×2 matrices to show that for $n \times n$ matrices A and B, $AB \neq BA$ but $|AB| = |BA|$.

79. If A and B are matrices and $|AB| = 0$, must $|A| = 0$ or $|B| = 0$? Explain.

80. If A and B are matrices and $|AB| = 0$, must $A = 0$ or $B = 0$? Explain.

Use a graphing calculator to evaluate each determinant.

81. $\begin{vmatrix} 2.3 & 5.7 & 6.1 \\ 3.4 & 6.2 & 8.3 \\ 5.8 & 8.2 & 9.2 \end{vmatrix}$

82. $\begin{vmatrix} .32 & -7.4 & -6.7 \\ 3.3 & 5.5 & -0.27 \\ -8 & -0.13 & 5.47 \end{vmatrix}$

Review   *Factor each polynomial, if possible.*

83. $x^2 + 3x - 4$

84. $2x^2 - 5x - 12$

85. $9x^3 - x$

86. $x^2 + 3x + 5$

Add the fractions.

87. $\dfrac{1}{x - 2} + \dfrac{2}{2x - 1}$

88. $\dfrac{2}{x} + \dfrac{3}{x^2} - \dfrac{1}{x - 1}$

89. $\dfrac{2}{x^2 + 1} + \dfrac{1}{x}$

90. $\dfrac{2x}{x^2 + 1} + \dfrac{1}{x}$

6.6 Partial Fractions

Objectives

1. Decompose a Fraction When the Denominator Has Distinct Linear Factors
2. Decompose a Fraction When the Denominator Has Distinct Quadratic Factors
3. Decompose a Fraction When the Denominator Has Repeated Linear Factors
4. Decompose a Fraction When the Denominator Has Repeated Quadratic Factors
5. Decompose a Fraction When the Degree of $P(x)$ Is Equal to or Greater Than the Degree of $Q(x)$

In this section, we will discuss how to write a complicated fraction as a sum of simpler fractions. This skill is used in calculus to solve problems such as modeling a population when there is a maximum population (called the **carrying capacity**) that an environment can sustain.

Suppose there are P gorillas in a specific region in Africa. If there is a carrying capacity of 300 gorillas in that region, the fraction $\dfrac{1}{P^2 - 300P}$ will occur in the model of population growth of gorillas in that region. To solve this modeling problem, we need to rewrite this fraction as a sum of two simpler fractions. In this section, we will learn to do that. The process is called **partial fraction decomposition**.

You will be asked to decompose this fraction in Problem 10 in the exercises.

We begin the discussion by reviewing how to add fractions. For example, to find the sum

$$\frac{2}{x} + \frac{6}{x+1} + \frac{-1}{(x+1)^2}$$

we write each fraction with an LCD of $x(x+1)^2$, add the fractions by adding their numerators and keeping the common denominator, and simplify.

$$\frac{2}{x} + \frac{6}{x+1} + \frac{-1}{(x+1)^2} = \frac{2(x+1)^2}{x(x+1)^2} + \frac{6x(x+1)}{(x+1)x(x+1)} + \frac{-1x}{(x+1)^2 x}$$

$$= \frac{2x^2 + 4x + 2 + 6x^2 + 6x - x}{x(x+1)^2}$$

$$= \frac{8x^2 + 9x + 2}{x(x+1)^2}$$

Johann Bernoulli
(1667–1748)
Bernoulli was a Swiss mathematician and teacher of Euler. His method of decomposition by partial fractions was a major contribution to the calculus.

To reverse the addition process and write a fraction as the sum of simpler fractions with denominators of smallest possible degree, we must **decompose a fraction into partial fractions.** To decompose the fraction

$$\frac{8x^2 + 9x + 2}{x(x+1)^2}$$

into partial fractions, we will assume that there are constants A, B, and C such that

$$\frac{8x^2 + 9x + 2}{x(x+1)^2} = \frac{A}{x} + \frac{B}{x+1} + \frac{C}{(x+1)^2}$$

After writing the terms on the right side as fractions with an LCD of $x(x+1)^2$, we add the fractions to get

$$\frac{8x^2 + 9x + 2}{x(x+1)^2} = \frac{A(x+1)^2}{x(x+1)^2} + \frac{Bx(x+1)}{x(x+1)(x+1)} + \frac{Cx}{(x+1)^2 x}$$

$$= \frac{Ax^2 + 2Ax + A + Bx^2 + Bx + Cx}{x(x+1)^2}$$

(1) $\quad \dfrac{8x^2 + 9x + 2}{x(x+1)^2} = \dfrac{(A+B)x^2 + (2A+B+C)x + A}{x(x+1)^2}$ Factor x^2 from $Ax^2 + Bx^2$.
Factor x from $2Ax + Bx + Cx$.

Since the fractions on the left and right sides of Equation 1 are equal and their denominators are equal, the coefficients of their polynomial numerators are equal.

$$\begin{cases} A + B = 8 & \text{These are the coefficients of } x^2. \\ 2A + B + C = 9 & \text{These are the coefficients of } x. \\ A = 2 & \text{These are the constants.} \end{cases}$$

We can solve this system of equations to find that $A = 2$, $B = 6$, and $C = -1$. Then we know that

$$\frac{8x^2 + 9x + 2}{x(x+1)^2} = \frac{2}{x} + \frac{6}{x+1} + \frac{-1}{(x+1)^2}$$

To check the result, we can add the previous fractions and show that the sum is the original fraction.

To find the partial-fraction decomposition of a fraction, we will follow these steps:

Decomposing a Fraction into Partial Fractions	1. Set up the decomposition with unknown constants A, B, C, ... in the numerator of the composition.
	2. Write each fraction with a common denominator and add the fractions and simplify.
	3. Set the coefficients of the corresponding terms of each numerator equal to each other.
	4. Solve the resulting system of linear equations.
	5. Use the values of A, B, C, ... to write the decomposition.

1. Decompose a Fraction When the Denominator Has Distinct Linear Factors

EXAMPLE 1 Decompose $\dfrac{9x + 2}{(x + 2)(3x - 2)}$ into partial fractions.

Solution **Step 1.** Since each factor in the denominator is linear, there are constants A and B such that

$$\frac{9x + 2}{(x + 2)(3x - 2)} = \frac{A}{x + 2} + \frac{B}{3x - 2}$$

Step 2. After writing the terms on the right side as fractions with an LCD of $(x + 2)(3x - 2)$, we add the fractions to get

(2) $\qquad \dfrac{9x + 2}{(x + 2)(3x - 2)} = \dfrac{A(3x - 2)}{(x + 2)(3x-2)} + \dfrac{B(x + 2)}{(x + 2)(3x - 2)}$

$\qquad \dfrac{9x + 2}{(x + 2)(3x - 2)} = \dfrac{3Ax - 2A + Bx + 2B}{(x + 2)(3x - 2)}$

(3) $\qquad \dfrac{9x + 2}{(x + 2)(3x - 2)} = \dfrac{(3A + B)x - 2A + 2B}{(x + 2)(3x - 2)}$ Factor x from $3Ax + Bx$.

Step 3. Since the fractions in Equation 3 are equal, the coefficients of their polynomial numerators are equal, and we have

$$\begin{cases} 9 = 3A + B & \text{These are the coefficients of } x. \\ 2 = -2A + 2B & \text{These are the constants.} \end{cases}$$

Step 4. The solution of the system of linear equations in Step 3 is $A = 2$ and $B = 3$.

Step 5. Finally, we will substitute the values of A and B and write the decomposition.

$$\frac{9x + 2}{(x + 2)(3x - 2)} = \frac{2}{x + 2} + \frac{3}{3x - 2}$$

Self Check 1 Decompose $\dfrac{2x + 2}{x(x + 2)}$ into partial fractions.

Comment

The values of A and B in Example 1 also can be found in the following way. From Equation 2, we know that $9x + 2 = A(3x - 2) + B(x + 2)$. In this equation, we can then let $x = -2$ to eliminate the term involving B and solve for A. Then we can let $x = \frac{2}{3}$ to eliminate the term involving A and solve for B. As before, we will find that $A = 2$ and $B = 3$.

2.   Decompose a Fraction When the Denominator Has Distinct Quadratic Factors

We can use the following theorem to decompose a fraction with linear and nonlinear factors in the denominator. We begin with an algebraic fraction—such as $\frac{P(x)}{Q(x)}$, the quotient of two polynomials—and write the fraction as the sum of two or more fractions with simpler denominators. By the theorem, we know that they will be either first-degree or irreducible second-degree polynomials, or powers of those.

Polynomial Factorization Theorem

The factorization of any polynomial $Q(x)$ with real coefficients is the product of polynomials of the forms

$$(ax + b)^n \quad \text{and} \quad (ax^2 + bx + c)^n$$

where n is a positive integer and $ax^2 + bx + c$ is irreducible over the real numbers.

If $Q(x)$ has a prime quadratic factor of the form $ax^2 + bx + c$, the partial fraction decomposition of $\frac{P(x)}{Q(x)}$ will have a corresponding term of the form

$$\frac{Bx + C}{ax^2 + bx + c}$$

Comment

Be sure that the quadratic factor is not factorable over the real numbers—check it with the discriminant. If $b^2 - 4ac$ is negative, the factor is prime.

EXAMPLE 2   Decompose $\dfrac{2x^2 + x + 1}{x^3 + x}$ into partial fractions.

Solution   **Step 1.**   Since the denominator can be written as a product of a linear factor and a prime quadratic factor, the partial fractions have the form

$$\frac{2x^2 + x + 1}{x(x^2 + 1)} = \frac{A}{x} + \frac{Bx + C}{x^2 + 1}$$

Step 2.   We add the fractions and simplify.

$$\frac{2x^2 + x + 1}{x(x^2 + 1)} = \frac{A(x^2 + 1)}{x(x^2 + 1)} + \frac{(Bx + C)x}{x(x^2 + 1)}$$

$$= \frac{Ax^2 + A + Bx^2 + Cx}{x(x^2 + 1)}$$

$$= \frac{(A + B)x^2 + Cx + A}{x(x^2 + 1)} \qquad \text{Factor } x^2 \text{ from } Ax^2 + Bx^2.$$

Step 3.   We can equate the corresponding coefficients of the numerators $2x^2 + 1x + 1$ and $(A + B)x^2 + Cx + A$ to get the system

$$\begin{cases} A + B = 2 & \text{These are the coefficients of } x^2. \\ C = 1 & \text{These are the coefficients of } x. \\ A = 1 & \text{These are the constants.} \end{cases}$$

Step 4. This system has solutions of $A = 1$, $B = 1$, and $C = 1$.

Step 5. Substituting the values of A, B, and C, we find that the partial fraction decomposition is

$$\frac{2x^2 + x + 1}{x(x^2 + 1)} = \frac{A}{x} + \frac{Bx + C}{x^2 + 1}$$

$$= \frac{1}{x} + \frac{1x + 1}{x^2 + 1}$$

$$= \frac{1}{x} + \frac{x + 1}{x^2 + 1}$$

Self Check 2 Decompose $\dfrac{2x^2 + 1}{x^3 + x}$ into partial fractions.

3. Decompose a Fraction When the Denominator Has Repeated Linear Factors

If $Q(x)$ has n linear factors of $ax + b$, then $(ax + b)^n$ is a factor of $Q(x)$. Each factor of the form $(ax + b)^n$ generates the following sum of n partial fractions:

$$\frac{A}{ax + b} + \frac{B}{(ax + b)^2} + \frac{C}{(ax + b)^3} + \cdots + \frac{D}{(ax + b)^n}$$

EXAMPLE 3 Decompose $\dfrac{3x^2 - x + 1}{x(x - 1)^2}$ into partial fractions.

Solution **Step 1.** Here, each factor in the denominator is linear. The linear factor x appears once, and the linear factor $x - 1$ appears twice. Thus, there are constants A, B, and C such that

$$\frac{3x^2 - x + 1}{x(x - 1)^2} = \frac{A}{x} + \frac{B}{x - 1} + \frac{C}{(x - 1)^2}$$

Comment

Note that $x - 1$ is a repeated factor. It would be incorrect to write

$$\frac{3x^2 - x + 1}{x(x - 1)^2} = \frac{A}{x} + \frac{B}{x - 1} + \frac{C}{x - 1}$$

Step 2. After writing the terms on the right side as fractions with an LCD of $x(x - 1)^2$, we combine them to get

$$\frac{3x^2 - x + 1}{x(x - 1)^2} = \frac{A(x-1)^2}{x(x-1)^2} + \frac{Bx(x - 1)}{x(x - 1)(x - 1)} + \frac{Cx}{(x - 1)^2 x}$$

$$= \frac{Ax^2 - 2Ax + A + Bx^2 - Bx + Cx}{x(x - 1)^2}$$

$$= \frac{(A + B)x^2 + (-2A - B + C)x + A}{x(x - 1)^2} \qquad \begin{array}{l}\text{Factor } x^2 \text{ from } Ax^2 + Bx^2.\\ \text{Factor } x \text{ from } -2Ax - Bx + Cx.\end{array}$$

Step 3.   Since the fractions are equal, the coefficients of the polynomial numerators are equal, and we have

$$\begin{cases} A + B = 3 & \text{These are the coefficients of } x^2. \\ -2A - B + C = -1 & \text{These are the coefficients of } x. \\ A = 1 & \text{These are the constants.} \end{cases}$$

Step 4.   We can solve this system to find that $A = 1$, $B = 2$, and $C = 3$.

Step 5.   Substituting the values of A, B, and C, we find that the partial fraction decomposition is

$$\frac{3x^2 - x + 1}{x(x - 1)^2} = \frac{A}{x} + \frac{B}{x - 1} + \frac{C}{(x - 1)^2}$$

$$= \frac{1}{x} + \frac{2}{x - 1} + \frac{3}{(x - 1)^2}$$

Self Check 3   Decompose $\dfrac{3x^2 + 7x + 1}{x(x + 1)^2}$ into partial fractions.

4.   Decompose a Fraction When the Denominator Has Repeated Quadratic Factors

If $Q(x)$ has n prime factors of $ax^2 + bx + c$, then $(ax^2 + bx + c)^n$ is a factor. Each factor of the form $(ax^2 + bx + c)^n$ generates a sum of n partial fractions of the form

$$\frac{Ax + B}{ax^2 + bx + c} + \frac{Cx + D}{(ax^2 + bx + c)^2} + \cdots + \frac{Ex + F}{(ax^2 + bx + c)^n}$$

EXAMPLE 4   Decompose $\dfrac{3x^2 + 5x + 5}{(x^2 + 1)^2}$ into partial fractions.

Solution   **Step 1.**   Since the quadratic factor $x^2 + 1$ is used twice, we must find constants A, B, C, and D such that

$$\frac{3x^2 + 5x + 5}{(x^2 + 1)^2} = \frac{Ax + B}{x^2 + 1} + \frac{Cx + D}{(x^2 + 1)^2}$$

Step 2.   We add the fractions on the right side to get

$$\frac{3x^2 + 5x + 5}{(x^2 + 1)^2} = \frac{(Ax + B)(x^2 + 1)}{(x^2 + 1)(x^2 + 1)} + \frac{Cx + D}{(x^2 + 1)^2}$$

$$= \frac{Ax^3 + Ax + Bx^2 + B + Cx + D}{(x^2 + 1)^2}$$

$$= \frac{Ax^3 + Bx^2 + (A + C)x + B + D}{(x^2 + 1)^2} \qquad \text{Factor } x \text{ from } Ax + Cx.$$

Step 3.   If we add the term $0x^3$ to the numerator on the left side, we can equate the corresponding coefficients in the numerators to get

$A = 0$ These are the coefficients of x^3.

$B = 3$ These are the coefficients of x^2.

$A + C = 5$ These are the coefficients of x.

$B + D = 5$ These are the constants.

Step 4. The solution to this system is $A = 0$, $B = 3$, $C = 5$, and $D = 2$.

Step 5. Substituting the values of A, B, C, and D, we find that the partial fraction decomposition is

$$\frac{3x^2 + 5x + 5}{(x^2 + 1)^2} = \frac{Ax + B}{x^2 + 1} + \frac{Cx + D}{(x^2 + 1)^2}$$

$$= \frac{0x + 3}{x^2 + 1} + \frac{5x + 2}{(x^2 + 1)^2}$$

$$= \frac{3}{x^2 + 1} + \frac{5x + 2}{(x^2 + 1)^2}$$

Self Check 4 Decompose $\dfrac{x^3 + 2x + 3}{(x^2 + 2)^2}$ into partial fractions.

5. Decompose a Fraction When the Degree of $P(x)$ Is Equal to or Greater Than the Degree of $Q(x)$

When the degree of $P(x)$ is equal to or greater than the degree of $Q(x)$ in the fraction $\frac{P(x)}{Q(x)}$, we do a long division before decomposing the fraction into partial fractions.

EXAMPLE 5 Decompose $\dfrac{x^2 + 4x + 2}{x^2 + x}$ into partial fractions.

Solution Because the degree of the numerator and denominator are the same, we must do a long division and express the fraction in quotient $+ \frac{\text{remainder}}{\text{divisor}}$ form:

$$
\begin{array}{r}
1 \\
x^2 + x \overline{\smash{)}x^2 + 4x + 2} \\
\underline{x^2 + x} \\
3x + 2
\end{array}
$$

So we can write

$$(3) \qquad \frac{x^2 + 4x + 2}{x^2 + x} = 1 + \frac{3x + 2}{x^2 + x}$$

Because the degree of the numerator of the fraction on the right side of Equation 3 is less than the degree of the denominator, we can find its partial fraction decomposition:

$$\frac{3x + 2}{x^2 + x} = \frac{3x + 2}{x(x + 1)}$$

$$= \frac{A}{x} + \frac{B}{x + 1}$$

$$= \frac{A(x + 1) + Bx}{x(x + 1)}$$

$$= \frac{(A + B)x + A}{x(x + 1)}$$

We equate the corresponding coefficients in the numerator and solve the resulting system of equations to find that the solution is $A = 2$, $B = 1$. So we have

$$\frac{x^2 + 4x + 2}{x^2 + x} = 1 + \frac{2}{x} + \frac{1}{x + 1}$$

Self Check 5 Decompose $\dfrac{x^2 + x - 1}{x^2 - x}$ into partial fractions.

Self Check Answers

1. $\dfrac{1}{x} + \dfrac{1}{x + 2}$ **2.** $\dfrac{1}{x} + \dfrac{x}{x^2 + 1}$ **3.** $\dfrac{1}{x} + \dfrac{2}{x + 1} + \dfrac{3}{(x + 1)^2}$

4. $\dfrac{x}{x^2 + 2} + \dfrac{3}{(x^2 + 2)^2}$ **5.** $1 + \dfrac{1}{x} + \dfrac{1}{x - 1}$

6.6 Exercises

Vocabulary and Concepts *Fill in the blanks.*

1. A polynomial with real coefficients factors as the product of _____ and _____ factors or powers of those.

2. The second-degree factors of a polynomial with real coefficients are _____, which means they don't factor further over the real numbers.

Practice *Decompose each fraction into partial fractions.*

3. $\dfrac{3x - 1}{x(x - 1)}$

4. $\dfrac{4x + 6}{x(x + 2)}$

5. $\dfrac{2x - 15}{x(x - 3)}$

6. $\dfrac{5x + 21}{x(x + 7)}$

7. $\dfrac{3x + 1}{(x + 1)(x - 1)}$

8. $\dfrac{9x - 3}{(x + 1)(x - 2)}$

9. $\dfrac{-4}{x^2 - 2x}$

10. $\dfrac{1}{P^2 - 300P}$

11. $\dfrac{-2x + 11}{x^2 - x - 6}$

12. $\dfrac{7x + 2}{x^2 + x - 2}$

13. $\dfrac{3x - 23}{x^2 + 2x - 3}$

14. $\dfrac{-x - 17}{x^2 - x - 6}$

15. $\dfrac{9x - 31}{2x^2 - 13x + 15}$

16. $\dfrac{-2x - 6}{3x^2 - 7x + 2}$

17. $\dfrac{4x^2 + 4x - 2}{x(x^2 - 1)}$

18. $\dfrac{x^2 - 6x - 13}{(x + 2)(x^2 - 1)}$

19. $\dfrac{x^2 + x + 3}{x(x^2 + 3)}$

20. $\dfrac{5x^2 + 2x + 2}{x^3 + x}$

21. $\dfrac{3x^2 + 8x + 11}{(x + 1)(x^2 + 2x + 3)}$

22. $\dfrac{-3x^2 + x - 5}{(x + 1)(x^2 + 2)}$

23. $\dfrac{5x^2 + 9x + 3}{x(x + 1)^2}$

24. $\dfrac{2x^2 - 7x + 2}{x(x - 1)^2}$

25. $\dfrac{-2x^2 + x - 2}{x^2(x - 1)}$

26. $\dfrac{x^2 + x + 1}{x^3}$

27. $\dfrac{3x^2 - 13x + 18}{x^3 - 6x^2 + 9x}$

28. $\dfrac{3x^2 + 13x + 20}{x^3 + 4x^2 + 4x}$

29. $\dfrac{x^2 - 2x - 3}{(x - 1)^3}$

30. $\dfrac{x^2 + 8x + 18}{(x + 3)^3}$

31. $\dfrac{x^3 + 4x^2 + 2x + 1}{x^4 + x^3 + x^2}$

32. $\dfrac{3x^3 + 5x^2 + 3x + 1}{x^2(x^2 + x + 1)}$

33. $\dfrac{4x^3 + 5x^2 + 3x + 4}{x^2(x^2 + 1)}$

34. $\dfrac{2x^2 + 1}{x^4 + x^2}$

35. $\dfrac{-x^2 - 3x - 5}{x^3 + x^2 + 2x + 2}$

36. $\dfrac{-2x^3 + 7x^2 + 6}{x^2(x^2 + 2)}$

37. $\dfrac{x^3 + 4x^2 + 3x + 6}{(x^2 + 2)(x^2 + x + 2)}$

38. $\dfrac{x^3 + 3x^2 + 2x + 4}{(x^2 + 1)(x^2 + x + 2)}$

39. $\dfrac{2x^4 + 6x^3 + 20x^2 + 22x + 25}{x(x^2 + 2x + 5)^2}$

40. $\dfrac{x^3 + 3x^2 + 6x + 6}{(x^2 + x + 5)(x^2 + 1)}$

41. $\dfrac{x^3}{x^2 + 3x + 2}$

42. $\dfrac{2x^3 + 6x^2 + 3x + 2}{x^3 + x^2}$

43. $\dfrac{3x^3 + 3x^2 + 6x + 4}{3x^3 + x^2 + 3x + 1}$

44. $\dfrac{x^4 + x^3 + 3x^2 + x + 4}{(x^2 + 1)^2}$

45. $\dfrac{x^3 + 3x^2 + 2x + 1}{x^3 + x^2 + x}$

46. $\dfrac{x^4 - x^3 + x^2 - x + 4}{(x^2 + 1)^2}$

47. $\dfrac{2x^4 + 2x^3 + 3x^2 - 1}{(x^2 - x)(x^2 + 1)}$

48. $\dfrac{x^4 - x^3 + 5x^2 + x + 6}{(x^2 + 3)(x^2 + 1)}$

Discovery and Writing

49. Is the polynomial $x^3 + 1$ prime?

50. Decompose $\dfrac{1}{x^3 + 1}$ into partial fractions.

Review *Simplify each radical expression. Use absolute value symbols if necessary.*

51. $\sqrt{8a^3b}$

52. $\sqrt{x^2 + 6x + 9}$

53. $\sqrt{18x^5} + x^2\sqrt{50x} - 5x^2\sqrt{2x}$

54. $\dfrac{\sqrt{x^2y^3}}{\sqrt{xy^5}}$

Solve each equation.

55. $\sqrt{x - 5} = x - 7$ **56.** $x - 5 = (x - 7)^2$

6.7 Graphs of Linear Inequalities

Objectives

1. Graph Inequalities
2. Graph Systems of Inequalities

A company builds 84-inch and 120-inch motorized-drive projection screens. If the company needs 2 hours to build an 84-inch screen, it will take $2x$ hours to build x of them. If the company needs 3 hours to build a 120-inch screen, it will take $3y$ hours to build y of them. If the company has 300 hours of labor available each day and cannot build a negative number of either screen, the following system of inequalities provides restrictions on x and y, the number of screens it can build.

$$\begin{cases} 2x + 3y \le 300 \\ x \ge 0 \\ y \ge 0 \end{cases}$$
The time it takes to build x small screens plus the time it takes to build y large screens must be less than or equal to 300. Both x and y must be greater than or equal to 0.

We will solve this system of inequalities in Example 6. In this section, we will use graphing to find the solution sets of many types of systems.

1.   Graph Inequalities

The **graph of an inequality** in x and y is the graph of all ordered pairs (x, y) that satisfy the inequality. We will start by considering graphs of **linear inequalities**— inequalities that can be written in one of the following forms:

$$Ax + By < C \quad Ax + By > C \quad Ax + By \le C \quad \text{or} \quad Ax + By \ge C$$

To graph the inequality $y > 3x + 2$, we note that one of the following statements is true:

$$y = 3x + 2 \quad y < 3x + 2 \quad \text{or} \quad y > 3x + 2$$

The graph of $y = 3x + 2$ is a line, as shown in Figure 6-11(a). The graphs of the inequalities are half-planes, one on each side of that line. We can think of the graph of $y = 3x + 2$ as a boundary separating the two half-planes. The graph of $y = 3x + 2$ is drawn with a dashed line to show that it is not part of the graph of $y > 3x + 2$.

To find which half-plane is the graph of $y > 3x + 2$, we can substitute the coordinates of any point on one side of the line—say, the origin $(0, 0)$—into the inequality and simplify:

$$y > 3x + 2$$
$$0 > 3(0) + 2$$
$$0 > 2 \qquad \text{false}$$

Since $0 > 2$ is false, the coordinates $(0, 0)$ do not satisfy the inequality, and the origin is not in the half-plane that is the graph of $y > 3x + 2$. Thus, the graph is the half-plane located on the other side of the dashed line. The graph of the inequality $y > 3x + 2$ is shown in Figure 6-11(b).

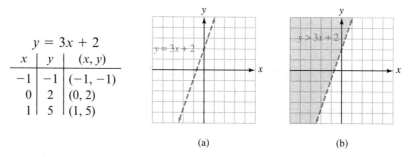

$y = 3x + 2$		
x	y	(x, y)
-1	-1	$(-1, -1)$
0	2	$(0, 2)$
1	5	$(1, 5)$

(a)

(b)

Figure 6-11

We can use the following steps to graph linear inequalities.

Graphing Linear Inequalities in Two Variables

1. Write the inequality as an equation by replacing the inequality symbol with an equal sign.
2. Graph the resulting equation to establish a boundary line. If the inequality is $<$ or $>$, draw a dashed line. If the inequality is $\leq$ or $\geq$, draw a solid line.
3. Select a test point that is not on the boundary.
4. Substitute the coordinates of the test point into the inequality.
5. Shade the region that contains the test point if the coordinates of the test point satisfy the inequality. If it doesn't, shade the region that does not contain the test point.

EXAMPLE 1   Graph the inequality: $2x - 3y \leq 6$.

Solution   **Step 1.**   This inequality is the combination of $2x - 3y < 6$ and $2x - 3y = 6$.

Step 2.   We start by graphing $2x - 3y = 6$ to establish the boundary line that separates the plane into two half-planes. We draw a solid line, because equality is permitted. See Figure 6-12(a).

Step 3.   Because the computations will be easy, we select the origin as a test point.

Step 4.   We will substitute the coordinates of the origin into the inequality and see whether the coordinates satisfy the inequality:

$$2x - 3y \leq 6$$
$$2(0) - 3(0) \leq 6$$
$$0 \leq 6 \quad \text{true}$$

Step 5.   Because $0 \leq 6$ is true, we shade the half-plane that contains the test point. The graph is shown in Figure 6-12(b).

$2x - 3y = 6$

x	y	(x, y)
0	-2	$(0, -2)$
3	0	$(3, 0)$

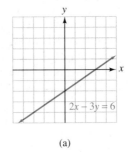 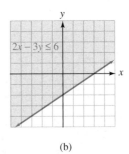

(a)   (b)

Figure 6-12

Self Check 1   Graph: $3x + 2y \leq 6$.

EXAMPLE 2   Graph the inequality: $y < 2x$.

Solution   **Step 1.**   We write the equation $y = 2x$.

Step 2.   Since the graph of $y = 2x$ is not part of the graph of the inequality, we graph the boundary with a dashed line, as in Figure 6-13(a).

Step 3.   We cannot use the origin as a test point, because the boundary line passes through the origin. So we choose some other point—say, $(3, 1)$.

Step 4. To determine which half-plane represents the graph $y < 2x$, we check whether the coordinates of the test point $(3, 1)$ satisfy the inequality.

$$y < 2x$$
$$1 < 2(3)$$
$$1 < 6 \qquad \text{true}$$

Step 5. Since $1 < 6$, the point $(3, 1)$ lies in the graph. The graph is shown in Figure 6-13(b).

$y = 2x$

x	y	(x, y)
0	0	$(0, 0)$
1	2	$(1, 2)$

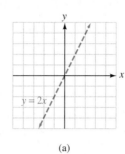

(a)

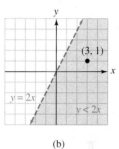

(b)

Figure 6-13

Self Check 2 Graph: $y > 3x$.

Accent on Technology

Graphing Inequalities

Many calculators (such as the TI-84 Plus) have a graphing-style icon in the $y =$ editor. See Figure 6-14(a). Some of the different graphing styles are as follows:

\	=	line	A straight line or curved graph is shown.
◥	=	above	Shading covers the area above the graph.
◣	=	below	Shading covers the area below the graph.

We can change the icon by placing the cursor on it and pressing Enter .

To graph the inequality of Example 1 using window settings of $x = [-10, 10]$ and $y = [-10, 10]$, we change the graphing-style icon to "above" (◥), enter the equation $2x - 3y = 6$ as $y = \frac{2}{3}x - 2$, and press GRAPH to obtain Figure 6-14(b).

To graph the inequality of Example 2 using window settings of $x = [-10, 10]$ and $y = [-10, 10]$, we change the graphing-style icon to "below" (◣), enter the equation $y = 2x$, and press GRAPH to obtain Figure 6-14(c).

If your calculator does not have a graphing-style icon, you can graph linear inequalities using a shade feature. To do so, consult your owner's manual.

Note that graphing calculators do not distinguish between solid and dashed lines to show whether or not the edge of a region is included in the graph.

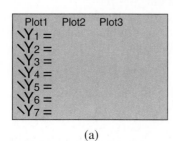

(a)

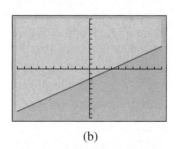

(b)

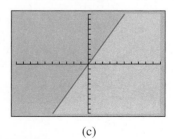

(c)

Figure 6-14

We can graph many inequalities that are not linear inequalities.

EXAMPLE 3   Graph the inequality: $x^2 + y^2 > 25$.

Solution   **Step 1.**   We form the equation $x^2 + y^2 = 25$.

Step 2.   The graph of $x^2 + y^2 = 25$ is a circle. Since the inequality is $>$, we draw the circle as a dashed circle, as in Figure 6-15(a).

Step 3.   Because the work is easy, we select the origin as the test point.

Step 4.   We substitute the coordinates of the origin into the inequality.

$$x^2 + y^2 > 25$$
$$0^2 + 0^2 > 25$$
$$0 > 25 \quad \text{false}$$

Step 5.   Since the coordinates of the origin do not satisfy the inequality, we shade the area outside the circle, as shown in Figure 6-15(b).

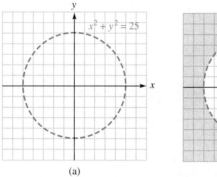

(a)   (b)

Figure 6-15

Self Check 3   Graph the inequality: $x^2 + y^2 \le 36$.

2.   Graph Systems of Inequalities

We now consider systems of inequalities. To graph the solution set of the system

$$\begin{cases} y < 5 \\ x \le 6 \end{cases}$$

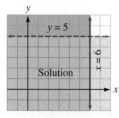

Figure 6-16

we graph each inequality on the same set of coordinate axes, as in Figure 6-16. The graph of the inequality $y < 5$ is the half-plane that lies below the line $y = 5$. The graph of the inequality $x \le 6$ includes the half-plane that lies to the left of the line $x = 6$ together with the line $x = 6$.

The portion of the xy-plane where the two graphs intersect is the graph of the system. Any point that lies in the doubly shaded region has coordinates that satisfy both inequalities in the system.

EXAMPLE 4   Graph the solution set of $\begin{cases} x + y \le 1 \\ 2x - y > 2 \end{cases}$.

Solution   On the same set of coordinate axes, we graph each inequality, as in Figure 6-17. The graph of $x + y \le 1$ includes the graph of $x + y = 1$ and all points below it. Because the boundary line is included, we draw it as a solid line.

$x + y = 1$		
x	y	(x, y)
0	1	$(0, 1)$
1	0	$(1, 0)$

$2x - y = 2$		
x	y	(x, y)
0	-2	$(0, -2)$
1	0	$(1, 0)$

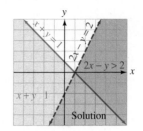

Figure 6-17

Comment

To be sure you graph the correct side of the boundary line, always use a test point.

The graph of $2x - y > 2$ contains only those points below the line graph of $2x - y = 2$. Because the boundary line is not included, we draw it as a dashed line.

The area that is shaded twice represents the solution of the system of inequalities. Any point in the doubly-shaded region has coordinates that satisfy both inequalities.

Self Check 4   Graph the solution set of $\begin{cases} x + y < 2 \\ x - 2y \geq 2 \end{cases}$.

EXAMPLE 5   Graph the solution set of $\begin{cases} y < x^2 \\ y > \dfrac{x^2}{4} - 2 \end{cases}$.

Solution   The graph of $y = x^2$ is a parabola opening upward with vertex at the origin, as shown in Figure 6-18. The points with coordinates that satisfy the inequality are the points below the parabola.

The graph of $y = \dfrac{x^2}{4} - 2$ is also a parabola opening upward. The points that satisfy the inequality are the points above the parabola. The graph of the solution set is the shaded area between the two parabolas.

$y = x^2$		
x	y	(x, y)
0	0	$(0, 0)$
1	1	$(1, 1)$
-1	1	$(-1, 1)$
2	4	$(2, 4)$
-2	4	$(-2, 4)$

$y = \dfrac{x^2}{4} - 2$		
x	y	(x, y)
0	-2	$(0, -2)$
2	-1	$(2, -1)$
-2	-1	$(-2, -1)$
4	2	$(4, 2)$
-4	2	$(-4, 2)$

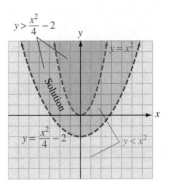

Figure 6-18

Self Check 5   Graph the solution set of $\begin{cases} y \leq 4 - x^2 \\ y > \dfrac{x^2}{4} \end{cases}$.

EXAMPLE 6   Graph the solution set of the system that began the section:

$$\begin{cases} 2x + 3y \le 300 \\ x \ge 0 \\ y \ge 0 \end{cases}$$

Solution   We graph each line, as shown in Figure 6-19. The graph of $2x + 3y \le 300$ includes the line $2x + 3y = 300$ and all points below it. Because the boundary line is included, we will draw it as a solid line.

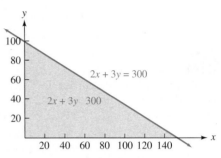

The graph of $x \ge 0$ includes the y-axis and all points to the right of it. The graph of $y \ge 0$ includes the x-axis and all points above it.

The solution set of the system is the set of points in the shaded region.

Figure 6-19

Self Check 6   Graph the solution set of the system $\begin{cases} x + 2y \le 50 \\ x \ge 0 \\ y \ge 0 \end{cases}$.

EXAMPLE 7   Graph the solution set of $\begin{cases} x + y \le 4 \\ x - y \le 6. \\ x \ge 0 \end{cases}$

Solution   We graph each inequality, as in Figure 6-20. The graph of $x + y \le 4$ includes the line $x + y = 4$ and all points below it. Because the boundary line is included, we draw it as a solid line. The graph of $x - y \le 6$ contains the line $x - y = 6$ and all points above it.

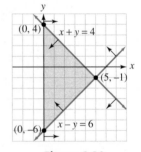

The graph of the inequality $x \ge 0$ contains the y-axis and all points to the right of the y-axis. The solution of the system of inequalities is the shaded area in the figure.

The coordinates of the corner points of the shaded area are $(0, 4)$, $(0, -6)$, and $(5, -1)$.

Figure 6-20

Self Check 7   Graph the solution set of $\begin{cases} x + y \le 5 \\ x - 3y \le -3. \\ x \ge 0 \end{cases}$

EXAMPLE 8   Graph the solution set of the system $\begin{cases} x \ge 1 \\ y \ge x \\ 4x + 5y < 20 \end{cases}$.

Solution   The graph of the solution set of $x \ge 1$ includes those points on the graph of $x = 1$ and to the right. See Figure 6-21(a).

The graph of the solution set of $y \ge x$ includes those points on the graph of $y = x$ and above it. See Figure 6-21(b).

The graph of the solution set of $4x + 5y < 20$ includes those points below the graph of $4x + 5y = 20$. See Figure 6-21(c).

If these graphs are merged onto a single set of coordinate axes, as in Figure 6-21(d), the graph of the original system of inequalities includes those points within the shaded triangle together with the points on the sides of the triangle drawn as solid lines. The coordinates of the corner points are $\left(1, \frac{16}{5}\right)$, $(1, 1)$, and $\left(\frac{20}{9}, \frac{20}{9}\right)$.

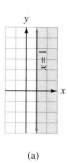
(a)

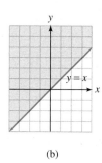
(b)

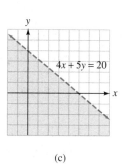
(c)

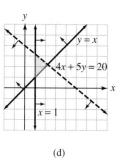
(d)

Figure 6-21

Self Check Answers

1.

2.

3.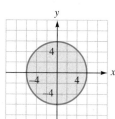

4.

5.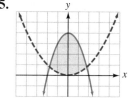

6.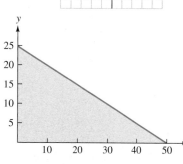

7.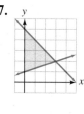

6.7 Exercises

Vocabulary and Concepts *Fill in the blanks.*

1. The graph of $Ax + By = C$ is a line. The graph of $Ax + By \leq C$ is a _____. The line is its _____.

2. The boundary of the graph $Ax + By < C$ is _____ (included, excluded) from the graph.

3. The origin _____ (is, is not) included in the graph of $3x - 4y > 4$.

4. The origin _____ (is, is not) included in the graph of $4x + 3y \leq 5$.

Practice *Graph each inequality.*

5. $2x + 3y < 12$

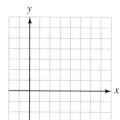

6. $4x - 3y > 6$

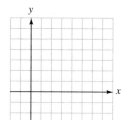

7. $x < 3$

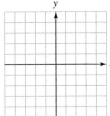

8. $y > -1$

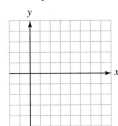

17. $y < x^2$

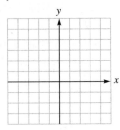

18. $y \geq |x|$

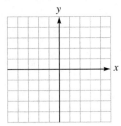

9. $4x - y > 4$

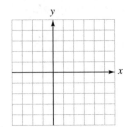

10. $x - 2y < 5$

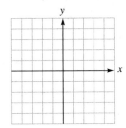

19. $x^2 + y^2 \leq 4$

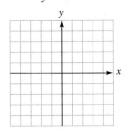

20. $x^2 + y^2 > 4$

11. $y > 2x$

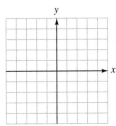

12. $y < 3x$

Graph the solution set of each system.

21. $\begin{cases} y < 3 \\ x \geq 2 \end{cases}$

22. $\begin{cases} y \geq -2 \\ x < 0 \end{cases}$

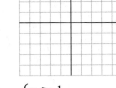

13. $y \leq \dfrac{1}{2}x + 1$

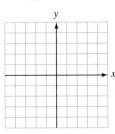

14. $y \geq \dfrac{1}{3}x - 1$

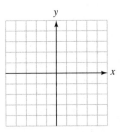

23. $\begin{cases} y \geq 1 \\ x < 2 \end{cases}$

24. $\begin{cases} y \leq -1 \\ x > -1 \end{cases}$

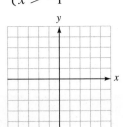

15. $2y \geq 3x - 2$

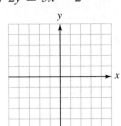

16. $3y \leq 2x + 3$

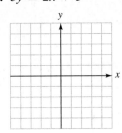

25. $\begin{cases} y \leq x - 2 \\ y \geq 2x + 1 \end{cases}$

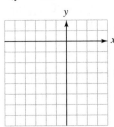

26. $\begin{cases} y < 3x + 2 \\ y < -2x + 3 \end{cases}$

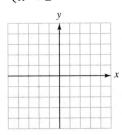

27. $\begin{cases} x + y < 2 \\ x + y \le 1 \end{cases}$

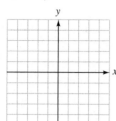

28. $\begin{cases} 3x + 2y \ge 6 \\ x + 3y \le 2 \end{cases}$

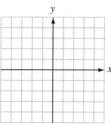

37. $\begin{cases} 2x - y \le 0 \\ x + 2y \le 10 \\ y \ge 0 \end{cases}$

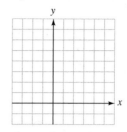

38. $\begin{cases} 3x - 2y \ge 5 \\ 2x + y \ge 8 \\ x \le 5 \end{cases}$

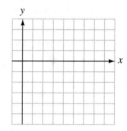

29. $\begin{cases} x + 2y < 3 \\ 2x - 4y < 8 \end{cases}$

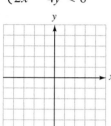

30. $\begin{cases} 3x + y \le 1 \\ -x + 2y \ge 9 \end{cases}$

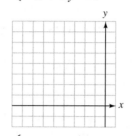

39. $\begin{cases} x - 2y \ge 0 \\ x - y \le 2 \\ x \ge 0 \end{cases}$

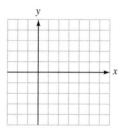

40. $\begin{cases} 2x + 3y \le 6 \\ x - y \ge 4 \\ y \ge -4 \end{cases}$

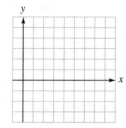

31. $\begin{cases} 2x - 3y \ge 6 \\ 3x + 2y < 6 \end{cases}$

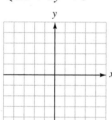

32. $\begin{cases} 4x + 2y \le 6 \\ 2x - 4y \ge 10 \end{cases}$

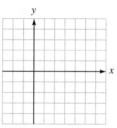

41. $\begin{cases} x + y \le 4 \\ x - y \le 4 \\ x \ge 0 \\ y \ge 0 \end{cases}$

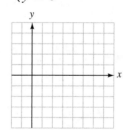

42. $\begin{cases} 2x + 3y \ge 12 \\ 2x - 3y \le 6 \\ x \ge 0 \\ y \le 4 \end{cases}$

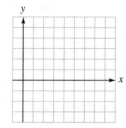

33. $\begin{cases} y \ge x^2 - 4 \\ y \le \dfrac{1}{2}x \end{cases}$

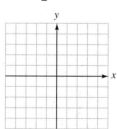

34. $\begin{cases} y \le -x^2 + 4 \\ y > -x - 1 \end{cases}$

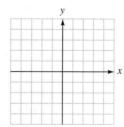

43. $\begin{cases} 3x - 2y \le 6 \\ x + 2y \le 10 \\ x \ge 0 \\ y \ge 0 \end{cases}$

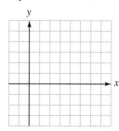

44. $\begin{cases} 3x + 2y \ge 12 \\ 5x - y \le 15 \\ x \ge 0 \\ y \le 4 \end{cases}$

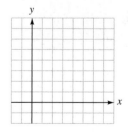

35. $\begin{cases} y \ge x^2 \\ y < 4 - x^2 \end{cases}$

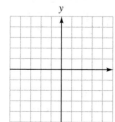

36. $\begin{cases} x^2 + y \le 1 \\ y - x^2 \ge -1 \end{cases}$

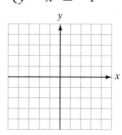

Applications

45. Building furniture   A furniture maker has 60 hours of labor to make sofas (s) and loveseats (l). It takes 6 hours to make a sofa and 4 hours to make a loveseat. Write a system of inequalities that provides the restrictions on the variables. (*Hint:* Remember that a negative number of pieces cannot be made.)

46. Installing video   Each week, Prime Time Video and Audio has 90 hours of labor to install satellite dishes (d) and home theater systems (t). On average, it takes 5 hours to install a satellite dish and 6 hours to install a home theater system. Write a system of inequalities that provides the restrictions on the variables.   (*Hint:* Remember that a negative number of units cannot be installed.)

Discovery and Writing

47. When graphing a linear inequality, explain how to determine the boundary.

48. When graphing a linear inequality, explain how to decide whether the boundary is included.

49. When graphing a linear inequality, explain how to decide which side of the boundary to shade.

50. Does the method you describe in Exercise 49 work for the inequality $3x - 2y \le 0$? Explain.

Review

51. State the remainder theorem.

52. State the factor theorem.

Fill in the blanks.

53. According to Descartes' rule of signs, $53x^4 + 13x^2 - 12 = 0$ has ___ positive solution(s) and ___ negative solution(s).

54. A root of the polynomial equation $P(x) = 0$ is a ____ of the polynomial $P(x)$.

55. If $x^{37} + 3x^{19} - 4$ is divided by $x - 1$, the remainder is __.

56. Is $x - 2$ a factor of $x^4 - 7x - 2$?

6.8   Linear Programming

Objectives

1. Solve Linear Programming Problems
2. Solve Applications of Linear Programming

Linear programming is a mathematical technique used to find the optimal allocation of resources in the military, business, telecommunications, and other fields. It got its start during World War II when it became necessary to move huge quantities of people, materials, and supplies as efficiently and economically as possible.

A simple linear programming problem might involve a television program director who wants to schedule comedy skits and musical acts for a prime-time variety show. Of course, the director wants to do this in a way that earns the maximum possible income for her network. Such an example is provided in Example 5 in this section.

1.   Solve Linear Programming Problems

To solve linear programming problems, we must maximize (or minimize) a function (called the **objective function**) subject to given restrictions on its variables.

Charles Babbage
(1792–1871)
In 1823, Babbage built a steam-powered digital calculator, which he called a difference engine. Thought to be a crackpot by his London neighbors, Babbage was a visionary. His machine embodied principles still used in modern computers.

These restrictions (called **constraints**) usually are given as a system of linear inequalities. For example, suppose that the annual profit (in millions of dollars) earned by a business is given by the equation $P = y + 2x$ and that x and y are subject to the following constraints:

$$\begin{cases} 3x + y \le 120 \\ x + y \le 60 \\ x \ge 0 \\ y \ge 0 \end{cases}$$

To find the maximum profit P that can be earned by the business, we solve the system of inequalities as shown in Figure 6-22(a) and find the coordinates of each corner point of the region R. This region is often called a **feasibility region.** We can then write the profit equation

$$P = y + 2x \qquad \text{in the form} \qquad y = -2x + P$$

The equation $y = -2x + P$ is the equation of a set of parallel lines, each with a slope of -2 and a y-intercept of P. The graph of $y = -2x + P$ for three values of P is shown as red lines in Figure 6-22(b). To find the red line that passes through region R and provides the maximum value of P, we locate the red line with the greatest y-intercept. Since line l has the greatest y-intercept and intersects region R at the corner point $(30, 30)$, the maximum value of P (subject to the given constraints) is

$$\begin{aligned} P &= y + 2x \\ &= 30 + 2(30) \\ &= 90 \end{aligned}$$

Thus, the maximum profit P that can be earned is \$90 million. This profit occurs when $x = 30$ and $y = 30$.

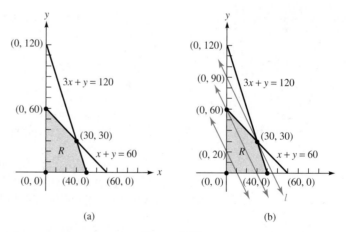

(a)   (b)

Figure 6-22

The preceding discussion illustrates the following important fact.

Maximum or Minimum of an Objective Function   If a linear function, subject to the constraints of a system of linear inequalities in two variables, attains a maximum or a minimum value, that value will occur at a corner point or along an entire edge of the region R that represents the solution of the system.

EXAMPLE 1   If $P = 2x + 3y$, find the maximum value of P subject to the following constraints:

$$\begin{cases} x + y \le 4 \\ 2x + y \le 6 \\ x \ge 0 \\ y \ge 0 \end{cases}$$

Solution   We solve the system of inequalities to find the feasibility region R shown in Figure 6-23. The coordinates of its corner points are $(0, 0)$, $(3, 0)$, $(0, 4)$, and $(2, 2)$.

Since the maximum value of P will occur at a corner of R, we substitute the coordinates of each corner point into the objective function $P = 2x + 3y$ and find the one that gives the maximum value of P.

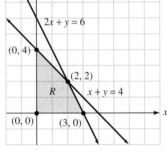

Point	$P = 2x + 3y$
$(0, 0)$	$P = 2(0) + 3(0) = 0$
$(3, 0)$	$P = 2(3) + 3(0) = 6$
$(2, 2)$	$P = 2(2) + 3(2) = 10$
$(0, 4)$	$P = 2(0) + 3(4) = 12$

Figure 6-23   The maximum value $P = 12$ occurs when $x = 0$ and $y = 4$.

Self Check 1   Find the maximum value of $P = 4x + 3y$ subject to the constraints of Example 1.

EXAMPLE 2   If $P = 3x + 2y$, find the minimum value of P subject to the following constraints:

$$\begin{cases} x + y \ge 1 \\ x - y \le 1 \\ x - y \ge 0 \\ x \le 2 \end{cases}$$

Solution   We refer to the feasibility region shown in Figure 6-24 with corner points at $\left(\frac{1}{2}, \frac{1}{2}\right)$, $(2, 2)$, $(2, 1)$, and $(1, 0)$.

Since the minimum value of P occurs at a corner point of region R, we substitute the coordinates of each corner point into the objective function $P = 3x + 2y$ and find the one that gives the minimum value of P.

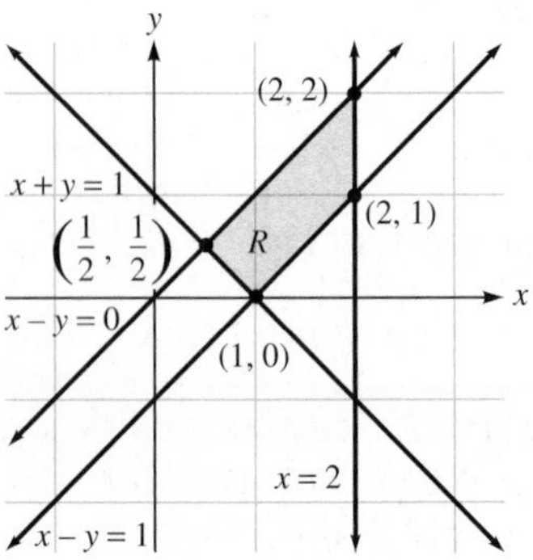

Point	$P = 3x + 2y$
$\left(\frac{1}{2}, \frac{1}{2}\right)$	$P = 3\left(\frac{1}{2}\right) + 2\left(\frac{1}{2}\right) = \frac{5}{2}$
$(2, 2)$	$P = 3(2) + 2(2) = 10$
$(2, 1)$	$P = 3(2) + 2(1) = 8$
$(1, 0)$	$P = 3(1) + 2(0) = 3$

Figure 6-24   The minimum value $P = \frac{5}{2}$ occurs when $x = \frac{1}{2}$ and $y = \frac{1}{2}$.

Self Check 2   Find the minimum value of $P = 2x + y$ subject to the constraints of Example 2.

2. Solve Applications of Linear Programming

Linear programming problems can be very complex and involve hundreds of variables. In this section, we will consider only a few simple problems. Since they involve only two variables, we can solve them using graphical methods.

    To solve a linear programming problem, we will follow these steps.

Solving Linear Programming Problems

1. Find the objective function and constraints.
2. Find the feasibility region by graphing the system of inequalities and identifying the coordinates of its corner points.
3. Find the maximum (or minimum) solution by substituting the coordinates of the corner points into the objective function.

EXAMPLE 3   An accountant prepares tax returns for individuals and for small businesses. On average, each individual return requires 3 hours of her time and 1 hour of computer time. Each business return requires 4 hours of her time and 2 hours of computer time. Because of other business considerations, her time is limited to 240 hours, and the computer time is limited to 100 hours. If she earns a profit of $80 on each individual return and a profit of $150 on each business return, how many returns of each type should she prepare to maximize her profit?

Solution   First, we organize the given information into a table.

	Individual tax return	Business tax return	Time available
Accountant's time	3	4	240 hours
Computer time	1	2	100 hours
Profit	$80	$150	

Then we solve the problem using the following steps.

Find the objective function   Suppose that x represents the number of individual returns to be completed and y represents the number of business returns to be completed. Since each of the x individual returns will earn an $80 profit, and each of the y business returns will earn a $150 profit, the total profit is given by the equation

$$P = 80x + 150y$$

Find the feasibility region   Since the number of individual returns and business returns cannot be negative, we know that $x \geq 0$ and $y \geq 0$.

    Since each of the x individual returns will take 3 hours of her time, and each of the y business returns will take 4 hours of her time, the total number of hours she will work will be $(3x + 4y)$ hours. This amount must be less than or equal to her available time, which is 240 hours. Thus, the inequality $3x + 4y \leq 240$ is a constraint on the accountant's time.

Since each of the *x* individual returns will take 1 hour of computer time, and each of the *y* business returns will take 2 hours of computer time, the total number of hours of computer time will be $(x + 2y)$ hours. This amount must be less than or equal to the available computer time, which is 100 hours. Thus, the inequality $x + 2y \le 100$ is a constraint on the computer time.

We have the following constraints on the values of *x* and *y*.

$$\begin{cases} x \ge 0 & \text{The number of individual returns is nonnegative.} \\ y \ge 0 & \text{The number of business returns is nonnegative.} \\ 3x + 4y \le 240 & \text{The accountant's time must be less than or equal to 240 hours.} \\ x + 2y \le 100 & \text{The computer time must be less than or equal to 100 hours.} \end{cases}$$

To find the feasibility region, we graph each of the constraints to find region *R*, as in Figure 6-25. The four corner points of this region have coordinates of (0, 0), (80, 0), (40, 30), and (0, 50).

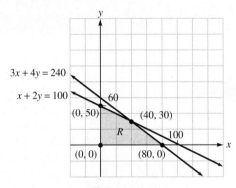

Figure 6-25

Find the maximum profit   To find the maximum profit, we substitute the coordinates of each corner point into the objective function $P = 80x + 150y$.

Point	$P = 80x + 150y$
(0, 0)	$P = 80(0) + 150(0) = 0$
(80, 0)	$P = 80(80) + 150(0) = 6{,}400$
(40, 30)	$P = 80(40) + 150(30) = 7{,}700$
(0, 50)	$P = 80(0) + 150(50) = 7{,}500$

From the table, we can see that the accountant will earn a maximum profit of $7,700 if she prepares 40 individual returns and 30 business returns.

EXAMPLE 4   Vigortab and Robust are two diet supplements. Each Vigortab tablet costs 50¢ and contains 3 units of calcium, 20 units of vitamin C, and 40 units of iron. Each Robust tablet costs 60¢ and contains 4 units of calcium, 40 units of vitamin C, and 30 units of iron. At least 24 units of calcium, 200 units of vitamin C, and 120 units of iron are required for the daily needs of one patient. How many tablets of each supplement should be taken daily for a minimum cost? Find the daily minimum cost.

Solution   First, we organize the given information into a table.

	Vigortab	Robust	Amount required
Calcium	3	4	24
Vitamin C	20	40	200
Iron	40	30	120
Cost	50¢	60¢	

Find the objective function   We can let x represent the number of Vigortab tablets to be taken daily and y the corresponding number of Robust tablets. Because each of the x Vigortab tablets will cost 50¢, and each of the y Robust tablets will cost 60¢, the total cost will be given by the equation

$$C = 0.50x + 0.60y \qquad \text{50¢ = \$0.50 and 60¢ = \$0.60.}$$

Find the feasibility region   Since there are requirements for calcium, vitamin C, and iron, there is a constraint for each. Note that neither x nor y can be negative.

$$\begin{cases} 3x + 4y \geq 24 & \text{The amount of calcium must be greater than or equal to 24 units.} \\ 20x + 40y \geq 200 & \text{The amount of vitamin C must be greater than or equal to 200 units.} \\ 40x + 30y \geq 120 & \text{The amount of iron must be greater than or equal to 120 units.} \\ x \geq 0, y \geq 0 & \text{The number of tablets taken must be greater than or equal to 0.} \end{cases}$$

We graph the inequalities to find the feasibility region and the coordinates of its corner points, as in Figure 6-26.

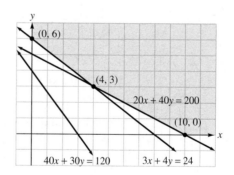

Figure 6-26

Find the minimum cost   In this case, the feasibility region is not bounded on all sides. The coordinates of the corner points are (0, 6), (4, 3), and (10, 0). To find the minimum cost, we substitute each pair of coordinates into the objective function.

Point	$C = 0.50x + 0.60y$
(0, 6)	$C = 0.50(0) + 0.60(6) = 3.60$
(4, 3)	$C = 0.50(4) + 0.60(3) = 3.80$
(10, 0)	$C = 0.50(10) + 0.60(0) = 5.00$

A minimum cost will occur if no Vigortab and 6 Robust tablets are taken daily. The minimum daily cost is \$3.60.

EXAMPLE 5   A television program director must schedule comedy skits and musical numbers for prime-time variety shows. Each comedy skit requires 2 hours of rehearsal time, costs $3,000, and brings in $20,000 from the show's sponsors. Each musical number requires 1 hour of rehearsal time, costs $6,000, and generates $12,000. If 250 hours are available for rehearsal, and $600,000 is budgeted for comedy and music, how many segments of each type should be produced to maximize income? Find the maximum income.

Solution   First, we organize the given information into a table.

	Comedy	Musical	Available
Rehearsal time (hours)	2	1	250
Cost (in $1,000s)	3	6	600
Generated income (in $1,000s)	20	12	

Find the objective function   We can let x represent the number of comedy skits and y the number of musical numbers to be scheduled. Since each of the x comedy skits generates $20 thousand, the income generated by the comedy skits is $20x$ thousand. The musical numbers produce $12y$ thousand. The objective function to be maximized is

$$I = 20x + 12y$$

Find the feasibility region   Since there are limits on rehearsal time and budget, there is a constraint for each. Note that neither x nor y can be negative.

$$\begin{cases} 2x + y \le 250 & \text{The total rehearsal time must be less than or equal to 250 hours.} \\ 3x + 6y \le 600 & \text{The total cost must be less than or equal to \$600 thousand.} \\ x \ge 0, y \ge 0 & \text{The numbers of skits and musical numbers must be greater than or equal to 0.} \end{cases}$$

We graph the inequalities to find the feasibility region as shown in Figure 6-27 and find the coordinates of each corner point.

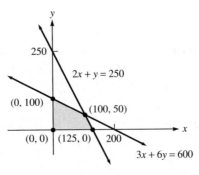

Figure 6-27

Find the maximum income   The coordinates of the corner points of the feasible region are $(0, 0)$, $(0, 100)$, $(100, 50)$, and $(125, 0)$. To find the maximum income, we substitute each pair of coordinates into the objective function.

Corner point	$I = 20x + 12y$
$(0, 0)$	$I = 20(0) + 12(0) = 0$
$(0, 100)$	$I = 20(0) + 12(100) = 1,200$
$(100, 50)$	$I = 20(100) + 12(50) = 2,600$
$(125, 0)$	$I = 20(125) + 12(0) = 2,500$

Maximum income will occur if 100 comedy skits and 50 musical numbers are scheduled. The maximum income will be 2,600 thousand dollars, or $2,600,000.

Self Check Answers   1. 14   2. $\frac{3}{2}$

6.8   Exercises

Vocabulary and Concepts   *Fill in the blanks.*

1. In a linear program, the inequalities are called _____.

2. Ordered pairs that satisfy the constraints of a linear program are called _____ solutions.

3. The function to be maximized (or minimized) in a linear program is called the _____ function.

4. The objective function of a linear program attains a maximum (or minimum), subject to the constraints, at a _____ or along an ____ of the feasibility region.

Practice   *Maximize P subject to the following constraints.*

5. $P = 2x + 3y$
$\begin{cases} x \geq 0 \\ y \geq 0 \\ x + y \leq 4 \end{cases}$

6. $P = 3x + 2y$
$\begin{cases} x \geq 0 \\ y \geq 0 \\ x + y \leq 4 \end{cases}$

7. $P = y + \frac{1}{2}x$
$\begin{cases} x \geq 0 \\ y \geq 0 \\ 2y - x \leq 1 \\ y - 2x \geq -2 \end{cases}$

8. $P = 4y - x$
$\begin{cases} x \geq 2 \\ y \geq 0 \\ x + y \geq 1 \\ 2y - x \leq 1 \end{cases}$

9. $P = 2x + y$
$\begin{cases} y \geq 0 \\ y - x \leq 2 \\ 2x + 3y \leq 6 \\ 3x + y \leq 3 \end{cases}$

10. $P = x - 2y$
$\begin{cases} x + y \leq 5 \\ y \leq 3 \\ x \leq 2 \\ x \geq 0 \\ y \geq 0 \end{cases}$

11. $P = 3x - 2y$
$\begin{cases} x \leq 1 \\ x \geq -1 \\ y - x \leq 1 \\ x - y \leq 1 \end{cases}$

12. $P = x - y$
$\begin{cases} 5x + 4y \leq 20 \\ y \leq 5 \\ x \geq 0 \\ y \geq 0 \end{cases}$

Minimize P subject to the following constraints.

13. $P = 5x + 12y$
$\begin{cases} x \geq 0 \\ y \geq 0 \\ x + y \leq 4 \end{cases}$

14. $P = 3x + 6y$
$\begin{cases} x \geq 0 \\ y \geq 0 \\ x + y \leq 4 \end{cases}$

15. $P = 3y + x$
$\begin{cases} x \geq 0 \\ y \geq 0 \\ 2y - x \leq 1 \\ y - 2x \geq -2 \end{cases}$

16. $P = 5y + x$
$\begin{cases} x \leq 2 \\ y \geq 0 \\ x + y \geq 1 \\ 2y - x \leq 1 \end{cases}$

17. $P = 6x + 2y$
$$\begin{cases} y \geq 0 \\ y - x \leq 2 \\ 2x + 3y \leq 6 \\ 3x + y \leq 3 \end{cases}$$

18. $P = 2y - x$
$$\begin{cases} x \geq 0 \\ y \geq 0 \\ x + y \leq 5 \\ x + 2y \geq 2 \end{cases}$$

19. $P = 2x - 2y$
$$\begin{cases} x \leq 1 \\ x \geq -1 \\ y - x \leq 1 \\ x - y \leq 1 \end{cases}$$

20. $P = y - 2x$
$$\begin{cases} x + 2y \leq 4 \\ 2x + y \leq 4 \\ x + 2y \geq 2 \\ 2x + y \geq 2 \end{cases}$$

Applications *Write the objective function and the inequalities that describe the constraints in each problem. Graph the feasibility region, showing the corner points. Then find the maximum or minimum value of the objective function.*

21. Making furniture Two woodworkers, Tom and Carlos, get $100 for making a table and $80 for making a chair. On average, Tom must work 3 hours and Carlos 2 hours to make a chair. Tom must work 2 hours and Carlos 6 hours to make a table. If neither wants to work more than 42 hours per week, how many tables and how many chairs should they make each week to maximize their income? Find the maximum income.

	Table	Chair	Time available
Income ($)	100	80	
Tom's time (hr)	2	3	42
Carlos's time (hr)	6	2	42

22. Making crafts Two artists, Nina and Rob, make yard ornaments. They get $80 for each wooden snowman they make and $64 for each wooden Santa Claus. On average, Nina must work 4 hours and Rob 2 hours to make a snowman. Nina must work 3 hours and Rob 4 hours to make a Santa Claus. If neither wants to work more than 20 hours per week, how many of each ornament should they make each week to maximize their income? Find the maximum income.

	Snowman	Santa Claus	Time available
Income ($)	80	64	
Nina's time (hr)	4	3	20
Rob's time (hr)	2	4	20

23. Inventories An electronics store manager stocks from 20 to 30 IBM-compatible computers and from 30 to 50 Apple computers. There is room in the store to stock up to 60 computers. The manager receives a commission of $50 on the sale of each IBM-compatible computer and $40 on the sale of each Apple computer. If the manager can sell all of the computers, how many should she stock to maximize her commissions? Find the maximum commission.

Inventory	IBM	Apple
Minimum	20	30
Maximum	30	50
Commission	$50	$40

24. Diet problems A diet requires at least 16 units of vitamin C and at least 34 units of vitamin B complex. Two food supplements are available that provide these nutrients in the amounts and costs shown in the table. How much of each should be used to minimize the cost?

Supplement	Vitamin C	Vitamin B	Cost
A	3 units/g	2 units/g	3¢/g
B	2 units/g	6 units/g	4¢/g

25. Production Manufacturing DVD players and TVs requires the use of the electronics, assembly, and finishing departments of a factory, according to the following schedule:

	Hours for DVD players	Hours for TV	Hours available per week
Electronics	3	4	180
Assembly	2	3	120
Finishing	2	1	60

Each DVD player has a profit of $40, and each TV has a profit of $32. How many DVD players and TVs should be manufactured weekly to maximize profit? Find the maximum profit.

26. **Production problems** A company manufactures one type of computer chip that runs at 2.0 GHz and another that runs at 2.8 GHz. The company can make a maximum of 50 fast chips per day and a maximum of 100 slow chips per day. It takes 6 hours to make a fast chip and 3 hours to make a slow chip, and the company's employees can provide up to 360 hours of labor per day. If the company makes a profit of $20 on each 2.8-GHz chip and $27 on each 2.0-GHz chip, how many of each type should be manufactured to earn the maximum profit?

27. **Financial planning** A stockbroker has $200,000 to invest in stocks and bonds. She wants to invest at least $100,000 in stocks and at least $50,000 in bonds. If stocks have an annual yield of 9% and bonds have an annual yield of 7%, how much should she invest in each to maximize her income? Find the maximum return.

28. **Production** A small country exports soybeans and flowers. Soybeans require 8 workers per acre, flowers require 12 workers per acre, and 100,000 workers are available. Government contracts require that there be at least 3 times as many acres of soybeans as flowers planted. It costs $250 per acre to plant soybeans and $300 per acre to plant flowers, and there is a budget of $3 million. If the profit from soybeans is $1,600 per acre and the profit from flowers is $2,000 per acre, how many acres of each crop should be planted to maximize profit? Find the maximum profit.

29. **Band trips** A high school band trip will require renting buses and trucks to transport no fewer than 100 students and 18 or more large instruments. Each bus can accommodate 40 students plus three large instruments and costs $350 to rent. Each truck can accommodate 10 students plus 6 large instruments and costs $200 to rent. How many of each type of vehicle should be rented for the cost to be minimum? Find the minimum cost.

30. **Making ice cream** An ice cream store sells two new flavors: Fantasy and Excess. Each barrel of Fantasy requires 4 pounds of nuts and 3 pounds of chocolate and has a profit of $500. Each barrel of Excess requires 4 pounds of nuts and 2 pounds of chocolate and has a profit of $400. There are 16 pounds of nuts and 18 pounds of chocolate in stock, and the owner does not want to buy more for this batch. How many barrels of each should be made for a maximum profit? Find the maximum profit.

Discovery and Writing

31. Does the objective function attain a maximum at the corners of a region defined by following nonlinear inequalities? Attempt to maximize $P(x) = x + y$ on the region and write a paragraph on your findings.

$$\begin{cases} x \geq 0 \\ y \geq 0 \\ y \leq 4 - x^2 \end{cases}$$

32. Attempt to minimize the objective function of Exercise 31.

Review *Write each matrix in reduced row echelon form. Problem 33 cannot be done with a calculator.*

33. $\begin{bmatrix} 1 & 2 & 3 \\ 1 & -2 & 3 \\ 0 & 2 & -3 \\ 2 & 0 & 6 \end{bmatrix}$

34. $\begin{bmatrix} 1 & 3 & -2 & 1 \\ 3 & 9 & -3 & 2 \end{bmatrix}$

35. The matrix in Exercise 34 is the system matrix of a system of equations. Find the general solution of the system.

36. Find the inverse of $\begin{bmatrix} 7 & 2 & 5 \\ 3 & 1 & 2 \\ 3 & 1 & 3 \end{bmatrix}$.

6.1   Systems of Linear Equations

Definitions and Concepts	Examples
To solve a system of equations by graphing, graph each equation in the system and find the coordinates of the point where all of the graphs intersect. If the graphs have no point in common, the system has no solutions. If the graphs coincide, the system has infinitely many solutions.	Solve $\begin{cases} x - 2y = -9 \\ 3x + y = 1 \end{cases}$ by graphing. We graph each equation as shown in the illustration and find the coordinates of the point where the graphs intersect. From the illustration, we see that the solution is $(-1, 4)$. 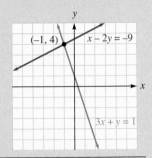
To solve a system by substitution, solve one equation for one variable, and substitute that result into the other equation. Then solve for the other variable.	Solve $\begin{cases} x - 2y = -9 \\ 3x + y = 1 \end{cases}$ by substitution. We can solve the first equation for x and substitute the result for x in the other equation. $x - 2y = -9 \rightarrow x = 2y - 9$ $3(2y - 9) + y = 1$ \quad Substitute $2y - 9$ for x. $6y - 27 + y = 1$ $7y = 28$ $y = 4$ To find x, we can substitute 4 for y in the equation $x = 2y - 9$ to get $x = -1$. The solution is $(-1, 4)$.
To solve a system by addition, multiply one or both equations by suitable constants so that when the results are added, one variable will be eliminated. Then back substitute to find the other variable.	Solve $\begin{cases} x - 2y = -9 \\ 3x + y = 1 \end{cases}$ by addition. We can multiply the second equation by 2 and add the equations to eliminate y. $\begin{cases} x - 2y = -9 \\ 3x + y = 1 \end{cases} \rightarrow \begin{cases} x - 2y = -9 \\ \underline{6x + 2y = \;\;\; 2} \\ 7x \qquad\;\; = -7 \\ \qquad\quad x = -1 \end{cases}$ We can substitute -1 for x in any equation and find that $y = 4$. The solution is $(-1, 4)$.

Exercises

Solve each system by graphing.

1. $\begin{cases} 2x - y = -1 \\ x + y = 7 \end{cases}$ **2.** $\begin{cases} 5x + 2y = 1 \\ 2x - y = -5 \end{cases}$

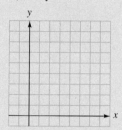 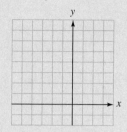

3. $\begin{cases} y = 5x + 7 \\ x = y - 7 \end{cases}$

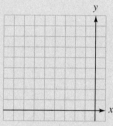

4. $\begin{cases} 3x + 2y = 6 \\ y = -\dfrac{3}{2}x + 3 \end{cases}$ **5.** $\begin{cases} 4x - y = 4 \\ y = 4(x - 2) \end{cases}$

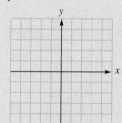 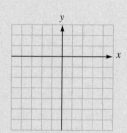

Solve by substitution.

6. $\begin{cases} 2y + x = 0 \\ x = y + 3 \end{cases}$ **7.** $\begin{cases} 2x + y = -3 \\ x - y = 3 \end{cases}$

8. $\begin{cases} \dfrac{x + y}{2} + \dfrac{x - y}{3} = 1 \\ y = 3x - 2 \end{cases}$

9. $\begin{cases} y = 3x - 4 \\ 9x - 3y = 12 \end{cases}$ **10.** $\begin{cases} x = -\dfrac{3}{2}y + 3 \\ 2x + 3y = 4 \end{cases}$

Solve by addition.

11. $\begin{cases} x + 5y = 7 \\ 3x + y = -7 \end{cases}$ **12.** $\begin{cases} 2x + 3y = 11 \\ 3x - 7y = -41 \end{cases}$

13. $\begin{cases} 2(x + y) - x = 0 \\ 3(x + y) + 2y = 1 \end{cases}$ **14.** $\begin{cases} 8x + 12y = 24 \\ 2x + 3y = 4 \end{cases}$

15. $\begin{cases} 3x - y = 4 \\ 9x - 3y = 12 \end{cases}$

Solve by any method.

16. $\begin{cases} 3x + 2y - z = 2 \\ x + y - z = 0 \\ 2x + 3y - z = 1 \end{cases}$ **17.** $\begin{cases} 5x - y + z = 3 \\ 3x + y + 2z = 2 \\ x + y = 2 \end{cases}$

18. $\begin{cases} 2x - y + z = 1 \\ x - y + 2z = 3 \\ x - y + z = 1 \end{cases}$

19. The buyer for a large department store must order 40 coats, some fake fur and some leather. He is unsure of the expected sales. He can buy 25 fake-fur coats and the rest leather for $9,300, or 10 fake-fur coats and the rest leather for $12,600. How much does he pay if he decides to split the order evenly?

20. Adult tickets for the championship game are usually $5, but on Seniors' Day, seniors paid $4. Children's tickets were $2.50. Sales of 1,800 tickets totaled $7,425, and children and seniors accounted for one-half of the tickets sold. How many of each were sold?

6.2 Gaussian Elimination and Matrix Methods

Definitions and Concepts	Examples
There are three matrices associated with a system of equations, a *coefficient matrix,* a *matrix of constants,* and an *augmented* or *systems matrix.*	There are three matrices associated with the following system of equations $\begin{cases} x + 3y + z = 2 \\ 3x - 2y - z = 5. \\ 4x + 2y + z = 9 \end{cases}$

Coefficient matrix   **Constants**   **System matrix**

$$\begin{bmatrix} 1 & 3 & 1 \\ 3 & -2 & -1 \\ 4 & 2 & 1 \end{bmatrix} \qquad \begin{bmatrix} 2 \\ 5 \\ 9 \end{bmatrix} \qquad \left[\begin{array}{ccc:c} 1 & 3 & 1 & 2 \\ 3 & -2 & -1 & 5 \\ 4 & 2 & 1 & 9 \end{array}\right]$$

Elementary row operations:

1. In a type 1 row operation, any two rows of a matrix can be interchanged.

2. In a type 2 row operation, the elements of any row of a matrix can be multiplied by any nonzero constant.

3. In a type 3 row operation, any row of a matrix can be changed by adding a multiple of another row to it.

To solve the previous system, we can use row operations to write the system matrix in row echelon form.

$$\left[\begin{array}{ccc:c} 1 & 3 & 1 & 2 \\ 3 & -2 & -1 & 5 \\ 4 & 2 & 1 & 9 \end{array}\right] \begin{array}{l} \\ (-3)R_1 + R_2 \rightarrow \\ (-4)R_1 + R_3 \rightarrow \end{array} \left[\begin{array}{ccc:c} 1 & 3 & 1 & 2 \\ 0 & -11 & -4 & -1 \\ 0 & -10 & -3 & 1 \end{array}\right]$$

$$(-1)R_3 + R_2 \rightarrow \left[\begin{array}{ccc:c} 1 & 3 & 1 & 2 \\ 0 & -1 & -1 & -2 \\ 0 & -10 & -3 & 1 \end{array}\right]$$

Row echelon form of a matrix:

1. The first nonzero entry in each row is 1.

2. Lead entries appear farther to the right as you move down the rows of the matrix.

3. Any rows containing only 0's are at the bottom of the matrix.

$$(-1)R_2 \rightarrow \left[\begin{array}{ccc:c} 1 & 3 & 1 & 2 \\ 0 & 1 & 1 & 2 \\ 0 & -10 & -3 & 1 \end{array}\right]$$

$$(10)R_2 + R_3 \rightarrow \left[\begin{array}{ccc:c} 1 & 3 & 1 & 2 \\ 0 & 1 & 1 & 2 \\ 0 & 0 & 7 & 21 \end{array}\right]$$

$$\left(\tfrac{1}{7}\right)R_3 \rightarrow \left[\begin{array}{ccc:c} 1 & 3 & 1 & 2 \\ 0 & 1 & 1 & 2 \\ 0 & 0 & 1 & 3 \end{array}\right]$$

From the last line, we see that $z = 3$. From the middle line, we see that $y + z = 2$. Since $z = 3$, we know that $y = -1$. From the top equation, we know that $x + 3y + z = 2$. Substituting the values for y and z, we can determine that $x = 2$. The solution is $(2, -1, 3)$.

Exercises

Solve by matrix methods, if possible.

21. $\begin{cases} 2x + 5y = 7 \\ 3x - y = 2 \end{cases}$

22. $\begin{cases} 3x - y = -4 \\ -6x + 2y = 8 \end{cases}$

23. $\begin{cases} x + 3y - z = 8 \\ 2x + y - 2z = 11 \\ x - y + 5z = -8 \end{cases}$

24. $\begin{cases} x + 3y + z = 3 \\ 2x - y + z = -11 \\ 3x + 2y + 3z = 2 \end{cases}$

25. $\begin{cases} x + y + z = 4 \\ 3x - 2y - 2z = -3 \\ 4x - y - z = 0 \end{cases}$

6.3 Matrix Algebra

Definitions and Concepts	Examples
Matrices are equal if and only if they are the same size and have the same corresponding entries.	The following matrices are equal because they are the same size and their corresponding entries are equal. $$\begin{bmatrix} 2 & \sqrt{9} \\ \frac{8}{4} & 7 \end{bmatrix} = \begin{bmatrix} \sqrt{4} & 3 \\ 2 & \sqrt{49} \end{bmatrix}$$
Two $m \times n$ matrices are added by adding the corresponding elements of those matrices.	$\begin{bmatrix} 2 & -3 \\ -7 & 5 \end{bmatrix} + \begin{bmatrix} -3 & 4 \\ -2 & 2 \end{bmatrix} = \begin{bmatrix} -1 & 1 \\ -9 & 7 \end{bmatrix}$ Add the corresponding entries
Two $m \times n$ matrices are subtracted by the rule $A - B = A + (-B)$.	$\begin{bmatrix} 2 & -3 \\ -7 & 5 \end{bmatrix} - \begin{bmatrix} -3 & 4 \\ -2 & 2 \end{bmatrix} = \begin{bmatrix} 5 & -7 \\ -5 & 3 \end{bmatrix}$ Subtract the corresponding entries
The product AB of the $m \times n$ matrix A and the $n \times p$ matrix B is the $m \times p$ matrix C. The ith-row, jth-column entry of C is found by keeping a running total of the products of the elements in the ith row of A with the corresponding elements in the jth column of B.	$\begin{bmatrix} 2 & -3 \\ -7 & 5 \end{bmatrix}\begin{bmatrix} -3 & 4 \\ -2 & 2 \end{bmatrix}$ $= \begin{bmatrix} 2(-3) + (-3)(-2) & 2(4) + (-3)(2) \\ (-7)(-3) + 5(-2) & -7(4) + (5)(2) \end{bmatrix}$ $= \begin{bmatrix} 0 & 2 \\ 11 & -18 \end{bmatrix}$

Exercises

26. Solve for x and y. $\begin{bmatrix} 1 & -4 \\ x & 2 \\ 0 & x+7 \end{bmatrix} = \begin{bmatrix} 1 & x \\ -4 & 2 \\ x+4 & y \end{bmatrix}$

Perform the matrix operations, if possible.

27. $\begin{bmatrix} 3 & 2 & 1 \\ 3 & 2 & 1 \end{bmatrix} + \begin{bmatrix} -2 & 1 & 3 \\ 1 & -2 & 1 \end{bmatrix}$

28. $\begin{bmatrix} 2 & 3 & 5 \\ 1 & -2 & 4 \\ 2 & 1 & -2 \end{bmatrix} - \begin{bmatrix} 0 & -2 & 1 \\ 3 & 4 & -2 \\ 6 & -4 & 1 \end{bmatrix}$

29. $\begin{bmatrix} 1 & -2 \\ -3 & 1 \end{bmatrix}\begin{bmatrix} 2 & 3 \\ -1 & 2 \end{bmatrix}$

30. $\begin{bmatrix} -2 & 3 & 5 \\ 1 & -2 & -3 \end{bmatrix}\begin{bmatrix} 2 & 1 \\ -1 & 2 \\ -2 & 3 \end{bmatrix}$

31. $\begin{bmatrix} 1 & -3 & 2 \end{bmatrix}\begin{bmatrix} 2 \\ 1 \\ 3 \end{bmatrix}$

32. $\begin{bmatrix} 1 \\ 2 \\ 1 \\ 5 \end{bmatrix}\begin{bmatrix} 2 & -1 & 1 & 3 \end{bmatrix}$

33. $\begin{bmatrix} 1 & -5 & 3 \\ 2 & 1 & -1 \end{bmatrix}\begin{bmatrix} 2 \\ -2 \\ 3 \end{bmatrix}\begin{bmatrix} 1 & -1 \\ -1 & 3 \end{bmatrix}\begin{bmatrix} 1 \\ -2 \end{bmatrix}$

34. $\begin{bmatrix} 1 & -3 & 2 \end{bmatrix}\begin{bmatrix} 2 \\ 1 \\ -5 \end{bmatrix} + \begin{bmatrix} 1 & -3 \end{bmatrix}\begin{bmatrix} 2 \\ 5 \end{bmatrix}$

35. $\left(\begin{bmatrix} 1 & -3 \\ 3 & 1 \end{bmatrix} + \begin{bmatrix} -1 & 3 \\ 1 & 1 \end{bmatrix}\right)\begin{bmatrix} 1 \\ -5 \end{bmatrix}$

| | 6.4 | **Matrix Inversion** |

Definitions and Concepts | **Examples**

The inverse of an $n \times n$ matrix A, if it exists, is A^{-1}, where

$$AA^{-1} = A^{-1}A = I$$

Use elementary row operations to transform $[A \mid I]$ into $[I \mid A^{-1}]$, where I is an identity matrix. If A cannot be transformed into I, then A is singular.

If A is invertible, then the solution of $AX = B$ is $X = A^{-1}B$.

To find the inverse of matrix $A = \begin{bmatrix} 1 & 4 \\ 3 & 2 \end{bmatrix}$, we form

the matrix $\begin{bmatrix} 1 & 4 & | & 1 & 0 \\ 3 & 2 & | & 0 & 1 \end{bmatrix}$ and use row operations to

transform matrix A to the left of the dashed line into the identity matrix. These row operations performed on the identity matrix to the right of the dashed line will transform it into the inverse.

$$\begin{bmatrix} 1 & 4 & | & 1 & 0 \\ 3 & 2 & | & 0 & 1 \end{bmatrix} \xrightarrow{(-3)R_1 + R_2} \begin{bmatrix} 1 & 4 & | & 1 & 0 \\ 0 & -10 & | & -3 & 1 \end{bmatrix}$$

$$\xrightarrow{\left(-\frac{1}{10}\right)R_2} \begin{bmatrix} 1 & 4 & | & 1 & 0 \\ 0 & 1 & | & \frac{3}{10} & -\frac{1}{10} \end{bmatrix}$$

$$\xrightarrow{-4R_2 + R_1} \begin{bmatrix} 1 & 0 & | & -\frac{1}{5} & \frac{2}{5} \\ 0 & 1 & | & \frac{3}{10} & -\frac{1}{10} \end{bmatrix}$$

The inverse is $\begin{bmatrix} -\frac{1}{5} & \frac{2}{5} \\ \frac{3}{10} & -\frac{1}{10} \end{bmatrix}$. Check by showing that

the product of matrix A and the inverse is the identity matrix.

Exercises

Find the inverse of each matrix, if possible.

36. $\begin{bmatrix} 2 & 3 \\ 3 & 5 \end{bmatrix}$

37. $\begin{bmatrix} 1 & 0 & 0 \\ 2 & 0 & -2 \\ 1 & 2 & 2 \end{bmatrix}$

38. $\begin{bmatrix} 1 & 0 & 8 \\ 3 & 7 & 6 \\ 1 & 2 & 3 \end{bmatrix}$

39. $\begin{bmatrix} -6 & 4 \\ -3 & 2 \end{bmatrix}$

40. $\begin{bmatrix} 4 & 4 & 1 \\ 1 & 1 & 1 \\ -1 & -1 & 0 \end{bmatrix}$

Use the inverse of the coefficient matrix to solve each system of equations.

41. $\begin{cases} 4x - y + 2z = 0 \\ x + y + 2z = 1 \\ x + z = 0 \end{cases}$

42. $\begin{cases} w + 3x + y + 3z = 1 \\ w + 4x + y + 3z = 2 \\ x + y = 1 \\ w + 2x - y + 2z = 1 \end{cases}$

6.5   Determinants

Definitions and Concepts	**Examples**

The determinant:

$$\det(A) = |A| = \begin{vmatrix} a & b \\ c & d \end{vmatrix} = ad - bc$$

The determinant of an $n \times n$ matrix A is the sum of the products of the entries of any row (or column) and the cofactors of those entries.

$$\begin{vmatrix} 2 & -3 \\ -4 & 5 \end{vmatrix} = 2(5) - (-3)(-4) = 10 - 12 = -2$$

Evaluate the following determinant by expanding along the first row.

$$\begin{vmatrix} 2 & 3 & -4 \\ 3 & 2 & 1 \\ -2 & 5 & -1 \end{vmatrix}$$

$$= 2\begin{vmatrix} 2 & 1 \\ 5 & -1 \end{vmatrix} - 3\begin{vmatrix} 3 & 1 \\ -2 & -1 \end{vmatrix} + (-4)\begin{vmatrix} 3 & 2 \\ -2 & 5 \end{vmatrix}$$

$$= 2(-2 - 5) - 3[-3 - (-2)] - 4[15 - (-4)]$$

$$= 2(-7) - 3(-1) - 4(19)$$

$$= -14 + 3 - 76$$

$$= -87$$

Properties of determinants:

If two rows (or columns) of a matrix are interchanged, the sign of its determinant is reversed.

$$\begin{vmatrix} 2 & 3 \\ 4 & 5 \end{vmatrix} = -\begin{vmatrix} 4 & 5 \\ 2 & 3 \end{vmatrix} \qquad \begin{vmatrix} 2 & 3 \\ 4 & 5 \end{vmatrix} = -\begin{vmatrix} 3 & 2 \\ 5 & 4 \end{vmatrix}$$

If a row (or column) of a matrix is multiplied by a constant k, the value of its determinant is multiplied by k.

$$\begin{vmatrix} 3 & 4 \\ -2 & 1 \end{vmatrix} = 11, \text{ but}$$

$$5\begin{vmatrix} 3 & 4 \\ -2 & 1 \end{vmatrix} = \begin{vmatrix} 5(3) & 5(4) \\ -2 & 1 \end{vmatrix} = \begin{vmatrix} 15 & 20 \\ -2 & 1 \end{vmatrix} = 55$$

and $55 = 5(11)$. We have seen that

$$\begin{vmatrix} 3 & 4 \\ -2 & 1 \end{vmatrix} = 11$$

If a row (or column) of a matrix is altered by adding to it a multiple of another row (or column), the value of its determinant is unchanged.

If we multiply row 2 by 5 and add it to row 1, the value of the determinant is still 11.

$$\begin{vmatrix} 3 & 4 \\ -2 & 1 \end{vmatrix}$$

$$= \begin{vmatrix} 3 + 5(-2) & 4 + 5(1) \\ -2 & 1 \end{vmatrix} = \begin{vmatrix} -7 & 9 \\ -2 & 1 \end{vmatrix} = 11$$

Cramer's rule:

Form quotients of two determinants. The denominator is the determinant of the coefficient matrix A. The numerator is the determinant of a modified coefficient matrix; when solving for the ith variable, replace the ith column of A with a column of constants B.

Use determinants to solve the following system for x.

$$\begin{cases} 2x + 3y - 4z = 1 \\ 3x + 2y + z = 6 \\ -2x + 5y - z = 2 \end{cases}$$

$$x = \frac{\begin{vmatrix} 1 & 3 & -4 \\ 6 & 2 & 1 \\ 2 & 5 & -1 \end{vmatrix}}{\begin{vmatrix} 2 & 3 & -4 \\ 3 & 2 & 1 \\ -2 & 5 & -1 \end{vmatrix}}$$

$$= \frac{1 \begin{vmatrix} 2 & 1 \\ 5 & -1 \end{vmatrix} - 3 \begin{vmatrix} 6 & 1 \\ 2 & -1 \end{vmatrix} - 4 \begin{vmatrix} 6 & 2 \\ 2 & 5 \end{vmatrix}}{-87} \quad \text{(See above.)}$$

$$= \frac{1(-7) - 3(-8) - 4(26)}{-87}$$

$$= \frac{-87}{-87}$$

$$= 1$$

Exercises

Evaluate each determinant.

43. $\begin{vmatrix} 3 & -2 \\ 1 & -3 \end{vmatrix}$

44. $\begin{vmatrix} 1 & -2 & 3 \\ 2 & -1 & 3 \\ 1 & -1 & 0 \end{vmatrix}$

45. $\begin{vmatrix} 1 & 3 & -1 \\ 1 & 2 & 1 \\ 1 & 0 & 2 \end{vmatrix}$

46. $\begin{vmatrix} 1 & 2 & 3 & 4 \\ -1 & 3 & -3 & 2 \\ 0 & 0 & 0 & -1 \\ 3 & 3 & 4 & 3 \end{vmatrix}$

Use Cramer's rule to solve each system.

47. $\begin{cases} x + 3y = -5 \\ -2x + y = -4 \end{cases}$

48. $\begin{cases} x - y + z = -1 \\ 2x - y + 3z = -4 \\ x - 3y + z = -1 \end{cases}$

49. $\begin{cases} x - 3y + z = 7 \\ x + y - 3z = -9 \\ x + y + z = 3 \end{cases}$

50. $\begin{cases} w + x - y + z = 4 \\ 2w + x + z = 4 \\ x + 2y + z = 0 \\ w + y + z = 2 \end{cases}$

If $\begin{vmatrix} a & b & c \\ d & e & f \\ g & h & i \end{vmatrix} = 7$, *evaluate each determinant.*

51. $\begin{vmatrix} 3a & 3b & 3c \\ d & e & f \\ g & h & i \end{vmatrix}$

52. $\begin{vmatrix} a & b & c \\ d+g & e+h & f+i \\ g & h & i \end{vmatrix}$

6.6   Partial Fractions

Definitions and Concepts	Examples
The fraction $\frac{P(x)}{Q(x)}$ can be written as the sum of simpler fractions with denominators determined by the prime factors of $Q(x)$.	Decompose $\frac{4x-1}{2x^2+3x+1}$ into partial fractions. We factor the denominator to get $\frac{4x-1}{(2x+1)(x+1)}$.

Since each factor in the denominator is linear, there are constants A and B such that

$$\frac{4x - 1}{(2x + 1)(x + 1)} = \frac{A}{2x + 1} + \frac{B}{x + 1}$$

Adding the fractions on the right, we get

$$\frac{4x - 1}{(2x + 1)(x + 1)}$$
$$= \frac{A(x + 1)}{(2x + 1)(x + 1)} + \frac{B(2x + 1)}{(x + 1)(2x + 1)}$$
$$= \frac{Ax + A + 2Bx + B}{(2x + 1)(x + 1)}$$
$$= \frac{(A + 2B)x + A + B}{(2x + 1)(x + 1)}$$

Setting the coefficients of the polynomial numerators equal, we have the following system of equations:

$$\begin{cases} A + 2B = 4 \\ A + B = -1 \end{cases}$$

whose solution is $A = -6$ and $B = 5$. Thus,

$$\frac{4x - 1}{(2x + 1)(x + 1)} = \frac{-6}{2x + 1} + \frac{5}{x + 1}$$

Exercises

Decompose into partial fractions.

53. $\dfrac{7x + 3}{x^2 + x}$

54. $\dfrac{4x^3 + 3x + x^2 + 2}{x^4 + x^2}$

55. $\dfrac{x^2 + 5}{x^3 + x^2 + 5x}$

56. $\dfrac{x^2 + 1}{(x + 1)^3}$

6.7   Graphs of Inequalities

Definitions and Concepts	**Examples**
Systems of inequalities in two variables can be solved by graphing. The solution is represented by a plane region with boundaries determined by graphing the inequalities as if they were equations.	To use graphing to solve the system $\begin{cases} y \le -x + 1 \\ 2x > y + 2 \end{cases}$, we graph each inequality on the same coordinate axes as shown below. 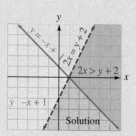 The solution is the overlapping region shown in purple.

Exercises

57. Graph: $y \geq -2x - 1$.

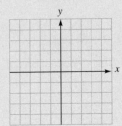

58. Graph: $x^2 + y^2 > 4$.

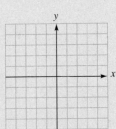

Solve each system by graphing.

59. $\begin{cases} 3x + 2y \leq 6 \\ x - y > 3 \end{cases}$

60. $\begin{cases} y \leq x^2 + 1 \\ y \geq x^2 - 1 \end{cases}$

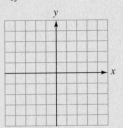

6.8 Linear Programming

Definitions and Concepts	Examples
The maximum and the minimum values of a linear function in two variables, subject to the constraints of a system of linear inequalities, are attained at a corner or along an entire edge of the region determined by the system of inequalities.	See Example 1 in this section.

Exercises

Maximize P subject to the given conditions.

61. $P = 2x + y$

$\begin{cases} x \geq 0 \\ y \geq 0 \\ x + y \leq 3 \end{cases}$

62. $P = 2x - 3y$

$\begin{cases} x \geq 0 \\ y \leq 3 \\ x - y \leq 4 \end{cases}$

63. $P = 3x - y$

$\begin{cases} y \geq 1 \\ y \leq 2 \\ y \leq 3x + 1 \\ x \leq 1 \end{cases}$

64. $P = y - 2x$

$\begin{cases} x + y \geq 1 \\ x \leq 1 \\ y \leq \dfrac{x}{2} + 2 \\ x + y \leq 2 \end{cases}$

65. A company manufactures two fertilizers, x and y. Each 50-pound bag of fertilizer requires three ingredients, which are available in the limited quantities shown in the table. The profit on each bag of fertilizer x is \$6 and on each bag of y, \$5. How many bags of each product should be produced to maximize the profit?

Ingredient	Number of pounds in fertilizer x	Number of pounds in fertilizer y	Total number of pounds available
Nitrogen	6	10	20,000
Phosphorus	8	6	16,400
Potash	6	4	12,000

CHAPTER TEST

Solve each system of equations by the graphing method.

1. $\begin{cases} x - 3y = -5 \\ 2x - y = 0 \end{cases}$
2. $\begin{cases} x = 2y + 5 \\ y = 2x - 4 \end{cases}$

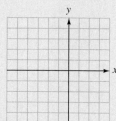

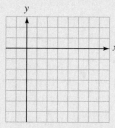

Solve each system of equations by the substitution or addition method.

3. $\begin{cases} 3x + y = 0 \\ 2x - 5y = 17 \end{cases}$

4. $\begin{cases} \dfrac{x + y}{2} + x = 7 \\ \dfrac{x - y}{2} - y = -6 \end{cases}$

5. **Mixing solutions** A chemist has two solutions, one has a 20% concentration and the other a 45% concentration. How many liters of each must she mix to obtain 10 liters of 30% concentration?

6. **Wholesale distribution** Ace Electronics, Hi-Fi Stereo, and CD World buy a total of 175 VCR/DVD players from the same distributor each month. Because CD World buys 25 more units than the other two stores combined, CDW's cost is only $160 per unit. The players cost Hi-Fi $165 each and Ace $170 each. How many players does each retailer buy each month if the distributor receives $28,500 each month from the sale of the players to the three stores?

Write each system of equations as a matrix and solve it by Gaussian elimination.

7. $\begin{cases} 3x - 2y = 4 \\ 2x + 3y = 7 \end{cases}$

8. $\begin{cases} x + 3y - z = 6 \\ 2x - y - 2z = -2 \\ x + 2y + z = 6 \end{cases}$

Write each system of equations as a matrix and solve it by Gauss–Jordan elimination. If the system has infinitely many solutions, show how each can be determined.

9. $\begin{cases} x + 2y + 3z = -5 \\ 3x + y - 2z = 7 \\ y - z = 2 \end{cases}$
10. $\begin{cases} x + 2y + z = 0 \\ 3x - 2y - 2z = 7 \\ 4x - z = 7 \end{cases}$

Perform the operations.

11. $3\begin{bmatrix} 2 & -3 & 5 \\ 0 & 3 & -1 \end{bmatrix} - 5\begin{bmatrix} -2 & 1 & -1 \\ 0 & 3 & 2 \end{bmatrix}$

12. $[1 \quad 2 \quad 3]\begin{bmatrix} 2 & -2 \\ -2 & 2 \\ 1 & 0 \end{bmatrix}\begin{bmatrix} 3 \\ -2 \end{bmatrix}$

Find the inverse of each matrix, if possible.

13. $\begin{bmatrix} 5 & 19 \\ 2 & 7 \end{bmatrix}$
14. $\begin{bmatrix} -1 & 3 & -2 \\ 4 & 1 & 4 \\ 0 & 3 & -1 \end{bmatrix}$

Use the inverses found in Questions 13 and 14 to solve each system.

15. $\begin{cases} 5x + 19y = 3 \\ 2x + 7y = 2 \end{cases}$
16. $\begin{cases} -x + 3y - 2z = 1 \\ 4x + y + 4z = 3 \\ 3y - z = -1 \end{cases}$

Evaluate each determinant.

17. $\begin{vmatrix} 3 & -5 \\ -3 & 1 \end{vmatrix}$
18. $\begin{vmatrix} 3 & 5 & -1 \\ -2 & 3 & -2 \\ 1 & 5 & -3 \end{vmatrix}$

Use Cramer's rule to solve each system for y.

19. $\begin{cases} 3x - 5y = 3 \\ -3x + y = 2 \end{cases}$

20. $\begin{cases} 3x + 5y - z = 2 \\ -2x + 3y - 2z = 1 \\ x + 5y - 3z = 0 \end{cases}$

25. Maximize $P = 3x + 2y$ subject to

$\begin{cases} y \geq 0 \\ x \geq 0 \\ 2x + y \leq 4 \\ y \leq 2 \end{cases}$

Decompose each fraction into partial fractions.

21. $\dfrac{5x}{2x^2 - x - 3}$

22. $\dfrac{3x^2 + x + 2}{x^3 + 2x}$

26. Minimize $P = y - x$ subject to

$\begin{cases} x \geq 0 \\ y \geq 0 \\ x + y \leq 8 \\ 2x + y \geq 2 \end{cases}$

Graph the solution set of each system.

23. $\begin{cases} x - 3y \geq 3 \\ x + 3y \leq 3 \end{cases}$

24. $\begin{cases} 3x + 4y \leq 12 \\ 3x + 4y \geq 6 \\ x \geq 0 \\ y \geq 0 \end{cases}$

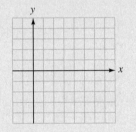

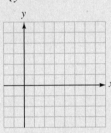

7

Conic Sections and Quadratic Systems

Careers and Mathematics

Astronomer   Astronomers use the principles of physics and mathematics to learn about the fundamental nature of the universe, including the Sun, moon, planets, stars, and galaxies. They also apply their knowledge to solve problems in navigation, space flight, and satellite communications, and to develop the instrumentation and techniques used to observe and collect astronomical data.

Almost all astronomers do research. Some are theoreticians, working on the laws governing the structure and evolution of astronomical objects. Others analyze large quantities of data gathered by observatories and satellites, and write scientific papers or reports on their findings. Some astronomers actually operate large space- or ground-based telescopes, usually as part of a team. However, astronomers may spend only a few weeks each year making observations with optical telescopes, radio telescopes, and other instruments. A small number of astronomers work in museums housing planetariums.

Astronomers held about 1,700 jobs in 2006.

Education   Because most jobs are in basic research and development, a doctoral degree is the usual educational requirement for astronomers. Master's degree holders qualify for some jobs in applied research and development, whereas bachelor's degree holders often qualify as research assistants.

Job Outlook   Employment of astronomers is expected to increase by 6 percent through 2016. Median annual earnings of astronomers were $95,740 in 2006.

For a sample application, see Example 7 in Section 7.2. For more information, see www.bls.gov/oco/ocoa052.htm.

In this chapter, we will study second-degree equations in x and y. They have many applications in such fields as navigation, astronomy, and satellite communications.

© AP Photo/Gail Burton

7.1 The Circle and the Parabola

Objectives

1. Write the Equation of a Circle
2. Graph Circles
3. Write the Equation of a Parabola in Standard Form
4. Graph Equations of Parabolas
5. Solve Applied Problems Involving Parabolas

The suspension bridge is one of the oldest types of bridges. An example is the Golden Gate Bridge that spans the opening of San Francisco Bay into the Pacific Ocean. Completed in 1937, it is currently the second longest suspension bridge in the United States. As the photograph shows, the main cables that hold up the bridge hang in the shape of a parabola. A parabola is an example of a *conic section*.

Second-degree equations in x and y have the general form

$$Ax^2 + Bxy + Cy^2 + Dx + Ey + F = 0$$

where at least one of the coefficients A, B, and C is not zero. The graphs of these equations fall into one of several categories: a point, a pair of lines, a circle, a parabola, an ellipse, an hyperbola, or no graph at all. These graphs are called **conic sections,** because each one is the intersection of a plane and a right-circular cone, as shown in Figure 7-1.

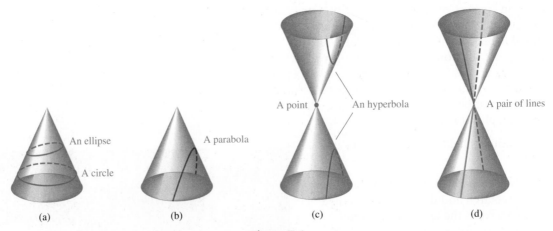

| (a) | (b) | (c) | (d) |

Figure 7-1

Although these shapes have been known since the time of the ancient Greeks, it wasn't until the 17th century that René Descartes (1596–1650) and Blaise Pascal (1623–1662) developed the mathematics needed to study them in detail.

In this section, we will consider two conic sections, the circle and the parabola.

1. Write the Equation of a Circle

In Section 2.4, we developed the following standard equations for circles:

The Standard Equation of a Circle with Center at (h, k)	The graph of any equation that can be written in the form $$(x - h)^2 + (y - k)^2 = r^2$$ is a circle with radius r and center at point (h, k).

If $h = 0$ and $k = 0$, we have this result.

The Standard Equation of a Circle with Center at $(0, 0)$	The graph of any equation that can be written in the form $$x^2 + y^2 = r^2$$ is a circle with radius r and center at the origin.

The graphs of two circles, one with center at (h, k) and one with center at $(0, 0)$, are shown in Figure 7-2.

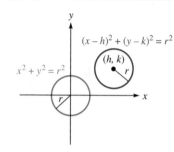

Figure 7-2

EXAMPLE 1   Find the equation of the circle shown in Figure 7-3.

Solution   We will use distance formula to find the radius of the circle and then substitute the coordinates of the center and radius into the standard equation of the circle.

To find the radius of the circle, we substitute the coordinates of the points $(1, 3)$ and $(-2, -1)$ into the distance formula and simplify:

$$r = \sqrt{(x_2 - x_1)^2 + (y_2 - y_1)^2}$$
$$r = \sqrt{(-2 - 1)^2 + (-1 - 3)^2} \quad \text{Substitute 1 for } x_1, \text{3 for } y_1, -2 \text{ for } x_2, \text{ and } -1 \text{ for } y_2.$$
$$= \sqrt{(-3)^2 + (-4)^2}$$
$$= \sqrt{9 + 16}$$
$$= \sqrt{25}$$
$$= 5$$

To find the equation of a circle with radius 5 and center at $(1, 3)$, we substitute 1 for h, 3 for k, and 5 for r in the standard equation of the circle.

$$(x - h)^2 + (y - k)^2 = r^2$$
$$(x - 1)^2 + (y - 3)^2 = 5^2$$

To write the equation in general form, we square the binomials and simplify.

$$x^2 - 2x + 1 + y^2 - 6y + 9 = 25$$
$$x^2 + y^2 - 2x - 6y - 15 = 0 \quad \text{Subtract 25 from both sides and simplify.}$$

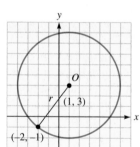

Figure 7-3

Self Check 1   Find the general equation of a circle with center at $(-2, 1)$ and radius of 4.

The final equation in Example 1 can be written as

$$1x^2 + 0xy + 1y^2 - 2x - 6y - 15 = 0$$

which illustrates that the graph of

$$Ax^2 + Bxy + Cy^2 + Dx + Ey + F = 0$$

is a circle whenever $B = 0$ and $A = C$.

2. Graph Circles

EXAMPLE 2   Graph the circle whose equation is $2x^2 + 2y^2 + 4x + 3y = 3$.

Solution   We will convert the general form of the circle into standard form so we can identify the center and radius. Then we will graph the circle.
   To find the coordinates of the center and the radius, we complete the square
on x and y. We begin by dividing both sides of the equation by 2 and rearranging
terms to get

$$x^2 + 2x + y^2 + \frac{3}{2}y = \frac{3}{2}$$

To complete the square on x and y, we add 1 and $\frac{9}{16}$ to both sides.

$$x^2 + 2x + 1 + y^2 + \frac{3}{2}y + \frac{9}{16} = \frac{3}{2} + 1 + \frac{9}{16}$$

We can factor $x^2 + 2x + 1$ and $y^2 + \frac{3}{2}y + \frac{9}{16}$ and simplify on the right side to get

$$(x + 1)^2 + \left(y + \frac{3}{4}\right)^2 = \frac{49}{16}$$

$$[x - (-1)]^2 + \left[y - \left(-\frac{3}{4}\right)\right]^2 = \left(\frac{7}{4}\right)^2$$

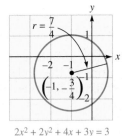

$2x^2 + 2y^2 + 4x + 3y = 3$

Figure 7-4

   From the equation, we see that the coordinates of the center of the circle are
$h = -1$ and $k = -\frac{3}{4}$ and that the radius of the circle is $\frac{7}{4}$. The graph is shown in
Figure 7-4.

Self Check 2   Graph the circle whose equation is $x^2 + y^2 - 6x - 2y = -6$.

Accent on   Graphing Circles
Technology

Since the graphs of circles fail the vertical line test, their equations do not represent functions. It is somewhat more difficult to use a graphing calculator to graph
equations that are not functions. For example, to graph the circle described by
$(x - 1)^2 + (y - 2)^2 = 4$, we must split the equation into two functions and graph
each one separately. To do this, we begin by solving the equation for y.

$$(x - 1)^2 + (y - 2)^2 = 4$$

$$(y - 2)^2 = 4 - (x - 1)^2 \qquad \text{Subtract } (x - 1)^2 \text{ from both sides.}$$

$$y - 2 = \pm\sqrt{4 - (x - 1)^2} \qquad \text{Take the square root of both sides.}$$

$$y = 2 \pm \sqrt{4 - (x - 1)^2} \qquad \text{Add 2 to both sides.}$$

   This equation defines two functions. If we use window settings of $[-3, 5]$ for
x and $[-3, 5]$ for y and graph the functions

$$y = 2 + \sqrt{4 - (x - 1)^2} \quad \text{and} \quad y = 2 - \sqrt{4 - (x - 1)^2}$$

we get the distorted circle shown in Figure 7-5(a). To get a better circle, we can use the graphing calculator's squaring feature, which gives an equal unit distance on both the x- and y-axes. Using this feature, we get the circle shown in Figure 7-5(b). Sometimes the two arcs will not join because of approximations made by the calculator at each endpoint.

Comment

Note that in this case, it is easier to graph the circle by hand.

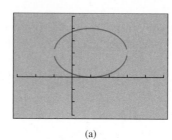 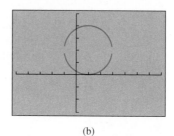

(a)   (b)

Figure 7-5

EXAMPLE 3   The effective broadcast area of a radio station is bounded by the circle

$$x^2 + y^2 = 2{,}500$$

where x and y are measured in miles. Another station's broadcast area is bounded by the circle

$$(x - 100)^2 + (y - 100)^2 = 900$$

Can any location receive both stations?

Solution   It is possible to receive both stations only if their circular broadcast areas overlap. (See Figure 7-6.) This happens when the sum of the radii, r and r', of the two circles is greater than the distance, d, between their centers. That is, $r + r' > d$.

The center of the circle $x^2 + y^2 = 2{,}500$ is $(x_1, y_1) = (0, 0)$, and its radius r is 50 miles. The center of the circle $(x - 100)^2 + (y - 100)^2 = 900$ is $(x_2, y_2) = (100, 100)$, and its radius r' is 30 miles.

We can use the distance formula to find the distance d between the centers.

$$d = \sqrt{(x_2 - x_1)^2 + (y_2 - y_1)^2}$$
$$d = \sqrt{(100 - 0)^2 + (100 - 0)^2}$$
$$= \sqrt{100^2 + 100^2}$$
$$= \sqrt{100^2 \cdot 2}$$
$$= 100\sqrt{2}$$
$$\approx 141 \text{ miles}$$

Figure 7-6

The sum of the radii, $r + r'$, of the two circles is $(50 + 30)$ miles, or 80 miles. Since this is less than the distance d between their centers (141 miles), there is no location where both stations can be received.

Self Check 3   In Example 3, if the station at the origin boosted its power to cover the area within $x^2 + y^2 = 9{,}000$, would the coverage overlap?

3. Write the Equation of a Parabola in Standard Form

In Chapter 3, we saw that the graphs of quadratic functions of the form

$$y = ax^2 + bx + c \quad \text{or} \quad y = a(x - h)^2 + k \qquad \text{provided } a \neq 0.$$

were parabolas that open up or down. We now discuss parabolas that open to the left or to the right and examine the properties of all parabolas in greater detail.

| The Parabola | A **parabola** is the set of all points in a plane equidistant from a line l (called the **directrix**) and a fixed point F (called the **focus**) that is not on line l. (See Figure 7-7.) |

The point on the parabola that is closest to the directrix is called the **vertex**, and the line passing through the vertex and the focus is called the **axis of the parabola.**

Hypatia
(370?–415)

Hypatia was one of the earliest known women in mathematics. She is also known as an astronomer and philosopher. Perhaps her most important contribution was her editing of the work *On the Conics of Apollonius,* which divides a cone into parts when cut by a plane. This concept led to the study of parabolas, ellipses, and hyperbolas.

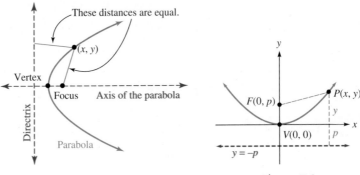

Figure 7-7　　　　　　**Figure 7-8**

A basic parabola is one whose vertex occurs at the origin and opens upward, as shown in Figure 7-8. Consider such a parabola with the following characteristics:

Vertex: $V(0, 0)$　　　　**Focus:** $F(0, p)$

Directrix: $y = -p$　　**Point on the parabola:** $P(x, y)$

To find the equation of the parabola, we will find the distance from point $P(x, y)$ on the parabola to the focus and find the distance from point $P(x, y)$ to the directrix and set the two distances equal. Using the distance formula, we can find the distance from $P(x, y)$ to $F(0, p)$.

$$d(PF) = \sqrt{x^2 + (y - p)^2}$$

From the figure, we see that the distance from $P(x, y)$ to the directrix $y = -p$ is:

$$|y - (-p)| = |y + p|$$

We can equate these distances and simplify.

$$\sqrt{x^2 + (y - p)^2} = |y + p|$$
$$x^2 + (y - p)^2 = |y + p|^2 \qquad \text{Square both sides.}$$
$$x^2 + (y - p)^2 = (y + p)^2 \qquad |y + p|^2 = (y + p)^2$$
$$x^2 + y^2 - 2yp + p^2 = y^2 + 2py + p^2 \qquad \text{Expand the binomials.}$$
$$x^2 - 2py = 2py \qquad \text{Subtract } y^2 + p^2 \text{ from both sides.}$$
(1)$$\qquad x^2 = 4py \qquad \text{Add } 2py \text{ to both sides.}$$

Equation 1 is one of the **standard equations of a parabola** with vertex at the origin. If $p > 0$ in Equation 1, the graph of the equation will be a parabola that opens upward. If $p < 0$, the graph of the equation will be a parabola that opens downward.

It can be shown that a parabola whose vertex is at the origin and opens to the left or to the right has an equation of the form $y^2 = 4px$.

The standard equations of a parabola with vertex $V(0, 0)$ and some of the characteristics of the parabola are summarized below.

Pierre de Fermat
(1601–1665)

Pierre de Fermat shares the honor of discovering analytic geometry (with Descartes) and of developing the theory of probability (with Pascal). But to Fermat alone goes credit for founding number theory. He is probably most famous for a theorem called Fermat's last theorem. It states that if n represents a number greater than 2, there are no whole numbers a, b, and c that satisfy the equation $a^n + b^n = c^n$. This theorem was proved by Andrew Wiles in 1995.

Standard Form of the Equation of a Parabola with Vertex (0, 0)		
Equation	$x^2 = 4py$	$y^2 = 4px$
Focus	$(0, p)$	$(p, 0)$
Directrix	$y = -p$	$x = -p$
Axis of Symmetry	Vertical, y-axis	Horizontal, x-axis
$p > 0$	Opens up	Opens right
$p < 0$	Opens down	Opens left

Consider the parabola with equation $x^2 = 8y$ that is shown in Figure 7-9. Since the equation of the parabola is written in standard form, $x^2 = 4py$, we know that $4p = 8$ and $p = 2$. Thus, the focus of the parabola is $F(0, p) = F(0, 2)$ and the directrix is $y = -p$ or $y = -2$. Because $p = 2$ and $2 > 0$, the parabola opens upward. The axis of symmetry is a vertical line, which is the y-axis.

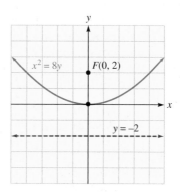

Figure 7-9

EXAMPLE 4   Find the equation of the parabola with vertex at the origin and focus at $(3, 0)$.

Solution   A sketch of the parabola is shown in Figure 7-10. Because the focus is to the right of the vertex, the parabola opens to the right, and because the vertex is the origin, the standard equation is $y^2 = 4px$. The distance between the focus and the vertex is $p = 3$. We can substitute 3 for p in the standard equation to get

$$y^2 = 4px$$
$$y^2 = 4(3)x$$
$$y^2 = 12x$$

The equation of the parabola is $y^2 = 12x$.

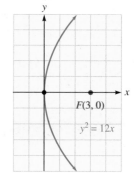

Figure 7-10

Self Check 4   Find the equation of the parabola with vertex at the origin and focus at $(-3, 0)$.

In Section 3.4, we used translations to sketch graphs of functions. Translations also can be applied to graphs that are parabolas. The following table summarizes the characteristics of parabolas with vertex at $V(h, k)$.

Standard Form of the Equation of a Parabola with Vertex (h, k)		
Equation	$(x - h)^2 = 4p(y - k)$	$(y - k)^2 = 4p(x - h)$
Focus	$(h, k + p)$	$(h + p, k)$
Directrix	$y = -p + k$	$x = -p + h$
Axis of Symmetry	Vertical $x = h$	Horizontal $y = k$
$p > 0$	Opens up	Opens right
$p < 0$	Opens down	Opens left

EXAMPLE 5   Find the equation of the parabola that opens up, has vertex at the point (4, 5), and passes through the point (0, 7).

Solution   Because the parabola opens upward, we will substitute the coordinates of the given vertex and the point into the equation $(x - h)^2 = 4p(y - k)$ and solve for p. Then we will substitute the coordinates of the vertex and the value of p into the previous equation.

Since $(h, k) = (4, 5)$ and the point (0, 7) is on the curve, we can substitute 4 for h, 5 for k, 0 for x, and 7 for y in the standard equation and solve for p.

$$(x - h)^2 = 4p(y - k)$$
$$(0 - 4)^2 = 4p(7 - 5)$$
$$16 = 8p$$
$$2 = p$$

To find the equation of the parabola, we substitute 4 for h, 5 for k, and 2 for p in the standard equation and simplify:

$$(x - h)^2 = 4p(y - k)$$
$$(x - 4)^2 = 4 \cdot 2(y - 5)$$
$$(x - 4)^2 = 8(y - 5)$$

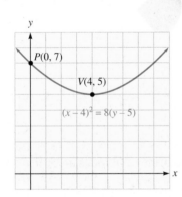

Figure 7-11

The graph of the equation appears in Figure 7-11.

Self Check 5   Find the equation of the parabola that opens up, has vertex at (4, 5), and passes through (0, 9).

EXAMPLE 6   Find the equations of two parabolas with a vertex at (2, 4) that pass through (0, 0).

Solution   The two parabolas are shown in Figure 7-12.

Part 1:   To find the parabola that opens to the left, we use the equation $(y - k)^2 = 4p(x - h)$. Since the curve passes through the point $(x, y) = (0, 0)$ and the vertex $(h, k) = (2, 4)$, we substitute 0 for x, 0 for y, 2 for h, and 4 for k in the standard equation and solve for p:

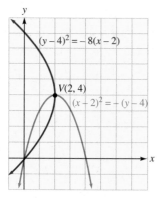

Figure 7-12

$$(y - k)^2 = 4p(x - h)$$
$$(0 - 4)^2 = 4p(0 - 2)$$
$$16 = -8p$$
$$-2 = p$$

Since $h = 2$, $k = 4$, $p = -2$, and the parabola opens to the left, its equation is

$$(y - k)^2 = 4p(x - h)$$
$$(y - 4)^2 = 4(-2)(x - 2)$$
$$(y - 4)^2 = -8(x - 2)$$

Part 2:   To find the equation of the parabola that opens down, we use the equation $(x - h)^2 = 4p(y - k)$ and substitute 2 for h, 4 for k, 0 for x, and 0 for y and solve for p:

$$(x - h)^2 = 4p(y - k)$$
$$(0 - 2)^2 = 4p(0 - 4)$$
$$4 = -16p$$
$$p = -\frac{1}{4}$$

Since $h = 2$, $k = 4$, $p = -\frac{1}{4}$, the equation is

$$(x - h)^2 = 4p(y - k)$$
$$(x - 2)^2 = 4\left(-\frac{1}{4}\right)(y - 4) \quad \text{Substitute 2 for } h, -\frac{1}{4} \text{ for } p, \text{ and 4 for } k.$$
$$(x - 2)^2 = -(y - 4)$$

Self Check 6   Find the equations of two parabolas with vertex at $(2, 4)$ and that pass through $(0, 8)$.

4.   Graph Equations of Parabolas

EXAMPLE 7   Find the vertex and y-intercepts of the parabola with the following equation and graph it: $y^2 + 8x - 4y = 28$.

Solution   To identify the coordinates of the vertex, we will complete the square on y and write the equation in standard form. To find the y-intercepts, we will let $x = 0$ and solve for y. Finally, we will graph the parabola by plotting the vertex and y-intercepts and drawing a smooth curve through the points.

Step 1: Complete the square and write the equation in standard form.

$$y^2 + 8x - 4y = 28$$
$$y^2 - 4y = -8x + 28 \qquad \text{Subtract } 8x \text{ from both sides.}$$
$$y^2 - 4y + 4 = -8x + 28 + 4 \qquad \text{Add 4 to both sides.}$$
$$(2) \qquad (y - 2)^2 = -8(x - 4) \qquad \text{Factor both sides.}$$

Equation 2 represents a parabola opening to the left with vertex at $(4, 2)$.

Step 2: Find the y-intercept.
To find the y-intercepts, we substitute 0 for x in Equation 2 and solve for y.

$$(y - 2)^2 = -8(x - 4)$$
$$(y - 2)^2 = -8(0 - 4) \qquad \text{Substitute 0 for } x.$$

$$y^2 - 4y + 4 = 32 \qquad \text{Remove parentheses.}$$
$$(3) \quad y^2 - 4y - 28 = 0$$

We can use the quadratic formula to find that the roots of Equation 3 are $y \approx 7.7$ and $y \approx -3.7$. So the points with coordinates of approximately $(0, 7.7)$ and $(0, -3.7)$ lie on the graph of the parabola and are the y-intercepts.

Step 3: Graph the parabola.
We can use the information above and the knowledge that the graph opens to the left and has vertex at $(4, 2)$ to draw the graph, as shown in Figure 7-13.

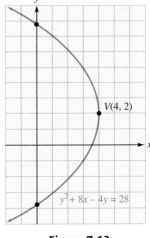

$V(4, 2)$

$y^2 + 8x - 4y = 28$

Figure 7-13

Self Check 7 Find the vertex and y-intercepts of the parabola with an equation of $y^2 - x + 2y = 3$. Then graph it.

To graph the equation $y^2 + 8x - 4y = 28$ (Example 7) with a graphing calculator, we first solve the equation for y.

$$y^2 + 8x - 4y = 28$$
$$y^2 - 4y = -8x + 28$$
$$y^2 - 4y + 4 = -8x + 28 + 4$$
$$(y - 2)^2 = -8x + 32$$
$$y - 2 = \pm \sqrt{-8x + 32}$$
$$y = 2 \pm \sqrt{-8x + 32}$$

If we use window settings of $[-2, 7]$ for x and $[-5, 9]$ for y and graph the functions

$$y = 2 + \sqrt{-8x + 32} \qquad \text{and} \qquad y = 2 - \sqrt{-8x + 32}$$

we will get a graph similar to the one shown in Figure 7-13.

5. Solve Applied Problems Involving Parabolas

Parabolas arise in many real-world settings. A few examples are:

- As shown in the photograph, a bouncing ball travels in a series of parabolic paths. The ball in the image was captured with a camera with a stroboscopic flash at 25 images per second. Air resistance causes the ball to deviate slightly from a perfect parabola shape.

- A satellite dish is a type of parabolic antenna designed with the specific purpose of transmitting signals to and/or receiving signals from satellites.

- Objects extended in space often follow parabolic paths, such as a diver jumping from a diving board or cliff. The diver may follow a complex motion, but his center of mass forms a parabola.

EXAMPLE 8   The Gateway Arch in St. Louis has a shape that approximates a parabola. (See Figure 7-14.) The vertex of the parabola is $V(0, 630)$ and the x-intercepts are $(315, 0)$ and $(-315, 0)$. Find the equation of the parabola that models the arch.

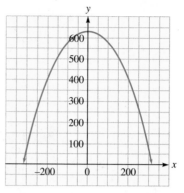

Figure 7-14

Solution   To find the equation of the parabola that opens downward, we use the equation $(x - h)^2 = 4p(y - k)$ and substitute the vertex $V(0, 630)$ and the x-intercept $(315, 0)$ and solve for p. Then we write the equation of the parabola.

We substitute 0 for h, 630 for k, 315 for x, and 0 for y and solve for p:

$$(x - h)^2 = 4p(y - k)$$
$$(315 - 0)^2 = 4p(0 - 630)$$
$$99{,}225 = -2{,}520p$$
$$p = -\frac{99{,}225}{2{,}520}$$
$$= -\frac{315}{8}$$

Since $h = 0$, $k = 630$, $p = -\frac{315}{8}$, the equation is

$$(x - h)^2 = 4p(y - k)$$
$$(x - 0)^2 = 4\left(-\frac{315}{8}\right)(y - 630) \qquad \text{Substitute 0 for } h, -\frac{315}{8} \text{ for } p, \text{ and 630 for } k.$$
$$x^2 = -\frac{315}{2}(y - 630) \qquad \text{Simplify.}$$

Self Check 8   What is the height of the arch 115 feet from the base?

EXAMPLE 9   If water is propelled straight up by a "super nozzle" during a water fountain show in Las Vegas, the equation $s = 128t - 16t^2$ expresses the water's height, s (in feet), t seconds after it is propelled. Find the maximum height reached by the water.

Solution   The graph of $s = 128t - 16t^2$, which expresses the height s of the water t seconds after it is propelled, is the parabola shown in Figure 7-15. To find the maximum height reached by the water, we find the s-coordinate k of the vertex of the parabola. To find k, we write the equation of the parabola in standard form by completing the square on t.

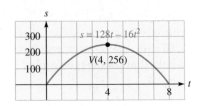

Figure 7-15

$$s = 128t - 16t^2$$

$$16t^2 - 128t = -s \qquad \text{Multiply both sides by } -1.$$

$$t^2 - 8t = \frac{-s}{16} \qquad \text{Divide both sides by 16.}$$

$$t^2 - 8t + 16 = \frac{-s}{16} + 16 \qquad \text{Add 16 to both sides to complete the square.}$$

$$(t - 4)^2 = \frac{-s + 256}{16} \qquad \text{Factor } t^2 - 8t + 16 \text{ and combine terms.}$$

$$(t - 4)^2 = -\frac{1}{16}(s - 256) \qquad \text{Factor out } -\frac{1}{16}.$$

This equation indicates that the maximum height is 256 feet.

Comment

The parabola shown in Figure 7-15 is not the path of the water. The water goes straight up and straight down.

Self Check 9   At what time will the water strike the ground?   (*Hint:* Find the *t*-intercept.)

Self Check Answers   **1.** $x^2 + y^2 + 4x - 2y - 11 = 0$ **2.**   **3.** no   **4.** $y^2 = -12x$
5. $(x - 4)^2 = 4(y - 5)$
6. $(y - 4)^2 = -8(x - 2)$,
$(x - 2)^2 = (y - 4)$

7. vertex $(-4, -1)$; *y*-intercepts $(0, 1)$, $(0, -3)$ **8.** approximately 376 feet
9. $t = 8$ seconds

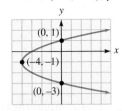

7.1 Exercises

Vocabulary and Concepts   *Give the coordinates of the circle's center and its radius.*

1. $(x - 2)^2 + (y + 5)^2 = 9$: center (__, __); radius __

2. $x^2 + y^2 - 36 = 0$: center (__, __); radius __

3. $x^2 + y^2 = 5$: center (__, __); radius ___

4. $2(x - 9)^2 + 2y^2 = 7$: center (__, __); radius ___

Determine whether the parabolic graph of the equation opens up, down, to the left, or to the right.

5. $y^2 = -4x$: opens

_____

6. $y^2 = 10x$: opens

_____

7. $x^2 = -8(y - 3)$:
opens _____

8. $(x - 2)^2 = (y + 3)$:
opens ___

Fill in the blanks.

9. A parabola is the set of all points in a plane equidistant from a line, called the _____, and a fixed point not on the line, called the _____.

10. The general form of a second-degree equation in the variables *x* and *y* is
Ax^2_____ $= 0$.

Practice   *Write the equation of each circle.*

11.

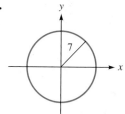

12.

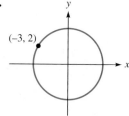

13.

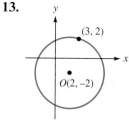

14.

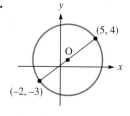

15. Radius of 6; center at the intersection of $3x + y = 1$ and $-2x - 3y = 4$

16. Radius of 8; center at the intersection of $x + 2y = 8$ and $2x - 3y = -5$

Graph each equation.

17. $x^2 + y^2 = 4$

18. $x^2 - 2x + y^2 = 15$

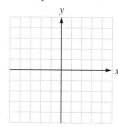

19. $3x^2 + 3y^2 - 12x - 6y = 12$

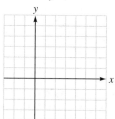

20. $2x^2 + 2y^2 + 4x - 8y + 2 = 0$

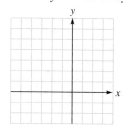

Find the vertex, focus, and directrix of each parabola.

21. $x^2 = 12y$

22. $y^2 = -12x$

23. $(y - 3)^2 = 20x$

24. $x^2 = -\dfrac{1}{2}(y + 5)$

25. $(x + 2)^2 = -24(y - 1)$

26. $(y + 1)^2 = 28(x - 2)$

Find the equation of each parabola.

27. Vertex at $(0, 0)$; focus at $(0, 3)$

28. Vertex at $(0, 0)$; focus at $(0, -3)$

29. Vertex at $(0, 0)$; focus at $(-3, 0)$

30. Vertex at $(0, 0)$; focus at $(3, 0)$

31. Vertex at $(3, 5)$; focus at $(3, 2)$

32. Vertex at $(3, 5)$; focus at $(-3, 5)$

33. Vertex at $(3, 5)$; focus at $(3, -2)$

34. Vertex at $(3, 5)$; focus at $(6, 5)$

35. Vertex at $(0, 2)$; directrix at $y = 3$

36. Vertex at $(-3, 4)$; directrix at $y = 2$

37. Vertex at $(1, -5)$; directrix at $x = -1$

38. Vertex at $(3, 5)$; directrix at $x = 6$

39. Vertex at $(2, 2)$; passes through $(0, 0)$

40. Vertex at $(-2, -2)$; passes through $(0, 0)$

41. Vertex at $(-4, 6)$; passes through $(0, 3)$

42. Vertex at $(-2, 3)$; passes through $(0, -3)$

43. Vertex at $(6, 8)$; passes through $(5, 10)$ and $(5, 6)$

44. Vertex at $(2, 3)$; passes through $\left(1, \dfrac{13}{4}\right)$ and $\left(-1, \dfrac{21}{4}\right)$

45. Vertex at $(3, 1)$; passes through $(4, 3)$ and $(2, 3)$

46. Vertex at $(-4, -2)$; passes through $(-3, 0)$ and $\left(\dfrac{9}{4}, 3\right)$

Change each equation to standard form and graph it.

47. $y = x^2 + 4x + 5$

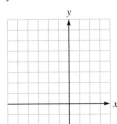

48. $2x^2 - 12x - 7y = 10$

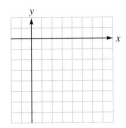

49. $y^2 + 4x - 6y = -1$

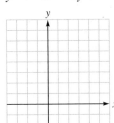

50. $x^2 - 2y - 2x = -7$

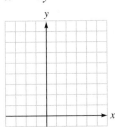

51. $y^2 + 2x - 2y = 5$

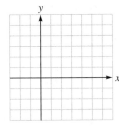

52. $y^2 - 4y = 4x - 8$

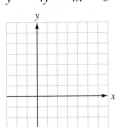

53. $y^2 - 4y = -8x + 20$

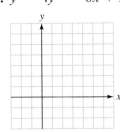

54. $y^2 - 2y = 9x + 17$

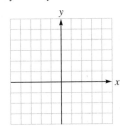

55. $x^2 - 6y + 22 = -4x$

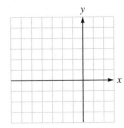

56. $4y^2 - 4y + 16x = 7$

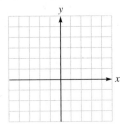

57. $4x^2 - 4x + 32y = 47$

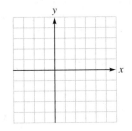

58. $4y^2 - 16x + 17 = 20y$

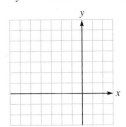

Applications

59. Broadcast range   A television tower broadcasts a signal with a circular range, as shown in the illustration. Can a city 50 miles east and 70 miles north of the tower receive the signal?

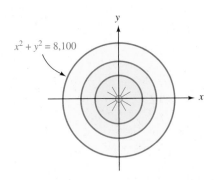

$x^2 + y^2 = 8,100$

60. Warning sirens   A tornado warning siren can be heard in the circular range shown in the illustration. Can a person 4 miles west and 5 miles south of the siren hear its sound?

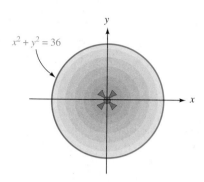

$x^2 + y^2 = 36$

61. Radio translators Some radio stations extend their broadcast range by installing a translator—a remote device that receives the signal and retransmits it. A station with a broadcast range given by $x^2 + y^2 = 1{,}600$, where x and y are in miles, installs a translator with a broadcast area bounded by $x^2 + y^2 - 70y + 600 = 0$. Find the greatest distance from the main transmitter that the signal can be received.

62. Ripples in a pond When a stone is thrown into the center of a pond, the ripples spread out in a circular pattern, moving at a rate of 3 feet per second. If the stone is dropped at the point $(0, 0)$ in the illustration, when will the ripple reach the seagull floating at the point $(15, 36)$?

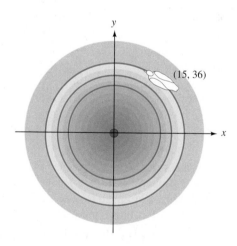

$(15, 36)$

63. Writing equations of circles Find the equation of the outer rim of the circular arch shown in the illustration.

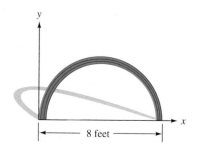

8 feet

64. Writing equations of circles The shape of the window shown is a combination of a rectangle and a semicircle. Find the equation of the circle of which the semicircle is a part.

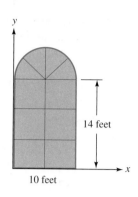

14 feet

10 feet

65. Meshing gears For design purposes, the large gear is described by the circle $x^2 + y^2 = 16$. The smaller gear is a circle centered at $(7, 0)$ and tangent to the larger circle. Find the equation of the smaller gear.

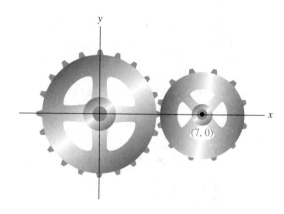

$(7, 0)$

66. Walkways The walkway shown is bounded by the two circles $x^2 + y^2 = 2{,}500$ and $(x - 10)^2 + y^2 = 900$, measured in feet. Find the largest and the smallest width of the walkway.

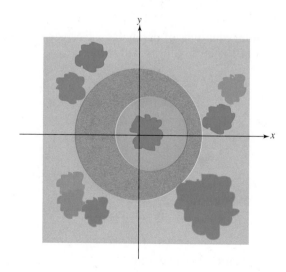

67. Solar furnaces A parabolic mirror collects rays of the sun and concentrates them at its focus. In the illustration, how far from the vertex of the parabolic mirror will it get the hottest? (All measurements are in feet.)

$y^2 = 8x$

68. Searchlight reflectors A parabolic mirror reflects light in a beam when the light source is placed at its focus. In the illustration, how far from the vertex of the parabolic reflector should the light source be placed? (All measurements are in feet.)

$y^2 = 12x$

69. Writing equations of parabolas Derive the equation of the parabolic arch shown.

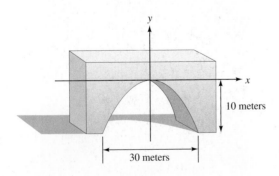

30 meters

10 meters

70. Projectiles The cannonball in the illustration follows the parabolic trajectory $y = 30x - x^2$. How far short of the castle does it land?

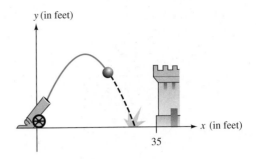

y (in feet)

x (in feet)

35

71. Satellite antennas The cross-section of the satellite antenna in the illustration is a parabola given by the equation $y = \frac{1}{16}x^2$, with distances measured in feet. If the dish is 8 feet wide, how deep is it?

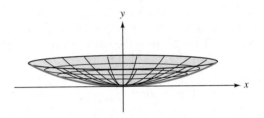

y

x

72. Design of a satellite antenna The cross-section of the satellite antenna shown is a parabola with the pickup at its focus. Find the distance d from the pickup to the center of the dish.

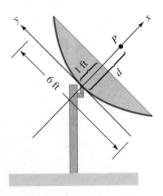

P

1 ft

6 ft

d

73. Operating a resort A resort owner plans to build and rent n cabins for d dollars per week. The price d that she can charge for each cabin depends on the number of cabins she builds, where $d = -45\left(\frac{n}{32} - \frac{1}{2}\right)$. Find the number of cabins she should build to maximize her weekly income.

74. Toy rockets A toy rocket is s feet above the Earth at the end of t seconds, where $s = -16t^2 + 80\sqrt{3}t$. Find the maximum height of the rocket.

75. Design of a parabolic reflector Find the outer diameter (the length $\overline{AB}$) of the parabolic reflector shown.

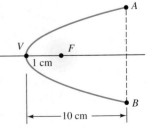

A

V F

1 cm

10 cm

B

76. Design of a suspension bridge The cable between the towers of the suspension bridge shown in the illustration has the shape of a parabola with vertex 15 feet above the roadway. Find the equation of the parabola.

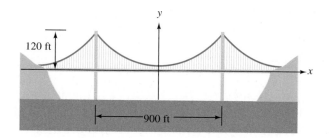

77. Gateway Arch The Gateway Arch in St. Louis has a shape that approximates a parabola. (See the illustration.) Find the width w of the arch 200 feet above the ground.

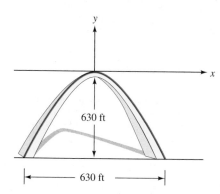

78. Building tunnels A construction firm plans to build a tunnel whose arch is in the shape of a parabola. (See the illustration.) The tunnel will span a two-lane highway 8 meters wide. To allow safe passage for vehicles, the tunnel must be 5 meters high at a distance of 1 meter from the tunnel's edge. Find the maximum height of the tunnel.

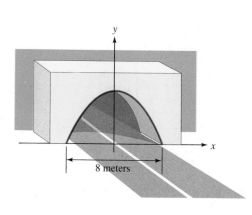

Discovery and Writing

79. Show that the standard form of the equation of a parabola $(y - 2)^2 = 8(x - 1)$ is a special case of the general form of a second-degree equation in two variables.

80. Show that the standard form of the equation of a circle $(x + 2)^2 + (y - 5)^2 = 36$ is a special case of the general form of a second-degree equation in two variables.

Find the equation, in the form $(x - h)^2 + (y - k)^2 = r^2$, of the circle passing through the given points.

81. $(0, 8), (5, 3),$ and $(4, 6)$

82. $(-2, 0), (2, 8),$ and $(5, -1)$

Find the equation of the parabola passing through the given points. Give the equation in the form $y = ax^2 + bx + c$.

83. $(1, 8), (-2, -1),$ and $(2, 15)$

84. $(1, -3), (-2, 12),$ and $(-1, 3)$

85. Ballistics A stone tossed upward is s feet above the Earth after t seconds, where $s = -16t^2 + 128t$. Show that the stone's height s seconds after it is thrown is equal to its height s seconds before it hits the ground.

86. Ballistics Show that the stone in Exercise 85 reaches its greatest height in one-half of the time it takes until it strikes the ground.

Review *Find the number that must be added to make each binomial a perfect square trinomial.*

87. $x^2 + 4x +$ __

88. $y^2 - 12y +$ __

89. $x^2 - 7x +$ __

90. $y^2 + 11y +$ __

Solve each equation.

91. $x^2 + 4x = 5$

92. $y^2 - 12y = 13$

93. $x^2 - 7x - 18 = 0$

94. $y^2 + 11y = -18$

7.2 The Ellipse

Objectives

1. Understand the Definition of an Ellipse
2. Write the Equation of an Ellipse
3. Graph Ellipses
4. Solve Problems Using Ellipses

Washington, D.C. has many historical sites. Many people touring the United States capitol visit Statuary Hall, originally the chamber of the House of Representatives. It is said that because of the oval shape of the ceiling, politicians on one side of the hall could eavesdrop on politicians talking on the opposite side of the hall. Such rooms often are called *whispering galleries*. Problem 47 in the Exercises discusses some properties of whispering galleries.

Today, Statuary Hall houses a collection of statues donated by individual states. Sam Houston, Daniel Webster, and Will Rogers are among those honored with a statue.

A third important conic is an oval-shaped curve, called an *ellipse*.

1. Understand the Definition of an Ellipse

The Ellipse An **ellipse** is the set of all points P in a plane such that the sum of the distances from P to two other fixed points F and F' is a positive constant.

We can illustrate this definition by showing how to construct an ellipse. To do so, we place two thumbtacks fairly close together, as in Figure 7-16. We then tie each end of a piece of string to a thumbtack, catch the loop with the point of a pencil, and (while keeping the string stretched tightly) draw the ellipse. Note that $d(F_1P) + d(PF_2)$ will be a positive constant.

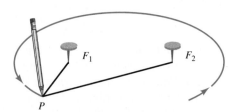

Figure 7-16

As we see in the figure, the graph of an ellipse is egg-shaped. The graphs of two ellipses are shown in Figure 7-17.

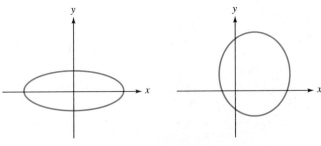

Ellipse with a horizontal axis

(a)

Ellipse with a vertical axis

(b)

Figure 7-17

Ellipses occur in many real-life settings.

- Rotating an ellipse about its axis produces a shape like a football.
- Planets such as Mars revolve around the Sun in an elliptical path.

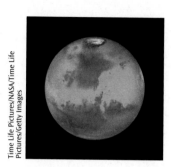

- A workout machine, called an *elliptic trainer,* uses elliptical movements for exercise without causing damage to the joints.
- An ellipse has an interesting property. Any light or sound that starts at one focus will be reflected through the other. This property is the basis of a medical procedure for treating kidney stones, called **lithotripsy.** The patient with the kidney stone is placed at one focus in an elliptical tank. Shock waves from a controlled explosion at the other focus are concentrated on the stone, pulverizing it. (See Figure 7-18.) Although some machines are still in use today, most are being replaced and patients lie down on a soft water-filled cushion.

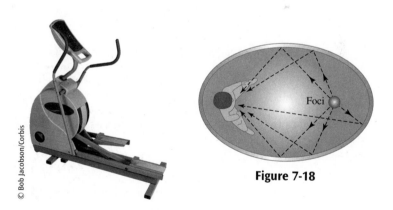

Figure 7-18

2. Write the Equation of an Ellipse

In the ellipse shown in Figure 7-19(a) on the next page, the fixed points F and F' are called **foci** (each is a **focus**), the midpoint of the chord FF' is called the **center,** and the chord VV' is called the **major axis.** Each of the endpoints V and V' of the major axis is called a **vertex.** The chord BB', perpendicular to the major axis and passing through the center C of the ellipse, is called the **minor axis.**

To derive the equation of the ellipse shown in Figure 7-19(b), we note that point O is the midpoint of chord FF' and let $d(OF) = d(OF') = c$, where $c > 0$. Then the coordinates of point F are $(c, 0)$, and the coordinates of F' are $(-c, 0)$. We also let $P(x, y)$ be any point on the ellipse.

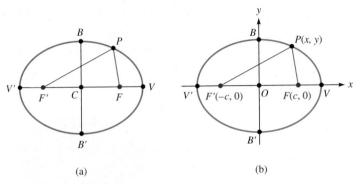

Figure 7-19

By the definition of an ellipse, $d(F'P) + d(PF)$ must be a positive constant, which we will call $2a$. Thus,

(1) $d(F'P) + d(PF) = 2a$

We can use the distance formula to compute the lengths of $F'P$ and PF,

$$d(F'P) = \sqrt{[x - (-c)]^2 + y^2} \quad \text{and} \quad d(PF) = \sqrt{(x - c)^2 + y^2}$$

and substitute these values into Equation 1 to obtain

$$\sqrt{[x - (-c)]^2 + y^2} + \sqrt{(x - c)^2 + y^2} = 2a$$

or

(2) $\sqrt{[x + c]^2 + y^2} = 2a - \sqrt{(x - c)^2 + y^2}$ (Subtract $\sqrt{(x - c)^2 + y^2}$ from both sides.)

We can square both sides of Equation 2 and simplify to get

$$(x + c)^2 + y^2 = 4a^2 - 4a\sqrt{(x - c)^2 + y^2} + [(x - c)^2 + y^2]$$
$$x^2 + 2cx + c^2 + y^2 = 4a^2 - 4a\sqrt{(x - c)^2 + y^2} + x^2 - 2cx + c^2 + y^2$$
$$4cx = 4a^2 - 4a\sqrt{(x - c)^2 + y^2}$$
$$cx = a^2 - a\sqrt{(x - c)^2 + y^2}$$
$$cx - a^2 = -a\sqrt{(x - c)^2 + y^2}$$

We square both sides again and simplify to get

$$c^2x^2 - 2a^2cx + a^4 = a^2[(x - c)^2 + y^2]$$
$$c^2x^2 - 2a^2cx + a^4 = a^2(x^2 - 2cx + c^2 + y^2)$$
$$c^2x^2 - 2a^2cx + a^4 = a^2x^2 - 2a^2cx + a^2c^2 + a^2y^2$$
$$c^2x^2 + a^4 = a^2x^2 + a^2c^2 + a^2y^2$$
$$a^4 - a^2c^2 = a^2x^2 - c^2x^2 + a^2y^2$$

(3) $a^2(a^2 - c^2) = (a^2 - c^2)x^2 + a^2y^2$

Because the shortest distance between two points is a line segment, $d(F'P) + d(PF) > d(F'F)$. Therefore, $2a > 2c$. Thus, $a > c$, and $a^2 - c^2$ is a positive number, which we will call b^2. Letting $b^2 = a^2 - c^2$ and substituting into Equation 3, we have

(4) $a^2b^2 = b^2x^2 + a^2y^2$

Dividing both sides of Equation 4 by a^2b^2 gives the equation

$$\frac{x^2}{a^2} + \frac{y^2}{b^2} = 1 \qquad \text{where } a > b > 0$$

This is the standard form of the equation of an ellipse centered at the origin and major axis on the x-axis.

To find the coordinates of the vertices V and V', we substitute 0 for y and solve for x:

$$\frac{x^2}{a^2} + \frac{y^2}{b^2} = 1$$

$$\frac{x^2}{a^2} + \frac{0^2}{b^2} = 1$$

$$\frac{x^2}{a^2} = 1$$

$$x^2 = a^2$$

$$x = a \quad \text{or} \quad x = -a$$

Since the coordinates of V are $(a, 0)$ and the coordinates of V' are $(-a, 0)$, a is the distance between the center of the ellipse and either of its vertices. Thus, the center of the ellipse is the midpoint of the major axis.

To find the coordinates of B and B', we substitute 0 for x and solve for y:

$$\frac{x^2}{a^2} + \frac{y^2}{b^2} = 1$$

$$\frac{0^2}{a^2} + \frac{y^2}{b^2} = 1$$

$$y^2 = b^2$$

$$y = b \quad \text{or} \quad y = -b$$

Since the coordinates of B are $(0, b)$ and the coordinates of B' are $(0, -b)$, the distance between the center of the ellipse and either endpoint of the minor axis is b. We have the following results.

The Ellipse: Major Axis on x-Axis, Center at (0, 0)

The standard equation of an ellipse with center at the origin and the major axis (horizontal) on the x-axis is

$$\frac{x^2}{a^2} + \frac{y^2}{b^2} = 1 \qquad \text{where } a > b > 0$$

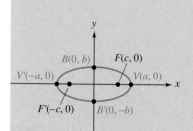

Vertices (ends of the major axis): $V(a, 0)$ and $V'(-a, 0)$

Length of major axis: $2a$

Ends of the minor axis: $B(0, b)$ and $B'(0, -b)$

Length of minor axis: $2b$

Foci: $F(c, 0)$ and $F'(-c, 0)$ where $c^2 = a^2 - b^2$

A similar equation results if the ellipse has a major axis on the y-axis and center at the origin.

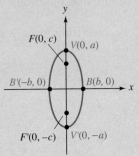

The Ellipse: Major Axis on y-Axis, Center at (0, 0)

The standard equation of an ellipse with center at the origin and the major axis (vertical) on the y-axis is

$$\frac{x^2}{b^2} + \frac{y^2}{a^2} = 1 \qquad \text{where } a > b > 0$$

Vertices (ends of the major axis): $V(0, a)$ and $V'(0, -a)$

Length of major axis: $2a$

Ends of the minor axis: $B(b, 0)$ and $B'(-b, 0)$

Length of minor axis: $2b$

Foci: $F(0, c)$ and $F'(0, -c)$ where $c^2 = a^2 - b^2$

EXAMPLE 1 Find the equation of the ellipse with center at the origin, major axis of length 6 units located on the x-axis, and minor axis of length 4 units.

Solution To find the equation, we determine a and b and substitute into the standard equation of an ellipse with center at the origin and major axis on the x-axis

$$\frac{x^2}{a^2} + \frac{y^2}{b^2} = 1$$

The center of the ellipse is given to be the origin and the length of the major axis is 6. Since the length of the major axis of an ellipse centered at the origin is $2a$, we have $2a = 6$ or $a = 3$. The coordinates of the vertices are $(3, 0)$ and $(-3, 0)$, as shown in Figure 7-20.

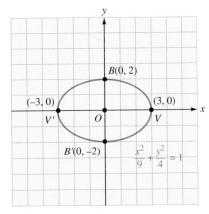

Figure 7-20

The length of the minor axis is given to be 4. Since the length of the minor axis of an ellipse centered at the origin is $2b$, we have $2b = 4$ or $b = 2$. The coordinates of B and B' are $(0, 2)$ and $(0, -2)$.

To find the equation of the ellipse, we substitute 3 for a and 2 for b in the standard equation of an ellipse with center at the origin and major axis on the x-axis.

$$\frac{x^2}{a^2} + \frac{y^2}{b^2} = 1$$

$$\frac{x^2}{3^2} + \frac{y^2}{2^2} = 1$$

$$\frac{x^2}{9} + \frac{y^2}{4} = 1$$

Self Check 1 Find the equation of the ellipse with center at the origin, major axis of length 10 on the y-axis, and minor axis of length 8.

EXAMPLE 2 Find the equation of the ellipse with center at the origin, focus at $\left(2\sqrt{3}, 0\right)$, and vertex at $(4, 0)$.

Solution Because the vertex and focus are on the x-axis, the major axis of the ellipse is on the x-axis. (See Figure 7-21.) To find the equation of the ellipse, we will determine a^2 and b^2 and substitute into the standard equation

$$\frac{x^2}{a^2} + \frac{y^2}{b^2} = 1$$

The distance between the center of the ellipse and the vertex is $a = 4$. The distance between the focus and the center is $c = 2\sqrt{3}$.

Since $b^2 = a^2 - c^2$, we can substitute 4 for a and $2\sqrt{3}$ for c and solve for b^2:

$$b^2 = a^2 - c^2$$
$$b^2 = 4^2 - \left(2\sqrt{3}\right)^2$$
$$= 16 - 12$$
$$= 4$$

To find the equation of the ellipse, we substitute 16 for a^2 and 4 for b^2 in the standard equation.

$$\frac{x^2}{a^2} + \frac{y^2}{b^2} = 1$$
$$\frac{x^2}{16} + \frac{y^2}{4} = 1$$

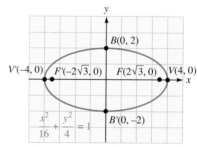

Figure 7-21

The graph of the ellipse $\dfrac{x^2}{16} + \dfrac{y^2}{4} = 1$ is shown in Figure 7-21.

Self Check 2   Find the equation of the ellipse with center at the origin, focus at $\left(0, 2\sqrt{3}\right)$, and vertex at $(0, 4)$.

To translate an ellipse to a new position centered at the point (h, k) instead of the origin, we replace x and y in the standard equations with $x - h$ and $y - k$, respectively, to get the following results.

The Ellipse: Major Axis Horizontal, Center at (h, k)

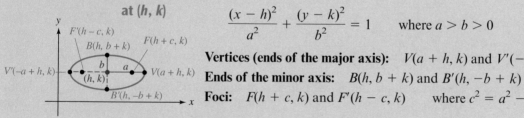

The standard equation of an ellipse with center at (h, k) and major axis horizontal is

$$\frac{(x - h)^2}{a^2} + \frac{(y - k)^2}{b^2} = 1 \qquad \text{where } a > b > 0$$

Vertices (ends of the major axis): $V(a + h, k)$ and $V'(-a + h, k)$
Ends of the minor axis: $B(h, b + k)$ and $B'(h, -b + k)$
Foci: $F(h + c, k)$ and $F'(h - c, k)$ where $c^2 = a^2 - b^2$

There is a similar result when the major axis is vertical.

The Ellipse: Major Axis Vertical, Center at (h, k)

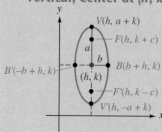

The standard equation of an ellipse with center at (h, k) and major axis vertical is

$$\frac{(x - h)^2}{b^2} + \frac{(y - k)^2}{a^2} = 1 \qquad \text{where } a > b > 0$$

Vertices (ends of the major axis): $V(h, a + k)$ and $V'(h, -a + k)$
Ends of the minor axis: $B(b + h, k)$ and $B'(-b + h, k)$
Foci: $F(h, k + c)$ and $F'(h, k - c)$ where $c^2 = a^2 - b^2$

In each case, the length of the major axis is $2a$, and the length of the minor axis is $2b$.

EXAMPLE 3 Find the equation of the ellipse with focus at $(-1, 7)$ and vertices at $V(-1, 8)$ and $V'(-1, -2)$.

Solution We will use the coordinates of the focus and the vertices to determine the center of the ellipse and a and b. We then will substitute these coordinates into the appropriate standard equation of an ellipse. Because the major axis passes through points V and V', we see that the major axis is a vertical line that is parallel to the y-axis. So the standard equation to use is

$$\frac{(x - h)^2}{b^2} + \frac{(y - k)^2}{a^2} = 1 \qquad \text{where } a > b > 0$$

Since the midpoint of the major axis is the center of the ellipse, the coordinates of the center are $(-1, 3)$, as in Figure 7-22.

From Figure 7-22, we see that the distance between the center of the ellipse and either vertex is $a = 5$. We also see that the distance between the focus and the center is $c = 4$.

Since $b^2 = a^2 - c^2$ is an ellipse, we can substitute 5 for a and 4 for c and solve for b^2:

$$b^2 = a^2 - c^2$$
$$= 5^2 - 4^2$$
$$= 9$$

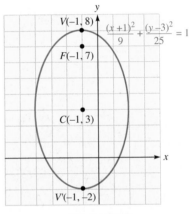

$$\frac{(x+1)^2}{9} + \frac{(y-3)^2}{25} = 1$$

Figure 7-22

To find the equation of the ellipse, we substitute -1 for h, 3 for k, 25 for a^2, and 9 for b^2 in the standard equation and simplify:

$$\frac{(x - h)^2}{b^2} + \frac{(y - k)^2}{a^2} = 1$$

$$\frac{[x - (-1)]^2}{9} + \frac{(y - 3)^2}{25} = 1$$

$$\frac{(x + 1)^2}{9} + \frac{(y - 3)^2}{25} = 1$$

Self Check 3 Find the equation of the ellipse with focus at $(3, 1)$ and vertices at $V(5, 1)$ and $V'(-5, 1)$.

3. Graph Ellipses

EXAMPLE 4 Graph the ellipse: $\dfrac{x^2}{49} + \dfrac{y^2}{4} = 1$

Solution To graph the ellipse, we will first determine the coordinates of the vertices and endpoints of the minor axis. We then will plot these points and draw an ellipse through them.

Because the equation is the standard form of an ellipse centered at the origin with major axis on the x-axis, we know that the center of the ellipse is at $(0, 0)$ and that the vertices lie on the x-axis. Because $a = 7$, the vertices are 7 units to the right and left of the origin at points $(7, 0)$ and $(-7, 0)$. Because $b^2 = 4$, $b = 2$ and the endpoints of the minor axis are 2 units above and below the origin at points $(0, 2)$ and $(0, -2)$. Using these points, we can sketch the ellipse shown in Figure 7-23.

Comment

In Example 4, note that $49 > 4$. Because 49 is the denominator of the x^2 term, the axis of the ellipse is horizontal and is the x-axis.

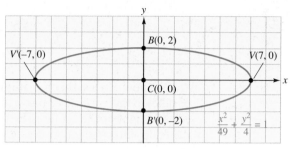

Figure 7-23

Self Check 4   Graph the ellipse: $\dfrac{x^2}{4} + \dfrac{y^2}{9} = 1$

EXAMPLE 5   Graph the ellipse: $\dfrac{(x + 2)^2}{4} + \dfrac{(y - 2)^2}{9} = 1$

Solution   The equation is the standard form of the equation of an ellipse whose major axis is vertical and whose center is at $(-2, 2)$. We will use translations to graph the ellipse.

Because $a^2 = 9$, we have $a = 3$ and the vertices are 3 units above and 3 units below the center at points $(-2, 5)$ and $(-2, -1)$. Because $b^2 = 4$, we have $b = 2$ and the endpoints of the minor axis are 2 units to the right and 2 units to the left of the center at points $(0, 2)$ and $(-4, 2)$. Using these points, we can sketch the ellipse as shown in Figure 7-24.

Comment

In Example 5, note that $9 > 4$. Because 9 is the denominator of the y^2 term, the axis of the ellipse is vertical.

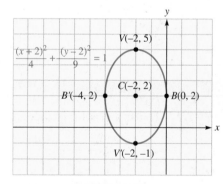

Figure 7-24

Self Check 5   Graph the ellipse: $\dfrac{(x - 2)^2}{9} + \dfrac{(y + 1)^2}{25} = 1$

EXAMPLE 6   Graph: $4x^2 + 9y^2 - 16x - 18y = 11$

Solution   We write the equation in standard form by completing the square on x and y and then use translations to graph the ellipse.

$$4x^2 + 9y^2 - 16x - 18y = 11$$
$$4x^2 - 16x + 9y^2 - 18y = 11$$
$$4(x^2 - 4x) + 9(y^2 - 2y) = 11$$
$$4(x^2 - 4x + 4) + 9(y^2 - 2y + 1) = 11 + 16 + 9$$
$$4(x - 2)^2 + 9(y - 1)^2 = 36$$
$$\frac{(x - 2)^2}{9} + \frac{(y - 1)^2}{4} = 1$$

We now can see that the graph is an ellipse with center at (2, 1) and major axis parallel to the x-axis. Because $a^2 = 9$, we have $a = 3$ and the vertices 3 units to the left and right of the center at $(-1, 1)$ and $(5, 1)$. Because $b = 4$, we have $b = 2$ and the endpoints of the minor axis are 2 units above and below the center at $(2, -1)$ and $(2, 3)$. Using these points, we can sketch the ellipse as shown in Figure 7-25.

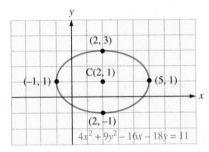

Figure 7-25

Self Check 6 Graph: $4x^2 + 9y^2 - 8x + 36y = -4$

Comment

If the coefficients of x^2 and y^2 in the equation of an ellipse are equal, the ellipse is a circle. If the coefficients are not equal, the ellipse is not a circle.

Accent on Technology **Graphing Ellipses**

To use a graphing calculator to graph

$$\frac{(x + 2)^2}{4} + \frac{(y - 1)^2}{25} = 1$$

we first clear the equation of fractions by multiplying both sides by 100 and solving for y.

$25(x + 2)^2 + 4(y - 1)^2 = 100$	**Multiply both sides by 100.**
$4(y - 1)^2 = 100 - 25(x + 2)^2$	**Subtract $25(x + 2)^2$ from both sides.**
$(y - 1)^2 = \dfrac{100 - 25(x + 2)^2}{4}$	**Divide both sides by 4.**
$y - 1 = \pm\dfrac{\sqrt{100 - 25(x + 2)^2}}{2}$	**Take the square root of both sides.**
$y = 1 \pm\dfrac{\sqrt{100 - 25(x + 2)^2}}{2}$	**Add 1 to both sides.**

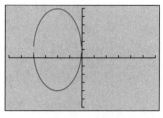

Figure 7-26

If we use window settings $[-6, 6]$ for x and $[-6, 6]$ for y and graph the functions

$$y = 1 + \frac{\sqrt{100 - 25(x + 2)^2}}{2} \quad \text{and} \quad y = 1 - \frac{\sqrt{100 - 25(x + 2)^2}}{2}$$

we will obtain the ellipse shown in Figure 7-26.

Everyday Connections

Eccentricity of an Ellipse

The eccentricity of an ellipse provides a measure of how much the curve resembles a true circle. Specifically, the eccentricity of a true circle equals 0.

When analyzing planetary orbits, astronomers plot the relationship between the length of the orbit's semimajor axis (measured in Astronomical Units, where 1 AU = 149,598,000 km) and the eccentricity of the orbit.

Use the given data plot to estimate how many of the 75 planets shown follow orbits that are true circles.

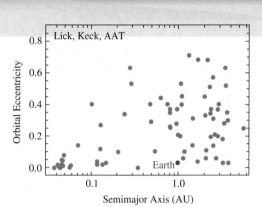

Source: Eccentricity vs. Semimajor axis for extrasolar planets. The 75 planets shown were found in a Doppler survey of 1300 FGKM main sequence stars using the Lick, Keck, and AAT telescopes. The survey was carried out by the California-Carnegie planet search team.

http://exoplanets.org/newsframe.html

4. Solve Problems Using Ellipses

EXAMPLE 7 The orbit of the Earth is approximately an ellipse, with the Sun at one focus. The ratio of c to a (called the **eccentricity** of the ellipse) is about $\frac{1}{62}$, and the length of the major axis is approximately 186,000,000 miles. How close does the Earth get to the Sun?

Solution We will assume that the ellipse has its center at the origin and vertices V' and V at $(-93,000,000, 0)$ and $(93,000,000, 0)$, as shown in Figure 7-27. Because the eccentricity $\frac{c}{a}$ is given to be $\frac{1}{62}$ and $a = 93,000,000$, we have

$$\frac{c}{a} = \frac{1}{62}$$

$$c = \frac{1}{62}a$$

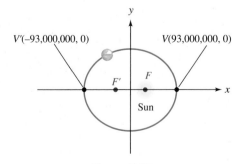

Figure 7-27

$$c = \frac{1}{62}(93{,}000{,}000)$$

$$= 1{,}500{,}000$$

The distance $d(FV)$ is the shortest distance between the Earth and the Sun. (You'll be asked to prove this in the exercises.) Thus,

$$d(FV) = a - c = 93{,}000{,}000 - 1{,}500{,}000 = 91{,}500{,}000 \text{ mi}$$

The Earth's point of closest approach to the Sun (called the **perihelion**) is approximately 91.5 million miles.

Self Check 7   Find the eccentricity of the ellipse: $\dfrac{(x+2)^2}{9} + \dfrac{(y-5)^2}{25} = 1$

We can use the eccentricity of an ellipse to judge its shape. If the eccentricity is close to 1, the ellipse is relatively flat, as in the ellipse on the left. If the eccentricity is close to 0, the ellipse is more circular, as in the ellipse on the right. Since the eccentricity of the Earth's orbit is $\frac{1}{62}$, the Earth's orbit is almost a circle.

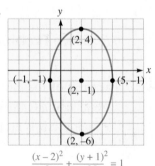

Self Check Answers   **1.** $\dfrac{x^2}{16} + \dfrac{y^2}{25} = 1$ **2.** $\dfrac{x^2}{4} + \dfrac{y^2}{16} = 1$ **3.** $\dfrac{x^2}{25} + \dfrac{(y-1)^2}{16} = 1$

4.

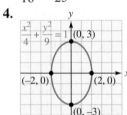

5.

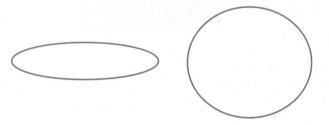

6.

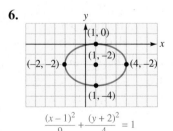

7. $\dfrac{4}{5}$

7.2   **Exercises**

Vocabulary and Concepts   *Fill in the blanks.*

1. An ellipse is the set of all points in the plane such that the ____ of the distances from two fixed points is a positive _____.
2. Each of the two fixed points in the definition of an ellipse is called a ____ of the ellipse.
3. The chord that joins the _____ is called the major axis of the ellipse.
4. The chord through the center of an ellipse and perpendicular to the major axis is called the _____ axis.
5. In the ellipse $\dfrac{x^2}{a^2} + \dfrac{y^2}{b^2} = 1$ $(a > b)$, the vertices are $V(\_,\_)$ and $V'(\_,\_)$.
6. In an ellipse, the relationship between a, b, and c is _____.
7. To draw an ellipse that is 26 inches wide and 10 inches tall, how long should the piece of string be, and how far apart should the two thumbtacks be?
8. To draw an ellipse that is 20 centimeters wide and 12 centimeters tall, how long should the piece of string be, and how far apart should the two thumbtacks be?

Practice   *Write the equation of the ellipse that has its center at the origin.*

9. Major axis of length 8 units located on the x-axis and minor axis of length 6 units.
10. Major axis of length 14 units located on the y-axis and minor axis of length 10 units.

11. Focus at $(3, 0)$; vertex at $(5, 0)$
12. Focus at $(0, 4)$; vertex at $(0, 7)$
13. Focus at $(0, 1)$; $\frac{4}{3}$ is one-half the length of the minor axis
14. Focus at $(1, 0)$; $\frac{4}{3}$ is one-half the length of the minor axis

15. Focus at $(0, 3)$; major axis equal to 8
16. Focus at $(5, 0)$; major axis equal to 12

Write the equation of each ellipse.

17. Center at $(3, 4)$; $a = 3$, $b = 2$; major axis parallel to the y-axis
18. Center at $(3, 4)$; passes through $(3, 10)$ and $(3, -2)$; $b = 2$
19. Center at $(3, 4)$; $a = 3$, $b = 2$; major axis parallel to the x-axis
20. Center at $(3, 4)$; passes through $(8, 4)$ and $(-2, 4)$; $b = 2$
21. Foci at $(-2, 4)$ and $(8, 4)$; $b = 4$
22. Foci at $(8, 5)$ and $(4, 5)$; $b = 3$
23. Vertex at $(6, 4)$; foci at $(-4, 4)$ and $(4, 4)$
24. Center at $(-4, 5)$; $\dfrac{c}{a} = \dfrac{1}{3}$; vertex at $(-4, -1)$
25. Foci at $(6, 0)$ and $(-6, 0)$; $\dfrac{c}{a} = \dfrac{3}{5}$
26. Vertices at $(2, 0)$ and $(-2, 0)$; $\dfrac{2b^2}{a} = 2$

Graph each ellipse.

27. $\dfrac{x^2}{25} + \dfrac{y^2}{9} = 1$

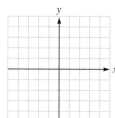

28. $\dfrac{x^2}{36} + \dfrac{y^2}{25} = 1$

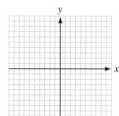

29. $\dfrac{x^2}{25} + \dfrac{y^2}{49} = 1$

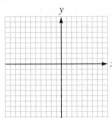

30. $4x^2 + y^2 = 4$

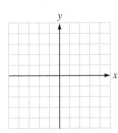

31. $\dfrac{x^2}{16} + \dfrac{(y+2)^2}{36} = 1$

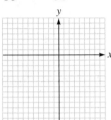

32. $(x-1)^2 + \dfrac{4y^2}{25} = 4$

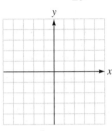

33. $\dfrac{(x-4)^2}{49} + \dfrac{(y-2)^2}{9} = 1$ **34.** $\dfrac{(x-1)^2}{25} + \dfrac{y^2}{4} = 1$

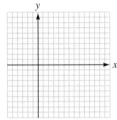

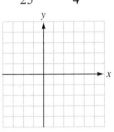

Write each ellipse in standard form.

35. $4x^2 + y^2 - 2y = 15$

36. $4x^2 + 25y^2 + 8x - 96 = 0$

37. $9x^2 + 4y^2 + 18x + 16y - 11 = 0$

38. $x^2 + 4y^2 - 10x - 8y = -13$

Graph each ellipse.

39. $x^2 + 4y^2 - 4x + 8y + 4 = 0$

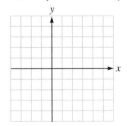

40. $x^2 + 4y^2 - 2x - 16y = -13$

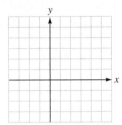

41. $16x^2 + 25y^2 - 160x - 200y + 400 = 0$

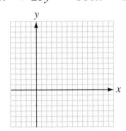

42. $3x^2 + 2y^2 + 7x - 6y = -1$

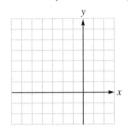

Applications

43. Pool tables   Find the equation of the outer edge of the elliptical pool table shown below.

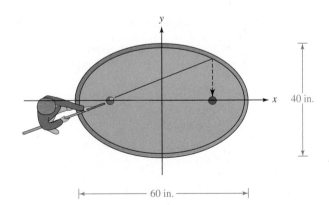

44. Astronomy   The moon has an orbit that is an ellipse, with the Earth at one focus. If the major axis of the orbit is 378,000 miles and the ratio of c to a is approximately $\dfrac{11}{200}$, how far does the moon get from the Earth? (This farthest point in an orbit is called the **apogee**.)

45. Equation of an arch An arch is a semiellipse 12 meters wide and 5 meters high. Write the equation of the ellipse if the ellipse is centered at the origin.

46. Design of a track A track is built in the shape of an ellipse with a maximum length of 100 meters and a maximum width of 60 meters. Write the equation of the ellipse and find its **focal width.** That is, find the length of a chord that is perpendicular to the major axis and passes through either focus of the ellipse.

47. Whispering galleries Any sound from one focus of an ellipse reflects off the ellipse directly back to the other focus. This property explains whispering galleries such as Statuary Hall in Washington, D.C. The ceiling of the whispering gallery shown has the shape of a semiellipse. Find the distance sound travels as it leaves focus F and returns to focus F'.

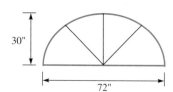

48. Finding the width of a mirror The oval mirror shown is in the shape of an ellipse. Find the width of the mirror 12 inches above its base.

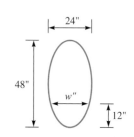

49. Finding the height of a window The window shown has the shape of an ellipse. Find the height of the window 20 inches from one end.

50. Area of an ellipse The area A of the ellipse

$$\frac{x^2}{a^2} + \frac{y^2}{b^2} = 1$$

is given by $A = \pi ab$. Find the area of the ellipse $9x^2 + 16y^2 = 144$.

Discovery and Writing

51. If F is a focus of the ellipse shown and B is an endpoint of the minor axis, use the distance formula to prove that the length of segment FB is a. (*Hint:* In an ellipse, $a^2 - c^2 = b^2$.)

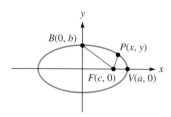

52. If F is a focus of the ellipse shown and P is any point on the ellipse, use the distance formula to show that the length of FP is $a - \frac{c}{a}x$. (*Hint:* In an ellipse, $a^2 - c^2 = b^2$.)

53. Finding the focal width In the ellipse shown, chord AA' passes through the focus F and is perpendicular to the major axis. Show that the length of AA' (called the **focal width**) is $\frac{2b^2}{a}$.

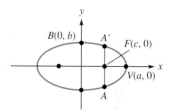

54. Prove that segment FV in Example 7 is the shortest distance between the Earth and the Sun. (*Hint:* Refer to Exercise 52.)

55. Constructing an ellipse The ends of a piece of string 6 meters long are attached to two thumbtacks that are 2 meters apart. A pencil catches the loop and draws it tight. As the pencil is moved about the thumbtacks (always keeping the tension), an ellipse is produced, with the thumbtacks as foci. Write the equation of the ellipse. (*Hint:* You'll have to establish a coordinate system.)

56. The distance between point $P(x, y)$ and the point $(0, 2)$ is $\frac{1}{3}$ of the distance of point P from the line $y = 18$. Find the equation of the curve on which point P lies.

57. Prove that $a > b$ in the development of the standard equation of an ellipse.

58. Show that the expansion of the standard equation of an ellipse is a special case of the general second-degree equation in two variables.

61. $5B - 2C$

62. $AC + B$

63. The additive inverse of B

64. The multiplicative inverse of A

Review   *Let* $A = \begin{bmatrix} 3 & -1 & 2 \\ 0 & 2 & -1 \\ 3 & 1 & 1 \end{bmatrix}$, $B = \begin{bmatrix} 1 & 2 \\ 2 & 0 \\ -1 & 1 \end{bmatrix}$,

and $C = \begin{bmatrix} 0 & 1 \\ -1 & 1 \\ -2 & 0 \end{bmatrix}$. *Find each of the following, if possible.*

59. AB

60. $B + C$

7.3   The Hyperbola

Objectives

1. Write Equations of Hyperbolas
2. Graph Hyperbolas

When a military plane exceeds the speed of sound, it is said to break the sound barrier. When this happens, a visible vapor cloud appears and an explosion-like sound that we call a *sonic boom* results.

As the plane moves faster beyond the speed of sound, a cone shaped shock wave intersects the ground forming one branch of a curve called an *hyperbola,* a fourth type of conic section. The sonic boom is heard at points inside this hyperbola.

Hyperbolas appear in many real-life applications.

- Hyperbolas are the basis of a navigational system known as LORAN (LOng RAnge Navigation). It is a radio navigation system that uses time intervals between radio signals to determine the location of a ship or aircraft.

- Hyperbolas are the basis for the design of hypoid and spiral bevel gears that are used in motor vehicles.

- Hyperbolas describe the orbits of some comets.

The definition of an hyperbola is similar to the definition of an ellipse, except that we require a constant difference of $2a$ instead of a constant sum.

The Hyperbola   An **hyperbola** is the set of all points P in a plane such that the absolute value of the difference of the distances from point P to two other points in the plane is a positive constant.

Although the definition of an ellipse and an hyperbola are quite similar, their graphs are very different. The graphs of two hyperbolas are shown in Figure 7-28.

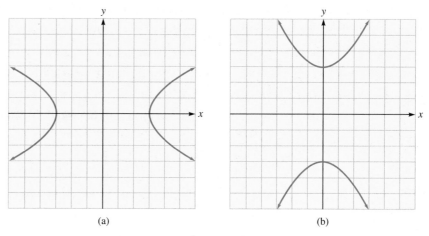

Figure 7-28

1.   Write Equations of Hyperbolas

In the graph of the hyperbola shown in Figure 7-29, points F and F' are called the **foci** of the hyperbola and the midpoint of chord FF' is called the **center.** The points V and V', where the hyperbola intersects FF', are called **vertices.** The segment VV' is called the **transverse axis.**

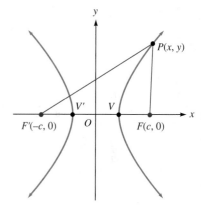

Figure 7-29

To develop the equation of the hyperbola centered at the origin, we note that the origin is the midpoint of chord FF', and we let $d(F'O) = d(OF) = c$, where $c > 0$. Then F is at $(c, 0)$, and F' is at $(-c, 0)$. The definition of an hyperbola requires that $d(F'P) - d(PF) = 2a$, where $2a$ is a positive constant. We use the distance formula to compute the lengths of $F'P$ and PF:

$$d(F'P) = \sqrt{[x - (-c)]^2 + y^2} \qquad d(PF) = \sqrt{(x - c)^2 + y^2}$$

Substituting these values into the equation $|d(F'P) - d(PF)| = 2a$ gives

$$\sqrt{(x + c)^2 + y^2} - \sqrt{(x - c)^2 + y^2} = 2a$$

or

$$\sqrt{(x + c)^2 + y^2} = 2a + \sqrt{(x - c)^2 + y^2}$$

After squaring, we have

$$(x + c)^2 + y^2 = 4a^2 + 4a\sqrt{(x - c)^2 + y^2} + (x - c)^2 + y^2$$
$$x^2 + 2cx + c^2 + y^2 = 4a^2 + 4a\sqrt{(x - c)^2 + y^2} + x^2 - 2cx + c^2 + y^2$$
$$4cx = 4a^2 + 4a\sqrt{(x - c)^2 + y^2}$$
$$cx - a^2 = a\sqrt{(x - c)^2 + y^2}$$

Squaring both sides again and simplifying gives

$$c^2x^2 - 2a^2cx + a^4 = a^2(x^2 - 2cx + c^2 + y^2)$$
$$c^2x^2 - 2a^2cx + a^4 = a^2x^2 - 2a^2cx + a^2c^2 + a^2y^2$$
$$c^2x^2 + a^4 = a^2x^2 + a^2c^2 + a^2y^2$$
$$(1) \qquad (c^2 - a^2)x^2 - a^2y^2 = a^2(c^2 - a^2)$$

Because $c > a$ (you will be asked to prove this in the exercises), $c^2 - a^2$ is a positive number. So we can let $b^2 = c^2 - a^2$ and substitute b^2 for $c^2 - a^2$ in Equation 1 to get

$$b^2x^2 - a^2y^2 = a^2b^2 \qquad \text{Substitute } c^2 - a^2 \text{ for } b^2.$$

If we divide both sides of the previous equation by a^2b^2, we will obtain the following equation.

$$\frac{x^2}{a^2} - \frac{y^2}{b^2} = 1$$

This equation is the standard form of the equation of an hyperbola with center at the origin and horizontal axis the x-axis.

To find the x-intercepts of the graph, we let $y = 0$ and solve for x. We get

$$\frac{x^2}{a^2} = 1$$
$$x^2 = a^2$$
$$x = a \quad \text{or} \quad x = -a$$

We now know that the x-intercepts are the vertices $V(a, 0)$ and $V'(-a, 0)$. The distance between the center of the hyperbola and either vertex is a, and the center of the hyperbola is the midpoint of the segment $V'V$ as well as that of the segment FF'.

We attempt to find the y-intercepts by letting $x = 0$. Then the equation becomes

$$\frac{-y^2}{b^2} = 1 \qquad \text{or} \qquad y^2 = -b^2$$

Because $-b^2$ represents a negative number, and y^2 cannot be negative, the equation has no real solutions. Since there are no y-values corresponding to $x = 0$, the hyperbola does not intersect the y-axis.

This discussion suggests the following results.

Hyperbola: Foci on **x-Axis, Center at (0, 0)**	The standard equation of an hyperbola with center at the origin and foci on the x-axis is $$\frac{x^2}{a^2} - \frac{y^2}{b^2} = 1$$ where $a^2 + b^2 = c^2$. **Vertices:** $V(a, 0)$ and $V'(-a, 0)$ **Foci:** $F(c, 0)$ and $F'(-c, 0)$

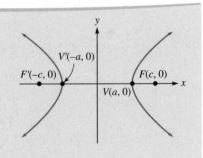

If the foci are on the y-axis, a similar equation results.

Hyperbola: Foci on y-Axis, Center at (0, 0)

The standard equation of an hyperbola with center at the origin and foci on the y-axis is

$$\frac{y^2}{a^2} - \frac{x^2}{b^2} = 1$$

where $a^2 + b^2 = c^2$.

Vertices: $V(0, a)$ and $V'(0, -a)$

Foci: $F(0, c)$ and $F'(0, -c)$

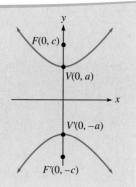

EXAMPLE 1 Write the equation of the hyperbola with vertices $V(4, 0)$ and $V'(-4, 0)$ and a focus at $F(5, 0)$.

Solution Because the foci lie on a the x-axis, we use the standard equation $\frac{x^2}{a^2} - \frac{y^2}{b^2} = 1$. We will find a^2 and b^2 and substitute the results into the standard equation.

The center of the hyperbola is midway between the vertices V and V'. Thus, the center is the origin $(0, 0)$. The distance between the vertex and the center is $a = 4$, and the distance between the focus and the center is $c = 5$. We can find b^2 by substituting 4 for a and 5 for c in the following equation to get

$$b^2 = c^2 - a^2 \quad \text{In an hyperbola, } b^2 = c^2 - a^2.$$
$$b^2 = 5^2 - 4^2$$
$$b^2 = 9$$

Substituting the values for a^2 and b^2 in the standard equation gives the equation of the hyperbola:

$$\frac{x^2}{a^2} - \frac{y^2}{b^2} = 1$$
$$\frac{x^2}{16} - \frac{y^2}{9} = 1$$

The graph of $\frac{x^2}{16} - \frac{y^2}{9} = 1$ is shown in Figure 7-30.

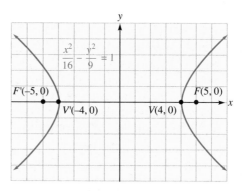

Figure 7-30

Self Check 1 Write the equation of the hyperbola with vertices $(0, 4)$ and $(0, -4)$ and a focus at $(0, 5)$.

To translate the hyperbola to a new position centered at the point (h, k) instead of the origin, we replace x and y with $x - h$ and $y - k$, respectively. We get the following results.

Hyperbola: Transverse Axis Horizontal, Center at (h, k)

The standard equation of an hyperbola with center at the (h, k) and foci on a line parallel to the x-axis is

$$\frac{(x - h)^2}{a^2} - \frac{(y - k)^2}{b^2} = 1$$

where $a^2 + b^2 = c^2$.

Vertices: $V(a + h, k)$ and $V'(-a + h, k)$

Foci: $F(c + h, k)$ and $F'(-c + h, k)$

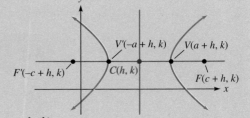

Hyperbola: Transverse Axis Vertical, Center at (h, k)

The standard equation of an hyperbola with center at (h, k) and foci on a line parallel to the y-axis is

$$\frac{(y - k)^2}{a^2} - \frac{(x - h)^2}{b^2} = 1$$

where $a^2 + b^2 = c^2$.

Vertices: $V(h, a + k)$ and $V'(h, -a + k)$

Foci: $F(h, c + k)$ and $F'(h, -c + k)$

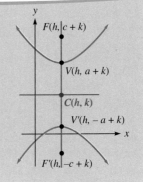

EXAMPLE 2   Write the equation of the hyperbola with vertices $(3, -3)$ and $(3, 3)$ and a focus at $(3, 5)$.

Solution   Because the foci lie on a vertical line as shown in Figure 7-31, we will use the standard form of the equation

$$\frac{(y - k)^2}{a^2} - \frac{(x - h)^2}{b^2} = 1$$

and determine h, k, a^2, and b^2, and substitute the results into the standard equation.

The center of the hyperbola is midway between the vertices V and V'. Thus, the center is point $(3, 0)$; $h = 3$; and $k = 0$. The distance between the vertex and the center is $a = 3$, and the distance between the focus and the center is $c = 5$. We can find b^2 by substituting 3 for a and 5 for c in the following equation to get

$$b^2 = c^2 - a^2 \qquad \text{In an hyperbola, } b^2 = c^2 - a^2.$$
$$b^2 = 5^2 - 3^2$$
$$b^2 = 16$$

Substituting the values for h, k, a^2, and b^2 in the standard equation gives the equation of the hyperbola:

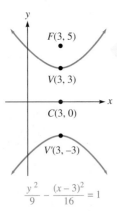

Figure 7-31

$$\frac{(y - k)^2}{a^2} - \frac{(x - h)^2}{b^2} = 1$$

$$\frac{(y - 0)^2}{9} - \frac{(x - 3)^2}{16} = 1$$

$$\frac{y^2}{9} - \frac{(x - 3)^2}{16} = 1$$

Self Check 2   Write the equation of the hyperbola with vertices $(3, 1)$ and $(-3, 1)$ and a focus at $(5, 1)$.

2.   Graph Hyperbolas

The graph of the hyperbola with equation $\dfrac{x^2}{25} - \dfrac{y^2}{4} = 1$ is shown in Figure 7-32. From the equation, we can see that $a^2 = 25$ and $b^2 = 4$ and that $a = 5$ and $b = 2$. The blue dashed lines in the figure are asymptotes of the graph.

The values of a and b can be used to find the equations of the asymptotes. In fact, the equations of the asymptotes are

$$y = \frac{b}{a}x \quad \text{and} \quad y = -\frac{b}{a}x$$

$$y = \frac{2}{5}x \qquad\qquad y = -\frac{2}{5}x$$

Andrew Wiles
(1953–   )
Wiles first learned of Fermat's last theorem (see page 605) in 1963, when he was 10 years old. He was so intrigued by the problem that he decided to study mathematics. While on the faculty at Princeton University, he would isolate himself in the attic of his home and work on the problem. After seven years of hard work, he announced a solution in June 1993. After another year of work to fix some gaps in the solution, his 130-page proof was published in 1995. After 350 years, Fermat's last theorem has been proved.

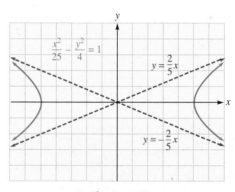

Figure 7-32

The values of a and b are also aids in drawing a special rectangle that can be used to help graph hyperbolas.

To show that this is true, we consider the hyperbola with equation of $\dfrac{x^2}{a^2} - \dfrac{y^2}{b^2} = 1$. This hyperbola has its center at the origin and vertices at $V(a, 0)$ and $V'(-a, 0)$. We can plot points V, V', $B(0, b)$, and $B'(0, -b)$ and form rectangle RSQP, called the **fundamental rectangle,** as shown in Figure 7-33. The extended diagonals of this rectangle are the asymptotes of the hyperbola.

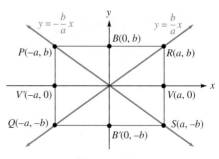

Figure 7-33

To show that the extended diagonals are asymptotes of the hyperbola, we solve the equation $\dfrac{x^2}{a^2} - \dfrac{y^2}{b^2} = 1$ for y and modify its form as follows:

$$\frac{x^2}{a^2} - \frac{y^2}{b^2} = 1$$

$$b^2x^2 - a^2y^2 = a^2b^2 \qquad \text{Multiply both sides by } a^2b^2.$$

$$y^2 = \frac{b^2x^2 - a^2b^2}{a^2} \qquad \text{Subtract } b^2x^2 \text{ from both sides and divide both sides by } -a^2.$$

$$y^2 = \frac{b^2x^2}{a^2}\left(1 - \frac{a^2}{x^2}\right) \qquad \text{Factor out the fraction } \frac{b^2x^2}{a^2}.$$

$$(2) \qquad y = \pm\frac{bx}{a}\sqrt{1 - \frac{a^2}{x^2}} \qquad \text{Take the square root of both sides.}$$

In Equation 2, if a is constant and $|x|$ approaches ∞, then $\dfrac{a^2}{x^2}$ approaches 0, and $\sqrt{1 - \dfrac{a^2}{x^2}}$ approaches 1. Thus, the hyperbola approaches the lines

$$y = \frac{b}{a}x \qquad \text{and} \qquad y = -\frac{b}{a}x$$

Using the fundamental rectangle, asymptotes, and vertices as guides, we can sketch the hyperbola shown in Figure 7-34. The Figure segment BB' is called the **conjugate axis** of the hyperbola.

If the hyperbola is not in standard form, we simply convert it to standard form by completing the square and making use of translations to graph the hyperbola. This information is summarized as follows:

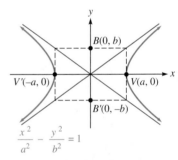

Figure 7-34

Graphing a Hyperbola
1. Write the hyperbola in standard form.
2. Identify the center and a and b.
3. Draw the fundamental rectangle and asymptotes.
4. Sketch the hyperbola beginning with the vertices and using the asymptotes as guides.

EXAMPLE 3   Graph the hyperbola: $\dfrac{y^2}{9} - \dfrac{x^2}{25} = 1$

Solution   We will apply the four steps listed above to graph the hyperbola.

Step 1: Write the hyperbola in standard form.   The equation of the hyperbola $\dfrac{y^2}{9} - \dfrac{x^2}{25} = 1$ is already written in the standard form $\dfrac{y^2}{a^2} - \dfrac{x^2}{b^2} = 1$.

Step 2: Identify the center and a and b.   From the standard equation of an hyperbola, we see that the center is $(0, 0)$, that $a = 3$ and $b = 5$, and that the vertices are on the y-axis.

Step 3: Draw the fundamental rectangle and asymptotes. The vertices V and V' are 3 units above and below the origin and have coordinates of $(0, 3)$ and $(0, -3)$. Points B and B' are 5 units right and left of the origin and have coordinates of $(5, 0)$ and $(-5, 0)$. We use the points V, V', B, and B' to construct the fundamental rectangle and extend the diagonals to graph the asymptotes as shown in Figure 7-35.

Step 4: Sketch the hyperbola. We begin with the vertices and sketch the graph of the hyperbola, as shown in Figure 7-35, using the asymptotes as guides.

Comment

The graphs of hyperbolas look very similar to the graphs of parabolas, but hyperbolas are not parabolas.

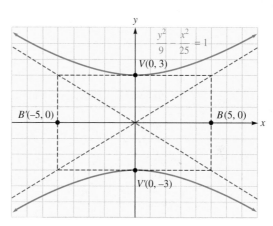

Figure 7-35

Self Check 3   Graph the hyperbola: $\dfrac{y^2}{9} - \dfrac{(x + 1)^2}{25} = 1$

EXAMPLE 4   Graph the hyperbola: $x^2 - y^2 - 2x + 4y = 12$

Solution   We will use the four steps listed above to graph the hyperbola.

Step 1: Write the hyperbola in standard form. We complete the square on x and y to write the equation into standard form:

$$x^2 - y^2 - 2x + 4y = 12$$
$$x^2 - 2x - (y^2 - 4y) = 12$$
$$x^2 - 2x + 1 - (y^2 - 4y + 4) = 12 + 1 - 4$$
$$(x - 1)^2 - (y - 2)^2 = 9$$
$$\frac{(x - 1)^2}{9} - \frac{(y - 2)^2}{9} = 1$$

Step 2: Identify the center and a and b. From the standard equation of an hyperbola, we see that the center is $(1, 2)$, that $a = 3$ and $b = 3$, and that the vertices are on a line segment parallel to the x-axis, as shown in Figure 7-36.

Step 3: Draw the fundamental rectangle and asymptotes. The vertices V and V' are 3 units to the right and left of the center and have coordinates of $(4, 2)$ and $(-2, 2)$. Points B and B', 3 units above and below the center, have coordinates of $(1, 5)$ and $(1, -1)$. We use the points V, V', B, and B' to construct the fundamental rectangle and extend the diagonals to draw the asymptotes as shown in Figure 7-36.

Step 4: Sketch the hyperbola. We begin with the vertices and sketch the graph of the hyperbola, as shown in Figure 7-36, using the asymptotes as guides.

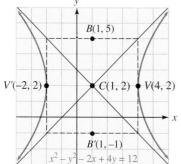

Figure 7-36

Self Check 4   Graph the hyperbola: $x^2 - y^2 + 6x + 2y = 8$

To graph the equation $x^2 - y^2 - 2x + 4y = 12$ with a graphing calculator, we first solve the equation for y.

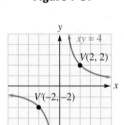

Figure 7-37

$$x^2 - y^2 - 2x + 4y = 12$$
$$-x^2 + y^2 + 2x - 4y = -12$$
$$y^2 - 4y = x^2 - 2x - 12$$
$$y^2 - 4y + 4 = x^2 - 2x - 12 + 4$$
$$(y - 2)^2 = x^2 - 2x - 8$$
$$y - 2 = \pm\sqrt{x^2 - 2x - 8}$$
$$y = 2 \pm\sqrt{x^2 - 2x - 8}$$

If we use window settings of $[-4, 6]$ for x and $[-3, 7]$ for y and graph the functions

$$y = 2 + \sqrt{x^2 - 2x - 8} \qquad \text{and} \qquad y = 2 - \sqrt{x^2 - 2x - 8}$$

we will get a graph shown in Figure 7-37.

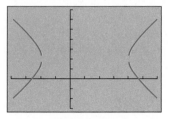

Figure 7-38

We have considered only those hyperbolas with a major axis that is horizontal or vertical. However, some hyperbolas have nonhorizontal or nonvertical major axes. For example, the graph of the equation $xy = 4$ is an hyperbola with vertices at $(2, 2)$ and $(-2, -2)$, as shown in Figure 7-38.

Self Check Answers   1. $\dfrac{y^2}{16} - \dfrac{x^2}{9} = 1$ 2. $\dfrac{x^2}{9} - \dfrac{(y - 1)^2}{16} = 1$

3. 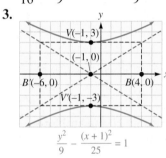   4.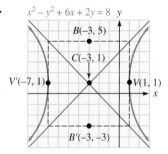

7.3   Exercises

Vocabulary and Concepts   *Fill in the blanks.*

1. An hyperbola is the set of all points in the plane such that the absolute value of the _____ of the distances from two fixed points is a positive _____.

2. Each of the two fixed points in the definition of an hyperbola is called a _____ of the hyperbola.

3. The vertices of the hyperbola $\dfrac{x^2}{a^2} - \dfrac{y^2}{b^2} = 1$ are

   $V(\_, \_)$ and $V'(\_, \_)$.

4. The vertices of the hyperbola $\dfrac{y^2}{a^2} - \dfrac{x^2}{b^2} = 1$ are

   $V(\_, \_)$ and $V'(\_, \_)$.

5. The chord that joins the vertices is called the _____ of the hyperbola.

6. In an hyperbola, the relationship between a, b, and c is _____.

Practice   *Write the equation of each hyperbola.*

7. Vertices $(5, 0)$ and $(-5, 0)$; focus $(7, 0)$

8. Focus (3, 0); vertex (2, 0); center (0, 0)

9. Center (2, 4); $a = 2$, $b = 3$; transverse axis is horizontal

10. Center $(-1, 3)$; vertex (1, 3); focus (2, 3)

11. Center (5, 3); vertex (5, 6); passes through (1, 8)

12. Foci (0, 10) and $(0, -10)$; $\dfrac{c}{a} = \dfrac{5}{4}$

13. Vertices (0, 3) and $(0, -3)$; $\dfrac{c}{a} = \dfrac{5}{3}$

14. Focus (4, 0); vertex (2, 0); center (0, 0)

15. Center $(1, -3)$; $a^2 = 4$; $b^2 = 16$

16. Center (1, 4); focus (7, 4); vertex (3, 4)

17. Center at the origin; passes through (4, 2) and $(8, -6)$

18. Center $(3, -1)$; y-intercept -1; x-intercept $3 + \dfrac{3\sqrt{5}}{2}$

Find the area of the fundamental rectangle of each hyperbola.

19. $4(x - 1)^2 - 9(y + 2)^2 = 36$
20. $x^2 - y^2 - 4x - 6y = 6$
21. $x^2 + 6x - y^2 + 2y = -11$
22. $9x^2 - 4y^2 = 18x + 24y + 63$

Write the equation of each hyperbola.

23. Center $(-2, -4)$; $a = 2$; area of fundamental rectangle is 36 square units

24. Center $(3, -5)$; $b = 6$; area of fundamental rectangle is 24 square units

25. Vertex (6, 0); one end of conjugate axis at $\left(0, \dfrac{5}{4}\right)$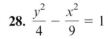

26. Vertex (3, 0); focus $(-5, 0)$; center (0, 0)

Graph each hyperbola.

27. $\dfrac{x^2}{9} - \dfrac{y^2}{4} = 1$

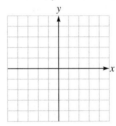

28. $\dfrac{y^2}{4} - \dfrac{x^2}{9} = 1$

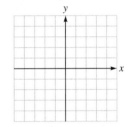

29. $4x^2 - 3y^2 = 36$

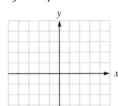

30. $3x^2 - 4y^2 = 36$

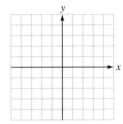

31. $y^2 - x^2 = 1$

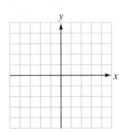

32. $x^2 - \dfrac{y^2}{4} = 1$

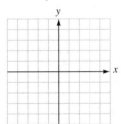

33. $\dfrac{(x + 2)^2}{9} - \dfrac{y^2}{4} = 1$

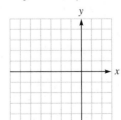

34. $\dfrac{y^2}{9} - \dfrac{(x - 2)^2}{36} = 1$

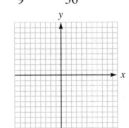

35. $4(y - 2)^2 - 9(x + 1)^2 = 36$

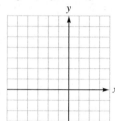

36. $9(y + 2)^2 - 4(x - 1)^2 = 36$

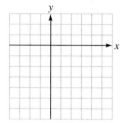

37.
$4x^2 - 2y^2 + 8x - 8y = 8$

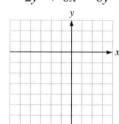

38.
$x^2 - y^2 - 4x - 6y = 6$

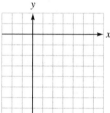

39.
$y^2 - 4x^2 + 6y + 32x = 59$

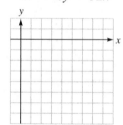

40.
$x^2 + 6x - y^2 + 2y = -11$

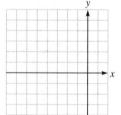

41. $-xy = 6$

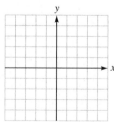

42. $xy = 20$

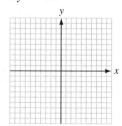

Find the equation of the curve on which point P lies.

43. The difference of the distances between $P(x, y)$ and the points $(-2, 1)$ and $(8, 1)$ is 6.

44. The difference of the distances between $P(x, y)$ and the points $(3, -1)$ and $(3, 5)$ is 5.

45. The distance between point $P(x, y)$ and the point $(0, 3)$ is $\frac{3}{2}$ of the distance between P and the line $y = -2$.

46. The distance between point $P(x, y)$ and the point $(5, 4)$ is $\frac{5}{3}$ of the distance between P and the line $x = -3$.

Applications

47. Fluids See the illustration below. Two glass plates in contact at the left, and separated by about 5 millimeters on the right, are dipped in beet juice, which rises by capillary action to form an hyperbola. The hyperbola is modeled by an equation of the form $xy = k$. If the curve passes through the point $(12, 2)$, what is k?

48. Astronomy Some comets have an hyperbolic orbit, with the Sun as one focus. When the comet shown in the illustration is far away from Earth, it appears to be approaching Earth along the line $y = 2x$. Find the equation of its orbit if the comet comes within 100 million miles of the Earth.

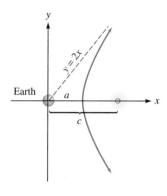

49. Alpha particles The particle in the illustration approaches the nucleus at the origin along the path $9y^2 - x^2 = 81$ in the coordinate system shown. How close does the particle come to the nucleus?

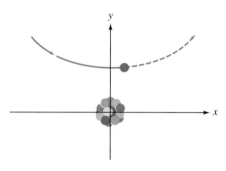

50. Physics Parallel beams of similarly charged particles are shot from two atomic accelerators 20 meters apart, as shown in the illustration. If the particles were not deflected, the beams would be 2.0×10^{-4} meter apart. However, because the charged particles repel each other, the beams follow the hyperbolic path $y = \dfrac{k}{x}$, for some k. Find k.

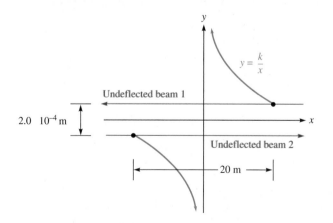

51. Navigation The LORAN system (LOng RAnge Navigation) in the illustration uses two radio transmitters 26 miles apart to send simultaneous signals. The navigator on a ship at $P(x, y)$ receives the closer signal first, and determines that the difference of the distances between the ship and each transmitter is 24 miles. That places the ship on a certain curve. Identify the curve and find its equation.

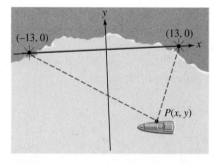

52. Navigation By determining the difference of the distances between the ship in the illustration and two radio transmitters, the LORAN navigation system places the ship on the hyperbola $x^2 - 4y^2 = 576$ in the coordinate system shown. If the ship is 5 miles out to sea, find its coordinates.

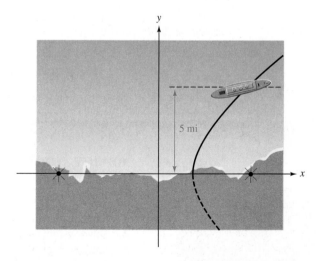

53. Wave propagation Stones dropped into a calm pond at points A and B create ripples that propagate in widening circles. In the illustration, points A and B are 20 feet apart, and the radii of the circles differ by 12 feet. The point $P(x, y)$ where the circles intersect moves along a curve. Identify the curve and find its equation.

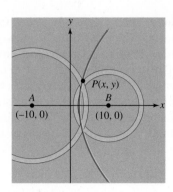

54. Sonic boom The position of a sonic boom caused by the faster-than-sound aircraft is one branch of the hyperbola $y^2 - x^2 = 25$ in the coordinate system shown. How wide is the hyperbola 5 miles from its vertex?

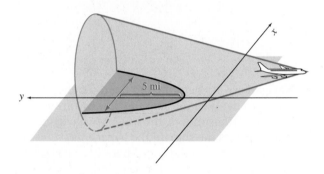

Discovery and Writing

55. Prove that $c > a$ for an hyperbola with center at $(0, 0)$ and line segment FF' on the x-axis.

56. Show that the extended diagonals of the fundamental rectangle of the hyperbola $\dfrac{x^2}{a^2} - \dfrac{y^2}{b^2} = 1$ are $y = \dfrac{b}{a}x$ and $y = -\dfrac{b}{a}x$.

57. Show that the expansion of the standard equation of an hyperbola is a special case of the general equation of second degree with $B = 0$.

58. Write a paragraph describing how you can tell from the equation of an hyperbola whether the transverse axis is vertical or horizontal.

Review *Find the inverse of each function. Write the answer in $y = f^{-1}(x)$ form.*

59. $f(x) = 3x - 2$

60. $f(x) = \dfrac{x + 1}{x}$

61. $f(x) = \dfrac{5x}{x + 2}$

62. $f(x) = x$

Let $f(x) = x^2 + 1$ and $g(x) = (x + 1)^2$. Find each composite function.

63. $f(g(x))$

64. $g(f(x))$

65. $f(f(x))$

66. $g(g(x))$

7.4 Solving Nonlinear Systems of Equations

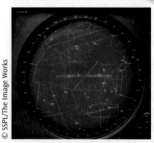

Objectives

1. Solve Systems by Graphing
2. Solve Systems by Substitution
3. Solve Systems by Addition
4. Solve Problems Using Systems of Nonlinear Equations

Air-traffic controllers maintain an orderly flow of air traffic. Since air-traffic controllers keep planes a safe distance apart, the paths of two planes will not intersect. This guarantees that there will be no mid-air collisions. Knowing where the graphs of two or more equations intersect can be very important. For example, the intersection point of the graphs of the flight paths of two planes will be a potential point of collision.

© SSPL/The Image Works

The photograph shows the flow of air traffic around a city on a typical afternoon.

We now will discuss techniques for solving systems of two equations in two variables, where at least one of the equations is nonlinear. We will use three methods to solve such systems: graphing, substitution, and addition.

1. Solve Systems by Graphing

To solve systems by the graphing method, we can graph each equation and identify the point(s) of intersection of the two graphs.

EXAMPLE 1 Solve $\begin{cases} x^2 + y^2 = 25 \\ 2x + y = 10 \end{cases}$ by graphing.

Solution To solve this system, we will graph each equation and identify the point(s) of intersection of the two graphs.

The graph of $x^2 + y^2 = 25$ is a circle with center at the origin and radius of 5. The equation of $2x + y = 10$ is a line with x-intercept at $(5, 0)$ and y-intercept at $(0, 10)$.

After graphing the circle and the line, as shown in Figure 7-39, we see that there are two intersection points, $P(3, 4)$ and $P'(5, 0)$. The solutions to the system are

$$\begin{cases} x = 3 \\ y = 4 \end{cases} \quad \text{and} \quad \begin{cases} x = 5 \\ y = 0 \end{cases}$$

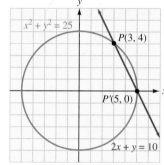

The line $2x + y = 10$ is a called a **secant line** because it intersects the graph of the circle at two points. In this case, the system has two solutions. If a line intersects the graph of a circle at one point, the line is called a **tangent line.** In this case, there is one solution. If the line does not intersect the circle, there is no solution.

Comment

The solutions can be verified by substituting into the two equations and checking to see that both equations are satisfied.

Figure 7-39

Self Check 1 Solve: $\begin{cases} x^2 + y^2 = 25 \\ 2x - y = 5 \end{cases}$

Accent on Technology

Solving Systems of Equations

To solve the system of equations in Example 1 using a graphing calculator, we must solve each of the equations for y. From the equation $x^2 + y^2 = 25$, we get two equations to graph:

$$y_1 = \sqrt{25 - x^2} \quad \text{and} \quad y_2 = -\sqrt{25 - x^2}$$

From the equation $2x + y = 10$, we get a third equation to graph:

$$y_3 = 10 - 2x$$

The graphs of these equations will be similar to those shown in Figure 7-40(a) on the next page. To find the top solution, we use ZOOM and TRACE to read the coordinates of the point of intersection, as shown in Figure 7-40(b): $x \approx 3$ and $y \approx 4$. Because the calculator graph is not complete at the other point of intersection shown in Figure 7-40(c), we must use our best judgment. We read the coordinates to be $x \approx 5$ and $y \approx 0$.

We also can find the coordinates of the intersection point by using the INTERSECT feature.

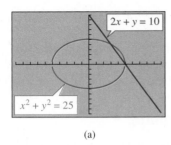 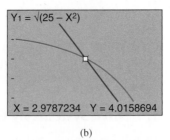 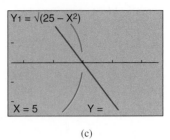

(a)   (b)   (c)

Figure 7-40

2. Solve Systems by Substitution

The substitution method can be used to find exact solutions. By making an appropriate substitution, we can convert a nonlinear system of two equations in two variables into one equation in one variable. We then can solve the resulting equation for one variable and back substitute to find the other variable.

EXAMPLE 2 Solve $\begin{cases} x^2 + y^2 = 25 \\ 2x + y = 10 \end{cases}$ by substitution.

Solution We will solve the linear equation for y to obtain an equation called the **substitution equation**. We then will substitute the expression that represents y into the second-degree equation to obtain an equation in the one variable x. After solving the equation for x, we can find y by back substituting the resulting values of x into the substitution equation and simplifying.

We begin by solving the linear equation for y.

$$2x + y = 10 \qquad \text{This equation is linear.}$$
$$y = -2x + 10 \qquad \text{This is the substitution equation.}$$

We now can substitute $-2x + 10$ for y in the second-degree equation and solve the resulting quadratic equation for x:

$$x^2 + y^2 = 25 \qquad \text{This equation is nonlinear.}$$
$$x^2 + (-2x + 10)^2 = 25 \qquad \text{Substitute } -2x + 10 \text{ for } y.$$
$$x^2 + 4x^2 - 40x + 100 = 25 \qquad \text{Square the binomial.}$$
$$5x^2 - 40x + 75 = 0 \qquad \text{Subtract 25 from both sides and combine like terms.}$$
$$x^2 - 8x + 15 = 0 \qquad \text{Divide both sides by 5.}$$
$$(x - 5)(x - 3) = 0 \qquad \text{Factor } x^2 - 8x + 15.$$
$$x - 5 = 0 \quad \text{or} \quad x - 3 = 0$$
$$x = 5 \quad | \quad x = 3$$

Because $y = -2x + 10$, if $x = 5$, then $y = 0$; and if $x = 3$, then $y = 4$. The two solutions are

$$\begin{cases} x = 5 \\ y = 0 \end{cases} \quad \text{and} \quad \begin{cases} x = 3 \\ y = 4 \end{cases}$$

Comment

Note that the solutions we obtained by graphing and by substitution are the same.

Self Check 2 Solve: $\begin{cases} x^2 + y^2 = 25 \\ 2x - y = 5 \end{cases}$ by substitution.

EXAMPLE 3   Solve: $\begin{cases} 4x^2 + 9y^2 = 5 \\ y = x^2 \end{cases}$ by substitution.

Solution   In this example, both equations are nonlinear. To solve it, we will substitute y for x^2 in the top equation to obtain one equation in the variable y. Then we will solve the resulting equation for y and use back substitution to find x.

$$4x^2 + 9y^2 = 5$$
$$4y + 9y^2 = 5 \qquad \text{Substitute } y \text{ for } x^2.$$
$$9y^2 + 4y - 5 = 0 \qquad \text{Add } -5 \text{ to both sides.}$$
$$(9y - 5)(y + 1) = 0 \qquad \text{Factor } 9y^2 + 4y - 5.$$
$$9y - 5 = 0 \quad \text{or} \quad y + 1 = 0$$
$$y = \frac{5}{9} \qquad \qquad y = -1$$

Because $y = x^2$, we can find x by solving the equations

$$x^2 = \frac{5}{9} \qquad \text{and} \qquad x^2 = -1$$

The solutions of $x^2 = \dfrac{5}{9}$ are

$$x = \frac{\sqrt{5}}{3} \qquad \text{and} \qquad x = -\frac{\sqrt{5}}{3}$$

Since the equation $x^2 = -1$ has no real solutions, the only solutions of the system are

$$\left(\frac{\sqrt{5}}{3}, \frac{5}{9} \right) \qquad \text{and} \qquad \left(-\frac{\sqrt{5}}{3}, \frac{5}{9} \right)$$

Comment
The substitution method is a good choice when one variable in the system is raised to the first power.

Self Check 3   Solve: $\begin{cases} x^2 + 3y^2 = 13 \\ x = y^2 - 1 \end{cases}$ by substitution.

3. Solve Systems by Addition

When we have two second-degree equations of the form $ax^2 + by^2 = c$, we can solve the system by using the addition method and eliminating one of the variables.

EXAMPLE 4   Solve: $\begin{cases} 3x^2 + 2y^2 = 36 \\ 4x^2 - y^2 = 4 \end{cases}$ by addition.

Solution   In this example, we have two second-degree equations of the form $ax^2 + by^2 = c$. In such cases, we can solve the system by eliminating one of the variables by addition, solving the resulting equation, and using back substitution to solve for the second variable.

To eliminate the terms involving y^2, we copy the first equation and multiply the second equation by 2 to obtain the following equivalent system.

$$\begin{cases} 3x^2 + 2y^2 = 36 \\ 8x^2 - 2y^2 = 8 \end{cases}$$

We then can add the equations and solve the resulting equation for x:

$$11x^2 = 44$$
$$x^2 = 4$$
$$x = 2 \quad \text{or} \quad x = -2$$

To find y, we substitute 2 for x and then -2 for x in the first equation.

For $x = 2$

$$3x^2 + 2y^2 = 36$$
$$3(2)^2 + 2y^2 = 36$$
$$12 + 2y^2 = 36$$
$$2y^2 = 24$$
$$y^2 = 12$$
$$y = \sqrt{12} \quad \text{or} \quad y = -\sqrt{12}$$
$$y = 2\sqrt{3} \quad | \quad y = -2\sqrt{3}$$

For $x = -2$

$$3x^2 + 2y^2 = 36$$
$$3(-2)^2 + 2y^2 = 36$$
$$12 + 2y^2 = 36$$
$$2y^2 = 24$$
$$y^2 = 12$$
$$y = \sqrt{12} \quad \text{or} \quad y = -\sqrt{12}$$
$$y = 2\sqrt{3} \quad | \quad y = -2\sqrt{3}$$

The four solutions of this system are

$$\left(2, 2\sqrt{3}\right), \quad \left(2, -2\sqrt{3}\right), \quad \left(-2, 2\sqrt{3}\right), \quad \text{and} \quad \left(-2, -2\sqrt{3}\right)$$

Self Check 4   Solve: $\begin{cases} 2x^2 + y^2 = 23 \\ 3x^2 - 2y^2 = 17 \end{cases}$ by addition.

4.   Solve Problems Using Systems of Nonlinear Equations

EXAMPLE 5   The area of a tennis court for singles matches is 2,106 square feet, and the perimeter is 210 feet. Find the dimensions of the court.

Solution   We will set up a system of two equations that models the problem and then use the substitution method to solve it.

Step 1: Write a system of two equations that models the problem.   We let x represent the length of the tennis court and y represent its width as shown in Figure 7-41.

Simon Bruty/Sports Illustrated/Getty Images
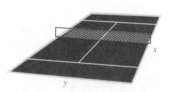

Figure 7-41

Because the area of the tennis court (2,106 square feet) is the product of its length and width, we can form the equation $xy = 2{,}106$. Because the perimeter of the court (210 feet) is the sum of twice its length and twice its width, we can write the equation $2x + 2y = 210$, or $x + y = 105$. This gives the following system of equations that models the problem.

$$\begin{cases} xy = 2{,}106 & \text{This equation is nonlinear.} \\ x + y = 105 & \text{This equation is linear.} \end{cases}$$

Sophie Germain
(1776–1831)

Sophie Germain was 13 years old during the French Revolution. Because of dangers caused by the insurrection in Paris, she was kept indoors and spent most of her time reading about mathematics in her father's library. Since interest in mathematics was considered inappropriate for a woman at that time, much of her work was written under the pen name of M. LeBlanc.

Step 2: Use substitution to solve the system.   First we solve the linear equation for y.

$$x + y = 105$$

$$y = -x + 105 \quad \text{This is the substitution equation.}$$

We now can substitute $-x + 105$ for y in the nonlinear equation and solve the resulting equation for x:

$$xy = 2{,}106$$

$$x(-x + 105) = 2{,}106 \quad \text{Substitute } -x + 105 \text{ for } y.$$

$$-x^2 + 105x = 2{,}106 \quad \text{Remove parentheses.}$$

$$-x^2 + 105x - 2{,}106 = 0 \quad \text{Subtract 2,106 from both sides.}$$

$$x^2 - 105x + 2{,}106 = 0 \quad \text{Multiply both sides by } -1.$$

$$(x - 78)(x - 27) = 0 \quad \text{Factor } x^2 - 105x + 2{,}106.$$

$$x - 78 = 0 \quad \text{or} \quad x - 27 = 0$$

$$x = 78 \quad \quad \quad x = 27$$

Because $y = -x + 105$, if $x = 78$, then $y = 27$; and if $x = 27$, then $y = 78$. The two solutions are

$$\begin{cases} x = 78 \\ y = 27 \end{cases} \quad \text{and} \quad \begin{cases} x = 27 \\ y = 78 \end{cases}$$

The dimensions of the tennis court are 78 feet by 27 feet.

Self Check 5   If the area of a tennis court for doubles matches is 2,808 square feet and the perimeter is 228 feet, find the dimensions of the court.

Self Check Answers

1. 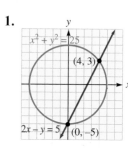 $\begin{cases} x = 4 \\ y = 3 \end{cases}$ and $\begin{cases} x = 0 \\ y = -5 \end{cases}$

2. $\begin{cases} x = 4 \\ y = 3 \end{cases}$ and $\begin{cases} x = 0 \\ y = -5 \end{cases}$

3. $\left(2, \sqrt{3}\right)$ and $\left(2, -\sqrt{3}\right)$

4. $\left(3, \sqrt{5}\right), \left(3, -\sqrt{5}\right),$ $\left(-3, \sqrt{5}\right), \left(-3, -\sqrt{5}\right)$

5. 78 ft by 36 ft

7.4   **Exercises**

Vocabulary and Concepts   *Fill in the blanks.*

1. Solutions of nonlinear systems of equations are the points of intersection of the _____ of conic sections.

2. Approximate solutions of nonlinear systems can be found _____, and exact solutions can be found algebraically using the methods of _____ or _____.

Practice   *Solve each system of equations by graphing.*

3. $\begin{cases} 8x^2 + 32y^2 = 256 \\ x = 2y \end{cases}$

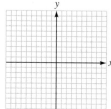

4. $\begin{cases} x^2 + y^2 = 2 \\ x + y = 2 \end{cases}$

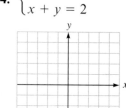

5. $\begin{cases} x^2 + y^2 = 90 \\ y = x^2 \end{cases}$

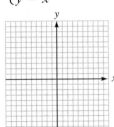

6. $\begin{cases} x^2 + y^2 = 5 \\ x + y = 3 \end{cases}$

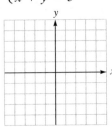

Solve each system of equations using substitution or addition for real values of x and y.

17. $\begin{cases} 25x^2 + 9y^2 = 225 \\ 5x + 3y = 15 \end{cases}$

18. $\begin{cases} x^2 + y^2 = 20 \\ y = x^2 \end{cases}$

7. $\begin{cases} x^2 + y^2 = 25 \\ 12x^2 + 64y^2 = 768 \end{cases}$

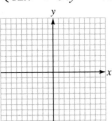

8. $\begin{cases} x^2 + y^2 = 13 \\ y = x^2 - 1 \end{cases}$

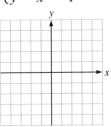

19. $\begin{cases} x^2 + y^2 = 2 \\ x + y = 2 \end{cases}$

20. $\begin{cases} x^2 + y^2 = 36 \\ 49x^2 + 36y^2 = 1{,}764 \end{cases}$

21. $\begin{cases} x^2 + y^2 = 5 \\ x + y = 3 \end{cases}$

22. $\begin{cases} x^2 - x - y = 2 \\ 4x - 3y = 0 \end{cases}$

9. $\begin{cases} x^2 - 13 = -y^2 \\ y = 2x - 4 \end{cases}$

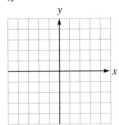

10. $\begin{cases} x^2 + y^2 = 20 \\ y = x^2 \end{cases}$

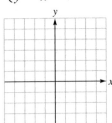

23. $\begin{cases} x^2 + y^2 = 13 \\ y = x^2 - 1 \end{cases}$

24. $\begin{cases} x^2 + y^2 = 25 \\ 2x^2 - 3y^2 = 5 \end{cases}$

25. $\begin{cases} x^2 + y^2 = 30 \\ y = x^2 \end{cases}$

26. $\begin{cases} 9x^2 - 7y^2 = 81 \\ x^2 + y^2 = 9 \end{cases}$

27. $\begin{cases} x^2 + y^2 = 13 \\ x^2 - y^2 = 5 \end{cases}$

28. $\begin{cases} 2x^2 + y^2 = 6 \\ x^2 - y^2 = 3 \end{cases}$

29. $\begin{cases} x^2 + y^2 = 20 \\ x^2 - y^2 = -12 \end{cases}$

30. $\begin{cases} xy = -\dfrac{9}{2} \\ 3x + 2y = 6 \end{cases}$

11. $\begin{cases} x^2 - 6x - y = -5 \\ x^2 - 6x + y = -5 \end{cases}$

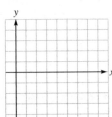

12. $\begin{cases} x^2 - y^2 = -5 \\ 3x^2 + 2y^2 = 30 \end{cases}$

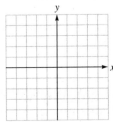

31. $\begin{cases} y^2 = 40 - x^2 \\ y = x^2 - 10 \end{cases}$

32. $\begin{cases} x^2 - 6x - y = -5 \\ x^2 - 6x + y = -5 \end{cases}$

33. $\begin{cases} y = x^2 - 4 \\ x^2 - y^2 = -16 \end{cases}$

34. $\begin{cases} 6x^2 + 8y^2 = 182 \\ 8x^2 - 3y^2 = 24 \end{cases}$

35. $\begin{cases} x^2 - y^2 = -5 \\ 3x^2 + 2y^2 = 30 \end{cases}$

36. $\begin{cases} \dfrac{1}{x} + \dfrac{1}{y} = 5 \\ \dfrac{1}{x} - \dfrac{1}{y} = -3 \end{cases}$

📟 *Use a graphing calculator to solve each system of equations.*

13. $\begin{cases} y = x + 1 \\ y = x^2 + x \end{cases}$

14. $\begin{cases} y = 6 - x^2 \\ y = x^2 - x \end{cases}$

15. $\begin{cases} 6x^2 + 9y^2 = 10 \\ 3y - 2x = 0 \end{cases}$

16. $\begin{cases} x^2 + y^2 = 68 \\ y^2 - 3x^2 = 4 \end{cases}$

37. $\begin{cases} \dfrac{1}{x} + \dfrac{2}{y} = 1 \\ \dfrac{2}{x} - \dfrac{1}{y} = \dfrac{1}{3} \end{cases}$

38. $\begin{cases} \dfrac{1}{x} + \dfrac{3}{y} = 4 \\ \dfrac{2}{x} - \dfrac{1}{y} = 7 \end{cases}$

39. $\begin{cases} 3y^2 = xy \\ 2x^2 + xy - 84 = 0 \end{cases}$ **40.** $\begin{cases} x^2 + y^2 = 10 \\ 2x^2 - 3y^2 = 5 \end{cases}$

41. $\begin{cases} xy = \dfrac{1}{6} \\ y + x = 5xy \end{cases}$ **42.** $\begin{cases} xy = \dfrac{1}{12} \\ y + x = 7xy \end{cases}$

Applications

43. Geometry   The area of a rectangle is 63 square centimeters, and its perimeter is 32 centimeters. Find the dimensions of the rectangle.

44. Dimensions of a whiteboard   The area of a SMART Board interactive whiteboard is 2,880 square inches, and its perimeter is 216 inches. Find the dimensions of the whiteboard.

45. Fencing pastures   The rectangular pasture shown below is to be fenced in along a riverbank. If 260 feet of fencing is to enclose an area of 8,000 square feet, find the dimensions of the pasture.

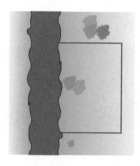

46. Investments   Grant receives $225 annual income from one investment. Jeff invested $500 more than Grant, but at an annual rate of 1% less. Jeff's annual income is $240. Find the amount and rate of Grant's investment.

47. Investments   Carol receives $67.50 annual income from one investment. John invested $150 more than Carol at an annual rate of $1\frac{1}{2}\%$ more. John's annual income is $94.50. Find the amount and rate of Carol's investment.   (*Hint:* There are two answers.)

48. Finding the rate and time   Jim drove 306 miles. Jim's brother made the same trip at a speed 17 miles per hour slower than Jim did and required an extra $1\frac{1}{2}$ hours. Find Jim's rate and time.

49. Paintball   See the illustration. A liquid-filled paintball is shot from the base of an incline and follows the parabolic path $y = -\frac{1}{300}x^2 + \frac{1}{5}x$, with distances measured in feet. The incline has a slope of $\frac{1}{10}$. Find the coordinates of the point of intersection of the paintball and the ground at impact.

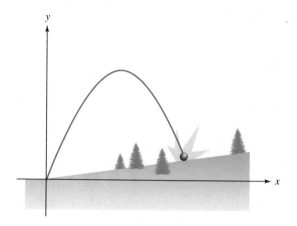

50. Artillery   See the illustration for Exercise 49. A shell fired from the base of a hill follows the parabolic path $y = -\frac{1}{6}x^2 + 2x$, with distances measured in miles. The hill has a slope of $\frac{1}{3}$. How far from the cannon is the point of impact? (*Hint:* Find the coordinates of the point and then the distance.)

51. Air traffic control   A plane is flying over an airport on a path whose equation is $y = x^2$. If a second plane, flying at the same altitude, is traveling on a path whose equation is $x + y = 2$. Is there any danger of a mid-air collision?

52. Ship traffic   One ship is steaming on a path whose equation is $y = x^2 + 1$ and another is steaming on a path whose equation is $x + y = -4$. Is there any danger of collision?

53. Radio reception   A radio station located 120 miles due east of Collinsville has a listening radius of 100 miles. A straight road joins Collinsville with Harmony, a town 200 miles to the east and 100 miles north. See the illustration on the next page. If a driver leaves Collinsville and heads toward Harmony, how far from Collinsville will the driver pick up the station?

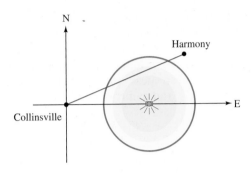

Collinsville

Harmony

N

E

54. Listening ranges   For how many miles will the driver in Exercise 53 continue to receive the signal?

Discovery and Writing

55. Is it possible for a system of second-degree equations to have no common solution? If so, sketch the graphs of the equations of such a system.

56. Can a system have exactly one solution? If so, sketch the graphs of the equations of such a system.

57. Can a system have exactly two solutions? If so, sketch the graphs.

58. Can a system have three solutions? If so, sketch the graphs.

59. Can it have exactly four solutions? If so, sketch the graphs.

60. Can a system have more than four solutions? If so, sketch the graphs.

Review   *Find the vertical, horizontal, or slant asymptotes of each of the following rational functions.*

61. $y = \dfrac{3x + 1}{x - 1}$

62. $y = \dfrac{x^2 + 3}{x - 1}$

63. $y = \dfrac{3x + 1}{x^2 - 1}$

64. $y = \dfrac{x^2 + 3}{x^2 + 1}$

Determine whether the graph of each function has x-axis, y-axis, or origin symmetry, or none of these symmetries.

65. $f(x) = \dfrac{3x^2 + 5}{5x^2 + 3}$

66. $f(x) = \dfrac{3x^3 + 5}{5x^2 + 3}$

67. $f(x) = \dfrac{3x^3}{5x^2 + 3}$

68. $f(x) = \dfrac{3x^4 + 8}{2x^2 - 5}$

CHAPTER REVIEW

7.1   The Circle and the Parabola

Definitions and Concepts	Examples
Standard equation of a circle with center at (0, 0) and radius r: $$x^2 + y^2 = r^2$$	The standard equation of the circle with center $(0, 0)$ and radius 5 is: $x^2 + y^2 = 5^2$ or $x^2 + y^2 = 25$. The graph of the circle is: 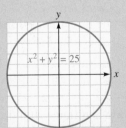
Standard equation of a circle with center at (h, k) and radius r: $$(x - h)^2 + (y - k)^2 = r^2$$	The standard equation of the circle with center $(4, 6)$ and radius 3 is: $$(x - 4)^2 + (y - 6)^2 = 9$$

General form of a second-degree equation in x and y:

$$Ax^2 + Bxy + Cy^2 + Dx + Ey + F = 0$$

The second-degree equation is a circle if $A = C$ and $B = 0$.

The general form of an equation of a circle is $x^2 + y^2 + 4x - 10y + 25 = 0$. To write the equation in standard form, we first subtract 25 from both sides to obtain

$$x^2 + y^2 + 4x - 10y = -25$$

and then complete the square on x and y.

$$x^2 + 4x + y^2 - 10y = -25 \quad \text{Rearrange terms.}$$

To complete the square, we add 4 and 25 to both sides.

$$x^2 + 4x + 4 + y^2 - 10y + 25 = -25 + 4 + 25$$

We can factor on the left side and simplify on the right side to get

$$(x + 2)^2 + (y - 5)^2 = 4$$

The circle is now written in standard form. From the equation, we can see that the coordinates of the center of the circle are $(-2, 5)$ and that the radius is 2.

Parabolas:

A **parabola** is the set of all points in a plane equidistant from a line l (called the **directrix**) and fixed point F (called the **focus**) that is not on line l.

Parabola opening	Vertex at origin
Right	$y^2 = 4px \quad (p > 0)$
Left	$y^2 = 4px \quad (p < 0)$
Up	$x^2 = 4py \quad (p > 0)$
Down	$x^2 = 4py \quad (p < 0)$

For a parabola that opens right or left with vertex at the origin, the directrix is $x = -p$ and the focus is $(p, 0)$.

For a parabola that opens up or down with vertex at the origin, the directrix is $y = -p$ and the focus is $(0, p)$.

Parabola opening	Vertex at $V(h, k)$
Right	$(y - k)^2 = 4p(x - h) \quad (p > 0)$
Left	$(y - k)^2 = 4p(x - h) \quad (p < 0)$
Up	$(x - h)^2 = 4p(y - k) \quad (p > 0)$
Down	$(x - h)^2 = 4p(y - k) \quad (p < 0)$

For a parabola that opens right or left with vertex at (h, k), the directrix is $x = -p + h$ and the focus is $(h + p, k)$.

For a parabola that opens up or down with vertex at (h, k), the directrix is $y = -p + k$ and the focus is $(h, k + p)$.

Find the equation of the parabola with vertex at the origin and focus at $(4, 0)$.

Because the focus is to the right of the vertex, the parabola opens to the right, and because the vertex is the origin, the standard equation is $y^2 = 4px$. Since the distance between the focus and the vertex is $p = 4$, we can substitute 4 for p in the standard equation to get

$$y^2 = 4px$$
$$y^2 = 4(4)x$$
$$y^2 = 16x$$

The graph of the parabola is shown in the figure.

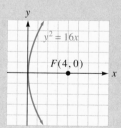

The ellipse with the major axis horizontal and center at (h, k):
The standard equation of an ellipse with center at (h, k) and major axis horizontal is

$$\frac{(x - h)^2}{a^2} + \frac{(y - k)^2}{b^2} = 1 \qquad \text{where } a > b > 0$$

Vertices (ends of the major axis): $V(a + h, k)$ and $V'(-a + h, k)$

Ends of the minor axis: $B(h, b + k)$ and $B'(h, -b + k)$

Foci: $F(h + c, k)$ and $F'(h - c, k)$ where $c^2 = a^2 - b^2$

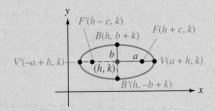

The ellipse with the major axis vertical and center at (h, k):
The standard equation of an ellipse with center at (h, k) and major axis vertical is

$$\frac{(x - h)^2}{b^2} + \frac{(y - k)^2}{a^2} = 1 \qquad \text{where } a > b > 0$$

Vertices (ends of the major axis): $V(h, a + k)$ and $V'(h, -a + k)$

Ends of the minor axis: $B(b + h, k)$ and $B'(-b + h, k)$

Foci: $F(h, k + c)$ and $F'(h, k - c)$ where $c^2 = a^2 - b^2$

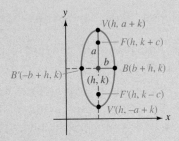

The major axis of the ellipse is vertical and the center of the ellipse is $(1, -2)$. The graph of the ellipse is shown below.

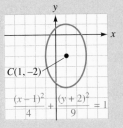

$$\frac{(x - 1)^2}{4} + \frac{(y + 2)^2}{9} = 1$$

Exercises

14. Write the equation of the ellipse with center at the origin, major axis that is horizontal and 12 units long, and minor axis 8 units long.

15. Write the equation of the ellipse with center at the origin, major axis that is vertical and 10 units long, and minor axis 4 units long.

16. Write the equation of the ellipse with center at point $(-2, 3)$ and curve passing through points $(-2, 0)$ and $(2, 3)$.

17. Write the equation in standard form and graph it.
$$4x^2 + y^2 - 16x + 2y = -13$$

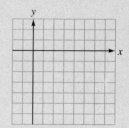

<table>
<tr><td colspan="2" style="background:black;color:white">7.3</td><td colspan="2">The Hyperbola</td></tr>
</table>

Definitions and Concepts	**Examples**

Definitions and Concepts

An **hyperbola** is the set of all points P in a plane such that the absolute value of the difference of the distances from point P to two other points in the plane is a positive constant.

Hyperbola with the foci on the x-axis and center at (0, 0):
The standard equation of an hyperbola with center at the origin and foci on the x-axis is

$$\frac{x^2}{a^2} - \frac{y^2}{b^2} = 1 \quad \text{where } a^2 + b^2 = c^2$$

Vertices: $V(a, 0)$ and $V'(-a, 0)$

Foci: $F(c, 0)$ and $F'(-c, 0)$

Asymptotes: $y = \pm \frac{b}{a}x$

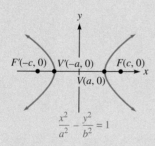

Examples

Write the equation of the hyperbola with vertices $V(3, 0)$ and $V'(-3, 0)$ and a focus at $F(5, 0)$.

Because the foci lie on the x-axis, we use the standard equation

$$\frac{x^2}{a^2} - \frac{y^2}{b^2} = 1$$

We will find a^2 and b^2 and substitute the results into the standard equation.

The center of the hyperbola is midway between the vertices V and V'. Thus, the center is the origin $(0, 0)$. The distance between the vertex and the center is $a = 3$, and the distance between the focus and the center is $c = 5$. We can find b^2 by substituting 3 for a and 5 for c in the following equation to get

$$b^2 = c^2 - a^2 \quad \text{In an hyperbola, } b^2 = c^2 - a^2.$$
$$b^2 = 5^2 - 3^2$$
$$b^2 = 16$$

Substituting the values for a^2 and b^2 in the standard equation gives the equation of the hyperbola:

$$\frac{x^2}{a^2} - \frac{y^2}{b^2} = 1$$
$$\frac{x^2}{9} - \frac{y^2}{16} = 1$$

Hyperbola with the foci on the y-axis and center at $(0, 0)$:

The standard equation of an hyperbola with center at the origin and foci on the y-axis is

$$\frac{y^2}{a^2} - \frac{x^2}{b^2} = 1 \qquad \text{where } a^2 + b^2 = c^2$$

Vertices: $V(0, a)$ and $V'(0, -a)$

Foci: $F(0, c)$ and $F'(0, -c)$

Asymptotes: $y = \pm\frac{a}{b}x$

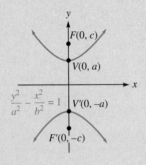

Hyperbola with the transverse axis horizontal and center at (h, k):

The standard equation of an hyperbola with center at (h, k) and foci on a line parallel to the x-axis is

$$\frac{(x - h)^2}{a^2} - \frac{(y - k)^2}{b^2} = 1 \qquad \text{where } a^2 + b^2 = c^2$$

Vertices: $V(a + h, k)$ and $V'(-a + h, k)$

Foci: $F(c + h, k)$ and $F'(-c + h, k)$

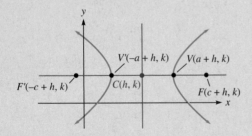

Graph the hyperbola: $\dfrac{y^2}{4} - \dfrac{x^2}{25} = 1$

Step 1: Write the hyperbola in standard form. The equation of the hyperbola $\dfrac{y^2}{4} - \dfrac{x^2}{25} = 1$ is already written in the standard form $\dfrac{y^2}{a^2} - \dfrac{x^2}{b^2} = 1$.

Step 2: Identify the center and a and b. From the standard equation of an hyperbola, we see that the center is $(0, 0)$, that $a = 2$ and $b = 5$, and that the vertices are on the y-axis.

Step 3: Draw the fundamental rectangle and asymptotes. The vertices V and V' are 2 units above and below the origin and have coordinates of $(0, 2)$ and $(0, -2)$. Points B and B' are 5 units right and left of the origin and have coordinates of $(5, 0)$ and $(-5, 0)$. We use the points V, V', B, and B' to construct the fundamental rectangle and extend the diagonals to graph the asymptotes as shown in the figure below.

Step 4: Sketch the hyperbola. We begin with the vertices and sketch the graph of the hyperbola, as shown in the figure below using the asymptotes as guides.

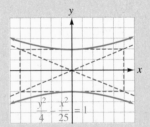

Hyperbola with the transverse axis vertical and center at (h, k):

The standard equation of a hyperbola with center at (h, k) and foci on a line parallel to the y-axis is

$$\frac{(y - k)^2}{a^2} - \frac{(x - h)^2}{b^2} = 1 \qquad \text{where } a^2 + b^2 = c^2$$

Vertices: $V(h, a + k)$ and $V'(h, -a + k)$

Foci: $F(h, c + k)$ and $F'(h, -c + k)$

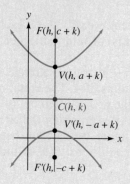

Graphing hyperbolas:

1. Write the hyperbola in standard form.
2. Identify the center and a and b.
3. Draw the fundamental rectangle and asymptotes.
4. Sketch the hyperbola beginning with the vertices and using the asymptotes as guides.

Exercises

18. Write the equation of the hyperbola with center at the origin, passing through points $(-2, 0)$ and $(2, 0)$, and having a focus at $(4, 0)$.

19. Write the equation of the hyperbola with center at the origin, one focus at $(0, 5)$, and one vertex at $(0, 3)$.

20. Write the equation of the hyperbola with vertices at points $(-3, 3)$ and $(3, 3)$ and a focus at point $(5, 3)$.

21. Write the equation of the hyperbola with vertices at points $(3, -3)$ and $(3, 3)$ and a focus at point $(3, 5)$.

22. Write the equation of the asymptotes of the hyperbola $\dfrac{x^2}{25} - \dfrac{y^2}{16} = 1$.

23. Write the equation in standard form and graph it.

$$9x^2 - 4y^2 - 16y - 18x = 43$$

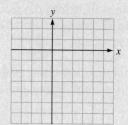

24. Graph: $4xy = 1$

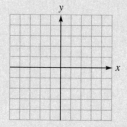

55. Use addition to solve $\begin{cases} x + 2y = -2 \\ 2x - y = 6 \end{cases}$.

56. Use any method to solve $\begin{cases} \dfrac{x}{10} + \dfrac{y}{5} = \dfrac{1}{2} \\ \dfrac{x}{2} - \dfrac{y}{5} = \dfrac{13}{10} \end{cases}$.

57. Evaluate: $\begin{vmatrix} 3 & -2 \\ 1 & -1 \end{vmatrix}$

58. Use Cramer's rule and solve for y only:
$$\begin{cases} 4x - 3y = -1 \\ 3x + 4y = -7 \end{cases}$$

59. Solve: $\begin{cases} x + y + z = 1 \\ 2x - y - z = -4 \\ x - 2y + z = 4 \end{cases}$

60. Solve: $\begin{cases} x + 2y + 3z = 6 \\ 3x + 2y + z = 6 \\ 2x + 3y + z = 6 \end{cases}$

Natural-Number Functions and Probability

8

In this chapter, we introduce a way to expand powers of binomials of the form $(x + y)^n$. This method leads to ideas needed when working with probability and statistics.

Careers and Mathematics

Statistician Statistics is the scientific application of mathematical principles to the collection, analysis, and presentation of numerical data. Statisticians apply their mathematical knowledge to the design of surveys and experiments; collection, processing, and analysis of data; and interpretation of the results.

Statisticians held about 22,000 jobs in 2006. Twenty percent of these were in the federal government. Most of the remaining jobs were in private industry. In addition, many professionals with a background in statistics were among the 54,000 full-time mathematical science faculty in colleges and universities.

Education A master's degree in statistics or mathematics is usually the minimum educational requirement for most statistician jobs. Research and academic positions usually require a Ph.D. in statistics. Beginning positions in industrial research often require a master's degree combined with several years of experience. Jobs with the federal government require at least a bachelor's degree.

Job Outlook Employment of statisticians is expected to increase by 9 percent through 2016. Individuals with a bachelor's or master's degree in statistics with a strong background in an allied field, such as finance, biology, or computer science, should have the best prospects of finding jobs in their field of study.

Median annual earnings of statisticians were $65,720 in 2006.

For a sample application, see Example 6 in Section 8.4. For more information, see www.bls.gov/oco/ocos045.htm.

Dave Martin/AFP/Getty Images

663

8.1 The Binomial Theorem

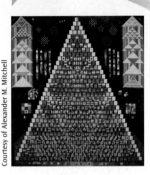

Courtesy of Alexander M. Mitchell

Objectives

1. Use Pascal's Triangle to Expand a Binomial
2. Define and Use Factorial Notation
3. Use the Binomial Theorem to Expand Binomials
4. Find a Specific Term of a Binomial Expansion

We begin this chapter by discussing a triangle, called **Pascal's triangle,** that is named after the French mathematician Blaise Pascal. Although the Western world gives credit for the triangle to Pascal, other mathematicians in other parts of the world studied it centuries before he did. Around 1068, the Indian mathematician Bhattotpala presented the first seventeen rows of the triangle, and about the same time, the great Persian mathematician Omar Khayyám discussed it. Today, in Iran, the triangle is referred to as the Khayyám triangle. Around 1250, the Chinese mathematician Yang Hui developed the triangle, and in China the triangle is known as Yang Hui's triangle.

Over the years, many people have searched the triangle for the many number patterns it contains. One such person, William H. Mitchell, was so impressed with the triangle, he created a needlepoint design of it.

We introduce the triangle by discussing a way to expand binomials of the form $(x + y)^n$, where n is a natural number. Consider the following expansions.

Blaise Pascal
(1623–1662)
Pascal was torn between religion and mathematics. Each surfaced at times in his life to dominate his interest. In mathematics, Pascal made contributions to the study of conic sections, probability, and differential calculus. At the age of 19, he invented a calculating machine. He is best known for a triangular array of numbers that bears his name.

$$(a + b)^0 = 1$$
$$(a + b)^1 = a + b$$
$$(a + b)^2 = a^2 + 2ab + b^2$$
$$(a + b)^3 = a^3 + 3a^2b + 3ab^2 + b^3$$
$$(a + b)^4 = a^4 + 4a^3b + 6a^2b^2 + 4ab^3 + b^4$$
$$(a + b)^5 = a^5 + 5a^4b + 10a^3b^2 + 10a^2b^3 + 5ab^4 + b^5$$
$$(a + b)^6 = a^6 + 6a^5b + 15a^4b^2 + 20a^3b^3 + 15a^2b^4 + 6ab^5 + b^6$$

Four patterns are apparent in the above expansions:

1. Each expansion has one more term than the power of the binomial.
2. The degree of each term in each expansion equals the exponent of the binomial.
3. The first term in each expansion is a raised to the power of the binomial.
4. The exponents on a decrease by 1 in each successive term, and the exponents on b, beginning with b^0 in the first term, increase by 1 in each successive term.

1. Use Pascal's Triangle to Expand a Binomial

To see another pattern, we write the coefficients of the above expansions in a triangular array:

$(a + b)^0 =$							1							Row 0
$(a + b)^1 =$						1		1						Row 1
$(a + b)^2 =$					1		2		1					Row 2
$(a + b)^3 =$				1		3		3		1				Row 3
$(a + b)^4 =$			1		4		6		4		1			Row 4
$(a + b)^5 =$		1		5		**10**		**10**		5		1		Row 5
$(a + b)^6 =$	1		6		15		**20**		15		6		1	Row 6

In this array, each entry other than the 1's is the sum of the closest pair of numbers in the line above it. For example, the first 3 in the third row is the sum of the 1 and 2 above it. The 20 in the sixth row is the sum of the 10's above it.

This array, called **Pascal's triangle** after Blaise Pascal (1623–1662), continues with the same pattern forever. The next two lines are

$$(a + b)^7 = \quad 1 \quad 7 \quad 21 \quad 35 \quad 35 \quad 21 \quad 7 \quad 1 \qquad \text{Row 7}$$
$$(a + b)^8 = 1 \quad 8 \quad 28 \quad 56 \quad 70 \quad 56 \quad 28 \quad 8 \quad 1 \qquad \text{Row 8}$$

EXAMPLE 1   Expand: $(x + y)^6$.

Solution   The first term is x^6, and the exponents on x will decrease by 1 in each successive term. A y will appear in the second term, and the exponents on y will increase by 1 in each successive term, concluding when the term y^6 is reached. The variables in the expansion are

$$x^6 \qquad x^5y \qquad x^4y^2 \qquad x^3y^3 \qquad x^2y^4 \qquad xy^5 \qquad y^6$$

We can use Pascal's triangle to find the coefficients of the variables. Because the binomial is raised to the sixth power, we choose Row 6 of Pascal's triangle. The coefficients of the variables are the numbers in that row.

$$1 \qquad 6 \qquad 15 \qquad 20 \qquad 15 \qquad 6 \qquad 1$$

Putting this information together gives the expansion:

$$(x + y)^6 = x^6 + 6x^5y + 15x^4y^2 + 20x^3y^3 + 15x^2y^4 + 6xy^5 + y^6$$

Self Check 1   Expand: $(p + q)^3$.

EXAMPLE 2   Expand: $(x - y)^6$.

Solution   We write the binomial as $[x + (-y)]^6$ and substitute $-y$ for y in the result of Example 1.

$$[x + (-y)]^6$$
$$= x^6 + 6x^5(-y) + 15x^4(-y)^2 + 20x^3(-y)^3 + 15x^2(-y)^4 + 6x(-y)^5 + (-y)^6$$
$$= x^6 - 6x^5y + 15x^4y^2 - 20x^3y^3 + 15x^2y^4 - 6xy^5 + y^6$$

Comment

In general, the signs in the expansion of $(x - y)^n$ alternate. The sign of the first term is $+$, the sign of the second term is $-$, and so on.

Self Check 2   Expand: $(p - q)^3$.

2.   Define and Use Factorial Notation

To use the binomial theorem to expand a binomial, we will need **factorial notation.**

Factorial Notation   If n is a natural number, the symbol $n!$ (read either as **"n factorial"** or as **"factorial n"**) is defined as

$$n! = n(n - 1)(n - 2)(n - 3) \cdot \, \cdots \, \cdot (3)(2)(1)$$

EXAMPLE 3 Evaluate: **a.** 3! **b.** 6! **c.** 10!

Solution We will apply the definition of factorial notation.

a. $3! = 3 \cdot 2 \cdot 1 = 6$
b. $6! = 6 \cdot 5 \cdot 4 \cdot 3 \cdot 2 \cdot 1 = 720$
c. $10! = 10 \cdot 9 \cdot 8 \cdot 7 \cdot 6 \cdot 5 \cdot 4 \cdot 3 \cdot 2 \cdot 1 = 3,628,800$

Self Check 3 Evaluate: **a.** 4! **b.** 7!

There are two fundamental properties of factorials.

Properties of Factorials **1.** By definition, $0! = 1$.
2. If n is a natural number, $n(n - 1)! = n!$.

EXAMPLE 4 Show that: **a.** $6 \cdot 5! = 6!$ **b.** $8 \cdot 7! = 8!$

Solution We will apply the definition of factorial notation.

a. $6 \cdot 5! = 6(5 \cdot 4 \cdot 3 \cdot 2 \cdot 1)$ **b.** $8 \cdot 7! = 8(7 \cdot 6 \cdot 5 \cdot 4 \cdot 3 \cdot 2 \cdot 1)$
$= 6 \cdot 5 \cdot 4 \cdot 3 \cdot 2 \cdot 1$ $= 8 \cdot 7 \cdot 6 \cdot 5 \cdot 4 \cdot 3 \cdot 2 \cdot 1$
$= 6!$ $= 8!$

Self Check 4 Show that: $4 \cdot 3! = 4!$

3. Use the Binomial Theorem to Expand Binomials

We can now state the binomial theorem.

The Binomial Theorem If n is any positive number, then

$$(a + b)^n = a^n + \frac{n!}{1!(n - 1)!}a^{n-1}b + \frac{n!}{2!(n - 2)!}a^{n-2}b^2$$

$$+ \frac{n!}{3!(n - 3)!}a^{n-3}b^3 + \cdots + \frac{n!}{r!(n - r)!}a^{n-r}b^r + \cdots + b^n$$

Comment

In the expansion of $(a + b)^n$, the term containing b^r is given by

$$\frac{n!}{r!(n - r)!}a^{n-r}b^r$$

A proof of the binomial theorem appears in Appendix I.

In the binomial theorem, the exponents on the variables in each term on the right side follow the familiar patterns:

1. The sum of the exponents on a and b in each term is n.

2. The exponents on a decrease by 1 in each successive term.

3. The exponents on b increase by 1 in each successive term.

However, the method of finding the coefficients is different. Except for the first and last terms, $n!$ is the numerator of each fractional coefficient. When the exponent on b is 2, the factors in the denominator of the fractional coefficient are 2! and $(n - 2)!$. When the exponent on b is 3, the factors in the denominator are 3! and $(n - 3)!$. When the exponent on b is r, the factors in the denominator are $r!$ and $(n - r)!$.

EXAMPLE 5   Use the binomial theorem to expand $(a + b)^5$.

Solution   We substitute directly into the binomial theorem.

$$(a + b)^5 = a^5 + \frac{5!}{1!(5 - 1)!}a^4b + \frac{5!}{2!(5 - 2)!}a^3b^2 + \frac{5!}{3!(5 - 3)!}a^2b^3 + \frac{5!}{4!(5 - 4)!}ab^4 + b^5$$

$$= a^5 + \frac{5 \cdot 4!}{1 \cdot 4!}a^4b + \frac{5 \cdot 4 \cdot 3!}{2 \cdot 1 \cdot 3!}a^3b^2 + \frac{5 \cdot 4 \cdot 3!}{3! \cdot 2 \cdot 1}a^2b^3 + \frac{5 \cdot 4!}{4! \cdot 1}ab^4 + b^5$$

$$= a^5 + 5a^4b + 10a^3b^2 + 10a^2b^3 + 5ab^4 + b^5$$

We note that the coefficients are the same numbers as in row 5 of Pascal's triangle.

Self Check 5   Use the binomial theorem to expand $(p + q)^4$.

EXAMPLE 6   Use the binomial theorem to expand $(2x - 3y)^4$.

Solution   We first find the expansion of $(a + b)^4$.

$$(a + b)^4 = a^4 + \frac{4!}{1!(4 - 1)!}a^3b + \frac{4!}{2!(4 - 2)!}a^2b^2 + \frac{4!}{3!(4 - 3)!}ab^3 + b^4$$

$$= a^4 + \frac{4 \cdot 3!}{1 \cdot 3!}a^3b + \frac{4 \cdot 3 \cdot 2!}{2 \cdot 1 \cdot 2!}a^2b^2 + \frac{4 \cdot 3!}{3! \cdot 1}ab^3 + b^4$$

$$= a^4 + 4a^3b + 6a^2b^2 + 4ab^3 + b^4$$

We then substitute $2x$ for a and $-3y$ for b in the result.

$$(a + b)^4 = a^4 + 4a^3b + 6a^2b^2 + 4ab^3 + b^4$$
$$[2x + (-3y)]^4 = (2x)^4 + 4(2x)^3(-3y) + 6(2x)^2(-3y)^2 + 4(2x)(-3y)^3 + (-3y)^4$$
$$(2x - 3y)^4 = 16x^4 - 96x^3y + 216x^2y^2 - 216xy^3 + 81y^4$$

Self Check 6   Use the binomial theorem to expand $(3x - 2y)^4$.

4.   Find a Specific Term of a Binomial Expansion

Suppose that we want to find the fifth term of the expansion of $(a + b)^{11}$. It would be tedious to raise the binomial to the 11th power and then look at the fifth term. The binomial theorem provides an easier way.

EXAMPLE 7   Find the fifth term of the expansion of $(a + b)^{11}$.

Solution   In the fifth term, the exponent on b is 4 (the exponent on b is always 1 less than the number of the term). Since the exponent on b added to the exponent on a equals 11, the exponent on a is 7. The variables of the fifth term are a^7b^4.

The number in the numerator of the fractional coefficient is $n!$, which in this case is 11!. The factors in the denominator are 4! and $(11 - 4)!$. The complete fifth term is

$$\frac{11!}{4!(11 - 4)!}a^7b^4 = \frac{11!}{4!7!}a^7b^4$$

$$= \frac{11 \cdot 10 \cdot 9 \cdot 8 \cdot 7!}{4 \cdot 3 \cdot 2 \cdot 1 \cdot 7!}a^7b^4$$

$$= 330a^7b^4$$

Comment

We can also use the formula for the term containing b^r, where $r = 4$ and $n = 11$.

$$\frac{n!}{r!(n-r)!}a^{n-r}b^r = \frac{11!}{4!(11-4)!}a^{11-4}b^4$$
$$= \frac{11!}{4!7!}a^7b^4$$
$$= 330a^7b^4$$

Self Check 7   Find the sixth term of the expansion in Example 7.

EXAMPLE 8   Find the sixth term of the expansion of $(a + b)^9$.

Solution   In the sixth term, the exponent on b is 5, and the exponent on a is $9 - 5$, or 4. The numerator of the fractional coefficient is 9!, and the factors in the denominator are 5! and $(9 - 5)!$. The sixth term of the expansion is

$$\frac{9!}{5!(9-5)!}a^4b^5 = \frac{9 \cdot 8 \cdot 7 \cdot 6 \cdot 5!}{5! \cdot 4!}a^4b^5$$
$$= \frac{9 \cdot 8 \cdot 7 \cdot 6}{4 \cdot 3 \cdot 2 \cdot 1}a^4b^5$$
$$= 126a^4b^5$$

Comment

We can also use the formula for the term containing b^r, where $r = 5$ and $n = 9$.

$$\frac{n!}{r!(n-r)!}a^{n-r}b^r = \frac{9!}{5!(9-5)!}a^{9-5}b^5$$
$$= \frac{9!}{5!4!}a^4b^5$$
$$= 126a^4b^5$$

Self Check 8   Find the fifth term of the expansion in Example 8.

EXAMPLE 9   Find the third term of the expansion of $(3x - 2y)^6$.

Solution   We begin by finding the third term of the expansion of $(a + b)^6$.

$$(1) \qquad \frac{6!}{2!(6-2)!}a^4b^2 = \frac{6 \cdot 5 \cdot 4!}{2 \cdot 1 \cdot 4!}a^4b^2 = 15a^4b^2$$

We then can substitute $3x$ for a and $-2y$ for b in Equation 1 to obtain the third term of the expansion of $(3x - 2y)^6$.

$$15a^4b^2 = 15(3x)^4(-2y)^2$$
$$= 15(3)^4(-2)^2x^4y^2$$
$$= 4{,}860x^4y^2$$

Self Check 9   Find the fourth term of the expansion in Example 9.

Self Check Answers   **1.** $p^3 + 3p^2q + 3pq^2 + q^3$ **2.** $p^3 - 3p^2q + 3pq^2 - q^3$ **3. a.** 24   **b.** 5,040
4. $4 \cdot (3 \cdot 2 \cdot 1) = 4!$ **5.** $p^4 + 4p^3q + 6p^2q^2 + 4pq^3 + q^4$
6. $81x^4 - 216x^3y + 216x^2y^2 - 96xy^3 + 16y^4$ **7.** $462a^6b^5$ **8.** $126a^5b^4$
9. $-4{,}320x^3y^3$

8.1 Exercises

Vocabulary and Concepts *Fill in the blanks.*

1. In the expansion of a binomial, there will be one more term than the _____ of the binomial.

2. The _____ of each term in a binomial expansion is the same as the exponent of the binomial.

3. The _____ term in a binomial expansion is the first term raised to the power of the binomial.

4. In the expansion of $(p + q)^n$, the _____ on p decrease by 1 in each successive term.

5. Expand 7!: _____

6. $0! =$ __

7. $n \cdot$ _____ $= n!$

8. In the seventh term of $(a + b)^{11}$, the exponent on a is __.

Practice *Evaluate each expression.*

9. $4!$
10. $-5!$
11. $3! \cdot 6!$
12. $0! \cdot 7!$
13. $6! + 6!$
14. $5! - 2!$
15. $\dfrac{9!}{12!}$
16. $\dfrac{8!}{5!}$
17. $\dfrac{5! \cdot 7!}{9!}$
18. $\dfrac{3! \cdot 5! \cdot 7!}{1!8!}$
19. $\dfrac{18!}{6!(18 - 6)!}$
20. $\dfrac{15!}{9!(15 - 9)!}$

Use Pascal's triangle to expand each binomial.

21. $(a + b)^5$
22. $(a + b)^7$

23. $(x - y)^3$
24. $(x - y)^7$

Use the binomial theorem to expand each binomial.

25. $(a + b)^3$
26. $(a + b)^4$
27. $(a - b)^5$
28. $(x - y)^4$
29. $(2x + y)^3$
30. $(x + 2y)^3$
31. $(x - 2y)^3$
32. $(2x - y)^3$
33. $(2x + 3y)^4$
34. $(2x - 3y)^4$
35. $(x - 2y)^4$
36. $(x + 2y)^4$
37. $(x - 3y)^5$

38. $(3x - y)^5$

39. $\left(\dfrac{x}{2} + y\right)^4$

40. $\left(x + \dfrac{y}{2}\right)^4$

Find the required term in each binomial expansion.

41. $(a + b)^4$; 3rd term
42. $(a - b)^4$; 2nd term

43. $(a + b)^7$; 5th term
44. $(a + b)^5$; 4th term

45. $(a - b)^5$; 6th term
46. $(a - b)^8$; 7th term

47. $(a + b)^{17}$; 5th term
48. $(a - b)^{12}$; 3rd term

49. $\left(a - \sqrt{2}\right)^4$; 2nd term
50. $\left(a - \sqrt{3}\right)^8$; 3rd term

51. $\left(a + \sqrt{3}b\right)^9$; 5th term
52. $\left(\sqrt{2}a - b\right)^7$; 4th term

53. $\left(\dfrac{x}{2} + y\right)^4$; 3rd term
54. $\left(m + \dfrac{n}{2}\right)^8$; 3rd term

55. $\left(\dfrac{r}{2} - \dfrac{s}{2}\right)^{11}$; 10th term
56. $\left(\dfrac{p}{2} - \dfrac{q}{2}\right)^9$; 6th term

57. $(a + b)^n$; 4th term
58. $(a - b)^n$; 5th term

59. $(a + b)^n$; rth term

60. $(a + b)^n$; $(r + 1)$th term

Discovery and Writing

61. Find the sum of the numbers in each row of the first ten rows of Pascal's triangle. Do you see a pattern?

62. Show that the sum of the coefficients in the binomial expansion of $(x + y)^n$ is 2^n. (*Hint:* Let $x = y = 1$.)

63. Find the constant term in the expansion of

$$\left(a - \frac{1}{a}\right)^{10}$$

64. Find the coefficient of x^5 in the expansion of

$$\left(x + \frac{1}{x}\right)^9$$

65. If we applied the pattern of coefficients to the coefficient of the first term in the binomial theorem, it would be $\frac{n!}{0!(n - 0)!}$. Show that this expression equals 1.

66. If we applied the pattern of coefficients to the coefficient of the last term in the binomial theorem, it would be $\frac{n!}{n!(n - n)!}$. Show that this expression equals 1.

67. Define factorial notation and explain how to evaluate 10!.

68. With a calculator, evaluate 69!. Explain why you cannot find 70! with a calculator.

69. Explain how the rth term of a binomial expansion is constructed.

70. Explain the four patterns apparent in a binomial expansion.

Review *Factor each expression.*

71. $3x^3y^2z^4 - 6xyz^5 + 15x^2yz^2$

72. $3z^2 - 15tz + 12t^2$

73. $a^4 - b^4$

74. $3r^4 - 36r^3 - 135r^2$

Simplify each complex fraction.

75. $\dfrac{\dfrac{1}{x} + \dfrac{1}{3}}{\dfrac{1}{x} - \dfrac{1}{3}}$

76. $\dfrac{x - \dfrac{1}{y}}{y - \dfrac{1}{x}}$

8.2 Sequences, Series, and Summation Notation

Objectives

1. Define and Use Sequences
2. Define a Sequence Recursively
3. Define Series
4. Define and Use Summation Notation

In this section, we will introduce a function whose domain is the set of natural numbers. This function, called a **sequence,** is a list of numbers in a specific order. Sequences can be seen all around us. For example, we can unlock a digital lock by entering the correct sequence of numbers, a pilot performs the same sequence of checks before taking off, and our lives are often changed by a sequence of events. We even use sequences as entertainment when we play the game of Sequence.

1. Define and Use Sequences

The meaning of the word *sequence* in mathematics is very similar to the meaning of the word in everyday English. It refers to a list of events in a particular order.

Sequences An **infinite sequence** is a function whose domain is the set of natural numbers.

A **finite sequence** is a function whose domain is the set of the first n natural numbers.

Since a sequence is a function whose domain is the set of natural numbers, we can write its terms as a list of numbers. For example, if n is a natural number, the function defined by $f(n) = 2n - 1$ generates the following infinite sequence

$$f(1) = 2(1) - 1 = 1$$
$$f(2) = 2(2) - 1 = 3$$
$$f(3) = 2(3) - 1 = 5$$
$$f(n) = 2(n) - 1 = 2n - 1$$

Writing the results horizontally, we have

$$1, 3, 5, \ldots, 2n - 1, \ldots$$

The number 1 is the first term, 3 is the second term, 5 is the third term, and $2n - 1$ is the **general,** or **nth term.**

If n is a natural number, the function $f(n) = 3n^2 + 1$ generates the infinite sequence

$$4, 13, 28, \ldots, 3n^2 + 1, \ldots$$

The number 4 is the first term, 13 is the second term, 28 is the third term, and $3n^2 + 1$ is the general term.

If n is one of the first four natural numbers, the function defined by $f(n) = n^2 + 1$ generates the following finite sequence

$$2, 5, 10, 17$$

A constant function such as $g(n) = 1$ is a sequence, because it generates the infinite sequence

$$1, 1, 1, \ldots$$

Comment

We seldom use function notation to denote a sequence, because it is often difficult or even impossible to write the general term. In such cases, if there is a pattern that is assumed to be continued, we simply list several terms of the sequence.

Some additional examples of infinite sequences are:

$$1^2, 2^2, 3^2, \ldots, n^2, \ldots$$
$$3, 9, 19, 33, \ldots, 2n^2 + 1, \ldots$$
$$1, 3, 6, 10, 15, 21, \ldots, \frac{n(n + 1)}{2}, \ldots$$
$$2, 3, 5, 7, 11, 13, 17, 19, 23, \ldots \quad \textbf{(prime numbers)}$$
$$1, 1, 2, 3, 5, 8, 13, 21, \ldots \quad \textbf{(Fibonacci sequence)}$$

The **Fibonacci sequence** is named after the 12th-century mathematician Leonardo of Pisa, also known as Fibonacci. After the two 1's in the Fibonacci sequence, each term is the sum of the two terms that immediately precede it. The Fibonacci sequence occurs in many fields, such as music, the growth patterns of plants, and the reproductive habits of bees. For example, in one octave of a piano

Leonardo Fibonacci
(late 12th and early 13th centuries)
Fibonacci, an Italian mathematician, is also known as Leonardo da Pisa. In his work *Liber abaci,* he advocated the adoption of Arabic numerals, the numerals that we use today.

keyboard, there are 2 black keys followed by 3 black keys for a total of 5 black keys. In one octave, there are 8 white keys and a total of 13 black and white keys. These are numbers in the Fibonacci sequence.

The fruitlets of a pineapple are arranged in interlocking spirals, 8 spirals in one direction and 13 in the other.

C D E F G A B C D E F G A B C D E F G A B

© The Natural History Museum/Alamy

2.   Define a Sequence Recursively

A sequence can be defined **recursively** by giving its first term and a rule showing how to obtain the $(n + 1)$th term from the nth term. For example, the information

$$a_1 = 5 \quad \text{(the first term)} \qquad \text{and} \qquad a_{n+1} = 3a_n - 2 \quad \text{(the rule showing how to get the } (n+1)\text{th term from the } n\text{th term)}$$

defines a sequence recursively. To find the first five terms of this sequence, we proceed as follows:

$a_1 = 5$

$a_2 = 3(a_1) - 2 = 3(5) - 2 = 13$ Substitute 5 for a_1 and simplify to get a_2.

$a_3 = 3(a_2) - 2 = 3(13) - 2 = 37$ Substitute 13 for a_2 and simplify to get a_3.

$a_4 = 3(a_3) - 2 = 3(37) - 2 = 109$ Substitute 37 for a_3 and simplify to get a_4.

$a_5 = 3(a_4) - 2 = 3(109) - 2 = 325$ Substitute 109 for a_4 and simplify to get a_5.

3.   Define Series

To add the terms of a sequence, we replace each comma between its terms with a + sign to form a **series.** If a sequence is infinite, the number of terms in the series associated with it is infinite also. Two examples of infinite series are

$$1^2 + 2^2 + 3^2 + \cdots + n^2 + \cdots$$

and

$$1 + 2 + 3 + 5 + 8 + 13 + 21 + \cdots$$

If a sequence is finite, the number of terms in the series associated with it is finite also. An example of a finite series is

$$3 + 7 + 11 + 15$$

If the signs between successive terms of a series alternate, the series is called an **alternating series.** Two examples of alternating infinite series are

$$-3 + 6 - 9 + 12 - 15 + 18 - \cdots + (-1)^n 3n + \cdots$$

Comment

The series $1 + \frac{1}{2} + \frac{1}{3} + \ldots$ is called a **harmonic series** and is important in calculus. The series $1 - \frac{1}{2} + \frac{1}{3} - \ldots$ is an alternating harmonic series.

and

$$2 - 4 + 8 - 16 + \cdots + (-1)^{n+1}2^n + \cdots$$

4.   Define and Use Summation Notation

Summation notation is a shorthand way to indicate the sum of the first n terms, or the **nth partial sum,** of a sequence. For example, the expression

$$\sum_{n=1}^{4} n \qquad \text{The symbol } \Sigma \text{ is the capital letter sigma in the Greek alphabet.}$$

indicates the sum of the four terms obtained when we successively substitute 1, 2, 3, and 4 for n.

$$\sum_{n=1}^{4} n = 1 + 2 + 3 + 4 = 10$$

The expression

$$\sum_{n=2}^{4} n^2$$

indicates the sum of the three terms obtained when we successively substitute 2, 3, and 4 for n.

$$\sum_{n=2}^{4} n^2 = (2)^2 + (3)^2 + (4)^2$$
$$= 4 + 9 + 16$$
$$= 29$$

The expression

$$\sum_{n=1}^{3} (2n^2 + 1)$$

indicates the sum of the three terms obtained if we successively substitute 1, 2, and 3 for n in the expression $2n^2 + 1$.

$$\sum_{n=1}^{3} (2n^2 + 1) = [2(1)^2 + 1] + [2(2)^2 + 1] + [2(3)^2 + 1]$$
$$= 3 + 9 + 19$$
$$= 31$$

EXAMPLE 1   Evaluate: $\displaystyle\sum_{n=1}^{4} (n^2 - 1)$.

Solution   Since n runs from 1 to 4, we substitute 1, 2, 3, and 4 for n in the expression $n^2 - 1$ and find the sum of the resulting terms:

$$\sum_{n=1}^{4} (n^2 - 1) = (1^2 - 1) + (2^2 - 1) + (3^2 - 1) + (4^2 - 1)$$
$$= 0 + 3 + 8 + 15$$
$$= 26$$

Self Check 1   Evaluate: $\displaystyle\sum_{n=1}^{5} (n^2 - 1)$.

EXAMPLE 2   Evaluate: $\displaystyle\sum_{n=3}^{5} (3n + 2)$.

Solution   Since n runs from 3 to 5, we substitute 3, 4, and 5 for n in the expression $3n + 2$ and find the sum of the resulting terms:

$$\sum_{n=3}^{5} (3n + 2) = [3(3) + 2] + [3(4) + 2] + [3(5) + 2]$$
$$= 11 + 14 + 17$$
$$= 42$$

Self Check 2   Evaluate: $\displaystyle\sum_{n=2}^{5} (3n + 2)$.

We can use summation notation to state the binomial theorem concisely. In Exercise 58, you will be asked to explain why the binomial theorem can be stated as

$$\sum_{r=0}^{n} \frac{n!}{r!(n - r)!} a^{n-r} b^r$$

There are three basic properties of summations. The first states that *the summation of a constant as k runs from 1 to n is n times the constant.*

Summation of a Constant   If c is a constant, then $\displaystyle\sum_{k=1}^{n} c = nc$.

Proof   Because c is a constant, each term is c for each value of k as k runs from 1 to n.

$$\sum_{k=1}^{n} c = \overbrace{c + c + c + c + \cdots + c}^{n \text{ number of } c\text{'s}} = nc$$

EXAMPLE 3   Evaluate: $\displaystyle\sum_{n=1}^{5} 13$.

Solution
$$\sum_{n=1}^{5} 13 = 13 + 13 + 13 + 13 + 13$$
$$= 5(13)$$
$$= 65$$

Self Check 3   Evaluate: $\displaystyle\sum_{n=1}^{6} 12$.

A second property states that *a constant factor can be brought outside a summation sign.*

Summation of a Product   If c is a constant, then $\displaystyle\sum_{k=1}^{n} cf(k) = c \sum_{k=1}^{n} f(k)$.

Proof
$$\sum_{k=1}^{n} cf(k) = cf(1) + cf(2) + cf(3) + \cdots + cf(n)$$

$$= c[f(1) + f(2) + f(3) + \cdots + f(n)] \qquad \text{Factor out } c.$$

$$= c\sum_{k=1}^{n} f(k)$$

EXAMPLE 4   Show that $\displaystyle\sum_{k=1}^{3} 5k^2 = 5\sum_{k=1}^{3} k^2$.

Solution

$$\sum_{k=1}^{3} 5k^2 = 5(1)^2 + 5(2)^2 + 5(3)^2 \qquad\qquad 5\sum_{k=1}^{3} k^2 = 5[(1)^2 + (2)^2 + (3)^2]$$

$$= 5 + 20 + 45 \qquad\qquad\qquad\qquad\qquad\quad\; = 5[1 + 4 + 9]$$

$$= 70 \qquad\qquad\qquad\qquad\qquad\qquad\qquad\;\; = 5(14)$$

$$\qquad\qquad\qquad\qquad\qquad\qquad\qquad\qquad\qquad\; = 70$$

The quantities are equal.

Self Check 4   Evaluate: $\displaystyle\sum_{k=1}^{4} 3k$.

The third property states that *the summation of a sum is equal to the sum of the summations.*

Summation of a Sum
$$\sum_{k=1}^{n} [f(k) + g(k)] = \sum_{k=1}^{n} f(k) + \sum_{k=1}^{n} g(k)$$

Proof
$$\sum_{k=1}^{n} [f(k) + g(k)] = [f(1) + g(1)] + [f(2) + g(2)] + [f(3) + g(3)] + \cdots + [f(n) + g(n)]$$

$$= [f(1) + f(2) + f(3) + \cdots + f(n)] + [g(1) + g(2) + g(3) + \cdots + g(n)]$$

$$= \sum_{k=1}^{n} f(k) + \sum_{k=1}^{n} g(k)$$

EXAMPLE 5   Show that $\displaystyle\sum_{k=1}^{3} (k + k^2) = \sum_{k=1}^{3} k + \sum_{k=1}^{3} k^2$.

Solution

$$\sum_{k=1}^{3} (k + k^2) = (1 + 1^2) + (2 + 2^2) + (3 + 3^2)$$

$$= 2 + 6 + 12$$

$$= 20$$

$$\sum_{k=1}^{3} k + \sum_{k=1}^{3} k^2 = (1 + 2 + 3) + (1^2 + 2^2 + 3^2)$$

$$= 6 + 14$$

$$= 20$$

Self Check 5   Evaluate: $\displaystyle\sum_{k=1}^{3} (k^2 + 2k)$.

EXAMPLE 6   Evaluate $\displaystyle\sum_{k=1}^{5}(2k-1)^2$ directly. Then expand the binomial, apply the previous properties, and evaluate the expression again.

Solution   **Part 1:** $\displaystyle\sum_{k=1}^{5}(2k-1)^2 = 1 + 9 + 25 + 49 + 81 = 165$

Part 2: $\displaystyle\sum_{k=1}^{5}(2k-1)^2 = \sum_{k=1}^{5}(4k^2 - 4k + 1)$

$$= \sum_{k=1}^{5}4k^2 + \sum_{k=1}^{5}(-4k) + \sum_{k=1}^{5}1$$ The summation of a sum is the sum of the summations.

$$= 4\sum_{k=1}^{5}k^2 - 4\sum_{k=1}^{5}k + \sum_{k=1}^{5}1$$ Bring the constant factors outside the summation signs.

$$= 4\sum_{k=1}^{5}k^2 - 4\sum_{k=1}^{5}k + 5$$ The summation of a constant as k runs from 1 to 5 is 5 times that constant.

$$= 4(1 + 4 + 9 + 16 + 25) - 4(1 + 2 + 3 + 4 + 5) + 5$$

$$= 4(55) - 4(15) + 5$$

$$= 220 - 60 + 5$$

$$= 165$$

Either way, the sum is 165.

Self Check 6   Evaluate: $\displaystyle\sum_{k=1}^{4}(2k-1)^2$.

Self Check Answers   **1.** 50   **2.** 50   **3.** 72   **4.** 30   **5.** 26   **6.** 84

8.2   **Exercises**

Vocabulary and Concepts   *Fill in the blanks.*

1. An infinite sequence is a function whose _____ is the set of natural numbers.

2. A finite sequence is a function whose _____ is the set of the first n natural numbers.

3. A _____ is formed when we add the terms of a sequence.

4. A series associated with a finite sequence is a _____ series.

5. A series associated with a infinite sequence is an _____ series.

6. If the signs between successive terms of a series alternate, the series is called an _____ series.

7. _____ is a shorthand way to indicate the sum of the first n terms of a sequence.

8. The symbol $\displaystyle\sum_{k=1}^{5}(k^2 - 3)$ indicates the ____ of the five terms obtained when we successively substitute 1, 2, 3, 4, and 5 for k.

9. $\displaystyle\sum_{k=1}^{5}6k^2 = \_\sum_{k=1}^{5}k^2$

10. $\displaystyle\sum_{k=1}^{5}(k^2 + 3k) = \sum_{k=1}^{5}k^2 + \_\_\_\_\_$

11. $\displaystyle\sum_{k=1}^{5}c$, where c is a constant, equals __.

12. The summation of a sum is equal to the ____ of the summations.

Practice   *Write the first six terms of the sequence defined by each function.*

13. $f(n) = 5n(n - 1)$

14. $f(n) = n\left(\dfrac{n-1}{2}\right)\left(\dfrac{n-2}{3}\right)$

Find the next term of each sequence.

15. $1, 6, 11, 16, \ldots$

16. $1, 8, 27, 64, \ldots$

17. $a, a + d, a + 2d, a + 3d, \ldots$

18. $a, ar, ar^2, ar^3, \ldots$

19. $1, 3, 6, 10, \ldots$

20. $20, 17, 13, 8, \ldots$

Find the sum of the first five terms of the sequence with the given general term.

21. n

22. $2k$

23. 3

24. $4k^0$

25. $2\left(\dfrac{1}{3}\right)^n$

26. $(-1)^n$

27. $3n - 2$

28. $2k + 1$

Assume that each sequence is defined recursively. Find the first four terms of each sequence.

29. $a_1 = 3$ and $a_{n+1} = 2a_n + 1$

30. $a_1 = -5$ and $a_{n+1} = -a_n - 3$

31. $a_1 = -4$ and $a_{n+1} = \dfrac{a_n}{2}$

32. $a_1 = 0$ and $a_{n+1} = 2a_n^2$

33. $a_1 = k$ and $a_{n+1} = a_n^2$

34. $a_1 = 3$ and $a_{n+1} = ka_n$

35. $a_1 = 8$ and $a_{n+1} = \dfrac{2a_n}{k}$

36. $a_1 = m$ and $a_{n+1} = \dfrac{a_n^2}{m}$

Determine whether each series is an alternating infinite series.

37. $-1 + 2 - 3 + \cdots + (-1)^n n + \cdots$

38. $a + \dfrac{a}{b} + \dfrac{a}{b^2} + \cdots + a\left(\dfrac{1}{b}\right)^{n-1} + \cdots; b = 4$

39. $a + a^2 + a^3 + \cdots + a^n + \cdots; a = 3$

40. $a + a^2 + a^3 + \cdots + a^n + \cdots; a = -2$

Evaluate each sum.

41. $\displaystyle\sum_{k=1}^{5} 2k$

42. $\displaystyle\sum_{k=3}^{6} 3k$

43. $\displaystyle\sum_{k=3}^{4} (-2k^2)$

44. $\displaystyle\sum_{k=1}^{100} 5$

45. $\displaystyle\sum_{k=1}^{5} (3k - 1)$

46. $\displaystyle\sum_{n=2}^{5} (n^2 + 3n)$

47. $\displaystyle\sum_{k=1}^{1,000} \dfrac{1}{2}$

48. $\displaystyle\sum_{x=4}^{5} \dfrac{2}{x}$

49. $\displaystyle\sum_{x=3}^{4} \dfrac{1}{x}$

50. $\displaystyle\sum_{x=2}^{6} (3k^2 + 2x) - 3\sum_{x=2}^{6} x^2$

51. $\displaystyle\sum_{x=1}^{4} (4x + 1)^2 - \sum_{x=1}^{4} (4x - 1)^2$

52. $\displaystyle\sum_{x=0}^{10} (2x - 1)^2 + 4\sum_{x=0}^{10} x(1 - x)$

53. $\displaystyle\sum_{x=6}^{8} (5x - 1)^2 + \sum_{x=6}^{8} (10x - 1)$

54. $\displaystyle\sum_{x=2}^{7} (3x + 1)^2 - 3\sum_{x=2}^{7} x(3x + 2)$

Discovery and Writing

55. Find a counterexample to disprove the proposition that the summation of a product is the product of the summations. In other words, prove that

$$\sum_{k=1}^{n} f(k)g(k) \neq \sum_{k=1}^{n} f(k) \sum_{k=1}^{n} g(k)$$

56. Find a counterexample to disprove the proposition that the summation of a quotient is the quotient of the summations. In other words, prove that

$$\sum_{k=1}^{n} \dfrac{f(k)}{g(k)} \neq \dfrac{\displaystyle\sum_{k=1}^{n} f(k)}{\displaystyle\sum_{k=1}^{n} g(k)}$$

57. Explain what it means to define something recursively.

58. Explain why the binomial theorem can be stated as

$$\sum_{r=0}^{n} \dfrac{n!}{r!(n - r)!} a^{n-r} b^r$$

Review *The triangles are similar. Find the value of x.*

59.

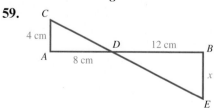

60.

Find the third side of each right triangle.

61.

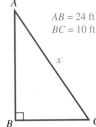

$AB = 24$ ft
$BC = 10$ ft

62.

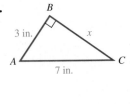

3 in. x

A 7 in. C

© Maseman

8.3   Arithmetic Sequences

Objectives

1. Define and Use Arithmetic Sequences
2. Find Arithmetic Means
3. Find the Sum of the First n Terms of an Arithmetic Series
4. Solve Problems Involving Sequences

The German mathematician Carl Friedrich Gauss (1777–1855), whose memorial is shown, was once a student in the class of a strict teacher. One day, the teacher asked the students to add together all of the natural numbers from 1 through 100. Gauss recognized that in the sum

$$1 + 2 + 3 + \cdots + 98 + 99 + 100$$

the first number (1) added to the last number (100) is 101, the second number (2) added to the second from the last number (99) is 101, and the third number (3) added to the third from the last number (98) is 101. He reasoned that there would be fifty pairs of such numbers, and that there would be fifty sums of 101. He multiplied 101 by 50 to get the correct answer of 5,050.

This story illustrates a problem involving the sum of the terms of a sequence, called an **arithmetic sequence,** in which each term except the first is found by adding a constant to the preceding term.

1.   Define and Use Arithmetic Sequences

Arithmetic Sequences   An **arithmetic sequence** is a sequence of the form

$$a, a + d, a + 2d, a + 3d, \ldots, a + (n - 1)d, \ldots$$

where a is the **first term,** $a + (n - 1)d$ is the **nth term,** and d is the **common difference.**

In this definition, the second term has an addend of d, the third term has an addend of $2d$, the fourth term has an addend of $3d$, and so on. This is why the nth term has an addend of $(n - 1)d$.

If an arithmetic sequence has infinitely many terms, it is called an **infinite arithmetic sequence.** If it has a finite number of terms, it is called a **finite arithmetic sequence.**

EXAMPLE 1   Write the first six terms and the 21st term of an arithmetic sequence with a first term of 7 and a common difference of 5.

Solution   Since the first term a is 7 and the common difference d is 5, the first six terms are

$$7, \quad 7 + 5, \quad 7 + 2(5), \quad 7 + 3(5), \quad 7 + 4(5), \quad 7 + 5(5)$$

or

$$7, \quad 12, \quad 17, \quad 22, \quad 27, \quad 32$$

To find the 21st term, we substitute 21 for n in the formula for the nth term:

$$n\text{th term} = a + (n - 1)d$$
$$21\text{st term} = 7 + (21 - 1)5$$
$$= 7 + (20)5$$
$$= 107$$

The 21st term is 107.

Self Check 1   Write the first five terms and the 18th term of an arithmetic sequence with a first term of 3 and a common difference of 6.

EXAMPLE 2   Find the 98th term of an arithmetic sequence whose first three terms are 2, 6, and 10.

Solution   Here $a = 2$, $n = 98$, and $d = 6 - 2 = 10 - 6 = 4$. Because we want to find the 98th term, we substitute these numbers into the formula for the nth term:

$$n\text{th term} = a + (n - 1)d$$
$$98\text{th term} = 2 + (98 - 1)4$$
$$= 2 + (97)4$$
$$= 390$$

Self Check 2   Write the 50th term of the arithmetic sequence whose first three terms are 3, 8, and 13.

2. Find Arithmetic Means

Numbers inserted between a first and last term to form a segment of an arithmetic sequence are called **arithmetic means.** When finding arithmetic means, we consider the last term, a_n, to be the nth term:

$$a_n = a + (n - 1)d$$

EXAMPLE 3   Insert three arithmetic means between -3 and 12.

Solution   Since we are inserting three arithmetic means between -3 and 12, the total number of terms is five. Thus, $a = -3$, $a_n = 12$, and $n = 5$. To find the common difference, we substitute -3 for a, 12 for a_n, and 5 for n in the formula for the last term and solve for d:

$$a_n = a + (n - 1)d$$
$$12 = -3 + (5 - 1)d$$
$$15 = 4d \qquad \text{Add 3 to both sides and simplify.}$$
$$\frac{15}{4} = d \qquad \text{Divide both sides by 4.}$$

Once we know d, we can find the other terms of the sequence:

$$a + d = -3 + \frac{15}{4} = \frac{3}{4}$$
$$a + 2d = -3 + 2\left(\frac{15}{4}\right) = -3 + \frac{30}{4} = 4\frac{1}{2}$$
$$a + 3d = -3 + 3\left(\frac{15}{4}\right) = -3 + \frac{45}{4} = 8\frac{1}{4}$$

The three arithmetic means are $\frac{3}{4}$, $4\frac{1}{2}$, and $8\frac{1}{4}$.

Self Check 3 Find three arithmetic means between -5 and 23.

3. Find the Sum of the First n Terms of an Arithmetic Series

If we replace the commas in an infinite arithmetic sequence with $+$ signs, we form an **infinite arithmetic series.**

$$a + (a + d) + (a + 2d) + (a + 3d) + \cdots + [a + (n - 1)d] + \cdots$$

If we replace the commas in a finite arithmetic sequence with $+$ signs, we form a **finite arithmetic series.**

$$a + (a + d) + (a + 2d) + (a + 3d) + \cdots + [a + (n - 1)d]$$

To find the sum of the first n terms of an arithmetic series, we use the following formula.

Sum of the First n Terms of an Arithmetic Series	The formula $$S_n = \frac{n(a + a_n)}{2}$$ gives the sum of the first n terms of an arithmetic series. In this formula, a is the first term, a_n is the last (or nth) term, and n is the number of terms.

Proof We write the first n terms of an arithmetic series (letting S_n represent their sum). Then we write the same sum in reverse order and add the equations term by term:

$$S_n = a + (a + d) + \cdots + [a + (n - 2)d] + [a + (n - 1)d]$$
$$S_n = [a + (n - 1)d] + [a + (n - 2)d] + \cdots + (a + d) + a$$
$$\overline{2S_n = [2a + (n - 1)d] + [2a + (n - 1)d] + \cdots + [2a + (n - 1)d] + [2a + (n - 1)d]}$$

Because there are n equal terms on the right side of the previous equation,

$$2S_n = n[2a + (n - 1)d]$$

or

(1) $\quad 2S_n = n\{a + [a + (n - 1)d]\}$ $\quad$ **Write $2a$ as $a + a$.**

We can substitute a_n for $a + (n - 1)d$ on the right side of Equation 1 and divide both sides by 2 to get

$$S_n = \frac{n(a + a_n)}{2}$$

EXAMPLE 4 $\quad$ Find the sum of the first 30 terms of the arithmetic series $5 + 8 + 11 + \cdots$.

Solution $\quad$ Here $a = 5$, $n = 30$, $d = 3$, and $a_n = 5 + 29(3) = 92$. Substituting these values into the formula for the sum of the first n terms of an arithmetic series gives

$$S_n = \frac{n(a + a_n)}{2}$$

$$S_{30} = \frac{30(5 + 92)}{2}$$

$$= 15(97)$$

$$= 1,455$$

The sum of the first 30 terms is 1,455.

Self Check 4 $\quad$ Find the sum of the first 50 terms of the arithmetic series $-2 + 5 + 12 + \cdots$.

4. Solve Problems Involving Sequences

EXAMPLE 5 $\quad$ A student deposits \$50 in a non-interest-bearing account and plans to add \$7 a week. How much will she have in the account one year after her first deposit?

Solution $\quad$ Her weekly balances form an arithmetic sequence:

$\quad$ 50, 57, 64, 71, 78, . . .

with a first term of 50 and a common difference of 7. To find her balance in one year (52 weeks), we substitute 50 for a, 7 for d, and 52 for n in the formula for the last term.

$$a_n = a + (n - 1)d$$

$$= 50 + (52 - 1)7$$

$$= 50 + (51)7$$

$$= 407$$

After one year, the balance will be \$407.

Self Check 5 $\quad$ How much will she have in the account after 60 weeks?

EXAMPLE 6 $\quad$ The equation $s = 16t^2$ represents the distance s (in feet) that an object will fall in t seconds.

- In 1 second, the object will fall 16 feet.
- In 2 seconds, the object will fall 64 feet.
- In 3 seconds, the object will fall 144 feet.

The object fell 16 feet during the first second, 48 feet during the next second, and 80 feet during the third second. Find the distance the object will fall during the 12th second.

Solution The sequence 16, 48, 80, . . . is an arithmetic sequence with $a = 16$ and $d = 32$. To find the 12th term, we substitute these values into the formula for the last term.

$$a_n = a + (n - 1)d$$
$$= 16 + (12 - 1)32$$
$$= 16 + 11(32)$$
$$= 368$$

During the 12th second, the object falls 368 feet.

Self Check 6 How far will the object fall during the 20th second?

Self Check Answers **1.** 3, 9, 15, 21, 27; 105 **2.** 248 **3.** 2, 9, 16 **4.** 8,475 **5.** $463 **6.** 624 ft

8.3 Exercises

Vocabulary and Concepts *Fill in the blanks.*

1. An arithmetic sequence is a sequence of the form

   $a, a + d, a + 2d, a + 3d, . . . , a + \underline{\quad\quad}d, . . .$

2. An arithmetic series is a series of the form

   $a + (a + d) + (a + 2d) + (a + 3d)$
   $+ \cdots [a + (n - 1)\underline{\quad}] + \cdots$

3. If an arithmetic series has infinitely many terms, it is called an _____ arithmetic series.

4. In an arithmetic sequence, a is the ____ term, d is the common _____, and n is the _____ of terms.

5. The nth term of an arithmetic sequence is given by the formula _____.

6. The formula for the sum of the first n terms of an arithmetic series is given by the formula

   _____.

7. _____ are numbers inserted between a first and last term of a sequence to form an arithmetic sequence.

8. The formula _____ gives the distance (in feet) that an object will fall in t seconds.

Practice *Write the first six terms of the arithmetic sequences with the given properties.*

9. $a = 1; d = 2$

10. $a = -12; d = -5$

11. $a = 5$; 3rd term is 2

12. $a = 4$; 5th term is 12

13. 7th term is 24; common difference is $\dfrac{5}{2}$

14. 20th term is -49; common difference is -3

Find the missing term in each arithmetic sequence.

15. Find the 40th term of an arithmetic sequence with a first term of 6 and a common difference of 8.

16. Find the 35th term of an arithmetic sequence with a first term of 50 and a common difference of -6.

17. The 6th term of an arithmetic sequence is 28, and the first term is -2. Find the common difference.

18. The 7th term of an arithmetic sequence is -42, and the common difference is -6. Find the first term.

19. Find the 55th term of an arithmetic sequence whose first three terms are $-8, -1$, and 6.

20. Find the 37th term of an arithmetic sequence whose second and third terms are -4 and 6.

21. If the fifth term of an arithmetic sequence is 14 and the second term is 5, find the 15th term.

22. If the fourth term of an arithmetic sequence is 13 and the second term is 3, find the 24th term.

Find the required means.

23. Insert three arithmetic means between 10 and 20.

24. Insert five arithmetic means between 5 and 15.

25. Insert four arithmetic means between -7 and $\frac{2}{3}$.

26. Insert three arithmetic means between -11 and -2.

Find the sum of the first n terms of each arithmetic series.

27. $5 + 7 + 9 + \cdots$ (to 15 terms)

28. $-3 + (-4) + (-5) + \cdots$ (to 10 terms)

29. $\sum_{n=1}^{20} \left(\frac{3}{2}n + 12 \right)$ **30.** $\sum_{n=1}^{10} \left(\frac{2}{3}n + \frac{1}{3} \right)$

Solve each problem.

31. Find the sum of the first 30 terms of an arithmetic sequence with 25th term of 10 and a common difference of $\frac{1}{2}$.

32. Find the sum of the first 100 terms of an arithmetic sequence with 15th term of 86 and first term of 2.

33. Find the sum of the first 200 natural numbers.

34. Find the sum of the first 1,000 natural numbers.

Applications

35. Interior angles   The sums of the angles of several polygons are given in the table. Assuming that the pattern continues, complete the table.

Figure	Number of sides	Sum of angles
Triangle	3	180°
Quadrilateral	4	360°
Pentagon	5	540°
Hexagon	6	720°
Octagon	8	
Dodecagon	12	

36. Borrowing money   To pay for college, a student borrows $5,000 interest-free from his father. If he pays his father back at the rate of $200 per month, how much will he still owe after 12 months?

37. Borrowing money   If Juanita borrows $5,500 interest-free from her mother to buy a new car and agrees to pay her mother back at the rate of $105 per month, how much will she still owe after 4 years?

38. Jogging   One day, some students jogged $\frac{1}{2}$ mile. Because it was fun, they decided to increase the jogging distance each day by a certain amount. If they jogged $6\frac{3}{4}$ miles on the 51st day, how much was the distance increased each day?

39. Sales   The year it incorporated, a company had sales of $237,500. Its sales were expected to increase by $150,000 annually for the next several years. If the forecast was correct, what will sales be in 10 years?

40. Falling objects   Find how many feet a brick will travel during the 10th second of its fall.

41. Falling objects   If a rock is dropped from the Golden Gate Bridge, how far will it fall in the third second?

42. Designing patios   Each row of bricks in the following triangular patio is to have one more brick than the previous row, ending with the longest row of 150 bricks. How many bricks will be needed?

43. Pile of logs   Several logs are stored in a pile with 20 logs on the bottom layer, 19 on the second layer, 18 on the third layer, and so on. If the top layer has one log, how many logs are in the pile?

44. Theater seating   The first row in a movie theater contains 24 seats. As you move toward the back, each row has 1 additional seat. If there are 30 rows, what is the capacity of the theater?

Discovery and Writing

45. Can an arithmetic sequence have a first term of 4, a 25th term of 126, and a common difference of $4\frac{1}{4}$? Explain.

46. In an arithmetic sequence, can a and d be negative, but a_n positive?

47. Can an arithmetic sequence be an alternating sequence? Explain.

48. Between 5 and $10\frac{1}{3}$ are three arithmetic means. One of them is 9. Find the other two.

Review *Solve each equation.*

49. $x + \sqrt{x + 3} = 9$

50. $\dfrac{x + 3}{x} + \dfrac{x - 3}{5} = 2$

51. $\dfrac{1}{x} + \dfrac{2}{x^2} + \dfrac{1}{x^3} = 0$

52. $x + 3\sqrt{x} - 10 = 0$

53. $x^4 - 1 = 0$

54. $x^4 - 29x^2 + 100 = 0$

8.4 Geometric Sequences

Objectives

1. Define and Use Geometric Sequences
2. Find Geometric Means
3. Find the Sum of the First n Terms of a Geometric Series
4. Define and Find the Sum of Infinite Geometric Series
5. Solve Problems Involving Geometric Sequences

Bungee jumping is a sport invented by daredevil A. J. Hackett. His first jump was off the 43-meter-high Kawarau Bridge in New Zealand. Since he was attached to a long rubber cord, he bounced up and down for a considerable time, giving him a thrilling ride. Today, this sport is gaining in popularity, but it is not for the faint of heart.

Suppose a jumper is attached to a cord that stretches to a length of 100 feet. Also suppose that on each bounce, he rebounds to a height that is 60% of the distance from which he fell. If we list the distances that he falls, we will get the following sequence:

$100, 60, 36, \ldots$ **Each number in the list is 60% of the number before it.**

If we list the distances that he rebounds, we will get the following sequence:

$60, 36, 21.6, \ldots$ **Each number in the list is 60% of the number before it.**

Each of these sequences is an example of a common sequence called a *geometric sequence*.

1. Define and Use Geometric Sequences

A **geometric sequence** is a sequence in which each term, except the first, is found by multiplying the preceding term by a constant.

Geometric Sequences

A **geometric sequence** is a sequence of the form

$$a, \quad ar, \quad ar^2, \quad ar^3, \ldots, ar^{n-1}, \ldots$$

where a is the **first term**, ar^{n-1} is the **nth term**, and r is the **common ratio**.

Comment

The nth term of a geometric sequence is given by

$$a_n = ar^{n-1}$$

In this definition, the second term of the sequence has a factor of r^1, the third term has a factor of r^2, the fourth term has a factor of r^3, and so on. This explains why the nth term has a factor of r^{n-1}.

If a geometric sequence has infinitely many terms, it is called an **infinite geometric sequence**. If it has a finite number of terms, it is called a **finite geometric sequence**.

EXAMPLE 1   Write the first six terms and the 15th term of the geometric sequence whose first term is 3 and whose common ratio is 2.

Solution   We first write the first six terms of the geometric sequence:

$$3, \quad 3(2), \quad 3(2)^2, \quad 3(2)^3, \quad 3(2)^4, \quad 3(2)^5$$

or

$$3, \quad 6, \quad 12, \quad 24, \quad 48, \quad 96$$

To find the 15th term, we substitute 15 for n, 3 for a, and 2 for r in the formula for the nth term:

$$n\text{th term} = \boldsymbol{ar^{n-1}}$$
$$15\text{th term} = \boldsymbol{3(2)^{15-1}}$$
$$= 3(2)^{14}$$
$$= 3(16{,}384)$$
$$= 49{,}152$$

Self Check 1   Write the first five terms of the geometric sequence whose first term is 2 and whose common ratio is 3. Find the 10th term.

EXAMPLE 2   Find the eighth term of a geometric sequence whose first three terms are 9, 3, and 1.

Solution   Here $a = 9$, $r = \frac{1}{3}$, and $n = 8$. To find the eighth term, we substitute these values into the formula for the nth term.

$$n\text{th term} = \boldsymbol{ar^{n-1}}$$
$$8\text{th term} = \boldsymbol{9\left(\frac{1}{3}\right)^{8-1}}$$
$$= 9\left(\frac{1}{3}\right)^7$$
$$= \frac{1}{243}$$

Self Check 2   Find the eighth term of a geometric sequence whose first three terms are $\frac{1}{3}$, 1, and 3.

2. Find Geometric Means

Numbers inserted between a first and last term to form a segment of a geometric sequence are called **geometric means.** When finding geometric means, we consider the last term, a_n, to be the nth term.

EXAMPLE 3 Insert two geometric means between 4 and 256.

Solution The first term is $a = 4$, and because 256 is the fourth term, $n = 4$ and $a_n = 256$. To find the common ratio, we substitute these values into the formula for the nth term and solve for r:

$$ar^{n-1} = a_n$$
$$4r^{4-1} = 256$$
$$r^3 = 64$$
$$r = 4$$

The common ratio is 4. The two geometric means are the second and third terms of the geometric sequence:

$$ar = 4 \cdot 4 = 16$$
$$ar^2 = 4 \cdot 4^2 = 4 \cdot 16 = 64$$

The first four terms of the geometric sequence are 4, 16, 64, and 256. The two geometric means between 4 and 256 are 16 and 64.

Self Check 3 Insert two geometric means between -3 and 192.

3. Find the Sum of the First n Terms of a Geometric Series

If we replace the commas in an infinite geometric sequence with $+$ signs, we form an **infinite geometric series.**

$$a + ar + ar^2 + ar^3 + \cdots + ar^{n-1} + \cdots$$

If we replace the commas in a finite geometric sequence with $+$ signs, we form a **finite geometric series.**

$$a + ar + ar^2 + ar^3 + \cdots + ar^{n-1}$$

To find the sum of the first n terms of a geometric series, we can use the following formula.

Sum of the First n Terms of a Geometric Series

The formula

$$S_n = \frac{a - ar^n}{1 - r} \quad (r \neq 1)$$

gives the sum of the first n terms of a geometric series. In the formula, S_n is the sum, a is the first term, r is the common ratio, and n is the number of terms.

Proof We write the sum of the first n terms of the geometric series:

(1) $\qquad S_n = a + ar + ar^2 + \cdots + ar^{n-3} + ar^{n-2} + ar^{n-1}$

Multiplying both sides of this equation by r gives

(2) $\qquad S_n r = ar + ar^2 + \cdots + ar^{n-2} + ar^{n-1} + ar^n$

We now subtract Equation 2 from Equation 1 and solve for S_n:

$$S_n - S_n r = a - ar^n$$

$$S_n(1 - r) = a - ar^n \quad \text{Factor out } S_n.$$

$$S_n = \frac{a - ar^n}{1 - r} \quad \text{Divide both sides by } 1 - r.$$

EXAMPLE 4   Find the sum of the first six terms of the geometric series $8 + 4 + 2 + \cdots$.

Solution   Here $a = 8$, $n = 6$, and $r = \frac{1}{2}$. Substituting these values in the formula for the sum of the first n terms of a geometric series gives

$$S_n = \frac{a - ar^n}{1 - r}$$

$$S_6 = \frac{8 - 8\left(\dfrac{1}{2}\right)^6}{1 - \dfrac{1}{2}}$$

$$= 2\left(\frac{63}{8}\right)$$

$$= \frac{63}{4}$$

The sum of the first six terms is $\frac{63}{4}$.

Self Check 4   Find the sum of the first eight terms of the geometric series $81, 27, 9, \ldots$.

4.   Define and Find the Sum of Infinite Geometric Series

Under certain conditions, we can find the sum of all of the terms in an **infinite geometric series.** To define this sum, we consider the infinite geometric series

$$a + ar + a^2 + \cdots$$

- The first partial sum, S_1, of the series is $S_1 = a$.
- The second partial sum, S_2, of the series is $S_2 = a + ar$.
- The nth partial sum, S_n, of the series is $S_n = a + ar + ar^2 + \cdots + ar^{n-1}$.

If the nth partial sum, S_n, of an infinite geometric series approaches some number S as n approaches ∞, then S is called the **sum of the infinite geometric series.** The following symbol denotes the sum, S, of an infinite geometric sequence, provided the sum exists.

$$S = \sum_{n=1}^{\infty} ar^{n-1}$$

To develop a formula for finding the sum of all the terms in an infinite geometric series, we consider the formula

$$(3) \quad S_n = \frac{a - ar^n}{1 - r} \quad (r \neq 1)$$

If $|r| < 1$ and a is a constant, then as n approaches ∞, ar^n approaches 0, and the term ar^n in Equation 3 can be dropped. As an illustration, suppose that $a = 1$

n	$1\left(\dfrac{1}{2}\right)^n$
0	1
1	$\dfrac{1}{2}$
2	$\dfrac{1}{4}$
3	$\dfrac{1}{8}$
4	$\dfrac{1}{16}$
10	$\dfrac{1}{1,024}$

and $r = \frac{1}{2}$. We can see from the table on the previous page that as n continues to increase, $ar^n = 1\left(\frac{1}{2}\right)^n$ approaches 0.

This argument gives the following formula.

| **Sum of the Terms of an Infinite Geometric Series** | If $|r| < 1$, the sum of the terms of an infinite geometric series is given by $$S_\infty = \frac{a}{1 - r}$$ where a is the first term and r is the common ratio. |

EXAMPLE 5   Change $0.\overline{4}$ to a common fraction.

Solution   We write the decimal as an infinite geometric series and find its sum:

$$S_\infty = \frac{4}{10} + \frac{4}{100} + \frac{4}{1,000} + \frac{4}{10,000} + \cdots$$

$$= \frac{4}{10} + \frac{4}{10}\left(\frac{1}{10}\right) + \frac{4}{10}\left(\frac{1}{10}\right)^2 + \frac{4}{10}\left(\frac{1}{10}\right)^3 + \cdots$$

Comment
If $|r| \geq 1$, the terms get larger and larger, and the sum does not approach a number. In this case, the previous formula does not apply.

Since the common ratio r equals $\frac{1}{10}$ and $\left|\frac{1}{10}\right| < 1$, we can use the formula for the sum of an infinite geometric series:

$$S_\infty = \frac{a}{1 - r} = \frac{\dfrac{4}{10}}{1 - \dfrac{1}{10}} = \frac{\dfrac{4}{10}}{\dfrac{9}{10}} = \frac{4}{9}$$

Comment
Using a generalization of the method used in Example 5, we can write any repeating decimal as a fraction.

Long division will verify that $\frac{4}{9} = 0.\overline{4}$.

Self Check 5   Change $0.\overline{7}$ to a common fraction.

5. Solve Problems Involving Geometric Sequences

Many of the exponential-growth problems discussed in Chapter 4 can be solved using the concepts of geometric sequences.

EXAMPLE 6   A statistician knows that a town with a population of 3,500 people has a predicted growth rate of 6% per year for the next 20 years. What should she predict the population to be 20 years from now?

Solution   Let p_0 be the initial population of the town. After 1 year, the population p_1 will be the initial population (p_0) plus the growth (the product of p_0 and the rate of growth, r).

$$p_1 = p_0 + p_0 r$$
$$= p_0(1 + r) \qquad \text{Factor out } p_0.$$

The population p_2 at the end of 2 years will be

$$p_2 = p_1 + p_1 r$$
$$= p_1(1 + r) \qquad \text{Factor out } p_1.$$
$$= p_0(1 + r)(1 + r) \qquad \text{Substitute } p_0(1 + r) \text{ for } p_1.$$
$$= p_0(1 + r)^2$$

The population at the end of the third year will be $p_3 = p_0(1 + r)^3$. Writing the terms as a sequence gives

$$p_0, \qquad p_0(1 + r), \qquad p_0(1 + r)^2, \qquad p_0(1 + r)^3, \qquad p_0(1 + r)^4, \ldots$$

This is a geometric sequence with p_0 as the first term and $1 + r$ as the common ratio. In this example, $p_0 = 3{,}500$, $1 + r = 1.06$, and (since the population after 20 years will be the value of the 21st term of the geometric sequence) $n = 21$. We can substitute these values into the formula for the last term of a geometric sequence to get

$$a_n = ar^{n-1}$$
$$= 3{,}500(1.06)^{21-1}$$
$$= 3{,}500(1.06)^{20}$$
$$\approx 11{,}224.97415 \qquad \text{Use a calculator.}$$

The population after 20 years will be approximately 11,225.

EXAMPLE 7   A student deposits $2,500 in a bank at 7% annual interest, compounded daily. If the investment is left untouched for 60 years, how much money will be in the account?

Solution   We let the initial amount in the account be a_0 and r be the rate. At the end of the first day, the account is worth

$$a_1 = a_0 + a_0\left(\frac{r}{365}\right) = a_0\left(1 + \frac{r}{365}\right)$$

After the second day, the account is worth

$$a_2 = a_1 + a_1\left(\frac{r}{365}\right) = a_1\left(1 + \frac{r}{365}\right) = a_0\left(1 + \frac{r}{365}\right)^2$$

The daily amounts form the following geometric sequence

$$a_0, a_0\left(1 + \frac{r}{365}\right), a_0\left(1 + \frac{r}{365}\right)^2, a_0\left(1 + \frac{r}{365}\right)^3, \ldots$$

where a_0 is the initial deposit and r is the annual rate of interest.

Because interest is compounded daily for 60 years (21,900 days), the amount at the end of 60 years will be the 21,901th term of the sequence.

$$a_{21{,}901} = 2{,}500\left(1 + \frac{0.07}{365}\right)^{21{,}900}$$

We can use a calculator to find that $a_{21{,}901} \approx \$166{,}648.71$.

EXAMPLE 8   A pump can remove 20% of the gas in a container with each stroke. Find the percentage of gas that remains in the container after six strokes.

Solution   We let V represent the volume of the container. Because each stroke of the pump removes 20% of the gas, 80% remains after each stroke, and we have the geometric sequence

$$V, \qquad 0.80V, \qquad 0.80(0.80V), \qquad 0.80[0.80(0.80V)], \ldots$$

or

$$V, \quad 0.80V, \quad (0.8)^2 V, \quad (0.8)^3 V, \quad (0.8)^4 V, \ldots$$

The amount of gas remaining after six strokes is the seventh term, a_n, of the sequence:

$$a_n = ar^{n-1}$$
$$= V(0.8)^{7-1}$$
$$= V(0.8)^6$$

We can use a calculator to find that approximately 26% of the gas remains after six strokes.

Self Check Answers   **1.** 2, 6, 18, 54, 162; 39,366   **2.** 729   **3.** 12, -48 **4.** $\frac{3,280}{27}$ **5.** $\frac{7}{9}$

8.4 Exercises

Vocabulary and Concepts   *Fill in the blanks.*

1. A geometric sequence is a sequence of the form a, ar, ar^2, ar^3, ... The nth term is $a(\underline{\quad})$.

2. In a geometric sequence, a is the _____ term, r is the common _____, and n is the _____ of terms.

3. The nth term of a geometric sequence is given by the formula $a_n = $ _____.

4. A geometric _____ is the sum of the terms of a geometric sequence.

5. A geometric series with infinitely many terms is called an _____ geometric series.

6. The formula for the sum of the first n terms of a geometric series is given by

   _____

7. _____ are numbers inserted between a first and a last term to form a geometric sequence.

8. If $|r| < 1$, the formula _____ gives the sum of the terms of an infinite geometric series.

Practice   *Write the first four terms of each geometric sequence with the given properties.*

9. $a = 10$; $r = 2$

10. $a = -3$; $r = 2$

11. $a = -2$ and $r = 3$

12. $a = 64$; $r = \frac{1}{2}$

13. $a = 3$; $r = \sqrt{2}$

14. $a = 2$; $r = \sqrt{3}$

15. $a = 2$; 4th term is 54

16. 3rd term is 4; $r = \frac{1}{2}$

Find the requested term of each geometric sequence.

17. Find the sixth term of the geometric sequence whose first three terms are $\frac{1}{4}$, 1, and 4.

18. Find the eighth term of the geometric sequence whose second and fourth terms are 0.2 and 5.

19. Find the fifth term of a geometric sequence whose second term is 6 and whose third term is -18.

20. Find the sixth term of a geometric sequence whose second term is 3 and whose fourth term is $\frac{1}{3}$.

Solve each problem.

21. Insert three positive geometric means between 10 and 20.

22. Insert five geometric means between -5 and 5, if possible.

23. Insert four geometric means between 2 and 2,048.

24. Insert three geometric means between 162 and 2. (There are two possibilities.)

Find the sum of the indicated terms of each geometric series.

25. $4 + 8 + 16 + \cdots$ (to 5 terms)

26. $9 + 27 + 81 + \cdots$ (to 6 terms)

27. $2 + (-6) + 18 + \cdots$ (to 10 terms)

28. $\dfrac{1}{8} + \dfrac{1}{4} + \dfrac{1}{2} + \cdots$ (to 12 terms)

29. $\displaystyle\sum_{n=1}^{6} 3\left(\dfrac{3}{2}\right)^{n-1}$ **30.** $\displaystyle\sum_{n=1}^{6} 12\left(-\dfrac{1}{2}\right)^{n-1}$

Find the sum of each infinite geometric series.

31. $6 + 4 + \dfrac{8}{3} + \cdots$ **32.** $8 + 4 + 2 + 1 + \cdots$

33. $\displaystyle\sum_{n=1}^{\infty} 12\left(-\dfrac{1}{2}\right)^{n-1}$ **34.** $\displaystyle\sum_{n=1}^{\infty} \left(\dfrac{1}{3}\right)^{n-1}$

Change each decimal to a common fraction.

35. $0.\overline{5}$ **36.** $0.\overline{6}$

37. $0.\overline{25}$ **38.** $0.\overline{37}$

Applications   *Use a calculator to help solve each problem.*

39. Staffing a department   The number of students studying algebra at State College is 623. The department chair expects enrollment to increase 10% each year. How many professors will be needed in 8 years to teach algebra if one professor can handle 60 students?

40. Bouncing balls   On each bounce, the rubber ball in the illustration rebounds to a height one-half of that from which it fell. Find the total vertical distance the ball travels.

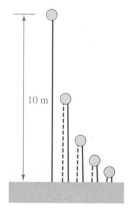

10 m

41. Bungee jumping   A bungee jumper is attached to a cord that stretches to a length of 100 feet. If he rebounds to 60% of the height jumped, how far will he fall on his fifth descent? How far will he travel when he comes to rest?

42. Bungee jumping   A bungee jumper is attached to a cord that stretches to a length of 100 feet. If she rebounds to 70% of the height jumped, how far will she travel upward on the fifth rebound? How far will she have traveled when she comes to rest?

43. Bouncing balls   A Super Ball rebounds to approximately 95% of the height from which it is dropped. If the ball is dropped from a height of 10 meters, how high will it rebound after the 13th bounce?

44. Genealogy   The following family tree spans 3 generations and lists 7 people. How many names would be listed in a family tree that spans 10 generations?

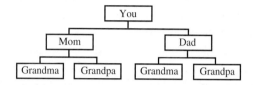

45. Investing money   If a married couple invests $1,000 in a 1-year certificate of deposit at $6\frac{3}{4}\%$ annual interest, compounded daily, how much interest will be earned during the year?

46. Biology   If a single cell divides into two cells every 30 minutes, how many cells will there be at the end of 10 hours?

47. Depreciation   A lawn tractor, costing c dollars when new, depreciates 20% of the previous year's value each year. How much is the lawn tractor worth after 5 years?

48. Financial planning   Maria can invest $1,000 at $7\frac{1}{2}\%$, compounded annually, or at $7\frac{1}{4}\%$, compounded daily. If she invests the money for a year, which is the better investment?

49. Population study   If the population of the Earth were to double every 30 years, approximately how many people would there be in the year 3020? (Consider the population in 2000 to be 6 billion and use 2000 as the base year.)

50. Investing money   If Linda deposits $1,300 in a bank at 7% interest, compounded annually, how much will be in the bank 17 years later? (Assume that there are no other transactions on the account.)

51. Real estate appreciation   If a house purchased for $50,000 in 1988 appreciates in value by 6% each year, how much will the house be worth in the year 2010?

52. Compound interest   Find the value of $1,000 left on deposit for 10 years at an annual rate of 7%, compounded annually.

53. Compound interest Find the value of $1,000 left on deposit for 10 years at an annual rate of 7%, compounded quarterly.

54. Compound interest Find the value of $1,000 left on deposit for 10 years at an annual rate of 7%, compounded monthly.

55. Compound interest Find the value of $1,000 left on deposit for 10 years at an annual rate of 7%, compounded daily.

56. Compound interest Find the value of $1,000 left on deposit for 10 years at an annual rate of 7%, compounded hourly.

57. Saving for retirement When John was 20 years old, he opened an individual retirement account by investing $2,000 at 11% interest, compounded quarterly. How much will his investment be worth when he is 65 years old?

58. Biology One bacterium divides into two bacteria every 5 minutes. If two bacteria multiply enough to completely fill a petri dish in 2 hours, how long will it take one bacterium to fill the dish?

59. Pest control To reduce the population of a destructive moth, biologists release 1,000 sterilized male moths each day into the environment. If 80% of these moths alive one day survive until the next, then after a long time the population of sterile males is the sum of the infinite geometric series

$$1,000 + 1,000(0.8) + 1,000(0.8)^2 + 1,000(0.8)^3 + \cdots$$

Find the long-term population.

60. Pest control If mild weather increases the day-to-day survival rate of the sterile male moths in Exercise 59 to 90%, find the long-term population.

61. Mathematical myth A legend tells of a king who offered to grant the inventor of the game of chess any request. The inventor said, "Simply place one grain of wheat on the first square of a chessboard, two grains on the second, four on the third, and so on, until the board is full. Then give me the wheat." The king agreed. How many grains did the king need to fill the chessboard?

62. Mathematical myth Estimate the size of the wheat pile in Exercise 61. (*Hint:* There are about one-half million grains of wheat in a bushel.)

Discovery and Writing

63. Does $0.999999 = 1$? Explain.

64. Does $0.999\ldots = 1$? Explain.

Review   *Perform each operation.* $\left(i = \sqrt{-1}.\right)$

65. $(3 + 2i) + (2 - 5i)$ **66.** $(7 + 8i) - (2 - 5i)$

67. $(3 + i)(3 - i)$ **68.** $(3 + \sqrt{3}i)(3 - \sqrt{3}i)$

69. $\dfrac{2 + 3i}{2 - i}$ **70.** $(7 + 3i)^2$

71. i^{127} **72.** i^{-127}

Image99/Jupiter Images

8.5   Mathematical Induction

Objectives

1. State the Principle of Mathematical Induction
2. Use Mathematical Induction to Prove Formulas

Suppose you are waiting in line to get into a movie and are wondering if you will be admitted. Even if the first person in line gets in, your worries will still be justified. Perhaps there is room in the theater for only a few more people.

It would be good news to hear the manager say, "If anyone gets in, the next person in line will also get in." However, this promise does not guarantee that anyone will be admitted. Perhaps the theater is full, and no one will get in.

However, when you see the first person in line walk in, you know that everyone will be admitted because you know two things:

- The first person was admitted.
- Because of the manager's promise, if the first person is admitted, then so is the second, and when the second person is admitted, then so is the third, and so on until everyone gets in.

This situation is similar to a game played with dominoes. Suppose that some dominoes are placed on end, as in Figure 8-1. When the first domino is knocked over, it knocks over the second. The second domino, in turn, knocks over the third, which knocks over the fourth, and so on until all of the dominoes fall. Two things must happen to guarantee that all of the dominoes will fall:

- The first domino must be knocked over.
- Every domino that falls must knock over the next one.

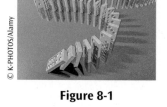

© K-PHOTOS/Alamy

Figure 8-1

When both conditions are met, it is certain that all of the dominoes will fall.

These two examples illustrate the basic principle of mathematical induction.

1. State the Principle of Mathematical Induction

In Section 8.3, we developed the following formula for the sum of the first n terms of an arithmetic series:

$$S_n = \frac{n(a + a_n)}{2}$$

If we apply this formula to the arithmetic series $1 + 2 + 3 + \cdots + n$, we have

$$S_n = 1 + 2 + 3 + \cdots + n = \frac{n(1 + n)}{2}$$

To see that this formula is true, we can check it for some positive numbers n:

For $n = 1$: $1 = \dfrac{1(1 + 1)}{2}$ is a true statement, because $1 = 1$.

For $n = 2$: $1 + 2 = \dfrac{2(1 + 2)}{2}$ is a true statement, because $3 = 3$.

For $n = 3$: $1 + 2 + 3 = \dfrac{3(1 + 3)}{2}$ is a true statement, because $6 = 6$.

For $n = 6$: $1 + 2 + 3 + 4 + 5 + 6 = \dfrac{6(1 + 6)}{2}$ is a true statement, because $21 = 21$.

Because the set of positive numbers is infinite, it is impossible to prove the formula by verifying it for all positive numbers. To verify this series formula for all positive numbers n, we must have a method of proof called **mathematical induction,** a method first used extensively by Giuseppe Peano (1858–1932).

Mathematical Induction	If a statement involving the natural number n has the following two properties

1. The statement is true for $n = 1$, and

2. If the statement is true for $n = k$, then it is true for $n = k + 1$,

then the statement is true for all natural numbers.

Mathematical induction provides a way to prove many formulas. Any proof by induction involves two parts. First, we must show that the formula is true for the number 1. Second, we must show that, if the formula is true for any natural number k, then it also is true for the natural number $k + 1$. A proof by induction is complete only when both of these properties are established.

2. Use Mathematical Induction to Prove Formulas

EXAMPLE 1   Use induction to prove that the following formula is true for every natural number n.

$$1 + 2 + 3 + \cdots + n = \frac{n(n + 1)}{2}$$

Solution   **Proof by induction:**

Part 1:   Verify that the formula is true for $n = 1$. When $n = 1$, there is a single term, the number 1, on the left side of the equation. Substituting 1 for n on the right side, we have

$$1 = \frac{n(n + 1)}{2}$$

$$1 = \frac{(1)(1 + 1)}{2}$$

$$1 = 1$$

The formula is true when $n = 1$. Part 1 of the proof is complete.

Part 2:   We assume that the given formula is true when $n = k$. By this assumption, called the **induction hypothesis,** we accept that

$$(1) \qquad 1 + 2 + 3 + \cdots + k = \frac{k(k + 1)}{2}$$

is a true statement. We must show that the induction hypothesis forces the given formula to be true when $n = k + 1$. We can show this by verifying the statement

$$(2) \qquad 1 + 2 + 3 + \cdots + k + (k + 1) = \frac{(k + 1)[(k + 1) + 1]}{2}$$

which is obtained from the given formula by replacing n with $k + 1$.

Comparing the left sides of Equations 1 and 2 shows that the left side of Equation 2 contains an extra term of $k + 1$. Thus, we add $k + 1$ to both sides of Equation 1 (which was assumed to be true) to obtain the equation

$$1 + 2 + 3 + \cdots + k + (k + 1) = \frac{k(k + 1)}{2} + (k + 1)$$

Because both terms on the right side of this equation have a common factor of $k + 1$, the right side factors, and the equation can be written as follows:

$$1 + 2 + 3 + \cdots + k + (k + 1) = (k + 1)\left(\frac{k}{2} + 1\right)$$

$$= (k + 1)\left(\frac{k + 2}{2}\right)$$

$$= \frac{(k + 1)(k + 2)}{2}$$

$$= \frac{(k + 1)[(k + 1) + 1]}{2}$$

This final result is Equation 2. Because the truth of Equation 1 implies the truth of Equation 2, part 2 of the proof is complete. Parts 1 and 2 together establish that the formula is true for any natural number n.

Self Check 1   Verify that the formula holds for $n = 4$.

EXAMPLE 2   Use induction to prove the following formula for all natural numbers n.

$$1 + 5 + 9 + \cdots + (4n - 3) = n(2n - 1)$$

Solution   **Proof by induction:**

Part 1:   First we verify the formula for $n = 1$. When $n = 1$, there is a single term, the number 1, on the left side of the equation. Substituting 1 for n on the right side, we have

$$1 = \mathbf{1}[2(\mathbf{1}) - 1]$$
$$1 = 1$$

The formula is true for $n = 1$. Part 1 of the proof is complete.

Part 2:   We assume that the formula is true for $n = k$. Hence,

(3) $$1 + 5 + 9 + \cdots + (4k - 3) = k(2k - 1)$$

is a true statement. To show that the induction hypothesis guarantees the truth of the formula for $k + 1$ terms, we add the $(k + 1)$th term to both sides of Equation 3. Because the terms on the left side increase by 4, the $(k + 1)$th term is $(4k - 3) + 4$, or $4k + 1$. Adding $4k + 1$ to both sides of Equation 3 gives

$$1 + 5 + 9 + \cdots + (4k - 3) + (4k + 1) = k(2k - 1) + (4k + 1)$$

We can simplify the right side and write the previous equation as follows:

$$1 + 5 + 9 + \cdots + (4k - 3) + [4(k + 1) - 3] = 2k^2 + 3k + 1$$
$$= (k + 1)(2k + 1)$$
$$= (k + 1)[2(k + 1) - 1]$$

Since this result has the same form as the given formula, except that $k + 1$ replaces n, the truth of the formula for $n = k$ implies the truth of the formula for $n = k + 1$. Part 2 of the proof is complete.

Because both of the induction requirements are true, the formula is true for all natural numbers n.

Self Check 2   Verify that the formula holds for $n = 6$.

EXAMPLE 3   Prove that $\dfrac{1}{2} + \dfrac{1}{4} + \dfrac{1}{8} + \cdots + \dfrac{1}{2^n} < 1$.

Solution   **Proof by induction:**

Part 1:   We verify the formula for $n = 1$. When $n = 1$, there is a single term, the fraction $\frac{1}{2}$, on the left side of the equation. Substituting 1 for n on the right side, we have the following true statement:

$$\frac{1}{2} < 1$$

The formula is true for $n = 1$. Part 1 of the proof is complete.

Part 2: We assume that the inequality is true for $n = k$. Thus,

$$\frac{1}{2} + \frac{1}{4} + \frac{1}{8} + \cdots + \frac{1}{2^k} < 1$$

We can multiply both sides of the above inequality by $\frac{1}{2}$ to get

$$\frac{1}{2}\left(\frac{1}{2} + \frac{1}{4} + \frac{1}{8} + \cdots + \frac{1}{2^k}\right) < 1\left(\frac{1}{2}\right)$$

or

$$\frac{1}{4} + \frac{1}{8} + \frac{1}{16} + \cdots + \frac{1}{2^{k+1}} < \frac{1}{2}$$

We now add $\frac{1}{2}$ to both sides of this inequality to get

$$\frac{1}{2} + \frac{1}{4} + \frac{1}{8} + \frac{1}{16} + \cdots + \frac{1}{2^{k+1}} < \frac{1}{2} + \frac{1}{2}$$

or

$$\frac{1}{2} + \frac{1}{4} + \frac{1}{8} + \frac{1}{16} + \cdots + \frac{1}{2^{k+1}} < 1$$

The resulting inequality is the same as the original inequality, except that $k + 1$ appears in place of n. Thus, the truth of the inequality for $n = k$ implies the truth of the inequality for $n = k + 1$. Part 2 of the proof is complete.

Because both of the induction requirements have been verified, this inequality is true for all natural numbers.

Self Check 3 Verify that the formula holds for $n = 8$.

Some statements are not true when $n = 1$ but are true for all natural numbers equal to or greater than some given natural number (say, q). In these cases, we verify the given statements for $n = q$ in part 1 of the induction proof. After establishing part 2 of the induction proof, the given statement is proved for all natural numbers that are greater than q.

Self Check Answers 1. $10 = 10$ 2. $66 = 66$ 3. $\frac{255}{256} < 1$

8.5 Exercises

Vocabulary and Concepts *Fill in the blanks.*

1. Any proof by induction requires ___ parts.
2. Part 1 is to show that the statement is true for

___.

3. Part 2 is to show that the statement is true for
 _____ whenever it is true for $n = k$.
4. When we assume that a formula is true for $n = k$, we call the assumption the induction _____.

Practice *Verify each formula for $n = 1, 2, 3,$ and 4.*

5. $5 + 10 + 15 + \cdots + 5n = \dfrac{5n(n + 1)}{2}$

6. $1^2 + 2^2 + 3^2 + \cdots + n^2 = \dfrac{n(n + 1)(2n + 1)}{6}$

7. $7 + 10 + 13 + \cdots + (3n + 4) = \dfrac{n(3n + 11)}{2}$

8. $1(3) + 2(4) + 3(5) + \cdots + n(n + 2) =$
$$\frac{n}{6}(n + 1)(2n + 7)$$

Prove each formula by induction, if possible.

9. $2 + 4 + 6 + \cdots + 2n = n(n + 1)$

10. $1 + 3 + 5 + \cdots + (2n - 1) = n^2$

11. $3 + 7 + 11 + \cdots + (4n - 1) = n(2n + 1)$

12. $4 + 8 + 12 + \cdots + 4n = 2n(n + 1)$

13. $10 + 6 + 2 + \cdots + (14 - 4n) = 12n - 2n^2$

14. $8 + 6 + 4 + \cdots + (10 - 2n) = 9n - n^2$

15. $2 + 5 + 8 + \cdots + (3n - 1) = \dfrac{n(3n + 1)}{2}$

16. $3 + 6 + 9 + \cdots + 3n = \dfrac{3n(n + 1)}{2}$

17. $1^2 + 2^2 + 3^2 + \cdots + n^2 = \dfrac{n(n + 1)(2n + 1)}{6}$

18. $1 + 2 + 3 + \cdots + (n - 1) + n + (n - 1) + \cdots$
$+ 3 + 2 + 1 = n^2$

19. $\dfrac{1}{3} + 2 + \dfrac{11}{3} + \cdots + \left(\dfrac{5}{3}n - \dfrac{4}{3}\right) = n\left(\dfrac{5}{6}n - \dfrac{1}{2}\right)$

20. $\dfrac{1}{1 \cdot 2} + \dfrac{1}{2 \cdot 3} + \dfrac{1}{3 \cdot 4} + \cdots + \dfrac{1}{n(n + 1)} = \dfrac{n}{n + 1}$

21. $\dfrac{1}{2} + \dfrac{1}{4} + \dfrac{1}{8} + \cdots + \left(\dfrac{1}{2}\right)^n = 1 - \left(\dfrac{1}{2}\right)^n$

22. $\dfrac{1}{3} + \dfrac{2}{9} + \dfrac{4}{27} + \cdots + \dfrac{1}{3}\left(\dfrac{2}{3}\right)^{n-1} = 1 - \left(\dfrac{2}{3}\right)^n$

23. $2^0 + 2^1 + 2^2 + 2^3 + \cdots + 2^{n-1} = 2^n - 1$

24. $1^3 + 2^3 + 3^3 + \cdots + n^3 = \left[\dfrac{n(n + 1)}{2}\right]^2$

25. Prove that $x - y$ is a factor of $x^n - y^n$. (*Hint:* Consider subtracting and adding xy^k to the binomial $x^{k+1} - y^{k+1}$.)

26. Prove that $n < 2^n$.

27. There are $180°$ in the sum of the angles of any triangle. Prove that $(n - 2)180°$ is the sum of the angles of any simple polygon when n is its number of sides. (*Hint:* If a polygon has $k + 1$ sides, it has $k - 2$ sides plus three more sides.)

28. Consider the equation
$$1 + 3 + 5 + \cdots + 2n - 1 = 3n - 2$$

  a. Is the equation true for $n = 1$?

  b. Is the equation true for $n = 2$?

  c. Is the equation true for all natural numbers n?

29. If $1 + 2 + 3 + \cdots + n = \dfrac{n}{2}(n + 1) + 1$ were true for $n = k$, show that it would be true for $n = k + 1$. Is it true for $n = 1$?

30. Prove that $n + 1 = 1 + n$ for each natural number n.

31. If n is any natural number, prove that $7^n - 1$ is divisible by 6.

32. Prove that $1 + 2n < 3^n$ for $n > 1$.

33. Prove that, if r is a real number where $r \neq 1$, then
$$1 + r + r^2 + \cdots + r^n = \frac{1 - r^{n+1}}{1 - r}$$

34. Prove the formula for the sum of the first n terms of an arithmetic series:
$$a + [a + d] + [a + 2d] + [a + (n - 1)d] = \frac{n(a + a_n)}{2}$$
where $a_n = a + (n - 1)d$.

Discovery and Writing

35. The expression a^m, where m is a natural number, was defined in Section 0.2. An alternative definition of a^m is (part 1) $a^1 = a$ and (part 2) $a^{m+1} = a^m \cdot a$. Use induction on n to prove the product rule for exponents, $a^m a^n = a^{m+n}$.

36. Use induction on n to prove the power rule for exponents, $(a^m)^n = a^{mn}$. (See Exercise 35.)

37. **Tower of Hanoi** A well-known problem in mathematics is "The Tower of Hanoi," first attributed to Edouard Lucas in 1883. In this problem, several disks, each of a different size and with a hole in the center, are placed on a board, with progressively smaller disks going up the stack. The object is to transfer the stack of disks to another peg by moving only one disk at a time and never placing a disk over a smaller one.

  a. Find the minimum number of moves required if there is only one disk.

  b. Find the minimum number of moves required if there are two disks.

  c. Find the minimum number of moves required if there are three disks.

  d. Find the minimum number of moves required if there are four disks.

38. Tower of Hanoi   The results in Exercise 37 suggest that the minimum number of moves required to transfer n disks from one peg to another is given by the formula $2^n - 1$. Use the following outline to prove that this result is correct using mathematical induction.

a. Verify the formula for $n = 1$.

b. Write the induction hypothesis.

c. How many moves are needed to transfer all but the largest of $k + 1$ disks to another peg?

d. How many moves are needed to transfer the largest disk to an empty peg?

e. How many moves are needed to transfer the first k disks back onto the largest one?

f. How many moves are needed to accomplish steps **c, d,** and **e**?

g. Show that part **f** can be written in the form $2^{(k+1)} - 1$.

h. Write the conclusion of the proof.

Review   *Graph each inequality.*

39. $3x + 4y \leq 12$

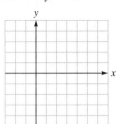

40. $4x - 3y < 12$

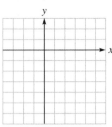

41. $2x + y > 5$

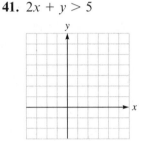

42. $x + 3y \geq 7$

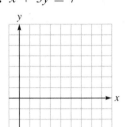

8.6   Permutations and Combinations

Objectives

1. Use the Multiplication Principle for Events
2. Solve Permutation Problems
3. Solve Combination Problems
4. Use Combination Notation to Write the Binomial Theorem
5. Find the Number of Permutations of Like Things

1.   Use the Multiplication Principle for Events

Lydia plans to go to dinner and attend a movie. If she has a choice of four restaurants and three movies, in how many ways can she spend her evening? There are four choices of restaurants and, for any one of these choices, there are three choices of movies, as shown in the tree diagram in Figure 8-2.

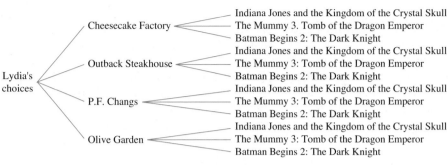

Figure 8-2

The diagram shows that Lydia has 12 ways to spend her evening. One possibility is to eat at the Cheesecake Factory and watch Indiana Jones. Another is to eat at the Olive Garden and watch Batman Begins 2.

Any situation that has several outcomes is called an **event.** Lydia's first event (choosing a restaurant) can occur in 4 ways. Her second event (choosing a movie) can occur in 3 ways. Thus, she has $4 \cdot 3$, or 12, ways to spend her evening. This example illustrates the **multiplication principle for events.**

Multiplication Principle for Events	Let E_1 and E_2 be two events. If E_1 can occur in a_1 ways, and if—after E_1 has occurred—E_2 can occur in a_2 ways, then the event "E_1 followed by E_2" can occur in $a_1 \cdot a_2$ ways.

The multiplication principle can be extended to n events.

EXAMPLE 1 If a traveler has 4 ways to go from New York to Chicago, 3 ways to go from Chicago to Denver, and 6 ways to go from Denver to San Francisco, in how many ways can she go from New York to San Francisco?

Solution We can let E_1 be the event "going from New York to Chicago," E_2 be the event "going from Chicago to Denver," and E_3 be the event "going from Denver to San Francisco." Since there are 4 ways to accomplish E_1, 3 ways to accomplish E_2, and 6 ways to accomplish E_3, the number of routes available is

$$4 \cdot 3 \cdot 6 = 72$$

Self Check 1 If a man has 4 sweaters and 5 pairs of slacks, how many different outfits can he wear?

2. Solve Permutation Problems

Suppose we want to arrange 7 books in order on a shelf. We can fill the first space with any one of the 7 books, the second space with any of the remaining 6 books, the third space with any of the remaining 5 books, and so on, until there is only one space left to fill with the last book. According to the multiplication principle, the number of ordered arrangements of the books is

$$7 \cdot 6 \cdot 5 \cdot 4 \cdot 3 \cdot 2 \cdot 1 = 5,040$$

When finding the number of possible ordered arrangements of books on a shelf, we are finding the number of **permutations.** The number of permutations of 7 books, using all the books, is 5,040. The symbol $P(n, r)$ is read as "the number of permutations of n things r at a time." Thus, $P(7, 7) = 5,040$.

EXAMPLE 2 Assume that there are 7 signal flags of 7 different colors to hang on a mast. How many different signals can be sent when 3 flags are used?

Solution We are asked to find $P(7, 3)$, the number of permutations (ordered arrangements) of 7 things using 3 of them. Any one of the 7 flags can hang in the top position on the mast. Any one of the 6 remaining flags can hang in the middle position, and any one of the remaining 5 flags can hang in the bottom position. By the multiplication principle, we have

$$P(7, 3) = 7 \cdot 6 \cdot 5 = 210$$

It is possible to send 210 different signals.

Self Check 2 How many signals can be sent if two of the 7 flags are missing?

Although it is correct to write $P(7, 3) = 7 \cdot 6 \cdot 5$, we will change the form of the answer to obtain a convenient formula. To derive this formula, we proceed as follows:

$$P(7, 3) = 7 \cdot 6 \cdot 5$$

$$= \frac{7 \cdot 6 \cdot 5 \cdot 4 \cdot 3 \cdot 2 \cdot 1}{4 \cdot 3 \cdot 2 \cdot 1} \qquad \textbf{Multiply numerator and denominator by } \mathbf{4 \cdot 3 \cdot 2 \cdot 1.}$$

$$= \frac{7!}{4!}$$

$$= \frac{7!}{(7 - 3)!}$$

The generalization of this idea gives the following formula.

Formula for $P(n, r)$ The number of permutations of n things r at a time is given by

$$P(n, r) = \frac{n!}{(n - r)!}$$

EXAMPLE 3 Find: **a.** $P(8, 4)$ **b.** $P(n, n)$ **c.** $P(n, 0)$

Solution We will substitute into the formula for $P(n, r)$ and simplify.

a. $P(8, 4) = \dfrac{8!}{(8 - 4)!} = \dfrac{8 \cdot 7 \cdot 6 \cdot 5 \cdot \cancel{4!}}{\cancel{4!}} = 1,680$

b. $P(n, n) = \dfrac{n!}{(n - n)!} = \dfrac{n!}{0!} = \dfrac{n!}{1} = n!$

c. $P(n, 0) = \dfrac{n!}{(n - 0)!} = \dfrac{\cancel{n!}}{\cancel{n!}} = 1$

Self Check 3 Find: **a.** $P(7, 5)$ **b.** $P(6, 0)$

Parts b and c of Example 3 establish the following formulas.

Formulas for $P(n, n)$ and $P(n, 0)$ The number of permutations of n things n at a time and n things 0 at a time are given by the formulas

$$P(n, n) = n! \qquad \text{and} \qquad P(n, 0) = 1$$

EXAMPLE 4   In how many ways can a baseball manager arrange a batting order of 9 players if there are 25 players on the team?

Solution   To find the number of permutations of 25 things 9 at a time, we substitute 25 for n and 9 for r in the formula for finding $P(n, r)$.

$$P(n, r) = \frac{n!}{(n - r)!}$$

$$P(25, 9) = \frac{25!}{(25 - 9)!}$$

$$= \frac{25!}{16!}$$

$$= \frac{25 \cdot 24 \cdot 23 \cdot 22 \cdot 21 \cdot 20 \cdot 19 \cdot 18 \cdot 17 \cdot 16!}{16!}$$

$$\approx 741,354,768,000$$

The number of permutations is approximately 741,354,768,000.

Self Check 4   In how many ways can the manager arrange a batting order if 2 players can't play?

EXAMPLE 5   In how many ways can 5 people stand in a line if 2 people refuse to stand next to each other?

Solution   The total number of ways that 5 people can stand in line is

$$P(5, 5) = 5! = 5 \cdot 4 \cdot 3 \cdot 2 \cdot 1 = 120$$

To find the number of ways that 5 people can stand in line if 2 people insist on standing together, we consider the two people as one person. Then there are 4 people to stand in line, and this can be done in $P(4, 4) = 4! = 24$ ways. However, because either of the two who are paired could be first, there are two arrangements for the pair who insist on standing together. Thus, there are $2 \cdot 4!$, or 48 ways that 5 people can stand in line if 2 people insist on standing together.

The number of ways that 5 people can stand in line if 2 people refuse to stand together is $5! = 120$ (the total number of ways to line up 5 people) minus $2 \cdot 4! = 48$ (the number of ways to line up the 5 people if 2 do stand together):

$$120 - 48 = 72$$

There are 72 ways to line up the people.

Self Check 5   In how many ways can 5 people stand in a line if one person demands to be first?

EXAMPLE 6   In how many ways can 5 people be seated at a round table?

Solution   If we were to seat 5 people in a row, there would be 5! possible arrangements. However, at a round table, each person has a neighbor to the left and to the right. If each person moves one, two, three, four, or five places to the left, everyone has the same neighbors and the arrangement has not changed. Thus, we must divide 5! by 5 to get rid of these duplications. The number of ways that 5 people can be seated at a round table is

$$\frac{5!}{5} = \frac{\cancel{5} \cdot 4!}{\cancel{5}} = 4! = 4 \cdot 3 \cdot 2 \cdot 1 = 24$$

Self Check 6   In how many ways can 6 people be seated at a round table?

The results of Example 6 suggest the following fact.

Circular Arrangements   There are $(n - 1)!$ ways to arrange n things in a circle.

3.   Solve Combination Problems

Suppose that a class of 12 students selects a committee of 3 people to plan a party. With committees, order is not important. A committee of John, Maria, and Raul is the same as a committee of Maria, Raul, and John. However, if we assume for the moment that order is important, we can find the number of permutations of 12 things 3 at a time.

$$P(12, 3) = \frac{12!}{(12 - 3)!} = \frac{12 \cdot 11 \cdot 10 \cdot \cancel{9!}}{\cancel{9!}} = 1{,}320$$

However, since we do not care about order, this result of 1,320 ways is too large. Because there are 6 ways (3! = 6) of ordering every committee of 3 students, the result of $P(12, 3) = 1{,}320$ is exactly 6 times too big. To get the correct number of committees, we must divide $P(12, 3)$ by 6:

$$\frac{P(12,3)}{6} = \frac{1{,}320}{6} = 220$$

In cases of selection where order is not important, we are interested in **combinations,** not permutations. The symbols $C(n, r)$ and $\binom{n}{r}$ both mean the number of combinations of n things r at a time.

If a committee of r people is chosen from a total of n people, the number of possible committees is $C(n, r)$, and there will be $r!$ arrangements of each committee. If we consider the committee as an ordered grouping, the number of orderings is $P(n, r)$. Thus, we have

(1) $r!\,C(n, r) = P(n, r)$

Comment

When discussing permutations, order counts. When discussing combinations, order doesn't count.

We can divide both sides of Equation 1 by $r!$ to obtain the formula for finding $C(n, r)$.

$$C(n, r) = \binom{n}{r} = \frac{P(n, r)}{r!} = \frac{n!}{r!(n - r)!}$$

Formula for C(n, r)   The number of combinations of n things r at a time is given by

$$C(n, r) = \binom{n}{r} = \frac{n!}{r!(n - r)!}$$

In the exercises, you will be asked to prove the following formulas.

Formulas for C(n, n)   If n is a whole number, then
and C(n, 0)

$$C(n, n) = 1 \quad \text{and} \quad C(n, 0) = 1$$

EXAMPLE 7   If Carla must read 4 books from a reading list of 10 books, how many choices does she have?

Solution   Because the order in which the books are read is not important, we find the number of combinations of 10 things 4 at a time:

$$C(10, 4) = \frac{10!}{4!(10 - 4)!} = \frac{10 \cdot 9 \cdot 8 \cdot 7 \cdot \cancel{6!}}{4 \cdot 3 \cdot 2 \cdot 1 \cdot \cancel{6!}}$$
$$= \frac{10 \cdot 9 \cdot 8 \cdot 7}{4 \cdot 3 \cdot 2}$$
$$= 210$$

Carla has 210 choices.

Self Check 7   How many choices would Carla have if she had to read 5 books?

EXAMPLE 8   A class consists of 15 men and 8 women. In how many ways can a debate team be chosen with 3 men and 3 women?

Solution   There are $C(15, 3)$ ways of choosing 3 men and $C(8, 3)$ ways of choosing 3 women. By the multiplication principle, there are $C(15, 3) \cdot C(8, 3)$ ways of choosing members of the debate team:

$$C(15, 3) \cdot C(8, 3) = \frac{15!}{3!(15 - 3)!} \cdot \frac{8!}{3!(8 - 3)!}$$
$$= \frac{15 \cdot 14 \cdot 13}{6} \cdot \frac{8 \cdot 7 \cdot 6}{6}$$
$$= 25,480$$

There are 25,480 ways to choose the debate team.

Self Check 8   In how many ways can the debate team be chosen if it is to have 4 men and 2 women?

4.   Use Combination Notation to Write the Binomial Theorem

The formula

$$C(n, r) = \frac{n!}{r!(n - r)!}$$

gives the coefficient of the $(r + 1)$th term of the binomial expansion of $(a + b)^n$. This implies that the coefficients of a binomial expansion can be used to solve problems involving combinations. The binomial theorem is restated below—this time listing the $(r + 1)$th term and using combination notation.

Binomial Theorem   If n is any positive integer, then

$$(a + b)^n = \binom{n}{0}a^n + \binom{n}{1}a^{n-1}b + \binom{n}{2}a^{n-2}b^2 + \cdots + \binom{n}{r}a^{n-r}b^r + \cdots + \binom{n}{n}b^n$$

EXAMPLE 9   Use Pascal's triangle to compute $C(7, 5)$.

Solution   Consider the eighth row of Pascal's triangle and the corresponding combinations:

$$\begin{array}{cccccccc} 1 & 7 & 21 & 35 & 35 & 21 & 7 & 1 \\ \binom{7}{0} & \binom{7}{1} & \binom{7}{2} & \binom{7}{3} & \binom{7}{4} & \binom{7}{5} & \binom{7}{6} & \binom{7}{7} \end{array}$$

Comment
In the expansion of $(a + b)^n$, the term containing b^r is given by

$$\binom{n}{r}a^{n-r}b^r$$

$$C(7, 5) = \binom{7}{5} = 21.$$

Self Check 9   Use Pascal's triangle to compute $C(6, 5)$.

5. Find the Number of Permutations of Like Things

A *word* is a distinguishable arrangement of letters. For example, six words can be formed with the letters a, b, and c if each letter is used exactly once. The six words are

abc,   acb,   bac,   bca,   cab,   and   cba

If there are n distinct letters and each letter is used once, the number of distinct words that can be formed is $n! = P(n, n)$. It is more complicated to compute the number of distinguishable words that can be formed with n letters when some of the letters are duplicates.

EXAMPLE 10   Find the number of "words" that can be formed if each of the 6 letters of the word *little* is used once.

Solution   For the moment, we assume that the letters of the word *little* are distinguishable: "LiTle." The number of words that can be formed using each letter once is $6! = P(6, 6)$. However, in reality we cannot tell the *l*'s or the *t*'s apart. Therefore, we must divide by a number to get rid of these duplications. Because there are 2! orderings of the two *l*'s and 2! orderings of the two *t*'s, we divide by $2! \cdot 2!$. The number of words that can be formed using each letter of the word *little* is

$$\frac{P(6, 6)}{2! \cdot 2!} = \frac{6!}{2! \cdot 2!} = \frac{6 \cdot 5 \cdot \overset{2}{\cancel{4}} \cdot 3 \cdot \cancel{2} \cdot 1}{\cancel{2} \cdot 1 \cdot \cancel{2} \cdot 1} = 180$$

Self Check 10   How many words can be formed if each letter of the word *balloon* is used once?

Example 10 illustrates the following general principle.

| **Permutations of Like Things** | The number of permutations of n things with a things alike, b things alike, and so on, is $$\frac{n!}{a!b!\cdots}$$ |

8.6 Exercises

Vocabulary and Concepts   *Fill in the blanks.*

1. If E_1 and E_2 are two events and E_1 can be done in 4 ways and E_2 can be done in 6 ways, then the event E_1 followed by E_2 can be done in ___ ways.

2. An arrangement of n objects is called a _____.

3. $P(n, r) =$ _____

4. $P(n, n) =$ ___

5. $P(n, 0) =$ ___

6. There are _____ ways to arrange n things in a circle.

7. $C(n, r) =$ _____

8. Using combination notation, $C(n, r) =$ ___.

9. $C(n, n) =$ ___

10. $C(n, 0) =$ ___

11. If a word with n letters has a of one letter, b of another letter, and so on, the number of different words that can be formed is _____ .

12. Where the order of selection is not important, we are interested in _____, not _____.

Practice   *Evaluate each expression.*

13. $P(7, 4)$

14. $P(8, 3)$

15. $C(7, 4)$

16. $C(8, 3)$

17. $P(5, 5)$

18. $P(5, 0)$

19. $\binom{5}{4}$

20. $\binom{8}{4}$

21. $\binom{5}{0}$

22. $\binom{5}{5}$

23. $P(5, 4) \cdot C(5, 3)$

24. $P(3, 2) \cdot C(4, 3)$

25. $\binom{5}{3}\binom{4}{3}\binom{3}{3}$

26. $\binom{5}{5}\binom{6}{6}\binom{7}{7}\binom{8}{8}$

27. $\binom{68}{66}$

28. $\binom{100}{99}$

Applications

29. **Choosing lunch**   A lunchroom has a machine with eight kinds of sandwiches, a machine with four kinds of soda, a machine with both white and chocolate milk, and a machine with three kinds of ice cream. How many different lunches can be chosen? (Consider a lunch to be one sandwich, one drink, and one ice cream.)

30. **Manufacturing license plates**   How many six-digit license plates can be manufactured if no license plate number begins with 0?

31. **Available phone numbers**   How many different seven-digit phone numbers can be used in one area code if no phone number begins with 0 or 1?

32. **Arranging letters**   In how many ways can the letters of the word *number* be arranged?

33. **Arranging letters with restrictions**   In how many ways can the letters of the word *number* be arranged if the e and r must remain next to each other?

34. **Arranging letters with restrictions**   In how many ways can the letters of the word *number* be arranged if the e and r cannot be side by side?

35. Arranging letters with repetitions   How many ways can five Scrabble tiles bearing the letters, F, F, F, L, and U be arranged to spell the word *fluff?*

36. Arranging letters with repetitions   How many ways can six Scrabble tiles bearing the letters B, E, E, E, F, and L be arranged to spell the word *feeble?*

37. Placing people in line   In how many arrangements can 8 women be placed in a line?

38. Placing people in line   In how many arrangements can 5 women and 5 men be placed in a line if the women and men alternate?

39. Placing people in line   In how many arrangements can 5 women and 5 men be placed in a line if all the men line up first?

40. Placing people in line   In how many arrangements can 5 women and 5 men be placed in a line if all the women line up first?

41. Combination locks   How many permutations does a combination lock have if each combination has 3 numbers, no two numbers of the combination are the same, and the lock dial has 30 notches?

42. Combination locks   How many permutations does a combination lock have if each combination has 3 numbers, no two numbers of the combination are the same, and the lock dial has 100 notches?

43. Seating at a table   In how many ways can 8 people be seated at a round table?

44. Seating at a table   In how many ways can 7 people be seated at a round table?

45. Seating at a table   In how many ways can 6 people be seated at a round table if 2 of the people insist on sitting together?

46. Seating at a table   In how many ways can 6 people be seated at a round table if 2 of the people refuse to sit together?

47. Arrangements in a circle   In how many ways can 7 children be arranged in a circle if Sally and John want to sit together and Martha and Peter want to sit together?

48. Arrangements in a circle   In how many ways can 8 children be arranged in a circle if Laura, Scott, and Paula want to sit together?

49. Selecting candy bars   In how many ways can 4 candy bars be selected from 10 different candy bars?

50. Selecting birthday cards   In how many ways can 6 birthday cards be selected from 24 different cards?

51. Circuit wiring   A wiring harness containing a red, a green, a white, and a black wire must be attached to a control panel. In how many different orders can the wires be attached?

52. Grading homework   A professor grades homework by randomly checking 7 of the 20 problems assigned. In how many different ways can this be done?

53. Forming words with distinct letters   How many words can be formed from the letters of the word *plastic* if each letter is to be used once?

54. Forming words with repeated letters   How many words can be formed from the letters of the word *banana* if each letter is to be used once?

55. Manufacturing license plates   How many license plates can be made using two different letters followed by four different digits if the first digit cannot be 0 and the letter O is not used?

56. Planning class schedules   If there are 7 class periods in a school day and a typical student takes 5 classes, how many different time patterns are possible for the student?

57. Selecting golf balls   From a bucket containing 6 red and 8 white golf balls, in how many ways can we draw 6 golf balls of which 3 are red and 3 are white?

58. Selecting a committee   In how many ways can you select a committee of 3 Republicans and 3 Democrats from a group containing 18 Democrats and 11 Republicans?

59. Selecting a committee   In how many ways can you select a committee of 4 Democrats and 3 Republicans from a group containing 12 Democrats and 10 Republicans?

60. Drawing cards   In how many ways can you select a group of 5 red cards and 2 black cards from a deck containing 10 red cards and 8 black cards?

61. Planning dinner   In how many ways can a husband and wife choose 2 different dinners from a menu of 17 dinners?

62. Placing people in line   In how many ways can 7 people stand in a row if 2 of the people refuse to stand together?

63. Geometry   How many lines are determined by 8 points if no 3 points lie on a straight line?

64. Geometry How many lines are determined by 10 points if no 3 points lie on a straight line?

65. Coaching basketball How many different teams can a basketball coach start if the entire squad consists of 10 players? (Assume that a starting team has 5 players and each player can play all positions.)

66. Managing baseball How many different teams can a manager start if the entire squad consists of 25 players? (Assume that a starting team has 9 players and each player can play all positions.)

67. Selecting job applicants There are 30 qualified applicants for 5 openings in the sales department. In how many different ways can the group of 5 be selected?

68. Sales promotions If a customer purchases a new stereo system during the spring sale, he may choose any 6 CDs from 20 classical and 30 jazz selections. In how many ways can the customer choose 3 of each?

69. Guessing on matching questions Ten words are to be paired with the correct 10 out of 12 possible definitions. How many ways are there of guessing?

70. Guessing on true-false exams How many possible ways are there of guessing on a 10-question true-false exam, if it is known that the instructor will have 5 true and 5 false responses?

71. Number of Wendy's® hamburgers Wendy's Old Fashioned Hamburgers offers eight toppings for their single hamburger. How many different single hamburgers can be ordered?

Discovery and Writing

72. Prove that $C(n, n) = 1$.

73. Prove that $C(n, 0) = 1$.

74. Prove that $\binom{n}{r} = \binom{n}{n-r}$.

75. Show that the binomial theorem can be expressed in the form

$$(a + b)^n = \sum_{k=0}^{n} \binom{n}{k} a^{n-k} b^k$$

76. Explain how to use Pascal's triangle to find $C(8, 5)$.

77. Explain how to use Pascal's triangle to find $C(10, 8)$.

Review *Find the value of x.*

78. $\log_x 16 = 4$

79. $\log_\pi x = \dfrac{1}{2}$

80. $\log_{\sqrt{7}} 49 = x$

81. $\log_x \dfrac{1}{2} = -\dfrac{1}{3}$

Determine whether the statement is true or false.

82. $\log_{17} 1 = 0$

83. $\log_5 0 = 1$

84. $\log_b b^b = b$

85. $\dfrac{\log_7 A}{\log_7 B} = \log_7 \dfrac{A}{B}$

8.7 Probability

Objectives

1. Compute Probabilities
2. Use the Multiplication Property of Probabilities

In some parts of the United States, casinos are a big part of the entertainment industry. Casinos design their games using the laws of probability to ensure they make huge amounts of money at their customers' expense.

Gerolamo Cardano
(1501–1576)

An Italian doctor and mathematician, Cardano's greatest mathematical work was his book, *Ars Magna*, in which he gave methods of solving cubic and quartic equations.

Gerolamo Cardano was an accomplished gambler who developed much of the theory of probability in the casino. His book *Liber de Ludo Aleae* (published in 1663 after his death) was the first book to develop the topics of probability logically. Cardano predicted the day of his death. On that day, when it looked as though he would live, it is said that he committed suicide to make his prediction come true.

1.   Compute Probabilities

The probability that an event will occur is a measure of the likelihood of that event. A tossed coin, for example, can land in two ways, either heads or tails. Because one of these two equally likely outcomes is heads, we expect that out of several tosses, about half will be heads. We say that the probability of obtaining heads in a single toss of the coin is $\frac{1}{2}$.

If records show that out of 100 days with weather conditions like today's, 30 have received rain, the weather service will report, "There is a $\frac{30}{100}$ or 30% probability of rain today."

An **experiment** is a process for which the outcome is uncertain. Tossing a coin, rolling a die, drawing a card, and predicting rain are examples of experiments. For any experiment, the set of all possible outcomes is called a **sample space.**

The sample space, S, for the experiment of tossing two coins is the set

$$S = \{(H, H), (H, T), (T, H), (T, T)\}$$

where the ordered pair (H, T) represents the outcome "heads on the first coin and tails on the second coin." Because there are two possible outcomes for the first coin and two for the second coin, we know (by the multiplication principle for events) that there are $2 \cdot 2 = 4$ possible outcomes. Since there are 4 elements in the sample space S, we write

$n(S) = 4$ **Read as "The number of elements in set S is 4."**

An **event** associated with an experiment is any subset of the sample space of that experiment. For example, if E is the event "getting at least one heads" in the experiment of tossing two coins, then

$$E = \{(H, H), (H, T), (T, H)\}$$

and $n(E) = 3$. Because the outcome of getting at least one heads can occur in 3 out of 4 possible ways, we say that the **probability** of a favorable outcome is $\frac{3}{4}$.

$$P(E) = P(\text{at least one heads}) = \frac{3}{4}$$

We define the probability of an event as follows.

Probability of an Event

If S is the sample space of an experiment with n distinct and equally likely outcomes, and E is an event that occurs in s of those ways, then the **probability of E** is

$$P(E) = \frac{n(E)}{n(S)} = \frac{s}{n}$$

Because $0 \le s \le n$, it follows that $0 \le \frac{s}{n} \le 1$. This implies that all probabilities have values from 0 to 1. An event that cannot happen has probability 0. An event that is certain to happen has probability 1.

To say that the probability of tossing heads on one toss of a coin is $\frac{1}{2}$ means that if a fair coin is tossed a large number of times, the ratio of the number of heads to the total number of tosses is nearly $\frac{1}{2}$.

To say that the probability of rolling 5 on one roll of a die is $\frac{1}{6}$ means that as the number of rolls approaches infinity, the ratio of the number of favorable outcomes (rolling a 5) to the total number of outcomes (rolling a 1, 2, 3, 4, 5, or 6) approaches $\frac{1}{6}$.

EXAMPLE 1   Show the sample space of the experiment "rolling two dice a single time."

Solution   We can list ordered pairs and let the first number be the result on the first die and the second number the result on the second die. The sample space, S, is the set with the following elements:

(1, 1)	(1, 2)	(1, 3)	(1, 4)	(1, 5)	(1, 6)
(2, 1)	(2, 2)	(2, 3)	(2, 4)	(2, 5)	(2, 6)
(3, 1)	(3, 2)	(3, 3)	(3, 4)	(3, 5)	(3, 6)
(4, 1)	(4, 2)	(4, 3)	(4, 4)	(4, 5)	(4, 6)
(5, 1)	(5, 2)	(5, 3)	(5, 4)	(5, 5)	(5, 6)
(6, 1)	(6, 2)	(6, 3)	(6, 4)	(6, 5)	(6, 6)

Since there are 6 possible outcomes with the first die and 6 possible outcomes with the second die, we expect $6 \cdot 6 = 36$ equally likely possible outcomes, and we have $n(S) = 36$.

Self Check 1   How many pairs in the above sample space have a sum of 7?

EXAMPLE 2   Find the probability of the event "rolling a sum of 7 on one roll of two dice."

Solution   The sample space is listed in Example 1. We let E be the set of favorable outcomes, those that give a sum of 7:

$$E = \{(1, 6), (2, 5), (3, 4), (4, 3), (5, 2), (6, 1)\}$$

Since there are 6 favorable outcomes among the 36 equally likely outcomes, $n(E) = 6$, and

$$P(E) = P(\text{rolling a 7}) = \frac{n(E)}{n(S)} = \frac{6}{36} = \frac{1}{6}$$

Self Check 2   Find the probability of rolling a sum of 4.

A standard playing deck of 52 cards has two red suits, hearts and diamonds, and two black suits, clubs and spades. Each suit has 13 cards, including an ace, a king, a queen, a jack, and cards numbered from 2 to 10. We will refer to a standard deck of cards in many examples and exercises.

EXAMPLE 3   Find the probability of drawing 5 cards, all hearts, from a standard deck of cards.

Solution   Since the number of ways to draw 5 hearts from the 13 hearts is $C(13, 5)$, we have $n(E) = C(13, 5)$. Since the number of ways to draw 5 cards from the deck is

$C(52, 5)$, we have $n(S) = C(52, 5)$. The probability of drawing 5 hearts is the ratio of the number of favorable outcomes to the number of possible outcomes.

$$P(5 \text{ hearts}) = \frac{C(13, 5)}{C(52, 5)}$$

$$P(5 \text{ hearts}) = \frac{\dfrac{13!}{5!8!}}{\dfrac{52!}{5!47!}}$$

$$= \frac{13!}{5!8!} \cdot \frac{5!47!}{52!}$$

$$= \frac{13 \cdot 12 \cdot 11 \cdot 10 \cdot 9 \cdot \cancel{8!}}{\cancel{8!}} \cdot \frac{\cancel{47!}}{52 \cdot 51 \cdot 50 \cdot 49 \cdot 48 \cdot \cancel{47!}}$$

$$= \frac{13 \cdot 12 \cdot 11 \cdot 10 \cdot 9}{52 \cdot 51 \cdot 50 \cdot 49 \cdot 48}$$

$$= \frac{33}{66,640}$$

The probability of drawing 5 hearts is $\frac{33}{66,640}$.

Self Check 3 Find the probability of drawing 6 cards, all diamonds, from the deck.

2. Use the Multiplication Property of Probabilities

There is a property of probabilities that is similar to the multiplication principle for events. In the following theorem, we read $P(A \cap B)$ as "the probability of A and B" and $P(B \mid A)$ as "the probability of A given B." If A and B are events, the set $A \cap B$ contains the outcomes that are in both A and B.

Multiplication Property of Probabilities If $P(A)$ represents the probability of event A, and $P(B \mid A)$ represents the probability that event B will occur after event A, then

$$P(A \cap B) = P(A) \cdot P(B \mid A)$$

EXAMPLE 4 A box contains 40 cubes of the same size. Of these cubes, 17 are red, 13 are blue, and the rest are yellow. If 2 cubes are drawn at random, without replacement, find the probability that 2 yellow cubes will be drawn.

Solution Of the 40 cubes in the box, 10 are yellow. The probability of getting a yellow cube on the first draw is

$$P(\text{yellow cube on the first draw}) = \frac{10}{40} = \frac{1}{4}$$

Because there is no replacement after the first draw, 39 cubes remain in the box, and 9 of these are yellow. The probability of drawing a yellow cube on the second draw is

$$P(\text{yellow cube on the second draw}) = \frac{9}{39} = \frac{3}{13}$$

The probability of drawing 2 yellow cubes in succession is the product of the probability of drawing a yellow cube on the first draw and the probability of drawing a yellow cube on the second draw.

$$P(\text{drawing two yellow cubes}) = \frac{1}{4} \cdot \frac{3}{13} = \frac{3}{52}$$

Self Check 4   Find the probability that 2 blue cubes will be drawn.

Everyday Connections

Office Pools Are Illegal in Many States

You might want to check your state's criminal code before you fill in the last teams in your predictions for the winning teams in your NCAA tournament bracket sheet.

David Stluka/Getty Images

In spite the efforts of some legislators to change statutes that some say are outdated and rarely enforced, office pools are illegal in many states.

Legislation was introduced recently that would have legalized most office pools in Wisconsin, but it never made it to a vote, even though no one spoke in opposition. There are plans to reintroduce the bill.

The NCAA Men's Basketball Tournament has a single-elimination format involving 64 teams. There are

63 games played: 32 games in the first round, 16 games in the second round, 8 games in the third round, 4 in the regional finals, 2 in the final four, and then the national championship game. The tournament champion must win 6 consecutive games.

1. If there are 63 tournament games played and you have 2 choices at each stage, how many ways can you fill out a bracket sheet?

2. If every U.S. citizen (300,000,000 people) fills out a bracket sheet, and every team has an equal chance of winning, what is the probability of someone filling out a perfect bracket sheet (i.e., correct choices for every possible game)?

Source: Adapted from http://sportsillustrated.cnn.com/

EXAMPLE 5   Repeat Example 3 using the multiplication property of probabilities.

Solution   The probability of drawing a heart on the first draw is $\frac{13}{52}$. The probability of drawing a heart on the second draw *given that we got a heart on the first draw* is $\frac{12}{51}$. The probability is $\frac{11}{50}$ on the third draw, $\frac{10}{49}$ on the fourth draw, and $\frac{9}{48}$ on the fifth draw. By the multiplication property of probabilities,

$$P(\text{5 hearts in a row}) = \frac{13}{52} \cdot \frac{12}{51} \cdot \frac{11}{50} \cdot \frac{10}{49} \cdot \frac{9}{48}$$

$$= \frac{33}{66,640}$$

Self Check 5   Find the probability of drawing 6 cards, all diamonds, from the deck.

EXAMPLE 6   In a school, 30% of the students are gifted in mathematics and 10% are gifted in art and mathematics. If a student is gifted in mathematics, find the probability that the student is also gifted in art.

Solution   Let $P(M)$ be the probability that a randomly chosen student is gifted in mathematics, and let $P(M \cap A)$ be the probability that the student is gifted in both art and mathematics. We must find $P(A \mid M)$, the probability that the student is gifted in art, given that he or she is gifted in mathematics. To do so, we substitute the given values

$$P(M) = 0.3 \quad \text{and} \quad P(M \cap A) = 0.1$$

in the formula for multiplication of probabilities and solve for $P(A \mid M)$:

$$P(M \cap A) = P(M) \cdot P(A \mid M)$$
$$0.1 = (0.3)P(A \mid M)$$
$$P(A \mid M) = \frac{0.1}{0.3}$$
$$= \frac{1}{3}$$

If a student is gifted in mathematics, there is a probability of $\frac{1}{3}$ that he or she is also gifted in art.

Self Check 6   If 40% of the students are gifted in art, find the probability that a student gifted in art is also gifted in mathematics.

Self Check Answers   **1.** 6   **2.** $\frac{1}{12}$ **3.** $\frac{33}{391,510}$ **4.** $\frac{1}{10}$ **5.** $\frac{33}{391,510}$ **6.** $\frac{1}{4}$

8.7   Exercises

Vocabulary and Concepts   *Fill in the blanks.*

1. An _____ is any process for which the outcome is uncertain.

2. A list of all possible outcomes for an experiment is called a _____.

3. The probability of an event E is defined as

$$P(E) = \underline{\quad\quad} = \frac{s}{n}$$

4. $P(A \cap B) = $ _____

Practice   *List the sample space of each experiment.*

5. Rolling one die and tossing one coin

6. Tossing three coins

7. Selecting a letter of the alphabet

8. Picking a one-digit number

An ordinary die is rolled. Find the probability of each event.

9. Rolling a 2

10. Rolling a number greater than 4

11. Rolling a number larger than 1 but less than 6

12. Rolling an odd number

Balls numbered from 1 to 42 are placed in a container and stirred. If one is drawn at random, find the probability of each result.

13. The number is less than 20.

14. The number is less than 50.

15. The number is a prime number.

16. The number is less than 10 or greater than 40.

If the spinner shown is spun, find the probability of each event. Assume that the spinner never stops on a line.

17. The spinner stops on red.

18. The spinner stops on green.

19. The spinner stops on orange.

20. The spinner stops on yellow.

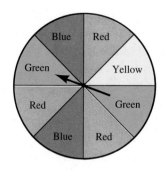

Find the probability of each event.

21. Rolling a sum of 4 on one roll of two dice

22. Drawing a diamond on one draw from a card deck

23. Drawing two aces in succession from a card deck if the card is replaced and the deck is shuffled after the first draw

24. Drawing two aces from a card deck without replacing the card after the first draw

25. Drawing a red egg from a basket containing 5 red eggs and 7 blue eggs

26. Getting 2 red eggs in a single scoop from a bucket containing 5 red eggs and 7 yellow eggs

27. Drawing a bridge hand of 13 cards, all of one suit

28. Drawing 6 diamonds from a card deck without replacing the cards after each draw

29. Drawing 5 aces from a card deck without replacing the cards after each draw

30. Drawing 5 clubs from the black cards in a card deck

31. Drawing a face card (king, queen, or jack) from a card deck

32. Drawing 6 face cards in a row from a card deck without replacing the cards after each draw

33. Drawing 5 orange cubes from a bowl containing 5 orange cubes and 1 beige cube

34. Rolling a sum of 4 with one roll of three dice

35. Rolling a sum of 11 with one roll of three dice

36. Picking, at random, 5 Republicans from a group containing 8 Republicans and 10 Democrats

37. Tossing 3 heads in 5 tosses of a fair coin

38. Tossing 5 heads in 5 tosses of a fair coin

Assume that the probability that an airplane engine will fail during a torture test is $\frac{1}{2}$ and that the aircraft in question has 4 engines.

39. Construct a sample space for the torture test.

40. Find the probability that all engines will survive the test.

41. Find the probability that exactly 1 engine will survive.

42. Find the probability that exactly 2 engines will survive.

43. Find the probability that exactly 3 engines will survive.

44. Find the probability that no engines will survive.

45. Find the sum of the probabilities in Exercises 40 through 44.

Assume that a survey of 282 people is taken to determine the opinions of doctors, teachers, and lawyers on a proposed piece of legislation, with the results as shown in the table. A person is chosen at random from those surveyed. Refer to the table to find each probability.

	Number that favor	Number that oppose	Number with no opinion	Total
Doctors	70	32	17	119
Teachers	83	24	10	117
Lawyers	23	15	8	46
Total	176	71	35	282

46. The person favors the legislation.

47. A doctor opposes the legislation.

48. A person who opposes the legislation is a lawyer.

49. **Quality control** In a batch of 10 tires, 2 are known to be defective. If 4 tires are chosen at random, find the probability that all 4 tires are good.

50. **Medicine** Out of a group of 9 patients treated with a new drug, 4 suffered a relapse. Find the probability that 3 patients of this group, chosen at random, will remain disease-free.

Use the multiplication property of probabilities.

51. If $P(A) = 0.3$ and $P(B \mid A) = 0.6$, find $P(A \cap B)$.

52. If $P(A \cap B) = 0.3$ and $P(B \mid A) = 0.6$, find $P(A)$.

53. Conditional probability The probability that a person owns a luxury car is 0.2, and the probability that the owner of such a car also owns a personal computer is 0.7. Find the probability that a person, chosen at random, owns both a luxury car and a computer.

54. Conditional probability If 40% of the population have completed college, and 85% of college graduates are registered to vote, what percent of the population are both college graduates and registered voters?

55. Conditional probability About 25% of the population watches the evening television news coverage as well as the morning soap operas. If 75% of the population watches the news, what percent of those who watch the news also watch the soaps?

56. Conditional probability The probability of rain today is 0.40. If it rains, the probability that Bill will forget his raincoat is 0.70. Find the probability that Bill will get wet.

Discovery and Writing

57. If $P(A \cap B) = 0.7$, is it possible that $P(B \mid A) = 0.6$? Explain.

58. Is it possible that $P(A \cap B) = P(A)$? Explain.

Review *Solve each equation.*

59. $|x + 3| = 7$

60. $|3x - 2| = |2x - 3|$

Graph each inequality.

61. $|x - 3| < 7$

62. $|3x - 2| \geq |2x - 3|$

CHAPTER REVIEW

| **8.1** | **The Binomial Theorem** |

Definitions and Concepts	**Examples**
Pascal's triangle is a triangular array of numbers showing the coefficients of the expansion of $(a + b)^n$. The example shows the first several rows of Pascal's triangle. Except for the 1's, each number in the triangle is the sum of the two numbers above it.	$\begin{matrix} & & & & 1 & & & & \\ & & & 1 & & 1 & & & \\ & & 1 & & 2 & & 1 & & \\ & 1 & & 3 & & 3 & & 1 & \\ 1 & & 4 & & 6 & & 4 & & 1 \\ \end{matrix}$ Row 0, Row 1, Row 2, Row 3, Row 4 $1 \quad 5 \quad 10 \quad 10 \quad 5 \quad 1$ Row 5
Factorial notation: $n! = n(n-1)(n-2) \cdot \cdots \cdot 3 \cdot 2 \cdot 1$ $0! = 1 \qquad n(n-1)! = n!$	$7! = 7 \cdot 6 \cdot 5 \cdot 4 \cdot 3 \cdot 2 \cdot 1 = 5{,}040$ $7 \cdot 6! = 7 \cdot 6 \cdot 5 \cdot 4 \cdot 3 \cdot 2 \cdot 1 = 7!$

The binomial theorem:
If n is any positive integer, then

$$(a + b)^n = a^n + \frac{n!}{1!(n-1)!}a^{n-1}b$$

$$+ \frac{n!}{2!(n-2)!}a^{n-2}b^2 + \frac{n!}{3!(n-3)!}a^{n-3}b^3$$

$$+ \frac{n!}{r!(n-r)!}a^{n-r}b^r + \cdots + b^n$$

$$(a + b)^3 = a^3 + \frac{3!}{1!(3-1)!}a^{3-1}b + \frac{3!}{2!(3-2)!}a^{3-2}b^2 + b^3$$

$$= a^3 + \frac{3 \cdot 2!}{1 \cdot 2!}a^2b + \frac{3 \cdot 2 \cdot 1}{2 \cdot 1 \cdot 1}ab^2 + b^3$$

$$= a^3 + 3a^2b + 3ab^2 + b^3$$

Note that the coefficients 1, 3, 3, and 1 are the numbers in the third row of Pascal's triangle.

Find the fourth term of $(a + b)^9$.

In the 4th term, the exponent on b is 3, and the exponent on a is $9 - 3$ or 6. The numerator of the fractional coefficient is 9! and the factors of the denominator are 3! and $(9 - 3)!$. Thus, the fourth is

$$\frac{9!}{3!(9-3)!}a^6b^3 = 84a^6b^2$$

Exercises

Find each value.

1. 6!

2. $7! \cdot 0! \cdot 1! \cdot 3!$

3. $\dfrac{8!}{7!}$

4. $\dfrac{5! \cdot 7! \cdot 8!}{6! \cdot 9!}$

Expand each expression.

5. $(x + y)^3$

6. $(p + q)^4$

7. $(a - b)^5$

8. $(2a - b)^3$

Find the required term of each expansion.

9. $(a + b)^8$; 4th term

10. $(2x - y)^5$; 3rd term

11. $(x - y)^9$; 7th term

12. $(4x + 7)^6$; 4th term

8.2   Sequences, Series, and Summation Notation

Definitions and Concepts	Examples
An **infinite sequence** is a function whose domain is the set of natural numbers.	2, 5, 8, 11, 14, 17, 20, ...
A **finite sequence** is a function whose domain is the set of the first n natural numbers.	2, 5, 8, 11, 14, 17, 20
If the commas in a sequence are replaced with + signs, the result is called a **series**.	**Infinite series:** $2 + 5 + 8 + 11 + 14 + 17 + 20 + \cdots$ **Finite series:** $2 + 5 + 8 + 11 + 14 + 17 + 20$
A sequence can be defined recursively by giving its first term and a rule showing how to obtain the $(n + 1)$th term from the nth term.	Write the first 4 terms of the sequence with a first term of 3 and a general term of $a_{n+1} = 2a_n + 1$. $a_1 = 3$ $a_2 = 2(3) + 1 = 7$ $a_3 = 2(7) + 1 = 15$ $a_4 = 2(15) + 1 = 31$ The first four terms are 3, 7, 15, and 31.

$\sum\limits_{k=1}^{4} n^3$ indicates the sum of the four terms obtained when 1, 2, 3, and 4 are substituted for n.

$$\sum_{k=1}^{4} n^3 = 1^3 + 2^3 + 3^3 + 4^3 = 1 + 8 + 27 + 64 = 100$$

If c is a constant, then

$$\sum_{k=1}^{n} c = nc$$

$$\sum_{k=1}^{5} 9 = 5 \cdot 9 = 45$$

If c is a constant, then

$$\sum_{k=1}^{n} cf(k) = c \sum_{k=1}^{n} f(k)$$

$$\sum_{k=1}^{n} [f(k) + g(k)] = \sum_{k=1}^{n} f(k) + \sum_{k=1}^{n} g(k)$$

$$\sum_{k=1}^{8} 4k^2 = 4 \sum_{k=1}^{8} k^2$$

$$\sum_{k=1}^{5} (k + k^2) = \sum_{k=1}^{5} k + \sum_{k=1}^{5} k^2$$

Exercises

Write the fourth term in each sequence.

13. $0, 7, 26, \ldots, n^3 - 1, \ldots$

14. $\dfrac{3}{2}, 3, \dfrac{11}{2}, \ldots, \dfrac{n^2 + 2}{2}, \ldots$

Find the first four terms of each sequence.

15. $a_1 = 5$ and $a_{n+1} = 3a_n + 2$

16. $a_1 = -2$ and $a_{n+1} = 2a_n^2$

Evaluate each expression.

17. $\sum\limits_{k=1}^{4} 3k^2$

18. $\sum\limits_{k=1}^{10} 6$

19. $\sum\limits_{k=5}^{8} (k^3 + 3k^2)$

20. $\sum\limits_{k=1}^{30} \left(\dfrac{3}{2}k - 12\right) - \dfrac{3}{2} \sum\limits_{k=1}^{30} k$

8.3   Arithmetic Sequences

Definitions and Concepts	Examples
An **arithmetic sequence** is of the form $\quad a, a + d, a + 2d, \ldots, a + (n - 1)d$ where a is the first term, d is the common difference, and $a_n = a + (n - 1)d$ is the nth term.	In the arithmetic sequence 2, 5, 8, 11, 14, 17, $\cdots$, 2 is the first term, 3 is the common difference, and the sixth term is 17. The twentieth term of the sequence above is given by $\quad\begin{aligned} a_n &= a + (n - 1)d \\ a_{20} &= 2 + (20 - 1)3 \\ &= 2 + (19)3 \\ &= 59 \end{aligned}$
Numbers inserted between a first and last term to form an arithmetic sequence are called **arithmetic means**.	To insert three arithmetic means between -4 and 16, we first determine the common difference d by substituting -4 for a, 16 for a_n, and 5 for n in the formula $a_n = a + (n - 1)d$ and solving for d. $\quad\begin{aligned} a_n &= a + (n - 1)d \\ 16 &= -4 + (5 - 1)d \\ 20 &= 4d \\ 5 &= d \end{aligned}$

Since the common difference is 5, the three arithmetic means are

$$-4 + 5, -4 + 2(5), -4 + 3(5) \quad \text{or} \quad 1, 6, 11$$

The formula

$$S_n = \frac{n(a + a_n)}{2}$$

gives the sum of the first n terms of an arithmetic series, where S_n is the sum, a is the first term, a_n is the last (or nth) term, and n is the number of terms.

The sum of the first 20 terms of the series

$$2 + 5 + 8 + 11 + 14 + 17 + 20 + \cdots$$

is given by

$$S_n = \frac{n(a + a_n)}{2}$$

$$S_{20} = \frac{20(2 + 59)}{2} \qquad a_{20} \text{ was found earlier.}$$

$$= 610$$

Exercises

Find the required term of each arithmetic sequence.

21. 5, 9, 13, . . .; 29th term

22. 8, 15, 22, . . .; 40th term

23. 6, -1, -8, . . .; 15th term

24. $\frac{1}{2}, -\frac{3}{2}, -\frac{7}{2}, \ldots$; 35th term

25. Find three arithmetic means between 2 and 8.

26. Find five arithmetic means between 10 and 100.

Find the sum of the first 40 terms in each sequence.

27. 5, 9, 13, . . .   **28.** 8, 15, 22

29. 6, -1, -8, . . .

30. $\frac{1}{2}, -\frac{3}{2}, -\frac{7}{2}, \ldots$

8.4   Geometric Sequences

Definitions and Concepts	Examples
A **geometric sequence** is of the form $$a, ar, ar^2, ar^3, \ldots, ar^{n-1}, \ldots$$ where a is the first term and r is the common ratio. The formula $a_n = ar^{n-1}$ gives the nth term of the sequence.	In the geometric sequence 2, 6, 18, 54, 162, $\cdots$, 2 is the first term, 3 is the common ratio, and the fifth term is 162. The tenth term of the sequence above is given by $$a_n = ar^{n-1}$$ $$a_{10} = 2 \cdot 3^{10-1}$$ $$= 2 \cdot 3^9$$ $$= 39{,}366$$
Numbers inserted between a first and last term that form a geometric sequence are called **geometric means.**	To insert a geometric mean between -1 and -16, we first determine the common ratio r by substituting -16 for a_n, -1 for a, and 3 for n in the formula $a_n = ar^{n-1}$ and then solving for r. $$a_n = ar^{n-1}$$ $$-16 = -1r^{3-1}$$ $$16 = r^2$$ $$\pm 4 = r$$ The geometric mean is either 4 or -4.

The formula

$$S_n = \frac{a - ar^n}{1 - r} \quad (r \neq 1)$$

gives the sum of the first n terms of a geometric series, where S_n is the sum, a is the first term, r is the common ratio, and n is the number of terms.

Find the sum of the first six terms of the geometric series $2 + 6 + 18 + 54 + 162 + \cdots$.

In this series, the first term is 2, the common ratio is 3, and the sixth term is $a_n = ar^{n-1} = 2(3)^5 = 486$.

$$S_n = \frac{a - ar^n}{1 - r}$$

$$S_6 = \frac{2 - 2(3)^6}{1 - 3}$$

$$= \frac{2 - 2(729)}{-2}$$

$$= 728$$

If $|r| < 1$, the formula

$$S_\infty = \frac{a}{1 - r}$$

gives the sum of the terms of an infinite geometric series, where S_∞ is the sum, a is the first term, and r is the common ratio.

Find the sum of the infinite series $162 + 54 + 18 + \cdots$.

In this infinite series, the first term is 162 and the common ratio is $\frac{1}{3}$. So the sum of the series is

$$S_\infty = \frac{a}{1 - r}$$

$$= \frac{162}{1 - \frac{1}{3}}$$

$$= 243$$

Exercises

Find the required term of each geometric sequence.

31. $81, 27, 9, \ldots$; 11th term

32. $2, 6, 18, \ldots$; 9th term

33. $9, \frac{9}{2}, \frac{9}{4}, \ldots$; 15th term

34. $8, -\frac{8}{5}, \frac{8}{25}, \ldots$; 7th term

35. Find three positive geometric means between 2 and 8.

36. Find four geometric means between -2 and 64.

37. Find the positive geometric mean between 4 and 64.

Find the sum of the first 8 terms in each sequence.

38. $81, 27, 9, \ldots$

39. $2, 16, 18, \ldots$

40. $9, \frac{9}{2}, \frac{9}{4}, \ldots$

41. $8, -\frac{8}{5}, \frac{8}{25}$

42. Find the sum of the first eight terms of the sequence $\frac{1}{3}, 1, 3, \ldots$.

43. Find the seventh term of the sequence $2\sqrt{2}, 4, 4\sqrt{2}, \ldots$.

Find the sum of each infinite sequence, if possible.

44. $\frac{1}{3}, \frac{1}{6}, \frac{1}{12}, \ldots$

45. $\frac{1}{5}, -\frac{2}{15}, \frac{4}{25}, \ldots$

46. $1, \frac{3}{2}, \frac{9}{4}, \ldots$

47. $0.5, 0.25, 0.125, \ldots$

Change each decimal into a common fraction.

48. $0.\overline{3}$

49. $0.\overline{9}$

50. $0.\overline{17}$

51. $0.\overline{45}$

52. **Investment problem** If Leonard invests $3,000 in a 6-year certificate of deposit at the annual rate of 7.75%, compounded daily, how much money will be in the account when it matures?

53. **College enrollments** The enrollment at Hometown College is growing at the rate of 5% over each previous year's enrollment. If the enrollment is currently 4,000 students, what will it be 10 years from now? What was it 5 years ago?

54. **House trailer depreciation** A house trailer that originally cost $10,000 depreciates in value at the rate of 10% per year. How much will the trailer be worth after 10 years?

8.5   Mathematical Induction

Definitions and Concepts	Examples

Mathematical induction: If a statement involving the natural number n has the two properties that
 	
1. the statement is true for $n = 1$, and
2. if the statement is true for $n = k$, then it is true for
 	
    $n = k + 1$,
 	
    then the statement is true for all natural numbers.
	See Examples 1, 2, and 3 in the section.

Exercises

55. Verify the following formula for $n = 1$, $n = 2$, $n = 3$, and $n = 4$:

$$1^3 + 2^3 + 3^3 + \cdots + n^3 = \frac{n^2(n + 1)^2}{4}$$

56. Prove the formula given in Exercise 55 by mathematical induction.

8.6   Permutations and Combinations

Definitions and Concepts	Examples
Multiplication principle for events: If event E_1 can occur in a_1 ways and event E_2 can occur in a_2 ways, then the event E_1 followed by E_2 can occur in $a_1 \cdot a_2$ ways.	If you choose one drink from **3** choices and one sandwich from **6** choices, in how many ways can you choose one drink and one sandwich? By the multiplication principle of events, the number of ways is $3 \cdot 6 = 18$.

Formulas for permutations and combinations:
 	
  $P(n, r) = \dfrac{n!}{(n - r)!}$
 	
  $P(n, n) = n!$
 	
  $P(n, 0) = 1$
 	
  $C(n, r) = \dbinom{n}{r} = \dfrac{n!}{r!(n - r)!}$
 	
  $C(n, n) = 1$
 	
  $C(n, 0) = 1$
	$P(7, 2) = \dfrac{7!}{(7 - 2)!} = \dfrac{7 \cdot 6 \cdot 5!}{5!} = 42$
	
	$P(7, 7) = \dfrac{7!}{(7 - 7)!} = \dfrac{7!}{0!} = \dfrac{7!}{1} = 7!$
	
	$P(7, 0) = \dfrac{7!}{(7 - 0)!} = \dfrac{7!}{7!} = 1$
	
	$C(7, 2) = \dfrac{7!}{2!(7 - 2)!} = \dfrac{7 \cdot 6 \cdot 5!}{2 \cdot 5!} = 7 \cdot 3 = 21$
	
	$C(7, 7) = \dfrac{7!}{7!(7 - 7)!} = \dfrac{7!}{7! \cdot 0!} = \dfrac{7!}{7! \cdot 1} = 1$
	
	$C(7, 0) = \dfrac{7!}{0!(7 - 0)!} = \dfrac{7!}{1 \cdot 7!} = \dfrac{7!}{1 \cdot 7!} = 1$

There are $(n - 1)!$ ways to place n things in a circle.	There are $(6 - 1)! = 5! = 120$ ways to seat 6 people at a round table.

Permutations of like objects: The number of permutations of n things with a things alike, b things alike, and so on, is $$\frac{n!}{a!b!\cdots}$$	The number of distinguishable words that can be formed from the letters of the word *happy* is $$\frac{5!}{1!\cdot 1!\cdot 2!\cdot 1!}$$ There are 5 letters. h, a, and y occur once; p occurs twice. $$=\frac{5\cdot 4\cdot 3\cdot 2!}{2!}=60$$

Exercises

Evaluate each expression.

57. $P(8, 5)$ **58.** $C(7, 4)$

59. $0!\cdot 1!$ **60.** $P(10, 2)\cdot C(10, 2)$

61. $P(8, 6)\cdot C(8, 6)$ **62.** $C(8, 5)\cdot C(6, 2)$

63. $C(7, 5)\cdot P(4, 0)$ **64.** $C(12, 10)\cdot C(11, 0)$

65. $\dfrac{P(8, 5)}{C(8, 5)}$ **66.** $\dfrac{C(8, 5)}{C(13, 5)}$

67. $\dfrac{C(6, 3)}{C(10, 3)}$ **68.** $\dfrac{C(13, 5)}{C(52, 5)}$

69. In how many ways can 10 teenagers be seated at a round table if 2 girls want to sit with their boyfriends?

70. How many distinguishable words can be formed from the letters of the word *casserole* if each letter is used exactly once?

8.7   Probability

Definitions and Concepts	**Examples**
An event that cannot happen has a probability of 0.	The probability that pigs fly is 0.
An event that is certain to happen has a probability of 1. All other events have probabilities between 0 and 1.	The probability that every pig will die is 1.
If S is the sample space of an experiment with n distinct and equally likely outcomes, and if E is an event that occurs in s of those ways, then the **probability of E** is $$P(E)=\frac{n(E)}{n(S)}=\frac{s}{n}$$	Since there are 28 days and 4 Fridays in February, the probability that a day chosen at random in February will be a Friday is $$P(\text{Friday})=\frac{4}{28}$$ the number of Fridays the number of days $$=\frac{1}{7}$$
Multiplication property of probabilities: $$P(A\cap B)=P(A)\cdot P(B\,|\,A)$$	A jar contains 25 marbles of the same size. Of these marbles, 10 are red, 10 are blue, and 5 are yellow. If 2 marbles are drawn (without replacement), find the probability that both marbles will be red. Of the 25 marbles in the jar, 10 are red. The probability of getting a red marble on the first draw is $$P(\text{red marble on the first draw})=\frac{10}{25}=\frac{2}{5}$$

Because there is no replacement after the first draw, 24 marbles remain in the jar and 9 of these are red. The probability of drawing a red marble on the second draw is

$$P(\text{red marble on the second draw}) = \frac{9}{24} = \frac{3}{8}$$

The probability of drawing 2 red marbles in a row is the product of the probabilities:

$$P(\text{drawing 2 red marbles}) = \frac{2}{5} \cdot \frac{3}{8} = \frac{3}{20}$$

Exercises

71. Make a tree diagram (like Figure 8-2 on page 699) to illustrate the possible results of tossing a coin four times.

72. In how many ways can you draw a 5-card poker hand of 3 aces and 2 kings?

73. Find the probability of drawing the hand described in Exercise 72.

74. Find the probability of not drawing the hand described in Exercise 72.

75. Find the probability of having a 13-card bridge hand consisting of 4 aces, 4 kings, 4 queens, and 1 jack.

76. Find the probability of choosing a committee of 3 men and 2 women from a group of 8 men and 6 women.

77. Find the probability of drawing a club or a spade on one draw from a standard card deck.

78. Find the probability of drawing a black card or a king on one draw from a standard card deck.

79. Find the probability of getting an ace-high royal flush in hearts (ace, king, queen, jack, and ten of hearts) in poker.

80. Find the probability of being dealt 5 cards of one suit in a poker hand.

81. Find the probability of getting 3 heads or fewer on 4 tosses of a fair coin.

CHAPTER TEST

Find each value.

1. $3! \cdot 0! \cdot 4! \cdot 1!$

2. $\dfrac{2! \cdot 4! \cdot 6! \cdot 8!}{3! \cdot 5! \cdot 7!}$

Find the required term in each expansion.

3. $(x + 2y)^5$; 2nd term

4. $(2a - b)^8$; 7th term

Find each sum.

5. $\displaystyle\sum_{k=1}^{3} (4k + 1)$

6. $\displaystyle\sum_{k=2}^{4} (3k - 21)$

Find the sum of the first ten terms of each sequence.

7. $2\ 5, 8, \ldots$

8. $5, 1, -3, \ldots$

9. Find three arithmetic means between 4 and 24.

10. Find two geometric means between -2 and -54.

Find the sum of the first ten terms of each sequence.

11. $\dfrac{1}{4}, \dfrac{1}{2}, 1, \ldots$

12. $6, 2, \dfrac{2}{3}, \ldots$

13. A car costing \$$c$ when new depreciates 25% of the previous year's value each year. How much is the car worth after 3 years?

14. A house costing \$$c$ when new appreciates 10% of the previous year's value each year. How much will the house be worth after 4 years?

15. Prove by induction:

$$3 + 4 + 5 + \cdots + (n + 2) = \tfrac{1}{2}n(n + 5)$$

16. How many six-digit license plates can be made if no plate begins with 0 or 1?

Find each value.

17. $P(7, 2)$

18. $P(4, 4)$

19. $C(8, 2)$

20. $C(12, 0)$

21. How many ways can 4 men and 4 women stand in line if all the women are first?

22. How many different ways can 6 people be seated at a round table?

23. How many different words can be formed from the letters of the word *bluff* if each letter is used once?

24. Show the sample space of the experiment: toss a fair coin three times.

Find each probability.

25. Rolling a 5 on one roll of a die

26. Drawing a jack or a queen from a standard card deck

27. Receiving 5 hearts for a 5-card poker hand

28. Rolling a sum of 9 on one roll of two dice.

29. A box contains 50 cubes of the same size. Of these cubes, 20 are red and 30 are blue. If 2 cubes are drawn at random, without replacement, find the probability that 2 blue cubes are drawn.

30. In a batch of 20 tires, 2 are known to be defective. If 4 tires are chosen at random, find the probability that all 4 tires are good.

The Mathematics of Finance

In this chapter, we will discuss the mathematics of finance—the rules that govern investing and borrowing money.

Careers and Mathematics

Actuary Actuaries use their broad knowledge of statistics, finance, and business to design insurance policies, pension plans, and other financial strategies, and ensure that these plans are maintained on a sound financial basis. They assemble and analyze data to estimate the probability and likely cost of an event such as death, sickness, injury, disability, or loss of property.

Most actuaries are employed in the insurance industry, specializing in either life and health insurance or property and casualty insurance. They produce probability tables or use modeling techniques that determine the likelihood that a potential event will generate a claim. From these, they estimate the amount a company can expect to pay in claims. Actuaries ensure that the premiums charged for such insurance will enable the company to cover claims and other expenses.

Actuaries held about 18,000 jobs in 2006.

Education Actuaries need a strong background in mathematics and general business. Actuaries usually earn an undergraduate degree in mathematics, statistics, or actuarial science, or a business-related field such as finance, economics or business. Actuaries must pass a series of examinations to gain full professional status.

Job Outlook Employment of actuaries is expected to increase by about 24 percent through 2016. Median annual earnings of actuaries were $82,800 in 2006.

For a sample application, see Example 3 in Section 9.3. For more information, see www.bls.gov.oco.ocos.041.htm.

Alexander Walter/Getty Images

723

9.1   Interest

Objectives

1. Compute Simple Interest
2. Compute Compound Interest
3. Borrow Money Using Bank Notes
4. Compute Effective Rate of Interest
5. Compute Present Value

Wages, rent, and interest are three common ways to earn money:

*   A wage refers to money received for letting someone use your labor.
*   Rent refers to money received for letting someone use your property, especially real estate.
*   Interest refers to money received for letting someone use your money.

Few people become wealthy by receiving wages. Unless you receive a large hourly rate of pay, there will not be enough left after daily living expenses to amass true wealth.

You will have a better chance of becoming wealthy by supplementing wages with rent. For example, if you borrow money to buy an apartment building, the rent received from tenants can pay off the loan, and eventually you will own the building without spending your own money.

Perhaps the easiest way to build wealth is to use money to earn interest. If you can earn a good rate of interest, compounded continuously, and keep the investment for a long time, it is amazing how large an investment can grow. In fact, it is said that *compound interest* is the eighth wonder of the world.

In this first section, we will discuss this important money-making tool: *interest.*

When money is borrowed, the lender expects to be paid back the amount of the loan plus an additional charge for the use of the money. This additional charge is called **interest.** When money is deposited in a bank, the bank pays the depositor for the use of the money. The money the deposit earns is also called *interest.*

Interest can be computed in two ways: either as *simple interest* or as *compound interest.*

1.   Compute Simple Interest

Simple interest is computed by finding the product of the **principal** (the amount of money on deposit), the **rate of interest** (usually written as a decimal), and the **time** (usually expressed in years).

Interest = principal · rate · time

This word equation suggests the following formula.

Simple Interest   The simple interest I earned on a principal P in an account paying an annual interest rate r for a length of time t is given by the formula

$$I = Prt$$

EXAMPLE 1 Find the simple interest earned on a deposit of $5,750 that is left on deposit for $3\frac{1}{2}$ years and earns an annual interest rate of $4\frac{1}{2}\%$.

Solution We write $3\frac{1}{2}$ and $4\frac{1}{2}\%$ as decimals and substitute the given values in the formula for simple interest.

$I = Prt$ $\qquad$ **This is the formula for simple interest.**

$I = 5{,}750 \cdot 0.045 \cdot 3.5$ $\quad$ **Substitute 5,750 for P, 0.045 for r, and 3.5 for t.**

$I = 905.625$ $\qquad$ **Perform the multiplications.**

In $3\frac{1}{2}$ years, the account will earn $905.63 in simple interest.

Self Check 1 Find the simple interest earned on a deposit of $12,275 that is left on deposit for $5\frac{1}{4}$ years and earns an annual interest rate of $3\frac{3}{4}\%$. ▬

EXAMPLE 2 Three years after investing $15,000, a retired couple received a check for $3,375 in simple interest. Find the annual interest rate their money earned during that time.

Solution The couple invested $15,000 (the principal) for 3 years (the time) and earned $3,375 (the simple interest). We must find the annual interest rate r. To do so, we substitute the given numbers into the simple interest formula and solve for r.

$I = Prt$

$3{,}375 = 15{,}000 \cdot r \cdot 3$ $\quad$ **Substitute 3,375 for I, 15,000 for P, and 3 for t.**

$3{,}375 = 45{,}000r$ $\qquad$ **Multiply.**

$\dfrac{3{,}375}{45{,}000} = \dfrac{45{,}000}{45{,}000}r$ $\quad$ **Divide both sides by 45,000.**

$0.075 = r$ $\qquad$ **Perform the divisions.**

$r = 7.5\%$ $\qquad$ **Write 0.075 as a percent.**

The couple received an annual rate of 7.5% for the 3-year period.

Self Check 2 Find the length of time it will take for the interest to grow to $9,000. ▬

2. Compute Compound Interest

When interest is left in an account and also earns interest, we say that the account earns **compound interest.**

EXAMPLE 3 A woman deposits $10,000 in a savings account paying 6% interest, compounded annually. Find the balance in her account after each of the first three years.

Solution At the end of the first year, the interest earned is 6% of the $10,000, or

$0.06(\$10{,}000) = \$600$

This interest is added to the $10,000 to get a new balance. After one year, this balance will be $10,600.

The second year's earned interest is 6% of $10,600, or

$0.06(\$10{,}600) = \$636$

This interest is added to $10,600, giving a second-year balance of $11,236.

The interest earned during the third year is 6% of $11,236, or

$$0.06(\$11{,}236) = \$674.16$$

This interest is added to $11,236 to give the woman a balance of $11,910.16, after three years.

Self Check 3   Find the balance in the woman's account after two more years.   ▬▬▬

We can generalize the method used in Example 3 to find a formula for compound interest calculations. Suppose that the original deposit in the account is A_0 dollars, that interest is paid at an **annual rate** r, and that the **accumulated amount** or the **future value** in the account at the end of the first year is A_1. Then the interest earned that year is $A_0 r$, and

| The amount after one year | equals | the original deposit | plus | the interest earned on the original deposit. |

$$A_1 = A_0 + A_0 r$$
$$\quad = A_0(1 + r) \quad \text{Factor out the common factor, } A_0.$$

The amount, A_1, at the end of the first year is the balance in the account at the beginning of the second year. So, the amount at the end of the second year, A_2, is

| The amount after two years | equals | the amount after one year | plus | the interest earned on the amount after one year. |

$$A_2 = A_1 + A_1 r$$
$$\quad = A_1(1 + r) \qquad \text{Factor out the common factor, } A_1.$$
$$\quad = A_0(1 + r)(1 + r) \quad \text{Substitute } A_0(1 + r) \text{ for } A_1.$$
$$\quad = A_0(1 + r)^2 \qquad \text{Simplify.}$$

By the end of the third year, the amount will be

$$A_3 = A_0(1 + r)^3$$

The pattern continues with the following result.

Compound Interest (Annual Compounding)   A single deposit A_0, earning compound interest for n years at an annual rate r, will grow to a **future value** A_n according to the formula

$$A_n = A_0(1 + r)^n$$

EXAMPLE 4   For their newborn child, parents deposit $10,000 in a college account that pays 8% interest, compounded annually. How much will be in the account on the child's 17th birthday?

Solution   We substitute $A_0 = 10{,}000$, $r = 0.08$, and $n = 17$ into the compound interest formula to find the future value A_{17}.

$$A_n = A_0(1 + r)^n$$
$$A_{17} = \mathbf{10{,}000}(1 + \mathbf{0.08})^{17}$$
$$\quad = 10{,}000(1.08)^{17}$$
$$\quad \approx 37{,}000.18054801 \qquad \text{Use a calculator.}$$

To the nearest cent, $37,000.18 will be available on the child's 17th birthday.

Self Check 4 If the parents leave the money on deposit for two more years, what amount will be available?

Interest compounded once each year is **compounded annually.** Many financial institutions compound interest more often. For example, instead of paying an annual rate of 8% once a year, a bank might pay 4% twice each year, or 2% four times each year. The annual rate, 8%, is also called the **nominal rate,** and the time between interest calculations is called the **conversion period.** If there are k periods each year, interest is paid at the **periodic rate** given by the following formula.

Periodic Rate

$$\text{Periodic rate} = \frac{\text{annual rate}}{\text{number of periods per year}}$$

This formula is often written as

$$i = \frac{r}{k}$$

where i is the periodic interest rate, r is the annual rate, and k is the number of times interest is paid each year.

If interest is calculated k times each year, in n years there will be kn conversions. Each conversion is at the periodic rate i. This leads to another form of the compound interest formula.

Compound Interest Formula An amount A_0, earning interest compounded k times a year for n years at an annual rate r, will grow to the future value A_n according to the formula

$$A_n = A_0(1 + i)^{kn}$$

where $i = \dfrac{r}{k}$ is the periodic interest rate.

Interest paid twice each year is called **semiannual** compounding, four times each year **quarterly** compounding, twelve times each year **monthly** compounding, and 360 or 365 times each year **daily** compounding.

EXAMPLE 5 If the parents of Example 4 invested that $10,000 in an account paying 8%, compounded quarterly, how much more money would they have after 17 years?

Solution We first calculate the periodic rate, i.

$$i = \frac{r}{k}$$

$$i = \frac{0.08}{4} \qquad \text{Substitute } r = 0.08 \text{ and } k = 4.$$

$$i = 0.02$$

We then substitute $A_0 = 10,000$, $i = 0.02$, $k = 4$, and $n = 17$ into the compound interest formula.

$$A_n = A_0(1 + i)^{kn}$$
$$A_{17} = 10,000(1 + 0.02)^{4 \cdot 17}$$
$$= 10,000(1.02)^{68}$$
$$\approx 38,442.50502546 \qquad \text{Use a calculator.}$$

To the nearest cent, $38,442.51 will be available, an increase of $1,442.33 over annual compounding.

Self Check 5   **a.** What would $10,000 become in 17 years if compounded monthly at a nominal rate of 8%?

   b. How does this compare with quarterly compounding?

Accent on Technology   **Growth of Money**

We can use a graphing calculator to find the time it would take a $10,000 investment to triple, assuming an 8% annual rate, compounded quarterly.

In n years, $10,000 earning 8% interest, compounded quarterly, will become the future value

$$10,000(1.02)^{4n}$$

To watch this value grow, we enter the function

$$Y_1 = 10000*1.02^{\wedge}(4*X)$$

in a graphing calculator, and set the window to $0 \le X \le 10$ (for 10 years) and $0 \le Y \le 40000$ (for the dollar amount).

The graph appears in Figure 9-1(a). To find the time it would take for the investment to triple, we use TRACE to move to the point with a Y-value close to 30,000. The X-value in Figure 9-1(b) shows that the investment would triple in about 13.9 years.

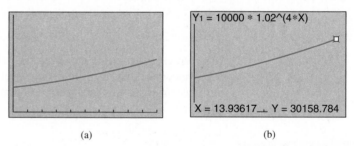

(a)   (b)

Figure 9-1

---

3.   Borrow Money Using Bank Notes

When a customer borrows money from a bank, the bank is making an investment in that person. The amount of the loan is the bank's deposit, and the bank expects to be repaid with interest in a single **balloon payment** at a later date. These loans, called **notes,** are based on a 360-day year, and they are usually written for 30 days, 90 days, or 180 days. We use the formula for compound interest to calculate the terms of the loan.

EXAMPLE 6   A student needs $4,000 for tuition. If his bank writes a 9%, 180-day note, with interest compounded daily, what will he owe at the end of 180 days?

Solution   In granting the loan, the bank invests $4,000. The amount to be repaid is the expected future value

$$A_n = A_0(1 + i)^{kn}$$

where A_0 is $4,000, the frequency of compounding k is 360, the periodic rate i is $\frac{0.09}{360} = 0.00025$, and the term n is 0.5 (180 days is one-half of 360 days). To determine what the student will owe, we substitute these numbers into the compound interest formula and solve for A_n.

$$A_n = A_0(1 + i)^{kn}$$
$$A_{0.5} = \mathbf{4{,}000}(1 + \mathbf{0.00025})^{360 \cdot 0.5}$$
$$= 4{,}000(1.00025)^{180}$$
$$\approx 4{,}184.087907996 \qquad \text{Use a calculator.}$$
$$\approx 4{,}184.09 \qquad \text{Round to the nearest cent.}$$

The student must repay $4,184.09.

Self Check 6   A woman borrows $7,500 for 90 days at 12%. If interest is compounded daily, how much will she owe at the end of 90 days?

4.   Compute Effective Rate of Interest

The true performance of an investment depends on both the frequency of compounding and the annual rate. To help investors compare different savings plans, financial institutions are required by law to provide the **effective rate**—the rate that, if compounded annually, would provide the same yield as a plan that is compounded more frequently.

To derive a formula for an effective rate, we assume that A_0 dollars are invested for n years at an annual rate r, compounded k times per year. That same investment of A_0 dollars, compounded annually at the effective rate R, would produce the same accumulated value. Since these amounts are to be equal, we have the equation

Accumulated amount at effective rate R, compounded annually	equals	accumulated amount at annual rate r, compounded k times per year.

$$A_0(1 + R)^n = A_0(1 + i)^{kn} \qquad i \text{ is the periodic rate, } i = \frac{r}{k}.$$

We can solve this equation for R.

$$A_0(1 + R)^n = A_0(1 + i)^{kn}$$
$$(1 + R)^n = (1 + i)^{kn} \qquad \text{Divide both sides by } A_0.$$
$$[(1 + R)^n]^{1/n} = [(1 + i)^{kn}]^{1/n} \qquad \text{Raise both sides to the } 1/n \text{ power.}$$
$$1 + R = (1 + i)^{k} \qquad \text{Multiply the exponents.}$$
$$R = (1 + i)^{k} - 1 \qquad \text{Subtract 1 from both sides.}$$

This result establishes the following formula.

Effective Rate of Interest The **effective rate of interest** R for an account paying a nominal rate r, compounded k times per year, is

$$R = (1 + i)^k - 1$$

where i is the periodic rate, $i = \dfrac{r}{k}$.

EXAMPLE 7 A bank offers the savings plans shown in the table. Calculate the effective interest rates for each investment.

	a. Money market fund	b. Certificate of deposit
Annual rate Compounding Effective rate	6.5% quarterly	7% monthly

Solution **a.** For the money market fund, $r = 0.065$ and $k = 4$, so $i = \frac{r}{k} = \frac{0.065}{4} = 0.01625$. To find the effective rate, we substitute $k = 4$ and $i = 0.01625$ in the formula for effective rate.

$$R = (1 + i)^k - 1$$
$$R = (1 + \mathbf{0.01625})^{\mathbf{4}} - 1$$

≈ 0.0666016088	Use a calculator.
≈ 0.0666	Round to the nearest ten thousandth.

As a percent, the effective rate is 6.66%, or approximately $6\frac{2}{3}\%$.

b. For the certificate of deposit, $r = 0.07$ and $k = 12$, so

$$i = \frac{0.07}{12} \approx 0.00583333$$

and

$$R = (1 + 0.00583333)^{12} - 1$$

≈ 0.0722900809	Use a calculator.
≈ 0.0723	Round to the nearest ten thousandth.

As a percent, the effective rate is 7.23%.

Self Check 7 A passbook savings account offers daily compounding (365 days per year) at an annual rate of 6%. Find the effective rate to the nearest hundredth.

5. Compute Present Value

For an initial deposit A_0, the compound interest formula gives the future value A_n of the account after n years. This is the situation suggested by Figure 9-2, where we know the beginning amount and need to find its future value.

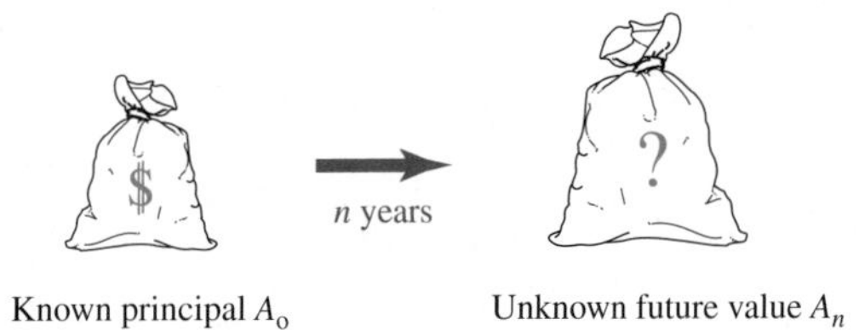

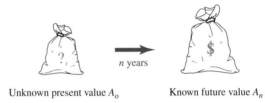

Known principal A_0 Unknown future value A_n

Figure 9-2

Often, the situation is reversed: We need to make a deposit now that will become a specific amount several years from now—perhaps enough to buy a car or pay tuition. As Figure 9-3 suggests, we need to know what single deposit *now* will accomplish that goal: What **present value** A_0 will yield a *specific* future value A_n?

Unknown present value A_0 Known future value A_n

Figure 9-3

To derive the formula for present value, we solve the compound interest formula for A_0.

$$A_n = A_0(1 + i)^{kn}$$ The original deposit is the present value, A_0.

$$\frac{A_n}{(1 + i)^{kn}} = \frac{A_0(1 + i)^{kn}}{(1 + i)^{kn}}$$ Divide both sides by $(1 + i)^{kn}$.

$$\frac{A_n}{(1 + i)^{kn}} = A_0$$ Divide on both sides: $\frac{(1 + i)^{kn}}{(1 + i)^{kn}} = 1$.

$$A_n(1 + i)^{-kn} = A_0$$ Use the definition of negative exponent: $\frac{1}{a^x} = a^{-x}$.

This result establishes the following formula.

Present Value   The **present value** A_0 is the amount that must be deposited now to provide a **future value** A_n after n years is given by the formula

$$A_0 = A_n(1 + i)^{-kn}$$

where interest is compounded k times per year at an annual rate r $\left(i \text{ is the periodic rate, } \frac{r}{k}\right)$.

EXAMPLE 8   When a medical student graduates in 8 years, she will need $25,700 to buy furniture for her medical office. What amount must she deposit now (at 8%, compounded twice per year) to meet this future obligation?

Solution   Use the annual rate ($r = 0.08$) and the frequency of compounding ($k = 2$) to find the periodic rate:

$$i = \frac{r}{k} = \frac{0.08}{2} = 0.04$$

In the present value formula, we substitute

the number of years, $n = 8$,

the periodic rate, $i = 0.04$,

the frequency of compounding, $k = 2$, and

the future value in 8 years, $A_8 = 25{,}700$.

$$A_0 = A_n(1 + i)^{-kn}$$
$$A_0 = 25{,}700(1 + 0.04)^{-2 \cdot 8} \qquad A_n = A_8 = 25{,}700$$
$$= 25{,}700(1.04)^{-16} \qquad \text{Use a calculator.}$$
$$\approx 13{,}721.44 \qquad \text{Round to the nearest cent.}$$

She must deposit \$13,721.44 now to have \$25,700 in 8 years.

Self Check 8 If the student decides to take two extra years to complete medical school, her obligation will be \$27,000. What present value will meet her goal?

Self Check Answers **1.** \$2,416.64 **2.** 8 yr **3.** \$13,382.26 **4.** \$43,157.01 **5. a.** \$38,786.48
b. \$343.97 more than with quarterly compounding **6.** \$7,728.37 **7.** 6.18%
8. \$12,322.45

9.1 Exercises

Vocabulary and Concepts *Fill in the blanks.*

1. A bank pays _____ for the privilege of using your money.

2. If interest is left on deposit to earn more interest, the account earns _____ interest.

3. Interest compounded once each year is called _____ compounding.

4. The initial deposit is called the _____ or the _____ value.

5. After a specific time, the principal grows to a _____ value.

6. Interest is calculated as a _____ of the amount on deposit.

7. Future value = principal + _____ earned

8. The annual rate also is called the _____ rate.

9. $\dfrac{\text{_____}}{} = \dfrac{\text{annual rate}}{\text{number of periods per year}}$

10. The time between interest calculations is the _____ period.

11. In the future value formula $A_n = A_0(1 + i)^{kn}$,
    A_0 is the _____,
    i is the _____,
    k is the _____, and
    n is the _____.

12. In the present value formula $A_0 = A_n(1 + i)^{-kn}$,
    A_n is the _____,
    i is the _____,
    k is the _____, and
    n is the _____.

13. To help consumers compare savings plans, banks advertise the _____ rate of interest.

14. If after one year, \$100 grows to \$110, the effective rate is ___%.

Practice

15. Find the simple interest earned in an account where \$4,500 is on deposit for 4 years at $3\frac{1}{4}\%$ annual interest.

16. Find the simple interest earned in an account where \$12,400 is on deposit for $8\frac{1}{4}$ years at $4\frac{1}{2}\%$ annual interest.

17. Find the principal necessary to earn \$814 in simple interest if the money is to be left on deposit for 4 years and earns $5\frac{1}{2}\%$ annual interest.

18. Find the time necessary for a deposit of \$11,500 to earn \$3,450 in simple interest if the money is to earn $3\frac{3}{4}\%$ annual interest.

19. Find the annual rate necessary for a deposit of $50,000 to earn $7,500 in simple interest if the money is to be left on deposit for $2\frac{1}{2}$ years.

20. Find the time necessary for a deposit of $5,000 to double in an account paying $6\frac{1}{4}\%$ simple interest.

Assume that $1,200 is deposited in an account in which interest is compounded annually at a rate of 8%. Find the accumulated amount after the given number of years.

21. 1 year
22. 3 years
23. 5 years
24. 20 years

Assume that $1,200 is deposited in an account in which interest is compounded annually at the given rate. Find the accumulated amount after 10 years.

25. 3%
26. 5%
27. 9%
28. 12%

Assume that $1,200 is deposited in an account in which interest is compounded at the given frequency, at an annual rate of 6%. Find the accumulated amount after 15 years.

29. $k = 2$
30. $k = 4$
31. $k = 12$
32. $k = 365$

Find the effective interest rate with the given annual rate r and compounding frequency k.

33. $r = 6\%, k = 4$
34. $r = 8\%, k = 12$

35. $r = 9\frac{1}{2}\%, k = 2$
36. $r = 10\%, k = 360$

Find the present value of $20,000 due in 6 years, at the given annual rate and compounding frequency.

37. 6%, semiannually
38. 8%, quarterly

39. 9%, monthly
40. 7%, daily (360 days/year)

Applications

41. **Small business**   To start a mobile dog-grooming service, a woman borrowed $2,500. If the loan was for 2 years and the amount of interest was $175, what simple interest rate was she charged?

42. **Banking**   Three years after opening an account that paid 6.45% simple interest, a depositor withdrew the $3,483 in interest earned. How much money was left in the account?

43. **Saving for college**   At the birth of their child, the Fieldsons deposited $7,000 in an account paying 6% interest, compounded quarterly. How much will be available when the child turns 18?

44. **Planning a celebration**   When the Fernandez family made reservations at the end of 2008 for the December 2014 New Year's celebration in Paris, they placed $5,700 into an account paying 8% interest, compounded monthly. What amount will be available at the time of the celebration?

45. **Planning for retirement**   When Jim retires in 12 years, he expects to live lavishly on the money in a retirement account that is earning $7\frac{1}{2}\%$ interest, compounded semiannually. If the account now contains $147,500, how much will be available at retirement?

46. **Pension fund management**   The managers of a pension fund invested $3 million in government bonds paying 8.73% annual interest, compounded semiannually. After 8 years, what will the investment be worth?

47. **Real estate investing**   Property values in the suburbs have been appreciating about 11% annually. If this trend continues, what will a $137,000 home be worth in four years? Give the result to the nearest dollar.

48. **Real estate investing**   Property in suburbs closer to the city is appreciating about 8.5% annually. If this trend continues, what will a $47,000 one-acre lot be worth in five years? Give the result to the nearest dollar.

49. **Gas consumption**   The gas utilities expect natural gas consumption to increase at 7.2% per year for the next decade. Monthly consumption for one county is currently 4.3 million cubic feet. What monthly demand for gas is expected in ten years?

50. **Comparing banks**   Bank One offers a passbook account with 4.35% annual rate, compounded quarterly. Bank Two offers a money market account at 4.3%, compounded monthly. Which account provides the better growth?   (*Hint:* Find the effective rates.)

51. Comparing accounts A savings and loan offers the two accounts shown in the table. Find the effective rates.

	Annual rate	Compounding	Effective rate
NOW account	7.2%	quarterly	
Money market	6.9%	monthly	

52. Comparing accounts A credit union offers the two accounts shown in the table. Find the effective rates.

	Annual rate	Compounding	Effective rate
Certificate of deposit	6.2%	semiannually	
Passbook	5.25%	quarterly	

53. Car repair Craig borrows $1,230 for unexpected car repair costs. His bank writes a 90-day note at 12%, with interest compounded daily. What will Craig owe?

54. Fly now, pay later For a 7-day Hawaii vacation, Beth borrowed $2,570 for 9 months at an annual rate of 11.4%, compounded monthly. What did she owe?

55. Buying a computer A man estimates that the computer he plans to buy in 18 months will cost $4,200. To meet this goal, how much should he deposit in an account paying 5.75%, compounded monthly?

56. Buying a copier An accounting firm plans to deposit enough money now in an account paying 7.6% interest, compounded quarterly, to finance the purchase of a $2,780 copier in 18 months. What should be the amount of that deposit?

Discovery and Writing

57. Adding to an investment To prepare for his retirement in 14 years, Jay deposited $12,000 in an account paying 7.5% annual interest, compounded monthly. Ten years later, he deposited another $12,000. How much will be available at retirement?

58. Changing rates Ten years ago, a man invested $1,100 in a 5-year certificate of deposit paying 10%, compounded monthly. When the CD matured, he invested the proceeds in another 5-year CD paying 8%, compounded semiannually. How much is available now?

59. The power of time "A young person's most powerful money-making scheme," said an investment advisor, "is *time.*" Write a paragraph explaining what the advisor meant.

60. **Watching money grow** $10,000 is invested at 10%, compounded annually. Use a graphing calculator to find how long it will take for the accumulated value to exceed $1 million.

61. Explain why the compound interest formula on page 727 is equivalent to the one on page 385.

62. Explain why the present value formula on page 731 is equivalent to the compound interest formula on page 727.

Review *Simplify each expression. Assume that all variables represent positive numbers.*

63. $\dfrac{x^2 - 2x - 15}{2x^2 - 9x - 5}$

64. $3x(x^2 - 5) - (x^3 - 2x)$

65. $\dfrac{(3 - x)(x + 3)}{-x^2 + 9}$

66. $-\sqrt{x^2 - 6x + 9}$ $(x \geq 3)$

9.2   Annuities and Future Value

Objectives

1. Find the Future Value of an Annuity
2. Work with Sinking Funds

Now that we understand how interest works, we will discuss how to use interest to build wealth to pay for retirement, a vacation, or some other special event. Two instruments for doing this are annuities and sinking funds.

1.   Find the Future Value of an Annuity

Financial plans that involve a series of payments are called **annuities.** Monthly mortgage payments, for example, are part of an annuity, as are regular contributions to a retirement plan.

Annuity	A plan involving payments made at regular intervals is called an **annuity.**
Future Value	The **future value** of an annuity is the sum of all the payments and the interest those payments earn.
Term	The time over which the payments are made is called the **term** of the annuity.
Ordinary Annuity	In an **ordinary annuity,** the payments are made at the *end* of each time interval.

In this book, we will consider only ordinary annuities with equal periodic payments made for a fixed term.

To understand how an annuity works, assume that a savings account pays 12% annual interest, compounded monthly. Its periodic rate is $\frac{12\%}{12}$, or 1%. Also assume that each month for the next year, $100 will be deposited in that account.

To determine the future value of this annuity, we think of each monthly payment as a one-time initial contribution to a compound-interest savings account. The future value of the annuity is the sum of the accumulated values of 12 individual accounts. As Figure 9-4 suggests, the first deposit earns interest for 11 months, the second for 10 months, and so on. Because the last deposit is made at the end of the year, it earns no interest. A few minute's work with a calculator will show that the total value of the account is $1,268.25.

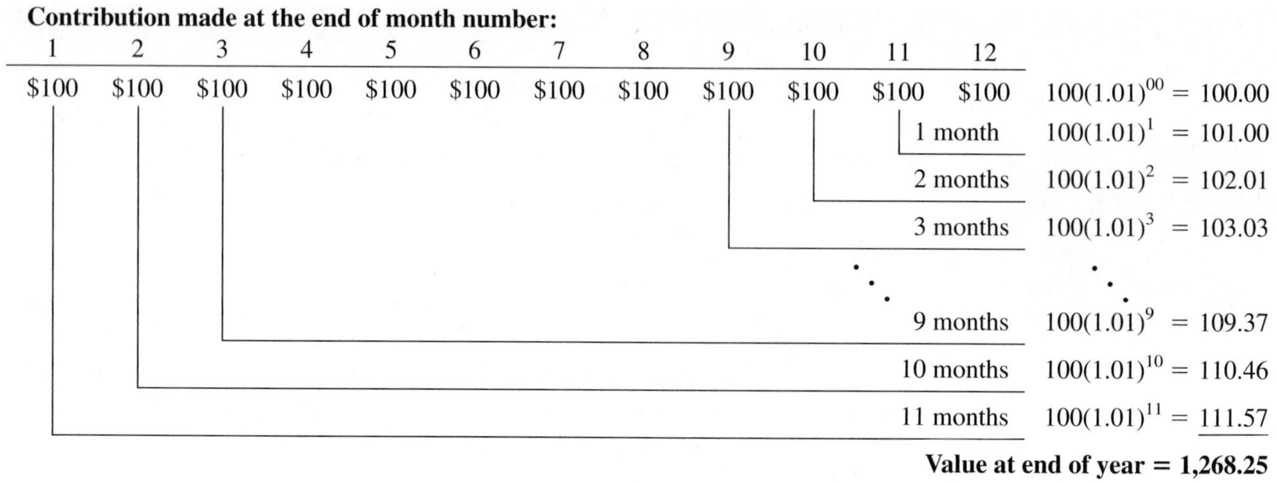

Contribution made at the end of month number:

1	2	3	4	5	6	7	8	9	10	11	12
\$100	\$100	\$100	\$100	\$100	\$100	\$100	\$100	\$100	\$100	\$100	\$100

$100(1.01)^{00} = 100.00$

1 month $\quad 100(1.01)^{1} = 101.00$

2 months $\quad 100(1.01)^{2} = 102.01$

3 months $\quad 100(1.01)^{3} = 103.03$

$\cdots$

9 months $\quad 100(1.01)^{9} = 109.37$

10 months $\quad 100(1.01)^{10} = 110.46$

11 months $\quad 100(1.01)^{11} = \underline{111.57}$

Value at end of year = 1,268.25

Figure 9-4

We will now generalize this example to derive a formula for the future value of an annuity. Consider an annuity with regular deposits of P dollars and with interest compounded k times per year for n years, at an annual rate r $\left(\text{periodic rate } i = \frac{r}{k}\right)$. During n years, there will be kn periods and kn deposits.

The last deposit, made at the end of the last period, earns no interest. The next-to-last deposit earns interest for one period, and so on. The first deposit, made at the end of the first period, earns compound interest for $kn - 1$ periods.

Future value of the annuity	=	future value of the last deposit	+	future value of the next-to-last deposit	+ $\cdots$ +	future value of the first deposit.

$$A_n = P + P(1 + i)^1 + P(1 + i)^2 + P(1 + i)^3 + \cdots + P(1 + i)^{kn-1}$$

This is a geometric sequence with first term P and a common ratio of $(1 + i)$. Its sum is given by

$$A_n = \frac{P[(1 + i)^{kn} - 1]}{i}$$

Recall that the sum of the terms of the geometric sequence $S_n = a + ar + ar^2 + ar^3 + \cdots + ar^{n-1}$ is $S_n = \frac{a(r^n - 1)}{r - 1}$.

We summarize this result.

Future Value of an Annuity

The future value A_n of an ordinary annuity with deposits of P dollars made regularly k times each year for n years, with interest compounded k times per year at an annual rate r, is

$$A_n = \frac{P[(1 + i)^{kn} - 1]}{i}$$

where i is the periodic rate, $i = \frac{r}{k}$.

EXAMPLE 1 Verify the future value of the annuity outlined in Figure 9-4.

Solution From Figure 9-4, we find

the term, in years: $\qquad n = 1$

the frequency of compounding: $\qquad k = 12$

the annual rate: $\qquad$ $r = 0.12$

the regular deposit: $\qquad$ $P = 100$

We then calculate the periodic interest rate: $i = \frac{r}{k} = \frac{0.12}{12} = 0.01$, and substitute these numbers in the formula for the future value of an annuity.

$$A_n = \frac{P[(1 + i)^{kn} - 1]}{i}$$

$$A_1 = \frac{100[(1 + 0.01)^{12 \cdot 1} - 1]}{0.01}$$

$$= \frac{100[(1.01)^{12} - 1]}{0.01}$$

$\approx 1{,}268.25030132$ $\qquad$ Use a calculator.

$\approx 1{,}268.25$ $\qquad$ Round to the nearest cent.

The annuity of Figure 9-4 will provide \$1,268.25 by the end of the year.

Self Check 1 $\quad$ Under the payroll savings plan, a student contributes \$50 a month to an ordinary annuity paying $7\frac{1}{2}\%$ annual interest, compounded monthly. How much will he have in 5 years?

Accent on Technology $\quad$ **Growth of Money**

The value of an annuity grows rapidly. To watch the value of the annuity of Example 1 grow, we graph the function

$$A_n = \frac{100[(1.01)^{12n} - 1]}{0.01}$$

$$= 10{,}000(1.01^{12n} - 1) \qquad \frac{100}{0.01} = 10{,}000$$

We enter the function

$$Y_1 = 10000*(1.01^{\wedge}(12*X) - 1)$$

on a graphing calculator, set the window values to be $0 \le X \le 50$ and $0 \le Y \le 1{,}000{,}000$, and graph it, as in Figure 9-5(a). Note how the amount increases more rapidly as the years go by. Using TRACE, as in Figure 9-5(b), we see that over \$1 million accumulates in just 39 years.

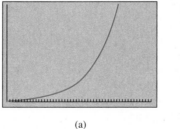
(a)

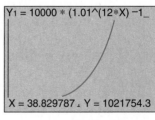
$Y_1 = 10000 * (1.01^{\wedge}(12*X) -1\_$

$X = 38.829787$ $Y = 1021754.3$

(b)

Figure 9-5

---

2. Work with Sinking Funds

If we know the amount of each deposit, we can calculate the future value of an annuity using the formula on page 736. This situation is often reversed: What regular deposits, made over time will provide a specific future amount? An annuity created to produce a fixed future value is called a **sinking fund.** To determine the required periodic payment P, we solve the future value formula for P.

$$A_n = \frac{P[(1 + i)^{kn} - 1]}{i}$$

$$A_n\frac{i}{(1 + i)^{kn} - 1} = \frac{P[(1 + i)^{kn} - 1]}{i} \cdot \frac{i}{(1 + i)^{kn} - 1}$$
To isolate P, multiply both sides by $\dfrac{i}{(1 + i)^{kn} - 1}$.

$$\frac{A_n i}{(1 + i)^{kn} - 1} = P$$
Simplify.

This result establishes the following formula.

Sinking Fund Payment For an annuity to provide a future value A_n, regular payments P are made k times per year for n years, with interest compounded k times per year at an annual rate n. The payment P is given by

$$P = \frac{A_n i}{(1 + i)^{kn} - 1}$$

where i is the periodic rate, $i = \dfrac{r}{k}$.

EXAMPLE 2 An accounting firm will need $17,000 in 5 years to replace its computer system. What periodic deposits to a sinking fund paying quarterly interest at a 9% annual rate will achieve that goal?

Solution The sinking fund will have the following characteristics:

future value:	$A_n = \$17{,}000$
annual rate:	$r = 0.09$
term:	$n = 5$
number of periods per year:	$k = 4$
periodic rate:	$i = \dfrac{r}{k} = \dfrac{0.09}{4} = 0.0225$

We substitute the given values in the formula to find the sinking fund payment.

$$P = \frac{A_n i}{(1 + i)^{kn} - 1}$$

$$= \frac{(17{,}000)(0.0225)}{(1 + 0.0225)^{4 \cdot 5} - 1}$$

$$\approx 682.415203056 \qquad \text{Use a calculator.}$$

$$\approx 682.42 \qquad \text{Round to the nearest cent.}$$

Quarterly payments of $682.42 will accumulate to $17,000 in 5 years.

Self Check 2 What quarterly deposits to the above account are required to raise the $50,000 startup cost of a branch office in 7 years?

Self Check Answers **1.** $3,626.36 **2.** $1,301.26

9.2 Exercises

Vocabulary and Concepts   *Fill in the blanks.*

1. Plans involving payments made at regular intervals are called _____.

2. In an ordinary annuity, payments are made at the ____ of each period.

3. The future value of an annuity is the sum of all the _____ and _____.

4. The time over which the payments are made is the ____ of the annuity.

5. In the future value formula, $A_n = \dfrac{P[(1 + i)^{kn} - 1]}{i}$

   P is the _____,
   i is the _____,
   k is the _____, and
   n is the _____.

6. An annuity created to fund a specific future obligation is a _____ fund.

Practice   *Assume that $100 is deposited at the end of each year in an account in which interest is compounded annually at a rate of 6%. Find the accumulated amount after the given number of years.*

7. 10 years
8. 5 years
9. 3 years
10. 20 years

Assume that $100 is deposited at the end of each year into an account in which interest is compounded annually at the given rate. Find the accumulated amount after 10 years.

11. 4%
12. 7%
13. 9.5%
14. 8.5%

Assume that $100 is deposited at the end of each period in an account in which interest is compounded at the given frequency, at an annual rate of 8%. Find the accumulated amount after 15 years.

15. $k = 2$
16. $k = 4$
17. $k = 12$
18. $k = 1$

Find the amount of each regular payment to provide $20,000 in 10 years, at the given annual rate and compounding frequency.

19. 4%, annually
20. 6%, quarterly
21. 9%, semiannually
22. 8%, monthly

Applications

23. **Saving for a vacation**   For next year's vacation, the Phelps family is saving $200 each month in an account paying 6% annual interest, compounded monthly. How much will be available a year from now?

24. **Planning for retirement**   Hank's regular $1,300 quarterly contributions to his retirement account have earned 6.5% annual interest, compounded quarterly, since he started 21 years ago. How much is in his account now?

25. **Pension fund management**   The managers of a company's pension fund invest the monthly employee contributions of $135,000 into a government fund paying 8.7%, compounded monthly. To what value will the fund grow in 20 years?

26. **Saving for college**   A mother has been saving regularly for her daughter's college—$25 each month for 11 years. The money has been earning $7\frac{1}{2}$% annual interest, compounded monthly. How much is now in the account?

27. **Buying office machines**   A company's new corporate headquarters will be completed in $2\frac{1}{2}$ years. At that time, $750,000 will be needed for office equipment. How much should be invested monthly to fund that expense? Assume 9.75% interest, compounded monthly.

28. **Retirement lifestyle**   A woman would like to receive a $500,000 lump-sum distribution from her retirement account when she retires in 25 years. She begins making monthly contributions now to an annuity paying 8.5%, compounded monthly. Find the amount of that monthly contribution.

29. **Comparing accounts**   Which account will require the lower annual contributions to fund a $10,000 obligation in 20 years?   (*Hint:* Compare the yearly total contributions.)

Bank A	5.5%; annually
Bank B	5.35%; monthly

30. Avoiding a balloon payment   The last payment of a home mortgage is a balloon payment of $47,000, which the owner is scheduled to pay in 12 years. How much *extra* should he start including in each monthly payment to eliminate the balloon payment? His mortgage is at 10.2%, compounded monthly.

Discovery and Writing

31. Retirement strategy   Jim will retire in 30 years. He will invest $100 each month for 15 years and then let the accumulated value continue to grow for the next 15 years. How much will be available at retirement? Assume 8%, compounded monthly.

32. Retirement strategy   (See Exercise 31.) Jim's brother Jack also will retire in 30 years. He plans on doing nothing during the first 15 years, then contributing twice as much—$200 monthly—to "catch up." How much will be available at retirement? Assume 8%, compounded monthly.

33. Changing plans   A woman needs $13,500 in 10 years. She would like to make regular annual contributions for the first 5 years and then let the amount grow at compound interest for the next 5 years. What should her contributions be? Assume 9%, compounded annually.

34. Talking financial sense   How would you explain to a friend who has just been hired for her first job that now is the time to start thinking about retirement?

Review   *Solve each equation.*

35. $\dfrac{2(5x - 12)}{x} = 8$

36. $\dfrac{2(5x - 12)}{x} = x$

37. $\sqrt{2x + 3} = 3$

38. $\sqrt{2x + 3} = x$

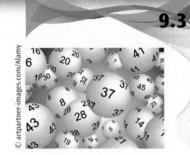

9.3   Present Value of an Annuity; Amortization

Objectives

1. Compute Present Value of an Annuity
2. Amortization

Suppose you are lucky enough to come into a great sum of money. Perhaps you will receive an inheritance, sell a business, or win the lottery. Would you spend the money wisely, or would you waste it as have so many lottery winners? In this section, we will discuss how to invest a portion of this money to guarantee that you will receive regular payments in the future.

1.   Compute Present Value of an Annuity

Instead of using an annuity to create a future value A_n, we might ask, "What *single* deposit made *now* would create that same future value?" The one deposit that gives the same final result as an annuity is called the **present value** of that annuity.

To find a formula for the present value of an annuity, we combine two previous formulas. A series of regular payments of P dollars for n years will grow to a future value A_n given by

(1) $$A_n = \frac{P[(1 + i)^{kn} - 1]}{i}$$

From a formula in Section 9.1, the present value of a future asset is given by

(2) $A_0 = A_n(1 + i)^{-kn}$

We can find the present value of a series of future payments by substituting the right side of Equation 1 into Equation 2.

$A_0 = A_n(1 + i)^{-kn}$ This is Equation 2.

$A_0 = \dfrac{P[(1 + i)^{kn} - 1]}{i}(1 + i)^{-kn}$ Substitute $\dfrac{P[(1 + i)^{kn} - 1]}{i}$ for A_n in

                                                                        Equation 2.

$\quad = \dfrac{P[(1 + i)^{kn}(1 + i)^{-kn} - 1(1 + i)^{-kn}]}{i}$ Use the distributive property.

$\quad = \dfrac{P[1 - (1 + i)^{-kn}]}{i}$ Simplify: $x^m x^{-m} = x^0 = 1.$

This establishes the following formula.

Present Value of an Annuity   The present value A_0 of an annuity with payments of P dollars made k times per year for n years, with interest compounded k times per year at an annual rate r, is

$$A_0 = \dfrac{P[1 - (1 + i)^{-kn}]}{i}$$

where i is the periodic rate, $i = \dfrac{r}{k}.$

EXAMPLE 1   To buy a boat in 2 years, the Higgins family plans to save $200 a month in an account that pays 12% interest, compounded monthly.

   a. Find the total amount of the payments.

   b. Find the value of the account in 2 years.

   c. Find the single deposit in that account that would give the same future value.

Solution   **a.** At $200 per month for 24 months, the total amount contributed is

$200(24) = $4,800$

   b. To find the value after 2 years, we use the formula for future value of an annuity found on page 736:

$$A_n = \dfrac{P[(1 + i)^{kn} - 1]}{i}$$

$$A_2 = \dfrac{200[1.01^{12 \cdot 2} - 1]}{0.01}$$

$\quad\quad \approx 5,394.692971$ Use a calculator.

$\quad\quad \approx 5,394.69$ Round to the nearest cent.

   c. To find the present value of the annuity, we substitute:

the term, in years: $n = 2$
the frequency of compounding: $k = 12$
the annual rate: $r = 0.12$

the payment: $\qquad$ $P = 200$

the periodic interest rate: $\qquad$ $i = \dfrac{r}{k} = \dfrac{0.12}{12} = 0.01$

in the present value formula.

$$A_0 = \frac{P[1 - (1 + i)^{-kn}]}{i}$$

$$A_0 = \frac{200[1 - (1 + 0.01)^{-12 \cdot 2}]}{0.01}$$

$$= \frac{200[1 - (1.01)^{-24}]}{0.01} \qquad \text{Simplify.}$$

$$\approx 4{,}248.677451 \qquad \text{Use a calculator.}$$

$$\approx 4{,}248.68 \qquad \text{Round to the nearest cent.}$$

The present value of the annuity is $4,248.68. That one deposit now will provide the same final amount, $5,394.69, as the annuity.

Self Check 1 For his retirement in 30 years, a man plans to make monthly contributions of $25 to an ordinary annuity paying $8\frac{1}{2}\%$ annually, compounded monthly.

a. Find the total amount of his contributions.

b. Find the single deposit now that will provide the same retirement benefit.

State lottery winnings are usually paid as a 20-year annuity. That is to the state's advantage, because it can fund the annuity with a single amount that is much smaller than the total prize.

EXAMPLE 2 Britta won the lottery. She will receive $75,000 per month for the next 20 years—a total of $18 million. What single deposit should the lottery commission make now to fund Britta's annuity? Assume 8.4% annual interest, compounded monthly.

Solution The lottery commission finds the present value of the annuity, with

the payment: $\qquad$ $P = 75{,}000$

the annual rate: $\qquad$ $r = 0.084$

the frequency of compounding: $\qquad$ $k = 12$

the periodic rate: $\qquad$ $i = \dfrac{r}{k} = \dfrac{0.084}{12} = 0.007$

the term, in years: $\qquad$ $n = 20$

These values are used in the formula for the present value of an annuity.

$$A_0 = \frac{P[1 - (1 + i)^{-kn}]}{i}$$

$$A_0 = \frac{75{,}000[1 - (1.007)^{-12 \cdot 20}]}{0.007}$$

$$\approx 8{,}705{,}700.365 \qquad \text{Use a calculator.}$$

$$\approx 8{,}705{,}700.37 \qquad \text{Round to the nearest cent.}$$

To fund the $18 million prize, the commission must deposit $8,705,700.37.

Self Check 2   The lottery pays a total prize of $120,000 in monthly installments, as a 10-year annuity. Assuming 8.4% interest, compounded monthly, what current deposit is needed to fund the annuity?

EXAMPLE 3   As a settlement in an automobile injury lawsuit, Robyn will receive $30,000 each year for the next 25 years, for a total of $750,000. The insurance company is offering a one-payment settlement of $300,000, now. Should she accept? Assume that the money can be invested at 9% annual interest.

Solution   Robyn should calculate the present value of an annuity with:

the payment:	$P = 30,000$
the annual rate:	$r = 0.09$
the frequency of compounding:	$k = 1$ (annual)
the periodic rate:	$i = \dfrac{r}{k} = \dfrac{0.09}{1} = 0.09$
the term, in years:	$n = 25$

She should use these values in the formula for the present value of an annuity.

$$A_0 = \frac{P[1 - (1 + i)^{-kn}]}{i}$$

$$A_0 = \frac{30,000[1 - (1.09)^{-1 \cdot 25}]}{0.09}$$

$\approx 294,677.3881$ **Use a calculator.**

$\approx 294,677.39$ **Round to the nearest cent.**

Since the annuity is worth $294,677.39 and the company is offering $300,000, Robyn should accept the $300,000.

Self Check 3   If Robyn could invest the settlement at 8% interest, should she still accept the lump-sum offer?

When a worker is employed, regular contributions are usually made to a retirement fund. After retirement, those funds are given back, either as an annuity or as a lump-sum distribution.

EXAMPLE 4   Carlos wants to fund an annuity to supplement his retirement income. How much should he deposit now to generate retirement income of $1,000 a month for the next 20 years? Assume that he can get $9\frac{3}{4}$% interest, compounded monthly.

Solution   Carlos must calculate the present value of a future stream of income, with:

the payment:	$P = 1,000$
the annual rate:	$r = 0.0975$
the frequency of compounding:	$k = 12$
the periodic rate:	$i = \dfrac{r}{k} = \dfrac{0.0975}{12} = 0.008125$
the term, in years:	$n = 20$

He should use these values in the formula for the present value of an annuity.

$$A_0 = \frac{P[1 - (1 + i)^{-kn}]}{i}$$

$$A_0 = \frac{1{,}000[1 - (1.008125)^{-12 \cdot 20}]}{0.008125}$$

$$\approx 105{,}428$$

If \$105,428 is deposited now, Carlos will receive \$1,000 per month in retirement income for 20 years.

Self Check 4 If Carlos can invest at $8\frac{3}{4}\%$, what deposit is needed now?

2. Amortization

Before a bank will lend money, you must sign a **promissory note** indicating that you will pay the money back. We discussed one-payment notes in Section 9.1. Most loans, however, are repaid in installments instead of all at once. Spreading the repayment over several equal payments is called **amortization.**

When a such a loan is made, the bank is buying an annuity from the borrower, and the bank pays the borrower a certain amount and expects regular payments in return. To calculate the amount of these regular installment payments, we solve the present value formula for P to get

$$A_0 = \frac{P[1 - (1 + i)^{-kn}]}{i}$$

$$A_0 i = P[1 - (1 + i)^{-kn}] \qquad \text{Multiply both sides by } i.$$

$$P = \frac{A_0 i}{1 - (1 + i)^{-kn}} \qquad \text{Divide both sides by } 1 - (1 + i)^{-kn}.$$

In this context, the present value A_0 is the amount of the loan.

Installment Payments The periodic payment P required to repay an amount A_0 is given by

$$P = \frac{A_0 i}{1 - (1 + i)^{-kn}}$$

where

r is the annual rate,

k is the frequency of compounding (usually monthly),

i is the periodic rate, $i = \dfrac{r}{k}$, and

n is the term of the loan.

EXAMPLE 5 The Almondi family takes a 15-year mortgage of \$200,000 for their new home, at 10.8%, compounded monthly.

a. Find their monthly payments.

b. Find the total of their payments over the full term.

Solution **a.** The mortgage has the following characteristics.

the amount:	$A_0 = 200{,}000$
the annual rate:	$r = 0.108$

the frequency of compounding: $k = 12$

the periodic rate: $i = \dfrac{r}{k} = \dfrac{0.108}{12} = 0.009$

the term, in years: $n = 15$

We substitute these values into the formula for installment payments to get

$$P = \frac{A_0 i}{1 - (1 + i)^{-kn}}$$

$$P = \frac{(200{,}000)(0.009)}{1 - (1.009)^{-12 \cdot 15}}$$

$$= \frac{1{,}800}{1 - 0.1993379912}$$

$$= 2{,}248.13964$$

Each monthly mortgage payment will be $2,248.14.

b. There are $12 \cdot 15 = 180$ payments of $2,248.14 each, for a total of $404,665—more than twice the amount borrowed!

Self Check 5   Instead of a 15-year mortgage, the Almondis considered a 30-year mortgage. Answer the previous two questions again.

Everyday Connections

Mortgage Rates

LIBOR is an abbreviation for "London Interbank Offered Rate," and is the interest rate offered by a specific group of London banks for U.S. dollar deposits of a stated maturity. LIBOR is used as a base index for setting rates of some adjustable rate financial instruments, including Adjustable Rate Mortgages (ARMs) and other loans.

One week in 2000, the average rate on a one-year adjustable mortgage surged to 6.51 percent, the highest since January 2001, from 5.84 percent the prior week. The rate also surpassed the cost of a 30-year fixed loan for the first time.

Suppose a prospective homeowner obtains a mortgage loan with the following terms:

- Mortgage amount $324,000
- Mortgage term   25 years
- Annual interest rate (fixed)   5.64%

Source: http://immobilienblasen.blogspot.com/2007_08_01_archive.html

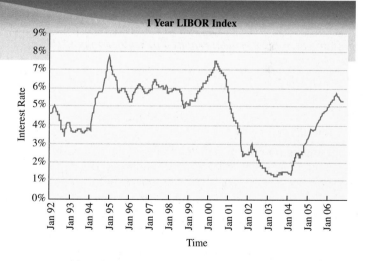

Calculate the annual monthly mortgage payment (principal and interest).

Self Check Answers   **1. a.** $9,000   **b.** $3,251.34   **2.** about $81,000   **3.** No; the annuity is now worth more than $320,243.   **4.** $113,159   **5. a.** $1,874.48   **b.** $674,814

9.3 Exercises

Vocabulary and Concepts *Fill in the blanks.*

1. The current worth of a future stream of income is the _____ of an annuity.

2. The amount required now to produce a future stream of income is the _____ of an annuity.

3. A loan is called a _____ because you promise to repay it.

4. Often, loan repayment is spread out over several _____.

5. Spreading repayment of a loan over several *equal* payments is called _____ the loan.

6. An amortized loan is also called a _____.

Practice *Find the present value of an annuity with the given terms.*

7. Annual payments of $3,500 at 5.25%, compounded annually for 25 years

8. Semiannual payments of $375 at a 4.92% annual rate, compounded semiannually for 10 years

Find the periodic payment required to repay a loan with the given terms.

9. $25,000 repaid over 15 years, with monthly payments at a 12% annual rate

10. $1,750 repaid in 18 monthly installments, at an annual rate of 19%

11. **Funding retirement** Instead of making quarterly contributions of $700 to a retirement fund for the next 15 years, a man would rather make only one contribution, now. How much should that be? Assume $6\frac{1}{4}\%$ annual interest, compounded quarterly.

12. **Funding a lottery** To fund Jamie's lottery winnings of $15,000 per month for the next 20 years, the lottery commission needs to make a single deposit now. Assuming 9.2% compounded monthly, what should the deposit be?

13. **Money up front** Instead of receiving an annuity of $12,000 each year for the next 15 years, a young woman would like a one-time payment, now. Assuming she could invest the proceeds at $8\frac{1}{2}\%$, what would be a fair amount?

14. **Funding retirement** What single amount deposited now into an account paying $7\frac{2}{3}\%$ annual interest, compounded quarterly, would fund an annuity paying $5,000 quarterly for the next 25 years?

15. **Buying a car** The Jepsens are buying a $21,700 car and financing it over the next 4 years. They secure an 8.4% loan. What will their monthly payments be?

16. **Total cost of buying a car** What will be the total amount the Jepsens will pay over the life of the loan? (See Exercise 15.)

17. **Choosing a mortgage** One lender offers two mortgages—a 15-year mortgage at 12%, and a 20-year mortgage at 11%. For each, find the monthly payment to repay $130,000.

18. **Total cost of a mortgage** For each of the mortgages in Exercise 17, find the total of the monthly payments.

Discovery and Writing

19. **Getting an early start** As Jorge starts working now at the age of 20, he decides to make regular contributions to a savings account. He wants to accumulate enough by age 55 to fund an annuity of $5,000 per month until age 80. What should his monthly contributions be? Assume that both accounts pay 8.75%, compounded monthly.

20. **Comparing annuities** Which of these 20-year plans is best, and why? All are at 8% annually.
    a. $1,000 each year for 10 years, and then let the accumulated amount grow for 10 years
    b. $500 each year for 20 years
    c. Do nothing for 10 years, and then contribute $2,000 each year for 10 years
    d. One payment of $8,000 now, and let it grow

21. **Changing the payment** A woman contributed $500 per quarter for the first 10 years of an annuity, but changed to quarterly payments of $1,500 for the last 10 years. Assuming $7\frac{1}{4}\%$ annual interest compounded quarterly, what is her accumulated value?

22. **Changing the rate** A woman contributed $150 per month for 10 years to an account that paid 5% for the first 5 years, but 6.5% for the last 5 years. How much has she saved?

Review *Simplify each expression. Assume that all variables represent positive numbers.*

23. $\dfrac{6\sqrt{30}}{3\sqrt{5}}$

24. $\dfrac{6}{\sqrt{7}-2}$

25. $3\sqrt{5x}+5\sqrt{20x}$

26. $\sqrt{\dfrac{x^3y^5}{x^5y^6}}$

CHAPTER REVIEW

9.1   Interest

Definitions and Concepts	Examples
If funds in a savings account earn **simple interest** at an **annual rate** r, the amount deposited is the **principal** P, and the length of **time** is t, the amount of **interest** I earned is given by the formula . $$I = Prt$$	Find the simple interest on a deposit of \$8,000 that is left on deposit for 15 years at an annual rate of 4.5%. $$I = Prt$$ $$I = 8{,}000 \cdot 0.045 \cdot 15$$ $$= 5{,}400$$ The interest earned is \$5,400 and the account will contain \$13,400.
Compound interest, annual compounding: A single deposit A_0 earning compound interest for n years at an annual rate r, will grow to a future value A_n according to the formula $$A_n = A_0(1 + r)^n$$	Find the amount in an account where \$8,000 is left on deposit for 15 years at an annual rate of 4.5%, compounded annually. $$A_n = A_0(1 + r)^n$$ $$A_{15} = 8{,}000(1 + 0.045)^{15}$$ $$= 15{,}482.25954$$ The amount will be \$15,482.26. This is \$2,082.26 more than when the money was deposited at simple interest.
Compound interest formulas: An amount A_0, earning interest compounded k times a year for n years at an annual rate r, will grow to a future value A_n according to the formula $$A_n = A_0(1 + i)^{kn}$$ where $i = \frac{r}{k}$ is the periodic interest rate.	Find the amount in an account in which \$8,000 is left on deposit for 15 years at an annual rate of 4.5%, compounded monthly. The periodic interest rate is $i = \frac{r}{k} = \frac{0.045}{12} = 0.00375$. $$A_n = A_0(1 + i)^{kn}$$ $$A_n = 8{,}000(1 + 0.00375)^{12 \cdot 15}$$ $$= 15{,}692.44007$$ The amount will be \$15,692.44, \$210.18 more than annual compounding.

The **effective rate** R is used to compare different savings plans.

$$R = (1 + i)^k - 1$$

Find the effective rate in the example above.

$$R = (1 + i)^k - 1$$
$$= (1 + 0.00375)^{12} - 1$$
$$= (1.00375)^{12} - 1$$
$$= 0.045939825$$

The effective rate is about 4.6%.

The **present value** A_0 is the single deposit *now* that will yield a specific future value, A_n.

$$A_0 = A_n(1 + i)^{-kn}$$

where interest is compounded k times a year at an annual rate r. $\left(i \text{ is the periodic rate } \frac{r}{k}. \right)$

Find the amount that must be deposited now to grow to be $15,692.44 in 15 years in an account earning 4.5%, compounded monthly.

As shown above, the periodic interest rate is 0.00375

$$A_0 = A_n(1 + i)^{-nk}$$
$$A_0 = 15,692.44(1 + 0.00375)^{-12 \cdot 15}$$
$$= 7,999.999967$$

The present value is $8,000. In the example above, we saw that $8,000 grew to be $15,692.44. So it is expected that the present value of $15,692.44 is $8,000.

Exercises

1. $2,000 is deposited in an account that earns 9% simple interest. Find the value of the account in 5 years.

2. $2,000 is deposited in an account in which interest is compounded annually at 9%. Find the value in 5 years.

3. Brian borrows $2,350 for medical bills. The bank writes a 60-day note at 14%, with interest compounded daily. What will Brian owe?

4. $2,000 earns interest, compounded quarterly, at an annual rate of 7.6% for 16 years. Find the future value.

5. BigBank advertises a savings account at a 6.3% rate, compounded quarterly. BestBank offers 6.21%, compounded daily. Calculate each effective rate and choose the better account.

6. What amount deposited now in an account paying 5.75% interest, compounded semiannually, will yield $7,900 in 6 years?

9.2   Annuities and Future Value

Definitions and Concepts

An **annuity** is a series of payments P made at regular intervals. Its **future value** A_n is the sum of all the payments and the interest those payments earn. The time over which the payments are made is called the **term** of the annuity. In an **ordinary annuity,** the payments are made at the *end* of each time interval.

The future value A_n of an ordinary annuity with deposits of P dollars made regularly k times each year for n years, with interest compounded k times per year at an annual rate r is

$$A_n = \frac{P[(1 + i)^{kn} - 1]}{i}$$

where i is the periodic rate, $i = \frac{r}{k}$.

Examples

Under a company savings plan, a worker contributes $100 a month to an ordinary annuity paying 8%, compounded monthly. How much will the annuity be worth in 50 years?

The periodic interest rate is $i = \frac{r}{k} = \frac{0.08}{12} \approx 0.0067$.

So,

$$A_n = \frac{P[(1 + i)^{kn} - 1]}{i}$$

$$A_{50} = \frac{100[(1 + 0.0067)^{12 \cdot 50} - 1]}{0.0067}$$

$$= 805,362.6379$$

The future value is $805,362.64.

An annuity with the purpose of funding a future obligation is a **sinking fund.** To yield a specific future value A_n, regular deposits of P dollars are made k times per year for n years, with interest compounded k times per year at an annual rate r. The payment P is

$$P = \frac{A_n i}{(1 + i)^{kn} - 1}$$

where i is the periodic rate, $i = \frac{r}{k}$.

What periodic deposit to a sinking fund paying 12% interest, compounded monthly, will amount to $50,000 in 30 years?

The periodic interest rate is $i = \frac{0.12}{12} = 0.01$.

So,

$$\begin{aligned}
P &= \frac{A_n i}{(1 + i)^{kn} - 1} \\
&= \frac{50{,}000(0.01)}{(1 + 0.01)^{12 \cdot 30} - 1} \\
&= 14.30629846 \\
&\approx 14.31
\end{aligned}$$

Exercises

7. $500 is deposited at the end of each year into an annuity in which interest is compounded annually at 5%. Find the accumulated amount after 13 years.

8. $150 is deposited monthly into an account that pays 8% annual interest, compounded monthly. Find the future value after 20 years.

9. The owners of a small dry cleaning shop will need $40,700 to open a second shop in 7 years. What monthly payments to a sinking fund earning 7.5% interest, compounded monthly, will meet that obligation?

9.3   Present Value of an Annuity; Amortization

Definitions and Concepts

The **present value** of an annuity is the current value of a future stream of income.

The present value A_0 of an annuity with payments of P dollars made k times per year for n years, with interest compounded k times per year at an annual rate r, is

$$A_0 = \frac{P[1 - (1 + i)^{-kn}]}{i}$$

where i is the periodic rate, $i = \frac{r}{k}$.

Examples

To buy a car in 2 years, a man intends to pay $300 each month into an account that pays 6% annual interest, compounded monthly. Find the present value of the annuity.

The periodic interest rate is $i = \frac{r}{k} = \frac{0.06}{12} = 0.005$.

So,

$$\begin{aligned}
A_0 &= \frac{P[1 - (1 + i)^{-kn}]}{i} \\
&= \frac{300[1 - (1 + 0.005)^{-12 \cdot 2}]}{0.005} \\
&= 6{,}768.859867
\end{aligned}$$

The present value of the annuity is $6,769.

Loans are often paid off in **installments.** If equal installments are paid over a fixed time, the payments are **amortized.**

The periodic payment P required to repay an amount A is given by

$$P = \frac{A_0 i}{1 - (1 + i)^{-kn}}$$

where

r is the annual rate,

k is the frequency of compounding,

i is the periodic rate, $i = \frac{r}{k}$, and

n is the term of the loan.

A family buys a new house for $250,000. To do so, they get a 30-year mortgage at 6% interest, compounded monthly. Find their monthly payment.

The periodic interest rate is $i = \frac{r}{k} = \frac{0.06}{12} = 0.005$.

So,

$$\begin{aligned} P &= \frac{A_0 i}{1 - (1 + i)^{-kn}} \\ &= \frac{250{,}000(0.005)}{1 - (1 + 0.005)^{-12 \cdot 30}} \\ &= 1{,}498.876313 \end{aligned}$$

The monthly payment for principal and interest will be $1,498.88. Taxes and insurance will be extra.

Exercises

10. An annuity pays $250 semiannually for 20 years. At a semiannually compounded rate of 6.5%, what is the present value?

11. The lottery must fund a 20-year annuity of $50,000 per year. At 9.6%, compounded annually, what must be invested now?

12. What are the monthly payments for a $150,500, 15-year, 10.75% mortgage? What is the total amount paid?

13. Answer the previous question, but for a 30-year mortgage.

CHAPTER TEST

Fill in the blanks.

1. When interest is left on deposit to earn more interest, the account earns _____ interest.

2. The annual rate of interest divided by the number of periods is called the _____ interest rate.

3. To compare different savings plans, compare the _____ rates of interest.

4. The nominal rate of interest is also called the _____ rate.

5. Plans involving regular periodic payments are called _____.

6. An annuity to fund a specific future obligation is a _____.

7. The current value of a series of future payments is the _____ of an annuity.

8. Repaying a loan over several regular, equal installments is called _____ the loan.

9. $1,300 is deposited in a new account that earns 5% simple interest. What will the account be worth in 10 years?

10. $1,300 is deposited in a new account that earns 5% interest, compounded annually. What will the account be worth in 10 years?

11. $1,300 is deposited in an account that earns 5% annual interest, compounded monthly. What will it be worth in 10 years?

12. What is the effective rate of the savings plan in Problem 11?

13. What single deposit now will yield $5,000 in 10 years? Assume 7% annual interest, compounded quarterly.

14. Each month for 5 years, a student made $700 payments to an account paying 7.3% annual interest, compounded monthly. Find the accumulated amount.

15. What monthly payment to a sinking fund will raise $8,000 in 5 years? Assume 6.5% annual interest, compounded monthly.

16. Find the present value of an annuity that pays $1,000 each month for 15 years. Assume 6.8% annual interest, compounded monthly.

17. What are the monthly payments for a 15-year, $90,000 mortgage at 8.95%?

CUMULATIVE REVIEW EXERCISES

Solve each system by graphing.

1. $\begin{cases} 2x + y = 8 \\ x - 2y = -1 \end{cases}$

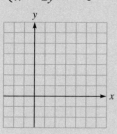

2. $\begin{cases} 3x = -y + 2 \\ y + x - 4 = -2x \end{cases}$

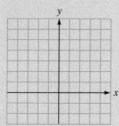

Solve each system.

3. $\begin{cases} 5x = 3y + 12 \\ 2x - 3y = 3 \end{cases}$

4. $\begin{cases} 2x + y - z = 7 \\ x - y + z = 2 \\ x + y - 3z = 2 \end{cases}$

Solve each system using matrices.

5. $\begin{cases} 2x + y - z = 0 \\ x - y + z = 3 \\ x + y - 3z = -5 \end{cases}$

6. $\begin{cases} 2x - 2y + 3z + t = 2 \\ x + y + z + t = 5 \\ -x + 2y - 3z + 2t = 2 \\ x + y + 2z - t = 4 \end{cases}$

Let $A = \begin{bmatrix} 2 & 1 \\ 1 & 4 \end{bmatrix}$, $B = \begin{bmatrix} -1 & 2 \\ 2 & 3 \end{bmatrix}$, *and*

$C = \begin{bmatrix} 2 & 0 & -1 \\ -1 & 2 & 2 \end{bmatrix}$. *Find each matrix.*

7. $A + B$

8. $B - A$

9. AC

10. $B^2 + 2A$

Find the inverse of each matrix, if possible.

11. $\begin{bmatrix} 2 & 6 \\ 2 & 4 \end{bmatrix}$

12. $\begin{bmatrix} 1 & -1 & 1 \\ 1 & 4 & 0 \\ 2 & 4 & 1 \end{bmatrix}$

Evaluate each determinant.

13. $\begin{vmatrix} -3 & 5 \\ 4 & 7 \end{vmatrix}$

14. $\begin{vmatrix} 2 & -3 & 2 \\ 0 & 1 & -1 \\ 1 & -2 & 1 \end{vmatrix}$

Set up the determinants to find x and y in the system $\begin{cases} 4x + 3y = 11 \\ -2x + 5y = 24 \end{cases}$. *Do not evaluate the determinants.*

15. $x =$

16. $y =$

Decompose each fraction into partial fractions.

17. $\dfrac{-x + 1}{(x + 1)(x + 2)}$

18. $\dfrac{x - 4}{(2x - 5)^2}$

Find each solution by graphing.

19. $y \le 2x + 6$

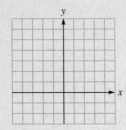

20. $\begin{cases} 2x + 3y \ge 6 \\ 2x - 3y \le 6 \end{cases}$

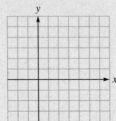

Write the equation of each circle with the given center O and radius r.

21. $O(0, 0); r = 4$

22. $O(2, -3); r = 11$

Complete the square on x and/or y and graph each equation.

23. $x^2 + y^2 - 4y = 12$

24. $x^2 - 2y - 2x = -7$

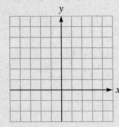

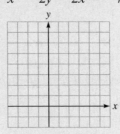

25. $x^2 + 4y^2 + 2x = 3$

26. $x^2 - 9y^2 - 4x = 5$

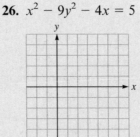

Write the equation of each ellipse.

27. Center $(0, 0)$; horizontal major axis of 12; minor axis of 8

28. Center $(2, 3); a = 5; c = 2$, major axis vertical

Write the equation of each hyperbola.

29. Center $(0, 0)$; focus $(3, 0)$; vertex $(2, 0)$

30. Center $(2, 4)$; area of fundamental rectangle is 36 square units; $a = b$; transverse axis parallel to y-axis

Find the required term of the expansion of $(x + 2y)^8$.

31. 2nd term

32. 6th term

Find each sum.

33. $\displaystyle\sum_{k=1}^{5} 2$

34. $\displaystyle\sum_{k=2}^{6} (3x + 1)$

Find the sum of the first six terms of each sequence.

35. $-2, 1, 4, \ldots$

36. $\dfrac{1}{9}, \dfrac{1}{3}, 1, \ldots$

Find each value.

37. $P(8, 4)$

38. $P(24, 0)$

39. $C(12, 10)$

40. $P(4, 4) \cdot C(6, 6)$

41. In how many ways can 6 men and 4 women be placed in a line if the women line up first?

42. In how many ways can a committee of 4 people be selected from a group of 12 people?

Find each probability.

43. Rolling 11 on one roll of two dice

44. Being dealt an all-red 5-card poker hand from a standard deck

45. If the probability that a person is married is 0.6 and the probability that a married person has children is 0.8, find the probability that a randomly chosen person is married with children.

46. Prove the formula by induction:

$$4 + 7 + 10 + \cdots + (3n + 1) = \frac{n(3n + 5)}{2}$$

47. What single deposit made now in an account that pays $8\frac{1}{2}\%$ interest, compounded annually, will grow to \$10,000 in 12 years?

48. A bank offers a \$110,000, 20-year mortgage at 8.75%. Find the monthly payment.

Appendix I

A Proof of the Binomial Theorem

The binomial theorem can be proved for positive-integer exponents using mathematical induction.

The Binomial Theorem   If n is a positive integer, then

$$(a + b)^n = a^n + \frac{n!}{1!(n - 1)!}a^{n-1}b^1 + \frac{n!}{2!(n - 2)!}a^{n-2}b^2 + \cdots$$

$$+ \frac{n!}{r!(n - r)!}a^{n-r}b^r + \cdots + b^n$$

Proof   As in all induction proofs, there are two parts.

Part 1:   Substituting the number 1 for n on both sides of the equation, we have

$$(a + b)^1 = a^1 + \frac{1!}{1!(1 - 1)!}a^{1-1}b^1$$

$$a + b = a + a^0 b$$

$$a + b = a + b$$

and the theorem is true when $n = 1$. Part 1 is complete.

Part 2:   We write expressions for two general terms in the statement of the induction hypothesis. We assume that the theorem is true for $n = k$:

$$(a + b)^k = a^k + \frac{k!}{1!(k - 1)!}a^{k-1}b + \frac{k!}{2!(k - 2)!}a^{k-2}b^2 + \cdots$$

$$+ \frac{k!}{(r - 1)!(k - r + 1)!}a^{k-r+1}b^{r-1}$$

$$+ \frac{k!}{r!(k - r)!}a^{k-r}b^r + \cdots + b^k$$

We multiply both sides of this equation by $a + b$ and hope to obtain a similar equation in which the quantity $k + 1$ replaces all of the n values in the binomial theorem:

$$(a + b)^k(a + b)$$

$$= (a + b)\left[a^k + \frac{k!}{1!(k - 1)!}a^{k-1}b + \frac{k!}{2!(k - 2)!}a^{k-2}b^2 + \cdots \right.$$

$$\left. + \frac{k!}{(r - 1)!(k - r + 1)!}a^{k-r+1}b^{r-1} + \frac{k!}{r!(k - r)!}a^{k-r}b^r + \cdots + b^k \right]$$

We distribute the multiplication first by a and then by b:

$$(a + b)^{k+1} = \left[a^{k+1} + \frac{k!}{1!(k-1)!}a^k b + \frac{k!}{2!(k-2)!}a^{k-1}b^2 + \cdots \right.$$

$$+ \frac{k!}{(r-1)!(k-r+1)!}a^{k-r+2}b^{r-1} + \frac{k!}{r!(k-r)!}a^{k-r+1}b^r + \cdots + ab^k \right]$$

$$+ \left[a^k b + \frac{k!}{1!(k-1)!}a^{k-1}b^2 + \frac{k!}{2!(k-2)!}a^{k-2}b^3 + \cdots \right.$$

$$+ \frac{k!}{(r-1)!(k-r+1)!}a^{k-r+1}b^r + \frac{k!}{r!(k-r)!}a^{k-r}b^{r+1} + \cdots + b^{k+1} \right]$$

Combining like terms, we have

$$(a + b)^{k+1} = a^{k+1} + \left[\frac{k!}{1!(k-1)!} + 1 \right]a^k b$$

$$+ \left[\frac{k!}{2!(k-2)!} + \frac{k!}{1!(k-1)!} \right]a^{k-1}b^2 + \cdots$$

$$+ \left[\frac{k!}{r!(k-r)!} + \frac{k!}{(r-1)!(k-r+1)!} \right]a^{k-r+1}b^r + \cdots + b^{k+1}$$

These results may be written as

$$(a + b)^{k+1} = a^{k+1} + \frac{(k+1)!}{1!(k+1-1)!}a^{(k+1)-1}b + \frac{(k+1)!}{2!(k+1-2)!}a^{(k+1)-2}b^2$$

$$+ \cdots + \frac{(k+1)!}{r!(k+1-r)!}a^{(k+1)-r}b^r + \cdots + b^{k+1}$$

This formula has precisely the same form as the binomial theorem, with the quantity $k + 1$ replacing all of the original n values. Therefore, the truth of the theorem for $n = k$ implies the truth of the theorem for $n = k + 1$. Because both parts of the axiom of mathematical induction are verified, the theorem is proved.

Appendix II

Tables

Table A Powers and Roots

n	n^2	$\sqrt{n}$	n^3	$\sqrt[3]{n}$	n	n^2	$\sqrt{n}$	n^3	$\sqrt[3]{n}$
1	1	1.000	1	1.000	51	2,601	7.141	132,651	3.708
2	4	1.414	8	1.260	52	2,704	7.211	140,608	3.733
3	9	1.732	27	1.442	53	2,809	7.280	148,877	3.756
4	16	2.000	64	1.587	54	2,916	7.348	157,464	3.780
5	25	2.236	125	1.710	55	3,025	7.416	166,375	3.803
6	36	2.449	216	1.817	56	3,136	7.483	175,616	3.826
7	49	2.646	343	1.913	57	3,249	7.550	185,193	3.849
8	64	2.828	512	2.000	58	3,364	7.616	195,112	3.871
9	81	3.000	729	2.080	59	3,481	7.681	205,379	3.893
10	100	3.162	1,000	2.154	60	3,600	7.746	216,000	3.915
11	121	3.317	1,331	2.224	61	3,721	7.810	226,981	3.936
12	144	3.464	1,728	2.289	62	3,844	7.874	238,328	3.958
13	169	3.606	2,197	2.351	63	3,969	7.937	250,047	3.979
14	196	3.742	2,744	2.410	64	4,096	8.000	262,144	4.000
15	225	3.873	3,375	2.466	65	4,225	8.062	274,625	4.021
16	256	4.000	4,096	2.520	66	4,356	8.124	287,496	4.041
17	289	4.123	4,913	2.571	67	4,489	8.185	300,763	4.062
18	324	4.243	5,832	2.621	68	4,624	8.246	314,432	4.082
19	361	4.359	6,859	2.668	69	4,761	8.307	328,509	4.102
20	400	4.472	8,000	2.714	70	4,900	8.367	343,000	4.121
21	441	4.583	9,261	2.759	71	5,041	8.426	357,911	4.141
22	484	4.690	10,648	2.802	72	5,184	8.485	373,248	4.160
23	529	4.796	12,167	2.844	73	5,329	8.544	389,017	4.179
24	576	4.899	13,824	2.884	74	5,476	8.602	405,224	4.198
25	625	5.000	15,625	2.924	75	5,625	8.660	421,875	4.217
26	676	5.099	17,576	2.962	76	5,776	8.718	438,976	4.236
27	729	5.196	19,683	3.000	77	5,929	8.775	456,533	4.254
28	784	5.292	21,952	3.037	78	6,084	8.832	474,552	4.273
29	841	5.385	24,389	3.072	79	6,241	8.888	493,039	4.291
30	900	5.477	27,000	3.107	80	6,400	8.944	512,000	4.309
31	961	5.568	29,791	3.141	81	6,561	9.000	531,441	4.327
32	1,024	5.657	32,768	3.175	82	6,724	9.055	551,368	4.344
33	1,089	5.745	35,937	3.208	83	6,889	9.110	571,787	4.362
34	1,156	5.831	39,304	3.240	84	7,056	9.165	592,704	4.380
35	1,225	5.916	42,875	3.271	85	7,225	9.220	614,125	4.397
36	1,296	6.000	46,656	3.302	86	7,396	9.274	636,056	4.414
37	1,369	6.083	50,653	3.332	87	7,569	9.327	658,503	4.431
38	1,444	6.164	54,872	3.362	88	7,744	9.381	681,472	4.448
39	1,521	6.245	59,319	3.391	89	7,921	9.434	704,969	4.465
40	1,600	6.325	64,000	3.420	90	8,100	9.487	729,000	4.481
41	1,681	6.403	68,921	3.448	91	8,281	9.539	753,571	4.498
42	1,764	6.481	74,088	3.476	92	8,464	9.592	778,688	4.514
43	1,849	6.557	79,507	3.503	93	8,649	9.644	804,357	4.531
44	1,936	6.633	85,184	3.530	94	8,836	9.695	830,584	4.547
45	2,025	6.708	91,125	3.557	95	9,025	9.747	857,375	4.563
46	2,116	6.782	97,336	3.583	96	9,216	9.798	884,736	4.579
47	2,209	6.856	103,823	3.609	97	9,409	9.849	912,673	4.595
48	2,304	6.928	110,592	3.634	98	9,604	9.899	941,192	4.610
49	2,401	7.000	117,649	3.659	99	9,801	9.950	970,299	4.626
50	2,500	7.071	125,000	3.684	100	10,000	10.000	1,000,000	4.642

Table B Base-10 Logarithms

N	0	1	2	3	4	5	6	7	8	9
1.0	.0000	.0043	.0086	.0128	.0170	.0212	.0253	.0294	.0334	.0374
1.1	.0414	.0453	.0492	.0531	.0569	.0607	.0645	.0682	.0719	.0755
1.2	.0792	.0828	.0864	.0899	.0934	.0969	.1004	.1038	.1072	.1106
1.3	.1139	.1173	.1206	.1239	.1271	.1303	.1335	.1367	.1399	.1430
1.4	.1461	.1492	.1523	.1553	.1584	.1614	.1644	.1673	.1703	.1732
1.5	.1761	.1790	.1818	.1847	.1875	.1903	.1931	.1959	.1987	.2014
1.6	.2041	.2068	.2095	.2122	.2148	.2175	.2201	.2227	.2253	.2279
1.7	.2304	.2330	.2355	.2380	.2405	.2430	.2455	.2480	.2504	.2529
1.8	.2553	.2577	.2601	.2625	.2648	.2672	.2695	.2718	.2742	.2765
1.9	.2788	.2810	.2833	.2856	.2878	.2900	.2923	.2945	.2967	.2989
2.0	.3010	.3032	.3054	.3075	.3096	.3118	.3139	.3160	.3181	.3201
2.1	.3222	.3243	.3263	.3284	.3304	.3324	.3345	.3365	.3385	.3404
2.2	.3424	.3444	.3464	.3483	.3502	.3522	.3541	.3560	.3579	.3598
2.3	.3617	.3636	.3655	.3674	.3692	.3711	.3729	.3747	.3766	.3784
2.4	.3802	.3820	.3838	.3856	.3874	.3892	.3909	.3927	.3945	.3962
2.5	.3979	.3997	.4014	.4031	.4048	.4065	.4082	.4099	.4116	.4133
2.6	.4150	.4166	.4183	.4200	.4216	.4232	.4249	.4265	.4281	.4298
2.7	.4314	.4330	.4346	.4362	.4378	.4393	.4409	.4425	.4440	.4456
2.8	.4472	.4487	.4502	.4518	.4533	.4548	.4564	.4579	.4594	.4609
2.9	.4624	.4639	.4654	.4669	.4683	.4698	.4713	.4728	.4742	.4757
3.0	.4771	.4786	.4800	.4814	.4829	.4843	.4857	.4871	.4886	.4900
3.1	.4914	.4928	.4942	.4955	.4969	.4983	.4997	.5011	.5024	.5038
3.2	.5051	.5065	.5079	.5092	.5105	.5119	.5132	.5145	.5159	.5172
3.3	.5185	.5198	.5211	.5224	.5237	.5250	.5263	.5276	.5289	.5302
3.4	.5315	.5328	.5340	.5353	.5366	.5378	.5391	.5403	.5416	.5428
3.5	.5441	.5453	.5465	.5478	.5490	.5502	.5514	.5527	.5539	.5551
3.6	.5563	.5575	.5587	.5599	.5611	.5623	.5635	.5647	.5658	.5670
3.7	.5682	.5694	.5705	.5717	.5729	.5740	.5752	.5763	.5775	.5786
3.8	.5798	.5809	.5821	.5832	.5843	.5855	.5866	.5877	.5888	.5899
3.9	.5911	.5922	.5933	.5944	.5955	.5966	.5977	.5988	.5999	.6010
4.0	.6021	.6031	.6042	.6053	.6064	.6075	.6085	.6096	.6107	.6117
4.1	.6128	.6138	.6149	.6160	.6170	.6180	.6191	.6201	.6212	.6222
4.2	.6232	.6243	.6253	.6263	.6274	.6284	.6294	.6304	.6314	.6325
4.3	.6335	.6345	.6355	.6365	.6375	.6385	.6395	.6405	.6415	.6425
4.4	.6435	.6444	.6454	.6464	.6474	.6484	.6493	.6503	.6513	.6522
4.5	.6532	.6542	.6551	.6561	.6571	.6580	.6590	.6599	.6609	.6618
4.6	.6628	.6637	.6646	.6656	.6665	.6675	.6684	.6693	.6702	.6712
4.7	.6721	.6730	.6739	.6749	.6758	.6767	.6776	.6785	.6794	.6803
4.8	.6812	.6821	.6830	.6839	.6848	.6857	.6866	.6875	.6884	.6893
4.9	.6902	.6911	.6920	.6928	.6937	.6946	.6955	.6964	.6972	.6981
5.0	.6990	.6998	.7007	.7016	.7024	.7033	.7042	.7050	.7059	.7067
5.1	.7076	.7084	.7093	.7101	.7110	.7118	.7126	.7135	.7143	.7152
5.2	.7160	.7168	.7177	.7185	.7193	.7202	.7210	.7218	.7226	.7235
5.3	.7243	.7251	.7259	.7267	.7275	.7284	.7292	.7300	.7308	.7316
5.4	.7324	.7332	.7340	.7348	.7356	.7364	.7372	.7380	.7388	.7396

Table B (continued)

N	0	1	2	3	4	5	6	7	8	9
5.5	.7404	.7412	.7419	.7427	.7435	.7443	.7451	.7459	.7466	.7474
5.6	.7482	.7490	.7497	.7505	.7513	.7520	.7528	.7536	.7543	.7551
5.7	.7559	.7566	.7574	.7582	.7589	.7597	.7604	.7612	.7619	.7627
5.8	.7634	.7642	.7649	.7657	.7664	.7672	.7679	.7686	.7694	.7701
5.9	.7709	.7716	.7723	.7731	.7738	.7745	.7752	.7760	.7767	.7774
6.0	.7782	.7789	.7796	.7803	.7810	.7818	.7825	.7832	.7839	.7846
6.1	.7853	.7860	.7868	.7875	.7882	.7889	.7896	.7903	.7910	.7917
6.2	.7924	.7931	.7938	.7945	.7952	.7959	.7966	.7973	.7980	.7987
6.3	.7993	.8000	.8007	.8014	.8021	.8028	.8035	.8041	.8048	.8055
6.4	.8062	.8069	.8075	.8082	.8089	.8096	.8102	.8109	.8116	.8122
6.5	.8129	.8136	.8142	.8149	.8156	.8162	.8169	.8176	.8182	.8189
6.6	.8195	.8202	.8209	.8215	.8222	.8228	.8235	.8241	.8248	.8254
6.7	.8261	.8267	.8274	.8280	.8287	.8293	.8299	.8306	.8312	.8319
6.8	.8325	.8331	.8338	.8344	.8351	.8357	.8363	.8370	.8376	.8382
6.9	.8388	.8395	.8401	.8407	.8414	.8420	.8426	.8432	.8439	.8445
7.0	.8451	.8457	.8463	.8470	.8476	.8482	.8488	.8494	.8500	.8506
7.1	.8513	.8519	.8525	.8531	.8537	.8543	.8549	.8555	.8561	.8567
7.2	.8573	.8579	.8585	.8591	.8597	.8603	.8609	.8615	.8621	.8627
7.3	.8633	.8639	.8645	.8651	.8657	.8663	.8669	.8675	.8681	.8686
7.4	.8692	.8698	.8704	.8710	.8716	.8722	.8727	.8733	.8739	.8745
7.5	.8751	.8756	.8762	.8768	.8774	.8779	.8785	.8791	.8797	.8802
7.6	.8808	.8814	.8820	.8825	.8831	.8837	.8842	.8848	.8854	.8859
7.7	.8865	.8871	.8876	.8882	.8887	.8893	.8899	.8904	.8910	.8915
7.8	.8921	.8927	.8932	.8938	.8943	.8949	.8954	.8960	.8965	.8971
7.9	.8976	.8982	.8987	.8993	.8998	.9004	.9009	.9015	.9020	.9025
8.0	.9031	.9036	.9042	.9047	.9053	.9058	.9063	.9069	.9074	.9079
8.1	.9085	.9090	.9096	.9101	.9106	.9112	.9117	.9122	.9128	.9133
8.2	.9138	.9143	.9149	.9154	.9159	.9165	.9170	.9175	.9180	.9186
8.3	.9191	.9196	.9201	.9206	.9212	.9217	.9222	.9227	.9232	.9238
8.4	.9243	.9248	.9253	.9258	.9263	.9269	.9274	.9279	.9284	.9289
8.5	.9294	.9299	.9304	.9309	.9315	.9320	.9325	.9330	.9335	.9340
8.6	.9345	.9350	.9355	.9360	.9365	.9370	.9375	.9380	.9385	.9390
8.7	.9395	.9400	.9405	.9410	.9415	.9420	.9425	.9430	.9435	.9440
8.8	.9445	.9450	.9455	.9460	.9465	.9469	.9474	.9479	.9484	.9489
8.9	.9494	.9499	.9504	.9509	.9513	.9518	.9523	.9528	.9533	.9538
9.0	.9542	.9547	.9552	.9557	.9562	.9566	.9571	.9576	.9581	.9586
9.1	.9590	.9595	.9600	.9605	.9609	.9614	.9619	.9624	.9628	.9633
9.2	.9638	.9643	.9647	.9652	.9657	.9661	.9666	.9671	.9675	.9680
9.3	.9685	.9689	.9694	.9699	.9703	.9708	.9713	.9717	.9722	.9727
9.4	.9731	.9736	.9741	.9745	.9750	.9754	.9759	.9763	.9768	.9773
9.5	.9777	.9782	.9786	.9791	.9795	.9800	.9805	.9809	.9814	.9818
9.6	.9823	.9827	.9832	.9836	.9841	.9845	.9850	.9854	.9859	.9863
9.7	.9868	.9872	.9877	.9881	.9886	.9890	.9894	.9899	.9903	.9908
9.8	.9912	.9917	.9921	.9926	.9930	.9934	.9939	.9943	.9948	.9952
9.9	.9956	.9961	.9965	.9969	.9974	.9978	.9983	.9987	.9991	.9996

Table C (continued)

N	0	1	2	3	4	5	6	7	8	9
5.5	1.7047	.7066	.7084	.7102	.7120	.7138	.7156	.7174	.7192	.7210
5.6	.7228	.7246	.7263	.7281	.7299	.7317	.7334	.7352	.7370	.7387
5.7	.7405	.7422	.7440	.7457	.7475	.7492	.7509	.7527	.7544	.7561
5.8	.7579	.7596	.7613	.7630	.7647	.7664	.7681	.7699	.7716	.7733
5.9	.7750	.7766	.7783	.7800	.7817	.7834	.7851	.7867	.7884	.7901
6.0	1.7918	.7934	.7951	.7967	.7984	.8001	.8017	.8034	.8050	.8066
6.1	.8083	.8099	.8116	.8132	.8148	.8165	.8181	.8197	.8213	.8229
6.2	.8245	.8262	.8278	.8294	.8310	.8326	.8342	.8358	.8374	.8390
6.3	.8405	.8421	.8437	.8453	.8469	.8485	.8500	.8516	.8532	.8547
6.4	.8563	.8579	.8594	.8610	.8625	.8641	.8656	.8672	.8687	.8703
6.5	1.8718	.8733	.8749	.8764	.8779	.8795	.8810	.8825	.8840	.8856
6.6	.8871	.8886	.8901	.8916	.8931	.8946	.8961	.8976	.8991	.9006
6.7	.9021	.9036	.9051	.9066	.9081	.9095	.9110	.9125	.9140	.9155
6.8	.9169	.9184	.9199	.9213	.9228	.9242	.9257	.9272	.9286	.9301
6.9	.9315	.9330	.9344	.9359	.9373	.9387	.9402	.9416	.9430	.9445
7.0	1.9459	.9473	.9488	.9502	.9516	.9530	.9544	.9559	.9573	.9587
7.1	.9601	.9615	.9629	.9643	.9657	.9671	.9685	.9699	.9713	.9727
7.2	.9741	.9755	.9769	.9782	.9796	.9810	.9824	.9838	.9851	.9865
7.3	.9879	.9892	.9906	.9920	.9933	.9947	.9961	.9974	.9988	2.0001
7.4	2.0015	.0028	.0042	.0055	.0069	.0082	.0096	.0109	.0122	.0136
7.5	2.0149	.0162	.0176	.0189	.0202	.0215	.0229	.0242	.0255	.0268
7.6	.0281	.0295	.0308	.0321	.0334	.0347	.0360	.0373	.0386	.0399
7.7	.0412	.0425	.0438	.0451	.0464	.0477	.0490	.0503	.0516	.0528
7.8	.0541	.0554	.0567	.0580	.0592	.0605	.0618	.0631	.0643	.0656
7.9	.0669	.0681	.0694	.0707	.0719	.0732	.0744	.0757	.0769	.0782
8.0	2.0794	.0807	.0819	.0832	.0844	.0857	.0869	.0882	.0894	.0906
8.1	.0919	.0931	.0943	.0956	.0968	.0980	.0992	.1005	.1017	.1029
8.2	.1041	.1054	.1066	.1078	.1090	.1102	.1114	.1126	.1138	.1150
8.3	.1163	.1175	.1187	.1199	.1211	.1223	.1235	.1247	.1258	.1270
8.4	.1282	.1294	.1306	.1318	.1330	.1342	.1353	.1365	.1377	.1389
8.5	2.1401	.1412	.1424	.1436	.1448	.1459	.1471	.1483	.1494	.1506
8.6	.1518	.1529	.1541	.1552	.1564	.1576	.1587	.1599	.1610	.1622
8.7	.1633	.1645	.1656	.1668	.1679	.1691	.1702	.1713	.1725	.1736
8.8	.1748	.1759	.1770	.1782	.1793	.1804	.1815	.1827	.1838	.1849
8.9	.1861	.1872	.1883	.1894	.1905	.1917	.1928	.1939	.1950	.1961
9.0	2.1972	.1983	.1994	.2006	.2017	.2028	.2039	.2050	.2061	.2072
9.1	.2083	.2094	.2105	.2116	.2127	.2138	.2148	.2159	.2170	.2181
9.2	.2192	.2203	.2214	.2225	.2235	.2246	.2257	.2268	.2279	.2289
9.3	.2300	.2311	.2322	.2332	.2343	.2354	.2364	.2375	.2386	.2396
9.4	.2407	.2418	.2428	.2439	.2450	.2460	.2471	.2481	.2492	.2502
9.5	2.2513	.2523	.2534	.2544	.2555	.2565	.2576	.2586	.2597	.2607
9.6	.2618	.2628	.2638	.2649	.2659	.2670	.2680	.2690	.2701	.2711
9.7	.2721	.2732	.2742	.2752	.2762	.2773	.2783	.2793	.2803	.2814
9.8	.2824	.2834	.2844	.2854	.2865	.2875	.2885	.2895	.2905	.2915
9.9	.2925	.2935	.2946	.2956	.2966	.2976	.2986	.2996	.3006	.3016

Table C Base-e Logarithms

N	0	1	2	3	4	5	6	7	8	9
1.0	.0000	.0100	.0198	.0296	.0392	.0488	.0583	.0677	.0770	.0862
1.1	.0953	.1044	.1133	.1222	.1310	.1398	.1484	.1570	.1655	.1740
1.2	.1823	.1906	.1989	.2070	.2151	.2231	.2311	.2390	.2469	.2546
1.3	.2624	.2700	.2776	.2852	.2927	.3001	.3075	.3148	.3221	.3293
1.4	.3365	.3436	.3507	.3577	.3646	.3716	.3784	.3853	.3920	.3988
1.5	.4055	.4121	.4187	.4253	.4318	.4383	.4447	.4511	.4574	.4637
1.6	.4700	.4762	.4824	.4886	.4947	.5008	.5068	.5128	.5188	.5247
1.7	.5306	.5365	.5423	.5481	.5539	.5596	.5653	.5710	.5766	.5822
1.8	.5878	.5933	.5988	.6043	.6098	.6152	.6206	.6259	.6313	.6366
1.9	.6419	.6471	.6523	.6575	.6627	.6678	.6729	.6780	.6831	.6881
2.0	.6931	.6981	.7031	.7080	.7129	.7178	.7227	.7275	.7324	.7372
2.1	.7419	.7467	.7514	.7561	.7608	.7655	.7701	.7747	.7793	.7839
2.2	.7885	.7930	.7975	.8020	.8065	.8109	.8154	.8198	.8242	.8286
2.3	.8329	.8372	.8416	.8459	.8502	.8544	.8587	.8629	.8671	.8713
2.4	.8755	.8796	.8838	.8879	.8920	.8961	.9002	.9042	.9083	.9123
2.5	.9163	.9203	.9243	.9282	.9322	.9361	.9400	.9439	.9478	.9517
2.6	.9555	.9594	.9632	.9670	.9708	.9746	.9783	.9821	.9858	.9895
2.7	.9933	.9969	1.0006	.0043	.0080	.0116	.0152	.0188	.0225	.0260
2.8	1.0296	.0332	.0367	.0403	.0438	.0473	.0508	.0543	.0578	.0613
2.9	.0647	.0682	.0716	.0750	.0784	.0818	.0852	.0886	.0919	.0953
3.0	1.0986	.1019	.1053	.1086	.1119	.1151	.1184	.1217	.1249	.1282
3.1	.1314	.1346	.1378	.1410	.1442	.1474	.1506	.1537	.1569	.1600
3.2	.1632	.1663	.1694	.1725	.1756	.1787	.1817	.1848	.1878	.1909
3.3	.1939	.1969	.2000	.2030	.2060	.2090	.2119	.2149	.2179	.2208
3.4	.2238	.2267	.2296	.2326	.2355	.2384	.2413	.2442	.2470	.2499
3.5	1.2528	.2556	.2585	.2613	.2641	.2669	.2698	.2726	.2754	.2782
3.6	.2809	.2837	.2865	.2892	.2920	.2947	.2975	.3002	.3029	.3056
3.7	.3083	.3110	.3137	.3164	.3191	.3218	.3244	.3271	.3297	.3324
3.8	.3350	.3376	.3403	.3429	.3455	.3481	.3507	.3533	.3558	.3584
3.9	.3610	.3635	.3661	.3686	.3712	.3737	.3762	.3788	.3813	.3838
4.0	1.3863	.3888	.3913	.3938	.3962	.3987	.4012	.4036	.4061	.4085
4.1	.4110	.4134	.4159	.4183	.4207	.4231	.4255	.4279	.4303	.4327
4.2	.4351	.4375	.4398	.4422	.4446	.4469	.4493	.4516	.4540	.4563
4.3	.4586	.4609	.4633	.4656	.4679	.4702	.4725	.4748	.4770	.4793
4.4	.4816	.4839	.4861	.4884	.4907	.4929	.4951	.4974	.4996	.5019
4.5	1.5041	.5063	.5085	.5107	.5129	.5151	.5173	.5195	.5217	.5239
4.6	.5261	.5282	.5304	.5326	.5347	.5369	.5390	.5412	.5433	.5454
4.7	.5476	.5497	.5518	.5539	.5560	.5581	.5602	.5623	.5644	.5665
4.8	.5686	.5707	.5728	.5748	.5769	.5790	.5810	.5831	.5851	.5872
4.9	.5892	.5913	.5933	.5953	.5974	.5994	.6014	.6034	.6054	.6074
5.0	1.6094	.6114	.6134	.6154	.6174	.6194	.6214	.6233	.6253	.6273
5.1	.6292	.6312	.6332	.6351	.6371	.6390	.6409	.6429	.6448	.6467
5.2	.6487	.6506	.6525	.6544	.6563	.6582	.6601	.6620	.6639	.6658
5.3	.6677	.6696	.6715	.6734	.6752	.6771	.6790	.6808	.6827	.6845
5.4	.6864	.6882	.6901	.6919	.6938	.6956	.6974	.6993	.7011	.7029

Use the properties of logarithms and ln 10 ≈ 2.3026 to find logarithms of numbers less than 1 or greater than 10.

Appendix III

Answers to Selected Exercises

1. set **3.** union **5.** decimal **7.** 2 **9.** composite
11. decimals **13.** negative **15.** $x + (y + z)$
17. $5m + 5 \cdot 2$ **19.** interval **21.** two **23.** positive
25. true **27.** false **29.** true **31.** {a, b, c, d, e, f, g}
33. {a, c, e} **35.** 1, 2, 6, 7 **37.** $-5, -4, 0, 1, 2, 6, 7$
39. $\sqrt{2}$ **41.** 6 **43.** $-5, 1, 7$
45.
47.
49.
51.
53. $(2, \infty)$
55. $(0, 5)$
57. $(-4, \infty)$
59. $[-2, 2)$
61. $(-\infty, 5]$
63. $(-5, 0]$
65. $[-2, 3]$
67. $[2, 6]$
69. $(-5, \infty) \cap (-\infty, 4)$
71. $[-8, \infty) \cap (-\infty, -3]$
73. $(-\infty, -2) \cup (2, \infty)$
75. $(-\infty, -1] \cup [3, \infty)$
77. 13 **79.** 0 **81.** -8 **83.** -32 **85.** $5 - \pi$
87. 0 **89.** $x + 1$ **91.** $-(x - 4)$ **93.** 5 **95.** 5
97. natural numbers **99.** integers

1. factor **3.** $3, 2x$ **5.** scientific, integer **7.** x^{m+n}
9. $x^n y^n$ **11.** 1 **13.** 169 **15.** -25 **17.** $4 \cdot x \cdot x \cdot x$
19. $(-5x)(-5x)(-5x)(-5x)$ **21.** $-8 \cdot x \cdot x \cdot x \cdot x$
23. $7x^3$ **25.** x^2 **27.** $-27t^3$ **29.** $x^3 y^2$ **31.** 10.648
33. -0.0625 **35.** x^5 **37.** z^6 **39.** y^{21} **41.** z^{26}
43. a^{14} **45.** $27x^3$ **47.** $x^6 y^3$ **49.** $\dfrac{a^6}{b^3}$ **51.** 1 **53.** 1
55. $\dfrac{1}{z^4}$ **57.** $\dfrac{1}{y^5}$ **59.** x^2 **61.** x^4 **63.** a^4 **65.** x
67. $\dfrac{m^9}{n^6}$ **69.** $\dfrac{1}{a^9}$ **71.** $\dfrac{a^{12}}{b^4}$ **73.** $\dfrac{1}{r^4}$ **75.** $\dfrac{x^{32}}{y^{16}}$ **77.** $\dfrac{9x^{10}}{25y^2}$
79. $\dfrac{4y^{12}}{x^{20}}$ **81.** $\dfrac{64z^7}{25y^6}$ **83.** 50 **85.** 4 **87.** -8 **89.** 216
91. -12 **93.** 20 **95.** $\dfrac{3}{64}$ **97.** 3.72×10^5
99. -1.77×10^8 **101.** 7×10^{-3} **103.** -6.93×10^{-7}
105. 1×10^{12} **107.** 937,000 **109.** 0.0000221 **111.** 3.2
113. -0.0032 **115.** 1.17×10^4 **117.** 7×10^4
119. 5.3×10^{19} **121.** 1.986×10^4 meters per min
123. 1.67248×10^{-15} g **125.** 9.3×10^7 mi, 1.48×10^8 mi
127. x^{n+2} **129.** x^{m-1} **131.** x^{m+4}
135.   **137.** $5 - \pi$

1. 0 **3.** not **5.** $a^{1/n}$ **7.** $\sqrt[n]{ab}$ **9.** $\neq$ **11.** 3
13. $\dfrac{1}{5}$ **15.** -3 **17.** 10 **19.** -4 **21.** not a real number
23. $4|a|$ **25.** $2|a|$ **27.** $-2a$ **29.** $-6b^2$ **31.** $\dfrac{4a^2}{5|b|}$
33. $-\dfrac{10x^2}{3y}$ **35.** 8 **37.** -64 **39.** -100 **41.** $\dfrac{1}{8}$
43. $\dfrac{1}{512}$ **45.** $-\dfrac{1}{27}$ **47.** $\dfrac{32}{243}$ **49.** $\dfrac{16}{9}$ **51.** $10s^2$
53. $\dfrac{1}{2y^2 z}$ **55.** $x^6 y^3$ **57.** $\dfrac{1}{r^6 s^{12}}$ **59.** $\dfrac{4a^4}{25b^6}$ **61.** $\dfrac{100s^8}{9r^4}$
63. a **65.** 7 **67.** 5 **69.** -5 **71.** $-\dfrac{1}{5}$ **73.** $6|x|$
75. $3y^2$ **77.** $2y$ **79.** $\dfrac{|x|y^2}{|z^3|}$ **81.** $\sqrt{2}$ **83.** $17x\sqrt{2}$

85. $2y^2\sqrt{3y}$ **87.** $12\sqrt[3]{3}$ **89.** $6z\sqrt[4]{3z}$ **91.** $6x\sqrt{2y}$

93. 0 **95.** $\sqrt{3}$ **97.** $\sqrt[3]{4}$ **99.** $\dfrac{2b\sqrt[4]{27a^2}}{3a}$

101. $\dfrac{u\sqrt[3]{6uv^2}}{3v}$ **103.** $\dfrac{1}{2\sqrt{5}}$ **105.** $\dfrac{1}{\sqrt[3]{3}}$ **107.** $\dfrac{b}{32a\sqrt[5]{2b^2}}$

109. $\dfrac{2\sqrt{3}}{9}$ **111.** $-\dfrac{\sqrt{2x}}{8}$ **113.** $\sqrt{3}$ **115.** $\sqrt[5]{4x^3}$

117. $\sqrt[6]{32}$ **119.** $\dfrac{\sqrt[4]{12}}{2}$ **125.** $(-2, 5]$ **127.** -5

129. 6.17×10^8

Exercises 0.4 (page 48)

1. monomial, variables   **3.** trinomial   **5.** one   **7.** like
9. coefficients, variables   **11.** yes, trinomial, 2nd degree
13. no   **15.** yes, binomial, 3rd degree   **17.** yes, monomial,
0th degree   **19.** yes, monomial, no defined degree
21. $6x^3 - 3x^2 - 8x$ **23.** $4y^3 + 14$ **25.** $-x^2 + 14$
27. $-28t + 96$ **29.** $-4y^2 + y$ **31.** $2x^2y - 4xy^2$

33. $8x^3y^7$ **35.** $\dfrac{m^4n^4}{2}$ **37.** $-4r^3s - 4rs^3$

39. $12a^2b^2c^2 + 18ab^3c^3 - 24a^2b^4c^2$ **41.** $a^2 + 4a + 4$
43. $a^2 - 12a + 36$ **45.** $x^2 - 16$ **47.** $x^2 + 2x - 15$
49. $3u^2 + 4u - 4$ **51.** $10x^2 + 13x - 3$
53. $9a^2 - 12ab + 4b^2$ **55.** $9m^2 - 16n^2$
57. $6y^2 - 16xy + 8x^2$ **59.** $9x^3 - yx^2 - 27xy + 3y^2$
61. $5z^3 - 5tz + 2tz^2 - 2t^2$ **63.** $27x^3 - 27x^2 + 9x - 1$
65. $6x^3 + 14x^2 - 5x - 3$ **67.** $6x^3 - 5x^2y + 6xy^2 + 8y^3$
69. $6y^{2n} + 2$ **71.** $-10x^{4n} - 15y^{2n}$ **73.** $x^{2n} - x^n - 12$
75. $6r^{2n} - 25r^n + 14$ **77.** $xy + x^{3/2}y^{1/2}$ **79.** $a - b$

81. $\sqrt{3} + 1$ **83.** $x(\sqrt{7} - 2)$ **85.** $\dfrac{x(x + \sqrt{3})}{x^2 - 3}$

87. $\dfrac{(y + \sqrt{2})^2}{y^2 - 2}$ **89.** $\dfrac{\sqrt{3} + 3 - \sqrt{2} - \sqrt{6}}{2}$

91. $\dfrac{x - 2\sqrt{xy} + y}{x - y}$ **93.** $\dfrac{1}{2(\sqrt{2} - 1)}$

95. $\dfrac{y^2 - 3}{y^2 + 2y\sqrt{3} + 3}$ **97.** $\dfrac{1}{\sqrt{x + 3} + \sqrt{x}}$ **99.** $\dfrac{2a}{b^3}$

101. $-\dfrac{2z^9}{3x^3y^2}$ **103.** $\dfrac{x}{2y} + \dfrac{3xy}{2}$ **105.** $\dfrac{2y^3}{5} - \dfrac{3y}{5x^3} + \dfrac{1}{5x^4y^3}$

107. $3x + 2$ **109.** $x - 7 + \dfrac{2}{2x - 5}$ **111.** $x - 3$

113. $x^2 - 2 + \dfrac{-x^2 + 5}{x^3 - 2}$ **115.** $x^4 + 2x^3 + 4x^2 + 8x + 16$

117. $6x^2 + x - 12$ **119.** $(x^2 + 3x - 10)$ ft^2
121. $(144x - 48x^2 + 4x^3)$ in.^{3} **129.** 27

131. $\dfrac{125x^3}{8y^6}$ **133.** $-b\sqrt[3]{2ab}$

Exercises 0.5 (page 58)

1. factor   **3.** $x(a + b)$ **5.** $(x + y)^2$
7. $(x + y)(x^2 - xy + y^2)$ **9.** $3(x - 2)$ **11.** $4x^2(2 + x)$
13. $7x^2y^2(1 + 2x)$ **15.** $(x + y)(a + b)$
17. $(4a + b)(1 - 3a)$ **19.** $(2x + 3)(2x - 3)$

21. $(2 + 3r)(2 - 3r)$ **23.** prime
25. $(x + z + 5)(x + z - 5)$ **27.** $(x + 4)^2$ **29.** $(b - 5)^2$
31. $(m + 2n)^2$ **33.** $(4x - 3y)(3x + 2y)$
35. $(x + 7)(x + 3)$ **37.** $(x - 6)(x + 2)$
39. $(2p + 3)(3p - 1)$ **41.** $(t + 7)(t^2 - 7t + 49)$
43. $(2z - 3)(4z^2 + 6z + 9)$ **45.** $3abc(a + 2b + 3c)$
47. $(x + 1)(3x^2 - 1)$ **49.** $t(y + c)(2x - 3)$
51. $(a + b)(x + y + z)$ **53.** $(x + y - z)(x - y + z)$
55. $-4xy$ **57.** $(x^2 + y^2)(x + y)(x - y)$
59. $3(x + 2)(x - 2)$ **61.** $2x(3y + 2)(3y - 2)$
63. prime   **65.** $(6a + 5)(4a - 3)$ **67.** $(3x + 7y)(2x + 5y)$
69. $2(6p - 35q)(p + q)$ **71.** $-(6m - 5n)(m - 7n)$
73. $-x(6x + 7)(x - 5)$ **75.** $x^2(2x - 7)(3x + 5)$
77. $(x^2 + 5)(x^2 - 3)$ **79.** $(a^n - 3)(a^n + 1)$
81. $(3x^n - 2)(2x^n - 1)$ **83.** $(2x^n + 3y^n)(2x^n - 3y^n)$
85. $(5y^n + 2)(2y^n - 3)$ **87.** $2(x + 10)(x^2 - 10x + 100)$
89. $(x + y - 4)(x^2 + 2xy + y^2 + 4x + 4y + 16)$
91. $(2a + y)(2a - y)(4a^2 - 2ay + y^2)(4a^2 + 2ay + y^2)$
93. $(a - b)(a^2 + ab + b^2 + 1)$
95. $(4x^2 + y^2)(16x^4 - 4x^2y^2 + y^4)$
97. $(x - 3 + 12y)(x - 3 - 12y)$
99. $(a + b - 5)(a + b + 2)$
101. $(x + 2)(x^2 - 2x + 4)(x - 1)(x^2 + x + 1)$
103. $(x^2 + x + 2)(x^2 - x + 2)$
105. $(x^2 + x + 4)(x^2 - x + 4)$
107. $(2a^2 + a + 1)(2a^2 - a + 1)$
109. $\frac{4}{3}\pi(r_1 - r_2)(r_1^2 + r_1r_2 + r_2^2)$ **115.** $2\left(\frac{3}{2}x + 1\right)$

117. $2\left(\frac{1}{2}x^2 + x + 2\right)$ **119.** $a\left(1 + \frac{b}{a}\right)$ **121.** $x^{1/2}(x^{1/2} + 1)$

123. $\sqrt{2}\left(\sqrt{2}x + y\right)$ **125.** $ab(b^{1/2} - a^{1/2})$
127. $(x - 2)(x + 3 + y)$ **129.** $(a + 1)(a^3 + a^2 + 1)$
131. 1 **133.** x^{20} **135.** 1 **137.** $-3\sqrt{5x}$

Exercises 0.6 (page 68)

1. numerator   **3.** $ad = bc$ **5.** $\dfrac{ac}{bd}$ **7.** $\dfrac{a + c}{b}$ **9.** equal

11. not equal   **13.** $\dfrac{a}{3b}$ **15.** $\dfrac{8x}{35a}$ **17.** $\dfrac{16}{3}$ **19.** $\dfrac{z}{c}$

21. $\dfrac{2x^2y}{a^2b^3}$ **23.** $\dfrac{2}{x + 2}$ **25.** $-\dfrac{x - 5}{x + 5}$ **27.** $\dfrac{3x - 4}{2x - 1}$

29. $\dfrac{x^2 + 2x + 4}{x + a}$ **31.** $\dfrac{x(x - 1)}{x + 1}$ **33.** $\dfrac{x - 1}{x}$

35. $\dfrac{x(x + 1)^2}{x + 2}$ **37.** $\dfrac{1}{2}$ **39.** $\dfrac{z - 4}{(z + 2)(z - 2)}$ **41.** 1

43. 1 **45.** $\dfrac{x(x + 3)}{x + 1}$ **47.** $\dfrac{x + 5}{x + 3}$ **49.** 4 **51.** $\dfrac{-1}{x - 5}$

53. $\dfrac{5x - 1}{(x + 1)(x - 1)}$ **55.** $\dfrac{2(a - 2)}{(a + 4)(a - 4)}$

57. $\dfrac{2}{(x + 2)(x - 2)}$ **59.** $\dfrac{2(x^2 - 3x + 1)}{(x + 1)^2(x - 1)}$ **61.** $\dfrac{3y - 2}{y - 1}$

63. $\dfrac{1}{x + 2}$ **65.** $\dfrac{2x - 5}{2x(x - 2)}$ **67.** $\dfrac{2x^2 + 19x + 1}{(x + 4)(x - 4)}$ **69.** 0

71. $\dfrac{-x^4 + 3x^3 - 43x^2 - 58x + 697}{(x + 5)(x - 5)(x + 4)(x - 4)}$ **73.** $\dfrac{b}{2c}$ **75.** $81a$

77. -1 **79.** $\dfrac{y + x}{x^2y^2}$ **81.** $\dfrac{y + x}{y - x}$ **83.** $\dfrac{a^2(3x - 4ab)}{ax + b}$

85. $\dfrac{x-2}{x+2}$ **87.** $\dfrac{3x^2y^2}{xy-1}$ **89.** $\dfrac{3x^2}{x^2+1}$ **91.** $\dfrac{x^2-3x-4}{x^2+5x-3}$

93. $\dfrac{x}{x+1}$ **95.** $\dfrac{5x+1}{x-1}$ **97.** $\dfrac{k_1k_2}{k_2+k_1}$ **99.** $\dfrac{3x}{3+x}$

101. $\dfrac{x+1}{2x+1}$ **107.** 6 **109.** $\dfrac{y^9}{x^{12}}$ **111.** $-\sqrt{5}$

Chapter Review (page 71)

1. 3, 6, 8 **2.** 0, 3, 6, 8 **3.** $-6, -3, 0, 3, 6, 8$
4. $-6, -3, 0, \frac{1}{2}, 3, 6, 8$ **5.** $\pi, \sqrt{5}$ **6.** $-6, -3, 0, \frac{1}{2}, 3, \pi,$
$\sqrt{5}, 6, 8$ **7.** 3 **8.** 6, 8 **9.** $-6, 0, 6, 8$
10. $-3, 3$ **11.** assoc. prop. of add. **12.** comm. prop. of
add. **13.** assoc. prop. of mult. **14.** distrib. prop.
15. comm. prop. of mult. **16.** comm. prop. of add.
17. double negative rule
18. ⊢────●───●────────●───●───⊢
   10 11 12 13 14 15 16 17 18 19 20
19. ⊢────────●────────●────⊢
   6 7 8 9 10 11 12 13 14
20. ⊢(────)──── **21.** ◄────────|────►
   -3 5 -1 0
22. ⊢(─────)──────
   -2 4
23. ⊢(─────)────────
   -5 2
24. ◄──────|────────
   -4 6

25. 6 **26.** 25 **27.** $\sqrt{2}-1$ **28.** $\sqrt{3}-1$ **29.** 12
30. $-5aaa$ **31.** $(-5a)(-5a)$ **32.** $3t^3$ **33.** $-6b^2$ **34.** n^6
35. p^6 **36.** $x^{12}y^8$ **37.** $\dfrac{a^{12}}{b^6}$ **38.** $\dfrac{1}{m^6}$ **39.** $\dfrac{q^6}{8p^6}$ **40.** $\dfrac{1}{a^3}$
41. $\dfrac{b^6}{a^4}$ **42.** $\dfrac{y^8}{9}$ **43.** $\dfrac{a^8}{b^{10}}$ **44.** $\dfrac{y^4}{9x^4}$ **45.** $-\dfrac{8m^{12}}{n^3}$
46. 18 **47.** 6.75×10^3 **48.** 2.3×10^{-4} **49.** 480
50. 0.00025 **51.** 1.5×10^{14} **52.** 11 **53.** $\dfrac{3}{5}$ **54.** $2x$
55. $3|a|$ **56.** $-10x^2$ **57.** not a real number **58.** $x^6|y|$
59. $\dfrac{y^2}{x^6}$ **60.** $-c$ **61.** a^2 **62.** 16 **63.** $\dfrac{1}{8}$ **64.** $\dfrac{8}{27}$
65. $\dfrac{4}{9}$ **66.** $\dfrac{9}{4}$ **67.** $\dfrac{125}{8}$ **68.** $36x^2$ **69.** $p^{2a/3}$ **70.** 6
71. -7 **72.** $\dfrac{3}{5}$ **73.** $\dfrac{3}{5}$ **74.** $|x|y^2$ **75.** x **76.** $\dfrac{m^2|n|}{p^4}$
77. $\dfrac{a^3b^2}{c}$ **78.** $7\sqrt{2}$ **79.** 0 **80.** $x\sqrt[3]{3x}$ **81.** $\dfrac{\sqrt{35}}{5}$
82. $2\sqrt{2}$ **83.** $\dfrac{\sqrt[3]{4}}{2}$ **84.** $\dfrac{2\sqrt[3]{5}}{5}$ **85.** $\dfrac{2}{5\sqrt{2}}$ **86.** $\dfrac{1}{\sqrt{5}}$
87. $\dfrac{2x}{3\sqrt{2x}}$ **88.** $\dfrac{21x}{2\sqrt[3]{49x^2}}$ **89.** 3rd degree, binomial
90. 2nd degree, trinomial **91.** 2nd degree, monomial
92. 4th degree, trinomial **93.** $5x-6$ **94.** $2x^3-7x^2-6x$
95. $9x^2+12x+4$ **96.** $6x^2-7xy-3y^2$
97. $8a^2-8ab-6b^2$ **98.** $3z^3+10z^2+2z-3$
99. $a^{2n}+a^n-2$ **100.** $2+2x\sqrt{2}+x^2$
101. $\sqrt{6}+\sqrt{2}+\sqrt{3}+1$ **102.** -5 **103.** $\sqrt{3}+1$

104. $-2\left(\sqrt{3}+\sqrt{2}\right)$ **105.** $\dfrac{2x\left(\sqrt{x}+2\right)}{x-4}$
106. $\dfrac{x-2\sqrt{xy}+y}{x-y}$ **107.** $\dfrac{x-4}{5\left(\sqrt{x}-2\right)}$ **108.** $\dfrac{1-a}{a\left(1+\sqrt{a}\right)}$
109. $\dfrac{y}{2x}$ **110.** $2a^2b+3ab^2$ **111.** x^2+2x+1
112. $x^3+2x-3-\dfrac{6}{x^2-1}$ **113.** $3t(t+1)(t-1)$
114. $5(r-1)(r^2+r+1)$ **115.** $(3x+8)(2x-3)$
116. $(3a+x)(a-1)$ **117.** $(2x-5)(4x^2+10x+25)$
118. $2(3x+2)(x-4)$ **119.** $(x+3+t)(x+3-t)$
120. prime **121.** $(2z+7)(4z^2-14z+49)$
122. $(7b+1)^2$ **123.** $(11z-2)^2$
124. $8(2y-5)(4y^2+10y+25)$ **125.** $(y-2z)(2x-w)$
126. $(x^2+1+x)(x^2+1-x)(x^4+1-x^2)$ **127.** $\dfrac{-1}{x-2}$
128. $\dfrac{a+3}{a-3}$ **129.** $(x-2)(x+3)$ **130.** $\dfrac{2y-5}{y-2}$
131. $\dfrac{t+1}{5}$ **132.** $\dfrac{p(p+4)}{(p^2+8p+4)(p-3)}$
133. $\dfrac{(x-2)(x+3)(x-3)}{(x-1)(x+2)^2}$ **134.** 1
135. $\dfrac{3x^2-10x+10}{(x-4)(x+5)}$ **136.** $\dfrac{2(2x^2+3x+6)}{(x+2)(x-2)}$
137. $\dfrac{3x^3-12x^2+11x}{(x-1)(x-2)(x-3)}$ **138.** $\dfrac{-5x-6}{(x+1)(x+2)}$
139. $\dfrac{-x^3+x^2-12x-15}{x^2(x+1)}$ **140.** $\dfrac{3x}{x+1}$ **141.** $\dfrac{20}{3x}$
142. $\dfrac{y}{2}$ **143.** $\dfrac{y+x}{xy(x-y)}$ **144.** $\dfrac{y+x}{x-y}$

Chapter Test (page 79)

1. $-7, 1, 3$ **2.** 3 **3.** comm. prop. of add. **4.** distrib.
prop. **5.** ⊢(─────)──► **6.** ◄────────|────►
        -4 2 -3 6
7. 17 **8.** $-(x-7)$ **9.** 16 **10.** 8 **11.** x^{11} **12.** $\dfrac{r}{s}$
13. a **14.** x^{24} **15.** 4.5×10^5 **16.** 3.45×10^{-4}
17. 3,700 **18.** 0.0012 **19.** $5a^2$ **20.** $\dfrac{216}{729}$ **21.** $\dfrac{9s^6}{4t^4}$
22. $3a^2$ **23.** $5\sqrt{3}$ **24.** $-4x\sqrt[3]{3x}$ **25.** $\dfrac{x\left(\sqrt{x}+2\right)}{x-4}$
26. $\dfrac{x-y}{x+2\sqrt{xy}+y}$ **27.** $-a^2+7$ **28.** $-6a^6b^6$
29. $6x^2+13x-28$ **30.** $a^{2n}-a^n-6$ **31.** x^4-16
32. $2x^3-5x^2+7x-6$ **33.** $6x+19+\dfrac{34}{x-3}$
34. x^2+2x+1 **35.** $3(x+2y)$ **36.** $(x+10)(x-10)$
37. $(5t-2w)(2t-3w)$ **38.** $3(a-6)(a^2+6a+36)$
39. $(x+2)(x-2)(x^2+3)$ **40.** $(3x^2-2)(2x^2+5)$
41. 1 **42.** $\dfrac{-2x}{(x+1)(x-1)}$ **43.** $\dfrac{(x+5)^2}{x+4}$
44. $\dfrac{1}{(x+1)(x-2)}$ **45.** $\dfrac{b+a}{a}$ **46.** $\dfrac{y}{y+x}$

Exercises 1.1 (page 88)

1. root, solution   **3.** no   **5.** linear   **7.** one
9. no restrictions   **11.** $x \neq 0$ **13.** $x \geq 0$ **15.** $x \neq 3$ and
$x \neq -2$ **17.** 5; conditional equation   **19.** no solution
21. 7; conditional equation   **23.** no solution   **25.** identity
27. 6; conditional equation   **29.** identity   **31.** 1   **33.** 9
35. 10   **37.** -3 **39.** 6   **41.** -2 **43.** $\frac{5}{2}$ **45.** 1
47. $-\frac{14}{11}$ **49.** identity   **51.** 2   **53.** 4   **55.** -4 **57.** no
solution   **59.** 17   **61.** $-\frac{2}{5}$ **63.** $\frac{2}{3}$ **65.** 3   **67.** 5
69. no solution   **71.** 2   **73.** $p = \frac{k}{2.2}$ **75.** $w = \frac{p - 2l}{2}$
77. $r^2 = \frac{3V}{\pi h}$ **79.** $s = \frac{f(P_n - L)}{i}$ **81.** $m = \frac{r^2 F}{Mg}$
83. $y = b\left(1 - \frac{x}{a}\right)$ **85.** $r = \frac{r_1 r_2}{r_1 + r_2}$ **87.** $n = \frac{l - a + d}{d}$
89. $n = \frac{360}{180 - a}$ **91.** $r_1 = \frac{-Rr_2 r_3}{Rr_3 + Rr_2 - r_2 r_3}$ **95.** $5|x|$
97. $\frac{4y^4}{25x^2}$ **99.** $5|y|$ **101.** $\frac{|ab^3|}{z^2}$

Exercises 1.2 (page 96)

1. add   **3.** amount   **5.** rate, time   **7.** 84   **9.** 94
11. 7   **13.** $2\frac{1}{2}$ ft   **15.** 20 ft by 8 ft   **17.** 20 ft
19. \$10,000 at 7%, \$12,000 at 6%   **21.** 327   **23.** \$17,500 at
8%, \$19,500 at $9\frac{1}{2}$%   **25.** \$79.95   **27.** 200 units
29. 21   **31.** $2\frac{6}{11}$ days   **33.** $1\frac{1}{3}$ hr   **35.** $21\frac{1}{9}$ hr
37. 10 oz   **39.** 1 liter   **41.** $\frac{1}{15}$ liter   **43.** about 4.5 gal
45. 4 liters   **47.** 50 lb   **49.** 30 gal   **51.** 39 mph going;
65 mph returning   **53.** $2\frac{1}{2}$ hr   **55.** 50 sec   **57.** 12 mph
59. 600 lb barley; 1,637 lb oats; 163 lb soybean meal
61. about 11.2 mm   **65.** $(x + 7)(x - 9)$
67. $(3x - 5)(3x + 1)$ **69.** $(x + 3)^2$
71. $(x + 2)(x^2 - 2x + 4)$

Exercises 1.3 (page 110)

1. $ax^2 + bx + c = 0$ **3.** $\sqrt{c}, -\sqrt{c}$ **5.** equal rational
numbers   **7.** 3, -2 **9.** 12, -12 **11.** 2, $-\frac{5}{2}$ **13.** 2, $\frac{3}{5}$
15. $\frac{3}{5}, -\frac{5}{3}$ **17.** $\frac{3}{2}, \frac{1}{2}$ **19.** 3, -3 **21.** $5\sqrt{2}, -5\sqrt{2}$
23. 3, -1 **25.** 2, -4 **27.** $x^2 + 6x + 9$
29. $x^2 - 4x + 4$ **31.** $a^2 + 5a + \frac{25}{4}$ **33.** $r^2 - 11r + \frac{121}{4}$
35. $y^2 + \frac{3}{4}y + \frac{9}{64}$ **37.** $q^2 - \frac{1}{5}q + \frac{1}{100}$ **39.** 5, 3
41. 2, -3 **43.** 0, 25   **45.** $\frac{2}{3}, -2$ **47.** $\frac{-5 \pm \sqrt{5}}{2}$
49. $\frac{-2 \pm \sqrt{7}}{3}$ **51.** $\pm 2\sqrt{3}$ **53.** 3, $-\frac{5}{2}$ **55.** 2, $-\frac{1}{5}$
57. 1, -2 **59.** $\frac{-5 \pm \sqrt{13}}{6}$ **61.** $\frac{-1 \pm \sqrt{61}}{10}$
63. $t = \pm \frac{\sqrt{2hg}}{g}$ **65.** $t = \frac{8 \pm \sqrt{64 - h}}{4}$
67. $y = \pm \frac{b\sqrt{a^2 - x^2}}{a}$ **69.** $a = \pm \frac{bx\sqrt{b^2 + y^2}}{b^2 + y^2}$

71. $x = \frac{-y \pm y\sqrt{5}}{2}$ **73.** rational and equal   **75.** not real
numbers   **77.** rational and unequal   **79.** not real numbers
81. yes   **83.** 2, 10   **85.** 3, -4 **87.** $\frac{3}{2}, -\frac{1}{4}$
89. $\frac{1}{2}, -\frac{4}{3}$ **91.** $\frac{5}{6}, -\frac{2}{5}$ **93.** $4 \pm 2\sqrt{2}$ **95.** 1, -1
97. $-\frac{1}{2}, 5$ **99.** -2 **101.** 3, $-\frac{8}{11}$ **111.** $2x^2 - 8x$
113. $12m$ **115.** $3x\sqrt{2x}$

Exercises 1.4 (page 118)

1. $A = lw$ **3.** 4 ft by 8 ft   **5.** 9 cm   **7.** width: $7\frac{1}{4}$ ft;
length: $13\frac{3}{4}$ ft   **9.** 2 in.   **11.** 40.1 in. by 22.6 in.
13. 20 mph going and 10 mph returning   **15.** 7 hr
17. 25 sec   **19.** about 9.5 sec   **21.** 1.6 sec   **23.** 10¢
25. 1,440   **27.** 24.3 ft   **29.** 4 hr   **31.** about 9.5 hr
33. no   **35.** Matilda at 8%; Maude at 7%   **37.** 10
39. 20   **41.** 10 m and 24 m   **43.** 1.70 in.   **47.** $\frac{x - 6}{x(x - 3)}$
49. $\frac{x + 3}{x(x + 2)}$ **51.** $\frac{y + x}{y - x}$

Exercises 1.5 (page 130)

1. imaginary   **3.** imaginary   **5.** $2 - 5i$ **7.** real
9. $x = 3; y = 5$ **11.** $x = \frac{2}{3}; y = -\frac{2}{9}$ **13.** $5 - 6i$
15. $-2 - 10i$ **17.** $4 + 10i$ **19.** $1 - i$ **21.** $-9 + 19i$
23. $-5 + 12i$ **25.** $52 - 56i$ **27.** $-6 + 17i$ **29.** $0 + i$
31. $4 + 0i$ **33.** $\frac{2}{5} - \frac{1}{5}i$ **35.** $\frac{1}{25} + \frac{7}{25}i$ **37.** $\frac{1}{2} + \frac{1}{2}i$
39. $-\frac{7}{13} - \frac{22}{13}i$ **41.** $-\frac{12}{17} + \frac{11}{34}i$ **43.** $\frac{6 + \sqrt{3}}{10} + \frac{3\sqrt{3} - 2}{10}i$
45. i **47.** -1 **49.** -1 **51.** -1 **53.** 5
55. $\sqrt{13}$ **57.** $7\sqrt{2}$ **59.** $\frac{\sqrt{2}}{2}$ **61.** 6   **63.** $\sqrt{2}$
65. $\frac{3\sqrt{5}}{5}$ **67.** 1   **69.** $-1 \pm i$ **71.** $-2 \pm i$
73. $1 \pm 2i$ **75.** $\frac{1}{3} \pm \frac{1}{3}i$ **77.** $(x + 2i)(x - 2i)$
79. $(5p + 6qi)(5p - 6qi)$ **81.** $2(y + 2zi)(y - 2zi)$
83. $2(5m + ni)(5m - ni)$ **85.** $21 + 12i$ **87.** $6.2 - 0.7i$
95. $4x^2\sqrt{2}$ **97.** $x - 4\sqrt{x + 1} + 5$ **99.** $\sqrt{5} + 1$

Exercises 1.6 (page 138)

1. equal   **3.** extraneous   **5.** 0, -5, -4 **7.** 0, $\frac{4}{3}$, $-\frac{1}{2}$
9. 5, -5, 1, -1 **11.** 6, -6, 1, -1 **13.** $3\sqrt{2}, -3\sqrt{2}$,
$\sqrt{5}, -\sqrt{5}$ **15.** 0, 1   **17.** $\frac{1}{8}, -8$ **19.** 1, 144   **21.** $\frac{1}{64}$
23. $\frac{1}{9}$ **25.** 27   **27.** $-\frac{1}{3}$ **29.** 2   **31.** $-\frac{3}{2}, -\frac{5}{2}$ **33.** 9
35. 20   **37.** 2   **39.** -2 **41.** 3, 4   **43.** $\frac{1}{5}, -1$ **45.** 3, 5
47. -2, 1   **49.** 2, $-\frac{5}{2}$ **51.** 0, 4   **53.** -2 **55.** no
solution   **57.** 3   **59.** 1   **61.** 400 ft   **63.** 8 ft
65. about \$3,109
69.

71. $[3, \infty)$ **73.** $[-2, 1)$

75. $(-\infty, 1) \cup [2, \infty)$

Exercises 1.7 (page 150)

1. right   **3.** $a < c$ **5.** $b - c$ **7.** $>$ **9.** linear
11. equivalent   **13.** $(-\infty, 1)$

15. $[1, \infty)$ **17.** $(-\infty, 1)$

19. $[1, \infty)$ **21.** $(-\infty, 3]$

23. $(-10/3, \infty)$ **25.** $(5, \infty)$

27. $[14, \infty)$ **29.** $(-\infty, 15/4)$

31. $(-44/41, \infty)$ **33.** $(6, 9]$

35. $(8, 22]$ **37.** $[-11, 4]$

39. $[5, 21]$ **41.** $(-\infty, 0)$

43. $(0, \infty)$ **45.** $(3, 14/3)$

47. $(2, \infty)$ **49.** $(-4, 5/6)$

51. $[-2, \infty)$ **53.** $(-4, -3)$

55. $(-\infty, 2] \cup [3, \infty)$ **57.** $(-3, -2)$

59. $(-\infty, -1/2] \cup [-1/3, \infty)$ **61.** $(1/3, 1/2)$

63. $(-\infty, -3/2] \cup [1, \infty)$ **65.** $(-\infty, -\sqrt{3}) \cup (\sqrt{3}, \infty)$

67. $(-\sqrt{11}, \sqrt{11})$ **69.** $(-3, 2)$

71. $(-\infty, -1) \cup (-1, 0) \cup (1, \infty)$ **73.** $(-\infty, -3) \cup (-3, -2] \cup (2, \infty)$

75. $(-\infty, -2) \cup (-2, -1/3) \cup (1/2, \infty)$ **77.** $(0, 3/2)$

79. $(-\infty, 0) \cup (3/2, \infty)$ **81.** $(-\infty, 2) \cup [13/5, \infty)$

83. $(-\infty, -\sqrt{7}) \cup (-1, 1) \cup (\sqrt{7}, \infty)$ **85.** 17 min   **87.** 12

89. $p \le \$1{,}124.12$ **91.** $a > \$50{,}000$ **93.** anything over $\$1{,}800$ **95.** $16\frac{2}{3}$ cm $< s < 20$ cm
97. $40 + 2w < P < 60 + 2w$ **101.** $0, -2, 2$
103. 2   **105.** All are real.

Exercises 1.8 (page 158)

1. x **3.** $x = k$ or $x = -k$ **5.** $-k < x < k$
7. $x \le -k$ or $x \ge k$ **9.** 7   **11.** 0   **13.** 2   **15.** $\pi - 2$
17. $x - 5$ **19.** x^3 if $x \ge 0$; $-x^3$ if $x < 0$ **21.** $0, -4$
23. $2, -\dfrac{4}{3}$ **25.** $\dfrac{14}{3}, -2$ **27.** $7, -3$ **29.** no solution
31. $\dfrac{2}{7}, 2$ **33.** $x \ge 0$ **35.** $-\dfrac{3}{2}$ **37.** $0, -6$ **39.** 0
41. $\dfrac{3}{5}, 3$ **43.** $-\dfrac{3}{13}, \dfrac{9}{5}$ **45.** $(-3, 9)$

47. $(-\infty, -9) \cup (3, \infty)$ **49.** $(-\infty, -7] \cup [3, \infty)$

51. $[-13/3, 1]$ **53.** $(-\infty, -3) \cup (-3, \infty)$

55. $(-1, 1/5)$ **57.** $(-\infty, -7/9) \cup (13/9, \infty)$

59. $(-5, 7)$ **61.** $(-2, -1/2) \cup (-1/2, 1)$

63. $(-7/3, -2/3) \cup (4/3, 3)$ **65.** $(-7, -1) \cup (11, 17)$

67. $(-18, -6) \cup (10, 22)$ **69.** $(-10, -7] \cup [5, 8)$

71. $\left[-\dfrac{1}{2}, \infty\right)$ **73.** $(-\infty, 0)$ **75.** $\left(-\infty, -\dfrac{1}{2}\right)$ **77.** $[0, \infty)$
79. $70° \le t \le 86°$ **81.** $|c - 0.6°| \le 0.5°$
83. $|h - 55| < 17$ **85. a.** 26.45%, 24.76%
b. It is less than or equal to 1%.   **91.** 3.725×10^4
93. 523,000   **95.** $-4xy$

Chapter Review (page 161)

1. no restrictions   **2.** $x \ne 0$ **3.** $x \ge 0$ **4.** $x \ne 2, x \ne 3$
5. $\dfrac{16}{27}$; conditional equation   **6.** -14; conditional equation
7. $\dfrac{16}{5}$; conditional equation   **8.** no solution
9. 7; conditional equation   **10.** identity   **11.** 7; conditional
equation   **12.** $\dfrac{1}{3}$; conditional equation   **13.** -2

14. 0   **15.** $F = \frac{9}{5}C + 32$ **16.** $f = \frac{is}{P_n - l}$

17. $f_1 = \frac{ff_2}{f_2 - f}$ **18.** $l = \frac{a - S + Sr}{r}$ **19.** 60%

20. 22.5 ft by 27.5 ft   **21.** 3 hr   **22.** 0.5 hr

23. 1.5 liters   **24.** about 3.9 hr   **25.** $5\frac{1}{7}$ hr

26. $3\frac{1}{3}$ oz   **27.** \$4,500 at 11%; \$5,500 at 14%   **28.** 10

29. $2, -\frac{3}{2}$ **30.** $\frac{1}{4}, -\frac{4}{3}$ **31.** $0, \frac{8}{5}$ **32.** $\frac{2}{3}, \frac{4}{9}$ **33.** 3, 5

34. $-4, -2$ **35.** $\frac{1 \pm \sqrt{21}}{10}$ **36.** $0, \frac{1}{5}$ **37.** $2, -7$

38. $9, -\frac{2}{3}$ **39.** $\frac{-1 \pm \sqrt{21}}{10}$ **40.** $-1 \pm 2i$ **41.** 1

42. unequal rational numbers   **43.** $\frac{1}{3}$ **44.** 10, 2   **45.** $-\frac{5}{2}$

46. $\frac{8}{5}, 5$ **47.** either 95 by 110 yd or 55 by 190 yd

48. 320 mph for prop plane; 440 mph for jet plane

49. 1 sec   **50.** $1\frac{1}{2}$ ft   **51.** $-2 - i$ **52.** $-2 - 5i$

53. $5 - 2i$ **54.** $21 - 9i$ **55.** $0 - 3i$ **56.** $0 - 2i$

57. $\frac{3}{2} - \frac{3}{2}i$ **58.** $-\frac{2}{5} + \frac{4}{5}i$ **59.** $\frac{4}{5} + \frac{3}{5}i$ **60.** $\frac{1}{2} - \frac{5}{2}i$

61. $0 + i$ **62.** $0 - i$ **63.** $\sqrt{10}$ **64.** 1

65. $\frac{1}{3} \pm \frac{\sqrt{2}}{3}i$ **66.** $\frac{1}{3} \pm \frac{\sqrt{11}}{3}i$ **67.** $2, -3$ **68.** $4, -2$

69. $1, 1, -1, -1$ **70.** $1, -1, 6, -6$ **71.** 9   **72.** $8, -27$

73. 5   **74.** 0   **75.** $4, -4$ **76.** no solution

77. $(-\infty, 7)$

78. $[-1/5, \infty)$

79. $(-\infty, 5/3)$

80. $(-\infty, -12/7)$

81. $[-3, 5)$

82. $(2, \infty)$

83. $(-\infty, -2) \cup (4, \infty)$

84. $(-4, 1)$

85. $(-1, 3)$

86. $(-\infty, -3/2) \cup (1, \infty)$

87. $(-\infty, -2] \cup (3, \infty)$

88. $(-4, 1]$

89. $[-2, 1] \cup (3, \infty)$

90. $(-\infty, 0) \cup (5/2, \infty)$

91. $5, -7$ **92.** 0   **93.** $-\frac{4}{3}, -6$ **94.** $-\frac{3}{8}, \frac{3}{10}$

95. $(-6, 0)$

96. $(-\infty, 2] \cup [8/3, \infty)$

97. $(-5, 1)$

98. $(-\infty, -29) \cup (35, \infty)$

99. $(-7/2, -2) \cup (-1, 1/2)$

100. $(-1, 4/3) \cup (4/3, 11/3)$

Chapter Test (page 171)

1. $x \neq 0, x \neq 1$ **2.** $x \geq 0$ **3.** $\frac{5}{2}$ **4.** 37   **5.** $x = \mu + z\sigma$

6. $a = \frac{bc}{c + b}$ **7.** 87.5   **8.** \$14,000   **9.** $\frac{1}{2}, \frac{3}{2}$ **10.** $\frac{3}{2}, -4$

11. $x = \frac{-b \pm \sqrt{b^2 - 4ac}}{2a}$ **12.** $\frac{5 \pm \sqrt{133}}{6}$ **13.** $5, -3$

14. 8 sec   **15.** $7 - 12i$ **16.** $-23 - 43i$ **17.** $\frac{4}{5} + \frac{2}{5}i$

18. $0 + i$ **19.** i **20.** 1   **21.** 13   **22.** $\frac{\sqrt{10}}{10}$

23. $2, -2, 3, -3$ **24.** $1, -\frac{1}{32}$ **25.** 139   **26.** -1

27. $(-\infty, 2]$

28. $(-\infty, 5)$

29. $(2, \infty)$

30. $[-2, 1)$

31. $2, -\frac{10}{3}$ **32.** 0   **33.** $(-\infty, 3/2) \cup (7/2, \infty)$

34. $[-9, 6]$

Cumulative Review Exercises (page 172)

1. $-2, 0, 2, 6$ **2.** 2, 5, 11   **3.** $[-4, 7)$

4. $(-\infty, 0) \cup [2, \infty)$ **5.** comm. prop. of addition

6. transitive prop.   **7.** $9a^2$ **8.** $81a^2$ **9.** a^6b^4 **10.** $\frac{9y^2}{x^4}$

11. $\frac{x^4}{16y^2}$ **12.** $\frac{4y^{10}}{9x^6}$ **13.** a^3b^3 **14.** abc^2 **15.** $\sqrt{3}$

16. $\frac{\sqrt[3]{2x^2}}{x}$ **17.** $\frac{3(y + \sqrt{3})}{y^2 - 3}$ **18.** $\frac{3x(\sqrt{x} + 1)}{x - 1}$

19. $5\sqrt{3} - 3\sqrt{5}$ **20.** $3\sqrt{2}$ **21.** $5 - 2\sqrt{6}$

22. 4   **23.** $-8x + 8$ **24.** $9x^4 - 4x^3$

25. $6x^2 + 11x - 35$ **26.** $z^3 + z^2 + 4$ **27.** $2x^2 - x + 1$

28. $3x^2 + 1 - \frac{x}{x^2 + 2}$ **29.** $3t(t - 2)$ **30.** $(3x + 2)(x - 4)$

31. $(x + 1)^2(x - 1)^2(x^2 + 1)^2$

32. $(x + 1)(x^2 - x + 1)(x - 1)(x^2 + x + 1)$ **33.** $\frac{x - 5}{x + 5}$

34. $(2x + 1)(x + 2)$ **35.** $\frac{5x^2 + 17x - 6}{(x + 3)(x - 3)}$

36. $\frac{-x^2 + x + 7}{(x + 2)(x + 3)}$ **37.** $b + a$ **38.** $-\frac{1}{xy}$ **39.** 0, 10

40. 34   **41.** $R = \frac{R_1 R_2}{R_1 + R_2}$ **42.** $r = \frac{a - S}{l - S}$ or $r = \frac{S - a}{S - l}$

43. either 8 ft by 24 ft or 12 ft by 16 ft   **44.** \$8,000

45. $\frac{3}{5} + \frac{4}{5}i$ **46.** $\frac{3}{2} - \frac{1}{2}i$ **47.** 5   **48.** $0 + 10i$ **49.** 3

50. $2, -2, 3, -3$ **51.** 2, 7   **52.** 64, 729

53. $\left(-\infty, \frac{11}{5}\right]$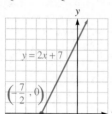

54. $(-\infty, 3) \cup (5, \infty)$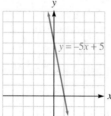

55. $[-3, -1] \cup (2, \infty)$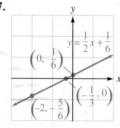

56. $(-\infty, -3) \cup (0, 3)$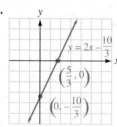

57. $(-\infty, -1] \cup [4, \infty)$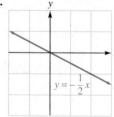

58. $\left(\frac{1}{3}, 3\right)$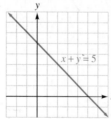

Exercises 2.1 (page 187)

1. quadrants **3.** to the right **5.** first **7.** linear
9. x-intercept **11.** horizontal **13.** midpoint
15. $A(2, 3)$ **17.** $C(-2, -3)$ **19.** $E(0, 0)$ **21.** $G(-5, -5)$
23. QI **25.** QIII **27.** QI **29.** positive x-axis

31. 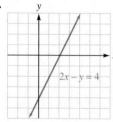 **33.**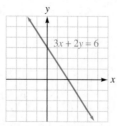

35. 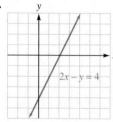 **37.**

39. 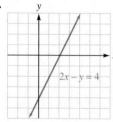 **41.**

43. 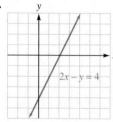 **45.**

47.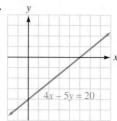

49.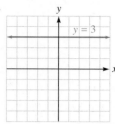

51.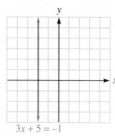

53.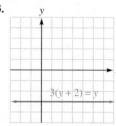

55.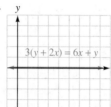

57. 1.22 **59.** 4.67 **61.** 5
63. $\sqrt{13}$ **65.** $\sqrt{2}$ **67.** 2
69. 5 **71.** 13 **73.** 25
75. $8\sqrt{2}$ **77.** 7 **79.** $(4, 6)$
81. $(0, 1)$ **83.** $(0, 0)$
85. $\left(\frac{\sqrt{5}}{2}, \frac{\sqrt{5}}{2}\right)$ **87.** $(5, 6)$

89. $(5, 15)$ **93.** $\sqrt{2}$ units **97.** \$312,500 **99.** 200
101. 100 rpm **103.** approx. 170 mi.
109. 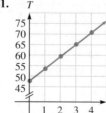 **111.** no graph; the intersection is the empty set **113.** -12

Exercises 2.2 (page 200)

1. divided **3.** run **5.** the change in **7.** vertical
9. perpendicular **11.** 1 **13.** 5 **15.** $-\frac{7}{4}$ **17.** -2
19. undefined **21.** $\frac{5}{3}$ **23.** -1 **25.** 3 **27.** $\frac{1}{2}$ **29.** $\frac{2}{3}$
31. 0 **33.** 0 **35.** undefined **37.** negative
39. positive **41.** undefined **43.** perpendicular
45. parallel **47.** perpendicular **49.** perpendicular
51. perpendicular **53.** parallel **55.** perpendicular
57. neither **59.** not on same line **61.** on same line
63. No two are perpendicular. **65.** PQ and PR are perpendicular. **67.** PQ and PR are perpendicular.
77. 3.5 students per yr **79.** \$642.86 per year
81. $\frac{\Delta T}{\Delta t}$ is the hourly rate of change in temperature.

83.

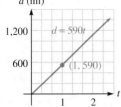

The slope is the speed of the plane.

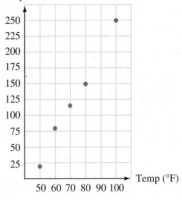

87. $y = -\frac{3}{7}x + 3$ **89.** $y = -\frac{2}{5}x + 2$

91. $(2p + 3)(3p - 4)$ **93.** $(m + n)(p + q)$

Exercises 2.3 (page 213)

1. $y - y_1 = m(x - x_1)$ **3.** slope–intercept   **5.** $-\frac{A}{B}$

7. $2x - y = 0$ **9.** $4x - 2y = -7$ **11.** $2x - 5y = -7$

13. $y = -3$ **15.** $\pi x - y = \pi^2$ **17.** $2x - 3y = -11$

19. $y = x$ **21.** $y = \frac{7}{3}x - 3$ **23.** $y = -\frac{9}{5}x + \frac{2}{5}$

25. $y = 3x - 2$ **27.** $y = 5x - \frac{1}{5}$ **29.** $y = ax + \frac{1}{a}$

31. $y = ax + a$ **33.** $3x - 2y = 0$ **35.** $3x + y = -4$

37. $\sqrt{2}x - y = -\sqrt{2}$

39. $1, (0, -1)$

41. $\frac{2}{3}, (0, 2)$

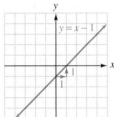

43. $-\frac{2}{3}, (0, 6)$

45. $\frac{3}{2}, (0, -4)$ **47.** $-\frac{1}{3}, \left(0, -\frac{5}{6}\right)$

49. $\frac{7}{2}, (0, 2)$ **51.** parallel

53. perpendicular

55. parallel

57. perpendicular

59. perpendicular

61. perpendicular

63. $y = 4x$

65. $y = 4x - 3$

67. $y = \frac{4}{5}x - \frac{26}{5}$ **69.** $y = -\frac{1}{4}x$ **71.** $y = -\frac{1}{4}x + \frac{11}{2}$

73. $y = -\frac{5}{4}x + 3$ **75.** $m = -\frac{4}{5}; (0, 4)$

77. $m = -\frac{2}{3}; (0, 4)$ **79.** $x = -2$ **81.** $x = 5$

83. $y = -3,200x + 24,300$ **85.** $y = 47,500x + 475,000$

87. $y = -\frac{710}{3}x + 1,900$ **89.** \$90   **91.** \$890   **93.** \$37,200

95. \$230   **97.** about 838   **99.** $C = \frac{5}{9}(F - 32)$

101. $y = -\frac{9}{10}x + 47; 11\%$ **103.** $y = 3.75x + 37.5; \$52\frac{1}{2}$

105. 1,655 barrels per day   **107. a.** See graph in the next

column. **b.** $y = \frac{23}{5}x - 210$ (answers may vary)

c. 204 (answers may vary)

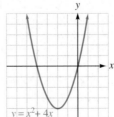

121. x^5 **123.** $\frac{125}{729}$ **125.** $-\sqrt{3}$ **127.** $\dfrac{5\left(\sqrt{x} - 2\right)}{x - 4}$

Exercises 2.4 (page 233)

1. x-intercept   **3.** axis of symmetry   **5.** x-axis   **7.** circle,

center   **9.** $x^2 + y^2 = r^2$ **11.** $(-2, 0), (2, 0); (0, -4)$

13. $(0, 0), \left(\frac{1}{2}, 0\right); (0, 0)$ **15.** $(-1, 0), (5, 0); (0, -5)$

17. $(1, 0), (-2, 0); (0, -2)$ **19.** $(-3, 0), (0, 0), (3, 0); (0, 0)$

21. $(-1, 0), (1, 0); (0, -1)$

23.

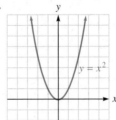

25.

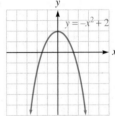

27.

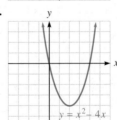

29.

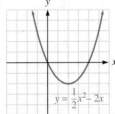

31. about the y-axis   **33.** about the x-axis   **35.** about the

x-axis, the y-axis, and the origin   **37.** about the y-axis

39. none   **41.** about the x-axis   **43.** about the y-axis

45. about the x-axis

47.

49.

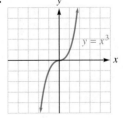

51.
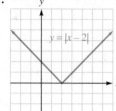
$y = |x - 2|$

53.
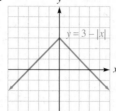
$y = 3 - |x|$

55.
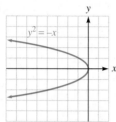
$y^2 = -x$

57.
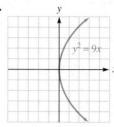
$y^2 = 9x$

59.
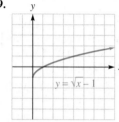
$y = \sqrt{x} - 1$

61.
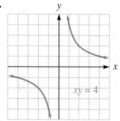
$xy = 4$

63. $(0, 0)$; 10   **65.** $(0, 5)$; 7   **67.** $(-6, 0)$; $\frac{1}{2}$ **69.** $(4, 1)$; 3

71. $\left(\frac{1}{4}, -2\right)$; $3\sqrt{5}$ **73.** $x^2 + y^2 = 25$

75. $x^2 + (y + 6)^2 = 36$ **77.** $(x - 8)^2 + y^2 = \frac{1}{25}$

79. $(x - 2)^2 + (y - 12)^2 = 169$ **81.** $x^2 + y^2 - 1 = 0$

83. $x^2 + y^2 - 12x - 16y + 84 = 0$

85. $x^2 + y^2 - 6x + 8y + 23 = 0$

87. $x^2 + y^2 - 6x - 6y - 7 = 0$

89. $x^2 + y^2 + 6x - 8y = 0$ **91.** $(x - 3)^2 + (y + 2)^2 = 9$

93. $(x - 5)^2 + (y - 6)^2 = 4$ **95.** $(x - 2)^2 + (y - 4)^2 = 9$

97.
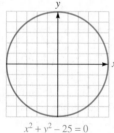
$x^2 + y^2 - 25 = 0$

99.
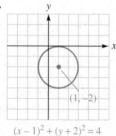
$(1, -2)$
$(x - 1)^2 + (y + 2)^2 = 4$

101.
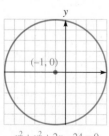
$(-1, 0)$
$x^2 + y^2 + 2x - 24 = 0$

103.
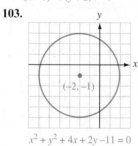
$(-2, -1)$
$x^2 + y^2 + 4x + 2y - 11 = 0$

105.
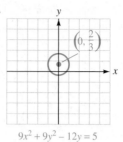
$\left(0, \frac{2}{3}\right)$
$9x^2 + 9y^2 - 12y = 5$

107.
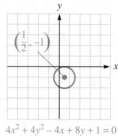
$\left(\frac{1}{2}, -1\right)$
$4x^2 + 4y^2 - 4x + 8y + 1 = 0$

109. $(0.25, 0.88)$ **111.** $(0.50, 7.25)$ **113.** ± 2.65

115. 1.44   **117.** 4 sec

119.
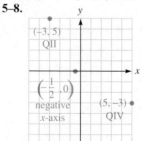
$(60, 342)$

121. $x^2 + y^2 - 14x - 8y + 40 = 0$ **125.** 6   **127.** -1
129. 20 oz

Exercises 2.5 (page 244)

1. quotient   **3.** means   **5.** extremes, means   **7.** inverse
9. joint   **11.** 14   **13.** $2, -3$ **15.** 18   **17.** $\frac{1}{2}$
19. 1,000   **21.** 1   **23.** $\frac{21}{4}$ **25.** 6   **27.** -8
29. direct variation   **31.** neither   **33.** 202, 172, 136
35. about 2 gal   **37.** $14\frac{6}{11}$ ft^3 **39.** 3 sec   **41.** 20 volts
43. 432 hertz   **45.** The force is multiplied by $\frac{9}{4}$.   **47.** $\frac{\sqrt{3}}{4}$
55. $\dfrac{3x + 5}{(x + 2)(x + 1)}$ **57.** $\dfrac{(x + 4)(x + 1)}{x - 1}$ **59.** $-\dfrac{1}{x}$

Chapter Review (page 246)

1. $(2, 0)$ **2.** $(-2, 1)$ **3.** $(0, -1)$ **4.** $(3, -1)$
5–8.
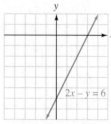
$(-3, 5)$
QII
$\left(-\frac{1}{2}, 0\right)$
negative x-axis
$(5, -3)$
QIV
$(0, -7)$
negative y-axis

9.
$2x - y = 6$

10.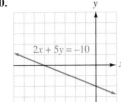

11.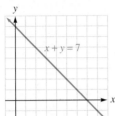

12.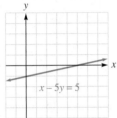

13.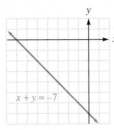

14.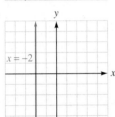

15.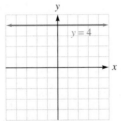

16.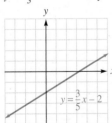

17. $12,150   **18.** $332,500
19. 10   **20.** $4\sqrt{2}$
21. 2   **22.** $2\sqrt{2}|a|$
23. $(0, 3)$ **24.** $\left(-6, \frac{15}{2}\right)$
25. $\left(\sqrt{3}, 8\right)$ **26.** $(0, 0)$
27. -6 **28.** 2   **29.** -1
30. 1   **31.** 3   **32.** 5
33. 0   **34.** undefined   **35.** negative   **36.** positive
37. perpendicular   **38.** neither   **39.** $y = 7$ **40.** $x = 3$
41. 200 ft/min   **42.** $48,750 per year   **43.** $7x + 5y = 0$
44. $4x + y = -7$ **45.** $x + 5y = -3$ **46.** $2x + y = 9$
47. $y = \frac{2}{3}x + 3$ **48.** $y = -\frac{3}{2}x - 5$
49. 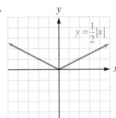   **50.**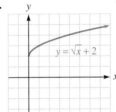

51. 2, $(0, 0)$ **52.** $-\frac{1}{2}$, $(0, 0)$ **53.** $\frac{3}{2}$, $(0, -5)$
54. $-\frac{1}{2}$, $(0, -2)$ **55.** slope $= -\frac{5}{2}$; $\left(0, \frac{7}{2}\right)$
56. slope $= \frac{3}{4}$; $\left(0, -\frac{7}{2}\right)$ **57.** $y = 17$ **58.** $x = -5$
59. $y = \frac{3}{4}x - \frac{3}{2}$ **60.** $y = -7x + 47$ **61.** $y = 3x + 5$
62. $y = \frac{1}{7}x - 3$ **63.** parallel   **64.** perpendicular
65. $(0, 0)$, $\left(\frac{1}{2}, 0\right)$; $(0, 0)$ **66.** $(12, 0)$, $(-2, 0)$; $(0, -24)$
67. about the x-axis   **68.** about the y-axis   **69.** about the y-axis   **70.** none

71.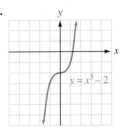

72.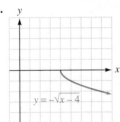

73.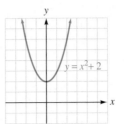

74.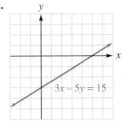

75.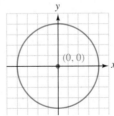

76.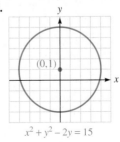

77.   **78.**   **79.**

80.   **81.** $(0, 0)$; 8   **82.** $(0, 6)$; 10
83. $(-7, 0)$; $\frac{1}{2}$ **84.** $(5, -1)$; 3
85. $x^2 + y^2 = 49$
86. $(x - 3)^2 + y^2 = \frac{1}{25}$ **87.** $(x + 2)^2 + (y - 12)^2 = 25$
88. $\left(x - \frac{2}{7}\right)^2 + (y - 5)^2 = 81$
89. $(x + 3)^2 + (y - 4)^2 = 144$, or
$x^2 + y^2 + 6x - 8y - 119 = 0$
90. $\left(x + \frac{1}{2}\right)^2 + \left(y - \frac{5}{2}\right)^2 = \frac{121}{2}$, or
$x^2 + y^2 + x - 5y - 54 = 0$
91. $(x + 3)^2 + (y - 2)^2 = 9$ **92.** $(x - 2)^2 + (y - 4)^2 = 25$
93. 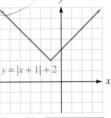   **94.**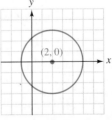

95. 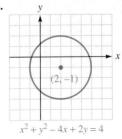   **96.**

$x^2 + y^2 - 2y = 15$ $x^2 + y^2 - 4x + 2y = 4$

97. 3.32, −3.32   **98.** −1, 0, 1   **99.** 1.73, −1.73, 1, −1
100. 4.19, −1.19   **101.** $x = 2, x = 5$ **102.** $x = -5, x = 5$
103. $\frac{9}{5}$ lb   **104.** $\frac{25}{9}$ **105.** $333\frac{1}{3}$ cc   **106.** 1
107. about 117 ohms   **108.** $385
109.   140 hr

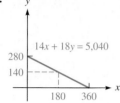

Chapter Test (page 258)

1. QII   **2.** negative y-axis
3.   **4.**

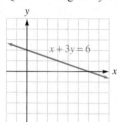   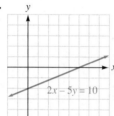

5.   **6.**

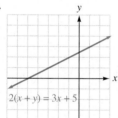   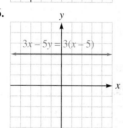

7.   **8.**

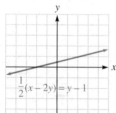   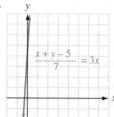

9. $\sqrt{41}$ **10.** approximately 4.44   **11.** (0, 0)

12. $\left(\sqrt{2}, 2\sqrt{2}\right)$ **13.** $-\frac{5}{4}$ **14.** $\frac{\sqrt{3}}{3}$ **15.** neither

16. perpendicular   **17.** $y = 2x - 11$ **18.** $y = 3x + \frac{1}{2}$
19. $y = 2x + 5$ **20.** $y = -\frac{1}{2}x + 5$ **21.** $y = 2x - \frac{11}{2}$
22. $x = 3$ **23.** (−4, 0), (0, 0), (4, 0); (0, 0)   **24.** (4, 0);
(0, 4)   **25.** about the x-axis   **26.** about the y-axis
27.   **28.**

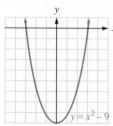   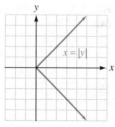

29.   **30.**

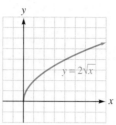   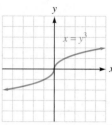

31. $(x - 5)^2 + (y - 7)^2 = 64$ **32.** $(x - 2)^2 + (y - 4)^2 = 32$
33.   **34.**

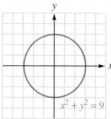   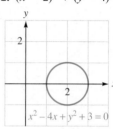

35. $y = kz^2$ **36.** $w = krs^2$ **37.** $P = \frac{35}{2}$ **38.** $x = \frac{27}{32}$
39. $x = 2.65$ **40.** $x = 5.85$

Exercises 3.1 (page 272)

1. function   **3.** domain   **5.** $y = f(x)$ **7.** x **9.** vertical,
once   **11.** a function   **13.** not a function   **15.** a function
17. not a function   **19.** not a function   **21.** a function
23. domain: $(-\infty, \infty)$ **25.** domain: $(-\infty, \infty)$ **27.** domain:
$(-\infty, -1) \cup (-1, \infty)$ **29.** domain: $[0, \infty)$ **31.** domain:
$(-\infty, -3) \cup (-3, \infty)$ **33.** domain: $(-\infty, -3) \cup (-3, \infty)$
35. domain: $(-\infty, -2) \cup (-2, 2) \cup (2, \infty)$
37. domain: $(-\infty, -1) \cup (-1, 5) \cup (5, \infty)$ **39.** 4; −11;
$3k - 2; 3k^2 - 5$ **41.** 4; $\frac{3}{2}; \frac{1}{2}k + 3; \frac{1}{2}k^2 + \frac{5}{2}$ **43.** 4; 9; k^2;
$k^4 - 2k^2 + 1$ **45.** 5, 10, $k^2 + 1$, $k^4 - 2k^2 + 2$ **47.** $\frac{1}{3}$; 2;
$\frac{2}{k + 4}; \frac{2}{k^2 + 3}$ **49.** $\frac{1}{3}; \frac{1}{8}; \frac{1}{k^2 - 1}; \frac{1}{k^4 - 2k^2}$
51. $\sqrt{5}; \sqrt{10}; \sqrt{k^2 + 1}; \sqrt{k^4 - 2k^2 + 2}$ **53.** 3
55. $2x + h$ **57.** $8x + 4h$ **59.** $2x + h + 3$
61. $4x + 2h - 4$ **63.** $3x^2 + 3xh + h^2$
65.   **67.**

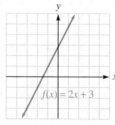   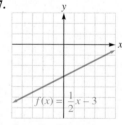

domain: $(-\infty, \infty)$ domain: $(-\infty, \infty)$
range: $(-\infty, \infty)$ range: $(-\infty, \infty)$

69.

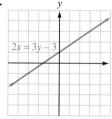

domain: $(-\infty, \infty)$
range: $(-\infty, \infty)$

71.

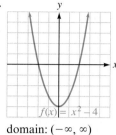

domain: $(-\infty, \infty)$
range: $[-4, \infty)$

73.

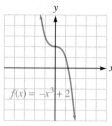

domain: $(-\infty, \infty)$
range: $(-\infty, \infty)$

75.

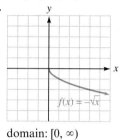

domain: $[0, \infty)$
range: $(-\infty, 0]$

77.

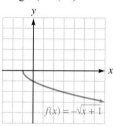

domain: $[-1, \infty)$
range: $(-\infty, 0]$

79.

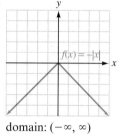

domain: $(-\infty, \infty)$
range: $(-\infty, 0]$

81.

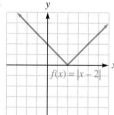

domain: $(-\infty, \infty)$
range: $[0, \infty)$

83.

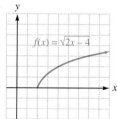

domain: $[2, \infty)$
range: $[0, \infty)$

85.

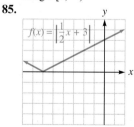

domain: $(-\infty, \infty)$
range: $[0, \infty)$

87.

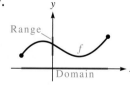

89. a function   **91.** a function   **93.** a function
95.   domain: $\left[\frac{5}{2}, \infty\right)$; range: $[0, \infty)$

97.   domain: $(-\infty, \infty)$; range: $(-\infty, \infty)$

99. a. $C(x) = 8x + 75$ **b.** \$755
101. a. $C(t) = 0.07t + 9.99$ **b.** \$11.39
103. a. $c = 95f + 14{,}000$ **b.** \$261,000
105. $c = 0.10x + 7$ **107.** $n = 1{,}692t + 6{,}400$ **109.** $-\frac{2}{3}$
115. 1, 7, 8   **117.** 7   **119.** $(-4, 7]$
121.

Exercises 3.2 (page 285)

1. $f(x) = ax^2 + bx + c$ **3.** $(3, 5)$ **5.** upward   **7.** $-\frac{b}{2a}$
9. up; minimum   **11.** down; maximum   **13.** down;
maximum   **15.** $(0, -1)$ **17.** $(3, 5)$ **19.** $(-6, -4)$
21. $(3, 0)$ **23.** $(2, 0)$ **25.** $(-3, -12)$ **27.** $(3, 1)$
29. $\left(\frac{2}{3}, \frac{11}{3}\right)$ **31.** $(-4, -11)$
33.

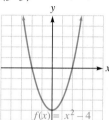

35.

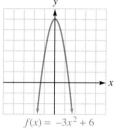

37.

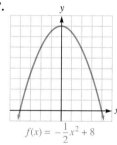

39.

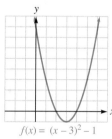

41.

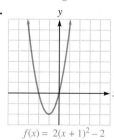

43.

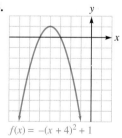

45.

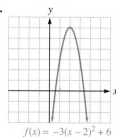

47.

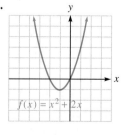

49.
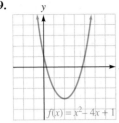
$f(x) = x^2 - 4x + 1$

51.
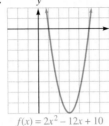
$f(x) = 2x^2 - 12x + 10$

53.
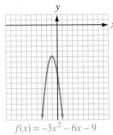
$f(x) = -3x^2 - 6x - 9$

55. 75 ft $\times$ 75 ft, 5,625 ft^2 **57.** 25 ft by 25 ft
59. $w = 12$ in.; $d = 6$ in.   **61.** 20 ft   **63.** 15.4 ft
65. 208 ft   **67.** 48 digital cameras; minimum cost $2,400
69. $95   **71.** $\frac{5}{2}$ sec   **73.** 100 ft   **75.** $(-2.25, -66.13)$
77. $(3.3, -68.5)$ **79.** 2, 3   **81.** 6 by $4\frac{1}{2}$ units   **83.** Both
numbers are 3.   **87.** $a^2 - 3a$; $a^2 + 3a$ **89.** $(5 - a)^2$;
$(5 + a)^2$ **91.** 7; 7

Exercises 3.3 (page 298)

1. 4   **3.** $n - 1$ **5.** odd   **7.** piecewise-defined   **9.** 3
11.
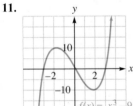
$f(x) = x^3 - 9x$

13.
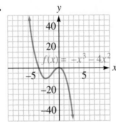
$f(x) = -x^3 - 4x^2$

15.
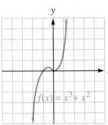
$f(x) = x^3 + x^2$

17.
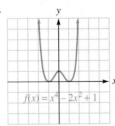
$f(x) = x^4 - 2x^2 + 1$

19.
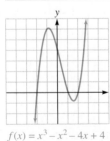
$f(x) = x^3 - x^2 - 4x + 4$

21.
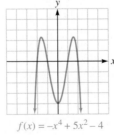
$f(x) = -x^4 + 5x^2 - 4$

23. even   **25.** neither   **27.** odd   **29.** odd
31. decreasing for $x < 0$; increasing for $x > 0$

33. increasing for $x < 0$; decreasing for $x > 0$
35. decreasing on $(-\infty, -2)$; constant on $(-2, 2)$; increasing
on $(2, \infty)$ **37.** decreasing for $x < 2$; increasing for $x > 2$
39. a. -2 **b.** 3   **41. a.** 2   **b.** 1   **c.** 3

43.
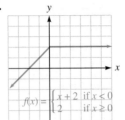
$f(x) = \begin{cases} x + 2 & \text{if } x < 0 \\ 2 & \text{if } x \geq 0 \end{cases}$

45.
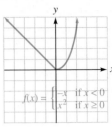
$f(x) = \begin{cases} -x & \text{if } x < 0 \\ x^2 & \text{if } x \geq 0 \end{cases}$

47.
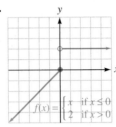
$f(x) = \begin{cases} x & \text{if } x \leq 0 \\ 2 & \text{if } x > 0 \end{cases}$

49.
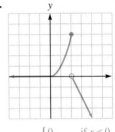
$f(x) = \begin{cases} 0 & \text{if } x < 0 \\ x^2 & \text{if } 0 \leq x \leq 2 \\ 4 - 2x & \text{if } x > 2 \end{cases}$

51. a. 3   **b.** -4 **c.** -3 **53. a.** 2   **b.** 3   **c.** 4
55.
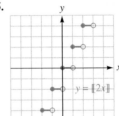
$y = \llbracket 2x \rrbracket$

57.
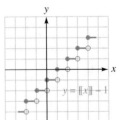
$y = \llbracket x \rrbracket - 1$

59. B
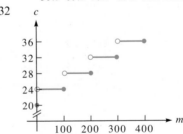

61. $32
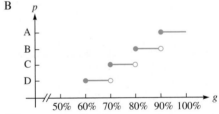

63. $1.60
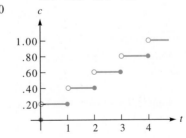

65.

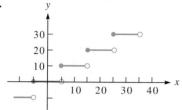

67. No; this is not defined at $x = 0$.

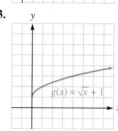

73. $3x + 5$; $3x + 3$ **75.** $\dfrac{3x - 8}{5}$; $\dfrac{3x + 1}{5} - 3$ or $\dfrac{3x - 14}{5}$

77. $-1, \dfrac{3}{2}$

Exercises 3.4 (page 313)

1. up   **3.** to the right   **5.** 2, down   **7.** y-axis
9. horizontally   **11.**

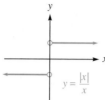

13.

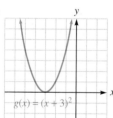

15.

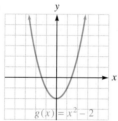

17.

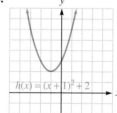

19.

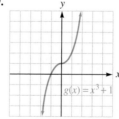

21.

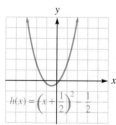

23.

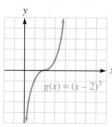

25.

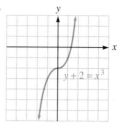

27.

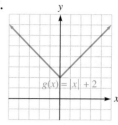

29.

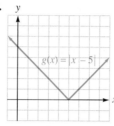

31.

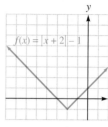

33.

35.

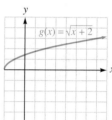

37.

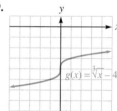

39.

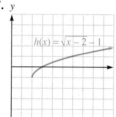

41.

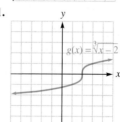

43.

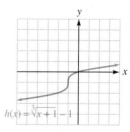

45.

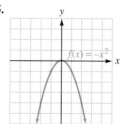

47.

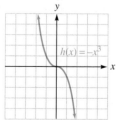

49.

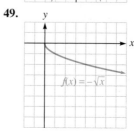

51.

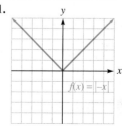

53.

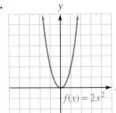

55.

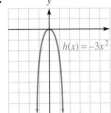

81.

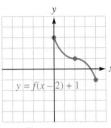

83.

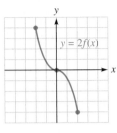

57.

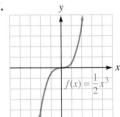

59.

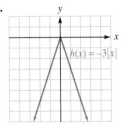

85.

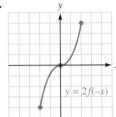

93. $\dfrac{x-2}{x+2}$

95. all real numbers except 3

97. $x + 2 + \dfrac{-2}{x+1}$

61.

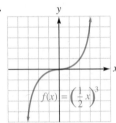

63.

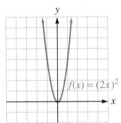

Exercises 3.5 (page 333)

1. asymptote **3.** vertical **5.** x-intercept **7.** the same
9. horizontal or slant, vertical **11.** vertical asymptote: $x = 2$;
horizontal asymptote: $y = 1$; domain: $(-\infty, 2) \cup (2, \infty)$.
range: $(-\infty, 1) \cup (1, \infty)$ **13.** 20 hr **15.** 12 hr
17. \$5,555.56 **19.** \$50,000 **21.** $(-\infty, 2) \cup (2, \infty)$
23. $(-\infty, -5) \cup (-5, 5) \cup (5, \infty)$
25. $(-\infty, -1) \cup (-1, 0) \cup (0, 1) \cup (1, \infty)$ **27.** $(-\infty, \infty)$
29. $x = 3$ **31.** $x = 1, x = -1$ **33.** $x = -2, x = 3$
35. none **37.** $y = 2$ **39.** $y = \frac{1}{2}$ **41.** $y = 0$ **43.** none
45. $y = x - 3$ **47.** $y = 2x + 3$ **49.** $y = x + 2$

65.

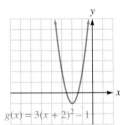

67.

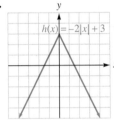

51.

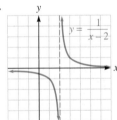

53.

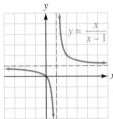

69.

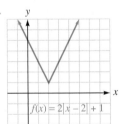

71.

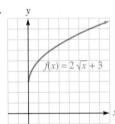

55.

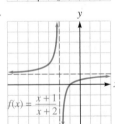

57.

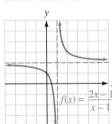

73.

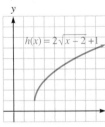

75.

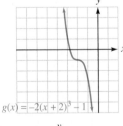

59.

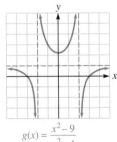

61.

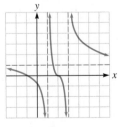

77.

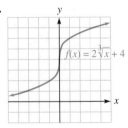

79.

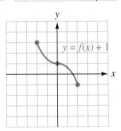

63.

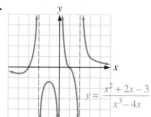

$$y = \frac{x^2 + 2x - 3}{x^3 - 4x}$$

65.

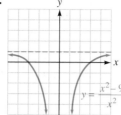

$$y = \frac{x^2 - 9}{x^2}$$

67.

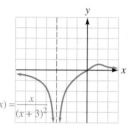

$$f(x) = \frac{x}{(x + 3)^2}$$

69.

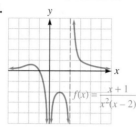

$$f(x) = \frac{x + 1}{x^2(x - 2)}$$

71.

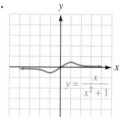

$$y = \frac{x}{x^2 + 1}$$

73.

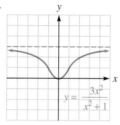

$$y = \frac{3x^2}{x^2 + 1}$$

75.

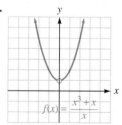

$$h(x) = \frac{x^2 - 2x - 8}{x - 1}$$

77.

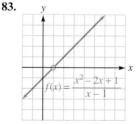

$$f(x) = \frac{x^3 + x^2 + 6x}{x^2 - 1}$$

79.

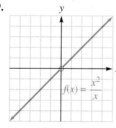

$$f(x) = \frac{x^2}{x}$$

81.

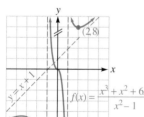

$$f(x) = \frac{x^3 + x}{x}$$

83.

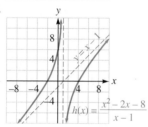

$$f(x) = \frac{x^2 - 2x + 1}{x - 1}$$

85.

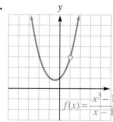

$$f(x) = \frac{x^3 - 1}{x - 1}$$

87. $c = f(x) = 1.25x + 700$

89. $\bar{c} = f(x) = \dfrac{1.25x + 700}{x}$

91. \$1,325

93. $c = f(n) = 0.09n + 7.50$

95. \$77.25   **97.** 9.75¢

99. a. $c(n) = 0.095n + 8.50$

b. $\bar{c}(n) = \dfrac{0.095n + 8.50}{n}$

c. 10.5¢   **111.** $3x^2 + x$ **113.** $10x^2 + 29x + 10$
115. $3x + 5$

Exercises 3.6 (page 348)

1. $f(x) + g(x)$ **3.** $f(x)g(x)$ **5.** intersection   **7.** $g(f(x))$
9. commutative   **11.** $(f + g)(x) = 5x - 1; (-\infty, \infty)$
13. $(f \cdot g)(x) = 6x^2 - x - 2; (-\infty, \infty)$
15. $(f - g)(x) = x + 1; (-\infty, \infty)$

17. $(f/g)(x) = \dfrac{x^2 + x}{x^2 - 1} = \dfrac{x}{x - 1}; (-\infty, -1) \cup (-1, 1) \cup (1, \infty)$

19. 7   **21.** 1   **23.** 12   **25.** no value   **27.** $f(x) = 3x^2$;
$g(x) = 2x$ **29.** $f(x) = 3x^2; g(x) = x^2 - 1$ **31.** $f(x) = 3x^3$;
$g(x) = -x$ **33.** $f(x) = x + 9; g(x) = x - 2$ **35.** 11
37. -17 **39.** 190   **41.** 145   **43.** $(-\infty, \infty)$;
$(f \circ g)(x) = 3x + 3$ **45.** $(-\infty, \infty); (f \circ f)(x) = 9x$
47. $(-\infty, \infty); (g \circ f)(x) = 2x^2$ **49.** $(-\infty, \infty); (g \circ g)(x) = 4x$
51. $[-1, \infty); (f \circ g)(x) = \sqrt{x + 1}$ **53.** $[0, \infty)$;
$(f \circ f)(x) = \sqrt[4]{x}$ **55.** $[-1, \infty); (g \circ f)(x) = x$
57. $(-\infty, \infty); (g \circ g)(x) = x^4 - 2x^2$
59. $(-\infty, 2) \cup (2, 3) \cup (3, \infty); (f \circ g)(x) = \dfrac{x - 2}{3 - x}$
61. $(-\infty, 1) \cup (1, 2) \cup (2, \infty); (f \circ f)(x) = \dfrac{x - 1}{2 - x}$
63. $f(x) = x - 2; g(x) = 3x$ **65.** $f(x) = x - 2; g(x) = x^2$
67. $f(x) = x^2; g(x) = x - 2$ **69.** $f(x) = \sqrt{x}; g(x) = x + 2$
71. $f(x) = x + 2; g(x) = \sqrt{x}$ **73.** $f(x) = x; g(x) = x$
75. 9   **77. a.** $A = 17w$ **b.** $w = \sqrt{d^2 - 289}$
c. $A = 17\sqrt{d^2 - 289}$ **79.** $P = 4\sqrt{A}$ **89.** $y = \dfrac{x + 7}{3}$
91. $y = \dfrac{3x}{1 - x}$

Exercises 3.7 (page 360)

1. one-to-one   **3.** identity   **5.** one-to-one
7. not one-to-one   **9.** not one-to-one   **11.** not one-to-one
13. not one-to-one   **15.** one-to-one   **17.** one-to-one
19. not a function   **25.** $f^{-1}(x) = \frac{1}{3}x$ **27.** $f^{-1}(x) = \dfrac{x - 2}{3}$
29. $f^{-1}(x) = \sqrt[3]{x - 2}$ **31.** $f^{-1}(x) = x^5$
33. $f^{-1}(x) = \frac{1}{x} - 3$ **35.** $f^{-1}(x) = \dfrac{1}{2x}$

37.

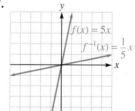

$f(x) = 5x$
$f^{-1}(x) = \dfrac{1}{5}x$

39.

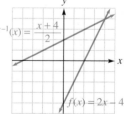

$f^{-1}(x) = \dfrac{x + 4}{2}$
$f(x) = 2x - 4$

41.

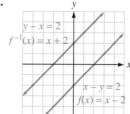

$y + x = 2$
$f^{-1}(x) = x + 2$
$x - y = 2$
$f(x) = x - 2$

43.

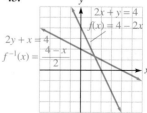

$2x + y = 4$
$f(x) = 4 - 2x$
$2y + x = 4$
$f^{-1}(x) = \dfrac{4 - x}{2}$

45.

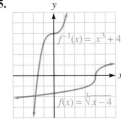

47.

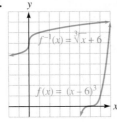

49.

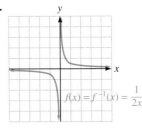

51.

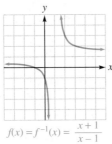

53. $f^{-1}(x) = -\sqrt{x+3}$ $(x \geq -3)$
55. $f^{-1}(x) = \sqrt[4]{x+8}$ $(x \geq -8)$
57. $f^{-1}(x) = \sqrt{4-x^2}$ $(0 \leq x \leq 2)$
59. domain: $(-\infty, 2) \cup (2, \infty)$; range: $(-\infty, 1) \cup (1, \infty)$
61. domain: $(-\infty, 0) \cup (0, \infty)$; range: $(-\infty, -2) \cup (-2, \infty)$
63. a. $y = 0.75x + 8.50$ **b.** \$11.50   **c.** $y = \frac{x - 8.50}{0.75}$
d. 2   **67.** 0   **69.** $a \geq 0$ **71.** 8   **73.** 4   **75.** $\frac{5}{4}$ **77.** $\frac{1}{7}$

Chapter Review (page 363)

1. a function   **2.** a function   **3.** not a function
4. a function   **5.** domain: $(-\infty, \infty)$
6. domain: $(-\infty, 5) \cup (5, \infty)$
7. domain: $[1, \infty)$ **8.** domain: $(-\infty, \infty)$ **9.** 8; -17; $5k - 2$
10. -2; $-\frac{3}{4}$; $\frac{6}{k-5}$ **11.** 0; 5; $|k - 2|$ **12.** $\frac{1}{7}$; $\frac{1}{2}$; $\frac{k^2 - 3}{k^2 + 3}$
13. 5   **14.** $4x + 2h - 7$

15.

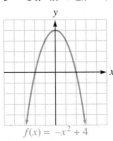

$f(x) = -x^2 + 4$
domain: $(-\infty, \infty)$
range: $(-\infty, 4]$

16.

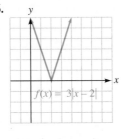

$f(x) = 3|x - 2|$
domain: $(-\infty, \infty)$
range: $[0, \infty)$

17.

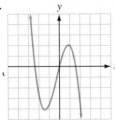

function

18.

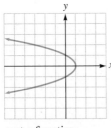

not a function

19. a. $I = 3.5h - 50$ **b.** \$650   **20.** $c = 0.15x + 5$
21. up; minimum   **22.** down; maximum   **23.** (1, 6)
24. $(-4, -5)$ **25.** $(-3, -13)$ **26.** $\left(\frac{1}{2}, -8\right)$

27.

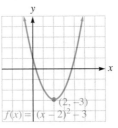

$f(x) = (x - 2)^2 - 3$
$(2, -3)$

28.

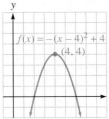

$f(x) = -(x - 4)^2 + 4$
$(4, 4)$

29.

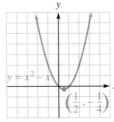

$y = x^2 - x$
$\left(\frac{1}{2}, -\frac{1}{4}\right)$

30.

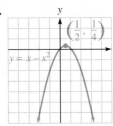

$\left(\frac{1}{2}, \frac{1}{4}\right)$
$y = x - x^2$

31.

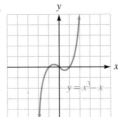

$y = x^2 - 3x - 4$
$\left(\frac{3}{2}, -\frac{25}{4}\right)$

32.

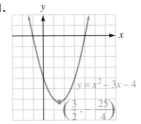

$y = 3x^2 - 8x - 3$
$\left(\frac{4}{3}, -\frac{25}{3}\right)$

33. 300 units   **34.** Both numbers are $\frac{1}{2}$.   **35.** 350 ft by 350 ft;
122,500 ft^2 **36.** 50 digital cameras; minimum cost \$1,100

37.

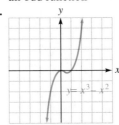

$y = x^3 - x$

an odd function

38.

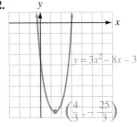

$y = x^2 - 4x$

neither even nor odd

39.

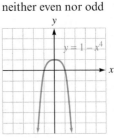

$y = x^3 - x^2$

neither even nor odd

40.

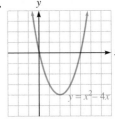

$y = 1 - x^4$

an even function

41. a. -4 **b.** 9   **42. a.** $\frac{1}{2}$ **b.** 3
43.

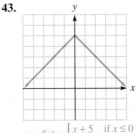

$y = f(x) = \begin{cases} x + 5 & \text{if } x \leq 0 \\ 5 - x & \text{if } x > 0 \end{cases}$

increasing for $x < 0$;
decreasing for $x > 0$

44.

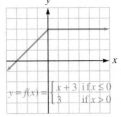

$y = f(x) = \begin{cases} x + 3 & \text{if } x \leq 0 \\ 3 & \text{if } x > 0 \end{cases}$

increasing for $x < 0$;
constant for $x > 0$

45. 3   **46.** −1

47.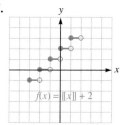

48.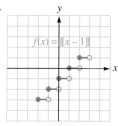

49. $44   **50.** $26

51.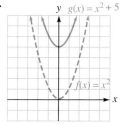

52.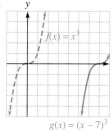

53.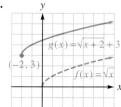

54.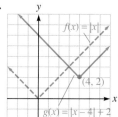

55.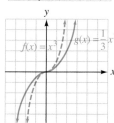

56.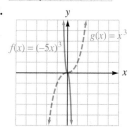

57.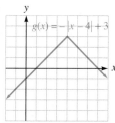

58.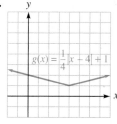

59.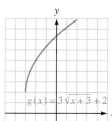

60.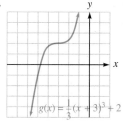

61.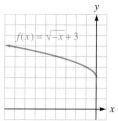

62.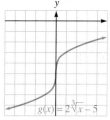

63. $(-\infty, -5) \cup (-5, 5) \cup (5, \infty)$ **64.** $(-\infty, \infty)$

65. $x = 1, x = -1$ **66.** $x = -7$ **67.** $x = 2, x = -3$

68. $x = 4, x = -1$ **69.** $y = \frac{1}{2}$ **70.** $y = -5$ **71.** $y = 0$

72. none   **73.** $y = 2x + 3$ **74.** none

75.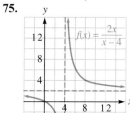

76.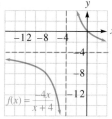

77.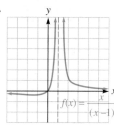

78.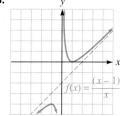

79.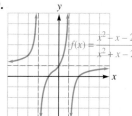

80.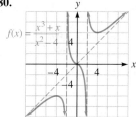

81. $(f + g)(x) = x^2 + 2x$, domain: $(-\infty, \infty)$

82. $(f \cdot g)(x) = 2x^3 + x^2 - 2x - 1$, domain: $(-\infty, \infty)$

83. $(f - g)(x) = x^2 - 2x - 2$, domain: $(-\infty, \infty)$

84. $(f/g)(x) = \dfrac{f(x)}{g(x)} = \dfrac{x^2 - 1}{2x + 1}$, domain: $\left(-\infty, -\frac{1}{2}\right) \cup \left(-\frac{1}{2}, \infty\right)$

85. 10   **86.** 60   **87.** 21   **88.** no value

89. $(f \circ g)(x) = f(g(x)) = 4x^2 + 4x$, domain: $(-\infty, \infty)$

90. $(g \circ f)(x) = g(f(x)) = 2x^2 - 1$, domain: $(-\infty, \infty)$

91. 20   **92.** −2   **93.** $f(x) = x^2; g(x) = x - 5$

94. $f(x) = x^3; g(x) = x + 6$ **95.** not one-to-one

96. is one-to-one

97.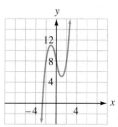

is one-to-one

98.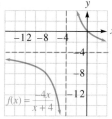

is not one-to-one

101. $f^{-1}(x) = \frac{x+1}{7}$ **102.** $f^{-1}(x) = 2 - \frac{1}{x}$

103. $f^{-1}(x) = \frac{x}{x+1}$ **104.** $f^{-1}(x) = \sqrt[3]{\frac{3}{x}}$

105. $f^{-1}(x) = \frac{x+5}{2}$

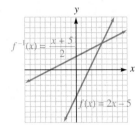

106. $\left(-\infty, \frac{2}{5}\right) \cup \left(\frac{2}{5}, \infty\right)$

Chapter Test (page 377)

1. domain: $(-\infty, 5) \cup (5, \infty)$ **2.** domain: $[-3, \infty)$ **3.** $\frac{1}{2}, 2$
4. $\sqrt{6}, 3$ **5.** $(7, -3)$ **6.** $(1, -4)$ **7.** $(4, -10)$
8. $(-2, 9)$
9.

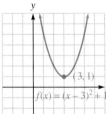

10.

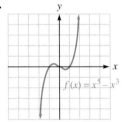

11. $\frac{25}{8}$ sec   **12.** $\frac{625}{4}$ ft   **13.** 10 ft   **14.** 110 ft
15.

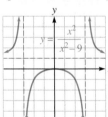

16.

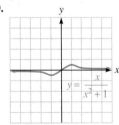

17. vertical asymptotes: $x = -3$ and $x = 3$; horizontal asymptote: $y = 0$ **18.** vertical asymptote: $x = 3$; horizontal asymptote: none; slant asymptote: $y = x - 2$
19.

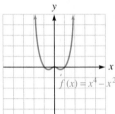

20.

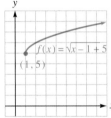

21.

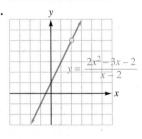

22.

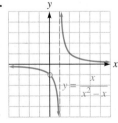

23. $(f + g)(x) = f(x) + g(x) = x^2 + 3x + 2$
24. $(g \circ f)(x) = g(f(x)) = 9x^2 + 2$
25. $(f/g)(x) = \frac{f(x)}{g(x)} = \frac{3x}{x^2 + 2}$
26. $(f \circ g)(x) = f(g(x)) = 3x^2 + 6$ **27.** $f^{-1}(x) = \frac{x+1}{x-1}$
28. $f^{-1}(x) = \sqrt[3]{x + 3}$ **29.** range: $(-\infty, -2) \cup (-2, \infty)$
30. range: $(-\infty, 3) \cup (3, \infty)$

Cumulative Review Exercises (page 378)

1.

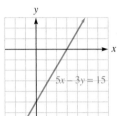

2.

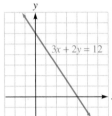

3. $\sqrt{41}; \left(\frac{1}{2}, \frac{3}{2}\right); -\frac{4}{5}$ **4.** $2\sqrt{29}; (-2, 5); \frac{2}{5}$
5. $y = -2x - 1$ **6.** $y = \frac{7}{2}x - \frac{11}{4}$ **7.** $y = \frac{3}{5}x + 6$
8. $y = -4x$
9.

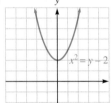

10.

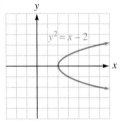

11.

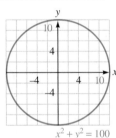

12.

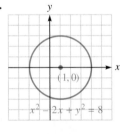

13. $x = 1, x = 10$ **14.** $x = 2, x = 8$ **15.** \$62.50   **16.** $\frac{25}{4}$
17. a function   **18.** a function   **19.** a function
20. not a function   **21.** domain: $(-\infty, \infty)$
22. domain: $(-\infty, -2) \cup (-2, \infty)$ **23.** domain: $[2, \infty)$
24. domain: $[-4, \infty)$ **25.** $\left(-\frac{5}{2}, -\frac{49}{4}\right)$ **26.** $\left(\frac{5}{2}, \frac{49}{4}\right)$
27.

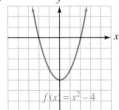

28.

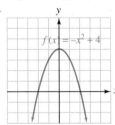

29.

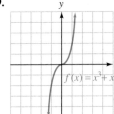

30.

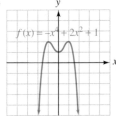

31.

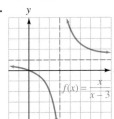

32.

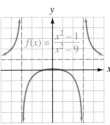

35.

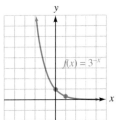

37.

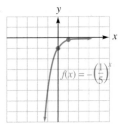

33. $(f + g)(x) = f(x) + g(x) = x^2 + 3x - 3$;
domain: $(-\infty, \infty)$
34. $(f - g)(x) = f(x) - g(x) = -x^2 + 3x - 5$;
domain: $(-\infty, \infty)$
35. $(f \cdot g)(x) = f(x) \cdot g(x) = 3x^3 - 4x^2 + 3x - 4$;

domain: $(-\infty, \infty)$ **36.** $(f/g)(x) = \dfrac{f(x)}{g(x)} = \dfrac{3x - 4}{x^2 + 1}$;

domain: $(-\infty, \infty)$ **37.** $(f \circ g)(2) = 11$ **38.** $(g \circ f)(2) = 5$
39. $(f \circ g)(x) = 3x^2 - 1$ **40.** $(g \circ f)(x) = 9x^2 - 24x + 17$
41. $f^{-1}(x) = \dfrac{x - 2}{3}$ **42.** $f^{-1}(x) = \dfrac{1}{x} + 3$

43. $f^{-1}(x) = \sqrt{x - 5}$ **44.** $f^{-1}(x) = \dfrac{x + 1}{3}$ **45.** $y = kwz$

46. $y = \dfrac{kx}{t^2}$

Exercises 4.1 (page 390)

1. exponential   **3.** $(-\infty, \infty)$ **5.** $(0, \infty)$ **7.** asymptote
9. 3   **11.** 2.72   **13.** increasing   **15.** 11.0357

17. 451.8079   **19.** $5^{2\sqrt{2}} = 25^{\sqrt{2}}$ **21.** a^4 **23.** 1, 25

25. 1, 9
27.

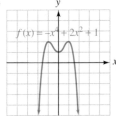

29.

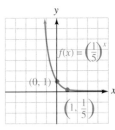

31.

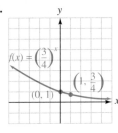

33.

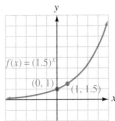

39. yes   **41.** no   **43.** $b = \dfrac{1}{2}$ **45.** no value of b
47. $b = 2$ **49.** $b = e$
51.

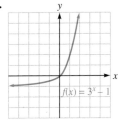

53.

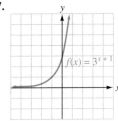

55.

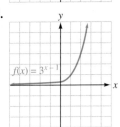

57.

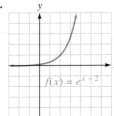

59.

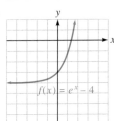

61.

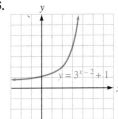

63.

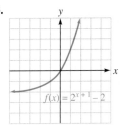

65.

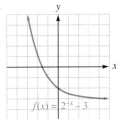

67.

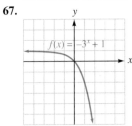

69.

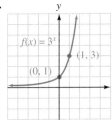

71.
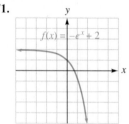
$f(x) = -e^x + 2$

73.

75.

77.

79.
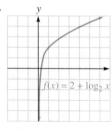

81. \$22,080.40 **83.** \$15.79 **85.** \$2,273,996.13
87. \$1,263.77 **89.** \$13,375.68 **91.** \$7,647.95 from continuous compounding, \$7,518.28 from annual compounding **93.** \$291.27 **95.** \$12,155.61 **97.** 13,228
101. 125 **103.** $x^2(1 + 9x^2)$ **105.** $(x + 4)(x - 3)$

Exercises 4.2 (page 398)

1. birth, death **3.** 0.1868 g **5.** about 10 kg
7. about 35.4% **9.** 0.1575 unit **11.** 2 lumens
13. about 8 lumens **15.** about 56,570 **17.** 61.9°C
19. 315 **21.** 13 **23.** 10.6 billion **25.** 2.6
27. about 0.07% **29.** 0 **31.** 18,394 **33.** about 24,060
35. about 492 **37.** 19.0 mm **39.** 49 mps
41. about 7 million **43.** about 72.2 years **49.** 8 **51.** 3
53. 2 **55.** 3

Exercises 4.3 (page 409)

1. $x = b^y$ **3.** range **5.** inverse **7.** exponent
9. $(b, 1), (1, 0)$ **11.** $\log_e x$ **13.** $(-\infty, \infty)$ **15.** 10
17. $3^4 = 81$ **19.** $\left(\frac{1}{2}\right)^3 = \frac{1}{8}$ **21.** $4^{-3} = \frac{1}{64}$ **23.** $\pi^1 = \pi$
25. $\log_8 64 = 2$ **27.** $\log_4 \frac{1}{16} = -2$ **29.** $\log_{1/2} 32 = -5$
31. $\log_x z = y$ **33.** 3 **35.** 3 **37.** 3 **39.** $\frac{1}{2}$ **41.** -3
43. 64 **45.** 7 **47.** 5 **49.** $\frac{1}{25}$ **51.** $\frac{1}{6}$ **53.** 5 **55.** $\frac{3}{2}$
57. 4 **59.** $\frac{2}{3}$ **61.** 5 **63.** 4 **65.** 0.5119 **67.** -2.3307
69. 3.8221 **71.** -0.4055 **73.** 3.5596 **75.** 2.0664
77. -0.2752 **79.** undefined **81.** 25.2522
83. 1.9498×10^{-4} **85.** 4.0645 **87.** 69.4079 **89.** 0.0245
91. 120.0719 **93.** 4 **95.** -3 **97.** 7 **99.** 4
101. $b = 2$ **103.** $b = 2$
105.
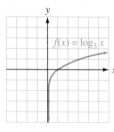
$f(x) = \log_3 x$

107.
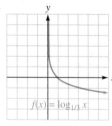
$f(x) = \log_{1/3} x$

109.
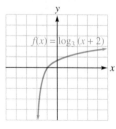
$f(x) = 2 + \log_2 x$

111.
$f(x) = \log_3 (x + 2)$

113.
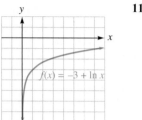
$f(x) = -3 + \ln x$

115.
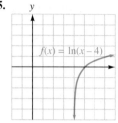
$f(x) = \ln(x - 4)$

117.

119.

121.

123.

125. b is larger.

129. $\left(-\frac{7}{2}, -\frac{37}{4}\right)$ **131.** 425 ft by 850 ft **133.** $y = 5x$

Exercises 4.4 (page 416)

1. $20 \log \dfrac{E_O}{E_I}$ **3.** $t = -\dfrac{1}{k} \ln\left(1 - \dfrac{C}{M}\right)$ **5.** $E = RT \ln\left(\dfrac{V_f}{V_i}\right)$
7. 55 db **9.** 29 db **11.** 49.5 db **13.** 4.4 **15.** 4
17. no **19.** 19.8 min **21.** about 5.8 yr **23.** about 9.2 yr
25. about 3,654 joules **27.** about 99% per year
29. 3 yr old **31.** about 10.8 yr **33.** about 208,000 V
35. 1 mi **39.** $y = 7x + 3$ **41.** $x = 2$
43. $\dfrac{1}{2x - 3}$ **45.** $\dfrac{x + 1}{3(x - 2)}$

Exercises 4.5 (page 426)

1. 0 **3.** M, N **5.** x, y **7.** x **9.** $\neq$ **11.** 0 **13.** 7
15. 10 **17.** 1 **25.** $\log_b 2 + \log_b x + \log_b y$
27. $\log_b 2 + \log_b x - \log_b y$ **29.** $2 \log_b x + 3 \log_b y$
31. $\frac{1}{3}(\log_b x + \log_b y)$ **33.** $\log_b x + \frac{1}{2} \log_b z$
35. $\frac{1}{3} \log_b x - \frac{1}{3} \log_b y - \frac{1}{3} \log_b z$ **37.** $7 \ln x + 8 \ln y$
39. $\ln x - 4 \ln y - \ln z$ **41.** $\log_b \dfrac{x + 1}{x}$ **43.** $\log_b x^2 \sqrt[3]{y}$
45. $\log_b \dfrac{\sqrt{z}}{x^3 y^2}$ **47.** $\log_b \dfrac{\dfrac{x}{z} + x}{\dfrac{y}{z} + y} = \log_b \dfrac{x}{y}$ **49.** $\ln \dfrac{x(x + 5)}{9}$
51. $\ln \dfrac{z}{x^6 y^2}$ **53.** true **55.** false **57.** true **59.** false
61. true **63.** false **65.** false **67.** true **69.** true
71. true **73.** false **75.** true **77.** 1.4472 **79.** 0.3521
81. 1.1972 **83.** 2.4014 **85.** 2.0493 **87.** 0.4682
89. 1.7712 **91.** 0.9597 **93.** 1.8928 **95.** 2.3219
97. 7.20 **99.** 4.77 **101.** from 5.01×10^{-4} to 1.26×10^{-3}
103. 19 db **105.** The original intensity must be raised to the 4th power. **107.** The volume V is squared. **121.** yes
123. no **125.** $(-\infty, \infty)$ **127.** $(-\infty, \infty)$

Exercises 4.6 (page 437)

1. exponential **3.** $A_0 2^{-t/h}$ **5.** 4 **7.** $-\frac{1}{3}$ **9.** 3, -1
11. $-2, -2$ **13.** 3 **15.** ± 5 **17.** 1.1610 **19.** 1.2702
21. 1.7095 **23.** 0 **25.** ± 1.0878 **27.** 0, 1.0566
29. 2.3026 **31.** 0.8959 **33.** 0 **35.** 0.2789 **37.** 1, 3

39. 0   **41.** 10, −10   **43.** 4   **45.** e^6 **47.** $\frac{1}{2}(e^4 + 7)$
49. 7   **51.** 50   **53.** 20   **55.** 10   **57.** 7   **59.** 6   **61.** 5
63. 4   **65.** 3, 4   **67.** 10^{10} **69.** no solution   **71.** 6
73. 9   **75.** 4   **77.** 7, 1   **79.** 20   **81.** 1.81
83. about 5.1 yr   **85.** about 42.7 days   **87.** about 2,900 yr
89. about 5.6 yr   **91.** about 5.4 yr   **93.** because
$\ln 2 \approx 0.70$ **95.** about 27 m   **97.** about 12 min
99. $k = \frac{\ln 0.75}{3}$ **101.** $\frac{\ln (2/3)}{5}$ **109.** $f^{-1}(x) = \frac{x - 2}{3}$
111. 19   **113.** $5x^2 - 1$

Chapter Review (page 439)

1. $5^{2\sqrt{2}}$ **2.** $2^{\sqrt{10}}$

3.
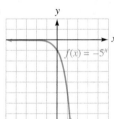

4.
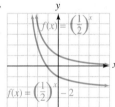

5. $p = 1, q = 7$ **6.** domain: $(-\infty, \infty)$; range: $(0, \infty)$

7.
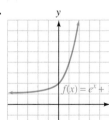

8.
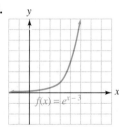

9.
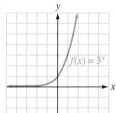

10.
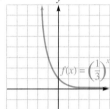

11.
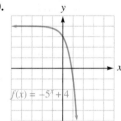

12.
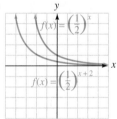

13. $2,189,703.45   **14.** $2,324,767.37   **15.** $\frac{2}{3}$
16. 0.19 lumen   **17.** about 635,000,000   **18.** about 2,708
19. $(0, \infty); (-\infty, \infty)$ **20.** $(0, \infty); (-\infty, \infty)$ **21.** 2   **22.** $-\frac{1}{2}$
23. 0   **24.** −2   **25.** $\frac{1}{2}$ **26.** $\frac{1}{3}$ **27.** 32   **28.** 9   **29.** 8
30. −1   **31.** $\frac{1}{8}$ **32.** 2   **33.** 4   **34.** 2   **35.** 10   **36.** $\frac{1}{25}$
37. 5   **38.** 3

39.
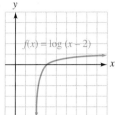

40.
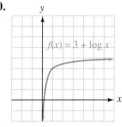

41.
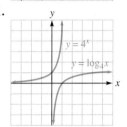

42.
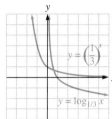

43. 6.1137   **44.** −0.1111   **45.** 10.3398   **46.** 2.5715

47.
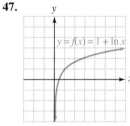

48.
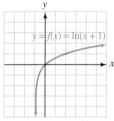

49. 12   **50.** $14x$ **51.** 53 db   **52.** 4.4   **53.** $9\frac{1}{2}$ min
54. 23 yr   **55.** 2,017 joules   **56.** 0   **57.** 1   **58.** 3
59. 4   **60.** 4   **61.** 0   **62.** 7   **63.** 3   **64.** 4   **65.** 9
66. $2 \log_b x + 3 \log_b y - 4 \log_b z$
67. $\frac{1}{2}(\log_8 x - \log_8 y - 2 \log_8 z)$
68. $4 \ln x - 5 \ln y - 6 \ln z$ **69.** $\frac{1}{3}(\ln x + \ln y + \ln z)$
70. $\log_b \frac{x^3 z^7}{y^5}$ **71.** $\log_b \frac{\sqrt{xy^3}}{z^7}$ **72.** $\ln \frac{x^4}{y^5 z^6}$ **73.** $\ln \frac{y^3 \sqrt{x}}{\sqrt[3]{z}}$
74. 3.36   **75.** 1.56   **76.** 2.64   **77.** −6.72   **78.** 1.7604
79. about 7.94×10^{-4} grams-ions per liter   **80.** $k \ln 2$ less
81. 2   **82.** −1, −3   **83.** 2   **84.** 3, −3
85. $\frac{\log 7}{\log 3} \approx 1.7712$ **86.** $\frac{\log 3}{\log 3 - \log 2} \approx 2.7095$
87. $\ln 8 \approx 2.079$ **88.** $\ln 7 \approx 1.9459$ **89.** −3   **90.** 8
91. 25, 4   **92.** 4   **93.** 2   **94.** 4, 3   **95.** 6   **96.** 31
97. $\frac{\ln 9}{\ln 2} \approx 3.1699$ **98.** no solution   **99.** $\frac{e}{e - 1} \approx 1.5820$
100. 1   **101.** about 3,300 yr

Chapter Test (page 451)

1.
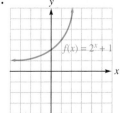

2.
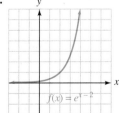

3. $\frac{3}{64}$ g   **4.** $1,060.90   **5.** $4,451.08   **6.** 3   **7.** −3
8. 17   **9.** 2   **10.** −3

11.

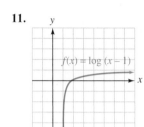

$f(x) = \log(x - 1)$

12.

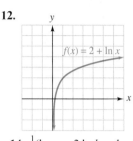

$f(x) = 2 + \ln x$

13. $2 \log a + \log b + 3 \log c$ **14.** $\frac{1}{2}(\ln a - 2 \ln b - \ln c)$

15. $\log \dfrac{b\sqrt{a+2}}{c^2}$ **16.** $\ln \dfrac{\sqrt[3]{\dfrac{a}{b^2}}}{c}$ **17.** 1.3801 **18.** 0.4259

19. $\dfrac{\log 3}{\log 7}$ or $\dfrac{\ln 3}{\ln 7}$ **20.** $\dfrac{\log e}{\log \pi}$ or $\dfrac{\ln e}{\ln \pi}$ or $\dfrac{1}{\ln \pi}$ **21.** true

22. false   **23.** 6.4 **24.** 46 db   **25.** $-1, 3$

26. $\dfrac{\log 3}{\log 3 - 2} \approx -0.3133$ **27.** $\ln 9$ **28.** 1 **29.** 10

30. 9

Exercises 5.1 (page 460)

1. whole   **3.** any   **5.** factor   **7.** $4x^2 + 2x + 1 + \dfrac{2}{x-1}$

9. $2x^3 - 3x^2 + 8x - 1 + \dfrac{-3}{x+2}$ **11.** -1 **13.** 9 **15.** 3

17. -3 **19.** 69 **21.** $70,249$ **23.** 128.3085938

25. true   **27.** true   **29.** false   **31.** true

33. $(x-1)(3x^2 + x - 5) - 9$

35. $(x-3)(3x^2 + 7x + 15) + 41$

37. $(x+1)(3x^2 - 5x - 1) - 3$

39. $(x+3)(3x^2 - 11x + 27) - 85$

41. $x^2 + 2x + 3$ **43.** $7x^2 - 10x + 5 + \dfrac{-4}{x+1}$

45. $4x^3 + 9x^2 + 27x + 80 + \dfrac{245}{x-3}$

47. $3x^4 + 12x^3 + 48x^2 + 192x$ **49.** 47 **51.** -569

53. $\frac{15}{8}$ **55.** 5 **57.** 0 **59.** 384 **61.** $16 + 2i$

63. $40 - 40i$ **65.** $\{-1, -5, 3\}$ **67.** $\left\{-\frac{1}{2}, 3, -3\right\}$

69. $\left\{1, 1, \sqrt{3}, -\sqrt{3}\right\}$ **71.** $\{2, 3, i, -i\}$

73. $x^2 - 9x + 20$ **75.** $x^3 - 3x^2 + 3x - 1$

77. $x^3 - 11x^2 + 38x - 40$ **79.** $x^4 - 3x^2 + 2$

81. $x^3 - \sqrt{2}x^2 + x - \sqrt{2}$ **83.** $x^3 - 2x^2 + 2x$ **85.** 0

89. QIV   **91.** QI   **93.** 10 **95.** 1

Exercises 5.2 (page 469)

1. zero   **3.** conjugate   **5.** 0 **7.** 2 **9.** lower bound

11. 10 **13.** 4 **15.** 4 **17.** $4, 4$ **19.** $5, 5$

21. $x^2 + 4 = 0$ **23.** $x^2 - 6x + 10 = 0$

25. $x^3 - 3x^2 + x - 3 = 0$ **27.** $x^3 - 6x^2 + 13x - 10 = 0$

29. $x^4 - 5x^3 + 7x^2 - 5x + 6 = 0$

31. $x^4 - 2x^3 + 3x^2 - 2x + 2 = 0$ **33.** 0 or 2 positive;

1 negative; 0 or 2 nonreal   **35.** 1 positive; 1 or 3 negative;

0 or 2 nonreal   **37.** 0 positive; 0 negative; 4 nonreal

39. 1 positive; 1 negative; 2 nonreal   **41.** 0 positive;

0 negative; 10 nonreal   **43.** 0 positive; 0 negative; 8 nonreal;

yes   **45.** 1 positive; 1 negative; 2 nonreal   **47.** $-2, 4$

49. $-1, 1$ **51.** $-4, 6$ **53.** $-4, 3$ **55.** $-2, 4$

61. k units to the right   **63.** reflected about the y-axis

65. stretched vertically by a factor of k

Exercises 5.3 (page 479)

1. -7 **3.** root   **5.** $\pm 1, \pm 2, \pm 3, \pm 4, \pm 6, \pm 12$

7. $\pm 1, \pm 2, \pm 3, \pm 6, \pm \frac{1}{2}, \pm \frac{3}{2}$

9. $\pm 1, \pm 2, \pm 5, \pm 10, \pm \frac{1}{2}, \pm \frac{5}{2}, \pm \frac{1}{4}, \pm \frac{5}{4}$ **11.** $1, -1, 5$

13. $1, 2, -1$ **15.** $1, 2, -2$ **17.** $3, -3, 2$ **19.** $1, -1, \frac{1}{2}$

21. $-1, -1, \frac{1}{3}$ **23.** $1, 2, 3, 4$ **25.** $2, -5$

27. $1, -1, 2, -2, -3$ **29.** $0, 2, 2, 2, -2, -2, -2$ **31.** $\frac{2}{3}$

33. $-1, 2, 3, \frac{2}{3}$ **35.** $-3, \frac{1}{3}, \frac{1}{2}, \frac{1}{2}$ **37.** $2, 2, -3, -\frac{1}{3}, \frac{1}{2}$

39. $\frac{3}{2}, \frac{2}{3}, -\frac{3}{5}$ **41.** $\frac{2}{3}, -\frac{3}{5}, 4$ **43.** $-\frac{1}{2}, \frac{3}{5}, \frac{6}{5}$

45. $3, \sqrt{2}, -\sqrt{2}$ **47.** $\frac{1}{2}, i, -i$

49. $2, -2, 1 + \sqrt{5}, 1 - \sqrt{5}$ **51.** $\frac{1}{2}, -1, 3i, -3i$

53. $1, 1, 1, 5i, -5i$ **55.** $3, 1 - i$ **57.** $3, -3, 1 - i$

59. $-\frac{2}{3}, 3, -1$ **61.** $\frac{1}{2}, \frac{1}{2}, \frac{1}{2}, 1, 1$ **63.** $10, 20, 60$ ohms

65. 13 in.   **69.** $(3, 7)$ or approx. $(1.54, 13.63)$

71. $6ab^2\sqrt{2abc}$ **73.** $8a\sqrt{2b}$ **75.** $\sqrt{3} + 1$

Exercises 5.4 (page 486)

1. $P(a)$ and $P(b)$ **3.** x_l and x_r **7.** $P(-2) = 3$;

$P(-1) = -2$, the signs of the results are opposites

9. $P(4) = -40$; $P(5) = 30$, the signs of the results are

opposites   **11.** $P(1) = 8$; $P(2) = -1$, the signs of the results

are opposites   **13.** $P(2) = -72$; $P(3) = 154$, the signs of the

results are opposites   **15.** $P(0) = 10$; $P(1) = -60$, the signs

of the results are opposites   **17.** 1.7 **19.** -2.2 **21.** 1.7

23. -1.2 **25.** $-2.2, 2.2$ **27.** $1, 2$; the bisection method

fails to find the solution 2.   **29.** yes   **31.** $(0.83, 0.56)$,

$(-0.83, -0.56)$ **35.** vertical: $x = 2, x = -2$;

horizontal: $y = 0$; slant: none   **37.** vertical: $x = 2$;

horizontal: none; slant: $y = x + 2$

Chapter Review (page 487)

1. 1 **2.** 66 **3.** 241 **4.** 34 **5.** false   **6.** false

7. true   **8.** true   **9.** $3x^3 + 9x^2 + 29x + 90 + \dfrac{277}{x-3}$

10. $2x^3 + 4x^2 + 5x + 13 + \dfrac{25}{x-2}$

11. $5x^4 - 14x^3 + 31x^2 - 64x + 129 + \dfrac{-259}{x+2}$

12. $4x^4 - 2x^3 + x^2 + 2x + \dfrac{1}{x+1}$ **13.** 151 **14.** -113

15. $\frac{13}{8}$ **16.** $-1 - 6i$ **17.** $\left\{3, \frac{1}{2}, -2\right\}$

18. $\left\{-2, -2, \sqrt{5}, -\sqrt{5}\right\}$ **19.** $2x^3 - 5x^2 - x + 6$

20. $2x^3 + 3x^2 - 8x + 3$ **21.** $x^4 + 3x^3 - 9x^2 + 3x - 10$

22. $x^4 + x^3 - 5x^2 + x - 6$ **23.** 6 **24.** 6 **25.** 65

26. $1,984$ **27.** $4, 4$ **28.** $40, 40$ **29.** $5, 5$ **30.** $3, 3$

31. $2 - i$ **32.** i **33.** $x^3 - 4x^2 + x - 4 = 0$

34. $x^3 + 5x^2 + x + 5 = 0$ **35.** 0 or 2 positive;

0 or 2 negative; 0, 2, or 4 nonreal   **36.** 1 or 3 positive;

1 negative; 0 or 2 nonreal   **37.** 1 positive; 0, 2, or 4 negative;

0, 2, or 4 nonreal   **38.** 1 or 3 positive; 0 or 2 negative; 2, 4, or

6 nonreal   **39.** 0 positive; 0 negative; 4 nonreal

40. 0 positive; 1 negative; 6 nonreal   **41.** $-1, 2$ **42.** $-5, 2$

43. $\pm 1, \pm 2, \pm 3, \pm 6, \pm \frac{1}{2}, \pm \frac{3}{2}$

44. $\pm 1, \pm 2, \pm 5, \pm 10, \pm \frac{1}{2}, \pm \frac{5}{2}, \pm \frac{1}{4}, \pm \frac{5}{4}$

45. $1, 4, 5$ **46.** $1, -1, 8$ **47.** $-5, -\frac{3}{2}, -2$ **48.** $\frac{1}{3}$

49. $2, -2, \frac{3}{2}, -\frac{3}{2}$ **50.** $4, 4, -2, -\frac{1}{2}$ **51.** $\frac{1}{3}, 4i, -4i$
52. $2, -2, 1 + \sqrt{6}, 1 - \sqrt{6}$ **53.** $P(-1) = -9; P(0) = 18$
54. $P(1) = -8; P(2) = 21$ **55.** 4.2   **56.** 2.5   **57.** 1.67
58. 0.67   **59.** 10 m by 6 m; 12 m by 5 m; 15 m by 4 m
60. 17.27 ft

Chapter Test (page 496)

1. yes   **2.** 1   **3.** -30 **4.** no
5. $(x - 2)(2x^2 + x - 2) - 5$
6. $(x + 1)(2x^2 - 5x + 1) - 2$
7. $2x + 3$ **8.** $3x^2 + x$ **9.** 5   **10.** -28 **11.** $\frac{11}{3}$
12. $6 - 3i$ **13.** $x^3 - 4x^2 - 5x$ **14.** $x^4 - 2x^2 - 3$
15. $x^3 - 2x^2 + x - 2 = 0$ **16.** $x^3 - 5x^2 + 9x - 5 = 0$
17. 3, 3   **18.** $3 + 2i$ **19.** 1 or 3 positive; 0 or 2 negative;
0, 2, or 4 nonreal   **20.** 1 positive; 0 or 2 negative; 0 or
2 nonreal   **21.** $-3, 3$ **22.** $-1, 6$ **23.** $\pm 1, \pm 2, \pm 4, \pm 8$
24. $2, -3, -\frac{1}{2}$ **25.** $2, i, -i$ **26.** not necessarily   **27.** 3.3

Cumulative Review Exercises (page 497)

1.

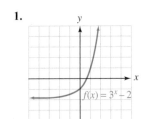

2.

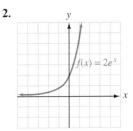

3.

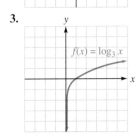

4.

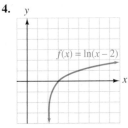

5. 6   **6.** -3 **7.** 3   **8.** 2   **9.** $\log a + \log b + \log c$
10. $2 \log a + \log b - \log c$ **11.** $\frac{1}{2}(\log a + \log b - 3 \log c)$
12. $\frac{1}{2} \ln a + \ln b - \ln c$ **13.** $\log \dfrac{a^3}{b^3}$ **14.** $\log \dfrac{\sqrt{a}\, b^3}{\sqrt[3]{c^2}}$
15. $x = \dfrac{\log 8}{\log 3} - 1$ **16.** $x = -1$ **17.** $x = 500$
18. $x = \sqrt{11}$ **19.** 9   **20.** -36 **21.** 4   **22.** $2 - i$
23. a factor   **24.** not a factor   **25.** a factor   **26.** a factor
27. 12   **28.** 2,000   **29.** 2 or 0 positive; 2 or 0 negative; 4, 2,
or 0 nonreal   **30.** 1 positive; 3 or 1 negative; 2 or 0 nonreal
31. $-1, -3, 3$ **32.** $-1, 1, 2$

Exercises 6.1 (page 509)

1. system   **3.** consistent   **5.** independent   **7.** consistent
9. dependent   **11.** is

13.

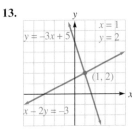

15.

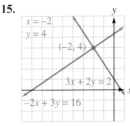

17. $(2.2, -4.7)$ **19.** $(1.7, 0.3)$ **21.** $(-1, -2)$
23. $(3, -2)$ **25.** $\left(\frac{1}{2}, \frac{1}{3}\right)$ **27.** no solution, inconsistent
system   **29.** dependent equations; a general solution is
$(x, 3x - 6)$ **31.** $(3, 1)$ **33.** $(3, 2)$ **35.** $(-3, 0)$
37. $\left(1, -\frac{1}{2}\right)$ **39.** dependent equations; a general solution is
$(x, 5 - 2x)$ **41.** no solution, inconsistent system
43. $(4, -7)$ **45.** $(2, -3)$ **47.** $(9, -1)$ **49.** $(1, 2, 0)$
51. $\left(0, -\frac{1}{3}, -\frac{1}{3}\right)$ **53.** $(1, 2, -1)$ **55.** $(1, 0, 5)$
57. no solution; inconsistent system   **59.** $(0, 1, 0)$
61. $\left(\frac{2}{3}, \frac{1}{4}, \frac{1}{2}\right)$ **63.** $\left(\frac{1}{2}, \frac{1}{2}, \frac{1}{2}\right)$ **65.** dependent equations; a
general solution is $(x, 2 - x, 1)$ **67.** 225 acres of corn;
125 acres of soybeans   **69.** 210,000 cm^2 **71.** 40 g and
20 g   **73.** 10 ft   **75.** $E(x) = 43.53x + 742.72$,
$R(x) = 89.95x$; 16 pairs per day   **77.** 15 hr cooking
hamburgers, 10 hr pumping gas, 5 hr janitorial
79. 1.05 million in 0–14 group, 1.56 million in 15–49 group,
0.39 million in 50-and-older group   **81.** 30°, 50°, 100°
89.

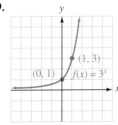

91. 8
93. $\log x - 2 \log y - \log z$
95. $\log \dfrac{xy^3}{\sqrt{z}}$

Exercise 6.2 (page 522)

1. matrix   **3.** coefficient   **5.** equation   **7.** row equivalent
9. interchanged   **11.** adding, multiple   **13.** $(2, 3)$
15. $(7, 6)$ **17.** $(1, 2, 3)$ **19.** $(1, 1, 3)$ **21.** row echelon
form   **23.** reduced row echelon form   **25.** $(2, -1)$
27. $(-2, 0)$ **29.** $(3, 1)$ **31.** no solution; inconsistent
system   **33.** $(0, -7)$ **35.** $(1, 0, 2)$ **37.** $(2, -2, 1)$
39. $(1, 1, 2)$ **41.** $(-1, 3, 1)$ **43.** $(1, -5, 3)$ **45.** $(13, 3)$
47. $(-13, 7, -2)$ **49.** $(10, 3)$ **51.** $(0, 0)$ **53.** $(3, 1, -2)$
55. $\left(\frac{1}{4}, 1, \frac{1}{4}\right)$ **57.** $(1, 2, 1, 1)$ **59.** $(1, 2, 0, 1)$
61. $\left(\frac{9}{4}, -3, \frac{3}{4}\right)$ **63.** $\left(0, \frac{20}{3}, -\frac{1}{3}\right)$ **65.** dependent equations;
a general solution is $\left(x, \frac{1}{2}x - 3, 0\right)$ **67.** $(1, -3)$
69. dependent equations; general solution is
$\left(\frac{8}{7} + \frac{1}{7}z, \frac{10}{7} - \frac{4}{7}z, z\right)$ **71.** dependent equations; general
solution is $(1 + z, -z, -1 - z, z)$ **73.** no solution;
inconsistent system   **75.** 1,300 mi   **77.** Dictionaries are
4.5 in. wide; atlases are 3.5 in. wide; thesauruses are 4 in. wide.
79. 2, 4, 6   **85.** $x = \pm 2, y = \pm 1, z = \pm 3$
87. $y = mx + b$ **89.** equal   **91.** $y = 2x + 7$ **93.** $x = 2$

Exercises 6.3 (page 535)

1. i, j **3.** corresponding   **5.** columns, rows   **7.** additive identity or zero   **9.** $x = 2, y = 5$ **11.** $x = 1, y = 2$

13. $\begin{bmatrix} -1 & 2 & 1 \\ -6 & 0 & 0 \end{bmatrix}$ **15.** $\begin{bmatrix} -6 & 5 & 0 \\ 1 & -1 & 0 \end{bmatrix}$

17. $\begin{bmatrix} -5 & 2 & -7 \\ 5 & 0 & -3 \\ 2 & -3 & 5 \end{bmatrix}$ **19.** $\begin{bmatrix} 15 & -15 \\ 0 & -10 \end{bmatrix}$

21. $\begin{bmatrix} 25 & 75 & -10 \\ -10 & -25 & 5 \end{bmatrix}$ **23.** $\begin{bmatrix} 18 & -1 & -4 \\ -35 & 0 & -1 \end{bmatrix}$

25. $\begin{bmatrix} 2 & -2 \\ 3 & 10 \end{bmatrix}$ **27.** $\begin{bmatrix} -22 & -22 \\ -105 & 126 \end{bmatrix}$ **29.** $\begin{bmatrix} 4 & 2 & 10 \\ 5 & -2 & 4 \\ 2 & -2 & 1 \end{bmatrix}$

31. $\begin{bmatrix} 32 \\ 2 \end{bmatrix}$ **33.** not possible   **35.** $\begin{bmatrix} 16 \\ 12 \\ 12 \end{bmatrix}$

37. not possible   **39.** $\begin{bmatrix} -36.29 \\ 16.2 \\ -19.26 \end{bmatrix}$

41. $\begin{bmatrix} -16.11 & 4.71 & 33.64 \\ -19.6 & 20.35 & 6.4 \\ -100.6 & 72.82 & 62.71 \end{bmatrix}$ **47.** $\begin{bmatrix} 4 & 5 \\ -7 & -1 \end{bmatrix}$

49. not possible   **51.** $\begin{bmatrix} 24 & 16 \\ 39 & 26 \end{bmatrix}$ **53.** $\begin{bmatrix} 47 \\ 81 \end{bmatrix}$

55. $QC = \begin{bmatrix} 2,000 \\ 1,700 \end{bmatrix}$ The cost to Supplier 1 is \$2,000. The cost to Supplier 2 is \$1,700.

57. $QP = \begin{bmatrix} 584.50 \\ 709.25 \\ 1,036.75 \end{bmatrix}$ adult males spent \$584.50 adult females spent \$709.25 children spent \$1,036.75

59. $\begin{bmatrix} 1 & 1 & 0 \\ 0 & 1 & 1 \\ 1 & 0 & 0 \end{bmatrix}$ **61.** $A^2 = \begin{bmatrix} 5 & 1 & 2 & 2 \\ 1 & 5 & 2 & 2 \\ 2 & 2 & 6 & 0 \\ 2 & 2 & 0 & 4 \end{bmatrix}$ indicates the

number of ways two cities can be linked with exactly one intermediate city to relay messages.

63. No; if $A = \begin{bmatrix} 1 & 1 \\ 1 & 1 \end{bmatrix}$ and $B = \begin{bmatrix} 1 & 0 \\ 0 & 0 \end{bmatrix}$, then $(AB)^2 \neq A^2B^2$.

65. Let $A = \begin{bmatrix} 1 & 2 \\ 1 & 2 \end{bmatrix}$ and $B = \begin{bmatrix} 2 & 2 \\ -1 & -1 \end{bmatrix}$. Neither is the zero matrix, yet $AB = 0$.   **67.** $6x^2 - 4x - 8$

69. $\dfrac{x + 1}{x - 1}$ **71.** $a = \dfrac{2s}{n} - l$

Exercises 6.4 (page 544)

1. $AB = BA = I$ **3.** $[I \mid A^{-1}]$ **5.** $\begin{bmatrix} 3 & 4 \\ 2 & 3 \end{bmatrix}$

7. $\begin{bmatrix} 5 & -7 \\ -2 & 3 \end{bmatrix}$ **9.** $\begin{bmatrix} -2 & 3 & -3 \\ -5 & 7 & -6 \\ 1 & -1 & 1 \end{bmatrix}$

11. $\begin{bmatrix} 4 & 1 & -3 \\ -5 & -1 & 4 \\ -1 & -1 & 1 \end{bmatrix}$ **13.** no inverse

15. $\begin{bmatrix} 1 & -2 & 1 \\ 0 & 1 & -2 \\ 0 & 0 & 1 \end{bmatrix}$ **17.** no inverse

19. $\begin{bmatrix} 1 & -2 & 1 & 0 \\ 0 & 1 & -2 & 1 \\ 0 & 0 & 1 & -2 \\ 0 & 0 & 0 & 1 \end{bmatrix}$ **21.** $\begin{bmatrix} 8 & -2 & -6 \\ -5 & 2 & 4 \\ 2 & 0 & -2 \end{bmatrix}$

23. $\begin{bmatrix} -2.5 & 5 & 3 & 5.5 \\ 5.5 & -8 & -6 & -9.5 \\ -1 & 3 & 1 & 3 \\ -5.5 & 9 & 6 & 10.5 \end{bmatrix}$ **25.** $x = 23, y = 17$

27. $x = 0, y = 0$ **29.** $x = 1, y = 2, z = 2$ **31.** $x = 54, y = -37, z = -49$ **33.** $x = 2, y = 1$ **35.** $x = 1, y = 2, z = 1$ **37.** 2 of model A, 3 of model B   **39.** Hi   **41.** no

45. $X = \begin{bmatrix} 0 \\ 0 \\ 0 \end{bmatrix}$ **51.** all reals except 2 and -2 **53.** all reals

55. $y \geq 0$ **57.** all reals

Exercises 6.5 (page 557)

1. $|A|$, det A **3.** 0   **5.** 0   **7.** 8   **9.** 1

11. $\begin{vmatrix} -2 & 3 \\ 8 & 9 \end{vmatrix}$ **13.** $\begin{vmatrix} 1 & -2 \\ 4 & 5 \end{vmatrix}$ **15.** $-\begin{vmatrix} -2 & 3 \\ 8 & 9 \end{vmatrix}$

17. $\begin{vmatrix} 1 & -2 \\ 4 & 5 \end{vmatrix}$ **19.** -54 **21.** -7 **23.** 86   **25.** -2

27. 2   **29.** 12   **31.** true   **33.** false   **35.** 3   **37.** 3

39. $(1, 2)$ **41.** $(3, 0)$ **43.** $(1, 0, 1)$ **45.** $(1, -1, 2)$

47. $(6, 6, 12)$ **49.** $\left(\frac{5}{6}, \frac{2}{3}, \frac{1}{2}, \frac{5}{2}\right)$ **51.** $3x - 2y = 0$

53. $6x + 7y = 9$ **55.** 30 sq. units   **57.** 73 sq. units

63. 8   **65.** -1 **67.** \$5,000 in HiTech, \$8,000 in SaveTel, \$7,000 in OilCo   **69.** 10   **71.** 24   **73.** 8   **77.** domain: $n \times n$ matrices; range: reals   **79.** yes   **81.** 21.468

83. $(x - 1)(x + 4)$ **85.** $x(3x + 1)(3x - 1)$

87. $\dfrac{4x - 5}{(x - 2)(2x - 1)}$ **89.** $\dfrac{x^2 + 2x + 1}{x(x^2 + 1)}$

Exercises 6.6 (page 567)

1. first-degree, second-degree   **3.** $\dfrac{1}{x} + \dfrac{2}{x - 1}$

5. $\dfrac{5}{x} - \dfrac{3}{x - 3}$ **7.** $\dfrac{1}{x + 1} + \dfrac{2}{x - 1}$ **9.** $\dfrac{2}{x} + \dfrac{-2}{x - 2}$

11. $\dfrac{1}{x - 3} - \dfrac{3}{x + 2}$ **13.** $\dfrac{8}{x + 3} - \dfrac{5}{x - 1}$

15. $\dfrac{5}{2x - 3} + \dfrac{2}{x - 5}$ **17.** $\dfrac{2}{x} + \dfrac{3}{x - 1} - \dfrac{1}{x + 1}$

19. $\dfrac{1}{x} + \dfrac{1}{x^2 + 3}$ **21.** $\dfrac{3}{x + 1} + \dfrac{2}{x^2 + 2x + 3}$

23. $\dfrac{3}{x} + \dfrac{2}{x + 1} + \dfrac{1}{(x + 1)^2}$ **25.** $\dfrac{1}{x} + \dfrac{2}{x^2} - \dfrac{3}{x - 1}$

27. $\dfrac{2}{x} + \dfrac{1}{x - 3} + \dfrac{2}{(x - 3)^2}$ **29.** $\dfrac{1}{x - 1} - \dfrac{4}{(x - 1)^3}$

31. $\dfrac{1}{x} + \dfrac{1}{x^2} + \dfrac{2}{x^2 + x + 1}$ **33.** $\dfrac{3}{x} + \dfrac{4}{x^2} + \dfrac{x + 1}{x^2 + 1}$

35. $-\dfrac{1}{x+1} - \dfrac{3}{x^2+2}$ **37.** $\dfrac{x+1}{x^2+2} + \dfrac{2}{x^2+x+2}$

39. $\dfrac{1}{x} + \dfrac{x}{x^2+2x+5} + \dfrac{x+2}{(x^2+2x+5)^2}$

41. $x - 3 - \dfrac{1}{x+1} + \dfrac{8}{x+2}$ **43.** $1 + \dfrac{2}{3x+1} + \dfrac{1}{x^2+1}$

45. $1 + \dfrac{1}{x} + \dfrac{x}{x^2+x+1}$ **47.** $2 + \dfrac{1}{x} + \dfrac{3}{x-1} + \dfrac{2}{x^2+1}$

49. No; it's the sum of two cubes.   **51.** $2|a|\sqrt{2ab}$

53. $3x^2\sqrt{2x}$ **55.** $x = 9$

Exercises 6.7 (page 575)

1. half-plane; boundary   **3.** is not

5. 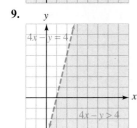   **7.**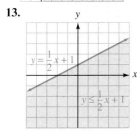

9. 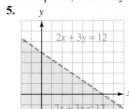   **11.**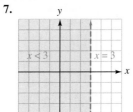

13. 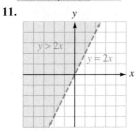   **15.**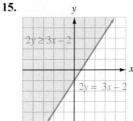

17. 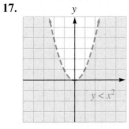   **19.**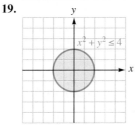

21. 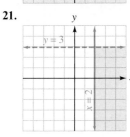   **23.**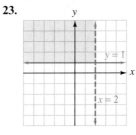

25. 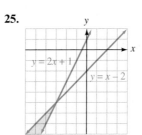   **27.**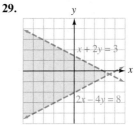

29. 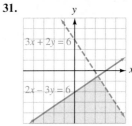   **31.**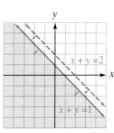

33. 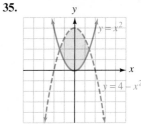   **35.**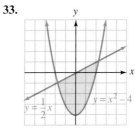

37. 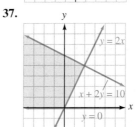   **39.**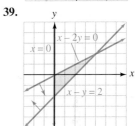

41. 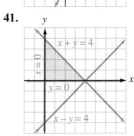   **43.**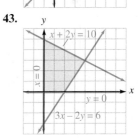

45. $\begin{cases} 6s + 4l \le 60 \\ s \ge 0 \\ l \ge 0 \end{cases}$ **53.** one, one   **55.** 0

Exercises 6.8 (page 585)

1. constraints   **3.** objective   **5.** $P = 12$ at $(0, 4)$

7. $P = \dfrac{13}{6}$ at $\left(\dfrac{5}{3}, \dfrac{4}{3}\right)$ **9.** $P = \dfrac{18}{7}$ at $\left(\dfrac{3}{7}, \dfrac{12}{7}\right)$ **11.** $P = 3$ at $(1, 0)$

13. $P = 0$ at $(0, 0)$ **15.** $P = 0$ at $(0, 0)$ **17.** $P = -12$ at $(-2, 0)$ **19.** $P = -2$ at the edge joining $(1, 2)$ and $(-1, 0)$

21. 3 tables and 12 chairs; $1,260   **23.** 30 IBMs and 30 Macintoshes; $2,700   **25.** 15 DVD players, 30 TVs; $1,560   **27.** $150,000 in stocks and $50,000 in bonds; $17,000   **29.** 2 buses, 2 trucks; $1,100

33. $\begin{bmatrix} 1 & 0 & 0 \\ 0 & 1 & 0 \\ 0 & 0 & 1 \\ 0 & 0 & 0 \end{bmatrix}$ **35.** $\left(\frac{1}{3} - 3y, y, -\frac{1}{3}\right)$

Chapter Review (page 588)

1.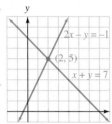

2.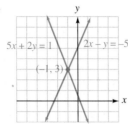

3.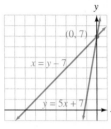

4.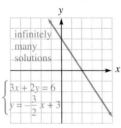

5.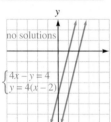

6. $x = 2, y = -1$
7. $x = 0, y = -3$
8. $x = 1, y = 1$
9. dependent equations; a general solution is $(x, 3x - 4)$
10. no solutions; inconsistent system **11.** $x = -3, y = 2$
12. $x = -2, y = 5$
13. $x = 2, y = -1$

14. no solutions; inconsistent system **15.** dependent equations; a general solution is $(x, 3x - 4)$ **16.** $x = 1$, $y = 0, z = 1$ **17.** $x = 1, y = 1, z = -1$ **18.** $x = 0$, $y = 1, z = 2$ **19.** \$10,400 **20.** 900 adult tickets, 450 senior tickets, 450 children's tickets **21.** $x = 1, y = 1$
22. dependent equations; a general solution is $(x, 3x + 4)$
23. $x = 3, y = 1, z = -2$ **24.** $x = -10, y = 1, z = 10$
25. no solution; inconsistent system **26.** $x = -4, y = 3$

27. $\begin{bmatrix} 1 & 3 & 4 \\ 4 & 0 & 2 \end{bmatrix}$ **28.** $\begin{bmatrix} 2 & 5 & 4 \\ -2 & -6 & 6 \\ -4 & 5 & -3 \end{bmatrix}$

29. $\begin{bmatrix} 4 & -1 \\ -7 & -7 \end{bmatrix}$ **30.** $\begin{bmatrix} -17 & 19 \\ 10 & -12 \end{bmatrix}$ **31.** $[5]$

32. $\begin{bmatrix} 2 & -1 & 1 & 3 \\ 4 & -2 & 2 & 6 \\ 2 & -1 & 1 & 3 \\ 10 & -5 & 5 & 15 \end{bmatrix}$ **33.** not possible **34.** $[-24]$

35. $\begin{bmatrix} 0 \\ -6 \end{bmatrix}$ **36.** $\begin{bmatrix} 5 & -3 \\ -3 & 2 \end{bmatrix}$ **37.** $\begin{bmatrix} 1 & 0 & 0 \\ -\frac{3}{2} & \frac{1}{2} & \frac{1}{2} \\ 1 & -\frac{1}{2} & 0 \end{bmatrix}$

38. $\begin{bmatrix} 9 & 16 & -56 \\ -3 & -5 & 18 \\ -1 & -2 & 7 \end{bmatrix}$ **39.** No inverse exists.

40. No inverse exists. **41.** $x = 1, y = 2, z = -1$
42. $w = 1, x = 1, y = 0, z = -1$ **43.** -7 **44.** -6
45. 3 **46.** -25 **47.** $x = 1, y = -2$ **48.** $x = 1, y = 0$, $z = -2$ **49.** $x = 1, y = -1, z = 3$ **50.** $w = 1, x = 0$, $y = -1, z = 2$ **51.** 21 **52.** 7 **53.** $\dfrac{3}{x} + \dfrac{4}{x + 1}$

54. $\dfrac{3}{x} + \dfrac{2}{x^2} + \dfrac{x - 1}{x^2 + 1}$ **55.** $\dfrac{1}{x} - \dfrac{1}{x^2 + x + 5}$

56. $\dfrac{1}{x + 1} - \dfrac{2}{(x + 1)^2} + \dfrac{2}{(x + 1)^3}$

57.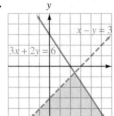

58.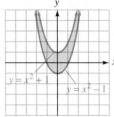

59.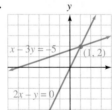

60.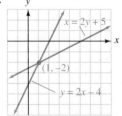

61. $P = 6$ at $(3, 0)$ **62.** $P = 12$ at $(0, -4)$
63. $P = 2$ at $(1, 1)$ **64.** $P = 3$ at $\left(-\frac{2}{3}, \frac{5}{3}\right)$
65. 1,000 bags of x, 1,400 bags of y

Chapter Test (page 597)

1.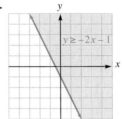

2.

3. $x = 1, y = -3$ **4.** $x = 3, y = 5$ **5.** 6 liters of 20% solution, 4 liters of 45% solution **6.** CD World 100 units, Ace 25 units, HiFi 50 units **7.** $x = 2, y = 1$
8. $x = 1, y = 2, z = 1$ **9.** $x = 1, y = 0, z = -2$
10. $x = -\frac{2}{5}y + \frac{7}{5}, z = -\frac{8}{5}y - \frac{7}{5}, y =$ any number

11. $\begin{bmatrix} 16 & -14 & 20 \\ 0 & -6 & -13 \end{bmatrix}$ **12.** $[-1]$ **13.** $\begin{bmatrix} -\frac{7}{3} & \frac{19}{3} \\ \frac{2}{3} & -\frac{5}{3} \end{bmatrix}$

14. $\begin{bmatrix} -13 & -3 & 14 \\ 4 & 1 & -4 \\ 12 & 3 & -13 \end{bmatrix}$ **15.** $x = \frac{17}{3}, y = -\frac{4}{3}$

16. $x = -36, y = 11, z = 34$ **17.** -12 **18.** -24
19. $-\frac{5}{4}$ **20.** 1 **21.** $\dfrac{3}{2x - 3} + \dfrac{1}{x + 1}$ **22.** $\dfrac{1}{x} + \dfrac{2x + 1}{x^2 + 2}$

23.

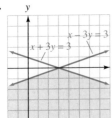

24.

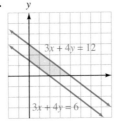

25. $P = 7$ at $(1, 2)$ **26.** $P = -8$ at $(8, 0)$

Exercises 7.1 (page 610)

1. $(2, -5), 3$ **3.** $(0, 0), \sqrt{5}$ **5.** to the left   **7.** down
9. directrix, focus   **11.** $x^2 + y^2 = 49$
13. $(x - 2)^2 + (y + 2)^2 = 17$
15. $(x - 1)^2 + (y + 2)^2 = 36$
17.

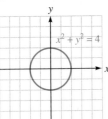

19.

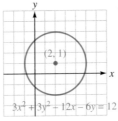

21. $(0, 0), (0, 3), y = -3$ **23.** $(0, 3), (5, 3), x = -5$
25. $(-2, 1), (-2, -5), y = 7$ **27.** $x^2 = 12y$
29. $y^2 = -12x$ **31.** $(x - 3)^2 = -12(y - 5)$
33. $(x - 3)^2 = -28(y - 5)$ **35.** $x^2 = -4(y - 2)$
37. $(y + 5)^2 = 8(x - 1)$ **39.** $(y - 2)^2 = -2(x - 2)$ or
$(x - 2)^2 = -2(y - 2)$ **41.** $(x + 4)^2 = -\frac{16}{3}(y - 6)$ or
$(y - 6)^2 = \frac{9}{4}(x + 4)$ **43.** $(y - 8)^2 = -4(x - 6)$
45. $(x - 3)^2 = \frac{1}{2}(y - 1)$ **47.**

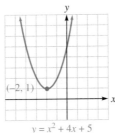

$y = x^2 + 4x + 5$
or
$y - 1 = (x + 2)^2$

49.

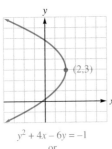

$y^2 + 4x - 6y = -1$
or
$(y - 3)^2 = -4(x - 2)$

51.

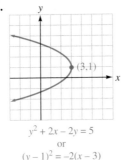

$y^2 + 2x - 2y = 5$
or
$(y - 1)^2 = -2(x - 3)$

53.

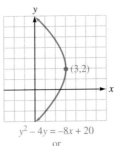

$y^2 - 4y = -8x + 20$
or
$(y - 2)^2 = -8(x - 3)$

55.

$x^2 - 6y + 22 = -4x$
or
$(x + 2)^2 = 6(y - 3)$

57.

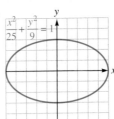

$4x^2 - 4x + 32y = 47$
or
$\left(x - \frac{1}{2}\right)^2 = -8\left(y - \frac{3}{2}\right)$

59. yes   **61.** 60 mi
63. $(x - 4)^2 + y^2 = 16$
65. $(x - 7)^2 + y^2 = 9$
67. 2 ft   **69.** $x^2 = \frac{-45}{2}y$
71. 1 ft   **73.** 8
75. about 12.6 cm
77. about 520 ft

79. $0x^2 + 0xy + y^2 - 8x - 4y + 12 = 0$
81. $x^2 + (y - 3)^2 = 25$ **83.** $y = x^2 + 4x + 3$ **87.** 4
89. $\frac{49}{4}$ **91.** $x = 1, -5$ **93.** $x = -2, 9$

Exercises 7.2 (page 627)

1. sum, constant   **3.** vertices   **5.** $(a, 0); (-a, 0)$
7. 26-in. string; tacks 24 in. apart   **9.** $\frac{x^2}{16} + \frac{y^2}{9} = 1$
11. $\frac{x^2}{25} + \frac{y^2}{16} = 1$ **13.** $\frac{9x^2}{16} + \frac{9y^2}{25} = 1$ **15.** $\frac{x^2}{7} + \frac{y^2}{16} = 1$
17. $\frac{(x - 3)^2}{4} + \frac{(y - 4)^2}{9} = 1$ **19.** $\frac{(x - 3)^2}{9} + \frac{(y - 4)^2}{4} = 1$
21. $\frac{(x - 3)^2}{41} + \frac{(y - 4)^2}{16} = 1$ **23.** $\frac{x^2}{36} + \frac{(y - 4)^2}{20} = 1$
25. $\frac{x^2}{100} + \frac{y^2}{64} = 1$ **27.**

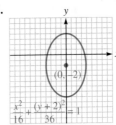

$\frac{x^2}{25} + \frac{y^2}{9} = 1$

29.

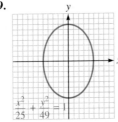

$\frac{x^2}{25} + \frac{y^2}{49} = 1$

31.

$\frac{x^2}{16} + \frac{(y + 2)^2}{36} = 1$

33.

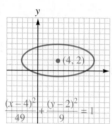

$\dfrac{(x-4)^2}{49} + \dfrac{(y-2)^2}{9} = 1$

35. $\dfrac{x^2}{4} + \dfrac{(y-1)^2}{16} = 1$

37. $\dfrac{(x+1)^2}{4} + \dfrac{(y+2)^2}{9} = 1$ **39.**

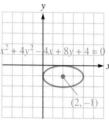

$x^2 + 4y^2 - 4x + 8y + 4 = 0$

$(2,-1)$

41.

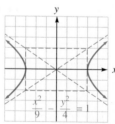

$(5,4)$

$16x^2 + 25y^2 - 160x - 200y + 400 = 0$

43. $\dfrac{x^2}{900} + \dfrac{y^2}{400} = 1$

45. $\dfrac{x^2}{36} + \dfrac{y^2}{25} = 1$

47. 26 m

49. about 26.9 in.

55. $\dfrac{x^2}{9} + \dfrac{y^2}{8} = 1$

59. $\begin{bmatrix} -1 & 8 \\ 5 & -1 \\ 4 & 7 \end{bmatrix}$ **61.** $\begin{bmatrix} 5 & 8 \\ 12 & -2 \\ -1 & 5 \end{bmatrix}$ **63.** $\begin{bmatrix} -1 & -2 \\ -2 & 0 \\ 1 & -1 \end{bmatrix}$

Exercises 7.3 (page 638)

1. difference, constant **3.** $(a,0); (-a,0)$ **5.** transverse axis

7. $\dfrac{x^2}{25} - \dfrac{y^2}{24} = 1$ **9.** $\dfrac{(x-2)^2}{4} - \dfrac{(y-4)^2}{9} = 1$

11. $\dfrac{(y-3)^2}{9} - \dfrac{(x-5)^2}{9} = 1$ **13.** $\dfrac{y^2}{9} - \dfrac{x^2}{16} = 1$

15. $\dfrac{(x-1)^2}{4} - \dfrac{(y+3)^2}{16} = 1$ or $\dfrac{(y+3)^2}{4} - \dfrac{(x-1)^2}{16} = 1$

17. $\dfrac{x^2}{10} - \dfrac{3y^2}{20} = 1$ **19.** 24 sq. units **21.** 12 sq. units

23. $\dfrac{(x+2)^2}{4} - \dfrac{4(y+4)^2}{81} = 1$ or $\dfrac{(y+4)^2}{4} - \dfrac{4(x+2)^2}{81} = 1$

25. $\dfrac{x^2}{36} - \dfrac{16y^2}{25} = 1$

27.

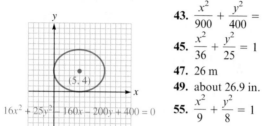

$\dfrac{x^2}{9} - \dfrac{y^2}{4} = 1$

29.

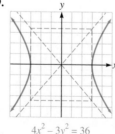

$4x^2 - 3y^2 = 36$

31.

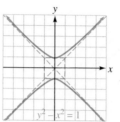

$y^2 - x^2 = 1$

33.

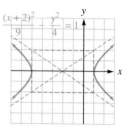

$\dfrac{(x+2)^2}{9} - \dfrac{y^2}{4} = 1$

35.

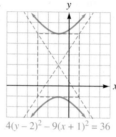

$4(y-2)^2 - 9(x+1)^2 = 36$

37.

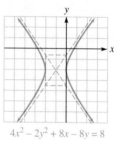

$4x^2 - 2y^2 + 8x - 8y = 8$

39.

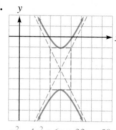

$y^2 - 4x^2 + 6y + 32x = 59$

41.

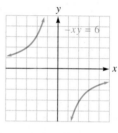

$-xy = 6$

43. $\dfrac{(x-3)^2}{9} - \dfrac{(y-1)^2}{16} = 1$ **45.** $4x^2 - 5y^2 - 60y = 0$

47. 24 **49.** 3 units **51.** hyperbola; $\dfrac{x^2}{144} - \dfrac{y^2}{25} = 1$

53. hyperbola; $\dfrac{x^2}{36} - \dfrac{y^2}{64} = 1$ **59.** $f^{-1}(x) = \dfrac{x+2}{3}$

61. $f^{-1}(x) = \dfrac{2x}{5-x}$ **63.** $f(g(x)) = (x+1)^4 + 1$

65. $f(f(x)) = (x^2+1)^2 + 1$

Exercises 7.4 (page 647)

1. graphs **3.**

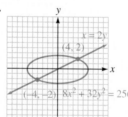

$x = 2y$

$(4,2)$

$(-4,-2)$ $8x^2 + 32y^2 = 256$

5.

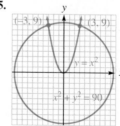

$(-3,9)$ $(3,9)$

$y = x^2$

$x^2 + y^2 = 90$

7.

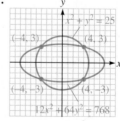

$x^2 + y^2 = 25$

$(-4,3)$ $(4,3)$

$(-4,-3)$ $(4,-3)$

$12x^2 + 64y^2 = 768$

9.

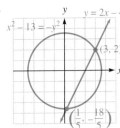

11.

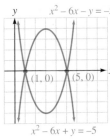

17. $\dfrac{(x-2)^2}{1} + \dfrac{(y+1)^2}{4} = 1$

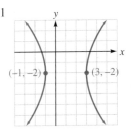

13. $(1, 2), (-1, 0)$ **15.** $(1, 0.67), (-1, -0.67)$ **17.** $(3, 0)$, $(0, 5)$ **19.** $(1, 1)$ **21.** $(1, 2), (2, 1)$ **23.** $(-2, 3), (2, 3)$ **25.** $\left(\sqrt{5}, 5\right), \left(-\sqrt{5}, 5\right)$ **27.** $(3, 2), (3, -2), (-3, 2), (-3, -2)$
29. $(2, 4), (2, -4), (-2, 4), (-2, -4)$
31. $\left(-\sqrt{15}, 5\right), \left(\sqrt{15}, 5\right), (-2, -6), (2, -6)$
33. $(0, -4), (-3, 5), (3, 5)$ **35.** $(-2, 3), (2, 3), (-2, -3)$, $(2, -3)$ **37.** $(3, 3)$ **39.** $(6, 2), (-6, -2), \left(\sqrt{42}, 0\right)$, $\left(-\sqrt{42}, 0\right)$ **41.** $\left(\frac{1}{2}, \frac{1}{3}\right), \left(\frac{1}{3}, \frac{1}{2}\right)$ **43.** 7 cm by 9 cm
45. 80 ft by 100 ft or 50 ft by 160 ft   **47.** either \$750 at 9% or \$900 at 7.5%   **49.** $(30, 3)$ **51.** yes, there are potential collision points at $(-2, 4)$ and $(1, 1)$.   **53.** about 23 mi
61. vertical: $x = 1$; horizontal: $y = 3$ **63.** vertical: $x = 1$, $x = -1$; horizontal: $y = 0$ **65.** y-axis   **67.** origin

18. $\dfrac{x^2}{4} - \dfrac{y^2}{12} = 1$ **19.** $\dfrac{y^2}{9} - \dfrac{x^2}{16} = 1$
20. $\dfrac{x^2}{9} - \dfrac{(y-3)^2}{16} = 1$ **21.** $\dfrac{y^2}{9} - \dfrac{(x-3)^2}{16} = 1$
22. $y = \pm\dfrac{4}{5}x$
23. $\dfrac{(x-1)^2}{4} - \dfrac{(y+2)^2}{9} = 1$

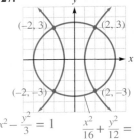

Chapter Review (page 650)

1. $x^2 + y^2 = 16$ **2.** $x^2 + y^2 = 100$
3. $(x-3)^2 + (y+2)^2 = 25$ **4.** $(x+2)^2 + (y-4)^2 = 25$
5. $(x-5)^2 + (y-10)^2 = 85$ **6.** $(x-2)^2 + (y-2)^2 = 89$

7.

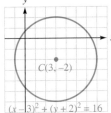

8.

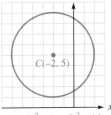

24.

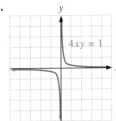

25.

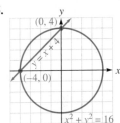

9. $y^2 = -2x$ **10.** $x^2 = 16y$ **11.** $(x+2)^2 = -\dfrac{4}{11}(y-3)$

12.

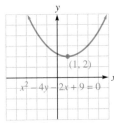

13.

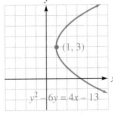

26.

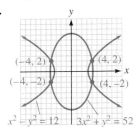

27.

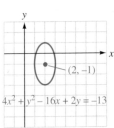

28. $(-4, 2), (-4, -2), (4, 2), (4, -2)$ **29.** $(0, 4), \left(2\sqrt{3}, -2\right)$
30. $(-2, 3), (-2, -3), (2, 3), (2, -3)$

14. $\dfrac{x^2}{36} + \dfrac{y^2}{16} = 1$ **15.** $\dfrac{y^2}{25} + \dfrac{x^2}{4} = 1$
16. $\dfrac{(x+2)^2}{16} + \dfrac{(y-3)^2}{9} = 1$

Chapter Test (page 660)

1. $(x-2)^2 + (y-3)^2 = 9$ **2.** $(x-2)^2 + (y-3)^2 = 41$
3. $(x-2)^2 + (y+5)^2 = 169$ **4.**

5. $(x - 3)^2 = 16(y - 2)$ **6.** $(y + 6)^2 = -4(x - 4)$

7. $(x - 2)^2 = \frac{4}{3}(y + 3)$ or $(y + 3)^2 = -\frac{9}{2}(x - 2)$

8.

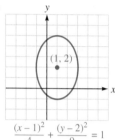

9. $\dfrac{x^2}{100} + \dfrac{y^2}{64} = 1$

10. $\dfrac{x^2}{169} + \dfrac{y^2}{144} = 1$

11. $\dfrac{x^2}{4} + \dfrac{y^2}{36} = 1$

12.

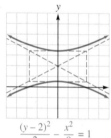

13. $\dfrac{x^2}{25} - \dfrac{y^2}{144} = 1$

14. $\dfrac{x^2}{36} - \dfrac{4y^2}{25} = 1$

15. $\dfrac{(x - 2)^2}{64} - \dfrac{(y + 1)^2}{36} = 1$

16.

17. $\left(\sqrt{7}, 4\right), \left(-\sqrt{7}, 4\right)$

18. $\left(3\sqrt{2}, 3\right), \left(-3\sqrt{2}, 3\right), \left(3\sqrt{2}, -3\right), \left(-3\sqrt{2}, -3\right)$

19. $(y - 2)^2 = 6(x + 3)$; parabola

20. $\dfrac{(x - 1)^2}{3} + \dfrac{(y + 2)^2}{2} = 1$; ellipse

Cumulative Review Exercises (page 661)

1. 16   **2.** $\frac{1}{2}$ **3.** y^2 **4.** $x^{17/12}$ **5.** $x^{4/3} - x^{2/3}$

6. $\frac{1}{x} + 2 + x$ **7.** $-3x$ **8.** $4t\sqrt{3t}$ **9.** $4x$ **10.** $x + 3$

11. $7\sqrt{2}$ **12.** $-12\sqrt[4]{2} + 10\sqrt[4]{3}$ **13.** $-18\sqrt{6}$

14. $\dfrac{5\sqrt[3]{x^2}}{x}$ **15.** $\dfrac{x + 3\sqrt{x} + 2}{x - 1}$ **16.** $\sqrt{xy}$ **17.** 2, 7

18. $\frac{1}{4}$ **19.** $1, -\frac{3}{2}$ **20.** $\dfrac{-2 \pm \sqrt{7}}{3}$ **21.** $7 + 2i$

22. $-5 - 7i$ **23.** $13 + 0i$ **24.** $12 - 6i$ **25.** $-12 - 10i$

26. $\frac{3}{2} + \frac{1}{2}i$ **27.** $\sqrt{13}$ **28.** $\sqrt{61}$ **29.** -2 **30.** $\left(1, \frac{1}{2}\right)$

31. $(-\infty, -2) \cup (3, \infty)$ **32.** $[-2, 3]$ **33.** 5   **34.** 27

35. $12x^2 - 12x + 5$ **36.** $6x^2 + 3$ **37.** $2^y = x$

38. $\log_3 a = b$ **39.** 5   **40.** 3   **41.** $\frac{1}{27}$ **42.** 1

43. $y = 2^x$ **44.** x **45.** 1.9912   **46.** 0.301   **47.** 1.6902

48. 0.1461   **49.** $\dfrac{2\log 2}{\log 3 - \log 2}$ **50.** 16   **51.** $2,848.31

52. 1.16056   **53.** (2, 1)   **54.** (1, 1)   **55.** $(-2, -2)$

56. (3, 1)   **57.** -1 **58.** -1 **59.** $(-1, -1, 3)$

60. (1, 1, 1)

Exercises 8.1 (page 669)

1. power   **3.** first   **5.** $7 \cdot 6 \cdot 5 \cdot 4 \cdot 3 \cdot 2 \cdot 1$ **7.** $(n - 1)!$

9. 24   **11.** 4,320   **13.** 1,440   **15.** $\frac{1}{1,320}$ **17.** $\frac{5}{3}$

19. 18,564   **21.** $a^5 + 5a^4b + 10a^3b^2 + 10a^2b^3 + 5ab^4 + b^5$

23. $x^3 - 3x^2y + 3xy^2 - y^3$ **25.** $a^3 + 3a^2b + 3ab^2 + b^3$

27. $a^5 - 5a^4b + 10a^3b^2 - 10a^2b^3 + 5ab^4 - b^5$

29. $8x^3 + 12x^2y + 6xy^2 + y^3$

31. $x^3 - 6x^2y + 12xy^2 - 8y^3$

33. $16x^4 + 96x^3y + 216x^2y^2 + 216xy^3 + 81y^4$

35. $x^4 - 8x^3y + 24x^2y^2 - 32xy^3 + 16y^4$

37. $x^5 - 15x^4y + 90x^3y^2 - 270x^2y^3 + 405xy^4 - 243y^5$

39. $\dfrac{x^4}{16} + \dfrac{x^3y}{2} + \dfrac{3x^2y^2}{2} + 2xy^3 + y^4$ **41.** $6a^2b^2$

43. $35a^3b^4$ **45.** $-b^5$ **47.** $2,380a^{13}b^4$ **49.** $-4\sqrt{2}a^3$

51. $1,134a^5b^4$ **53.** $\dfrac{3x^2y^2}{2}$ **55.** $-\dfrac{55r^2s^9}{2,048}$

57. $\dfrac{n!}{3!(n - 3)!}a^{n-3}b^3$

59. $\dfrac{n!}{(r - 1)!(n - r + 1)!}a^{n-r+1}b^{r-1}$ **63.** -252

71. $3xyz^2(x^2yz^2 - 2z^3 + 5x)$ **73.** $(a^2 + b^2)(a + b)(a - b)$

75. $\frac{3 + x}{3 - x}$

Exercises 8.2 (page 676)

1. domain   **3.** series   **5.** infinite   **7.** summation
notation   **9.** 6   **11.** $5c$ **13.** 0, 10, 30, 60, 100, 150

15. 21   **17.** $a + 4d$ **19.** 15   **21.** 15   **23.** 15   **25.** $\frac{242}{243}$

27. 35   **29.** 3, 7, 15, 31   **31.** $-4, -2, -1, -\frac{1}{2}$

33. k, k^2, k^4, k^8 **35.** $8, \dfrac{16}{k}, \dfrac{32}{k^2}, \dfrac{64}{k^3}$ **37.** an alternating

series   **39.** not an alternating series   **41.** 30   **43.** -50

45. 40   **47.** 500   **49.** $\frac{7}{12}$ **51.** 160   **53.** 3,725

59. 6 cm   **61.** 26 ft

Exercises 8.3 (page 682)

1. $(n - 1)$ **3.** infinite   **5.** $a_n = a + (n - 1)d$

7. arithmetic means   **9.** 1, 3, 5, 7, 9, 11   **11.** $5, \frac{7}{2}, 2,$
$\frac{1}{2}, -1, -\frac{5}{2}$ **13.** $9, \frac{23}{2}, 14, \frac{33}{2}, 19, \frac{43}{2}$ **15.** 318

17. 6   **19.** 370   **21.** 44   **23.** $\frac{25}{2}, 15, \frac{35}{2}$ **25.** $-\frac{82}{15}, -\frac{59}{15},$
$-\frac{12}{5}, -\frac{13}{15}$ **27.** 285   **29.** 555   **31.** $157\frac{1}{2}$ **33.** 20,100

35. $1,080°, 1,800°$ **37.** $460   **39.** $1,587,500   **41.** 80 ft

43. 210   **49.** 6   **51.** -1 **53.** $1, -1, i, -i$

Exercises 8.4 (page 690)

1. r^{n-1} **3.** ar^{n-1} **5.** infinite   **7.** geometric means

9. 10, 20, 40, 80   **11.** $-2, -6, -18, -54$ **13.** $3, 3\sqrt{2}, 6,$
$6\sqrt{2}$ **15.** 2, 6, 18, 54   **17.** 256   **19.** -162 **21.** $10\sqrt[4]{2},$
$10\sqrt{2}, 10\sqrt[4]{8}$ **23.** 8, 32, 128, 512   **25.** 124

27. $-29,524$ **29.** $\frac{1,995}{32}$ **31.** 18   **33.** 8   **35.** $\frac{5}{9}$ **37.** $\frac{25}{99}$

39. 23   **41.** 12.96 ft, 400 ft   **43.** 5.13 m   **45.** $69.82

47. about $\frac{1}{3}c$ **49.** about 1.03×10^{20} **51.** $180,176.87

53. $2,001.60   **55.** $2,013.62   **57.** $264,094.58

59. 5,000   **61.** 1.8447×10^{19} grains   **63.** no   **65.** $5 - 3i$
67. $10 + 0i$ **69.** $\frac{1}{5} + \frac{8}{5}i$ **71.** $-i$

Exercises 8.5 (page 696)

1. two   **3.** $n = k + 1$
5. $5 = 5; 15 = 15; 30 = 30; 50 = 50$
7. $7 = 7; 17 = 17; 30 = 30; 46 = 46$ **29.** no
37. a. 1   **b.** 3   **c.** 7   **d.** 15
39.   **41.**

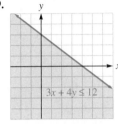   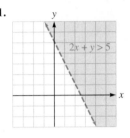

Exercises 8.6 (page 705)

1. 24   **3.** $\frac{n!}{(n-r)!}$ **5.** 1   **7.** $\frac{n!}{r!(n-r)!}$ **9.** 1

11. $\frac{n!}{a! \cdot b! \ldots}$ **13.** 840   **15.** 35   **17.** 120   **19.** 5
21. 1   **23.** 1,200   **25.** 40   **27.** 2,278   **29.** 144
31. 8,000,000   **33.** 240   **35.** 6   **37.** 40,320
39. 14,400   **41.** 24,360   **43.** 5,040   **45.** 48   **47.** 96
49. 210   **51.** 24   **53.** 5,040   **55.** 2,721,600   **57.** 1,120
59. 59,400   **61.** 272   **63.** 28   **65.** 252   **67.** 142,506
69. 66   **71.** 256   **79.** $\sqrt{\pi}$ **81.** 8   **83.** false   **85.** false

Exercises 8.7 (page 712)

1. experiment   **3.** $\frac{n(E)}{n(S)}$ **5.** {(1, H), (2, H), (3, H), (4, H),
(5, H), (6, H), (1, T), (2, T), (3, T), (4, T), (5, T), (6, T)}
7. {a, b, c, d, e, f, g, h, i, j, k, l, m, n, o, p, q, r, s, t, u, v, w, x, y, z}
9. $\frac{1}{6}$ **11.** $\frac{2}{3}$ **13.** $\frac{19}{42}$ **15.** $\frac{13}{42}$ **17.** $\frac{3}{8}$ **19.** 0
21. $\frac{1}{12}$ **23.** $\frac{1}{169}$ **25.** $\frac{5}{12}$ **27.** about 6.3×10^{-12} **29.** 0
31. $\frac{3}{13}$ **33.** $\frac{1}{6}$ **35.** $\frac{1}{8}$ **37.** $\frac{5}{16}$
39. (S = survive, F = fail){SSSS, SSSF, SSFS, SFSS, FSSS,
SSFF, SFSF, FSSF, SFFS, FSFS, FFSS, SFFF, FSFF, FFSF,
FFFS, FFFF}   **41.** $\frac{1}{4}$ **43.** $\frac{1}{4}$ **45.** 1   **47.** $\frac{32}{119}$ **49.** $\frac{1}{3}$
51. 0.18   **53.** 0.14   **55.** about 33%   **57.** no
59. $-10, 4$ **61.**

Chapter Review (page 714)

1. 720   **2.** 30,240   **3.** 8   **4.** $\frac{280}{3}$
5. $x^3 + 3x^2y + 3xy^2 + y^3$
6. $p^4 + 4p^3q + 6p^2q^2 + 4pq^3 + q^4$
7. $a^5 - 5a^4b + 10a^3b^2 - 10a^2b^3 + 5ab^4 - b^5$
8. $8a^3 - 12a^2b + 6ab^2 - b^3$ **9.** $56a^5b^3$ **10.** $80x^3y^2$
11. $84x^3y^6$ **12.** $439,040x^3$ **13.** 63   **14.** 9
15. 5, 17, 53, 161   **16.** $-2, 8, 128, 32,768$ **17.** 90
18. 60   **19.** 1,718   **20.** -360 **21.** 117   **22.** 281
23. -92 **24.** $-\frac{135}{2}$ **25.** $\frac{7}{2}, 5, \frac{13}{2}$ **26.** 25, 40, 55, 70, 85

27. 3,320   **28.** 5,780   **29.** $-5,220$ **30.** $-1,540$
31. $\frac{1}{729}$ **32.** 13,122   **33.** $\frac{9}{16,384}$ **34.** $\frac{8}{15,625}$
35. $2\sqrt{2}, 4, 4\sqrt{2}$ **36.** $4, -8, 16, -32$ **37.** 16
38. $\frac{3,280}{27}$ **39.** 6,560   **40.** $\frac{2,295}{128}$ **41.** $\frac{520,832}{78,125}$ **42.** $\frac{3,280}{3}$
43. $16\sqrt{2}$ **44.** $\frac{2}{3}$ **45.** $\frac{3}{25}$ **46.** no sum   **47.** 1
48. $\frac{1}{3}$ **49.** 1   **50.** $\frac{17}{99}$ **51.** $\frac{5}{11}$ **52.** \$4,775.81
53. 6,516; 3,134   **54.** \$3,486.78
55. $1 = 1; 9 = 9; 36 = 36; 100 = 100$ **57.** 6,720   **58.** 35
59. 1   **60.** 4,050   **61.** 564,480   **62.** 840   **63.** 21
64. 66   **65.** 120   **66.** $\frac{56}{1,287}$ **67.** $\frac{1}{6}$ **68.** $\frac{33}{66,640}$
69. 20,160   **70.** 90,720   **72.** 24   **73.** $\frac{1}{108,290}$
74. $\frac{108,289}{108,290}$ **75.** about 6.3×10^{-12} **76.** $\frac{60}{143}$ **77.** $\frac{1}{2}$
78. $\frac{7}{13}$ **79.** $\frac{1}{2,598,960}$ **80.** $\frac{33}{16,660}$ **81.** $\frac{15}{16}$

Chapter Test (page 722)

1. 144   **2.** 384   **3.** $10x^4y$ **4.** $112a^2b^6$ **5.** 27
6. -36 **7.** 155   **8.** -130 **9.** 9, 14, 19   **10.** $-6, -18$
11. 255.75   **12.** about 9   **13.** about \$0.42c
14. about \$1.46c   **16.** 800,000   **17.** 42   **18.** 24
19. 28   **20.** 1   **21.** 576   **22.** 120   **23.** 60
24. {(H, H, H), (H, H, T), (H, T, H), (H, T, T), (T, H, H),
(T, H, T), (T, T, H), (T, T, T)}   **25.** $\frac{1}{6}$ **26.** $\frac{1}{13}$ **27.** $\frac{33}{66,640}$
28. $\frac{1}{9}$ **29.** $\frac{87}{245}$ **30.** $\frac{12}{19}$

Exercises 9.1 (page 732)

1. interest   **3.** annual   **5.** future   **7.** interest
9. periodic rate   **11.** principal, periodic rate, frequency of
compounding, number of years   **13.** effective   **15.** \$585
17. \$3,700   **19.** 6%   **21.** \$1,296   **23.** \$1,763.19
25. \$1,612.70   **27.** \$2,840.84   **29.** \$2,912.71
31. \$2,944.91   **33.** 6.14%   **35.** 9.73%   **37.** \$14,027.60
39. \$11,678.47   **41.** $3\frac{1}{2}$%   **43.** \$20,448.10
45. \$356,867.13   **47.** \$207,976   **49.** 8.62 million ft^3
51. 7.4%, 7.12%   **53.** \$1,267.45   **55.** \$3,853.73
57. \$50,363.13   **63.** $\frac{x+3}{2x+1}$ **65.** 1

Exercises 9.2 (page 739)

1. annuities   **3.** payments, interest   **5.** regular payment,
periodic rate, frequency of compounding, number of years
7. \$1,318.08   **9.** \$318.36   **11.** \$1,200.61   **13.** \$1,556.03
15. \$5,608.49   **17.** \$34,603.82   **19.** \$1,665.81
21. \$637.52   **23.** \$2,467.11   **25.** \$86,803,923.58
27. \$22,177.71   **29.** Bank B   **31.** \$114,432.12
33. \$1,466.08   **35.** 12   **37.** 3

Exercises 9.3 (page 746)

1. present value   **3.** promissory note   **5.** amortizing
7. \$48,116.14   **9.** \$300.04   **11.** \$27,128.43
13. \$99,650.84   **15.** \$533.84   **17.** 15-yr: \$1,560.22;
20-yr: \$1,341.84   **19.** \$220.13   **21.** \$146,506.50
23. $2\sqrt{6}$ **25.** $13\sqrt{5x}$

Chapter Review (page 747)

1. $2,900   **2.** $3,077.25   **3.** $2,405.47   **4.** $6,670.80
5. BigBank, 6.45%; BestBank, 6.4%; BigBank   **6.** $5,622.23
7. $8,856.49   **8.** $88,353.06   **9.** $369.89   **10.** $5,552.11
11. $437,563.50   **12.** $1,687.03; $303,665.40
13. $1,404.89; $505,760.40

Chapter Test (page 750)

1. compound   **2.** periodic   **3.** effective   **4.** annual
5. annuities   **6.** sinking fund   **7.** present value
8. amortizing   **9.** $1,950   **10.** $2,117.56   **11.** $2,141.11
12. 5.116%   **13.** $2,498.00   **14.** $50,506.10   **15.** $113.20
16. $112,652.71   **17.** $910.16

Cumulative Review Exercises (page 751)

1.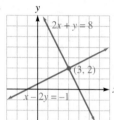

2.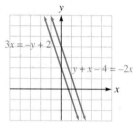

inconsistent system

3. $x = 3, y = 1$ **4.** $x = 3, y = 2, z = 1$
5. $x = 1, y = 0, z = 2$ **6.** $x = 1, y = 2, z = 1, t = 1$
7. $\begin{bmatrix} 1 & 3 \\ 3 & 7 \end{bmatrix}$ **8.** $\begin{bmatrix} -3 & 1 \\ 1 & -1 \end{bmatrix}$ **9.** $\begin{bmatrix} 3 & 2 & 0 \\ -2 & 8 & 7 \end{bmatrix}$

10. $\begin{bmatrix} 9 & 6 \\ 6 & 21 \end{bmatrix}$ **11.** $\begin{bmatrix} -1 & \frac{3}{2} \\ \frac{1}{2} & -\frac{1}{2} \end{bmatrix}$ **12.** $\begin{bmatrix} 4 & 5 & -4 \\ -1 & -1 & 1 \\ -4 & -6 & 5 \end{bmatrix}$

13. -41 **14.** -1 **15.** $\dfrac{\begin{vmatrix} 11 & 3 \\ 24 & 5 \end{vmatrix}}{\begin{vmatrix} 4 & 3 \\ -2 & 5 \end{vmatrix}}$ **16.** $\dfrac{\begin{vmatrix} 4 & 11 \\ -2 & 24 \end{vmatrix}}{\begin{vmatrix} 4 & 3 \\ -2 & 5 \end{vmatrix}}$

17. $\dfrac{2}{x + 1} - \dfrac{3}{x + 2}$ **18.** $\dfrac{\frac{1}{2}}{2x - 5} - \dfrac{\frac{3}{2}}{(2x - 5)^2}$

19. 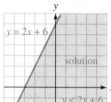   **20.**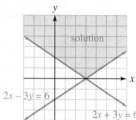

21. $x^2 + y^2 = 16$ **22.** $(x - 2)^2 + (y + 3)^2 = 121$

23. 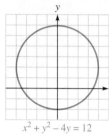   **24.**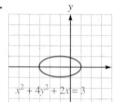

25.   **26.**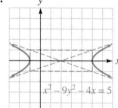

27. $\dfrac{x^2}{36} + \dfrac{y^2}{16} = 1$ **28.** $\dfrac{(x - 2)^2}{21} + \dfrac{(y - 3)^2}{25} = 1$
29. $\dfrac{x^2}{4} - \dfrac{y^2}{5} = 1$ **30.** $\dfrac{(y - 4)^2}{9} - \dfrac{(x - 2)^2}{9} = 1$
31. $16x^7y$ **32.** $1,792x^3y^5$ **33.** 10   **34.** 65   **35.** 33
36. $\frac{364}{9}$ **37.** 1,680   **38.** 1   **39.** 66   **40.** 24
41. 17,280   **42.** 495   **43.** $\frac{1}{18}$ **44.** $\frac{506}{19,992} = \frac{253}{9,996}$
45. 0.48   **47.** $3,757.02   **49.** $972.08

Index

Index of Applications

Examples that are applications are shown with boldface numbers.
Exercises that are applications are shown with lightface numbers.